Portable Spectroscopy and Spectrometry 1

Portable Spectroscopy and Spectrometry 1

Technologies and Instrumentation

Edited by

Richard A. Crocombe
Crocombe Spectroscopic Consulting
Winchester
MA, USA

Pauline E. Leary
Federal Resources
Stevensville
MD, USA

Brooke W. Kammrath
Department of Forensic Science
Henry C. Lee College of Criminal Justice and Forensic Sciences
University of New Haven
West Haven
CT, USA

and

Henry C. Lee Institute of Forensic Science
West Haven
CT, USA

WILEY

This edition first published 2021

Registered Offices
John Wiley & Sons, Inc., 111 River Street, Hoboken, NJ 07030, USA
John Wiley & Sons Ltd, The Atrium, Southern Gate, Chichester, West Sussex, PO19 8SQ, UK

Editorial Office
The Atrium, Southern Gate, Chichester, West Sussex, PO19 8SQ, UK

For details of our global editorial offices, customer services, and more information about Wiley products visit us at www.wiley.com.

Wiley also publishes its books in a variety of electronic formats and by print-on-demand. Some content that appears in standard print versions of this book may not be available in other formats.

Library of Congress Cataloging-in-Publication Data

Names: Crocombe, Richard A., editor. | Leary, Pauline E., editor. | Kammrath, Brooke W. editor.
Title: Portable spectroscopy and spectrometry 1: technologies and instrumentation / edited by Richard A. Crocombe, Pauline E. Leary, Brooke W. Kammrath.
Description: First edition. | Hoboken, NJ : Wiley, 2021. | Includes bibliographical references and indexes. | Contents: v. 1. Technologies and instrumentation.
Identifiers: LCCN 2020042686 (print) | LCCN 2020042687 (ebook) | ISBN 9781119636366 (v. 1 ; hardback) | ISBN 9781119636335 (v. 1 ; adobe pdf) | ISBN 9781119636410 (v. 1 ; epub)
Subjects: LCSH: Spectrometer. | Spectrum analysis.
Classification: LCC QC373.S7 P67 2021 (print) | LCC QC373.S7 (ebook) | DDC 621.36/1–dc23
LC record available at https://lccn.loc.gov/2020042686
LC ebook record available at https://lccn.loc.gov/2020042687

Cover Design: Wiley
Cover Image: Hand © Artem Furman/Shutterstock;
Infrared Spectrum: Courtesy of Pauline Leary and Brooke Kammrath

Set in 9.5/12.5pt STIXTwoText by SPi Global, Chennai, India
Printed and bound by CPI Group (UK) Ltd, Croydon, CR0 4YY

C9781119636366_240321

To my parents, my family, and all those in spectroscopy I've worked with, and learned from, over the years.

Richard A. Crocombe, Ph.D.

This book is dedicated to **John A. Reffner**. *Everyone needs someone in their life like you are to me: A* **teacher** *to provide them guidance when needed, a* **mentor** *to help them see what they can achieve, and a* **friend** *to talk to when they need to know it will all work out.*

Pauline E. Leary, Ph.D.

The dedication of this book is shared amongst all of the members of my family. To my mother Shirley and late father Milton, I am grateful for all of the many life lessons you taught me and your encouragement to pursue a career that I love. To my twin sister Lindsey, I appreciate having you as my forever best friend, partner in crime, and womb-mate. To my husband Matt, a choice in a life partner is the most important decision a person can make, and I chose very wisely. You are my biggest champion, and I am yours – together we can accomplish anything. And to my children Riley and Grayson, I love you so much. I hope you always stay curious because curiosity is the best teacher, the mother of invention, the cure for boredom, the key to creativity, the engine of achievement, and the beginning of science.

Brooke W. Kammrath, Ph.D.

Contents

List of Contributors

Jean Albert
Bluefield Technologies Inc.
San Francisco
CA, USA

Angie Ambers
Department of Forensic Science
Henry C. Lee College of Criminal Justice
and Forensic Sciences
University of New Haven
West Haven
CT, USA

Miquel Antoine
Johns Hopkins Applied Physics Laboratory
Laurel
MD, USA

Yotam Ariel
Bluefield Technologies Inc.
San Francisco
CA, USA

Dieter Bingemann
Wasatch Photonics, Inc.
Logan
UT, USA

Kenion H. Blakeman
908 Devices Inc.
Boston
MA, USA

Griffin Cassata
Department of Forensic Science
Henry C. Lee College of Criminal Justice
and Forensic Sciences
University of New Haven
West Haven
CT, USA

Sterling Cornaby
Moxtek, Inc.
Orem
UT, USA

David Creasey
Wasatch Photonics, Inc.
Logan
UT, USA

Richard A. Crocombe
Crocombe Spectroscopic Consulting
Winchester
MA, USA

Brian Damit
Johns Hopkins Applied Physics Laboratory
Laurel
MD, USA

Anshuman Das
Labby Inc.
Cambridge
MA, USA

David Day
Sciaps Incorporated
Woburn
MA, USA

Charlotte De Bleye
Department of Pharmacy
University of Liege
Liege, Belgium

Reno F. DeBono
Trace Detection Scientist
Formerly of Barringer Instruments
Warren
NJ, USA

and

Smiths Detection
Edgewood
MD, USA

Claire L. Glynn
Department of Forensic Science
Henry C. Lee College of Criminal Justice
and Forensic Sciences
University of New Haven
West Haven
CT, USA

Nathaniel R. Gomer
ChemImage Corporation
Pittsburgh
PA, USA

Heinrich Grüger
Fraunhofer Institute for
Photonic Microsystems
Dresden, Germany

Philippe Hubert
Department of Pharmacy
University of Liege
Liege, Belgium

Brooke W. Kammrath
Department of Forensic Science
Henry C. Lee College of Criminal Justice and
Forensic Sciences
University of New Haven
West Haven
CT, USA

and

Henry C. Lee Institute of Forensic Science
West Haven
CT, USA

Vassili Karanassios
Department of Chemistry and
Waterloo Institute for Nanotechnology
University of Waterloo
Waterloo
ON, Canada

Koby L. Kizzire
Department of Forensic Science
Henry C. Lee College of Criminal Justice
and Forensic Sciences
University of New Haven
West Haven
CT, USA

Richard L. Lachance
Bluefield Technologies Inc.
San Francisco
CA, USA

Hugo Lavoie
Defence R&D Canada – Valcartier Research Center
Quebec
QC, Canada

Pauline E. Leary
Federal Resources
Stevensville
MD, USA

Scott E. Miller
908 Devices Inc.
Boston
MA, USA

Matthew P. Nelson
ChemImage Corporation
Pittsburgh
PA, USA

Eldon Puckrin
Defence R&D Canada – Valcartier Research Center
Quebec
QC, Canada

Oliver Pust
Delta Optical Thin Film A/S
Hørsholm, Denmark

Cicely Rathmell
Wasatch Photonics, Inc.
Logan
UT, USA

John A. Reffner
John Jay College of Criminal Justice
City University of New York
New York
NY, USA

Pierre-Yves Sacré
Department of Pharmacy
University of Liege
Liege, Belgium

Terry Sauer
Presco Engineering
Woodbridge
CT, USA

Alexander Scheeline
Department of Chemistry
University of Illinois at Urbana-Champaign
Urbana
IL, USA

and

SpectroClick Inc.
Savoy
IL, USA

David W. Schiering
RedWave Technology
Danbury
CT, USA

Dalton T. Snyder
Resource for Native Mass Spectrometry Guided Structural Biology
The Ohio State University
Columbus
OH, USA

John T. Stein
JTS Consultancy
Ashburn
VA, USA

Andrei Stratilatov
Moxtek, Inc.
Orem
UT, USA

Jean-Marc Thériault
Defence R&D Canada – Valcartier Research Center
Quebec
QC, Canada

Alexandre Thibeault
Bluefield Technologies Inc.
San Francisco
CA, USA

Mark Zieg
Wasatch Photonics, Inc.
Logan
UT, USA

Eric Ziemons
Department of Pharmacy
University of Liege
Liege, Belgium

Foreword

When I first learned that Richard Crocombe, Pauline Leary, and Brooke Kammrath were editing a two-volume series of books that cover the development of field-portable analytical technologies and the numerous applications of these technologies, I was excited because I knew that these scientists had the experience, knowledge, and energy to produce a great product, books that I will immediately add to my library.

So, who am I to make such a bold evaluation of these books? My name is John A. Reffner and I am currently a tenured full Professor of Forensic Science at the City University of New York's John Jay College of Criminal Justice. I have also received several distinguished awards including some for developments that enabled portable spectroscopic instruments. After graduating from Akron University in 1956, I joined the "Works Technical Analytical Laboratory" at the B.F. Goodrich Tire and Rubber Company. This experience taught me the valuable lesson that chemistry is essential for a major corporation to be successful, a lesson which was continually reinforced throughout my almost 65 years of professional experiences. I have had the good fortune of working with many prominent scientists and business leaders. I have seen how science and chemistry change the world. I have also seen how consumer demands drive technology and innovation, leading us to where we are today, immersed in essential portable technologies that have changed the world.

A short story that exemplifies my passion for the field of portable instruments is the introduction of the DuraScope at the Pittsburgh Conference in 1998, and the subsequent development of the TravelIR portable infrared spectrometer. Our SensIR Technologies team, which included the likes of Don Sting, Jim Fitzpatrick, Don Wilks, and Bob Burch, introduced this new micro-ATR (Attenuated Total Reflection) accessory for FT-IR (Fourier transform infrared) spectrometers. While it might not seem that such an accessory could make a system portable, a scientist from a major supplier of chemicals was very excited about the product. In his work, he traveled to paper companies to resolve customer complaints. While he didn't need an ATR accessory, what he did need was a small FT-IR, possibly an ATR-based infrared system, that could fit in the overhead storage compartment of a commercial airplane. As a result of these conversations, the TravelIR was born. The TravelIR was the first portable infrared spectrometer delivered to the market, enabling the identification of an infinite number of samples at the sample site.

The novelty of the TravelIR attracted a lot of interest, but portability was low on the list of requirements by a majority of end users. That was, of course, until 11 September 2001. One week after the 9/11 terrorist attacks, letters containing anthrax spores were delivered to members of the news media, and to United States Senators Tom Daschle and Patrick Leahy. A total of five people died from exposure to these spores, and seventeen others were infected. These terrorist events had a significant impact on field-portable analytical instruments. There became an immediate need to identify dangerous chemicals, including white powders, quickly and reliably at the sample site. This need catalyzed the portable-spectroscopy market.

Infrared spectrometers like the TravelIR were well-suited as a chemical identifier and could meet the analytical needs of field users, but it was clear that simply having a small footprint capable of providing reliable answers to a trained scientist was not enough. Systems for deployment needed to endure the rough handling and environmental conditions required of a valid field-portable analytical device. In addition, it was necessary for these systems to collect spectral data and translate that data into actionable results in real time by a nonscientist operator with minimal training. SensIR's follow-up to the TravelIR was a product known as the HazMatID. This system was ruggedized to meet aggressive military specification standards including those for ruggedness and total immersion in decontamination solution. The "genie was out of the bottle" and the need for field-portable instruments exploded.

As you read the 45 chapters of these books, you will see the versatility in both the instrumentation and technologies, as well as the tremendous impact these systems have upon our society. Whether considering how portable spectrometers are used in hazmat and military operations to assess safety and defense concerns, by archaeologists and other cultural historians to help understand artwork and ancient civilizations, or the value these systems offer to practitioners of the forensic, pharmaceutical, and geological sciences, the reader will appreciate the challenges to their development, the breadth of their applicability, and the irreplaceable value they afford to the end user.

November 2020

John A. Reffner
Professor of Forensic Science
at the City University of New York's
John Jay College of Criminal Justice

Preface for Volume 1

There have been tremendous developments in portable spectroscopy and spectrometry instrumentation over the last several decades resulting in their use in a variety of applications. However, portable spectroscopic instruments have not had significant visibility within the scientific community when compared with the current generation of high-performance laboratory spectrometers. This is attributable to a number of factors, including their very applied (as opposed to research) nature, which has resulted in their predominant use by non-scientists and a dearth of publications in the refereed literature. The numbers of these instruments deployed today can therefore come as a surprise. For instance, as early as the 1990s over 60 000 ion mobility spectrometer (IMS) systems were deployed worldwide; in 2019, a single model of an IMS detector had deployments from the previous 14 years of over 90 000; cumulative handheld x-ray fluorescence (XRF) shipments today total more than 100 000; and portable Raman shipments number in the tens of thousands. For these, and some other portable spectrometers, there are more instruments in the field than there are in laboratories.

One of us (RAC) published a review article in late 2018 on portable spectroscopy, covering optical and elemental (XRF and laser-induced breakdown spectroscopy (LIBS)) instruments and applications. Just covering those areas required 50 printed pages. Two of us (RAC and PEL) have also taught day-long short courses on this topic, and realized that we were barely scratching the surface of this field. It therefore became apparent that this field could benefit from comprehensive coverage in a book. The editors bring different backgrounds to this: from instrumentation and applications development (RAC), training on instrumentation and applications in safety and security, the military, and pharmaceuticals (PEL), and teaching instrumentation and forensic applications (BWK). As a result of this, our original concept expanded from one book to two volumes. Here, in Volume 1, we focus on the instrumentation itself and enabling technologies. Volume 2 focuses on the myriad of applications of portable instrumentation.

Volume 1 starts with a chapter on engineering portable spectrometers in general, and then proceeds through the electromagnetic spectrum covering XRF, ultraviolet (UV)-visible, near-infrared, mid-infrared and Raman spectroscopies. Microplasmas and LIBS areas are addressed, nuclear magnetic resonance (NMR), and also a number of portable mass spectrometry instrument types are described. Some special topics include smartphone spectroscopy, optical-filter technology and micro-electromechanical systems/micro-opto-electromechanical systems (MEMS/MOEMS) (key to many very small spectrometers and hyperspectral imagers) and stand-off detection. And as we write this, under coronavirus lockdown, the volume also includes chapters on DNA instrumentation and biological analyzers.

We hope that the combination of Volumes 1 and 2 provides thorough coverage of the whole field, with chapters written by technical experts with hands-on expertise. The authors have a range of experiences, from working in instrument companies, universities, and research institutes, to those intimately involved with the military and hazardous-material teams, drawn from North America, Europe, Asia, and Australia. These authors have made efforts for their chapters to cover the field up to early 2020, and we thank all the authors for their contributions. In addition, most of the chapters were also reviewed by third-party experts to ensure their quality and completeness, and we thank the reviewers for their helpful suggestions.

June 2020

Richard A. Crocombe
Crocombe Spectroscopic Consulting
Winchester, MA, USA

Pauline E. Leary
Federal Resources
Stevensville, MD, USA

Brooke W. Kammrath
Department of Forensic Science
Henry C. Lee College of Criminal Justice and Forensic Sciences
University of New Haven
West Haven
CT, USA

and

Henry C. Lee Institute of Forensic Science
West Haven
CT, USA

Acknowledgements

The editors would like to thank and acknowledge the following scientists and subject matter experts for their willingness to review the chapters within these volumes.

Volume 1

Katherine Bakeev
Sheldon Brunk
Michael Burka
Cecil Dybowski
Peter Griffiths
Michael Kessler
Richard Lareau
Christopher Palmer
Alexander Scheeline
Ken Schreiber
Dalton Snyder
Mazdak Taghioskoui

Volume 2

Jose Almirall
Paul Bartholomew
Amy Bauer
Lucia Burgio
Craig Gardner
Richard Jackson
Lena Kim
Larry McDermott
Ellen Miseo
John A. Reffner
Luis Rodriguez-Saona
Shiv Sharma
Brandye Smith-Goettler

1

Introduction to Portable Spectroscopy

Pauline E. Leary[1], Richard A. Crocombe[2] and Brooke W. Kammrath[3,4]

[1] *Federal Resources, Stevensville, MD, USA*
[2] *Crocombe Spectroscopic Consulting, Winchester, MA, USA*
[3] *Department of Forensic Science, Henry C. Lee College of Criminal Justice and Forensic Sciences, University of New Haven, West Haven, CT, USA*
[4] *Henry C. Lee Institute of Forensic Science, West Haven, CT, USA*

1.1 Introduction

In this book, we regard a portable spectrometer as an analytical instrument, which generates clear answers for its operator, when it is carried to the sample, i.e., spectrometer to the sample rather than sample to the spectrometer. The instrument ideally will operate in point-and-shoot mode, or at least minimize sampling handling, and the primary output is not *a spectrum* but rather *a result*. In some cases, the result might be a sample identification; in others it may be a pass/fail visual or audible alarm (green light/red light). The operators of these instruments are rarely scientists, but may instead be hazardous-material technicians, armed-services personnel, or even scrap-metal dealers. These spectrometers may have to conform to regulatory standards such as Title 21 CFR Part 11, which governs electronic records and signatures within the pharmaceutical industry in the United States, or legal standards for the admissibility of scientific evidence, such as Daubert and Frye [1, 2], which govern the admissibility of scientific evidence in all United States Courts. The environments within which they may be used can be life threatening, such as when dealing with improvised explosive devices (IEDs), to analyze suspicious white powders, during kinetic military action, or chemical spills. When performing analytical testing in these situations, reliable, rapid results that are easy for the operator to interpret, understand, and act on are critical.

1.2 Defining Portable Spectrometers

The definition of a portable spectrometer can be quite variable. In some instances, portable simply means the instrument can be moved to the sample location, even if the instrument is large and cumbersome; in some cases even a laboratory-based system that is built into a fixed mobile laboratory is considered portable. In other instances, and what seems to be more the expectation from today's users, a portable spectrometer should be small, lightweight, and capable of running on battery power for a reasonable duration of time. These instruments are intended for use in the field and, therefore, must perform well outside the climate-controlled environment of a laboratory. They must operate and maintain their calibration over wide ranges of temperature and humidity. They are usually dust and water resistant, if not waterproof, and capable of withstanding physical shocks, for instance, being dropped onto a hard surface from a few feet off the ground. If used for the detection and identification of hazardous materials, they must be able to withstand decontamination procedures, which can involve being scrubbed down with a solution of bleach.

Portable Spectroscopy and Spectrometry 1: Technologies and Instrumentation, First Edition.
Edited by Richard A. Crocombe, Pauline E. Leary and Brooke W. Kammrath.

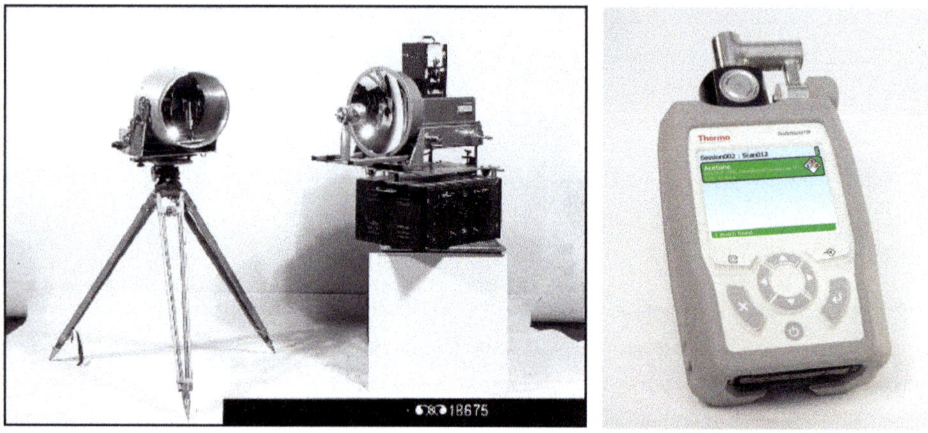

Figure 1.1 One of the earliest "portable" LOPAIR spectrometers, left (*Source*: Image courtesy of the United States Army), alongside a modern portable ATR FTIR (right). *Source*: Thermo Fisher Scientific.

Frequently, the first iterations of a portable spectrometer are not truly portable from a size, weight, and power (SWaP) perspective, but rather are considered portable because the spectrometer is taken to the sample. Over time, though, as the technology evolves, a greater degree of SWaP friendliness is usually achieved. As an example, Figure 1.1 shows one of the earliest (if not *the* earliest) "portable" infrared (IR) spectrometers, which was based upon long-path IR (LOPAIR) technology, alongside a modern portable IR attenuated total reflection (ATR) spectrometer. In 1954, the United States military began development of this LOPAIR system for the detection of chemical warfare agents (CWAs). The principle behind this device was that the G-series nerve agents absorb radiation in certain portions of the IR spectrum. Such a device would scan the atmosphere continuously in advance of troops and give a warning alarm when G-series agents were detected. The prototype model, designated the E33 Area Scanning G-Agent Alarm, performed satisfactorily up to about 300 yards, but weighed over 250 lbs and used too much electrical power [3]. In contrast, a large percentage of today's portable IR systems are Fourier transform IR (FT–IR) systems that weigh only a few pounds and operate for hours on an internal battery with no external computer or sampling accessories required. It is important to note that modern portable FT–IR instruments, using an ATR sample interface, are typically deployed to perform single-measurement analysis and not scan environments continuously as was the intention of the original E33 LOPAIR system. There are some modern portable FT–IR systems that scan continuously to monitor environments (see the Chapter 12 in this volume by Lavoie et al.), but these generally employ transmission sampling, rather than ATR, and tend to be larger and heavier than portable FT–IR ATR spectrometers. They also require significantly more computer processing power since they scan and provide results continuously, performing identifications in seconds, and running 24/7/365. Examples include the Gasmet Technologies Oy DX4000, which is used to monitor stack emissions and can monitor up to 50 gases simultaneously [4, 5], as well as the MKS Instruments, Inc. AIRGARD®, which is used to monitor environments for CWAs and toxic industrial chemicals (TICs) [6, 7]. Although individual chapters in this book may use slightly different definitions, the authors of this chapter have provided a general classification scheme of portable spectrometers in Table 1.1 based upon SWaP expectations.

1.3 Performance

When judged against its laboratory counterpart, a portable spectrometer, evaluated strictly on typical laboratory specifications for analytical performance, usually will not perform as well. Laboratory-based systems attempt to

Table 1.1 General classification of portable spectrometers.

Class	Size	Weight	Power	Examples
Vehicle mounted and transported				Military application (see chapter by Samuels [Volume 2])
Vehicle transportable	Much greater than 40 × 35 × 35 cm (16 × 14 × 14 in.)	>20 kg >44 lbs	May require shore power with no battery option and external computing accessories	First-generation portables; installs in fixed mobile laboratories
Person portable	<~40 × 35 × 35 cm (<~16 × 14 × 14 in.)	~3–20 kg ~7–44 lbs	May run for a limited duration on battery with hot-swappable battery replacement option	Portable GC–MS systems; IMS systems used in aviation security for passenger screening
Handheld	<~30 × 15 × 7 cm (<~12 × 6 × 3 in.)	~0.5–3 kg ~1–7 lbs	Self-contained, battery-powered, embedded data system	Portable XRF, FTIR, Raman, and LIBS
Wearable	<~10 × 20 × 5 cm (<~5 × 10 × 2 in.)	<~0.5 kg <~1 lb	Some visible region instruments	IMS-based CWA-monitoring devices; some visible-region instruments
Smartphone-based	<~4 × 4 × 4 cm (<~1.5 × 1.5 × 1.5 in.)	<~0.5 kg <~1 lb	Visible region only, using the smartphone camera as the sensor	See chapters by Scheeline (Volume 1), and by Peveler and Algar (Volume 2)

Sizes and weights are approximate, and some instruments in each class may be smaller and lighter.

achieve optimal performance in terms of criteria such as resolution, signal throughput, and signal-to-noise ratio (SNR). They frequently have great flexibility in their data acquisition parameters, in their sample introduction or sampling accessories, and even in the spectral range over which they operate. Portable systems tend to have poorer performance in terms of these types of criteria. The spectral range and resolution of portable instruments is usually either fixed, or predetermined by the manufacturer's hardware and engineering choices, and in many cases the measurement time is automatically determined as the spectral data are collected, ensuring that the spectrum has a sufficient SNR for the type of analysis, but not necessarily an optimal SNR. Frequently, only one sampling method is provided, with an emphasis toward point-and-shoot capability over versatility in sampling approach.

While all of this discussion of performance implies that portable instruments are inferior to their laboratory-based counterparts, raw spectroscopic performance is only part of the rationale for using portable spectrometers and, therefore, only part of the criteria that should be used to establish performance of a portable spectrometer. The fit-for-purpose operational requirements, including deployment potential, that are critical for a field instrument must also be considered. An instrument's deployment potential is impacted by items such as required accessories, SWaP, software interface, training requirements, ability to be used when the operator is encumbered by required personal protective equipment (PPE), and the ability of the operator to achieve results (not data) in the field. If these factors are not appropriate for use in the intended environment, the system would offer no value in the field.

With regard to deployment potential, requirements for storage and use in extreme environmental conditions are normal. It is common for portable spectrometers to rate themselves against aggressive standards such as MIL-STD-810, the Test Method Standard for Environmental Engineering Considerations and Laboratory Tests

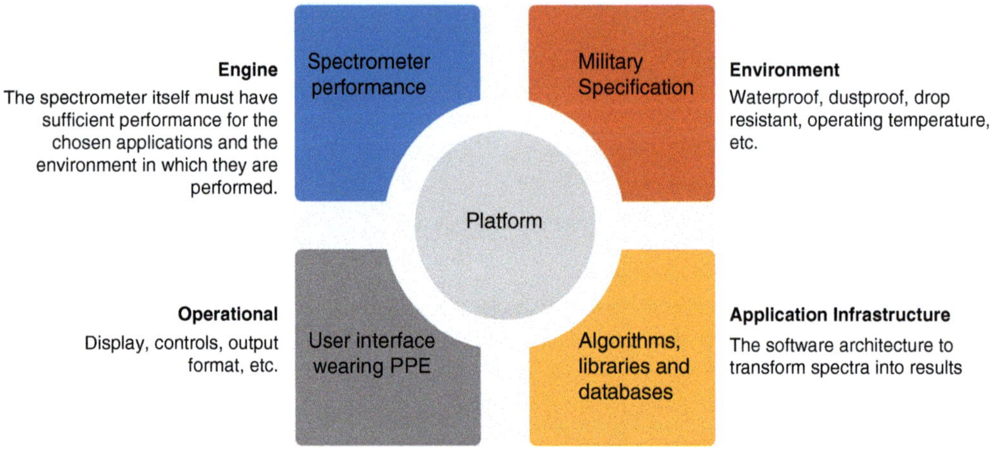

Figure 1.2 Main components of a successful portable spectroscopy platform.

of the United States Department of Defense. This Standard contains materiel acquisition program planning and engineering direction for considering the influences that environmental stresses have on materiel throughout all phases of its service life [8]. The ability of a system to function under variable environmental conditions including temperature, humidity, and altitude as well as endure and remain functional after exposure to drops and vibrations is evaluated. Even in situations where portable systems may not be intended for use by the military, comparison with this standard is sometimes made and referenced, so that an evaluation of overall ruggedness and deployment potential of a system can be easily assessed. When both analytical performance and operational requirements are considered, the true value of a portable spectrometer can be effectively assessed for a specific application. Figure 1.2 summarizes the main components for consideration during the design and development of a portable spectroscopy platform.

The ability of a portable system to meet these reconciled goals between analytical and operational performance is variable, and frequently dependent upon the technology. For instance, portable IR and Raman instruments generally achieve many of them, and instruments of this type that offer good analytical performance, and are easy to deploy, have been available from a number of vendors for the past 15 years. Portable gas chromatography–mass spectrometry (GC–MS) systems, on the other hand, frequently perform well from an analytical perspective, but offer deployment challenges due to SWaP, their need to perform under vacuum, use of compressed gases, and the complexity of data generated which is not always easily converted to results. These requirements add deployment challenges. The smallest portable GC–MS that is currently commercially available is significantly larger and heavier than even a combined portable IR/Raman system.

1.4 History and Availability

For many of the spectroscopic techniques, the availability of portable spectrometers was driven by safety, security, terrorism, and military concerns. The major exceptions to this are handheld X-ray fluorescence (XRF), laser-induced breakdown spectroscopy (LIBS), and near-infrared (NIR) spectroscopy systems, where field instrumentation has been driven almost exclusively by commercial reasons. Figure 1.3 shows a time line of critical events that have transpired over the past century and have contributed either critically or peripherally to the need for and subsequent development of portable spectrometers. Figure 1.4 details the initial portable-instrument deployments based upon technology for systems capable of being hand carried to the scene and operating at the sample site using battery power.

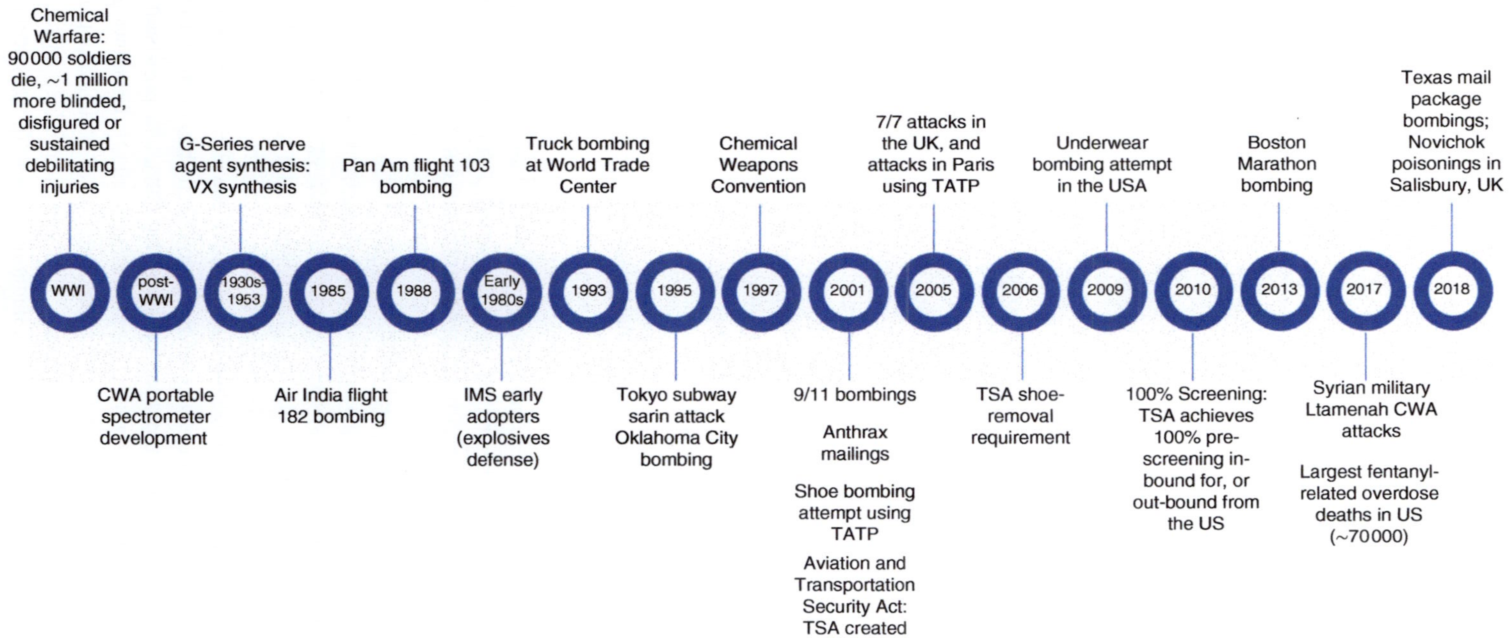

Figure 1.3 Time line of events of importance to the development of portable spectrometers.

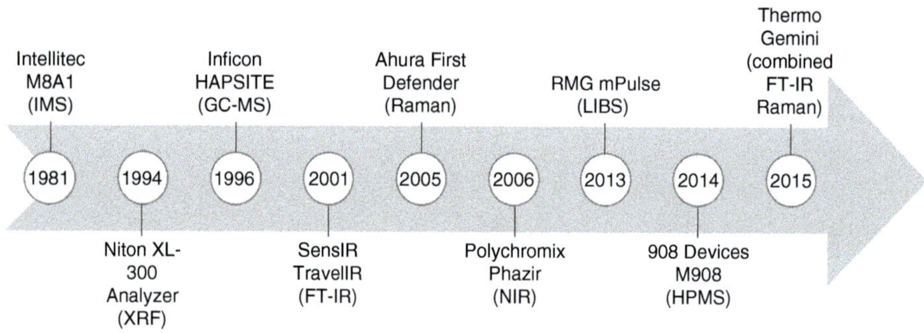

Figure 1.4 Initial technology deployments for portable spectrometers capable of being carried to the scene and operated on battery.

While the consequences of 9/11 have certainly been a driving force for innovation in technology, and were instrumental in the development of many modern portable platforms including IR, Raman, and high-pressure mass spectrometry (HPMS) systems, the use of portable instruments at the sample site, and the value this type of analysis provides, began much further back in our history. Jeffery Smart, Command Historian at Aberdeen Proving Ground, in his summary entitled "History of Chemical and Biological Detectors, Alarms, and Warning Systems" reviewed military-related chemical detectors used throughout the twentieth century [3]. The earliest detectors he describes are vapor field detectors used to replace the "sniff test" used during World War I to detect and identify dangerous chemicals. There, troops learned that German mustard agents indeed smelled like mustard, while the allied mustard agent smelled like garlic. These observations were used by trained gas scouts, positioned to provide advance warning to the main trench line of an incoming gas cloud, but had serious negative consequences for those with continued exposure to dangerous chemicals at low concentrations, and much more significant consequences when concentrations were high. Vapor field detectors did not require troops to remove their protective masks to identify these dangerous substances. Some of the vapor field detectors that served as sniff-test replacements included instruments based upon a copper flame test lantern, a selenious acid field detector, an iodine pentoxide test, an iodic acid test field detector, and a hydrogen sulfide field detector [3].

The military were not the only early users of portable spectrometers. Kammrath et al. in their chapter on the applications of portable spectrometers to the field of forensic science (Volume 2) discuss the use of a portable spectrometer called a Drunkometer in the 1930s to measure breath-alcohol levels. This system was developed at Indiana University and relied on the colorimetric reduction of potassium permanganate by ethyl alcohol. These authors report the most recent developments in breath-alcohol instruments combine electrochemical sensors and IR spectrometers in a single platform.

With regard to portable GC–MS, the first truly portable GC–MS system was introduced in 1996 for the analysis of hazardous air pollutants at the sample site [9]. Earlier attempts to take the GC–MS instrument to the sample, which happened to be on the planet Mars, were successful when GC–MS was deployed on the Viking voyage in 1976 [10], but the GC–MS technology used was not as much *portable* as it was *transportable*; the system was designed specifically for the mission and was, essentially, part of a fixed mobile laboratory [11].

Some of the first ion mobility spectrometers (IMSs) were deployed to the military in the early 1980s for detection of CWAs [3]. The United States Federal Bureau of Investigation (FBI) has deployed IMS systems, at least since 1991, for use in a wide variety of drug-interdiction missions, as well as for post-blast forensic analysis in the World Trade Center, Oklahoma City, and Atlanta Olympics incidents [12]. These systems were ultimately deployed extensively within the aviation security market for the detection of explosives (see Chapter 8 by Leary and Joshi in Volume 2). Military interest in the ability to detect CWAs in theater is long-standing, but interest in the detection of explosives within the aviation security market is more recent. In 1986, the United States Bureau of Alcohol, Tobacco and Firearms (ATF) reported that between the years 1977 and 1986, of the nearly 12 000 bombings reported, only about 100 considered either military bases or airports as the primary target [13]. Since this time, there has been significant interest in explosives detection within the aviation security market, which, as Leary and Joshi describe in their chapter on the applications of IMS, was critically impacted by terrorist bombings of Air India Flight 182 in 1982, Pan Am flight 103 in 1988, and American Airlines flights 11 and 77 as well as United Airlines flights 175 and 73 on 11 September 2001. The three days in history on which these acts of terrorism transpired mark the deadliest terror attacks in the histories of Canada, the United Kingdom, and the United States, respectively, and were integral to the development of particle explosives detection systems using IMS. During the late 1980s, the Canadian government funded development of a high-speed IMS-based explosive particle detection system, which formed the basis of the Barringer Instruments, Inc. (now Smiths Detection) IONSCAN 150 and subsequent Barringer IMS platforms. Prior to this, the focus for explosives detection capabilities using IMS had been on detecting volatile explosives such as nitroglycerin (NG), dinitrotoluene (DNT), and ethylene glycol dinitrate (EGDN) in the vapor phase. This vapor-phase approach did not allow for the detection of low-volatility explosives such as trinitrotoluene (TNT), pentaerythritol tetranitrate (PETN), and Research Department Explosive (RDX) [14]. In 1996 the Federal Aviation Administration (FAA) of the United States Department of Transportation established the qualification process for checkpoint screening [15]. In November 2001, the Aviation and Transportation Security Act was signed into law in the United States, and the Transportation Security Administration (TSA) was created to oversee security in all modes of transportation [16]. IMS has been an integral part of the aviation security market for detection of particulate residues of explosives since its initial introduction to it in the early 1990s.

As noted above, handheld XRF and LIBS developments have been commercially driven, primarily by the metals markets, from field geology and mining, through fabrication and inspection, to recycling. The history of portable XRF analyzers was summarized in a 2012 symposium [17] and started with an isotope-source lead paint analyzer in the early 1990s, followed by alloy analyzers in the mid-1990s, and later converting almost exclusively to instruments using miniature X-ray tube sources. Key enabling technologies specific to XRF were the development of thermoelectrically cooled silicon PIN-diode (p-type–intrinsic–n-type) detectors, [18] and miniature X-ray tubes [19, 20]. Handheld LIBS instruments are a much more recent development, and have been applied in essentially the same areas as handheld XRF with each having its advantages and limitations (e.g., LIBS is capable of lighter element analysis however is minimally destructive).

Although NIR spectroscopy is widely used in the food, feed, and agriculture sectors, and also in pharmaceutical manufacturing, the adoption of portable instruments has been much slower for NIR as compared with Raman instruments. This may be because there has not been a single "killer app" – there are a multitude of individual applications, each requiring its own validated database or calibration. This is due to the somewhat nonspecific nature of NIR spectra, especially at shorter wavelengths, in that there are no bands that are highly characteristic of individual compounds.

1.5 Instrument Design and Enabling Technologies

The design and manufacture of portable systems is not a simple task. When an outsider first peers into the world of portable spectrometers, they frequently consider them to be simplified versions of laboratory systems. This impression is misguided, though, because from their theoretical inception to their manifestation in the final product, these instruments are fundamentally different from their equivalent laboratory-based counterparts, in both hardware and software. Historically, and even in laboratories today, almost naturally, a spectroscopist's concern is optimizing analytical performance. What can be done to get the most signal with the least noise, improve the resolution, and offer the most versatile sampling options? When developing portable technologies, though, this approach must be abandoned. The question that needs to be answered is not, "What product offers the best performance at an affordable price?", but rather, "What product or technology successfully integrates sufficient certainty

Table 1.2 Enabling technologies for portable spectrometry.

All portables	Optical	XRF	MS and GC-MS
Mobile computing power: processors, memory	Optical telecom	Miniature X-ray sources	Miniature ion traps
Consumer electronics: user interfaces	Diode lasers	X-ray detectors (PIN, silicon drift detectors [SDD])	Pump technology
Communications: Wi-Fi, Bluetooth	Fiber optics	Pulse-counting electronics	Narrow-bore GC columns
MEMS and MOEMS technologies	Cameras, sensors, and array detectors	Fundamental parameter theory and calculations	MS sampling and sample introduction
Battery power and power management	Compact spectrographs		Ionization methods
Chemometrics and algorithms	Holographic filters (Raman)		
"The cloud"	ATR technology (IR)		

of results while meeting operational requirements, especially deployment potential, for the task at hand?" The answers to the different facets of this question are frequently at odds with each other, but a portable platform will thrive when the right balance has been struck among them. For this reason, it is critical that an instrument developer understands not only the technology, but also the application, the end user, and the deployment environment.

Design of a system is fundamentally limited by the availability of enabling components and technologies. All portable spectrometers have benefitted from the developments in computing power and consumer electronics, but there are other advances that have also been integral to the development of these systems. A short list of these includes low-power-consumption processors, low-cost/high-capacity memory, compact displays, mobile user interfaces, power management, battery capacity, Wi-Fi, Bluetooth, and cameras. Table 1.2 details some of the enabling technologies that have been important to the development of portable platforms.

1.6 Producing Results

While some portable spectrometers may be employed primarily to collect spectra for further analysis (e.g., see Chapters 21 by Pozzi et al. and 22 by Donais and Vandenabeele in Volume 2), the majority are designed to produce actionable results at the time of analysis. To generate these actionable answers, the developers of these instruments have put extensive resources into reliable identification algorithms, spectroscopic libraries or databases, and qualitative and quantitative calibrations. Real-life samples are most often mixtures, and this has been addressed with a combination of calibration, library, and algorithm development. These issues are addressed in the Chapters 2 and 3 by Zhang and Schreyer in Volume 2. This work has been done by the established large instrument companies, but there are new companies marketing devices in the visible-to-1000 nm region directly to the public, making wide claims about their analytical capabilities, and potentially relying on crowd-sourced data. This is a cause for concern in the scientific community and is explored in Chapter 23 by Crocombe in Volume 2. Figure 1.5 shows the primary contributing factors that enable a successful portable spectroscopy product.

Applications are critical for the success of portable spectrometers. An overview of the role applications development plays in the development and deployment of portable instruments is presented in the "Preface" for Volume 2.

Figure 1.5 Criteria for a successful portable spectroscopy product.

1.7 Outline of These Volumes

This book, in two volumes, seeks to cover today's portable spectroscopy instrumentation and markets. We have also included chapters on biological analyzers and DNA instrumentation, which, while not strictly spectroscopic in nature, complement the other technologies, and are of particular interest as this book is compiled in 2020, in the midst of the COVID-19 pandemic. Due to the size of this project, a small number of portable spectroscopic instruments is not covered, one of which is portable radiation detectors.

It seemed the best way to begin these volumes was with a discussion of the role engineering plays in the development of portable systems. A spectrometer is truly a system, including elements designed by mechanical, electrical, optical, and software engineers. To fit all of this together into a small, light, rugged, simple-to-operate instrument requires a conscious effort at systems engineering. In an optimal design, each of these elements will impact the others, as discussed by Sauer in Chapter 2 of this Volume. For instance, in an integrated handheld instrument, issues like noise pickup, power management, and heat dissipation are all greatly magnified as compared with laboratory instruments. In addition, the use of PPE limits the ability to use touch screens and small physical controls. This chapter describes these details and considers approaches to options that address these challenges. This discussion is then followed by in-depth reviews of both the established and emerging portable technologies, as well as some of the critical components that continue to enable them. Whether it be optical methods like IR and Raman that underwent a revolution after 9/11, or methods including ion mobility, MS, GC–MS, and HPMS that are capable of detecting and quickly identifying invisible threats in dangerous environments, these chapters provide in-depth coverage of these systems, components, and the critical applications that have driven their success. Volume 1 ends with discussions of both DNA (deoxyribonucleic acid) and biological technologies; these analyzers and their use in industries such as defense, medicine, food and agriculture, as well as the environmental and forensic sciences, are considered.

Volume 2 demonstrates why portable spectrometers have become so successful. Due to an inherent need for reliable detection and identification at the sample site, this volume begins with a discussion by these editors regarding the impact of applications on the development of portable spectrometers. This is followed by chapters describing algorithm and library development for portable systems; both are critical to the success of a portable platform and the applications that drive them. Portable systems must provide answers without requiring complex data interpretation or assessment on the part of the operator. As such, a significant amount of energy is spent developing proprietary search algorithms with the goal of requiring no input from the user to successfully identify the sample components, even at low concentrations. Algorithms designed to identify individual components in mixture samples are becoming more common and are described during this discussion. Electronic libraries for spectral matching are also critical. Instrument vendors will frequently reference the size of their onboard libraries, but it is

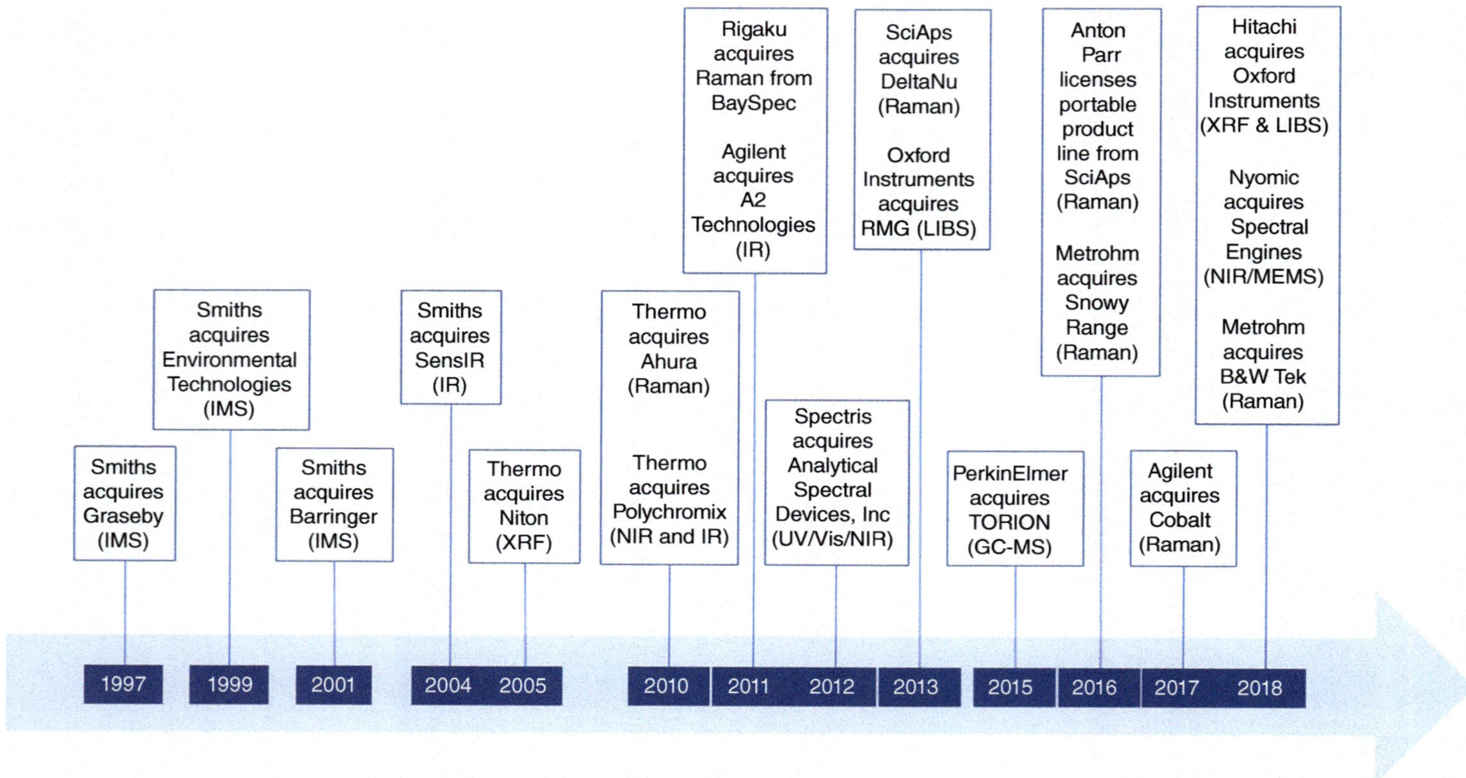

Figure 1.6 Acquisitions/consolidations in the portable spectroscopy industry.

clear that having the right library of reference data can be more important than having a large number of library entries. Target libraries focused on specific applications are important, and help improve automatic-identification accuracy.

Some of the major applications that have relied heavily upon portable spectroscopy are then reviewed from the perspective of the end user. These chapters are presented by expert users that not only understand the technologies, but the advantages and limitations when they are applied in the real world. There is an impressive diversity in these applications. When warranted based upon the extensive breadth of applications, some chapters cover a variety of applications for a specific technology; others approach the chapter from an industry or application field perspective, and demonstrate how different portable technologies are used within a specific industry. The reader will not only be able to appreciate how critical these portable systems have become within these industries, but also understand how they are successfully deployed. We end with a look to the future and how things will likely continue to evolve as we move ahead in a world of ever-changing needs. It is interesting to note that the growth in portable spectroscopy sales has spurred a wave of acquisition and consolidation in the industry (Figure 1.6). The first wave of acquisitions, between 1997 and 2010, appears to have been driven by a desire to access the safety, security, and military markets, and the second, from 2010 until now, to access the growing commercial/civilian markets. It will be interesting to see how this evolves in the coming years.

Acronyms and Abbreviations

ATF	Bureau of Alcohol, Tobacco and Firearms
ATR	Attenuated total reflection
CFR	Code of federal regulation
COVID-19	Coronavirus disease of 2019
CWA	Chemical warfare agent
DNA	Deoxyribonucleic acid
DNT	Dinitrotoluene
EGDN	Ethylene glycol dinitrate
FAA	Federal Aviation Administration
FBI	Federal Bureau of Investigation
FT–IR	Fourier transform infrared
GC–MS	Gas chromatography–mass spectrometry
HPMS	High-pressure mass spectrometry
IED	Improvised explosive device
IMS	Ion mobility spectrometry
IR	Infrared
kg	kilogram
lb	Pound
LIBS	Laser-induced breakdown spectroscopy
LOPAIR	Long-path infrared
MEMS	Microelectromechanical systems
MIL-STD-810	Test Method Standard for Environmental Engineering Considerations and Laboratory Tests of the United States Department of Defense
MOEMS	Micro-opto-electromechanical systems
MS	Mass spectrometry
NG	Nitroglycerin
NIR	Near infrared

nm	Nanometer
PETN	Pentaerythritol tetranitrate
PIN	p-type–intrinsic–n-type
PPE	Personal protective equipment
TIC	Toxic industrial chemical
TNT	Trinitrotoluene
TSA	Transportation Security Administration
SDD	Silicon drift detector
SNR	Signal-to-noise ratio
SWaP	Size, weight, and power
XRF	X-ray fluorescence

References

1 D.C. Circuit Court (1923). *Frye v. United States*. 293 F. 1013.

2 United States Supreme Court (1993). *Daubert v. Merrell Dow Pharmaceuticals, Inc.* 509 U.S. 579.

3 Smart, J. (2009). *History of U.S. Army Research & Development, Chemical and Biological Detectors, Alarms, and Warning Systems*. Aberdeen Proving Ground: U.S. Army Research, Development and Engineering Command.

4 Gasmet Technologies Oy. *Gasmet Portable Gas Analyzers, DX4000*. Gasmet [Online]. https://www.gasmet.com/products/category/portable-gas-analyzers/dx4000 (accessed 24 May 2020).

5 Gasmet Technologies Oy. *Gasmet Portable FTIR Gas Analyzer, DX4000*. Gasmet [Online]. https://www.gasmet.com/wp-content/uploads/2018/01/Product-Prochure-DX4000-2020-web.pdf (accessed 24 May 2020).

6 MKS Instrumnents, Inc (2010). *Application Note, CWA/TIC Detection (App Note 1/10-8/13)*. MKS Instruments [Online]. https://www.mksinst.com/mam/celum/celum_assets/resources/AIRGARD-CWA-TICDetection-AppNote.pdf?1 (accessed 16 September 2020).

7 MKS Instruments, Inc *Ambient FTIR Air Analyzer AIRGARD*. MKS Instruments [Online]. https://www.mksinst.com/p/AIRGARD-CWA (accessed 16 September 2020).

8 United States Department of Defense (2019). *Department of Defense Test Method Standard Environmental Engineering Considerations and Laboratory Tests*. United States Department of Defense.

9 Crume, C. (2009). *History of Inficon HAPSITE: Maintenance Management, Support, and Repair*. KD Aanlytical [Online]. http://www.kdanalytical.com/instruments/inficon-hapsite-history.aspx (accessed 16 September 2020).

10 Biemann, K., Oro, J., Toulmin, P. III, et al. (1977). The search for organic substances and inorganic volatile compounds in the surface of Mars. *Journal of Geophysical Research* 82: 4641–4658.

11 Leary, P.E., Kammrath, B.A., and Reffner, J.A. (2018). *Field-Portable Gas Chromatography-Mass Spectrometry*. Wiley Encyclopedia of Analytical Chemistry.

12 Wood, K. (1998). Statement Submitted to the House Committee on Transportation Subcommittee on Public Buildings and Economic Development Relative to Technological Solutions to Federal Building Security, June 4.

13 Nyden, M.R. (1990). *A Technical Assessment of Portable Explosives Vapor Detection Devices*. United States Department of Justice NIJ Report 300-89.

14 DeBono, R. (2020). Personal Communication, May 25.

15 Kraus, T.L. (2008). *The Federal Aviation Administration: A Historical Perspective, 1903–2008*. Washington, DC: United States Department of Transportation.

16 United States Transportation Security Administration. *Transportation Security Timeline*. Transportation Security Administration [Online]. https://www.tsa.gov/timeline (accessed 30 March 2020).

17 Bosco, G.L. (2013). Development and application of portable, hand-held X-ray fluorescence spectrometers. *TrAC Trends in Analytical Chemsitry* 45: 121–134.

18 Pantazis, T., Pantazis, J., Huber, A., and Redus, R. (2010). The historical development of the thermoelectrically cooled X-ray detector and its impact on the portable and hand-held XRF industries. *X-Ray Spectrometry* 39: 90–97.

19 Cornaby, S. and Kozaczek, K. (2016). X-ray sources for handheld X-ray fluorescence instruments. In: *Encyclopedia of Analytical Chemistry*. Wiley.

20 Crocombe, R.A. (2018). Portable spectroscopy. *Applied Spectroscopy* 72: 1701–1751.

2

Engineering Portable Instruments

Terry Sauer

Presco Engineering, Woodbridge, CT, USA

In the past two decades, spectrometers have moved out of the laboratory and into a wide variety of portable applications. Designing portable spectrometers requires a change in the mindset for product managers and engineers alike. Traditional spectrometer systems are judged on performance and on flexibility. They are designed to support a broad range of applications typically without a dedicated purpose in mind. This makes it difficult to specify how much performance is adequate, or "good enough," so the designers are tasked with making performance as good as possible, within the constraints of budget and size. Portables, by contrast, are typically designed for a fixed task, or, at most, a limited set of purposes, such as unknown material identification or quality control. The metric of instrument quality is likely to be something completely different from signal to noise or stability over time. The end user is likely not to be a spectroscopist, and cares only about whether the instrument works and gives the right answer. Users do not care about whether a portable Fourier transform infrared (FTIR) system supports four different methods of apodization; they want a simple user interface and consistent performance.

While the user desires a system that is simple to use, there are many features that require significant thought and design effort for such a system to be successful. Obvious concerns are size, weight, and power. We will explore these and many other considerations which are important for portable spectroscopy systems. The focus of this chapter will be on engineering aspects for FTIR and Raman systems. The chapter will address a range of aspects which are all necessary for the design and manufacture of a successful spectrometer. A spectrometer is truly a system, including elements designed by mechanical, electrical, optical, and software engineers. To fit all this together into a small, light, rugged, simple-to-operate instrument requires a conscious effort at systems engineering. In an optimal design, each of these elements will impact the others.

2.1 Size/Weight

For a system to be easily portable, it must be relatively small and lightweight. As with many electronic devices, systems continue to get ever smaller and faster. In 2001, SensIR Technologies won an R&D 100 award for the TravelIR (The Free Library 2019). This "portable" system was an FTIR instrument which by itself weighed 26 pounds. The instrument required a separate power supply and laptop computer. To operate it away from a wall socket required a "Portawattz" portable battery and inverter. The entire system with computer and Portawattz weighed over 50 pounds. By contrast, newer handheld FTIR instruments (Smiths' HazMatID Elite and ThermoFischer Scientific's TruDefender FTX) weigh in the range of 3–5 pounds complete with built-in computer and lithium-ion rechargeable battery. Handheld Raman systems can be even smaller, with the Metrohm Raman Mira DS weighing in at 1.5 pounds.

Portable Spectroscopy and Spectrometry 1: Technologies and Instrumentation, First Edition.
Edited by Richard A. Crocombe, Pauline E. Leary and Brooke W. Kammrath.

2.2 Sample Interface

Laboratory spectrometers have a wide variety of available accessories to support sampling of liquids, solids, and powders. A typical spectrometer has a sample compartment with a defined beam path. Accessories are mounted into this compartment and can be used to make a variety of measurements including transmission, diffuse or specular reflectance, grazing angle or ATR (attenuated total reflection). Samples may need to be prepared in particular ways in order to be measured.

Portable instruments will likely be used outside of a laboratory, without access to sample preparation equipment. The ability to analyze neat samples, or to provide specialized simple sample prep, is key to being useful in the field.

Many portable spectrometers provide a single, permanently mounted sample interface. For FTIR systems, a diamond ATR element is a popular choice as it is extremely durable and versatile. Raman systems typically have a single fixed-focus lens. There are exceptions to both of these generalizations. The Agilent 4300 portable FTIR instrument has clip-on front ends to provide diffuse reflectance, grazing angle, external reflectance options along with both diamond and germanium ATR elements. The Metrohm Raman Mira system has several different focal length lenses, calibration standards, and specialty sampling accessories which attach to the output aperture of the system.

External sampling accessories enable a reduced overall size for the instrument. There is no need for a compartment large enough to contain all sizes of accessory. Handheld instruments favor accessories which require no alignment after installation.

2.3 Embedded Computer vs. External Personal Computer (PC)

Laboratory spectrometers have traditionally been coupled to external computers. In the early eighties, these were proprietary devices built by the instrument makers. Nicolet Instruments developed a dedicated 20-bit computer, the Nicolet 1080, to allow higher precision arithmetic than the 16-bit computers prevalent at the time. The computer filled a 19-in. rack and had a small monochrome oscilloscope-type display.

In the mid-1980s, the IBM personal computer (PC) and its clones took over the laboratory. Instruments connected to the PC using proprietary interface cards. The PC enabled more standardized software, migrating over time from Microsoft disk operating system (MS-DOS) to Microsoft Windows and providing increasingly complex display and analysis capabilities.

TravelIR continued this path, connecting to a standard Windows laptop using a parallel interface provided by a third-party data acquisition company.

Despite the success of the TravelIR, separate PCs were not ideal for portable instruments, and designers started to integrate computers into the instruments themselves. HazMatID included a PC-on-a-card. It was an industrial computer in the PC-104 form factor, about 4 in. by 4 in. The portable device operated with Windows XP Embedded, and included a 6.4-in. touchscreen as a user interface.

Even this embedded computer proved too large for a true handheld instrument, and subsequent generations of the instrument moved to personal digital assistants, or PDAs, as the control and interface device. These devices often operated using versions of consumer-operating systems, such as Microsoft Windows CE.

The next step in size reduction required the design of specialized PC boards for embedded control. The embedded processors in these systems are designed for low power and to support advanced power management enabling extended battery life. Some instruments incorporated credit card-sized or smaller "systems on module" or SOMs. These SOMs are computing engines which include the tightly coupled elements required for a computer to operate. The main central processing unit (CPU), random access memory (RAM), permanent storage, and display drivers are all laid out efficiently. These high-speed signals can be tricky to arrange properly, as they may require low capacitance, controlled impedance, and path length matching for parallel signals. An instrument designer can purchase all of this demanding work in a predesigned module, reducing time and cost of getting to market.

The SOM has an additional advantage when it comes to creating the software to operate the instrument. The SOM manufacturer will generally provide a BSP, or Board Support Package. This pre-built software image will support the processor and set of peripherals on the SOM. BSPs can support a number of operating systems, including Linux and embedded Windows. Typical BSPs include features, such as display drivers for the integrated video electronics, touchscreen or keyboard support, a boot loader to start the board from power on, and management tools for volatile and nonvolatile memory.

SOMs are often designed to support easily connecting them to a custom-printed circuit board which has the dedicated components required to operate the spectrometer. This connection may take the form of mezzanine connectors which allow boards to stack in layers. SOMs often connect with a single inline memory module (SIMM) socket, which is familiar as the connector used to install RAM into personal computers.

For the ultimate in size reductions, designers place processors and support chips directly on a PC board which drives the instrument. This approach provides the most control over shape of the board as well as removing the need for connectors between a SOM and a carrier board. Removing these connectors reduces system volume and can improve system reliability.

The trade-offs among these approaches allow the designer to optimize for instrument size, development cost, reuse of existing software, and instrument bill of material cost.

2.4 Reduced Feature Set

Laboratory instruments are like Swiss Army knives. Their owners expect them to have a wide variety of uses. As capital equipment in an applications lab, they must justify their cost by being able to address a range of analyses. Bench space in labs is inevitably at a premium, so each instrument must provide flexibility. This can take the form of a sample compartment supporting many accessories, an external port for larger accessories, and flexible software with many choices to optimize performance for each desired analysis.

By contrast, portable instruments are typically targeted at a single problem or a small set of problems. Many of the instruments sold to Emergency Responders are designed simply to identify unknown materials. The operators of these identifiers may be under a great deal of stress as they work to find out whether the suspect material is a threat to life or health. It is critical for these operators that the instrument be simple to operate, with no unnecessary settings which could confuse the operator or be accidentally left in the wrong position, potentially impacting the results.

Laboratory instruments are often sold based on their broad feature set, with the thought that each additional feature or setting may be the one which causes a potential customer to select this instrument over a competitor. Portable instrument designers must adapt a new mindset which values simplicity and reliability over complexity and adaptability.

2.5 Target of Non-Spectroscopist

Users of benchtop instruments are often spectroscopists. In routine analysis, a technician may operate the system on a daily basis, but the analytical methods are generally developed and maintained by a supervisor with an understanding of spectroscopy. By contrast, nonscientists operate many handheld systems. This further reinforces the requirement that the software in particular be simple and intuitive to operate, including all maintenance and administrative functions as well as the execution of sample measurement and analysis. Wherever possible, software engineers should consider preselecting values or lumping them into groups. For example, instead of asking a method developer to enter a numeric value for the number of scans to collect, provide "sensitivity" options of low, medium and high, which the system will translate into appropriate values for the particular spectrometer design.

2.6 Power Budget

Portable systems are often expected to operate on a rechargeable battery. Battery technology continues to evolve, but it is almost always a limiting factor in the operation of a spectrometer. There are several concerns related to battery power as contrasted to an alternating current (AC) supply. The user is most concerned with operating lifetime as measured in hours. The designer must also plan for a limited amount of current, a safe shutdown when the battery charge runs out, and the ability to operate consistently over a range of voltages over the charge cycle of the battery.

Establishing a power budget for the system is important before selecting a battery. Every electronic component will consume power. The first step in the power budget is to add up the current used by each consumer. This may be more complicated than reading values from a data sheet, particularly with microprocessors whose current requirements may vary significantly depending on what they are doing. As an example, an NXP i.MX6ULL (NXP 2019) lists current draw ranging from 500 µA to 750 mA. The processor has four different power modes to allow the system to conserve energy whenever possible. Several microprocessor manufacturers provide tools to estimate power draw for their CPUs based on configuration and activity. An example is Silicon Labs' AppBuilder, part of their Precision32™ development suite. AppBuilder provides a Power Estimator function (Silicon Labs 2019).

Because the device may have components operating at a variety of voltages, it can be simpler to compute the power budget in Watts rather than in Amps. The power requirement for each component is the product of voltage and current.

$$P = IV$$

where:

P = Power in Watts
I = Current in Amps
V = Voltage

The current drawn from the battery is important to calculate using the battery voltage. This is likely to differ from the voltage required by the components. Every conversion to a different voltage is less than 100% efficient. Losses in voltage conversion can be significant. So, it is important to categorize them as well.

With information about the end power consumer's current requirements and the voltages of both battery and device, the total power for the device can be calculated by summing the power for each component and the loss for each voltage conversion.

In some situations, the power budget is too large for the preferred battery. Increasing the size of the battery can negatively impact the overall size and weight of the instrument. This can be mitigated by balancing the use of components, and not turning multiple power consumers on at the same time. An example would be a Raman spectrometer, where the laser consumes hundreds of milliwatts when it is on, and a search processor also requires hundreds of milliwatts. By powering the processor down until the laser is turned off, the system can avoid consuming too much current for a limited-capacity battery.

2.7 Voltage Conversion

The integrated circuits within a spectrometer may require a set of power supplies at different voltages, known as power rails. With a single battery as a power source, the designer must provide voltage conversion to generate the necessary rails. Logic circuitry generally operates with voltages which are positive relative to system ground. However, the first stage of analog circuits, such as the preamplifier for an optical detector, often requires symmetric voltages above and below ground. Charge coupled device (CCD) detectors, such as those used in Raman spectrometers, may require several voltage rails. The Hamamatsu S11850-1106 (Hamamatsu 2019) is a CCD array detector with 2060 × 70 pixels. It requires voltage supplies of +24 VDC, +12 VDC, +5 VDC, and −8 VDC.

These voltages are typically generated from the battery voltage within the instrument by the use of DC/DC converters, also known as switching power supplies. The detailed design of a DC/DC conversion circuit is beyond the scope of this work. However, there are some key attributes worth considering for the implementation of voltage conversion within a battery-powered device.

Despite the DC (for Direct Current) in the name, DC/DC converter circuits operate internally at high switching frequencies. These frequencies can be a source of electronic noise, which can cause problems with noise in the instrument's internal signal circuitry and can lead to radiated electric field emissions which may appear when the instrument is tested for regulatory compliance. Switching frequencies can range from hundreds of kHz to several MHz. Selection of operating frequencies for the DC/DC converters can have a significant impact on instrument size as well as performance.

DC/DC converters are available in chip packages. They generally require a few external components, typically capacitors, resistors and inductors. As the switching frequency increases, the size of the external components gets smaller. EE Times magazine performed a set of tests in which they found that moving the switching frequency from 350 kHz to 1.6 MHz reduced the board space of their DC/DC converter by 35% (Nowakowski and King 2006).

A downside to higher frequency conversion is a decrease in efficiency. The same EETimes experiment found that the efficiency of the converter dropped from 91% at 350 kHz to just under 87% at 1.6 MHz.

DC/DC converters are constant power output devices. As their input voltage lowers, they will pull more and more current at the supply voltage in order to provide the desired output power. It is important to consider this for battery-powered devices. As the battery discharges, the supplies will draw more current to satisfy their power needs. This becomes a downward spiral. Higher currents through a battery's internal resistance will lower the effective output voltage even more. It is wise to consider the battery's operational voltage range and not just its nominal rating. The design should include a means to monitor the battery voltage and to provide a graceful shutdown before the system enters the spiral caused by low voltage and increasing current.

2.8 Decon/Ingress Protection (IP) Rating

Portable systems are likely to be used outdoors in inclement weather. In addition, systems may be exposed to harmful materials, including toxic industrial chemicals, biological threats, or chemical weapons or their precursors. Systems must be capable of withstanding the sample materials as well as the cleaning solutions required to decontaminate them.

2.8.1 Ingress Protection

Systems which need to be decontaminated require a degree of liquid-tightness. This is often specified using an IP (Ingress Protection) rating. IP ratings are defined by IEC (International Electrotechnical Commission) Standard 60529 (IEC 1989). IP ratings are two-digit codes. The first digit refers to mechanical ingress, such as objects or dust. The second digit refers to liquid ingress. The letter X is often used as a wildcard to allow a single number to cover a range of ratings. For example, IP6X includes both IP67 and IP68 products. This is not an official designation per IEC 60529. Typical IP ratings for portable instruments are provided in Table 2.1.

2.8.2 Sealing an Enclosure

Achieving IP67 and IP68 is challenging. The number of seams must be minimized. Seals for X7 and X8 are typically achieved using O-rings. Compression of the O-ring should follow the manufacturer's recommendation. Compressing an O-ring over a large panel opening may require a large number of fasteners to ensure even compression. Using hollow O-ring cord can dramatically reduce the force needed to achieve a good seal. Screws holding the

Table 2.1 IP ratings.

Rating	Solid particle	Liquid
40	No particle or wire greater than 1 mm in size can enter enclosure.	No liquid protection.
50	Dust protected – ingress of dust is not completely eliminated, but not enough dust can enter to prevent system from operating.	No liquid protection.
51	Dust protected – ingress of dust is not completely eliminated, but not enough dust can enter to prevent system from operating.	Protected against dripping water. Water is dripped for 10 min vertically at the flow rate of 1 mm/min.
52	Dust protected – ingress of dust is not completely eliminated, but not enough dust can enter to prevent system from operating.	Protected against dripping water while tilted. Unit is tilted 15° with water dripping for 10 min at 3 mm/min.
67	Dust tight – no ingress of dust.	Immersion in water up to 1 m of depth for 30 min.
68	Dust tight – no ingress of dust.	Immersion in water of 1 m or more, duration specified by the manufacturer.

case parts together must be outside the seal and threaded into blind holes or must have a gasket under the head to prevent leaks along the threads. APM Hexseal Seelscrew™ screws (APM Hexseal, APMhexseal.com) have an integrated O-ring in a groove under the head.

2.9 Testing the Seal

Seal testing can be performed in several ways. On a new instrument design, it may require several iterations to attain a reliable seal. The simplest starting position is an (empty) instrument housing. Dunk the instrument at a defined distance for a defined time. After removing the instrument from the tank, all external water must be removed before opening the instrument for inspection.

In production, this dunk, remove, inspect test is not practical. How do we know that we have not introduced a new leak when we reseal the instrument after inspection? A more reliable means of testing for water ingress is to use an accurate balance to weigh the instrument before and after dunking. The outside of the instrument must be thoroughly dried, and the balance must both support the mass of the instrument and have the precision to detect the change in mass caused by a few drops of water.

Better yet, a pressure or vacuum tester can evaluate seals without the risk of filling a valuable instrument with water because an assembly technician missed torqueing a single screw. The design engineer must include a connection port in the housing; this can be as simple as a sealing screw which is installed after the test is complete. Commercial testers are available which can connect to this port. They create a pressure differential between the inside of the case and the ambient air. The test setup monitors the decay in this pressure differential over a defined period to identify any leaks. Manufacturers of pressure and vacuum decay testers include Cincinnati Test and Uson.

2.10 Gloved Operation

Instruments used by hazardous materials teams are subject to a special set of requirements for user interface. The suits worn by the teams include several layers of gloves. An example of an external glove is shown in Figure 2.1,

Figure 2.1 HazMat gloves.

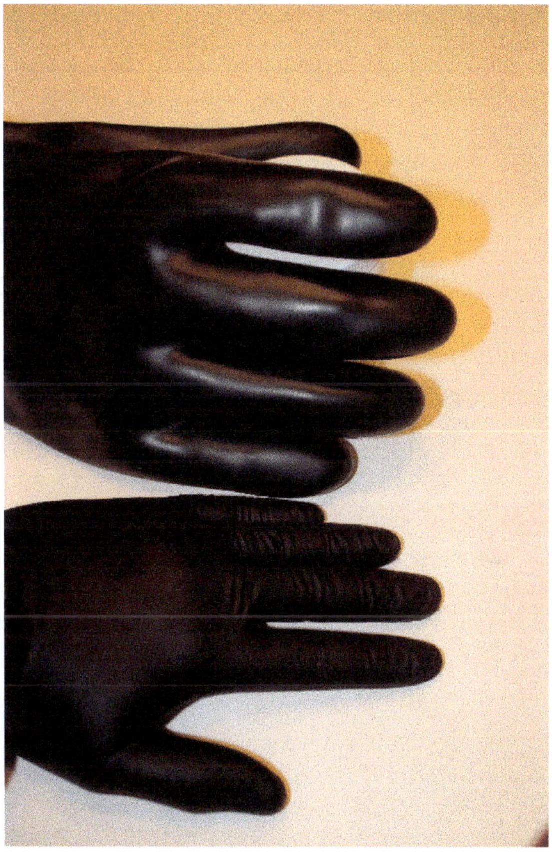

with a standard nitrile glove for comparison. These gloves obviously impair the dexterity of the operator. Systems for this type of use require large controls with good separation between buttons to reduce the chance of pressing more than one button at a time. Raised molded buttons on a keypad help to ensure the operator can easily locate the button through the glove.

Capacitive touch screens are popular on consumer handheld devices, but are incompatible with many gloves. Thin nitrile gloves used in medical applications will often work with capacitive screens. Resistive touch screens work with any glove material. Drawbacks to resistive screens are that they are typically not multitouch, and that they require a flexible membrane on the outside surface. These membranes are more susceptible to physical or chemical damage than the glass one can use with capacitive screens.

It can be difficult to touch a small icon with a bulky glove. Potential solutions include ensuring that icons and all software controls are scaled as large as possible. Using a stylus can help. Typical styli for electronic devices are too small to grasp easily in these gloves. HazMatID used a specially designed large diameter stylus with a tether to prevent loss. This stylus was turned from a solid piece of acetyl with a rounded point to activate the screen.

2.11 Display

Another key piece of the user interface is the display. There are many aspects of the display for a handheld instrument which deserve special consideration.

2.11.1 Content

Traditional spectrometers are all about the spectrum. Users want to be able to see and manipulate them and to perform analysis, such as baseline correction, smoothing, etc. For most smaller portable spectrometers, this is too much capability. Spectral display and manipulation are challenging to implement with a restricted user interface. Users are looking to answer, typically "what is this sample?" or "what are the components of this mixture?" Designers should consider limiting the display to information which is useful to the intended user. Method development and detailed spectral analysis can be migrated to a PC-based application.

2.11.2 Brightness

Sunlight legibility can be extremely important and challenging to achieve. Liquid crystal display (LCD) panels, commonly used for portable displays, are available in a range of brightnesses, measured in nits (a nit is 1 candela per square meter). A survey of LCD panels for Original Equipment Manufacturer (OEM) use shows a range from 300 to 850 nits (Sharp Electronics Corporation 2019). High-brightness LCD panels can reach as high as several thousand nits. Several manufacturers of LCD panels suggest that a minimum of 700 nits should be specified for a display to be sunlight readable.

Organic light-emitting diode (OLED) displays are an alternative which features daylight readability and requires no separate backlight, as the pixels are light-emitting. OLEDs are available in a much more limited range of sizes in the form factors required for portable equipment.

2.11.3 Legibility

HazMat operators may wear multiple face shields while operating equipment in their personal protective gear. The shields may cause distortion due to curvature or condensation. This can make it difficult to read the detailed text or graphics on small screens. User displays should be designed with this in mind, using large, highly contrasting text and controls.

2.11.4 Guidance

Operators of this equipment may not be trained spectroscopists and may not operate the equipment every day. Providing cartoons or videos illustrating the next step in the analysis process can assist the operator in getting better results.

2.11.5 Display Power

Backlights for LCD panels are often one of the larger consumers of power. The Sharp LQ035Q3DG05 display draws up to 238 mW (Sharp Electronics 2009). Power draw for backlights is generally proportional to the surface area of the display. A 6.4 in. display will draw approximately four times as much backlight power as a 3.1 in. display of the same technology. Transflective displays are able to reflect ambient light from the rear of the panel and can reduce the amount of light required from the backlight.

2.11.6 Temperature Range

LCD panels have operational ranges which may be limited relative to the environments in which portable equipment is operated. At cold temperatures, the response time of the liquid crystals slows down, and below a threshold, the display will cease to operate. Display manufacturers continue to extend these ranges. The Sharp LQ035Q3DG05 is a 3.5 in. diagonal display which can operate over a temperature range from −30 to +80 °C. In extreme situations, heaters can be added to displays to increase their low-temperature capability.

2.11.7 Ruggedness

Like a cell phone, handheld portables are prone to being dropped. Landing on a display or a corner of a case can place large sudden stresses on a display. Laminated displays can be custom-fabricated to greatly enhance the durability of LCD panels. Optically bonding a window to the front of the display using an index-matching adhesive can improve the display's apparent brightness as well as its mechanical robustness (General Digital 2019). This process must be started early in the development cycle, as lead times on these custom-assembled displays can be extensive.

2.12 Thermal Concerns

Temperature control is essential for operation and performance of optical spectrometers. Changes in temperature can affect optical alignment through expansion or contraction of support materials. Detectors have performance curves which vary with temperature and may stop working entirely if they get too warm. This is the case with DTGS (deuterated tryglycene sulfate) pyroelectric detectors, which have a Curie point above which they de-pole and no longer function. Newer versions of these pyroelectrics are doped with L-alanine to form deuterated L-alanine doped triglycine sulphate (DLATGS), which are self-poling, and recover when temperatures are returned below the Curie point.

Spectrometer instruments almost invariably contain sources of heat. The most obvious source is the optical energy source, particularly in the case of an incandescent filament or globar. In some laboratory instruments, the source generates enough heat to require liquid cooling. Less obvious sources of heat include the various electronic components, such as power supplies, batteries, and the motors or drive electronics used to control moving parts within the system. Some optical benches include heaters on the spectrometer's optical assembly or on components, such as gas cells. These heaters are used to provide a controlled, stable temperature. The heaters themselves and the control electronics driving them provide additional heat load to the instrument.

In a laboratory system, it is common to provide vents in the enclosure, and a cooling fan to ensure airflow through the system. In some systems, the speed of the fan is controlled electronically to assist with temperature control of the overall system. In these systems, as the interior temperature rises the fan's speed is increased to provide additional cooling.

2.12.1 Cooling Performance of a Ventilated System

Systems with vents and fans cool the internal components by transferring their heat to air, which is then exhausted from the enclosure by a fan or blower. As the warm air is forced out of the enclosure, it is replaced by fresh air at ambient temperature. With sufficient airflow, components inside the enclosure can be kept within a few degrees of ambient temperature. This cooling efficiency depends on the surface area available to transfer the component heat to the moving air. Hot components, such as power transistors can be coupled to extruded heatsinks with fins running parallel to the air motion.

2.12.2 Sealed Systems

A portable system faces further challenges with respect to remaining cool. These systems are typically designed to work in less well-controlled environments. The potential presence of dust, moisture, or even rain suggests that the enclosure should be sealed from the environment. This precludes the option of blowing air through the enclosure. At the same time, the instrument is likely to be moved outside of the climate-controlled laboratory, and may face temperatures ranging from below freezing to as high as 50 °C. Furthermore, portable instruments may be housed

Table 2.2 Detector performance at temperature.

Cooling	Operating temperature (°C)	D^*
Two-stage TEC	−40	≥4.0 E10
Three-stage TEC	−65	≥6.0 E10
Four-stage TEC	−75	≥6.0 E10
LN_2	−196	≥1.0 E11

in plastic enclosures. Molded enclosures provide advantages over metal enclosures with respect to cost, weight, and ergonomics. However, in general, plastics are far worse conductors of heat than metal, such as aluminum.

2.12.3 Detector Cooling

Detectors in both Raman and FTIR instruments are generally temperature sensitive.

- Dark current in CCD array detectors used in Raman systems doubles with every increase of 2–5 °C. (Hamamatsu 2015)
- DLATGS detectors are sensitive to temperature in a non-monotonic way. The D^* (a measure of performance for optical detectors) of Leonardo's Series 99 detector has a maximum value between 35 and 40 °C. By the time the temperature reaches 50 °C, the D^* is only 80% of the maximum (Leonardo 2017). The same detector's responsivity drops as temperature rises, resulting in a desired operating point on the lower side of the D^* peak.
- Mercury cadmium telluride (MCT) detectors operate best at liquid nitrogen temperatures, but can also be operated using multistage thermoelectric coolers. Some examples from Infrared Associates' website show some relative performance values for 1 mm by 1 mm square detectors (Infrared Associates n.d.) (Table 2.2):

Any of the thermoelectric means of stabilizing or lowering temperature requires a heatsink. The thermoelectric cooler (TEC) moves heat from the detector into its opposing side. This hot side of the TEC must be cooled, usually using ambient air.

2.12.4 Moving Heat to Ambient Air

In a sealed system, forced air is not an option for cooling local hot spots, such as the high temperature side of the TECs often used for controlling detector temperatures. With semiconductors, a designer can often locate the hot component on a wall of the enclosure, allowing direct conduction to fins on the outside of the enclosure for ambient convection cooling. The location of detectors is dictated by the optical design, and the TEC may not be conveniently located at the outside wall. Conduction is generally the most efficient means of moving heat within an enclosure, but there are several options for conduction. Once the heat reaches the outside wall, there are further considerations on how to dissipate it in a way which does not negatively impact the instrument design. Adding external fins may not fit with the instrument aesthetic, may provide a means for mechanical shock and/or electrical disturbances to be coupled directly to the sensitive detector, and, in extreme situations, may cause risk of operator discomfort or injury due to elevated temperatures. In the next section, we will calculate temperature rise at the hot side of a hypothetical TEC with a selection of thermal approaches. For our input, we will assume we are operating a four-stage TEC to cool an MCT detector in a TO66 can. The approximate power draw of such a TEC is 0.6 A at 8.25 VDC, for a power of about 5 W. (Note that far less power is actually removed from the detector itself.) We will explore three scenarios.

2.12.4.1 Direct Coupling to Heatsink Convection Cooling

For all the scenarios, we will work with a Wakefield–Vette 396-1AB heatsink. This is an extruded aluminum form with ten fins. Its base measures 127 × 76.2 mm, and the fins are 35.1 mm tall. Wakefield–Vette's data sheet lists a thermal resistance of 1.85 °C/W with natural convection, and 1.07 °C/W with forced air at 500 LFM (linear feet/minute).

For our first scenario, we will couple the hot side of the TEC directly to the heatsink. (For all the scenarios, we are assuming perfect coupling at junctions between components. In reality, there will be additional temperature rise at these interfaces. This rise can be minimized using thermal pads or compounds). We are injecting 5 W into the heatsink, with a thermal resistance of 1.85 °C/W. The temperature rise of the heatsink above ambient T_{hs} is then simply:

$$\Delta T_{\text{hs}} = QR_{\text{tc}} = 5(1.85) = 9.25°\text{C}$$

where

Q = Heat flow
R_{tc} = Thermal resistance of the heatsink under convection

If our ambient temperature T_{a} is 25.0 °C, the temperature T_{c} of the hot side of the TEC will be:

$$T_{\text{c}} = T_{\text{a}} + \Delta T = 25.0 + 9.25 = 34.25°\text{C}.$$

2.12.4.2 Direct Coupling to Heatsink-Forced Air Cooling

If we want to provide more efficient cooling, which may be required if the ambient air temperature is higher than 25 °C, we can compute the effect of adding a fan. This fan is forcing air over the heatsink, and can be outside our sealed enclosure. We simply substitute the forced air thermal resistance R_{tf} for the convection thermal resistance R_{tc}.

$$\Delta T_{\text{hs}} = QR_{\text{tf}} = 5(1.07) = 5.35°\text{C}$$

This change has reduced the temperature rise from 9.25 to 5.35 °C.

2.12.4.3 Remote Heatsink with Aluminum Bar

As described above, sometimes we do not have the option of locating the TEC directly on the wall of our enclosure. If the TEC is in the center of the instrument, we can couple it to an outside heatsink by means of an aluminum bar. Let us assume a bar which is a rectangular prism, 2 cm square in cross-section by 10 cm long. Fourier's Law of steady state conduction tells us that:

$$q = KA\Delta T/L$$

where

q = Heat flow
K = Material's conductivity
A = Surface area of conductor
i = Length of conductor
ΔT = Temperature difference from inside to outside enclosure

Solving for ΔT gives:

$$\Delta T = qL/KA$$

A representative value for K for aluminum is 236 W/(m-K). Our temperature rise through the length of the aluminum bar ΔT_{ab} can be calculated by substituting:

$$\Delta T_{ab} = \frac{qL}{KA} = \frac{5(\mathrm{W}) * 0.1(\mathrm{m})}{236\left(\frac{\mathrm{W}}{\mathrm{mK}}\right) * (0.02\ \mathrm{m}) * (0.02\ \mathrm{m})} = 5.3^{\circ}\mathrm{C}$$

This temperature rise is almost as much as we observe between the heatsink and ambient with forced air cooling. We can add this temperature rise to that of the heatsink to find T_c:

$$T_c = T_a + \Delta T_{ab} + \Delta T_{hc} = 25.0 + 5.3 + 9.25 = 39.5^{\circ}\mathrm{C}.$$

2.12.4.4 Remote Heatsink with Heatpipe

Between the heatsink–air interface and the aluminum bar's resistance, we have a temperature rise of almost 15 °C at the high side of the TEC. Transferring heat from one location to another can be done extremely efficiently by replacing our aluminum bar using a heatpipe. See Figure 2.2 for examples of heatpipes. Heatpipes are partially evacuated hollow tubes containing a heat transfer fluid and a wicking material. The fluid undergoes a phase change when one end of the tube is heated. The vapor moves to the other end of the tube where it condenses. The liquid then wicks back to the hot end to repeat the cycle. The resulting flow results in the equivalent of a thermal conductivity in the range of 30 000–40 000 W/mK, hundreds of times higher than that of aluminum. If we use 30 000 as our value for a 6 mm-diameter heatpipe in the equation above, our new temperature rise becomes:

$$\Delta T_{hp} = \frac{qL}{KA} = \frac{5(\mathrm{W}) * 0.1(\mathrm{m})}{30\,000\left(\frac{\mathrm{W}}{\mathrm{mK}}\right) * \pi * (0.003\ \mathrm{m})^2} = 0.59^{\circ}\mathrm{C}$$

The TEC hotside temperature T_c is now:

$$T_c = T_a + \Delta T_{hp} + \Delta T_{hc} = 25.0 + 0.59 + 9.25 = 34.8^{\circ}\mathrm{C}.$$

Thus, we see that a heatpipe of 5 mm diameter can replace a 2 cm square bar and reduce the thermal rise by a factor of 10. The heatpipe requires less space and weighs significantly less than the aluminum bar.

Heatpipes can be bent following the manufacturer's guidelines. Advanced Thermal Solutions (ATS) recommends a minimum bend radius at least three times the heatpipe's diameter.

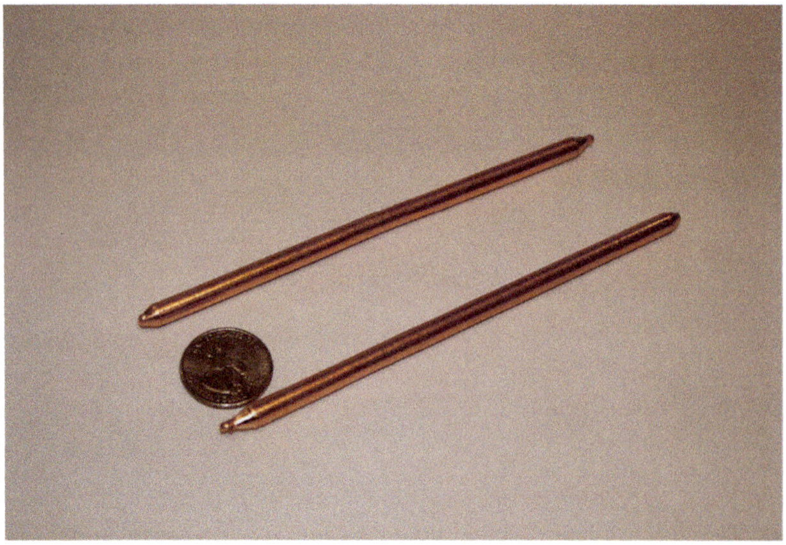

Figure 2.2 Heatpipes.

2.13 Optical Elements

Laboratory instruments are able to take advantage of infrastructure within the facility. As well as AC power sources, this can include a source of dry gas for purging optical systems. This constant supply of desiccated gas protects optics from moisture, allowing the use of hydrophilic materials, such as KBr, a standard material for beamsplitters and optical windows used in the mid-infrared. These materials are easily damaged by the presence of moisture in the atmosphere, and can quickly become cloudy or completely destroyed if they are exposed to excess water vapor. KBr is frequently used in general-purpose instruments because it has a wide transmission window, allowing spectroscopy over the range of 48 800–345 cm^{-1} (Spectral Systems n.d.).

In portable instruments, KBr is not a suitable choice. Other materials are more appropriate, though they can limit the available spectral window. Given the general assumption that the portable instrument is more limited in purpose, it is often possible to select a material which is transparent over an adequate spectral range for the application. Zinc Selenide, ZnSe, is a frequent choice for portable FTIR systems. It has a transmission range of 18 000–461 cm^{-1}.

Zinc Selenide is commonly available as a window material with two flat, parallel surfaces. It can also be diamond-turned into relatively complex shapes to be used as focusing elements where appropriate.

2.14 Interferometer Optical Design

Portable FTIR spectrometers face several restrictions with respect to their laboratory counterparts. These constraints should be considered when selecting the optical layout of the interferometer itself. This section will review three interferometer topologies and provide comments regarding the tradeoffs made for portability.

2.14.1 Traditional Michelson

A standard Michelson interferometer consists of two flat reflective surfaces and a beamsplitter. See Figure 2.3. The angle between the two beams may be 90°, though other angles are also frequently used. One mirror is fixed, and the second is translated with motion parallel to the optical beams. These mirrors must be aligned to be perpendicular to their respective beams. In particular, the moving mirror must maintain its alignment to within a fraction of a wavelength over its entire range of motion. The allowed deviation from perpendicularity is a function of the required wavelength resolution of the instrument. The offset in beampath caused by a given angular misalignment

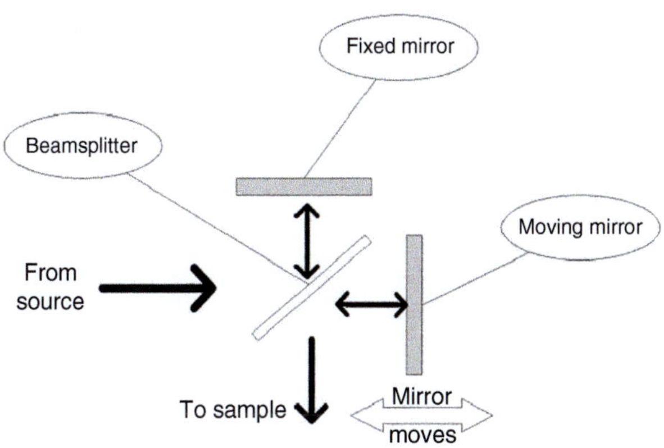

Figure 2.3 Michelson interferometer.

is proportional to the diameter of the beam. As the beam travels out and back, the beampath error is double that of the mirror error. Smaller beams are more forgiving of angular misalignment than larger beams. The mechanism used to align these assemblies generally tilts either the fixed mirror or the beamsplitter itself. HazMatID and GasID are examples of portable FTIR systems using flat mirrors in Michelson interferometers.

2.14.2 Michelson with Cube-Corner Reflectors

Cube corners are optical constructions with three orthogonal reflective surfaces. These structures are familiar from the "cat's eye" reflectors common on bicycles. Any beam of light which enters the acceptance angle of the cube corner will be returned parallel to the axis upon which it arrived. The beam will be translated by a distance which depends upon the distance from the optical center of the cube corner. The chief advantage of cube corners in interferometers is that they are insensitive to angular misalignment. Aligning a cube-corner interferometer does not require a tilt to any optical component; rather the fixed cube corner is translated in X and Y axes until its return beam is satisfactorily aligned to superpose itself over the beam returning from the moving mirror. TravelIR used a cube-corner Michelson interferometer.

2.14.3 Dual-Pendulum

A dual-pendulum interferometer uses two cube-corner reflectors mounted to a two-armed rotating yoke. See Figure 2.4. This type of interferometer was first described by Rippel and Jaacks (Rippel and Jaacks 1988). The entire yoke assembly is rotated over a small angle about an axis. The two reflectors move in opposite directions relative to the beamsplitter; as one moves closer, the other moves farther away. This has the effect of doubling the change in relative pathlengths, allowing a smaller displacement for a given resolution. The rotation is achieved by a magnet and coil assembly similar to that used to move the heads of a magnetic disk drive. A major advantage for a portable interferometer is that the entire rotating assembly, including mirrors, arms, and drive components, can be balanced with its center of mass aligned with the pivot point. This balance reduces the interferometer's sensitivity

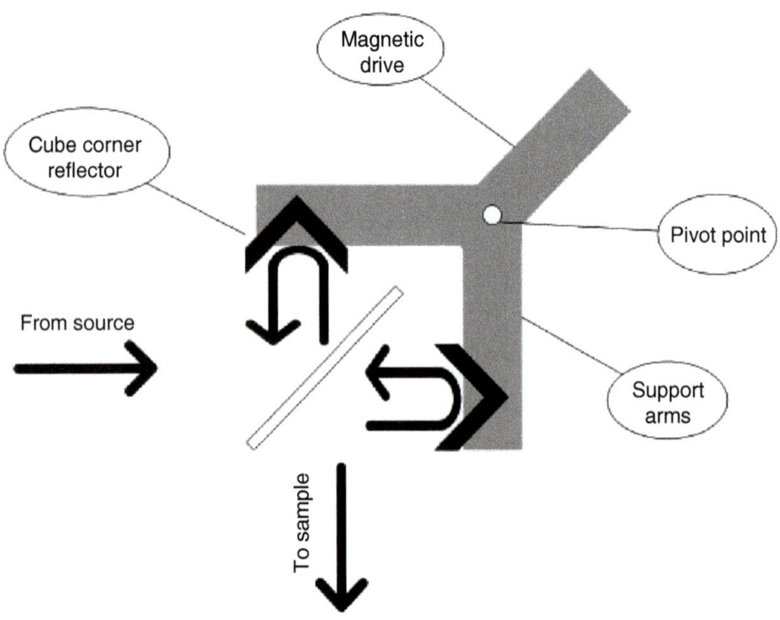

Figure 2.4 Dual-pendulum interferometer.

to shock or vibration, as any force imparted to the moving assembly by the axle will not impart any rotational moment to the assembly. A dual-pendulum design is used in the ABB-Bomem line of process instruments. It was also included in the Smiths Detection HazMatID Elite handheld FTIR system.

2.15 Interferometer Bearings

Moving mirrors and their supports must move accurately and reliably over the life of the instrument. As we have seen, the tolerances on this motion are measured in fractions of wavelengths. In many instruments, the mirror is moving anytime the instrument is powered on. This results in tens or hundreds of thousands of mirror strokes. The motion of the mirror must be constrained by bearings, but typical bearings used in mechanical devices are not adequate for our purposes. Tolerances on pins and leaves in a hinge mechanism are not typically tight enough to provide rotation while not allowing sideways motion.

Many interferometers rely on flexing steel members for joints. Nicolet's 5-series interferometer used a "porch swing" mechanism including four rigid arms with a pivot at each end of each arm. See Figure 2.5. Each of the pivot points was a small piece of spring steel which would bend to allow the arm to move like a classic porch swing. See Figure 2.5. The assembly maintained the alignment of the carriage, which, in turn, supported the moving mirror in a constant orientation throughout the stroke. The flexible elements, known as pivot bearings, can be purchased in a cylindrical format. The assembly includes two spring steel leaves coupling two aluminum cylinders. One cylinder is pressed into the fixed structure, and the second is pressed into the moving arm. As long as the movement is constrained to within the pivot's design limit, its life is essentially infinite. Typical angular limits range from ± 3.7 to $\pm 15°$, depending on the bearing's size, material of construction, and load limit (C-Flex Bearing Co 2015).

As instruments get smaller, the size and number of components in a porch swing become problematic. The design team at SensIR Technologies combined the flex pivots and rigid arms of a porch swing design into two flat spring-steel leaves, one at each end of the moving mirror assembly. See Figure 2.6. These leaves were cut as a stack of material to ensure the shape of all leaves is identical. The length of the leaf must be adequate compared with the desired stroke of the moving mirror, in order to ensure the steel is not flexed beyond its fatigue limit. This design is inherently self-centering. With no power applied, the moving assembly rests in the center of the stroke. The SensIR design included mechanical end-of-travel limits to prevent the spring steel from being over-flexed as the instrument is subjected to vibration or shock in transit.

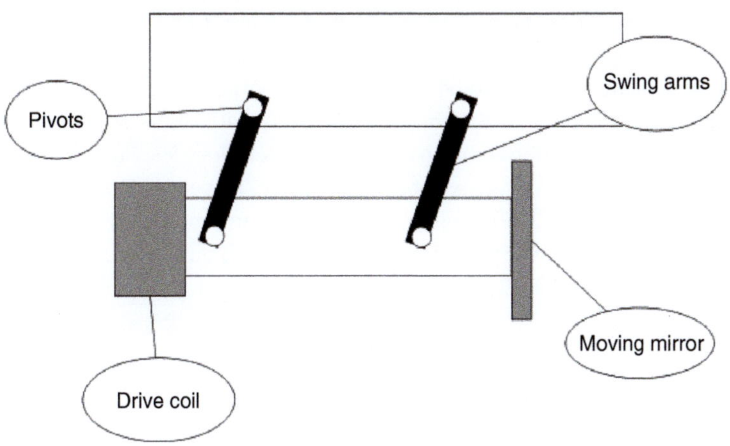

Figure 2.5 Porch swing bearing.

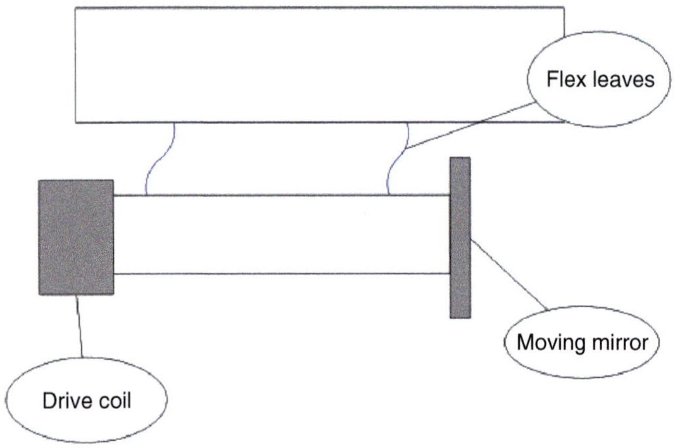

Figure 2.6 Flex-leaf bearing.

Instruments using cube-corner reflectors are more tolerant of angular movement during scanning. The TravelIR instrument took advantage of this by using a high-precision ball bearing linear slide to provide motion for the moving mirror. This slide provided low friction and was simple to assemble into the interferometer. Because the slide is not self-centering, TravelIR included a solenoid-operated brake which would lock the sliding mechanism whenever power was removed from the system. TravelIR was prone to dust or rough spots in the slides, sometimes caused by drops of the instrument. These small sticking points in the travel would cause nonconstant motion of the moving mirror, which resulted in distortions in the spectrum.

Rotational scanning interferometers, such as the dual-pendulum design described above, can take advantage of the flex pivot components to provide both movement and a self-centering design. Placing two flex pivots in the same line allows stability about the axis while providing for adequate rotation.

2.16 Vibration

Vibration is a concern for both bench and portable devices. Bench devices are often mounted on damping feet to reduce transmission of vibration from the bench into the device. FTIR systems are particularly susceptible as they collect data at frequencies in the audible spectrum and are, therefore, sensitive to noises. They also incorporate a moving mirror whose motion can be disturbed by impact or system vibration. This problem is worse for handheld devices as users cannot hold them perfectly still during sampling.

Various instrument designers have incorporated feedback loops from the FTIR metrology laser to the servo drive on the moving mirror within the Michelson interferometer. The Smiths Detection HazMatID Elite instrument uses a Phase Locked Loop (PLL) to tie the frequency of the interference pattern of the metrology laser to a crystal oscillator. The error signal from the PLL is fed back into the drive voltage for the servo drive. This ensures the mirror moves at a constant velocity. A paper presented at the May 2012 SPIE conference describes an experiment in which the resulting disturbance rejection dramatically reduces the noise present in a spectrum of caffeine acquired while the instrument is rocked at approximately 10 Hz (Arno et al. 2012).

2.17 Shock

Portable instruments are much more likely than benchtop systems to be subjected to rough handling or dropping. In addition to issues with data collection during a shock event, the systems are likely to be damaged. Shock

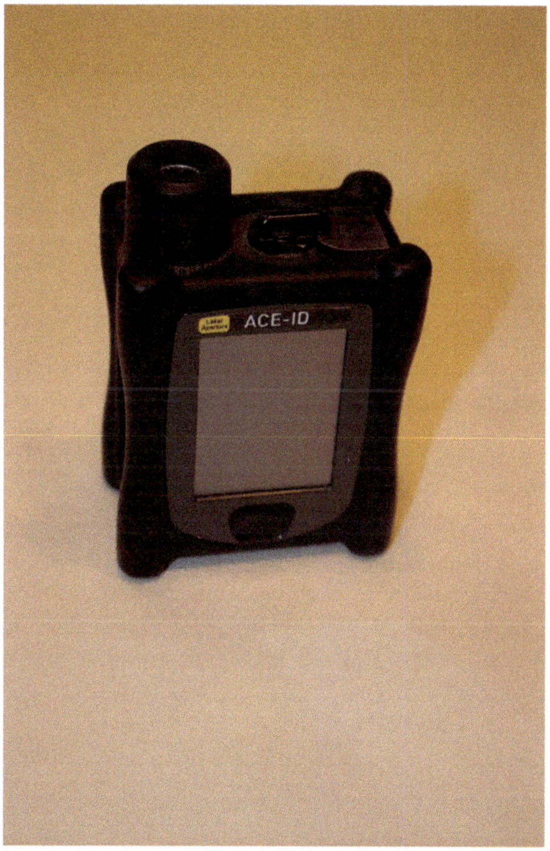

Figure 2.7 Smiths detection ACE–ID.

exposure can be minimized by the application of elastomeric bumpers on corners or surrounding the entire instrument. Arno et al. describe a means of designing "smart bumpers" with optimized durometer and shape to minimize the G forces experienced by the instrument when dropped. The Smiths Detection ACE–ID Raman spectrometer fits into a molded rubber jacket which cushions all corners of the instrument while leaving exposed metal surfaces to allow component cooling. See Figure 2.7.

2.18 Battery

Most portable instruments are battery-powered. Battery technology improves every year, increasing power density and battery lifetime. Specifying and maintaining the battery and its associated system is a complex task. In addition to engineering issues, batteries can impose regulatory restrictions with respect to travel or shipment. It is important to understand these implications early in the design cycle. Some larger capacity lithium-ion batteries require special paperwork to ship.

2.18.1 Primary vs. Rechargeable Batteries

In general, systems are designed for exclusive use of specialty rechargeable batteries. Some smaller handheld systems have requirements for using commonly available primary (non-rechargeable) cells, usually for ease of availability. When operating in the field, it is often easier to locate a pack of three cells in a local store than it is to find or recharge a special lithium-ion battery pack.

2.18.2 Charge Management

The user of an instrument will expect to have some information about how much charge is left in the battery. Over time, users have raised expectations due to their experience with devices, such as cell phones, which constantly display percent charge remaining. There are a number of ways to compute the remaining battery life, depending on the battery configuration. The most obvious means is to measure the voltage provided by the battery. As the battery is depleted, its output voltage drops following a curve which varies depending on battery chemistry. Issues with this means of measurement include the variation of voltage as a function of instrument current draw, and the shape of the curve for some chemistries. Lithium-ion batteries maintain an almost constant voltage for much of their discharge cycle. When the voltage starts to drop, it falls steeply. This makes it easy to judge when a battery is 90% discharged, but difficult to establish when it reaches 50 or 75%.

2.18.3 User-Replaceable Batteries

Replaceable batteries provide flexibility in use. They allow the user to maintain a stock of ready-to-use batteries and minimize the downtime when an instrument is discharged. Replaceable batteries are commonly used in cordless tools. Allowing the user to replace a battery can add significant complexity for instrument designers. From a mechanical perspective, the instrument needs a door or other means of inserting and removing the battery. It requires robust yet easy-to-operate connections, and the battery itself must be packaged in a way to survive outside the instrument. This adds extra volume and weight for the battery case. These are among the drivers in the consumer lines of phones and tablets which have manufacturers moving away from user-replaceable batteries. Electrical designers also have reason to avoid replaceable batteries. User-replaceable batteries make it harder for the system to know the state of charge through coulomb counting, as the system loses all known history whenever the battery is removed from the instrument. On the positive side for electrical design, if the battery is chargeable only outside the instrument, then the instrument does not need to provide support for charging circuitry, reducing the component count and worst-case heat load.

2.18.4 Smart Batteries

A means of delegating the battery state-of-charge calculation is to employ a "smart battery." These battery backs incorporate one or more integrated circuits specifically to manage the state of charge of the battery. Most of these smart battery packs comply with the System Management Bus, or SMBus protocols. SMBus is based on the I2C hardware definition. I2C is a two-wire bidirectional serial bus. The SMBus protocol was originally defined by Intel in 1995. The specification is maintained by the Smart Battery System Implementers Forum (System Management Interface Forum 2013).

The Smart Battery contains nonvolatile memory. The built-in battery circuitry uses this memory to log important parameters. The instrument can query the battery to learn about its present voltage, state of charge, time to empty at current discharge rates, and the number of charge/discharge cycles the battery has experienced. The instrument's internal charging circuitry can work with the smart battery to provide optimal control over charge rate. Specialized charger chips have been developed to support smart battery management. Chip families, such as TI's BQ25000 series, list over 250 different specialized parts for managing rechargeable batteries (TI 2019). External smart chargers are also commercially available to support charging smart batteries outside the instrument.

2.19 Electrostatic Discharge (ESD)

Electronic devices are subject to disturbance from Electrostatic Discharge, or ESD. Almost all devices sold in the European Union must be certified to meet CE requirements. CE certification requires that devices survive static discharges as specified in EN61000. ESD compliance is tested using an ESD simulator, a piece of test equipment which injects a spark at known voltage to critical parts of the device. The EUT, or Equipment Under Test, is placed on an insulted sheet on top of a grounded reference surface prior to the test. The ESD simulator has a pointed tip which is placed on or near the EUT. The tip is then pulsed to high voltage, typically in the range of 2–15 kV. The tip is moved to a number of locations selected on the outside surface of the EUT.

For traditional line-powered equipment, the standard defense against ESD is to provide a low-resistance path from any exterior surface to earth ground via the device's line cord. Handheld and portable devices often have no line cord or may have it only during the time they are being charged. This presents a different challenge to the portable equipment designer. Without the "infinite drain" of earth ground into which to dump the sudden influx of ESD-generated charge, what can we do to mitigate problems caused by ESD?

ESD damage in circuits is not caused by an elevated voltage over the entire circuit, but rather by the large difference in voltages across the circuit, or an individual component, caused by the many-kilovolt level discharge into a single point in the instrument. This near-instantaneous injection of charge can cause extremely high currents to flow for short periods of time. In a battery-powered device, we cannot shunt these currents around our circuits to earth ground. The charge in the end will equilibrate across the instrument, bringing the entire device to a new voltage potential. What we can do is to provide paths for the equilibrating currents which do not flow through our delicate electronic circuits.

Bill Jackson of TI published a white paper in 2008 describing means of mitigating ESD and radio frequency (RF) issues in battery packs themselves (Jackson 2008). These techniques are also valid for reducing the impact of ESD on our handheld devices.

ESD mitigation recommendations include:

- Minimize entry of charge via an insulated case. The ESD tester generates a 15 kV voltage at its tip. This voltage is looking for its easiest path to ground. A plastic case with adequate thickness can prevent the voltage from "breaking through" to conductors inside the instrument. The thickness required can be calculated from the breakdown voltage (also known as dielectric strength) of the material used for the case. Nylon 6 has a breakdown voltage of between 10 and 25 kV/mm depending on temperature and humidity (Unitika 2019). A case thickness of 2 mm would provide a minimum breakdown voltage of 20 kV, more than the 15 kV required for ESD. Different plastics have different breakdown voltages. The value is typically listed on a material data sheet from the plastic supplier. This strategy can defend most of the instruments; however, certain controls, screws, and connectors for power and communications can still be vulnerable.
- Provide voltage clamps on communication lines. Universal serial bus (USB) connectors are subject to frequent user interaction and are, therefore, likely areas for ESD events. Standard practice is to include a diode-based voltage clamp component as close as possible to the connector itself. These are simple six- or ten-lead devices which connect between the USB D+ and D− pins and ground. In use, they limit the ability of the data lines to move away from ground.
- Understand the surge path. The true key to minimizing ESD disturbance is to recognize the path which the surge in voltage will take as it translates to current. In a line-powered instrument, the obvious strategy is to create what Jackson calls a high-current ground path, which couples all exposed conductive surfaces to earth ground. This high-current ground path is intentionally separated from the low-current ground path, which is used for signals and more sensitive internal components. These two separate ground paths should be connected at a single point to prevent voltage gradients along the high-current path from being transmitted to the low-current path. Physical separation is also important, as the sudden current surge caused by an ESD event will create magnetic fields which can transmit the surge into adjacent traces. ESD current sources, such as the ground points in the USB transient suppressors, should be connected to the high-current ground path. The use of ground planes and ESD spark gaps within the printed circuit board layers can help minimize injection of ESD events into sensitive circuitry.

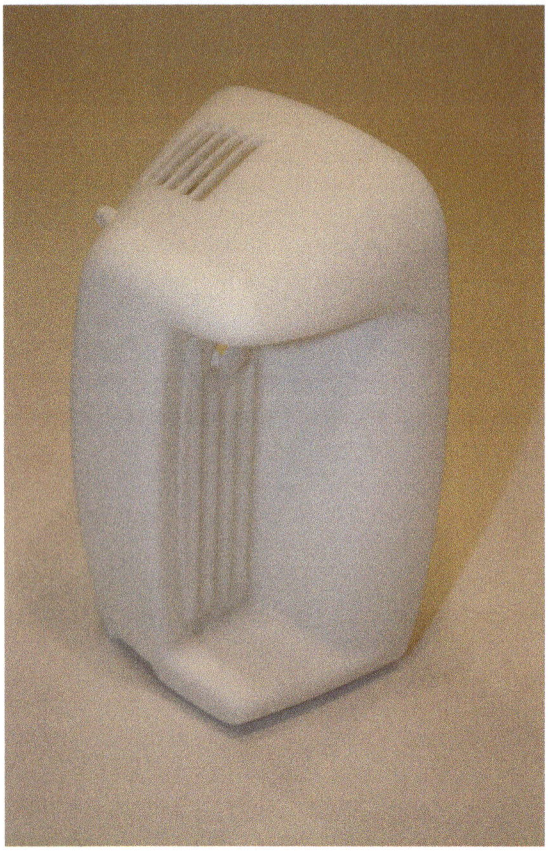

Figure 2.8 Printed handgrip.

2.20 Ergonomics

Portable equipment is by its nature handheld or at the minimum transportable. Transportable equipment must have well-defined lift points and be within weight limits for users to move and carry. Handheld equipment has much more stringent requirements. An instrument which will be carried and used throughout the day must be comfortable to operate for users across a wide range of sizes and strengths.

A key goal for the design of a handheld device is to provide grips and controls which fit comfortably in a range of hand sizes. The range of hand sizes has been extensively studied by a wide range of organizations. The National Institutes of Health (NIH) published a paper studying the hand size of firefighters (Hsiao 2015). The document includes several key dimensions, listing the range for each.

It is often extremely useful to create a three-dimensional model of the proposed handheld design, or at least of some key sections. These models may be sculpted from clay or foam. It is now easy to design the model using CAD (Computer Aided Design) and to 3D print an assortment of shapes. These shapes can be evaluated by users with a range of hand sizes to determine the best shape and size for comfort, ease of grip, and access to key controls. Figure 2.8 shows a sample 3D-printed handgrip used for user review.

2.21 Laser Safety

Many spectrometers contain lasers. For Raman systems, the laser is the excitation source for the measurement. FTIR systems almost always include a laser used as a reference to measure travel of the moving mirror.

In the US, laser safety is enforced by the CDRH, the Center for Devices and Radiologic Health, a division of the FDA (Food and Drug Administration). In Europe, laser products are certified according to EN-60825. Manufacturers of laser devices are required to file reports with the CDRH certifying the class and safety aspects of the lasers they produce (US FDA 2018).

Lasers are divided into safety classes, which are similar, if not identical, between the CDRH and EN-60825 nomenclature. Most handheld spectrometers include lasers in the classes between 1 and 3R or 3B. Class 1 lasers are considered eye-safe and need no special handling or safeguards. They can still be hazardous if viewed using optical aids, such as magnifiers or telescopes. Class 3R or 3B lasers are considered hazardous to eyes and/or skin depending on exposure time. The classification of a laser depends on several factors, including its wavelength, power, and angle of divergence (21 CFR 1000.40 2019).

The laser classification of a device is based on the amount of laser energy which is accessible to the user; thus, a device may contain a class 3 laser, but still be a class 1 device if the energy is prevented from leaving the confines of the device. Benchtop FTIR systems including Helium Neon lasers often provided beam-blocking mechanisms to prevent laser energy from escaping; thus, allowing the instrument to be listed as a Class 1 device despite its internal Class 3 laser. Portable FTIR systems are much more likely to use a diode laser as a reference. With proper design to limit power output, these diode lasers can be constructed as Class 1 modules in their own right. This ensures that laser safety is not compromised when the instrument cover is opened.

Lasers for Raman systems pose more issues. The power level required to excite a sample is generally high enough to prevent it from being a Class 1 laser. Typical Raman handheld systems deliver approximately 100 mW of laser power at wavelengths of 785 or 1064 nm. Handheld Raman systems frequently have lasers rated class 3B.

Bench Raman spectrometers generally require the user to place the sample into a light-tight compartment with a laser interlock switch. This system prevents the laser light from escaping to a point where it could cause harm to the user.

Some portable Raman systems have small vial holder compartments which allow the sampling of liquid samples in glass vials. These compartments have sensors to determine that they are closed, providing a safety interlock which allows the system to operate as a Class 1 device.

However, one of the key advantages of a handheld Raman system is that it allows a user to analyze a sample without removing it from its container. This container may be a large reservoir of unknown liquid found in a suspected clandestine drug lab. Opening the container may pose a different risk to the operator. It is safer and easier to measure the spectrum through the wall of the container. This requires that the laser be accessible outside the instrument.

A key value for determining eye safety is the Nominal Ocular Hazard Distance, or NOHD. The NOHD is the distance from the laser at which the beam no longer exceeds the Maximum Permissible Exposure, or MPE. The calculation of the NOHD is beyond our scope. The MPE is expressed in irradiance, or power density as W/m^2. The faster a laser beam diverges, the more rapidly its power density will decrease and the shorter the NOHD. Using a fast optic at the exit of the Raman system creates a focal point close to the laser exit, after which the beam will diverge rapidly. The Thermo Scientific TruScan RM's Statement of Compliance states that it is eye-safe at a distance greater than 14 cm (Thermo Scientific 2016).

2.22 Stability

Many spectroscopic techniques are ratiometric. They measure a sample spectrum by computing the ratio between measurements with and without a sample present. These two separate measurements are often known as single-beams. The measurement taken with no sample present, known as the "background" or "reference," is used to characterize the current line-shape of the spectrometer system, including the source, alignment, and all optics in the light path. The measurement with a sample present includes all the instrument information as well as new information provided by the sample itself. Dividing the sample measurement by the background measurement cancels the instrument line-shape, leaving only sample information in the form of a spectrum.

The preceding statement assumes that the instrument line-shape has not changed significantly in the time between measuring the background and the sample single-beams. If there has been a significant change, then that change will show up in the ratioed spectrum.

Stability for FTIR instruments is generally measured and specified by collecting two background single-beams separated by a defined time period, often measured in hours. These two single-beams are ratioed, resulting in a "100 percent line." In a perfect instrument with no noise and no instability, the result would be a straight line with the value of 1.00, or 100%. As time is increased in our imperfect real-world instruments, the line will become fuzzy as a result of noise, and will tilt or bend as a result of instability. Instrument makers specify stability as the maximum amount of deviation over a defined time at a particular wavelength or wavelengths. Agilent's Cary 630 FTIR Spectrometer Preventive Maintenance Checklist (Agilent Technologies 2016) specifies comparing the values at 1000 and 3000 cm^{-1} over a 30-minute period.

Changes in line-shape are principally results of four causes:

- Thermal expansion, which causes the optical elements to move relative to each other. If these movements happen within the interferometer, they will be wavelength dependent, seen as a tilt in the spectrum from one end to the other.
- Mechanical misalignment, caused by optical components shifting due to shock or vibration.
- Interferents in the beam path including changes in moisture or CO_2 content.
- Temperature changes in thermally sensitive components.

In benchtop instruments, designers use a number of techniques to ensure minimal change in instrument line-shape over time. Unfortunately, many of these techniques are not easily available to designers of portable instruments. Some of the common approaches and their applicability to portables include:

2.22.1 Mass

High-resolution, high-performance bench instruments use massive metal baseplates, often cast from aluminum. The large mass helps stabilize the optical path thermally and mechanically. The mass will be slow to respond to changes in temperature. The large mechanical size of the baseplate can also provide increased rigidity, reducing susceptibility to motion due to shock or vibration. Adding mass to a portable instrument is problematic as one of the primary requirements is typically to keep the weight as low as possible.

2.22.2 Active Thermal Control

Laboratory instruments often incorporate electric heating elements coupled to the spectrometer body. These heaters keep the instrument or subsystem at a constant temperature slightly above ambient. By minimizing temperature change with time, the instrument can minimize thermal motion and associated changes in optical alignment. This strategy is difficult to implement in a portable instrument because of the increased power requirement to operate the heaters, which significantly reduces battery life.

2.22.3 Long Warm-Up Time

When an instrument is powered on from a cold start, it undergoes significant thermal change. Every power-consuming element generates heat. This heat is not evenly distributed throughout the instrument. So, some parts will rise in temperature faster than others. The change in temperature causes thermal stresses and movement within the optical structures. In a benchtop instrument, the operator can "wait this out," leaving the instrument powered on for several hours to stabilize prior to making measurements. In a portable instrument, this is not an option for two reasons. First, the battery life is limited and the user does not want to waste a good portion of the battery just coming to thermal equilibrium. Second, and more importantly, the portable instrument is probably being deployed to generate time-sensitive information, such as the identity of a potentially hazardous material. Operators of portable spectrometers are almost always in a hurry, and cannot afford to wait for systems to stabilize thermally before making a measurement.

2.22.4 Dry-Gas Purge

As mentioned above, dry-gas purges can extend the life of hydrophilic optical elements. They can also reduce or eliminate the presence of light-absorbing moisture and CO_2 in the beam path. Dry gas is typically provided in the form of compressed Nitrogen, boil-off from a liquid Nitrogen dewar, or compressed air which has been desiccated using an air dryer. All of these options require some combination of volume, weight, or power which is not present in a portable system.

The above techniques are challenging or impossible to apply to portable spectrometers. We need to find some new approaches. Several successful techniques include:

2.22.5 Minimize Beam Path

Spectral components caused by moisture or CO_2 in the beam path are predictable using Beer's Law. The amount of signal is proportionate to three things:

1. The spectral absorptivity of the material.
2. The concentration of the material.
3. The path length of interaction between the material and the light beam.

We cannot control the absorptivity of the material, but we can minimize the concentration and the path length of interaction. Designing a system with a short beam path is consistent with our goal of making the instrument as small and light as possible. Because materials change size thermally as a percent of their original size, smaller components will exhibit less absolute movement for the same temperature change.

2.22.6 Seal and Desiccate

The other component of Beer's law that we can control is the concentration of interferents. The most common problems are seen from H_2O and CO_2, which are present in varying concentration in the atmosphere, and may change rapidly in the presence of human operators due to water vapor and carbon dioxide in their breath.

The obvious way to prevent change in the level of gas or vapor interferents is to seal the system, or as much of the optical path as is practical. It is important to provide a means to purge the system with dry gas before sealing it. A desiccant will absorb whatever moisture penetrates the seal. Dessicants are often considered user-replaceable, and the design should provide for this if it is felt to be necessary. A molecular sieve is a class of desiccant which can trap molecules other than water, including CO_2.

Because FTIR spectroscopy is a ratiometric technique, we can tolerate some degree of interference, as long as it is constant. The sealed design and presence of a dessicant or molecular sieve will stabilize the concentrations of CO_2 and water vapor, allowing them to ratio out of the sample spectrum.

2.22.7 Collect Backgrounds Frequently

Spectrometers change over time. This is particularly true at startup as the system temperature is changing rapidly. The more closely in time we can collect our background and sample single-beams, the less the instrument line-shape will change between the two measurements. Software in portable instruments is often structured to force the user to collect a background immediately prior to collecting the sample single-beam. This background single-beam can be scanned automatically for unusual spectral features and, thereby, be used to establish system cleanliness before allowing the user to place the sample.

2.23 Service

Portable instrumentation is subject to rough handling, from unintentional banging against furniture to operators who throw the instrument into the back of a pickup truck before heading out to the field. Despite every effort to design robustness into the system, it will eventually need to be serviced. Traditional laboratory instruments are usually serviced on site by a visiting service technician.

2.24 Communications/Wireless

As mentioned above, portable spectrometers often lack full-size display screens, keyboards, and mice. This can impose challenges in analyzing data on the device itself. It is often practical to move the data onto a PC for analysis or onto a network for archiving. Almost every modern portable device includes a USB port for this purpose. In a production or hot zone environment, having wireless communications can significantly reduce the amount of time and effort required to get data from the portable equipment to a fixed computer. In an office or lab environment, this may be as simple as adding a Bluetooth or Wi-Fi module to the system. For a hazardous materials team, the range of these technologies may not be adequate to reach from the operator to the command vehicle, which is typically outside the hot zone. A range of proprietary radio modems is available for this type of communications. Using a commercial module with Federal Communications Commission (FCC) approval can reduce the certification burden compared to designing a solution using component parts.

References

21 CFR 1000.40 (April 2019). Title 21 – Food and Drugs. *21 CFR 1000.40.*

Agilent Technologies (2016). *Cary 630 FTIR Spectrometer Preventive Maintenance Checklist – Standard*. Agilent Technologies.

Arno, J., Cardillo, L., Judge, K. et al. (2012). Advances in hand-held FT-IR instrumentation. In: *Proceedings of the SPIE 8374, Next-Generation Spectroscopic Technologies V, 837406* (17 May 2012). Baltimore, MD: SPIE.

C-Flex Bearing Co (2015). https://c-flex.com/pivot-bearings/life-projections (accessed 17 September 2019).

General Digital (2019). *Optical Bonding of Flat Panel Displays*. General Digital. https://www.generaldigital.com/optical-bonding-of-flat-panel-displays (accessed 17 September 2019).

Hamamatsu (2015). *Back-thinned TDI-CCD Technical Information*. Hamamatsu. https://www.hamamatsu.com/resources/pdf/ssd/tdi-ccd_kmpd9004e.pdf (accessed 20 September 2019).

Hamamatsu (2019). *Line Sensors for Spectrophotometry*. Hamamatsu. https://www.hamamatsu.com/resources/pdf/ssd/s11850-1106_kmpd1132e.pdf (accessed 20 September 2019).

Hsiao, H. (2015). Firefighter hand anthropometry and structural glove sizing: a new perspective. *Hum. Factors*: 1359–1377.

IEC (1989). *ANSI Web Store*. ansi.org. https://webstore.ansi.org/Standards/IEC/IEC60529Ed1989 (accessed 19 September 2019).

Infrared Associates (n.d.). *Infrared Associates HgCdTe (MCT) Detectors*. Infrared Associates. http://irassociates.com/index.php?page=hgcdte (accessed 20 September 2019).

Jackson, B. (2008). *ESD and RF Mitigation in Handheld Battery Pack Electronics*. TI.

Leonardo (2017). *Leonardo DLATGS IR Detector*. Leonardo. https://www.leonardocompany.com/documents/20142/3151790/DLATGS_IR_Detectors_for_Instrumentation_LQ_mm08096_b.pdf?t=1538987888860 (accessed 20 September 2019).

Nowakowski, R. and King, B. (2006). Choosing the optimum switching frequency of your DC/DC converter. *EE|Times*. https://www.eetimes.com/document.asp?doc_id=1272335# (accessed 17 September 2019).

NXP (2019). *NXP Processors and Microcontrollers*. NXP. https://www.nxp.com/docs/en/data-sheet/IMX6ULLIEC.pdf (accessed 15 September 2019).

Rippel, H. and Jaacks, R. (1988). Performance data of the double pendulum interferometer. *Mikrochim. Acta [Wien]* II: 303–306.

Sharp Electronics (2009). Device Specification for TFT-LCD Module Model No. LQ35Q3DG03 (30 October).

Sharp Electronics Corporation (2019). *Sharp Electonics Corporation Device Division*. Sharp Electronics Corporation. https://www.sharpsma.com/products?sharpCategory=Industrial%20LCD&p_p_parallel=0 (accessed 17 September 2019).

Silicon Labs (2019). *Precision32 32-bit Microcontroller Development Suite*. Silicon Labs. http://www.ti.com/tool/TIDA-00285?keyMatch=POWER%20TOOL&tisearch=Search-EN-everything&usecase=part-number (accessed 15 September 2019).

Spectral Systems (n.d.). *Properties of Materials for Spectroscopy*. Spectral Systems. https://www.spectral-systems.com/wp-content/uploads/2014/04/Properties-Chart_L.pdf (accessed 30 September 2019).

System Management Interface Forum (2013). *SMBus*. System Management Interface Forum. http://www.smbus.org (accessed 15 September 2019).

The Free Library (2019). 39th Annual R&D 100 Awards (October 2019). *The Free Library*. https://www.thefreelibrary.com/39th+Annual+R%26D+100+Awards.-a079197011 (accessed 15 September 2019).

Thermo Scientific (2016). *Thermo Scientific TruScan RMUnited States Pharmacopeia (USP)Chapter <1120> — Raman SpectroscopyStatement of Compliance*. Tewksbury, MA: Thermo Scientific.

TI (2019). *Battery Charger ICs*. TI. http://www.ti.com/power-management/battery-management/charger-ics/products.html (accessed 15 September 2019).

Unitika (2019). *Nylon Resin (Nylon 6) General Properties*. Unitika. https://www.unitika.co.jp/plastics/e/nylon/nylon6/03.html (accessed 15 September 2019).

US FDA (2018). *Radiation Safety Report*. US Food and Drug Administration. https://www.fda.gov/media/72606/download (accessed 13 September 2019).

3

Design Considerations for Portable Mid-Infrared FTIR Spectrometers Used for In-Field Identifications of Threat Materials

David W. Schiering[1] and John T. Stein[2]

[1] *RedWave Technology, Danbury, CT, USA*
[2] *JTS Consultancy, Ashburn, VA, USA*

3.1 Introduction and Background

The use of chemical weapons on civilians by the Iraqi military, the World Trade Center terrorist attacks of 11 September 2001, the anthrax letters sent through the US Postal Service mail system, and threats to military assets and civilian populations due to the use of improvised explosive devices (IEDs) have served to significantly change the capabilities of teams charged with ensuring safety and responding to chemical, biological, radiological, nuclear, and explosives (CBRNE) threats. The most sophisticated response teams, for example, the combating Weapons of Mass Destruction Civil Support Teams (WMD-CST), the New York City Fire Department (FDNY) Hazmat Team, the Chemical Biological Incident Response Force (CBIRF), or the US Army 20th CBRNE Command, are outfitted with mobile platforms containing the most advanced, in-field analyses technologies for detecting a broad range of CBRNE threats. The nature of the threat is continually changing due to the imagination and innovation of the terrorist adversaries. The evolving threat has created an environment for innovation in the scientific, technology development, and engineering communities charged with bringing improved capabilities to these sleuths and responders.

What are the needs of the responder or sleuth working in the field with instrumentation to identify threats and provide safety and security to the population? The responder desires solutions, not technology. The instrumentation must be easy to use and useable in the hot zone, where the threat is located. The operator wants actionable intelligence, not only a chemical or product name. The instrument should be lightweight, but rugged, durable, and dependable. It should identify multiple threats, have an easily expandable library, and the ability to analyze dirty samples. The instrument should be safe to use and useable by an operator encumbered by personal protective equipment (PPE). Preferably, there should be no consumables and the instrument should use commonly available, commercial-off-the-shelf batteries. Managing size, weight, and power (SWaP) while maintaining acceptable analytical performance is perhaps the dominant design driver for a portable device, one not appearing in the design trade space for the previous generation of laboratory, benchtop instruments.

Infrared (IR) spectroscopy is a mainstay analytical method for the identification and characterization of materials. The absorption bands in the IR spectrum arise from vibrational transitions in molecules and, therefore, the overall spectrum is very sensitive to changes in molecular structure. Coblentz [1] is generally credited with first understanding that the observed bands in the IR spectrum are associated with molecular bonds and atomic groupings, and the structural properties of molecules. Most of the IR absorption bands associated with fundamental vibrational modes are observed in the mid-IR region from 4000 to 400 cm^{-1}. Characteristic frequencies of IR absorption bands associated with molecular functional groups provide a basis for interpreting

Portable Spectroscopy and Spectrometry 1: Technologies and Instrumentation, First Edition.
Edited by Richard A. Crocombe, Pauline E. Leary and Brooke W. Kammrath.

IR spectra and classifying materials [2, 3]. Since the vast majority of covalently bonded molecules give rise to IR absorption, IR spectroscopy may be applied across many industries and disciplines, including criminalistics and forensic science. IR spectroscopy has been categorized as among "Category A" techniques, having "maximum potential discriminating power" by the Scientific Working Group for the Analysis of Seized Drugs (SWGDRUG) [4]. The IR absorption spectrum is related to the complex refractive index, a physical constant for a molecule. The IR spectrum is essentially a fingerprint of a molecule and can be used to identify unknowns when many candidates are possible. The identification is directed by digital comparison with a database of spectra of known materials. An example of identification from a portable spectrometer is illustrated in Figure 3.1, a screen capture of a threat analysis. Ammonium nitrate is a common oxidizer used by terrorists in homemade explosives (HMEs). In this software implementation, the threat spectrum can be viewed and overlaid with the spectral library data. A similarity score indicating the goodness of fit between the threat and library spectra, and threat chemical and safety data may also be displayed. Different portable spectrometers have different software implementations, but a common feature is the reliance on spectral matching. Spectral matching is challenged with samples that are mixtures or materials that are not represented in the library. Most commercial portable Fourier transform infrared (FTIR) instruments also employ some type of automated mixture analysis method.

The development of portable IR spectrometers initiated significant growth in the use and applications of spectroscopy systems, including both IR and Raman. Small, hardened, and tailored instruments found their way into the hands of many professionals who had no prior experience with sophisticated scientific instrumentation. This discussion shall be concerned with the instrumental aspects of portable FTIR spectrometers used in condensed-phase sample analyses, with particular focus placed on aspects that challenge portability and use in the field. We will confine the discussion to mid-IR FTIR instruments and fabrication methods currently employed for commercial instrumentation. A few groups have reported FTIR instrumental developments based on micro-electromechanical systems (MEMS) fabrication methods [5–7]; see the chapter in Volume 1 of this text by Grüger for a discussion of MEMS technology. A commercial MEMS FTIR spectrometer product is available that operates in the near-IR spectral region. While MEMS methods offer unparalleled potential for reducing the size and power requirements for the optical and mechanical elements of an FTIR platform, the current state of the art for IR sources, detectors and sample interfaces limit the reduction in overall instrument size that could be realized through the incorporation of MEMS technology. Also, current MEMS structures are so small that

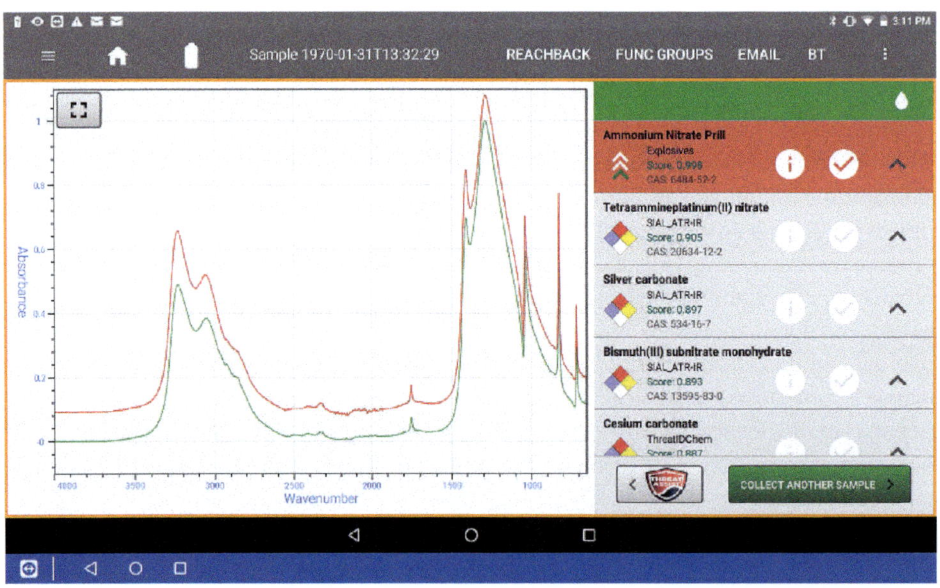

Figure 3.1 Analysis software screen for the identification of a homemade explosive (HME) component, ammonium nitrate. *Source*: Courtesy of RedWave Technology.

the reduction in the mid-IR signal is such that currently available, uncooled detectors do not provide sufficient signal-to-noise ratio (SNR) for practical use.

The first commercial portable FTIR spectrometer was the *TravelIR*™ (SensIR Technologies, LLC, Danbury, CT) introduced in 2000. One of the first applications of this spectrometer was in the mitigation of clandestine drug laboratories, collecting on-site evidence and chemical hazard data for site safety and remediation. The customers were US federal drug enforcement agencies. After the events of 2001, particularly the anthrax letters, field responders looked for better tools for responding to chemical incidents, providing immediate on-site intelligence and capability to address chemical, biological, and explosive threats. By 2002, the *TravelIR* was deployed into many response teams [8]. Early on, the limitations of the *TravelIR* became apparent. The size and weight were too large. The instrument was not waterproof nor could it be decontaminated, precluding its use where exposure to biological or chemical warfare agents (CWA) was a possibility. Although battery operation was possible, the instrument was almost always used with mains AC power as the battery was very heavy and cumbersome. The *SensIR* team immediately began development of a portable FTIR spectrometer that would address these shortcomings. In July 2003, the *HazMatID*™ (SensIR Technologies, LLC, Danbury, CT) was first shipped – more than 80 units to a Department of Defense (DoD) entity for use in base protection. The *HazMatID* was a small, suitcase-mounted FTIR spectrometer that was size reduced and hardened specifically for field response applications [9]. The instrument was submersible and fabricated of materials that would withstand corrosive decontamination fluids. Importantly, the instrument would operate greater than two hours using an integrated battery. These developments were firsts for any FTIR spectrometer. The *HazMatID* was marketed until 2018 by *Smiths Detection, Inc.* Several thousand were shipped and many are still in use.

Subsequent to these initial efforts, several companies have developed portable FTIR systems [10–12]. Current models of portable FTIR spectrometers are indicated in Figure 3.2, with some of their pertinent specifications. Suitcase-portable instruments provide a higher level of optical and computational performance, more software features, and larger screen size, while handhelds provide the ultimate in portability – size and weight reduction.

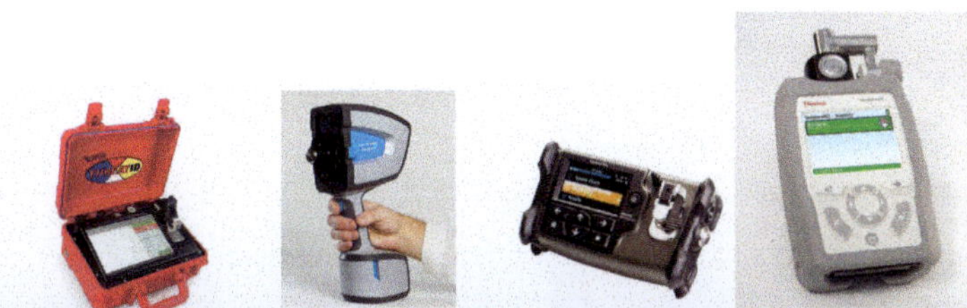

Year of introduction	2019	2014	2012	2010
Company	Redwave Technology	Agilent Technologies	Smiths Detection	Thermo Scientific
Size	14" × 10.6" × 6.12"	4" × 7.5" × 13.6"	10.6" × 5.6" × 3.1"	7.8" × 4.4" × 2.1"
Weight	15 lb	4.4 lbs	5 lbs	3 lbs
Spectral range	650 – 4000 cm^{-1}	650 – 4000 cm^{-1}	650 – 4000 cm^{-1}	650 – 4000 cm^{-1}
Resolution	4 cm^{-1}	4 cm^{-1}	4 cm^{-1}	4 cm^{-1}
Battery life	>4 h	3 h	>5 h	>2 h
Operational temperature	−10 to + 50°C	0 to + 50°C	−20 to + 50°C	−25 to + 40°C

Figure 3.2 Current models of portable FTIR spectrometers. *Source*: RedWave Technology, © Agilent Technologies, Inc., Image used with permission © Smiths Detection, and © Thermo Fisher Scientific.

Numerous in-field applications of portable FTIR spectrometers have been reported and include identification or analysis of CWA [13, 14], biological toxins [15], narcotic drugs [16], explosives [17], cultural heritage materials [18, 19], geologic materials [20], counterfeit pharmaceuticals [21], and regulated plasticizers in polymer products [22].

3.2 FTIR System Components

3.2.1 Interferometer Overview

The heart of an FTIR spectrometer is the interferometer. An absorption spectrum is not directly recorded. The recorded signal in an FTIR spectrometer is an optical interference signal called an interferogram. The IR spectrum is obtained by computation, transforming the time-based interferogram to a frequency-based spectrum, using a Fourier transform algorithm. Many of the details of this process are omitted in this discussion; the reader is directed towards several excellent descriptions [23–26]. We will be concerned primarily with instrumental aspects that impact miniaturization and field-use of FTIR systems and will discuss interferograms and spectra largely without connecting them.

A diagrammatical representation of a Michelson interferometer [27, 28] is shown in Figure 3.3. The IR source emission is directed towards and split by a beamsplitter into reflected and transmitted beams. The reflected and transmitted beams impinge onto fixed and moving mirrors, plane mirrors in this example. Ideally, the beamsplitter is 50% reflective and 50% transmissive. The beamsplitter is fabricated by coating an IR-transparent substrate that is very flat, typically <1/20 of the highest wavelength of interest. The beamsplitter coating can be multilayered and use different materials, but in the mid-IR region, germanium (Ge) is a significant component in the beamsplitter composition. To compensate for refraction and beam shifts in the beamsplitter substrate, an IR plate of equal

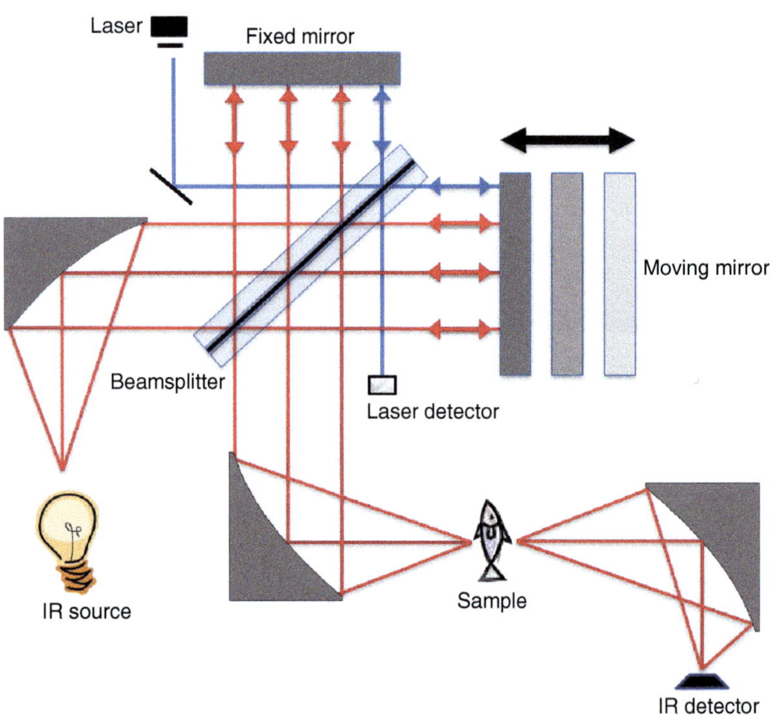

Figure 3.3 The Michelson interferometer configuration used for FTIR spectrometry.

thickness and flatness is placed over the beamsplitter. This is the compensator plate. The beamsplitter substrate, beamsplitter coating, and compensator plate comprise the beamsplitter assembly.

An optical interference signal is generated by varying the optical path difference (OPD) between the fixed and moving arms of the Michelson interferometer. The interferogram represents the intensity of this signal as a function of the OPD. The easiest way to visualize this is to first consider a monochromatic light source.

In an FTIR spectrometer, a laser reference system is used to ensure the sampling accuracy of the measured IR interferogram points. The laser has an essentially monochromatic output. The laser is directed to the same optics as the IR radiation, but uses a different beamsplitter coating and detector. The moving mirror is scanned continuously, but it is instructive to consider the laser interferogram construction in specific increments of the laser wavelength. When the distance between the fixed and moving mirror and beamsplitter is the same, the laser beam wavefronts at the beamsplitter are in phase and the two beams constructively interfere, yielding an intensity maximum measured at the detector. At a moving mirror position of ¼ the laser wavelength, the two beams are ½ wavelength out of phase, complete destructive interference results, and zero signal is measured at the detector. In this condition, all of the laser intensity is directed back towards the laser head. Since the beam travels twice the distance between the fixed and moving mirrors and the beamsplitter (forwards-backwards), the OPD is twice (2×) the difference between the distance between the beamsplitter and fixed and moving mirrors. At a moving mirror position of ½ the laser wavelength, the wavefronts at the beamsplitter are again in phase, constructive interference results, and a maximum signal is measured at the detector. At OPD that are integral multiples of the laser wavelength ($n * \lambda$), a signal maximum is measured; at OPD that are $n/2 * \lambda$, no signal is measured at the detector. The monochromatic signal intensity varies between zero and a maximum between $n/2 * \lambda$ and $n * \lambda$. Thus, the intensity of a monochromatic light source varies sinusoidally; it is a cosine wave. The interferometer modulates the beam at a frequency (in Hz), $f = 2V\nu$ where V is the mirror velocity in cm/s and ν is the spectral frequency in cm^{-1}. The situation is more complicated when the broadband IR source output is considered. When the OPD is zero, all IR wavelengths are in phase and constructively interfere; the IR interferogram signal intensity is the maximum. This is the centerburst of the interferogram, observed at zero optical path difference (ZPD). As the moving mirror is scanned, the IR wavelengths recombining at the beamsplitter undergo various degrees of constructive and destructive interference. This leads to a characteristic appearance of a polychromatic source with a high signal at ZPD and much lower signal on the wings of the interferogram away from ZPD. Each frequency in the IR spectrum is encoded by the interferometer in the audio frequency range by the relation $2V\nu$. The Fourier transformation process decodes this time-based frequency information into the spectrum.

In an FTIR spectrometer, a laser reference system is used to measure the OPD corresponding to each sample of the IR interferogram. The laser is directed to the same optics as the IR radiation, but uses a different beamsplitter coating and detector suitable for the laser wavelength, see Figure 3.3. As the laser has an essentially monochromatic output, the output produced by the laser detector is a cosine wave as described above. Sampling of the IR interferogram is typically performed at the zero crossings of the laser interferogram.

A benefit of an FTIR spectrometer is that the spectral resolution is constant over the spectral range and is determined by the maximum OPD. In the theoretical case, where the IR beam passing through the interferometer is perfectly collimated (beam divergence is zero), the achievable spectral resolution is $\mathbf{1/OPD_{max}}$, e.g. 0.25 cm OPD for 4 cm^{-1} spectral resolution. In this report, our interest is in condensed-phase applications where moderate spectral resolutions (4–8 cm^{-1}) will suffice. This is a benefit to the developer since the moving mirror will not be moved nor control required over larger distances, and allows some interferometer size reduction.

3.2.2 Optics

Brief mention can be made of the optics in the spectrometer. More detailed descriptions regarding the source, detector, metrology laser, and sample interface are made below. The intended field application for portable FTIR spectrometers limits the choice of transmission optics in the system. Most laboratory-based FTIR instruments operating in the mid-IR use potassium bromide (KBr) optics for the beamsplitter substrate, compensator plate, detector window, and sample compartment windows. KBr provides good spectral range coverage; however, the material is hygroscopic and must be protected from atmospheric humidity either by flushing the instrument with dry air or nitrogen, or by sealing and desiccating the spectrometer enclosure. Due to high humidity that can be encountered outdoors, KBr optics are not suitable for portable FTIR systems used for chemical defense or field forensics. Zinc selenide (ZnSe), a non-hygroscopic material, is used for the beamsplitter assembly and IR detector window. ZnSe suffers from high reflection losses; therefore, the surfaces are coated with a broadband antireflection coating to improve efficiency. ZnSe optics yield a mid-IR spectral range of 4000–650 cm^{-1}. Aspheric reflective focusing optics in the system, ellipsoids and paraboloids, are typically fashioned in aluminum (Al) by diamond turning machining, leading to a very precise optical figure, then overcoated with gold (Au) or Al for the reflective surface. The sample interface optics are discussed below.

Most portable FTIR interferometers use a Michelson configuration very much as illustrated in Figure 3.3. The *HazMatID* and *ThreatID*™ (RedWave Technology, Danbury, CT) employ a 60° beamsplitter orientation that provides greater throughput. Maintaining optical alignment is an important factor particularly for in-field units that may experience severe random vibration and shock environments. Software-actuated precision alignment is afforded by suitcase models. The motors used to align the fixed mirror require some space to integrate and contribute to the power budget, but can ensure that the interferometer is always aligned. The operator is alerted when alignment is needed and may be required to intervene.

Handheld units are permanently aligned by bonding or mechanically fixing the optics. Eliminating provisions for in-field alignment simplifies the optical system and aids miniaturization but at the expense of spectroscopy performance as the instrument must operate with some level of misalignment present. The *HazMatID Elite*™ (Smiths Detection, Inc., Edgewood, MD) design utilizes a double pendulum interferometer with cube corner optics [29, 30]. The double pendulum interferometer is a symmetric rotary mechanism and provides some optical compensation for misalignment during the course of a scan [23, 24, 31].

3.2.3 Interferometer Scanning System

Precise control of the interferometer moving mirror motion and sampling of interferogram points is a significant contributor to acquiring good-quality analytical data. The interferometer scanning control and signal digitization require significant engineering design and development expertise and effort, to realize the flawless integration of optics, electronics, and software. FTIR spectrometers universally use a voice coil actuator to drive the moving mirror. Most of the systems in Figure 3.2 utilize a blade flexure attached to the moving mirror assembly, the assembly then is attached to the actuator. These flexures are very durable and reliable in the field. The *HazMatID Elite* rotary interferometer utilizes crossed pair of flexures, a flex pivot, that constrain translation while providing the desired single rotational degree of freedom. Another type of interferometer moving mirror assembly is the tubular, "syringe" type [32]. In the syringe-type assembly, a circular moving mirror, attached to the voice coil actuator, moves within an outer tube that can be constructed of glass. This design inhibits tilting of the moving mirror during a scan. To reduce friction, the moving mirror is attached to a graphite piston that slides within the tube. Apparently, Thermo Scientific (previously Ahura Scientific) uses an interferometer of this type in its handheld spectrometers, even proposing a liquid material like silicon oil to reduce friction [33].

As discussed above, the interferogram of a monochromatic light source, a laser, is a sinusoidal signal, a cosine wave. As with their laboratory-based counterparts, portable FTIR spectrometers use a laser fringe-referencing method to control scanning and digitization of the IR interferogram. It is instructive to reference Figure 3.4 to understand the process. On the left of the figure are full-scale representations of the laser interferogram, the voice coil drive current, and a double-sided IR interferogram for a single stroke of the moving mirror assembly. The sloping, increasing coil current is the shape observed for a drive with a restoring force as with spring flexures or flex pivots. At the right in Figure 3.4 is a magnified view of a portion of the scan that allows visualization of the

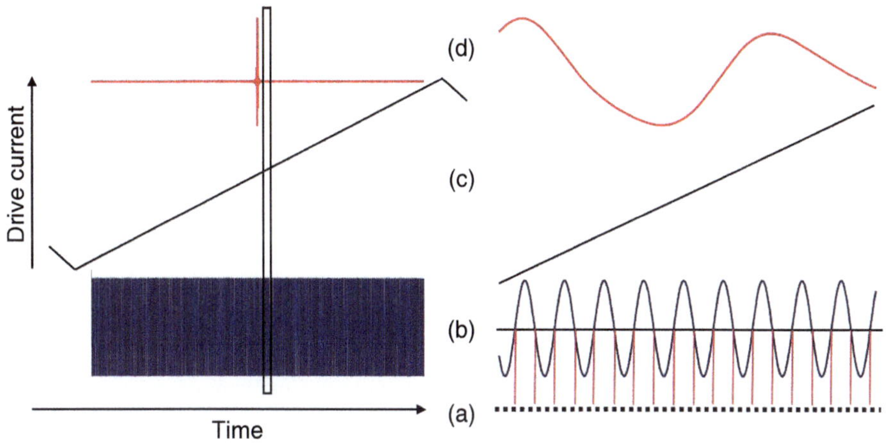

Figure 3.4 Interferometer control and interferogram digitization. (a) Delta sigma digitizer clock. (b) Laser interferogram with zero crossings indicated. (c) Voice coil drive current. (d) IR interferogram trace.

laser fringe spacing, zero crossings of the laser fringes, and the IR interferogram. In this illustration, every laser fringe zero crossing is indicated. One method of sampling the interferogram is to use the laser zero crossings to trigger the analog-to-digital converter (ADC) sample and hold circuit to digitize an IR interferogram datum. In this method, the laser interferogram is a continuous analog signal and only the IR signal points are digitized. In portable spectrometers, the wavelength of a 1310 nm diode laser yields a Nyquist frequency of 7633 cm^{-1}, if the interferogram is sampled at every fringe. The *HazMatID* and *HazMatID Elite* instruments generate multiples of the laser zero crossings by using electrical circuits and sample the IR interferogram at fractions of zero crossings using the ADC triggering mode. As suggested by Minami and Kawata [34], portable FTIR spectrometers developed by RedWave Technology and Agilent Technologies (formerly A2 Technologies) use delta-sigma ADCs and high-precision digital methods to sample the IR interferogram.

Delta-sigma ADCs were developed for digital audio applications and are relatively inexpensive since they are used in consumer products. Additionally, they are 24-bit precision and have multiple signal inputs. However, they do not incorporate sample-and-hold circuits so there is no triggering of the ADC; a different method of fringe referencing is required. The laser interferogram is digitized using one of the delta-sigma ADC inputs. These converters incorporate clocks with speeds >100 kHz. The bandwidth of a portable FTIR spectrometer is 3.8 kHz (Nyquist frequency) when running with a 1310 nm near-IR laser and OPD velocity of 0.5 cm/s. The delta sigma ADC provides significant oversampling of the laser interferogram, in equal time increments provided by the clock. Using digital processing techniques [35, 36], the laser zero crossing points are located very precisely by interpolation of clock sampling points on the interferogram. Delta sigma ADCs offer several advantages – high dynamic range, relaxed requirements for hardware band-limiting filters as the interferogram is sampled many times greater than the Nyquist limit, more forgiveness for velocity errors as the laser zero crossings are accurately located via interpolation and processing the interferogram, and changes and improvements can be made in software rather than necessitating hardware changes [37].

Another important function of laser fringe referencing is moving mirror velocity control. The control electronics of FTIR spectrometers include a proportional–integral–derivative (PID) servo controller that makes real-time adjustments to the voice coil drive current, keeping the laser fringes at constant separation in both OPD and time (constant frequency). This results in enhanced immunity to mechanical vibration as would occur in a moving vehicle and allows the bandwidth of the signal conditioning amplifier to be minimized without incurring phase errors. In early portable instruments, the PID servo was implemented as an analog circuit. In later instruments, the PID servo was coded in firmware in a field-programmable gate array (FPGA) integrated circuit. Current portable instruments integrate as much electronic function as feasible into FPGAs to reduce size by eliminating hardware circuits and because changes and improvements are easier to make, effectively requiring changes in software, not hardware changes.

3.2.4 Infrared Radiation Sources

Radiation sources for FTIR spectrometers emit broadband IR radiation when heated by passing a current through a resistive material; by definition, they are incandescent sources. These broadband IR radiation sources approximate the output of an idealized blackbody radiator where the spectral distribution of output brightness is determined by the temperature of the radiator, via Planck's Law, and the source emissivity. To realize high brightness and, therefore, higher SNR, increasing the source temperature as high as possible is warranted. In practice, there are limitations to increasing source temperature and brightness. Planck's Law, expressed in terms of spectral density as a function of frequency in cm^{-1} and temperature, is expressed in Eq. (3.1).

$$U_\nu(\nu, T) = \frac{2hc^2\nu^3}{\exp\left({}^{hc\nu}/_{kT}\right) - 1} \tag{3.1}$$

There is significant diversity in the availability of IR source components. Examples of IR sources that have been used or investigated for use in portable FTIR spectrometers are shown in Table 3.1 with their nominal operating temperatures and operating power requirement. The first three (3) sources in Table 3.1 have been used in commercial instruments and the fourth source has at the minimum been investigated for use. The computed blackbody spectral brightness of these four sources is shown in Figure 3.5.

The most common IR source in laboratory-based FTIR spectrometry is the air-cooled silicon-carbide (SiC) ceramic, sometimes called the *Globar*. When operated with the recommended voltage, the source temperature is 1035 °C (1308 K). It should be emphasized that the values in Table 3.1 are nominal. These sources can be operated within a voltage range. Higher voltages will result in higher color temperatures and higher spectral density at the expense of source lifetime. At the recommended voltages, the sources in Table 3.1 exceed 20,000 operating hours and rarely need replacement. *TravelIR*, the first portable FTIR spectrometer for condensed-phase threat assessment [8], was configured with a SiC source as in Table 3.1, operated at 10 V DC. This instrument was primarily powered with a mains AC power supply or a rather bulky, heavy DC battery power source.

The IR source is the most significant element in the power budget for a portable FTIR spectrometer and in subsequent instrument designs, focus was placed on this component for power reduction, reducing SWaP, and increasing battery charge life, albeit with a concomitant tradeoff in lower SNR. Reduction in power budgets for battery operation and reduced SWaP are not the greatest concern for laboratory instruments operating on mains power.

The *HazMatID* and subsequent commercial portable FTIR instruments utilized lower power wire sources fashioned from *Kanthal*® or other proprietary materials. As can be seen from Figure 3.5, there is as much as a 30%

Table 3.1 IR source choices for portable FTIR spectrometers.

Source type	Nominal operational temperature (°C)	Nominal operational power (W)
SiC ceramic (air-cooled)	1035	10.2
Kanthal® wire	950	4.2
Proprietary wire	900	2.1
2-D MEMS emitter	650	0.18

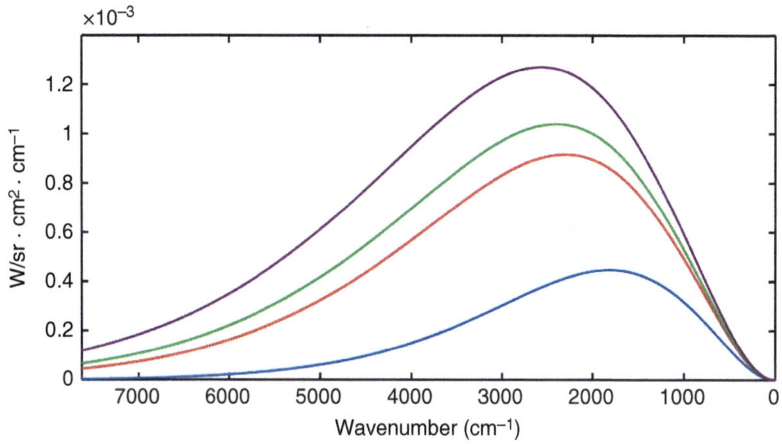

Figure 3.5 IR source brightness computed from Planck's Law (Eq. (3.1)) for sources in Table 3.1. Blue: Film 2D source; red: proprietary wire; green: Kanthal wire; purple: SiC ceramic.

reduction in peak signal with the wire sources compared to the *Globar*. However, the power reduction afforded by wire sources allowed further miniaturization and development of handheld FTIR spectrometers.

More reduction in source operating power may be gained by using thick film sources produced using MEMS manufacturing methods [38, 39]. Using such automated production methods, these components can be produced in great volumes and at low cost. These types of sources are available from several producers and their primary application has been for use in low-cost, single-purpose nondispersive IR (NDIR) analyzers. There are numerous designs available, but a common feature is the formation of a resistive element on a planar substrate. The resistive element is formed by thick film deposition techniques. These sources have low thermal mass, allowing the element to be heated and cooled rapidly to realize pulsed operation with an attendant reduction in power dissipation. The benefits accruing from this are manifold, as the reduced power dissipation reduces the burden placed on the thermal management structures required to reject the heat to the ambient as well as reduces the volume and weight of the battery. The performance cost is in SNR. Because they are operated at lower color temperatures, 2D sources generate less brightness and, therefore, less signal.

3.2.5 Infrared Detectors

IR radiation detectors convert photon energy to a voltage or current proportional to the IR energy flux. Detectors for FTIR spectroscopy fall into two (2) classes, thermal detectors or photon detectors. Typical detectors used for mid-IR spectroscopy are indicated in Table 3.2 with representative detector performance metrics. The specific detectivity, D^*, is a measure of the intrinsic sensitivity of the IR detector and is expressed in Eq. (3.2), below. D^* is the reciprocal of the noise equivalent power, NEP, normalized to the square root of the detector area, $A_\mathrm{d}^{1/2}$. NEP is a measured quantity that is the ratio of the noise spectral density (V/Hz$^{1/2}$) to the detector responsivity (V/W), a measure of the SNR due the detector element.

$$D^* = \frac{A_\mathrm{d}^{1/2}}{\mathrm{NEP}} \tag{3.2}$$

Mercury–Cadmium–Telluride (MCT) detectors are photon detectors that are commonly used in a photoconductive mode. These detectors operate by taking advantage of the intrinsic electro-optical properties of the MCT semiconductor. The band gap of this material is such that a mid-IR photon can promote an electron in the semiconductor from the valence band to the conduction band, increasing conductance and lowering resistance. A current

Table 3.2 Typical mid-IR detectors for FTIR spectrometry with operational metrics.

Detector element	Detector type	Spectral range[a),b)] (cm^{-1})	Operating temperature	D^* ($cm\ Hz^{1/2}/W$)
DLATGS	Pyroelectric	ca. 15 000–650	−20 to +55 °C	5.2×10^8
$PbZrO_3$–$PbTiO_3$ (PZT)	Pyroelectric	ca. 15 000–650	−20 to +70 °C	1×10^7
$LiTaO_3$	Pyroelectric	ca. 15 000–650	−20 to +70 °C	2×10^8
NB HgCdTe	Photoconductive	ca. 15 000–830	−78 °C	$\geq 4.0 \times 10^8$
NB HgCdTe	Photoconductive	ca. 15 000–750	77 K	4×10^{10}

a) Pyroelectric spectral response is limited by ZnSe optics.
b) Photoconductive detector high-frequency spectral response is limited by ZnSe optics.

is biased through the detector and the change in conductance induces a change in voltage developed across the detector that is proportional to IR flux. MCT detectors have the advantage of high intrinsic responsivity and low noise and are the most sensitive detectors (highest SNR) used in FTIR spectrometry for mid-IR applications. However, these detectors must be cooled and yield the best performance at cryogenic temperatures. MCT detectors can be cooled by using liquid N_2, a Stirling refrigerator, or a four-stage thermoelectric cooler (TEC). Although the SNR performance of MCT detectors cannot be approached with thermal detectors, the practical limitations imposed by cryogenic cooling, including access to cryogenic fluids, high power requirements and volume associated with mechanical refrigeration or multistage TECs, make MCT devices unsuitable for portable instruments.

To the knowledge of these authors, all commercial portable and handheld FTIR spectrometers employ deuterated L-alanine-doped triglycine sulfate (DLATGS) thermal detectors. DLATGS is a ferroelectric material that exhibits pyroelectric behavior. At temperatures below their Curie point, pyroelectric crystals are spontaneously electrically polarized. Under illumination by IR radiation, the extent of the electrical polarization of the crystal is thermally altered. This alteration of the electrical polarization state can be measured as an electrical signal when electrodes are attached to the faces of the pyroelectric crystal. The choice of DLATGS is due to its higher D^* and concomitant higher SNR performance than other thermal, pyroelectric detectors, such as lithium tantalate ($LiTaO_3$) or lead zirconate : lead titanate (PZT), see Table 3.2. DLATGS does present a challenge to the developer, however. The Curie point of DLATGS is 59 °C; at higher temperatures, the crystal "de-poles," and the detector does not respond to IR radiation. For portable in-field FTIR instruments, the operational temperature requirement can be as high as 50 °C ambient, and under a solar load with additional heat from the internal source and electronics, the internal temperature can be as high as 60 °C. Therefore, it becomes necessary to stabilize the DLATGS temperature using an integrated single-stage TEC. The power usage of the TEC must be considered in the context of heat rejection from the instrument, battery size and weight. In instrument designs employing thermoelectrically cooled detectors, the power demands of the TEC are often second only to those of the IR source. The temperature set point of the TEC can be programmed and in the *HazMatID Elite* instrument is held near 32 °C [30]. To reduce power consumption and battery size and mass, thermal management (heat dissipation) and its impact on the design of the spectrometer case and heat sinks have been considered and modeled using Finite Element Analysis [30].

The proximity of the TEC to the IR detector element allows mechanical noise produced within the TEC assembly to couple to the IR detector. As pyroelectric materials are typically piezoelectric as well, this results in the degradation of the signal developed by the IR detector. Strobing TEC operation in a manner synchronous with the interferometer scanning or other strategies may be effective in reducing this effect.

Detectors for FTIR are operated in an AC-coupled mode; the analog electronics that process the electrical signal only respond to the modulated component of the IR signal. As indicated above, the IR radiation is modulated in the audio frequency range by the interferometer. Ideally, the modulation frequency range should be matched to the optimum responsivity band of the detector. There are limitations to perfectly matching the optimum detector frequency response. Generally, the responsivity of the pyroelectric detector increases with decreasing frequency ($1/f$), with greater D^* at <100 Hz. To produce such a low modulation frequency, the moving mirror must scan extremely slowly, resulting in an undesired, long scan time. Slower optical velocities lead to poorer baseline stabilities in transformed spectra and the dynamic range of the ADC might be exceeded at the interferogram centerburst. A compromise in optimum responsivity is made for DLATGS detector operation. The interferometer scanning mechanism is generally operated at ~0.5 cm/s OPD velocity. This leads to a modulation frequency bandwidth of 325–2000 Hz in the spectral range of interest.

3.2.6 Metrology Laser

A key enabling technology for practical FTIR spectrometry is the reference laser metrology system and servo control of the mirror drive voltage, maintaining uniform velocity and fringe spacing during the scan. A separate but coincident laser interferometer is universal in FTIR spectrometers. As previously presented, the laser interferogram consists of a cosine wave whose period is precisely known. By sampling the IR interferogram at the zero crossings of this waveform, this precision is transferred to wavenumber axis of the spectrum derived from the IR interferogram.

Until the advent of portable FTIR spectrometers, the helium–neon (HeNe) gas laser was universally employed as the reference laser. The *TravelIR* used an HeNe laser, but, as mentioned above, did not provide usual operations under battery power. Gas lasers provide narrow output linewidths and wavelength stability. These properties are excellent benefits for a laser metrology system for FTIR spectrometry. However, HeNe lasers employ a high voltage power source for the discharge tube; therefore, their operating powers are >10 W for the models used in FTIR spectrometry. As with the IR source, the HeNe reference laser is a significant consumer of power and inhibitor to reducing SWaP. In addition, the HeNe laser and its associated power supply are quite bulky and heavy, adversely contributing to size and mass. In developing battery-integrated portable FTIR systems, a lower power reference laser was required.

The telecommunications industry has been a boon to the development of photonics systems, including low-power, rugged, small, solid state communications lasers. The *HazMatID* was the first commercial FTIR spectrometer to replace the HeNe reference laser with a solid state laser. Most of the commercial, portable FTIR spectrometers use a distributed feedback (DFB) telecommunications laser that operates in the near-infrared (NIR). This type of laser typically employs an InGaAsP/InP diode fabricated in quantum well structure. The power required to operate the diode laser is 0.024 W, a significant reduction in required operational power. Additionally, the size and mass of the diode assembly are miniscule in comparison to the HeNe laser and power supply. Performance compromises are made with diode lasers relative to HeNe. The output bandwidth of diode lasers is significantly greater than HeNe. NIR diodes have a specified output bandwidth of >5 cm^{-1}, much greater than HeNe which are better than 0.05 cm^{-1}. In practice, the NIR diode laser exhibits significantly narrower linewidths than specified, yielding measured IR spectral bandwidths of 2 cm^{-1}; however, they are only to be used for applications where moderate spectral resolution (4–8 cm^{-1}) will suffice, as in the case of condensed-phase unknown identification. The output wavelength will drift with temperature with the diode laser, causing a spectral shift in the recorded spectra. Manufacturers have employed temperature-stabilizing hardware in the diode assembly or spectral calibration methods to correct for these temperature-induced wavelength shifts.

3.2.7 Internal Reflection Sample Interface

The placement and removal of the unknown threat sample are important user-interface attributes in a portable FTIR system. Indeed, the first portable FTIR instrument was very much designed around its sample interface, which predated the spectrometer by a couple of years. A rugged, chemically resistant and optically transparent interface is required for portable FTIR instruments used in condensed-phase unknown analysis. Traditional IR analyses of liquids and solids were accomplished by transmission spectroscopy, requiring some degree of sample preparation and the use of alkali halide crystal materials, such as NaCl, KBr, or KCl. These materials are far less than optimum for field use as they are brittle, hygroscopic, and highly water soluble. Attenuated total reflection (ATR) IR spectroscopy is well known and has been used for analytical IR spectroscopy since the 1960s. ATR spectroscopy relies on the phenomenon of internal reflection. The theory and experimental aspects of ATR spectroscopy were developed by Harrick [40, 41] and Fahrenfort [42]. In this report, we shall not address many of the details of the theory and the reader is encouraged to consult the referenced literature. ATR spectroscopy requires an internal reflection element (IRE), an IR-transmitting crystal prism that is optically dense (higher refractive index than the samples analyzed). The IREs used for traditional ATR optics were most commonly Ge, ZnSe, and KRS-5 (ThI : ThBr). Incident IR radiation illuminates the sample through the crystal at an incident angle, θ_i, from normal. Radiation reflected from the internal surface of the prism exits at an equal reflection angle θ_r. If $\theta_i, \theta_r \geq \theta_c$, the critical angle as defined in Eq. (3.3) below, total internal reflection at the IRE/sample interface occurs. n_2 is the refractive index of the sample; and n_1 is the refractive index of the IRE. For a sample with a refractive index, n_2, of 1.5, a good estimate for many samples encountered, and an IRE refractive index, n_1, of 2.4, the critical angle, θ_c, is 39°. To satisfy this condition, most ATR interfaces are designed with a nominal angle of incidence, θ_i, that is 45° or slightly greater.

$$\theta_c = \sin^{-1}\left({}^{n_2}/{}_{n_1}\right) \tag{3.3}$$

At the reflection interface between the IRE and sample, the incident IR radiation electric field extends some distance beyond the interface and into the sample material. This nonzero electric field amplitude extending into the sample is called the *evanescent wave* and it is this evanescent wave that interacts with the molecules comprising the sample, leading to the observed IR absorption spectrum. Equation (3.4), below, describes the depth of penetration for the evanescent wave for a sample in contact with an IRE where radiation undergoes total internal reflection. The refractive indices are defined as above.

$$d_p = \frac{\lambda}{2\pi n_1 \sqrt{(\sin^2\theta - \left({}^{n_2}/{}_{n_1}\right)^2}} \tag{3.4}$$

θ is the angle of incidence relative to the normal of the IRE surface; λ is the wavelength of the incident IR radiation. For IR radiation in the mid-IR region (4000–650 cm^{-1} or 2.5–15.4 μm), the penetration depth, d_p, varies from ca. 0.5 to 3 μm across the spectral range. At these pathlengths, all IR bands are observed at <1 absorbance unit (AU), in the linear regime. As the total quantity of material sampled is ca. 2.3 μg, small quantities are required for high-quality spectra, and these instruments can be considered as micro-sampling tools.

ATR spectroscopy has practical advantages over transmission spectroscopy because (i) little or no sample preparation is required; the sample is measured by contact with the IRE; and (ii) the optical pathlength is reproducible and controlled for a given sample and is determined by the ratio the refractive index of a sample in good contact with the IRE and the refractive index of the IRE, both constants.

The traditional IRE crystals are not water soluble, but generally are not physically durable enough for use in austere environments in the field. Wilks [43] first reported the use of a diamond prism for use as an IRE. These early diamonds were natural Type-IIA, extremely expensive, and due to cost, generally not practical or commercially viable. Optical-grade, synthetic diamond, first produced in the mid-1990s [44], provided an optimum sample interface for instruments used in the field for condensed-phase sample analysis. The availability of highly pure, single crystal, Type-IIA synthetic diamonds has served to transform qualitative IR spectroscopy methods. Now, most qualitative IR analyses are performed with ATR methods using diamond IREs. All currently available, portable FTIR instruments use some type of diamond ATR sample interface.

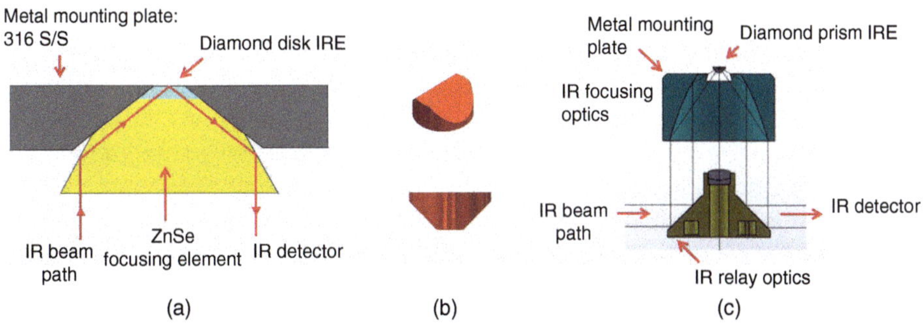

Figure 3.6 Single-reflection diamond ATR optics employed in portable FTIR systems. (a) Diamond disk IRE optically contacted with ZnSe. (b) "Rooftop" diamond IRE typical in prism configuration. (c) Diamond prism IRE with associated optics.

What makes diamond material, specifically Type-IIA diamonds, one of the best choices for IR spectroscopy and specifically in-field IR spectroscopic analyses? Type-IIA diamonds are free from nitrogen and other impurities that limit optical transmission. Type-IIA diamonds are optically transparent from the ultraviolet (UV) to the far-IR. Diamond is the hardest known substance. The Knoop hardness index for diamond is 9000 kg/mm^2. This hardness property makes diamond impervious to scratching and optical performance degradation, providing a rugged surface for hard, abrasive samples and ease in cleaning. Diamond is chemically resistant. Diamond will not react with encountered chemicals, including caustic materials and will withstand decontamination by common decontamination fluids including hypochlorite solutions. Finally, diamond has a relatively high refractive index, ca. 2.4, that makes it a good candidate for an IRE for IR spectroscopy.

There are two general ATR optical designs employed for field-portable FTIR spectrometers, as shown in Figure 3.6. All instruments used in threat identification use single-reflection ATR optics. Early instruments, including the *TravelIR* and *the HazMatID*, employed the configuration as shown in Figure 3.6a. This arrangement, referred to as a *DuraDisk*™ (SensIR Technologies, LLC, Danbury, CT), used a ZnSe-focusing crystal shaped with aspheric surfaces, optically contacted to a synthetic diamond disk that served as the sample interface [45, 46]. The ZnSe–diamond assembly is mechanically fixed into a stainless steel (in some cases Hastelloy®) disk that is integrated into the instrument top plate. As the refractive indices of ZnSe and diamond are nearly identical, nominally 2.4, there are no reflective losses at the ZnSe–diamond interface and the optical efficiency is very high. At the time, 1995–2002, this configuration was more cost-effective to produce than is the case today. Greater availability of cheaper, single-crystal synthetic diamonds caused designers to shift to prism IRE designs as shown in Figure 3.6c. Some variation of this design is employed in all portable ATR FTIR spectrometers today. The diamond IRE is an inverted "rooftop" that is fashioned from a truncated cylinder as illustrated in Figure 3.6b. The diamond is a single crystal with a circular sample-contacting surface that is 2–2.5 mm in diameter. The modulated IR source radiation is introduced into the crystal face by an aspheric-focusing reflector, either Al- or Au-coated. The detector-return-focusing optic is also an aspheric reflector. In either configuration (Figure 3.6a or c), a small amount of sample is brought into contact with the diamond and the IR spectrum is recorded in an ATR mode. For solids, an integrated press applies 15–50 lbs force to the sample, ensuring good contact with the diamond surface and spectra with maximized absorbances.

3.3 FTIR Spectrometer Performance Attributes

FTIR spectrometry has inherent advantages over other infrared instrumental methods, such as scanning monochromator or spectrograph instruments. For the most part, these advantages readily translate to portable or handheld applications. The first advantage to discuss is the *multiplex advantage*, or Fellgett's advantage. In an

FTIR spectrometer, all wavelengths are measured simultaneously. In comparison with a scanning monochromator, this leads to an SNR advantage or time advantage. All other parameters being equal, the FTIR instrument yields spectra with $M^{1/2}$ better SNR for equal scan times or $M^{1/2}$ faster acquisition for equal SNR, where M is the number of resolution elements, the spectral range divided by the spectral resolution, $\Delta\nu$. Spectrograph monochromator instruments are also multiplexing spectrometers. However, it is not possible to cover the mid-IR range with a single grating in a spectrograph without significantly degrading resolution. Spectrographs are not presently adequate for broad threat, in-field identifications based on spectral comparison methods.

The second advantage enjoyed by FTIR spectrometry is the *throughput advantage*, or the Jacquinot advantage. Throughput (or étendue), Θ, is a constant for radiometric optical systems and is the geometric light-gathering power, equal to the product of the area and solid angle, $A * \Omega$, at any focus. As FTIR spectrometers do not employ a slit, as do scanning or spectrograph monochromator systems, the throughput and, therefore, the signal, is higher for FTIR instruments operating at equal resolution. As we will discuss below, throughput is an important trade-off for portable instruments. Higher throughput translates to larger optics and larger instruments.

The last advantage of FTIR instruments that we will discuss is the *wavelength precision* or Conne's advantage. The wavelength accuracy and reproducibility of an FTIR instrument is far better than a dispersive spectrometer due to the laser-referencing interferometer control and accurate tracking of mirror position by using the laser interferogram fringes. This is quite important for the in-field identification of threats based on spectral comparison methods. Wavelength shifts between the sample and library spectra would degrade the comparison metrics and, therefore, the accuracy of identifications. Although the wavelength accuracy and reproducibility of portable FTIR spectrometers are not as high as HeNe-referenced instruments, it is good enough that spectral library reference spectra recorded on different models of FTIR spectrometers, including laboratory systems, can be used on the portables with good success in threat identifications in the field.

3.3.1 Trading Rules for FTIR Spectrometry

In an ideal scenario, the developer of an instrument would like to maximize all of performance attributes that are important for the intended applications of the instrument, SNR and spectral resolution frequently being the most important applications attributes. In a radiometric instrument, the signal amplitude is a function of IR source power and optical parameters, noise originates from the detector, associated electronics, and uncertainty in the OPD at which the interferogram is sampled.

$$\frac{\text{Signal}}{\text{Noise}} = \frac{U_\nu(T)\Theta\Delta\nu t^{1/2}\varepsilon}{\text{NEP}} = \frac{U_\nu(T)\Theta\Delta\nu t^{1/2}D^*\varepsilon}{A_d^{1/2}} \tag{3.5}$$

The SNR of an FTIR spectrometer can be estimated from Eq. (3.5) above. This relation ties together many of the physical attributes that we have discussed. For in-field threat identification, the SNR and wavelength precision are the two of the most important characteristics for analytical performance. Maximizing SNR allows more precise spectral comparisons and identification of minor components in mixtures, an important requirement, since many field samples are impure.

Equation (3.5) reflects the so-called "trading rules" for FTIR spectrometry. These physical parameters require balancing in design decisions for portable spectrometers, perhaps more so than in laboratory-based systems for which the management of SWaP is not a factor. The acquisition time, t, is the total integration time during data collection including signal averaging of scans. The acquisition time is an operator-selectable parameter, and of small consequence to design. The OPD velocity determines the time per scan and is usually selected to yield interferogram modulation frequencies that are compatible with the detector frequency response. Depending on the manner that interferograms are collected, the acquisition time per scan is typically in the range from one to four seconds. Typical default total acquisition times (signal averaging multiple individual scans) are 30 seconds to record the spectrum of the unknown threat.

Perhaps the easiest design parameter to balance is the best spectral resolution, $\Delta\nu$, expressed in cm^{-1}. The majority of portable and handheld FTIR spectrometers are targeted toward condensed-phase unknown identifications, not quantitative analyses of gas-phase molecules where higher spectral resolutions yield more accurate concentration determinations. For condensed-phase samples, the IR spectral linewidth is determined by intermolecular interactions and generally exceeds moderate instrumental spectral resolutions. Field-portable spectrometers have instrumental spectral resolutions no better than $4\,cm^{-1}$. Decreasing resolution to $8\,cm^{-1}$ would improve SNR and not degrade identification accuracy. Defining $4\,cm^{-1}$ as the best spectral resolution design specification somewhat simplifies the interferometer requirements, optical design, and size.

$U_\nu(T)$ is the IR source brightness, estimated by Planck's Law and expressed in Eq. (3.1) above. The design trade-off with higher source brightness (higher signal) is higher operating power and, therefore, larger SWaP. Optical throughput, Θ, should be conserved in the optical system from the source element to the detector. The typical DLATGS detector element size is 1 mm square. The diamond ATR element is ~2 mm at the sample. Beamsplitter diameters (apertures) are 25 mm for suitcase instruments and 12.5 mm for handheld instruments. The larger beamsplitters in suitcase-portable instruments, like *HazMatID* and *ThreatID*, yield significant SNR advantages over handhelds, albeit at the expense of SWaP. The final signal-related parameter, ε, is the optical efficiency. The detected IR signal amplitude is reduced by losses in the optical system. Fifty percent of the IR radiation is directed back towards the source by the beamsplitter, only 50% travels towards the sample. Most mirrors in the optical system are aluminum coated with reflectivity of 98.7% at $1000\,cm^{-1}$ resulting in 1.3% signal loss per reflection. The diamond ATR element has both reflection losses and absorptive losses. As refractive index of diamond is relatively high, 2.4, the reflection losses on the prism faces are 17% per face, leading to an efficiency of 69%. However, the transmission through the diamond prism can be increased as much as 30% through application of a broadband antireflection coating to the prism faces [47]. Two-phonon absorption of diamond is observed near $2000\,cm^{-1}$ [48] leading to significant absorption loss (and reduced signal) in this region of the spectrum. Fortunately, few chemical compounds exhibit fundamental IR absorption bands in this range, the exceptions being triple-bonded compounds, such as nitriles, cyanides, and alkynes, or cumulated double-bonded compounds, such as isocyanates and thioisocyanates. The beamsplitter substrate, compensator plate, and detector window are composed of ZnSe and comprise six (6) optical surfaces. Since the refractive index of ZnSe is also 2.4, the transmission efficiency through these optics is expected to be $(1 - 0.17)^6$, or 33%. Again, through the application of a broadband antireflection coating to the ZnSe surfaces, the transmission efficiency is significantly improved.

Equation (3.5) provides an SNR estimate presuming that the instrument is detector-noise limited. The NEP and corresponding D^* values are measured quantities for a given detector. In reality, there are noise sources that do not originate in the detector that can be manifested in the signal train and cause the real SNR performance to be less than detector-noise limited (the measured noise is greater than the detector noise). Potential noise sources include radiated energy picked up by the analog electronics on the preamplifier and filtering circuits, variations in the source intensity, and the TEC on the DLATGS detector. Through good engineering design and testing practice, these noise sources are minimized to the degree possible. Another noise source is due to sampling error caused by velocity jitter that can arise through external perturbations to the interferometer scanning mechanism. This is a particularly nettlesome problem for a handheld or portable field instrument used in uncontrolled environments. With modern FTIR technology, these scan errors may be compensated by signal-processing methods when the digitized laser interferogram is available for post processing with the IR interferogram [30].

3.4 Modeling and Simulation Guide Portable Instrument Design and Development

In this section, we discuss ways in which a simulation model can be used during the technology development and product definition phase of a new portable FTIR spectrometer development.

The first step in the design process for a new instrument is product definition, the object being to specify the instrument's capabilities in a way that best aligns with the perceived needs. The metrics used to specify these capabilities fall into two general categories: "performance" metrics, such as SNR and spectral resolution which the designer typically strives to maximize, and "cost" metrics which should be minimized. Cost metrics refer to parameters in limited supply; in the case of a portable instrument, the most expensive of these is frequently SWaP.

Performance metrics are generally interrelated by reciprocal functions where the designer may improve one aspect of instrument performance by accepting a reduction in another, an example being spectral resolution and SNR. The relationships linking performance and cost metrics are generally direct, i.e. improved performance implies a greater cost.

In the past, when the term "FTIR" implied a large instrument operating in the controlled environment of a laboratory and tethered to the power grid, the designer's task during product definition was to best balance the performance metrics. Cost metrics, particularly SWaP, were of little importance as the market placed no premium on minimizing these. To assist in balancing the performance metrics, "trading" rules, each of which defines the reciprocal relationship between a pair of performance metrics, were used.

The appeal of the trading rule lay in the insight it provided to the designer as the various aspects of instrument performance were balanced. The relationships were readily illustrated as curves in 2D spaces, and significant features, such as maxima or inflection points were readily identifiable.

3.4.1 The Need for a New Tool

As FTIR spectrometers shrunk from benchtop box to case-mounted portable units and handheld devices and the operating venue moved from the lab to the field, SWaP became a significant cost metric. Market demand for portability dictated smaller and lighter products, while the limited energy storage capacity of battery technology made reducing power consumption a critical factor.

Also accompanying the new generation of portable instruments were new performance metrics, such as run-time, robustness to thermal and vibratory environments, compatibility with decontamination procedures and user interface ergonomics. Many of these new metrics are strongly linked to SWaP.

The greater number of metrics implied more complex trading rules whose results could no longer be displayed in easily visualized two- or three-dimensional spaces. The inability to display the results in an intuitive format reduced the usefulness of the trade rule to the designer.

3.4.2 System Simulation as a Multiple-Dimension Trade Rule

A system-level math model may be viewed as the general case of a trading rule. The model represents the complete set of metrics, and the results set generated when one or more of the model parameters are varied represents the complete set of functional relationships present between the metrics. The model's usefulness during product definition lies in the ability it provides to the designer to extract subsets of the results suitable for display in an insightful format.

Such a model is extensible to accommodate any number of cost or performance metrics. Synthetic figures of merit (FOMs), e.g. run-time/watt-hour, may be defined and subsequently displayed and traded just as metrics corresponding to directly observable characteristics, e.g. spectral range.

3.4.3 Simulation Model Architecture

Because it provides easily visible linkages between the elements of a physical instrument and those of the simulation model, a model that is functionally partitioned by clearly defined interfaces and in a fashion similar to that of the physical instrument facilitates development, troubleshooting, and validation. Figure 3.7 illustrates a

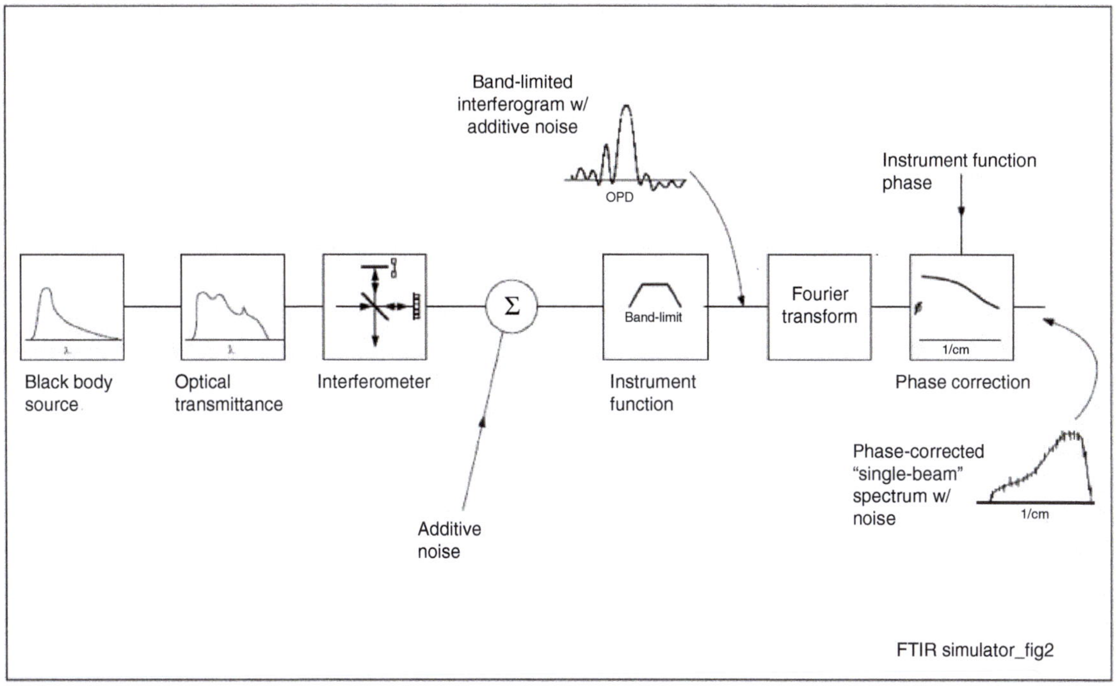

Figure 3.7 Functional partitioning of the modelling simulator. The system provides good visibility into the system operation by allowing observation of intermediate results, such as the interferogram or the power spectral density entering the interferometer.

model architecture that closely mirrors that of a physically realizable system. The presence of internal nodes corresponding to observable points in the actual instrument, e.g. the interferogram observed at the output of the IR preamplifier, facilitates validation of the model against actual hardware. This level of visibility also makes the model a useful tool for isolating faults occurring in an actual instrument.

The complexity of the simulation model and the level of fidelity of its results are functions of its intended application. If the model is to be used during requirements definition or anomaly investigation, it must provide accurate, precise results and provide good visibility into internal nodes implying a detailed, complex model. If the model will be used only as a guide during product definition, a lower level of performance and fewer inputs are required, and a simpler model may suffice.

Fortunately, the physics underlying FTIR is well understood and documented in the literature and the mathematics is general in nature and robust, factors which reduce the burden of developing a suitable model. A more detailed summary of the elements of an FTIR spectrometer system that can be modeled by the simulator and their pertinent inputs are presented in Table 3.3. The model is detailed, adhering to the parameters of Eq. (3.5), and includes signal elements as well as noise contributions from the detector and electronics, as well as phase contributions.

The design of the user interface is a strong determinant of the model's ease-of-use. A "dashboard" metaphor on which all user-definable parameters, salient results, and other "system-state" variables are displayed is an efficient interface design. Inclusion of some form of progress indicator to assure the user that the simulation has not "hung" is useful, particularly if the time required to execute a simulation exceeds a few seconds, which is often the case.

General-purpose script-based software tools, such as *MATLAB®*, are well-suited to performing the required mathematical operations. The requirement to generate textual scripts and the lack of a convenient method for incorporating blocks of explanatory text and comments are significant shortcomings of the tool.

Table 3.3 Simulation model elements and their corresponding elements of a physically realizable instrument.

Model element	Physical element	Domain	User-specified parameters
Foreoptics	IR source	Phys. optics	Color temp., Phys. dimensions
	Collimator	Geo optics	Phys. dimensions, Beam divergence
	Optical coatings	Phys. optics	Spec. transmittance
Sample	Sample	Phys. optics	Spec. transmittance
	ATR	Phys. optics	Spec. transmittance, dispersion
Interferometer	OPL modulator	Phys. optics	Stroke, pitch, Jitter, Mirror quality, Dispersion
Detector	Detector and FET	Complex freq.	Spec. responsivity, Freq. resp., Johnson & 1/f noise
Signal channel electronics	Preamplifier, freq. equalization, ADC	Complex freq.	Gain, Freq. resp., Johnson & quantization noise
Signal processing	Fourier transform, phase correct	Numerical processing	Phase correction strategy

The rightmost column of the table lists some of the parameters specified by the user during simulation.

Interpreted coding languages, such as *Python*, are also capable of performing the calculations but exhibit the same shortcomings. Also, as the graphics capability of most such languages is much less developed than that of *MATLAB®*, a user will most likely have to export the simulation results to another package for formatting and display.

Numerical word processors, such as *MathCAD®* and *SMath Studio*, produce "live" note sheets incorporating both text and familiar mathematical symbology. The word-processor paradigm makes incorporating textual notes and comments easy and the note sheet format executes as it is typed eliminating the need to develop scripts. The mathematical and graphics capabilities of *MathCAD* are sufficient for the task, although not as well-developed as those of *MATLAB*.

3.4.4 An Example – Optics and Source Trade Study

Figure 3.8 illustrates the use of a simulation model to identify which of the four candidate instrument configurations will meet a specified single scan, peak SNR requirement of 400 : 1 while minimizing unit size and power demands.

The simulation was set up to display single scan, single beam SNR as the performance metric. The diameter of the optical beam (beamsplitter) and the temperature of the IR source were defined as cost metrics serving as surrogates for instrument size and power, respectively. Two options for each of the cost metrics were assessed, resulting in the four simulation cases shown.

Three of the cases satisfy the SNR requirement. The large margin between the performance and the requirement seen in the high-power/large-beam example indicates a design significantly exceeding the stated target, and that the required performance could most likely be achieved at lower "cost."

This is borne out by the models that incorporate a larger beamsplitter with a lower power source and a smaller beamsplitter with a higher power source, Figure 3.8, middle, indicating that the requirement can be met with either of the cost metrics at the smaller of the possible values. The availability of two compliant configurations rather than just one provides the designer additional freedom during subsequent trades.

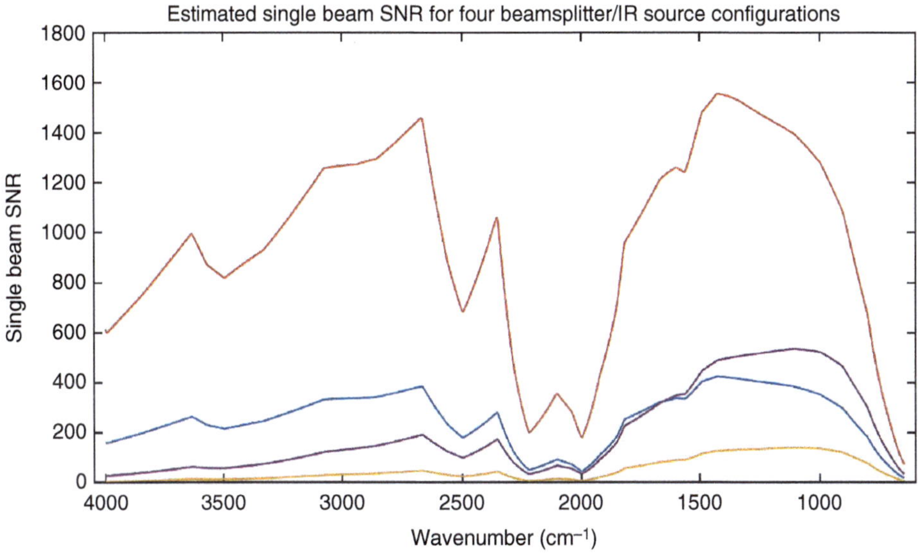

Figure 3.8 Single beam SNR, single scan simulation results for hypothetical miniaturized FTIR spectrometers with different beamsplitter diameters (throughput) and source color temperatures. Orange: 12.5 mm-diameter beamsplitter and 1176 K source color temperature; blue: 6.25 mm-diameter beamsplitter and 1176 K source color temperature; purple: 12.5 mm-diameter beamsplitter and 730 K source color temperature; yellow: 6.25 mm-diameter beamsplitter and 730 K source color temperature.

3.4.5 Power Budget Simulation

As mentioned previously, instrument size, weight and power are significant "cost" factors driving the design of a battery-powered portable FT instrument. The power demanded by the instrument will directly impact the size and weight of the battery, as both the volumetric power density (watt-h/in.^{3}) and the gravimetric power density (watt-hour/kg) of the battery are limited by the chemistry employed. As the battery is often the largest and heaviest single component in the instrument, its size and mass will significantly impact those of the instrument.

Increased power demand will also impact the instrument volume and weight indirectly through the increased complexity of the thermal management structures needed to reject the greater heat produced. Consequently, the power demand posed by a prospective instrument design can serve as a representative surrogate for all three elements of SWaP.

Estimating the power demanded by a prospective instrument design is readily accomplished through simulation. As the instrument performs a defined sequence of operations a defined number of times, while recording a spectrum, the instrument may be modeled as a set of discrete states, each representing a specific function, e.g. increment scanning mirror position, and search library for spectrum match. In operation, the instrument follows a path passing through all states, in some cases, revisiting a state repeatedly. The states are characterized by a common set of state variables, each of which indicates the condition assumed by a given hardware element and its power demand or the power required to execute a software process while in that state.

Summing the power values over the set of state variables and multiplying by the dwell time spent in the state yields an estimate of the total energy dissipated by the instrument while in the given state. Summing the energy values over the trajectory followed during the process yields an estimate of the total energy required to execute the process.

As the IR radiation source is frequently the single largest power dissipator in the instrument, the choice of technology utilized for this component will strongly affect the rate at which the battery is depleted, and, consequently, the number of spectra that may be acquired between battery replacement or re-charge.

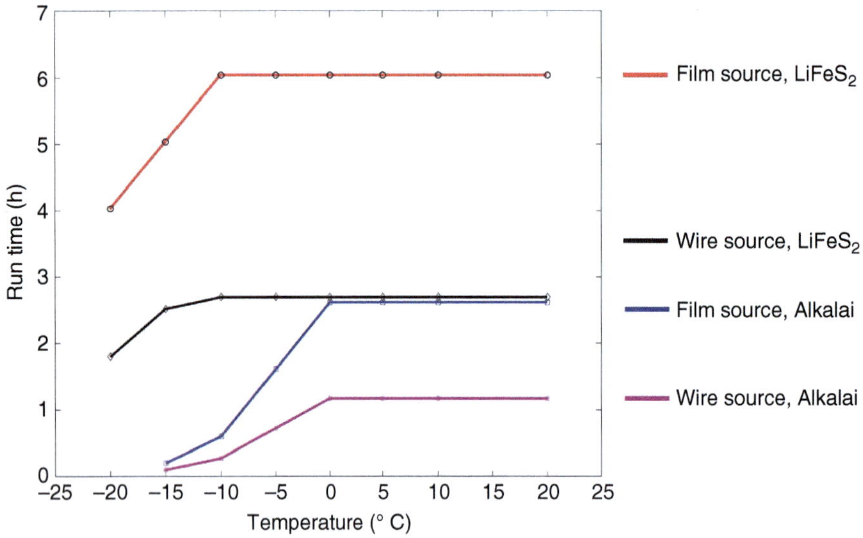

Figure 3.9 Simulator-estimated battery run-time vs. operating temperature for a hypothetical FTIR spectrometer configuration for two (2) IR source types and two (2) battery chemistries.

Shown in Figure 3.9 are the estimated battery-charge run-times for four hypothetical spectrometer hardware configurations – two IR source types and two battery technologies. The reduction in run-time with falling temperature is a consequence of the diminishing battery capacity in cold environments that is typical of most battery chemistries. As indicated by the red plot, the low power requirements posed by a power-strobed thick-film IR source coupled with the high capacity of lithium batteries result in an estimated run-time in a "warm" ambient environment of six hours. As a hypothetical FTIR instrument design matures and the power metrics are updated to better reflect the actual instrument design, the precision of the estimates improves.

It must be borne in mind that the selection of battery chemistry may not be made based solely on the run-time provided. The allowed thermal environments, the level of "care" demanded during recharge, possibility of spontaneous combustion, and transport limitations imposed on some battery chemistries are all factors that influence the selection of battery chemistry, in addition to device price.

3.5 Portable FTIR Performance Benchmarks

The SNR of a portable FTIR spectrometer is an important analytical performance attribute relative to the ability to identify unknowns. As we have discussed, SNR should be maximized, but achieving high SNR is counter to miniaturization. Representative measured absorbance noise for portable suitcase and handheld FTIR systems are shown in Figure 3.10. These data were recorded with DLATGS detectors in the mid-IR region from 4000 to 650 cm^{-1} for a one-minute acquisition using 4 cm^{-1} spectral resolution. The spectra are plotted on a common scale; the handheld data is offset for clarity. In both the cases, the absorbance noise is <0.001 AU across the mid-IR range (significantly less for the suitcase model) except in the diamond absorption region near 2000 cm^{-1} and in the low-frequency region where the source emission and optical efficiency fall off. The maximum sample absorbances for single-reflection diamond ATR measurements fall in the range from 0.15 to 0.5 AU; therefore, the peak-to-peak absorbance-to-noise ratios exceed 150 : 1. The noise for the suitcase-portable unit is ca. 4× less than the handheld, and even better than this in the high-frequency region. The major contributor is optical throughput. Also, the

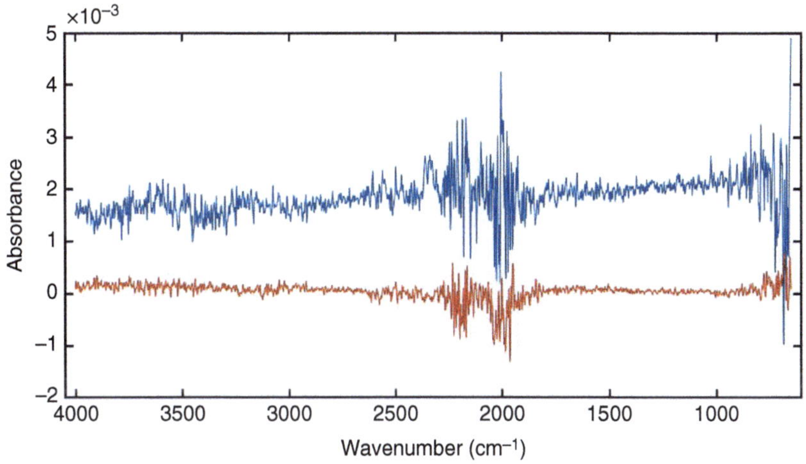

Figure 3.10 Representative absorbance noise recorded from handheld (blue) and suitcase-portable (orange) spectrometers. Measurement conditions – one minute acquisition, 4 cm^{-1} spectral resolution, single reflection diamond ATR, and DLATGS detector.

source color temperature is higher in the case of the suitcase-portable instrument resulting in higher signal in the high-frequency region of the spectrum in comparison with the handheld. Note that these observations are consistent with the SNR plots generated by simulation and presented in Figure 3.8.

The absorbance and wavenumber reproducibility are also important metrics for ATR spectroscopy systems. While providing ease in sampling and a rugged sample interface, there is more instrument-to-instrument variability in ATR spectroscopy when compared to transmission spectroscopy. Since library spectra are not always recorded on the same instrument as the unknown sample, this instrument-to-instrument variability can be important, particularly when analyzing for minor components in mixtures. It is possible that library spectra are recorded with completely different ATR optics than on the instrument used for analyses. As shown in Eq. (3.4), the penetration depth and, therefore, the absorbances of ATR spectra are dependent on the angle of incidence and the refractive index ratio of the sample and the IRE. Observed ATR spectra are sensitive to angle-of-incidence variations, particularly when close to the critical angle. Also, band shifts due to dispersion in the real part of the refractive index are observed in spectra when closer to the critical angle. The alignment of the ATR optics is not exact across instruments and this leads to small, effective incident-angle variations. These angle variations can contribute to both variation in absorbance intensity and shifts in the indicated wavenumber.

Representative instrument-to-instrument variation is shown in Figure 3.11 in which 2-propanol spectra were recorded in single-reflection mode with diamond ATR optics on 70 different instruments and are overlaid. It should be mentioned that there might also be some sample placement variation in these data; the diamond surface might not be completely covered in all cases. Also shown is the standard deviation over the spectral range from 4000 to 650 cm^{-1}. In this case, the variation appears to be intensity differences primarily. The ratio of the mean IR spectrum standard deviation to the standard deviation across all samples is 29 : 1. The instrument-to-instrument variation due to the ATR sample interface exceeds the instrumental noise contribution. This variation does not impact the major component identification, but can place a limit on fractional concentration limits of identification (LOI) for minor components in mixtures. Currently, minor components in mixtures can be identified at concentrations of ca. 10% or greater using automated software methods, lower if using expert operator-guided spectral interpretation skills.

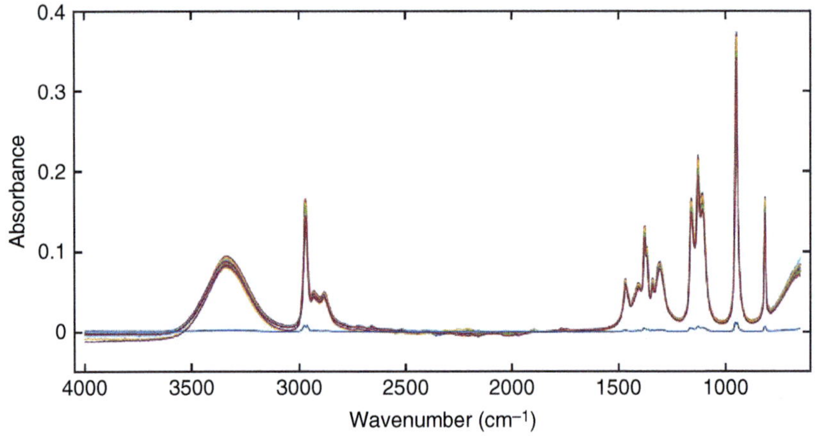

Figure 3.11 2-propanol (liquid) recorded on 70 different portable FTIR instruments. Also shown (blue) is the standard deviation at each spectral frequency. Measurement conditions: 30 seconds acquisition, 4 cm^{-1} spectral resolution, single reflection diamond ATR, and DLATGS detector.

3.6 Conclusion

The needs of the responder who is called to analyze chemical threats, for example, chemical warfare agents, toxic industrial chemicals, explosives, and narcotic drugs, drove the technology development of portable FTIR spectrometers. This led to a revolution in miniaturizing and hardening vibrational spectroscopy instrumentation in general. The footprint of laboratory systems has also been reduced, due to the performance and commercial success that portable spectrometers have enjoyed. In this report, we have discussed the technologies that have enabled the development of portable FTIR spectrometers and the performance tradeoffs faced in developing these instruments. The size reduction and performance of portable FTIR spectroscopy have relied on technology advances in numerous fields. Continued developments in the integrated circuit design and miniaturization of computers have provided smaller platforms while concomitantly increasing computational speed and solid-state memory capacity. Materials development, primarily made for abrasives, led to affordable, single crystal, synthetic diamond optics. The availability of embedded systems, e.g. FPGAs, digital signal processors (DSPs), has allowed the reduction in parts count and allotted volume for electronic hardware components. The development of Li-ion battery technology for portable computers provided a higher energy density needed to achieve acceptable run-time from a portable FTIR spectrometer. The telecommunications revolution provided small, low-power, solid state lasers, optical coatings, and fabrication methods. The digital electronics, multimedia industry provided low-cost, high-precision ADCs. Improvements in electronics-manufacturing methods, including multilayered printed circuit board (PCB) manufacture and flex circuits, also minimized the size contribution of electronics.

Abbreviations and Acronyms

2D	two-dimensional
ADC	analog-to-digital converter
ATR	attenuated total reflection
AU	absorbance unit
CBIRF	Chemical Biological Incident Response Force

CBRNE	Chemical, Biological, Radiological, Nuclear, and Explosives
CWA	Chemical Warfare Agents
DFB	distributed feedback
DLATGS	deuterated L-alanine-doped triglycine sulfate
DoD	Department of Defense (USA)
DSP	digital signal processor
FDNY	New York City Fire Department
FOM	figure of merit
FTIR	Fourier transform infrared
FPGA	field-programmable gate array
HME	homemade explosive
IED	improvised explosive device
IR	infrared
IRE	internal reflection element
LOI	limit of identification
MCT	mercury–cadmium–telluride
MEMS	micro-electromechanical systems
NDIR	nondispersive infrared
NIR	near-infrared
OPD	optical path difference
PCB	printed circuit board
PID	proportional-integral-differential
PZT	lead zirconate titanate ($PbZrO_3$-$PbTiO_3$)
SNR	signal-to-noise ratio
SWaP	size, weight, and power
SWGDRUG	Scientific Working Group for the Analysis of Seized Drugs
TEC	thermo electric cooler
WMD-CST	Weapons of Mass Destruction Civil Support Team (USA)
ZPD	zero-path difference

References

1 Coblentz, W.W. (1905). *Investigations of Infra-red Spectra, Publication No. 35*. Washington, DC: Carnegie Institution of Washington.

2 Colthup, N.B., Daly, L.H., and Wiberley, S.E. (1990). *Introduction to Infrared and Raman Spectroscopy*, 3e. St Louis, MO: Academic Press.

3 Bellamy, L.J. (1975). *The Infra-red Spectra of Complex Molecules*. Dordrecht: Springer.

4 Scientific Working Group for the Analysis of Seized Drugs (2019). *Scientific Working Group for the Analysis of Seized Drugs (SWGDRUG) Recommendations, Version 8*. http://www.swgdrug.org/Documents/SWGDRUG%20Recommendations%20Version%208_FINAL_ForPosting_092919.pdf (accessed 3 March 2020).

5 Khalil, D., Omran, H., Medhat, M., and Saadany, B. (2010). Miniaturized tunable integrated Mach-Zehnder MEMS interferometer for spectrometer applications. *Proc. SPIE* 7594: 75940T-1–75940T-13. https://doi.org/10.1117/12.843658.

6 Deutsch, E.R., Reyes, D., Schildkraut, E.R., and Kim, J. (2009). High resolution miniature FTIR spectrometer enabled by a large linear travel MEMS pop-up mirror. *Proc. SPIE* 7319: 73190J-1–73190J-8. https://doi.org/10.1117/12.818862.

7 Tortschanoff, A., Kenda, A., Kraft, M. et al. (2009). Improved MOEMS-based ultra-rapid Fourier transform infrared spectrometer. *Proc. SPIE* 7319: 73190I-1. https://doi.org/10.1117/12.818646.

8 Schiering, D.W. (2003). The TravelIR Haz-Mat Chemical Identifier. *Fire Eng.* 98: 98–99.

9 Norman, M.L., Gagnon, A.M., Reffner, J.A. et al. (2004). An FT-IR sensor for identifying chemical WMD and hazardous materials. *Proc. SPIE* 5269: 143–149.

10 Mukhopadhyay, R. (2004). Portable FTIR spectrometers get moving. *Anal. Chem.* 76: 369–372A.

11 Sorak, D., Herberholz, L., Iwascek, S. et al. (2012). New developments and applications of handheld Raman, mid-infrared, and near-infrared spectrometers. *Appl. Spectrosc. Rev.* 47 (2): 83–115. https://doi.org/10.1080/05704928.2011.625748.

12 Crocombe, R.A. (2018). Portable spectroscopy. *Appl. Spectrosc.* 72 (12): 1701–1751. https://doi.org/10.1177/0003702818809719.

13 Bryant, C.K., LaPuma, P.T., Hook, G.L., and Houser, E.J. (2007). Chemical agent identification by field-based attenuated total reflectance infrared detection and solid-phase microextraction. *Anal. Chem.* 79: 2334–2340.

14 Ong, K.Y., Baldauf, F.C., Carey, L.F., and St. Amant, D.C. (2003). Domestic preparedness program: evaluation of the TravelIR HCI™ HazMat chemical identifier. *ECBC-TR-355*. DIR, ECBC, ATTN: AMSSB-RRT-AT, APG, MD 21010-5424. https://apps.dtic.mil/dtic/tr/fulltext/u2/a421836.pdf (accessed 3 March 2020).

15 Schiering, D.W., Walton, R.B., Brown, C.W. et al. (2004). Towards the characterization of biological toxins using field-based FT-IR spectroscopic instrumentation. *Proc. SPIE* 5585: 21–32.

16 Manali, D. and Seelenbinder, J. (2016). Automated fast screening method for cocaine identification in seized drug samples using a portable fourier transform infrared (FT-IR) instrument. *Appl. Spectrosc.* 70 (5): 916–922.

17 Arnó, J., Frunzi, M., Weber, C., and Levy, D. (2013). Advanced sampling techniques for hand-held FT-IR instrumentation. *SPIE Proc. Vol. 8726, Next-Gen. Spectrosc. Technol.* VI: 87260U-1–87260U-8. https://doi.org/10.1117/12.2016148.

18 Vagnini, M., Gabrieli, F., Daveri, A., and Sali, D. (2017). Handheld new technology Raman and portable FT-IR spectrometers as complementary tools for the in situ identification of organic materials in modern art. *Spectrochim. Acta A* 176: 174–182.

19 Bell, J., Nel, P., and Stuart, B. (2019). Non-invasive identification of polymers in cultural heritage collections: evaluation, optimisation and application of portable FTIR (ATR and external refectance) spectroscopy to three-dimensional polymer-based objects. *Herit. Sci.* 7 (95): 1–18. https://doi.org/10.1186/s40494-019-0336-0.

20 Tang, P.L., Alqassim, M., Daé, N.N. et al. (2016). Nondestructive handheld Fourier Transform Infrared (FT-IR) analysis of spectroscopic changes and multivariate modeling of thermally degraded plain portland cement concrete and its slag and fly ash-based analogs. *Appl. Spectrosc.* 70 (5): 923–931.

21 Bansal, D., Malla, S., Gudala, K., and Tiwari, P. (2013). Anti-counterfeit technologies: a pharmaceutical industry perspective. *Sci. Pharm.* 81 (1): 1–14. https://doi.org/10.3797/scipharm.1202-03.

22 Higgins, F. (2013). *Rapid and reliable phthalate screening in plastics by portable FTIR spectroscopy – application note. Agilent Technologies Publication number: 5991-3649EN*, 1–8. Agilent Technologies.

23 Griffiths, P.R. and DeHaseth, J.A. (2007). *Fourier Transform Infrared Spectrometry*, 2e. New York: Wiley-Interscience.

24 Jackson, R.S. (2002). Continuous scanning interferometers for mid infrared spectrometry. In: *Handbook of Vibrational Spectroscopy*, vol. *I* (eds. J.M. Chalmers and P.R. Griffiths), 264–282. New York: Wiley.

25 Sumner, P.D., Abrams, M.C., and Brault, J.W. (2001). *Fourier Transform Spectrometry*. San Diego, CA: Academic Press.

26 Griffiths, P.R., Foskett, C.T., and Curbelo, R. (1972). Rapid-scan infrared Fourier transform spectroscopy. *Appl. Spectrosc. Rev.* 6 (1): 31–78.

27 Michelson, A.A. (1891). Visibility of interference-fringes in the focus of a telescope. *Philos. Mag., Ser. 5* 31: 256–259.

28 Michelson, A.A. (1903). *Light Waves and their Uses*. Chicago: University of Chicago Press.

29 Schiering, D.W., Arnó, J., Messerschmidt, R.G., and Zou, P. (2010). Performance trade-off modeling for a handheld FTIR spectrometric vapor identifier. *Proc. SPIE* 7680: 76800R1-10.

30 Arnó, J., Cardillo, L., Judge, K. et al. (2012). Advances in handheld FT-IR instrumentation. *Proc. SPIE* 8374: 837406–837414. https://doi.org/10.1117/12.921052.

31 Rippel, H. and Jaacks, R. (1988). Performance data of the double pendulum interferometer. *Mikrochem. Acta (Wein)* II: 303–306.

32 Coffin, J. (1999). *Interferometer Having Glass Graphite Bearing*, US Patent No. 5,896,197. Washington, DC: US Patent and Trademark Office.

33 Azimi, M., Bibby, A., Brown, C.D. et al. (2011). *Handheld Infrared and Raman Measurement Devices and Methods*, US Patent No. 7,928,391 B2. Washington, DC: US Patent and Trademark Office.

34 Minami, K. and Kawata, S. (1993). Dynamic range enhancement of Fourier transform infrared spectrum measurement using delta sigma modulation. *Appl. Optics* 32 (25): 4822–4827.

35 Brault, J.W. (1996). New approach to high precision Fourier transform spectrometer design. *Appl. Optics* 35: 2891–2896.

36 Turner, A., Hoult, R.A., and Forster, M.D. (1999). *Digitisation of Interferograms in Fourier Transform Spectrometry*, US Patent No. 5,914,780. Washington, DC: US Patent and Trademark Office.

37 Brault, J.W. (1997). New developments in FT-IR spectrometer design. *Mikrochem. Acta (Suppl.)* 14: 121–124.

38 Hildenbrand, J., Kürzinger, A., Peter, C., et al. (2008). Micromachined mid-infrared emitter for fast transient temperature operation for optical gas sensing systems. *IEEE SENSORS 2008 Conference Proceedings*, Lecce, Italy (26–29 October 2008), pp. 297–300.

39 Sebastian Weise, S., Bastian Steinbach, B., and Steffen Biermann, S. (2016). MEMS-based IR sources. *Proc. SPIE* 9752: 97521E-1–97521E-7. https://doi.org/10.1117/12.2208410.

40 Harrick, N.J. (1960). Surface chemistry from spectral analysis of totally internally reflected radiation. *J. Phys. Chem.* 64 (9): 1110–1114.

41 Harrick, N.J. (1967). *Internal Reflection Spectroscopy*. New York: Wiley.

42 Fahrenfort, J. (1961). Attenuated total reflection – a new principle for the production infra-red reflection spectra of organic compounds. *Spectrochim. Acta A* 17: 698–709.

43 Wilks, P.A. Jr., (September, 1977). Use of diamonds in internal reflection spectroscopy. *Am. Lab*: 79–83.

44 Harris, D.C. (1994). Development of chemical-vapour-deposited diamond for infrared optical applications – status report and summary. *Office of Naval Research: NAWCWPNS TP 8210*. https://apps.dtic.mil/sti/pdfs/ADA282999.pdf (accessed 3 March 2020).

45 Sting, D.W. and Milosevic, M. (1997). *Optical Sensing with Crystal Assembly Sensing Tip*, US Patent No. 5,703,366. Washington, DC: US Patent and Trademark Office.

46 Milosevic, M., Sting, D., and Rein, A. (1995). Diamond composite sensor for ATR spectroscopy. *Spectroscopy* 10: 44–49.

47 Spectral-Systems. *Application Note: Diamond Anti-Reflection Coating*. Hopewell Junction, NY: Spectral-Systems. www.spectral-systems.com (accessed 3 March 2020).

48 Wehner, R., Honk, H., and Kress, W. (1967). Lattice dynamics and infra-red absorption of diamond. *Solid State Commun.* 5: 307–309.

4

PAT Applications of NIR Spectroscopy in the Pharmaceutical Industry

Pierre-Yves Sacré, Charlotte De Bleye, Philippe Hubert and Eric Ziemons

Department of Pharmacy, University of Liege, Liege, Belgium

4.1 Introduction

This chapter covers applications of small, miniature, and portable near-infrared (NIR) spectrometers in process analytical technology (PAT) in pharmaceutical manufacturing. Many of the unit processes in secondary pharmaceutical manufacturing employ mobile, moving and/or high-vibration processing equipment, and, in addition, the space available for an analyzer is often very small. These factors lead to the use of miniature and portable spectrometers, sometimes battery-powered, and communicating via wireless.

The technologies employed in these spectrometers are described elsewhere in these volumes, for instance, in the chapter by Pust on optical filter technologies, and by Grüger et al. on micro-electro-mechanical systems (MEMS) and micro-opto-electro-mechanical systems (MOEMS) spectrometers. Other applications of portable NIR instruments are covered by Siesler and colleagues (materials analysis in general), Miseo, Meyer and Ryan (food) and Beć, Grabska and Huck (food, feed and agriculture).

4.2 Continuous Manufacturing and Real-Time Release Testing

4.2.1 Continuous Manufacturing

Fifteen years ago, in the launch of the PAT initiative by the Food and Drug Administration (FDA), the pharmaceutical industry was asked to continue to improve the quality of the manufactured pharmaceuticals [1]. The pharmaceutical industry mainly works in a batch release manner; that is to say that a final drug product is the result of several independent production steps. These may also take place in different geographical areas, leading to the shipping and storage of the different intermediates in containers to the next manufacturing facility. This increases the risk of degradation over time, or due to environmental conditions (light, humidity, etc.). Therefore, increasing the quality of pharmaceutical manufacturing will both decrease the risks for the patients, and the costs for the manufacturer linked to the destruction of out-of-specification batches.

One way to tackle this issue is to move from a batch release to a continuous release paradigm, and towards continuous manufacturing in the pharmaceutical industry, is supported by the emerging technology program of the FDA [2]. Emerging technologies are defined as innovative or novel: (i) product-manufacturing technology; (ii) manufacturing process; and/or (iii) testing and monitoring technology [3]. Continuous manufacturing is not a new concept, and is already implemented in other industries, but very few pharmaceutical processes are actually continuous [1]. This may be partially explained by the costs needed for a continuous manufacturing implementation, and the relatively low actual level of knowledge within the pharmaceutical industry of continuous manufacturing processes. A continuous process is one in which materials are ceaselessly charged

Portable Spectroscopy and Spectrometry 1: Technologies and Instrumentation, First Edition.
Edited by Richard A. Crocombe, Pauline E. Leary and Brooke W. Kammrath.

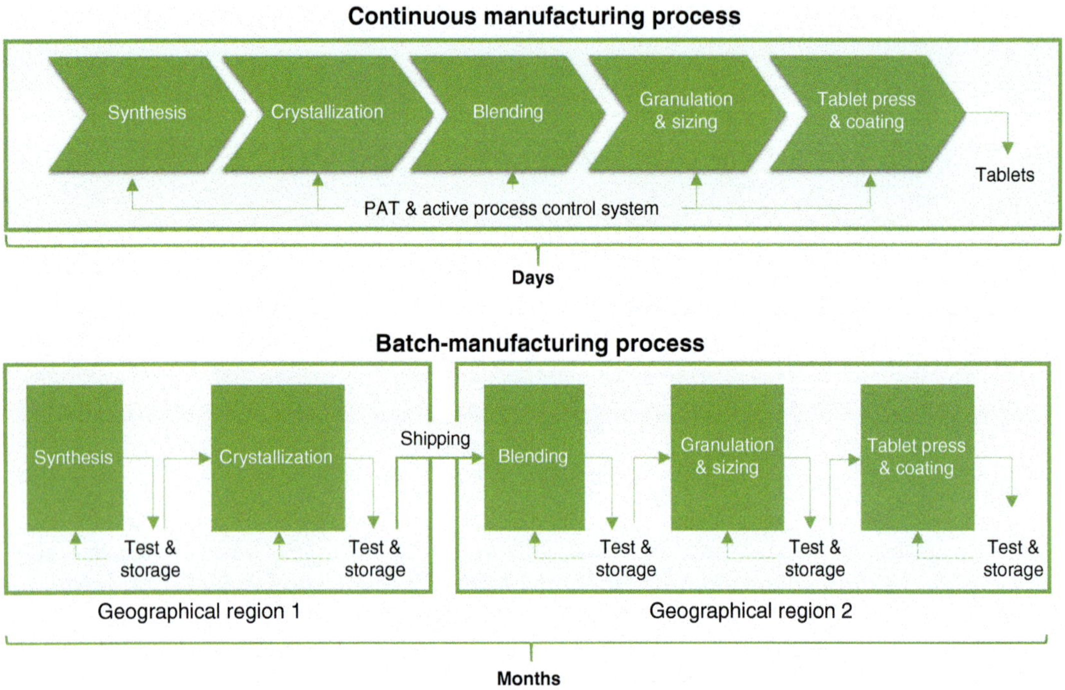

Figure 4.1 Comparison between continuous manufacturing and batch manufacturing.

into the system, while the final product is continuously discharged [4]. The opposite of batch release, this concept implies the full connection of the production units, with the use of PAT systems, together with process-control systems to monitor and control the integrated manufacturing plant (see Figure 4.1). The continuous process units are generally more efficient, more reactive, reduced in volumes, and have an augmented yield and fewer wastes compared to classical process units. These types of production units may, therefore, respond faster to a drug shortage or sudden changes in demand or need (such as in epidemic outbreak). Furthermore, their reduced size enables their transport directly to places where the drugs are needed [5].

However, continuous manufacturing needs the process to be designed and conceived as a single system. It is, therefore, necessary to have a thorough understanding of the process including the different connections between their process units. This understanding is achieved through the use of the concepts of Quality by Design (QbD) [6, 7]. The QbD concept implies the definition of the quality target drug product profile (QTPP) that identifies the critical quality attributes (CQAs) of the drug product (e.g. active pharmaceutical ingredient – API – content, pH, moisture, flowability, or particle size). Once these CQAs are defined, the critical process parameters (CPPs; e.g. screw speed, solid feed rate, and liquid feed rate), and the critical material attributes (CMAs; e.g. particle size, particle shape, cohesive/adhesive properties, and electrostatic properties) are identified, and the links between these CMAs and CPPs to the CQAs are elucidated using a design of experiments (DoE) (see Figure 4.2). This understanding of the process leads to the development of a control strategy to ensure that the CMAs and CPPs are within the appropriate (and previously defined) ranges of the design space (DS), ensuring the final product quality [8–11]. This real-time monitoring of the process, and the insurance of the quality of the final product through the design of the process itself, naturally lends itself to real-time release testing (RTRT) of the final product. Another important regulatory aspect associated with the definition of a DS is that modifying the operating conditions, while staying within the limits of the design space, is not considered as a change, and does not require a regulatory

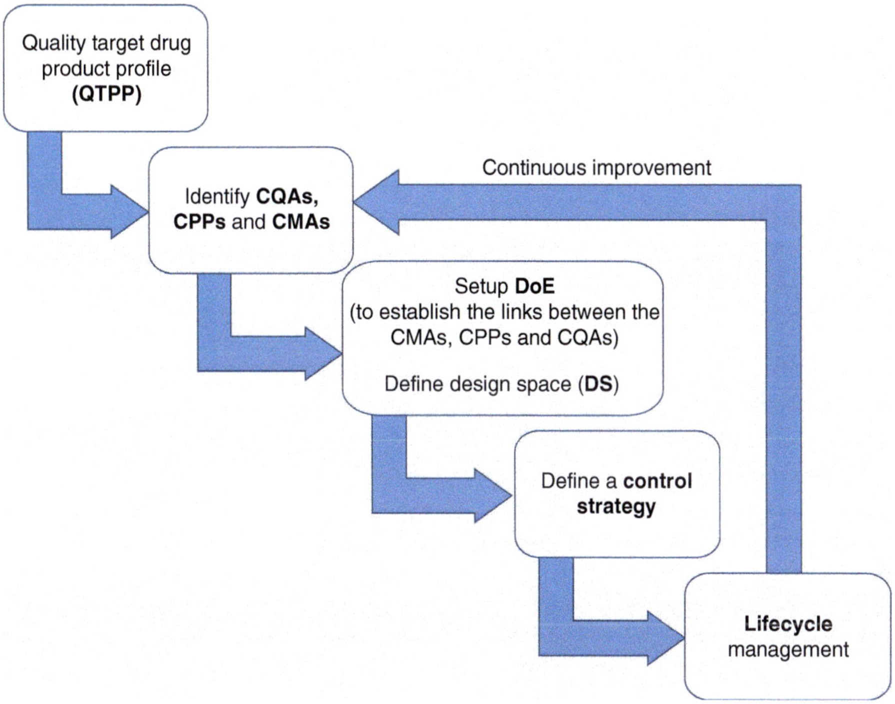

Figure 4.2 QbD typical flowchart. CQA: critical quality attributes; CMA: critical material attributes; CPP: critical process parameters; DoE: design of experiments.

post-approval change [12]. A nice example of the whole QbD continuous manufacturing in a good manufacturing practice (GMP) environment has recently been reported by Roggo et al. [13].

These concepts are already implemented in the International Council for Harmonisation of Technical Requirements for Pharmaceuticals for Human Use (ICH)[1] via quality guidance documents ICH Q8/Q9/Q10/Q11 and Q13 [14]. However, there are very few continuous manufacturing processes actually working. Indeed, only five drug products manufactured in this way are actually in the market: Orkambi® and Symdeko® (Vertex); Prezista® (J&J); Verzenio® (Eli Lilly); and Daurismo® (Pfizer) [14]. Many reasons may explain this, but one of these is that the transfer of active process control from development to manufacturing is not straightforward. Indeed, most units of production are still controlled one at a time. Therefore, once a disturbance is detected, the adjustment of the CPP variable takes a long time, resulting in a long period of generation of nonconforming products. Therefore, the recently introduced concept of quality by control (QbC) insists on the need for "an active process control system based on a high degree of quantitative and predictive product and process understanding," enabling real-time and automated adjustment of the CPPs [14]. This automated adjustment is part of the "smart factories" concept. This relatively new concept pushes forward all the previously described initiatives with a virtual twin of the factory using artificial intelligence to anticipate each adjustment of the process and its consequences. This virtualization of the manufacturing plant is the future of PAT strategies, called cyber-physical-based PAT [15]. Another issue is the definition of "batch," and the raw material traceability required by regulatory agencies. These items may be addressed by the use of residence time distribution (RTD). The RTD is used to describe how a material travels inside the unit operations of a continuous process system. It is also used to set the measurement frequency of sensors to ensure that any unacceptable content uniformity variations would be detected [16–18].

1 https://www.ich.org/page/quality-guidelines.

All the above presented concepts and initiatives are based on computations and predictions obtained through statistical modeling of the process. However, to be able to obtain accurate modeling and control of the process, and, therefore, high-quality pharmaceutical products, the statistical models have to be fed with high-quality data coming from sensors. These sensors are very different by nature depending on the CPPs to monitor (temperature, speed, etc.), or the CQAs of the by-products (assay, humidity, particle size, etc.) [19]. NIR spectroscopy is part of the family of soft sensors frequently used in continuous manufacturing, since it is able to assess both physical (particle size, tensile strength, and powder density) and chemical information (API content, blend homogeneity, and water content) in a single analysis [20–23]. However, to be efficient, the sensing device should be representative of both the material, the process, and the analyte [24–26].

4.2.2 Analytical Sensors and Their Positioning

To ensure this representativity, the analyst has several possibilities, among which are the selection of the appropriate location of the probe, and the selection of the adequate device properties.

The appropriate location will obviously depend on the considered process. For a tableting process, two main locations have been reported in the literature: the transition chute during the material discharge [27, 28], and the tablet-press feed frame before the dies [29–31]. The chute is the easiest location to place the probe, since there is a lot of room, and it allows the rejection of the blend if it is out of the specifications. The feed frame on the other side is more difficult to access, but is more representative of the final tablet. Indeed, Sierra-Vega et al., among others, showed that there was an additional mixing effect in the feed frame [32]. To do so, they used the miniature MicroNIR™ spectrophotometer, and compared both the chute and the feed-frame locations. Their results demonstrated that the drug concentration in the chute was not representative of the final drug concentration in the tablets, contrarily to the feed-frame-acquired spectra. Furthermore, the relative standard deviation in the feed frame was 60% lower than that in the chute. The feed frame seems, therefore, to be the best location for the assessment of drug concentration in the final product, while allowing rejecting (and reprocessing) the nonconforming blends. In this example, the authors used a dispersive miniature device. However, this choice is not obvious since many options exist: dispersive vs. Fourier transform (FT), transmission vs. reflection, single-point vs. multipoint measurements, etc.

The choice of dispersive vs. FT devices was investigated by Shi et al. [33] in a continuous tableting process. Contrary to the preconceived idea that dispersive is better than FT for monitoring blend homogeneity, because the movement of the mirror in the interferometer would lead to the detection of different flowing powder sections in the blending process, they showed a more contrasted picture. They tested several spectral resolutions with the FT device and different co-adds for both devices. The dispersive spectrophotometer outperformed the FT when the measurements were performed without co-adds. However, when performing co-adds, the FT outperformed the dispersive device for sensitivity and selectivity, while being equivalent for other tested parameters. They concluded that FT systems might be of interest depending on the process analyzed and the measurement time available.

Most PAT NIR measurements are performed in diffuse reflectance mode because of the ease of implementation and the various interfaces available: large scanning area (e.g. Bruker's Matrix Probe head), specific cleaning systems for blending monitoring (e.g. GEA's Lighthouse probe) or small size probes (e.g. Sentronic's SentroPAT). Another possibility offered by reflectance NIR measurements is the use of multifiber probes to perform real-time fast hyperspectral imaging measurements. One of these systems called spatially resolved spectroscopy (SRS) has been successfully used to detect and reject tablets with a wrong API content, or nonhomogenous distribution of API right after tableting. The 26 NIR spectra were collected simultaneously using an integration time of 1 ms [34].

Nevertheless, NIR transmittance measurements are also very interesting. Indeed, they allow measuring a larger and more representative volume of the sample. Sánchez-Paternina et al. [35] were able to measure through a 5.6 mm-thick sample that is considerably more than the 1.7–2.9 mm depth of penetration of NIR radiation in the diffuse reflectance mode reported by Iyer et al. [36]. The root mean square error of prediction (RMSEP) values obtained in transmittance mode were 2.35–5.61 times less than the ones obtained in reflectance. However, transmittance measurements also suffer from limitations. Alam et al. [37] investigated the implementation of NIR transmittance measurements in a chute of powder. The quantitation of the API in the flowing powder was affected by the flow rate changes and the thickness of the powder stream. The quantitation of the API was only possible through a powder stream thickness of 1–2 mm. This constitutes a challenge to implement NIR transmission measurements in continuous manufacturing since the powder flow rate is subject to variations depending on the production needs.

4.2.3 Chemometric Modeling of the Data

These changes in powder flow rate and, more generally, the fact that the materials to be investigated are flowing at relatively high speeds, constitute a major challenge to the modeling of NIR data. Indeed, traditionally, calibration and validation samples (powder mixes, granulates, tablets, etc.) are prepared off-line at several concentration levels to avoid a too high consumption of raw materials, and to be able to correctly correlate the NIR spectrum to the sample [38]. However, this way of calibrating NIR quantitative models often leads to discrepancies and bad results due to lack of correspondence between laboratory spectra and process spectra [39, 40]. To get around this difficulty, Blanco et al. tried to compute a process spectrum (S_p), and then multiply the laboratory spectra by a determined factor to increase their correspondence [41]. The main advantage of this strategy is that it does not need a reference method. Figure 4.3 shows the principal components score plot of tablets produced on an industrial manufacturing line, and powder blends mixed in laboratory used to build the regression model. As one can see on the left figure, the two blends are spectrally different, but once the laboratory blends are multiplied by the process spectrum, the two sets of spectra overlap, indicating a closer spectral similarity. The higher the similarity, the more accurate the regression model to predict manufactured tablets.

Other teams have used chemometrics tools to remove the physical influence of the samples on the NIR spectra. Pauli et al. [42] used independent component analysis (ICA) to dissociate information from the concentration and from the speed of the powder stream in a continuous manufacturing process. First, they reduced the spectral range to avoid high-noise spectral regions, and then they added a derivative preprocessing step to reduce the influence of speed. Using ICA, they were able to have an accurate qualitative monitoring of the API content. However, when quantitative values are needed, quantitative regression models are necessary, but are more difficult to set up, especially due to the different tablet speeds. They recommended developing quantitative models only for production in routine production, and not during the development of the process because of the huge workload. This consideration is a common limitation of multivariate statistical process control (MSPC) strategies. Indeed, MSPC strategies are based on the projection of the measured data (e.g. the NIR spectrum) onto a small number of new dimensions (e.g. principal components), and the assessment that they are below statistical control thresholds (e.g. Hoteling's T^2 or Q residuals) [43]. Nevertheless, these thresholds have to be set, and, generally, this is done with limited calibration sets that may not capture all the future variability of the process. This may lead to biased decisions and under optimal process control during the development stages.

Colón et al. [44] investigated the use of NIR in a continuous manufacturing process. They observed issues using NIR sensor due to solvate exchange with water in the environment that induced a change in NIR spectrum (but not in final product quality), and issues due to the relaxation of tablets. To handle these differences, they updated the calibration model using both laboratory tablets (for the complete set of concentration levels), pilot scale tablets at 100% of label claim, and tablets at 90, 100, and 110% of the label claim coming from the continuous manufacturing process. This augmentation of the calibration dataset allowed the incorporation of as much variability as possible to increase the robustness of the prediction to changes in the process. They also identified and removed the spectral ranges that were the most impacted by changes in humidity.

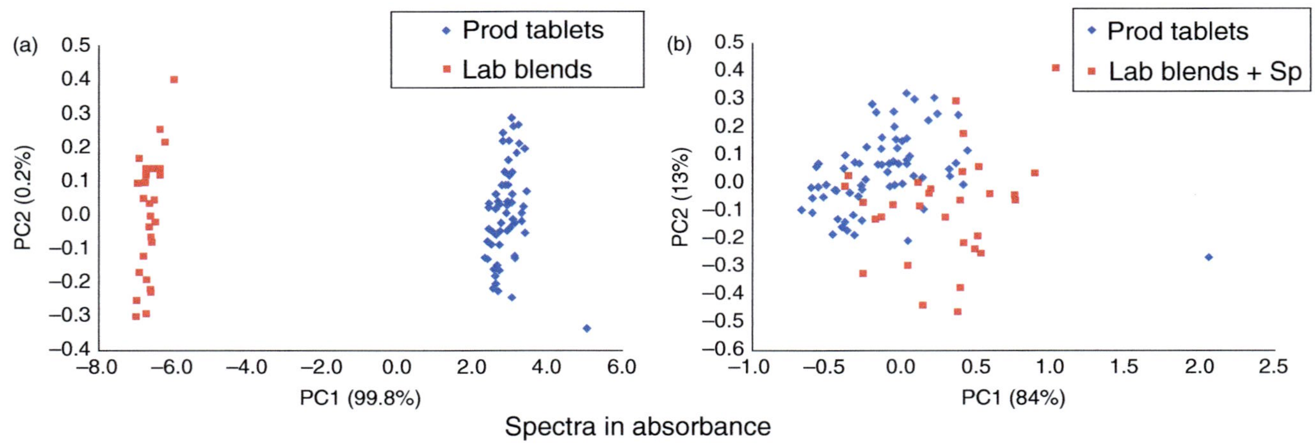

Figure 4.3 Scatter plot of PCA scores from absorbance spectra over the wavelength range 1100–2300 nm for: (a) laboratory blends and production tablets and (b) laboratory blends + S_p and production tablets. *Source*: Blanco et al. [40]© 2011, Elsevier.

Using spectroscopic sensors to monitor CQAs also affects the control of the process because of its inherent difference of uncertainty, compared to CPP measurements. Indeed, spectroscopic measurements are altered by the imperfect interfacing with the sample, and the need to use a statistical modeling of the measurement, while CPPs are often measured by mechanical or electrical sensors. Therefore, adjusting CPPs using noisy CQA measurements may potentially amplify variations and influence the final product quality. In this framework, Su et al. propose to use the concept of data reconciliation [45]. Data reconciliation explicitly uses process knowledge by combining measurement data with a process model. This "ensures that the reconciled measurements are consistent with the known relationships between process variables with minimum adjustment of the measurement data." This approach was shown to be superior to conventional MSPC approach in a rotary tablet press process. Another aspect to keep in mind while developing NIR spectroscopy methods to monitor processes is the maintenance of the chemometric model [46, 47]. Although some *qualitative* models are developed for each production and are *de facto* maintenance free [48], most of the *quantitative* models need to be maintained through time in order to ensure their accuracy from production to production. Several maintenance strategies have been proposed to tackle this issue [49, 50].

4.3 PAT Implementation of Near-Infrared Spectroscopy

Besides these challenges and limitations, NIR spectroscopy is one of the most used PAT techniques [51, 52]. The sections below will make a nonexhaustive demonstration of the possible implementations of NIR as PAT sensor in the manufacturing of solid pharmaceutical dosage forms. Although presented separately, the different process operations are intended to be interconnected in a continuous manufacturing line [53]. The monitoring of bioprocesses is out of the scope of the present chapter. The interested reader may find relevant information in the following papers [54, 55].

4.3.1 Raw Material Identification

Raw material identification (RMID) is a routine task in the pharmaceutical industry. These tests are performed before processing the materials, in order to avoid mistakes as much as possible, and, therefore, saving time and money. This material verification applies not only to both purchased materials (e.g. excipients), but also to some internal transfers of materials, for instance, APIs manufactured at another plant. In particular, if RMID can be performed on the loading dock, it means that samples do not have to be taken, and analyzed in a laboratory setting, while the materials are quarantined in a warehouse.

Originally, these tests were performed in laboratories using destructive tests, such as liquid chromatography. Nowadays, thanks to the recent developments of handheld and portable spectroscopic devices, these identification tests may be performed in the warehouse even directly through the packaging [56, 57]. Both Raman and NIR spectroscopies are used in this task for both active ingredients and excipients [58–60]. While Raman spectroscopy is gradually expanding its capabilities and applications, NIR spectroscopy still has its advantages when examining fluorescent materials, and many "small-molecule" APIs are conjugated molecules (aromatics), leading to the high probability of being fluorescent. NIR spectroscopy also presents a strong advantage for the characterization of physical parameters of pharmaceutical powders (e.g. particle size, moisture, etc.) [61] (see Figure 4.4). A detailed review of the NIR spectroscopy applications in pharmaceutical particle technology may be found elsewhere [62]. These physical parameters may also be very important to monitor since they may be key quality attributes for the subsequent processes. Therefore, a fast and nondestructive analysis of the incoming raw materials may ensure their future processability or even allow adjustments of the process parameters according to the defined QbD strategy.

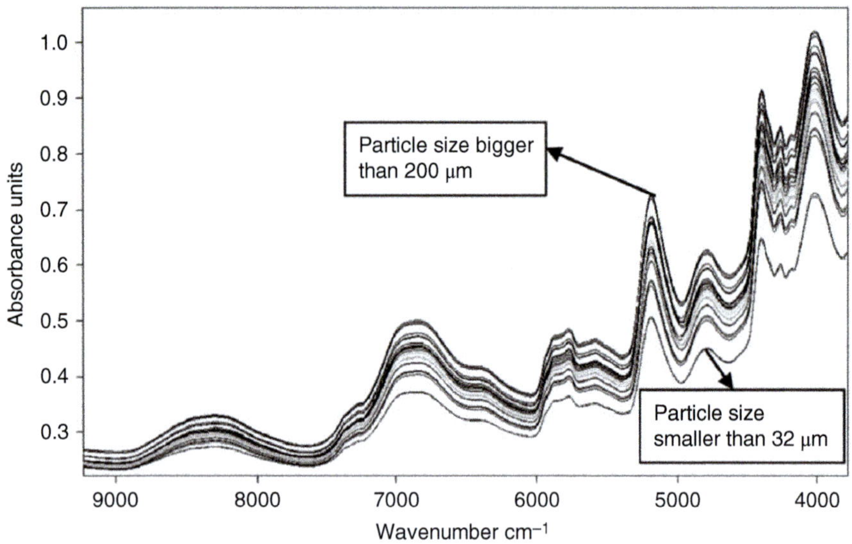

Figure 4.4 Original NIR spectra of different particle size fractions of HPMC. *Source*: Devjak Novak et al. [61] © 2012, Elsevier.

4.3.2 Blending

Once identified and weighed, the raw materials often require a blending step to ensure the homogenous mixing of the different powders. This is a critical step of the pharmaceutical solid-state products manufacturing since it has a direct impact on the final product quality and homogeneity. The blending process is affected not only by several physical properties of the powders, such as particle size, shape, and density, but also by process parameters, such as blender size, shape, fill level, and blend time [63–65]. Those parameters/properties may lead to nonhomogenous blends if not correctly monitored. Most of the actual techniques to assess blend homogeneity are based on the manufacturer's experience and on physical sampling, using thieves, despite the demonstrated drawbacks [66]. Indeed, physical sampling disturbs the powder flow, and the sampling is often nonrepresentative due to the uneven flow of particles of different shapes. Therefore, the possibility to monitor the blending process noninvasively is a natural implementation of the PAT technique. Among these, NIR has been the most used one because of the influence of both physical and chemical information [20, 67–69]. The sensors used to monitor blending processes are mostly wireless MEMS-based devices (e.g. SentroPAT™ BU from Sentronic or MicroNIR™ from Viavi). However, several practical issues may arise, such as the fouling of the measurement window due to powder deposition. In order to circumvent this issue, manufacturers developed more complex measurement systems (e.g. Lighthouse Probe™ from GEA) with an integrated cleaning system of the probe. Other kinds of sensors, such as chemical imaging devices, have also been successfully used to monitor and evaluate blending operations in order to increase the sampling representativity [64, 70, 71].

The analysis of the spectral data may be qualitative or quantitative. Quantitative monitoring of the blending process monitors content variations and assumes the end point as being a constant content. It may be the most accurate approach [72–74], although discrepancies between calibration and routine measurements may appear due to the powder flow [75]. El-Hagrasy and Drennen [76] found that multivariate linear regression (MLR) was more efficient to follow the blending process than principal component regression (PCR) or partial least squares (PLS), due to the fact that it considered a reduced spectral range and was, therefore, more robust to process variations. Similarly, Nakagawa et al. showed that the locally weighted PLS (LW-PLS) improved by 38.6% (in root mean squared error of prediction) the conventional PLS calibration [77, 78]. However, quantitative calibration is time and resource consuming, and, therefore, requires an important effort that is not feasible during the development stage of a new process. Moreover, the robustness, transfer and scale-up of the models are also important limitations.

On the other hand, qualitative analysis considers spectral variance, and the blend is said to be optimal once a stable or desired spectral profile is reached. Several statistical methods have been successfully applied among which are moving block standard deviation (MBSD) [67], principal component score distance analysis (PC-SDA) [63], soft independent modeling of class analogy (SIMCA) [79], or moving F-test [48, 80]. These approaches are easily deployed, and may be used during the development of a new process while providing information on the evolution of ingredient properties, process settings and environmental conditions, but no insight into the final quantitative composition of the blend. Nevertheless, the obtained information may be used to adapt subsequent operations, such as granulation.

4.3.3 Granulation and Sizing

Sometimes, due to their physical properties, the different ingredients of the pharmaceutical formulation do not mix well, and segregate during the blending operation. Therefore, it is appropriate to granulate powdered ingredients through compression, dry granulation, or in the presence of a binder under wet conditions; this is wet granulation. Granulation ensures the homogeneity and downstream processability of the blend (e.g. improved powder flow and compressibility). The granulated material is eventually dried, if wet granulation was used, blended, and compressed.

Although reviewing the different granulation possibilities is out of the scope of the present chapter, most of the spectroscopic investigations are focused on the determination of water during wet-granulation or post-granulation drying. Although our attention is centered on NIR applications, it is important to keep in mind that other kinds of sensors may also be envisaged. For example, microwave sensors may advantageously replace NIR probes for moisture determination, thanks to the highest penetration depth (2–5 cm compared to 0.5–2.5 mm for NIR reflection), with a significant reduction of effort for the development of the quantitative regression model [81].

Gavan et al. [82] developed a QbD strategy for a fluidized bed granulation process of two APIs, and used a portable NIR sensor to monitor the moisture content during the process with an orthogonal-PLS model in order to ensure the necessary robustness regarding changes in the process parameters. Fonteyne et al. [83, 84] used both NIR and Raman spectroscopy to follow up the polymorphic forms of theophylline during a continuous drying process together with the moisture content. Although the results were relatively accurate, they emphasized the fact that "care should be taken when evaluating the average moisture content and solid-state values obtained on the total granule load, since significant differences between the different sieve fractions of the granules were observed."

More recently, Pauli et al. [85] used NIR spectroscopy to monitor in-line the particle size distribution (PSD) of the granules during a GMP continuous granulation process. They used three PLS models to predict, respectively, the size range of 20–234 μm (X10), 98–1017 μm (X50), and 748–2297 μm (X90), respectively. They obtained RMSEP values of 17, 97, and 174 μm for PSD X10, X50, and X90, respectively that were sufficiently accurate to monitor changes and take adequate dispositions to adjust the process. This application is somewhat challenging since the correlation of particle size-related information to variances in NIR spectra is mainly based on changes in spectral slope and baseline offset, but water content and other sample variations due to the continuous processing context disturbed these factors.

4.3.4 Extrusion

Besides granulation operations, other processing tools may also be envisaged to ensure a homogenous mixture of the component while increasing the bioavailability of poorly soluble drugs. One of these tools is the hot melt extrusion (HME). Extrusion is the process of shaping a material (extrudate) by forcing the material through an orifice (die) using pressure. This process consists of the mixture of the active ingredients with a meltdown water-soluble polymer. It allows obtaining a molecular dispersion of the API within the matrix, increasing its solubilization in the gastro-enteral fluids. NIR spectroscopy has been extensively used in HME experiments to monitor both the API content, the solid state of the extrudates and elucidate the interactions between the ingredients [86–89].

Baronsky-Probst et al. [90] used NIR spectroscopy to monitor the homogeneity of the incoming materials, and the strand after cooling to assess the API content and the solid state in the final product. They preferred these locations to the more conventional position in the die to avoid pressure and temperature effects on the sample, and, therefore, obtain a more robust model. Wahl et al. [91] paid particular attention to the location of the NIR probe regarding the theory of sampling (TOS) precepts, and decided to place the NIR probe close to the die after the mixing screws. The connector to the die was modified to force the melt into an annular flow with high shear rates ensuring representative sampling. The developed PLS model was able to quantify the API, and changes in screw speed from 150 to 250 rpm did not affect the predictions much. Saerens et al. [92] used NIR spectroscopy in an HME process to get insights into the types of interactions between the API and the polymer together with the assay of the API. They elucidated that the interactions are manifested as hydrogen bonds between Kollidon® SR and metoprolol tartrate molecules. These findings were confirmed off-line using ATR-FTIR and Raman spectroscopy and differential scanning calorimetry (DSC). However, interpretation of NIR spectra remains complicated, and is often advantageously replaced by Raman spectroscopy when the purpose is the elucidation of interaction mechanisms.

4.3.5 Tableting

Once the homogenous mixing of ingredients is obtained (via direct blending, granulation or HME), the natural next step is tableting. This step may appear less important to monitor considering that homogeneity of powders has been ensured during the previous steps [93]. However, this step of the process is the closest one to the final product. Therefore, it is sometimes easier to monitor the quality of the product directly in the press, especially if there is a following coating step.

Dalvi et al. [94] compared the quantitative performances of a portable NIR spectrophotometer (Viavi MicroNIR™), placed directly in the feed frame with in-line measurements of the resulting tablets (Visiotec VisioNIR™), and at-line measurements with a benchtop spectrophotometer (Bruker MPA™). Good agreements were found between these three measurements. However, discrepancies were observed between NIR measurements and ultraviolet (UV) assay possibly due to nonrepresentative surface distribution of API, together with spectral variations resulting from dynamic sample presentation. The same team also combined results from NIR spectroscopy and NIR chemical imaging (NIR-CI) in the feed frame [95]. The main advantage was that the NIR-CI approach had a sampling volume five times higher than the NIR probe alone, and allowed an image of the distribution of the ingredients in the feed frame to be obtained. These insights might be valuable when developing a new process.

Durão et al. [96] monitored five components during the tableting process of a multivitamin formulation (31 components) using NIR spectroscopy, light-induced fluorescence (LIF) spectroscopy and red, green and blue (RGB) color imaging together. They used multiblock PLS (mb-PLS) that combines the scores obtained by separate PLS models for each block (one block per PAT sensor) in a single model. Therefore, they were able to use information coming from the different sensors, which allowed better predictions than the NIR sensor alone. Figure 4.5 shows the predicted concentration of ferrous fumarate in the blend using NIR only (Figure 4.5a), or all the sensors (Figure 4.5b). A sensible improvement may be observed when all sensors were used compared to NIR alone.

Pauli et al. [42] placed two NIR probes at different locations of a continuous tableting process: in the feed frame, and in the press just before tablet ejection. They were able to analyze 100% of the produced tablets (70 000 tablets/h), but the spectra were highly impacted by process parameters and, therefore, PLS models developed with off-line spectra were not suitable for the analysis of in-line measurements. They also observed a time-shift between the predictions of the two probes. This reflects the mixing phase inside the feed frame,

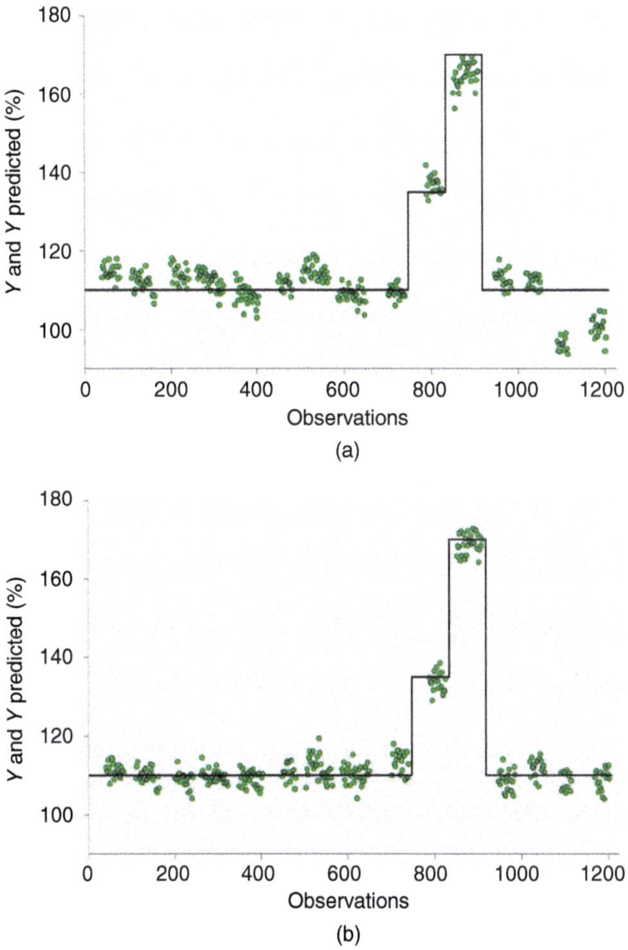

Figure 4.5 Predicted value of Ferrous Fumarate with a PLS model using data from (a) NIR and (b) all tools. The dark line represents the theoretical *Y* value. *Source*: Durão et al. [96] © 2017 Elsevier.

implying a recycling loop of the excessive granules, estimated at one minute in the present case. Therefore, they emphasized in the necessity to have several sensors at different locations to get better knowledge on the process and the different residence times. Indeed, these phenomena may have a direct impact in the management of process adjustments based on the information provided by the different sensors, especially if they are placed at different locations.

NIR spectroscopy may also be used to understand the powder behavior in the tablet press. He et al. [97] used quantitative NIR models to gain knowledge on powder segregation phenomenon inside a tablet press. The use of spectroscopic tools allowed fast, nondestructive analysis at several places of the process.

4.3.6 Coating

The coating process is a crucial step of oral solid preparations manufacturing. Indeed, the coating may act as a physical screen to avoid the effects of oxidation, humidity and illumination conditions in order to improve the stability of the final product or process intermediates. It is sometimes used to protect workers avoiding the generation of toxic dusts of intermediate products. The coating may also play an active role in the protection (gastro-resistant)

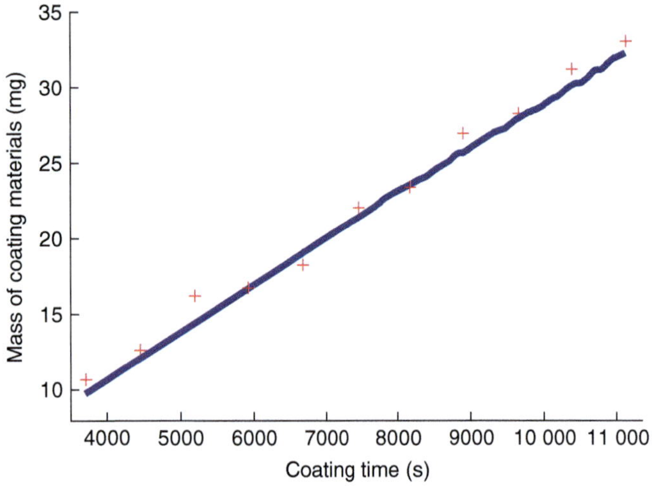

Figure 4.6 Real-time monitoring of coating deposit mass: (*) real-time NIR predictions during the pan-coating operation and (+) experimental results obtained by weighing. *Source*: Gendre et al. [100] © 2017, Elsevier.

and liberation (modified release) of the drug in vivo. The coating homogeneity and thickness are important to control the timing of the drug-release. Many off-line techniques exist to monitor the coating thickness, such as the changes in the weight, height or diameter of the pellet/tablet cores coated during the process [98]. Other more advanced techniques, such as Terahertz 3D imaging, are also used [99]. These techniques necessitate a sampling step during the process, and the measure of several samples to be representative, which is time consuming, and neither allows a fast and accurate adjustment of the process, nor an accurate end-point detection. Therefore, these off-line tools are advantageously replaced by in-line sensors among which NIR probes especially for aqueous-based coatings [100–103]. Figure 4.6 shows an example of coating deposition monitoring using NIR spectroscopy (blue stars) versus the weighing off-line reference method (red crosses). Compared to weighing, NIR spectroscopy allows a noninvasive (and nondisturbing) follow-up of the process at a higher sampling rate. Bogomolov et al. [104] compared NIR spectroscopy, Raman spectroscopy, and their combination, to follow up a coating process. Their findings suggest that both techniques may provide interesting insights, and offered comparable performances depending on the nature of ingredients to follow. Wahl et al. [105] decided to compare optical coherence tomography (OCT) and NIR spectroscopy as in-line PAT tools for a coating process. They used an NIR contact probe, highlighting the fact that the distance of the sample to the probe is crucial. Therefore, they applied a filter based on the absorbance ratio at $I_{1634\,\text{nm}}/I_{2030\,\text{nm}}$ ensuring that the spectra eventually analyzed with the PLS model are effectively tablet spectra. They compared off-line and in-line NIR measurements and obtained satisfactory results of RMSEP: 7.8 and 7.2 μm for off-line and in-line measurements, respectively and a final coating thickness of 70 μm. NIR and OCT results were consistent and well correlated to each other. However, the method development effort was significantly smaller for OCT since it directly provides a thickness measurement, while chemometrics are necessary for NIR spectra interpretation.

Igne et al. [106] studied the influence of the sampling frequency on the analysis of a coating process. They compared several combinations: with and without co-adds (1–64 co-adds), and spectral sampling over time (from 1 to 15 pan rotations). They found that the sampling frequency had a significant impact on the spectral data quality and the resulting quantitative analyses. A single co-add is noisier (smaller signal/noise ratio) than a 64 co-adds spectrum; however, it is less sensitive to the averaging of disturbances due to the dynamic nature of the process. In conclusion, they argued that a single co-add provides information on the process (bed height variations) and the tablet-to-tablet variations. The 64 co-adds-based predictions were more focused on the variability amongst the tablets; this time-dependent averaging improved up to 4 averages, but then stabilized. They concluded that if the sought information is the tablet-to-tablet variability, then averaging may be limited to one spectrum, and the number of co-adds adapted to the process variability. Another important finding was that the correlation between measured spectra and the reference data was not obvious. Indeed, the matching of physical sample collection and spectral acquisition is not the optimal modeling approach due to the slow phenomenon of the coating process. In their case, a temporal offset up to 60 seconds was less biased and had comparable RMSEP. This emphasizes once again the complexity of correlating spectral data of dynamic processes with reference values to build robust quantitative models.

Another important aspect associated with the coating process is the curing stage consisting of a post-coating drying step during which the polymer is exposed to a high temperature. This step changes the structure of the film influencing mechanical properties of the film, film–tablet adhesion, and the dissolution profiles of coated substrates [107].

Gendre et al. [108, 109] used X-ray micro-diffraction (XμCT), NIR and Raman spectroscopies to gain insights into the curing process and structures of the film coating. It allowed them to compare both dynamic and static curing conditions, and determine the end point of the processes, leading to similar dissolution profiles of the drugs. They were then able to demonstrate that a 24-hour static curing led to the same results as a 4-hour dynamic curing process. Korasa et al. [110] used NIR spectroscopy to monitor the curing process of pellets. They had reliable results predicting the coating thickness, but discrepancies appeared when predicting the drug-release rate because the pellet moisture content disturbed the PLS model at high moisture content.

4.3.7 End-Product Testing

Once the pharmaceutical product has completed all the processing steps, a last step of verification may ensure that only "good-quality" or conforming products are packaged and marketed. Application of NIR spectroscopy in the quality control of pharmaceutical preparations will not be extensively reviewed in this chapter, and detailed applications may be found in recent reviews [111–113].

Nevertheless, some interesting applications may be pointed out, such as the prediction of dissolution profiles. Pomerantsev et al. [114] used NIR spectroscopy, together with PLS and dynamic modeling, to predict the dissolution profiles of pH-sensitive coated pellets. Pawar et al. [115] were also able to predict the dissolution of immediate release tablets at-line, on a continuous manufacturing plant, using transmission NIR (see Figure 4.7). They included several variables, such as API concentration, blender speed, feed frame speed and compaction force to build a multilinear regression model using the principal component scores.

Other, more specific, applications are related to the new axes of research in the pharmaceutical field, such as the printing of tablets or oral films. These drugs may be printed on demand with a personalized composition [116, 117]. However, it is almost impossible to imagine coupling these manufacturing techniques with a destructive technique, such as high-pressure liquid chromatography (HPLC), in a drug store or even at the patient's home. Therefore, it might be interesting to use nondestructive portable systems, such as NIR spectrophotometers or NIR-chemical imagers, to perform the quality control of such drugs [118, 119].

Another important part of the end-product quality control covers the analysis of substandard and falsified medicines [120]. In this context, NIR spectroscopy has already proven its usefulness [121–124] and the development of low-cost handheld devices is already demonstrating its applicability [57, 125–128]. See the chapter by Leary, Crocombe and Kalyanaraman in these volumes for more details.

4.4 Conclusion

NIR spectroscopy is a light-based technique enabling the remote and nondestructive analysis of samples, providing both physical and chemical information. These characteristics may explain that NIR spectroscopy is one of

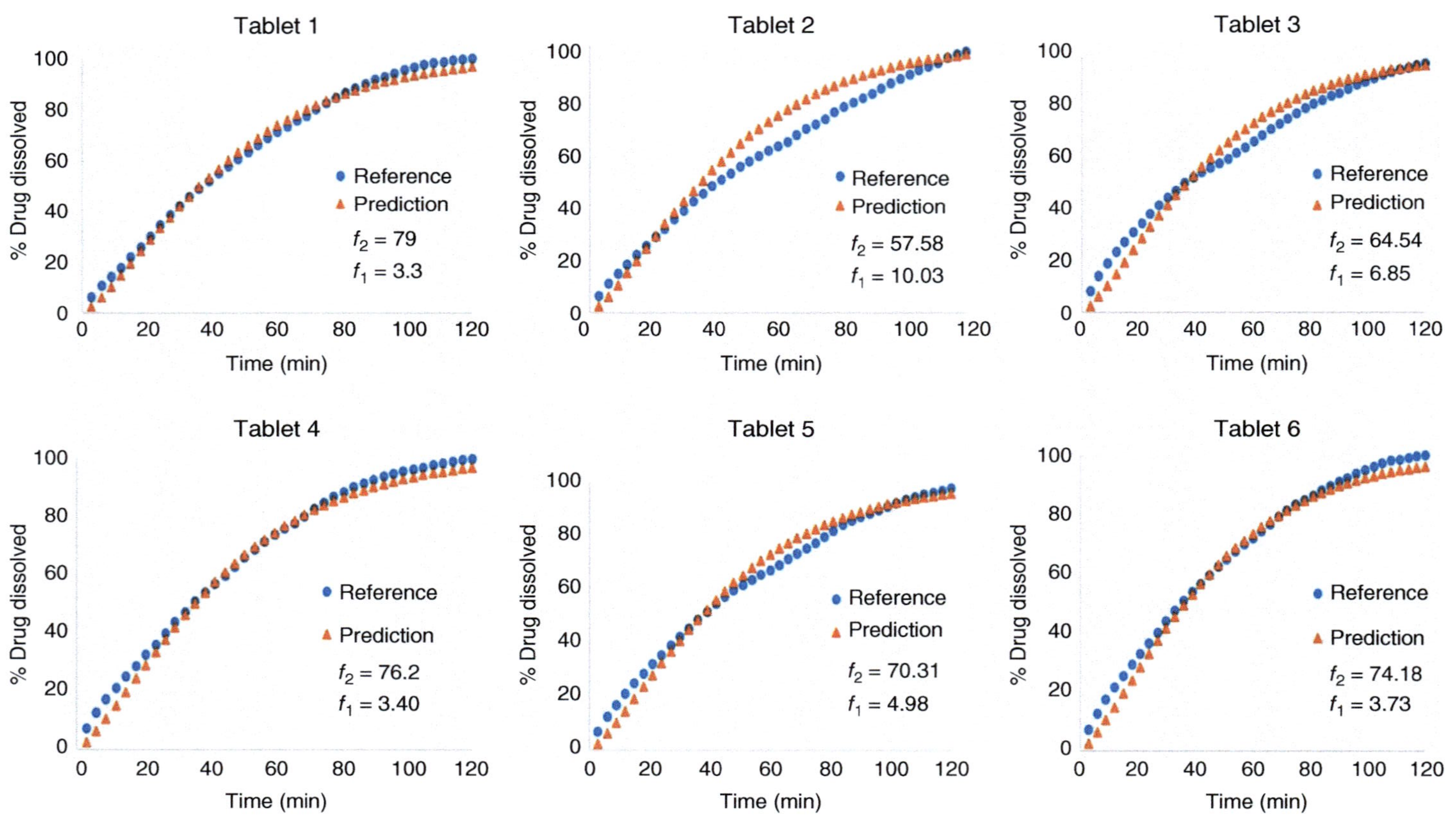

Figure 4.7 Predicting individual dissolution profile from model-dependent approach. The similarity factor is greater than 50 indicating a good agreement between the observed and predicted results. *Source*: Pawar et al. [115] © 2016, Elsevier.

the emerging technologies recommended by the regulatory agencies to monitor continuously the pharmaceutical processes. Nevertheless, its characteristics may also constitute a limitation to its implementation. Indeed, the spectra are hardly interpretable and must be processed by advanced machine learning tools in order to obtain meaningful usable information. The building of a robust quantitative model is not straightforward. It needs a lot of time and effort to be applicable in routine. One of the major gaps is the correlation of the spectra obtained in a dynamic environment with reference values obtained off-line with reference methods. Therefore, several strategies have been exposed in the beginning of the chapter to handle this issue.

Nevertheless, the amount of research and application papers in the pharmaceutical field demonstrates once more that NIR spectroscopy has a bright future ahead.

Glossary

Continuous Manufacturing (CM): CM of pharmaceutical drug products is a new approach within the pharmaceutical industry, opposing traditional batch-manufacturing processes based on its potential to increase manufacturing flexibility and efficiency. In CM, all process units are directly connected to each other. The starting material is continuously charged into the first process unit at the beginning of the line, while the final product is simultaneously discharged at the end [13].

Critical Process Parameter (CPP): CPPs are process inputs that have a direct and significant influence on critical quality attributes when they are varied within regular operation range [6].

Critical Quality Attribute (CQA): CQAs are physical, chemical, biological, or microbiological properties or characteristics that must be controlled directly or indirectly to ensure the quality of the product [6].

Design of Experiments (DoE): A DoE is a structured, organized set of experiments used to determine, model and quantify the relationship between factors affecting a process and the output of that process.

Design Space (DS): The DS is the multidimensional combination and interaction of input variables (e.g. material attributes) and process parameters that have been demonstrated to provide assurance of quality. Working within the DS is not considered as a change (see also ICH Q8 R2 [12]).

Dissolution profile: Graphical representation in terms of concentration vs. time of the release of an active pharmaceutical ingredient from a dosage form in an appropriate selected dissolution medium. It is used to verify that the drug will be correctly released in the patient's body ensuring its safety and biological activity.

High Pressure Liquid Chromatography (HPLC): HPLC is used to separate, identify and quantify each substance present in a mixture. It is composed of a solid stationary phase and a liquid mobile phase. The separation occurs because of differences of affinity of the components of the mixture towards the stationary and the mobile phase. Several types of detectors may be used to analyze the substances after their physical separation, such as ultraviolet spectroscopy, mass spectrometry, and fluorescence spectroscopy.

Partial Least Squares (PLS): PLS is a variable reduction technique which reduces the number of variables by making linear combinations of the original variables. These combinations are called PLS factors. The PLS factors are defined in such a way that they maximize the covariance with the response variable. This way latent variables are more directly related to the response variable than, for example, those obtained in PCA. PLS may be applied as a regression or a classification algorithm.

Polymorphism: In crystalline solids, the polymorphism is the ability of a substance to exist in different molecular arrangements and/or different molecular conformations. The polymorphic behavior of pharmaceutically active substances can affect their therapeutic efficacy and safety.

Principal Component Analysis (PCA): PCA is a variable reduction technique which reduces the number of variables by making linear combinations of the original variables. These combinations are called the principal components and are defined in such a way that they explain the highest (remaining) variability in the data and are by definition orthogonal. The importance of the original variables in the definition of a principal component is represented by its loading and the projections of the objects on to the principal components are called the scores of the objects.

Process Analytical Technology (PAT): PAT is defined by the US FDA as "a system for designing, analyzing, and controlling manufacturing through timely measurements (i.e., during processing) of critical quality and performance attributes of raw and in-process materials and processes, with the goal of ensuring final product quality. It is important to note that the term analytical in PAT is viewed broadly to include chemical, physical, microbiological, mathematical, and risk analysis conducted in an integrated manner. The goal of PAT is to enhance understanding and control the manufacturing process".

Quality by Design (QbD): The QbD concept implies that the final product quality is built in the conception (the design) of the process itself. It necessitates a deep understanding of the process and the product with the definition of CPPs influencing the CQAs of the product.

Quality by Testing (QbT): QbT is an actual system implemented in the pharmaceutical industry ensuring the quality of the final product by following a sequence of testing steps. These may include, but are not limited to, testing of incoming raw materials, in-process control and final product testing. The QbT approach is not dynamic by essence.

Residence Time Distribution (RTD): RTD is a term coming from the chemical engineering field. It is defined as "the probability of time that a solid or fluid material stays inside one or more unit operations in a continuous system." [17] It is a crucial information to better understand and model the material flow in the process. It is also used to define the "batch" notion in CM.

Root Mean Square Error (RMSE): This measure is frequently used in multivariate data analysis to evaluate the goodness of fit of a regression model. It may be applied on the calibration samples (RMSEC), on the cross-validation samples (RMSECV) or with the predictions of an external set of validation (RMSEP). It is computed as follows:

$$\text{RMSE} = \sqrt{\sum_{i=1}^{n} \frac{(\hat{y}_i - y_i)^2}{n}}$$

where $\hat{y}_i$ is the value of the sample predicted by the model, y_i is the actual value of the sample, and n is the total number of samples.

Target Product Quality Profile (TPQP): It consists of the characteristics that the drug product should possess in order to reproducibly deliver the therapeutic benefit promised. The TPQP guides formulation scientists to establish formulation strategies and keep the formulation effort focused and efficient. TPQP is related to identity, assay, dosage form, purity, and stability in the label [6].

Theory of Sampling (TOS): The theory of sampling is theorization of the sampling process enabling a representative sampling. It focuses only on the sampling process and not on the sample itself [129].

References

1 Yu, L.X. and Kopcha, M. (2017). The future of pharmaceutical quality and the path to get there. *Int. J. Pharm.* 528 (1–2): 354–359.

2 FDA and CDER (2017). Advancement of emerging technology applications for pharmaceutical innovation and modernization guidance for industry. *FDA-2015-D-4644*.

3 O'Connor, T. and Lee, S. (2017). Emerging technology for modernizing pharmaceutical production. In: *Developing Solid Oral Dosage Forms*, 2ee (eds. Y. Qiu, Y. Chen, G. Zhang, et al.), 1031–1046. Elsevier.

4 Lee, S.L., O'Connor, T.F., Yang, X. et al. (2015). Modernizing pharmaceutical manufacturing: from batch to continuous production. *J. Pharm. Innov.* 10 (3): 191–199.

5 Adamo, A., Beingessner, R.L., Behnam, M. et al. (2016). On-demand continuous-flow production of pharmaceuticals in a compact, reconfigurable system. *Science* 352 (6281): 61–67.

6 Yu, L.X. (2008). Pharmaceutical quality by design: product and process development, understanding, and control. *Pharm. Res.* 25 (4): 781–791.

7 Zhang, L. and Mao, S. (2017). Application of quality by design in the current drug development. *Asian J. Pharm. Sci.* 12 (1): 1–8.

8 Lebrun, P., Krier, F., Mantanus, J. et al. (2012). Design space approach in the optimization of the spray-drying process. *Eur. J. Pharm. Biopharm.* 80 (1): 226–234.

9 Singh, B.N. (2019). Product development, manufacturing, and packaging of solid dosage forms under QbD and PAT paradigm: DOE case studies for industrial applications. *AAPS PharmSciTech* 20 (8): 313.

10 Ziemons, E., Hubert, C., and Hubert, P. (2019). Process analysis – overview. In: *Encyclopedia of Analytical Science*, 3e, vol. 8, 396–402. Oxford, UK: Academic Press.

11 Yu, L.X., Amidon, G., Khan, M.A. et al. (2014). Understanding pharmaceutical quality by design. *AAPS J.* 16 (4): 771–783.

12 International Conference on Harmonization (2009). *ICH Q8(R2) Pharmaceutical Development.* https://database.ich.org/sites/default/files/Q8(R2) Guideline.pdf.

13 Roggo, Y., Pauli, V., Jelsch, M. et al. (2020). Continuous manufacturing process monitoring of pharmaceutical solid dosage form: a case study. *J. Pharm. Biomed. Anal.* 179: 112971.

14 Su, Q., Ganesh, S., Moreno, M. et al. (2019). A perspective on quality-by-control (QbC) in pharmaceutical continuous manufacturing. *Comput. Chem. Eng.* 125: 216–231.

15 Barenji, R.V., Akdag, Y., Yet, B., and Oner, L. (2019). Cyber-physical-based PAT (CPbPAT) framework for Pharma 4.0. *Int. J. Pharm.* 567: 118445.

16 Engisch, W. and Muzzio, F. (2016). Using residence time distributions (RTDs) to address the traceability of raw materials in continuous pharmaceutical manufacturing. *J. Pharm. Innov.* 11 (1): 64–81.

17 Gao, Y., Muzzio, F.J., and Ierapetritou, M.G. (2012). A review of the residence time distribution (RTD) applications in solid unit operations. *Powder Technol.* 228: 416–423.

18 Bhaskar, A. and Singh, R. (2019). Residence time distribution (RTD)-based control system for continuous pharmaceutical manufacturing process. *J. Pharm. Innov.* 14 (4): 316–331.

19 Ierapetritou, M., Muzzio, F., and Reklaitis, G. (2016). Perspectives on the continuous manufacturing of powder-based pharmaceutical processes. *AIChE J.* 62 (6): 1846–1862.

20 Nagy, B., Farkas, A., Magyar, K. et al. (2018). Spectroscopic characterization of tablet properties in a continuous powder blending and tableting process. *Eur. J. Pharm. Sci.* 123: 10–19.

21 Pestieau, A., Krier, F., Thoorens, G. et al. (2014). Towards a real time release approach for manufacturing tablets using NIR spectroscopy. *J. Pharm. Biomed. Anal.* 98: 60–67.

22 Rehrl, J., Karttunen, A.-P.P., Nicolaï, N. et al. (2018). Control of three different continuous pharmaceutical manufacturing processes: use of soft sensors. *Int. J. Pharm.* 543 (1–2): 60–72.

23 Rehrl, J., Kruisz, J., Sacher, S. et al. (2016). Optimized continuous pharmaceutical manufacturing via model-predictive control. *Int. J. Pharm.* 510 (1): 100–115.

24 Petersen, L., Minkkinen, P., and Esbensen, K.H. (2005). Representative sampling for reliable data analysis: theory of sampling. *Chemom. Intel. Lab. Syst.* 77 (1–2): 261–277.

25 Sánchez-Paternina, A., Sierra-Vega, N.O., Cárdenas, V. et al. (2019). Variographic analysis: a new methodology for quality assurance of pharmaceutical blending processes. *Comput. Chem. Eng.* 124: 109–123.

26 Esbensen, K.H., Friis-Petersen, H.H., Petersen, L. et al. (2007). Representative process sampling – in practice: variographic analysis and estimation of total sampling errors (TSE). *Chemom. Intel. Lab. Syst.* 88 (1): 41–59.

27 Vargas, J.M., Nielsen, S., Cárdenas, V. et al. (2018). Process analytical technology in continuous manufacturing of a commercial pharmaceutical product. *Int. J. Pharm.* 538 (1–2): 167–178.

28 Vargas, J.M., Roman-Ospino, A.D., Sanchez, E., and Romañach, R.J. (2017). Evaluation of analytical and sampling errors in the prediction of the active pharmaceutical ingredient concentration in blends from a continuous manufacturing process. *J. Pharm. Innov.* 12 (2): 155–167.

29 Li, Y., Anderson, C.A., Drennen, J.K. et al. (2019). Development of an in-line near-infrared method for blend content uniformity assessment in a tablet feed frame. *Appl. Spectrosc.* 73 (9): 1028–1040.

30 Harms, Z.D., Shi, Z., Kulkarni, R.A., and Myers, D.P. (2019). Characterization of near-infrared and raman spectroscopy for in-line monitoring of a low-drug load formulation in a continuous manufacturing process. *Anal. Chem.* 91 (13): 8045–8053.

31 Gosselin, R., Durão, P., Abatzoglou, N., and Guay, J.-M. (2017). Monitoring the concentration of flowing pharmaceutical powders in a tableting feed frame. *Pharm. Dev. Technol.* 22 (6): 699–705.

32 Sierra-Vega, N.O., Román-Ospino, A., Scicolone, J. et al. (2019). Assessment of blend uniformity in a continuous tablet manufacturing process. *Int. J. Pharm.* 560: 322–333.

33 Shi, Z., McGhehey, K.C., Leavesley, I.M., and Manley, L.F. (2016). On-line monitoring of blend uniformity in continuous drug product manufacturing process – the impact of powder flow rate and the choice of spectrometer: dispersive vs. FT. *J. Pharm. Biomed. Anal.* 118: 259–266.

34 Boiret, M. and Chauchard, F. (2017). Use of near-infrared spectroscopy and multipoint measurements for quality control of pharmaceutical drug products. *Anal. Bioanal. Chem.* 409 (3): 683–691.

35 Sánchez-Paternina, A., Román-Ospino, A.D., Martínez, M. et al. (2016). Near infrared spectroscopic transmittance measurements for pharmaceutical powder mixtures. *J. Pharm. Biomed. Anal.* 123: 120–127.

36 Iyer, M., Morris, H.R., and Drennen, J.K. (2002). Solid dosage form analysis by near infrared spectroscopy: comparison of reflectance and transmittance measurements including the determination of effective sample mass. *J. Near Infrared Spectrosc.* 10 (4): 233–245.

37 Alam, M.A., Shi, Z., Drennen, J.K., and Anderson, C.A. (2017). In-line monitoring and optimization of powder flow in a simulated continuous process using transmission near infrared spectroscopy. *Int. J. Pharm.* 526 (1–2): 199–208.

38 Dal Curtivo, C.P., Funghi, N.B., Tavares, G.D. et al. (2015). The critical role of NIR spectroscopy and statistical process control (SPC) strategy towards captopril tablets (25 mg) manufacturing process understanding: a case study. *Pharm. Dev. Technol.* 20 (3): 345–351.

39 Blanco, M. and Peguero, A. (2010). Influence of physical factors on the accuracy of calibration models for NIR spectroscopy. *J. Pharm. Biomed. Anal.* 52 (1): 59–65.

40 Blanco, M., Cueva-Mestanza, R., and Peguero, A. (2011). NIR analysis of pharmaceutical samples without reference data: improving the calibration. *Talanta* 85 (4): 2218–2225.

41 Blanco, M. and Peguero, A. (2010). Analysis of pharmaceuticals by NIR spectroscopy without a reference method. *TrAC – Trends Anal. Chem.* 29 (10): 1127–1136.

42 Pauli, V., Roggo, Y., Pellegatti, L. et al. (2019). Process analytical technology for continuous manufacturing tableting processing: a case study. *J. Pharm. Biomed. Anal.* 162: 101–111.

43 Bro, R. and Smilde, A.K. (2014). Principal component analysis. *Anal. Methods* 6 (9): 2812–2831.

44 Colón, Y.M., Vargas, J., Sánchez, E. et al. (2017). Assessment of robustness for a near-infrared concentration model for real-time release testing in a continuous manufacturing process. *J. Pharm. Innov.* 12 (1): 14–25.

45 Su, Q., Bommireddy, Y., Shah, Y. et al. (2019). Data reconciliation in the Quality-by-Design (QbD) implementation of pharmaceutical continuous tablet manufacturing. *Int. J. Pharm.* 563: 259–272.

46 European Medicines Agency (2014). Guideline on the use of Near Infrared Spectroscopy (NIRS) by the pharmaceutical industry and the data requirements for new submissions and variations. *Guideline-EMEA/CHMP/CVMP/QWP/17760/2009* 44: 1–28.

47 Food and Drug Administration (2015). *Development and Submission of Near Infrared Analytical Procedures Guidance for Industry*. Food and Drug Administration.

48 Fonteyne, M., Vercruysse, J., De Leersnyder, F. et al. (2016). Blend uniformity evaluation during continuous mixing in a twin screw granulator by in-line NIR using a moving F-test. *Anal. Chim. Acta* 935: 213–223.

49 Miyano, T., Nakagawa, H., Watanabe, T. et al. (2015). Operationalizing maintenance of calibration models based on near-infrared spectroscopy by knowledge integration. *J. Pharm. Innov.* 10 (4): 287–301.

50 Wise, B.M. and Roginski, R.T. (2015). Model maintenance: the unrecognized cost in PAT and QbD. *Chim. Oggi-Chem. Today* 33 (2): 38–43.

51 Jamrógiewicz, M. (2012). Application of the near-infrared spectroscopy in the pharmaceutical technology. *J. Pharm. Biomed. Anal.* 66: 1–10.

52 De Beer, T., Burggraeve, A., Fonteyne, M. et al. (2011). Near infrared and Raman spectroscopy for the in-process monitoring of pharmaceutical production processes. *Int. J. Pharm.* 417 (1–2): 32–47.

53 Singh, R., Velazquez, C., Sahay, A. et al. (2016). Advanced control of continuous pharmaceutical tablet manufacturing processes. In: *Process Simulation and Data Modeling in Solid Oral Drug Development and Manufacture. Methods in Pharmacology and Toxicology* (eds. M. Ierapetritou and R. Ramachandran), 191–224. New York, NY: Humana Press.

54 Classen, J., Aupert, F., Reardon, K.F. et al. (2017). Spectroscopic sensors for in-line bioprocess monitoring in research and pharmaceutical industrial application. *Anal. Bioanal. Chem.* 409: 651–666.

55 Rathore, A.S., Bhambure, R., and Ghare, V. (2010). Process analytical technology (PAT) for biopharmaceutical products. *Anal. Bioanal. Chem.* 398 (1): 137–154.

56 Mansouri, M.A., Sacré, P.-Y., Coïc, L. et al. (2020). Quantitation of active pharmaceutical ingredient through the packaging using Raman handheld spectrophotometers: a comparison study. *Talanta* 207: 120306.

57 Deidda, R., Sacre, P.-Y., Clavaud, M. et al. (2019). Vibrational spectroscopy in analysis of pharmaceuticals: critical review of innovative portable and handheld NIR and Raman spectrophotometers. *TrAC Trends Anal. Chem.* 114: 251–259.

58 Sun, L., Hsiung, C., Pederson, C.G. et al. (2016). Pharmaceutical raw material identification using miniature near-infrared (MicroNIR) spectroscopy and supervised pattern recognition using support vector machine. *Appl. Spectrosc.* 70 (5): 816–825.

59 Calvo, N.L., Alvarez, V.A., Lamas, M.C., and Leonardi, D. (2019). New approaches to identification and characterization of tioconazole in raw material and in pharmaceutical dosage forms. *J. Pharm. Anal.* 9 (1): 40–48.

60 Diehl, B., Chen, C., Grout, B. et al. (2012). Non-destructive material identification. *Eur. Pharm. Rev.* 17 (5): 3–8.

61 Devjak Novak, S., Šporar, E., Baumgartner, S., and Vrečer, F. (2012). Characterization of physicochemical properties of hydroxypropyl methylcellulose (HPMC) type 2208 and their influence on prolonged drug release from matrix tablets. *J. Pharm. Biomed. Anal.* 66: 136–143.

62 Razuc, M., Grafia, A., Gallo, L. et al. (2019). Near-infrared spectroscopic applications in pharmaceutical particle technology. *Drug Dev. Ind. Pharm.* 45 (10): 1565–1589.

63 Puchert, T., Holzhauer, C.-V., Menezes, J.C. et al. (2011). A new PAT/QbD approach for the determination of blend homogeneity: combination of on-line NIRS analysis with PC Scores Distance Analysis (PC-SDA). *Eur. J. Pharm. Biopharm.* 78 (1): 173–182.

64 Bakri, B., Weimer, M., Hauck, G., and Reich, G. (2015). Assessment of powder blend uniformity: comparison of real-time NIR blend monitoring with stratified sampling in combination with HPLC and at-line NIR Chemical Imaging. *Eur. J. Pharm. Biopharm.* 97: 78–89.

65 Anderson, C.A. and Velez, N.L. (2018). Blending and characterization of pharmaceutical powders. In: *Particles and Nanoparticles in Pharmaceutical Products. AAPS Advances in the Pharmaceutical Sciences Series*, vol. 29 (eds. H.G. Merkus, G.M.H. Meesters and W. Oostra), 233–275. Cham: Springer.

66 Muzzio, F.J., Robinson, P., Wightman, C., and Brone, D. (1997). Sampling practices in powder blending. *Int. J. Pharm.* 155 (2): 153–178.

67 Sekulic, S.S., Ward, H.W., Brannegan, D.R. et al. (1996). On-line monitoring of powder blend homogeneity by near-infrared spectroscopy. *Anal. Chem.* 68 (3): 509–513.

68 Ma, H. and Anderson, C.A. (2008). Characterization of pharmaceutical powder blends by NIR chemical imaging. *J. Pharm. Sci.* 97 (8): 3305–3320.

69 Chavez, P., Lebrun, P., Sacré, P.-Y. et al. (2015). Optimization of a pharmaceutical tablet formulation based on a design space approach and using vibrational spectroscopy as PAT tool. *Int. J. Pharm.* 486 (1–2): 13–20.

70 Wu, Z., Tao, O., Dai, X. et al. (2012). Monitoring of a pharmaceutical blending process using near infrared chemical imaging. *Vib. Spectrosc.* 63: 371–379.

71 Li, W., Woldu, A., Kelly, R. et al. (2008). Measurement of drug agglomerates in powder blending simulation samples by near infrared chemical imaging. *Int. J. Pharm.* 350 (1–2): 369–373.

72 Blanco, M., Cueva-Mestanza, R., and Cruz, J. (2012). Critical evaluation of methods for end point determination in pharmaceutical blending process. *Anal. Methods* 4 (9): 2694–2703.

73 Palmer, J., O'Malley, C.J., Wade, M.J. et al. (2018). Opportunities for process control and quality assurance using online NIR analysis to a continuous wet granulation tableting line. *J. Pharm. Innov.* 1–15.

74 Corredor, C.C., Lozano, R., Bu, X. et al. (2014). Analytical method quality by design for an on-line near-infrared method to monitor blend potency and uniformity. *J. Pharm. Innov.* 10 (1): 47–55.

75 Momose, W., Katz, J.M., Drennen, J.K., and Anderson, C.A. (2015). Development of NIR methods for blend analysis using small quantities of materials. *J. Pharm. Innov.* 10 (1): 36–46.

76 El-Hagrasy, A.S. and Drennen, J.K. (2006). A process analytical technology approach to near-infrared process control of pharmaceutical powder blending. Part III: quantitative near-infrared calibration for prediction of blend homogeneity and characterization of powder mixing kinetics. *J. Pharm. Sci.* 95 (2): 422–434.

77 Kim, S., Kano, M., Nakagawa, H., and Hasebe, S. (2011). Estimation of active pharmaceutical ingredients content using locally weighted partial least squares and statistical wavelength selection. *Int. J. Pharm.* 421 (2): 269–274.

78 Nakagawa, H., Kano, M., Hasebe, S. et al. (2014). Verification of model development technique for NIR-based real-time monitoring of ingredient concentration during blending. *Int. J. Pharm.* 471 (1–2): 264–275.

79 El-Hagrasy, A.S., Delgado-Lopez, M., and Drennen, J.K. (2006). A process analytical technology approach to near-infrared process control of pharmaceutical powder blending: part II: qualitative near-infrared models for prediction of blend homogeneity. *J. Pharm. Sci.* 95 (2): 407–421.

80 Besseling, R., Damen, M., Tran, T. et al. (2015). An efficient, maintenance free and approved method for spectroscopic control and monitoring of blend uniformity: the moving F-test. *J. Pharm. Biomed. Anal.* 114: 471–481.

81 Corredor, C.C., Bu, D., and Both, D. (2011). Comparison of near infrared and microwave resonance sensors for at-line moisture determination in powders and tablets. *Anal. Chim. Acta* 696 (1–2): 84–93.

82 Gavan, A., Iurian, S., Casian, T. et al. (2020). Fluidised bed granulation of two APIs: QbD approach and development of a NIR in-line monitoring method. *Asian J. Pharm. Sci.* 15 (4): 506–517.

83 Fonteyne, M., Vercruysse, J., Díaz, D.C. et al. (2013). Real-time assessment of critical quality attributes of a continuous granulation process. *Pharm. Dev. Technol.* 18 (1): 85–97.

84 Fonteyne, M., Gildemyn, D., Peeters, E. et al. (2014). Moisture and drug solid-state monitoring during a continuous drying process using empirical and mass balance models. *Eur. J. Pharm. Biopharm.* 87 (3): 616–628.

85 Pauli, V., Roggo, Y., Kleinebudde, P., and Krumme, M. (2019). Real-time monitoring of particle size distribution in a continuous granulation and drying process by near infrared spectroscopy. *Eur. J. Pharm. Biopharm.* 141: 90–99.

86 Vo, A.Q., He, H., Zhang, J. et al. (2018). Application of FT-NIR analysis for in-line and real-time monitoring of pharmaceutical hot melt extrusion: a technical note. *AAPS PharmSciTech* 19 (8): 3425–3429.

87 Kelly, A.L., Halsey, S.A., Bottom, R.A. et al. (2015). A novel transflectance near infrared spectroscopy technique for monitoring hot melt extrusion. *Int. J. Pharm.* 496 (1): 117–123.

88 Kelly, A.L., Gough, T., Dhumal, R.S. et al. (2012). Monitoring ibuprofen–nicotinamide cocrystal formation during solvent free continuous cocrystallization (SFCC) using near infrared spectroscopy as a PAT tool. *Int. J. Pharm.* 426 (1–2): 15–20.

89 Islam, M.T., Scoutaris, N., Maniruzzaman, M. et al. (2015). Implementation of transmission NIR as a PAT tool for monitoring drug transformation during HME processing. *Eur. J. Pharm. Biopharm.* 96: 106–116.

90 Baronsky-Probst, J., Möltgen, C.-V., Kessler, W., and Kessler, R.W. (2016). Process design and control of a twin screw hot melt extrusion for continuous pharmaceutical tamper-resistant tablet production. *Eur. J. Pharm. Sci.* 87: 14–21.

91 Wahl, P.R., Treffer, D., Mohr, S. et al. (2013). Inline monitoring and a PAT strategy for pharmaceutical hot melt extrusion. *Int. J. Pharm.* 455 (1–2): 159–168.

92 Saerens, L., Dierickx, L., Quinten, T. et al. (2012). In-line NIR spectroscopy for the understanding of polymer–drug interaction during pharmaceutical hot-melt extrusion. *Eur. J. Pharm. Biopharm.* 81 (1): 230–237.

93 Järvinen, K., Hoehe, W., Järvinen, M. et al. (2013). In-line monitoring of the drug content of powder mixtures and tablets by near-infrared spectroscopy during the continuous direct compression tableting process. *Eur. J. Pharm. Sci.* 48 (4–5): 680–688.

94 Dalvi, H., Langlet, A., Colbert, M.-J. et al. (2019). In-line monitoring of Ibuprofen during and after tablet compression using near-infrared spectroscopy. *Talanta* 195: 87–96.

95 Dalvi, H., Fauteux-Lefebvre, C., Guay, J.-M. et al. (2018). Concentration monitoring with near infrared chemical imaging in a tableting press. *J. Spectr. Imaging* 7 (1): a5.

96 Durão, P., Fauteux-Lefebvre, C., Guay, J.-M. et al. (2017). Using multiple process analytical technology probes to monitor multivitamin blends in a tableting feed frame. *Talanta* 164: 7–15.

97 He, X., Han, X., Ladyzhynsky, N., and Deanne, R. (2013). Assessing powder segregation potential by near infrared (NIR) spectroscopy and correlating segregation tendency to tabletting performance. *Powder Technol.* 236: 85–99.

98 Knop, K. and Kleinebudde, P. (2013). PAT-tools for process control in pharmaceutical film coating applications. *Int. J. Pharm.* 457 (2): 527–536.

99 Brock, D., Zeitler, J.A., Funke, A. et al. (2012). A comparison of quality control methods for active coating processes. *Int. J. Pharm.* 439 (1–2): 289–295.

100 Gendre, C., Genty, M., Boiret, M. et al. (2011). Development of a Process Analytical Technology (PAT) for in-line monitoring of film thickness and mass of coating materials during a pan coating operation. *Eur. J. Pharm. Sci.* 43 (4): 244–250.

101 Moltgen, C.V., Puchert, T., Menezes, J.C.C. et al. (2012). A novel in-line NIR spectroscopy application for the monitoring of tablet film coating in an industrial scale process. *Talanta* 92: 26–37.

102 Liu, R., Li, L., Yin, W. et al. (2017). Near-infrared spectroscopy monitoring and control of the fluidized bed granulation and coating processes – a review. *Int. J. Pharm.* 530 (1–2): 308–315.

103 Korasa, K. and Vrečer, F. (2018). Overview of PAT process analysers applicable in monitoring of film coating unit operations for manufacturing of solid oral dosage forms. *Eur. J. Pharm. Sci.* 111: 278–292.

104 Bogomolov, A., Engler, M., Melichar, M., and Wigmore, A. (2010). In-line analysis of a fluid bed pellet coating process using a combination of near infrared and Raman spectroscopy. *J. Chemometr.* 24 (7–8): 544–557.

105 Wahl, P.R., Peter, A., Wolfgang, M., and Khinast, J.G. (2019). How to measure coating thickness of tablets: method comparison of optical coherence tomography, near-infrared spectroscopy and weight-, height- and diameter gain. *Eur. J. Pharm. Biopharm.* 142: 344–352.

106 Igne, B., Arai, H., Drennen, J.K., and Anderson, C.A. (2016). Effect of sampling frequency for real-time tablet coating monitoring using near infrared spectroscopy. *Appl. Spectrosc.* 70 (9): 1476–1488.

107 Felton, L.A. (2013). Mechanisms of polymeric film formation. *Int. J. Pharm.* 457 (2): 423–427.

108 Gendre, C., Genty, M., da Silva, J.C. et al. (2012). Comprehensive study of dynamic curing effect on tablet coating structure. *Eur. J. Pharm. Biopharm.* 81 (3): 657–665.

109 Gendre, C., Genty, M., Fayard, B. et al. (2013). Comparative static curing versus dynamic curing on tablet coating structures. *Int. J. Pharm.* 453 (2): 448–453.

110 Korasa, K., Hudovornik, G., and Vrečer, F. (2016). Applicability of near-infrared spectroscopy in the monitoring of film coating and curing process of the prolonged release coated pellets. *Eur. J. Pharm. Sci.* 93: 484–492.

111 Görög, S. and Szántay, C. (2017). Spectroscopic methods in drug quality control and development. In: *Encyclopedia of Spectroscopy and Spectrometry*, 3e (eds. J.C. Lindon, G.E. Tranter and D.W. Koppenaal), 178–187. Academic Press.

112 Singh, R. (2018). Implementation of control system into continuous pharmaceutical manufacturing pilot plant (powder to tablet). In: *Computer Aided Chemical Engineering*, vol. 41, 447–469.

113 Román-Ospino, A.D., Cárdenas, V., Ortega-Zuñiga, C., and Singh, R. (2018). PAT for pharmaceutical manufacturing process involving solid dosages forms. In: *Computer Aided Chemical Engineerin.*, vol. 41, 293–315. Elsevier.

114 Pomerantsev, A.L., Rodionova, O.Y., Melichar, M. et al. (2011). In-line prediction of drug release profiles for pH-sensitive coated pellets. *Analyst* 136 (22): 4830.

115 Pawar, P., Wang, Y., Keyvan, G. et al. (2016). Enabling real time release testing by NIR prediction of dissolution of tablets made by continuous direct compression (CDC). *Int. J. Pharm.* 512 (1): 96–107.

116 Rantanen, J. and Khinast, J. (2015). The future of pharmaceutical manufacturing sciences. *J. Pharm. Sci.* 104 (11): 3612–3638.

117 Awad, A., Trenfield, S.J., Gaisford, S., and Basit, A.W. (2018). 3D printed medicines: a new branch of digital healthcare. *Int. J. Pharm.* 548 (1): 586–596.

118 Trenfield, S.J., Goyanes, A., Telford, R. et al. (2018). 3D printed drug products: non-destructive dose verification using a rapid point-and-shoot approach. *Int. J. Pharm.* 549 (1–2): 283–292.

119 Khorasani, M., Edinger, M., Raijada, D. et al. (2016). Near-infrared chemical imaging (NIR-CI) of 3D printed pharmaceuticals. *Int. J. Pharm.* 515 (1–2): 324–330.

120 WHO (2018). *Substandard and Falsified Medical Products – Fact Sheet*. WHO.

121 Rebiere, H., Guinot, P., Chauvey, D., and Brenier, C. (2017). Fighting falsified medicines: the analytical approach. *J. Pharm. Biomed. Anal.* 142: 286–306.

122 Been, F., Roggo, Y., Degardin, K. et al. (2011). Profiling of counterfeit medicines by vibrational spectroscopy. *Forensic Sci. Int.* 211 (1–3): 83–100.

123 Zontov, Y.V., Balyklova, K.S., Titova, A.V. et al. (2016). Chemometric aided NIR portable instrument for rapid assessment of medicine quality. *J. Pharm. Biomed. Anal.* 131: 87–93.

124 Mbinze, J.K., Sacré, P.-Y., Yemoa, A. et al. (2015). Development, validation and comparison of NIR and Raman methods for the identification and assay of poor-quality oral quinine drops. *J. Pharm. Biomed. Anal.* 111: 21–27.

125 Crocombe, R.A. (2018). Portable spectroscopy. *Appl. Spectrosc.* 72 (12): 1701–1751.

126 Ciza, P.H., Sacre, P.-Y., Waffo, C. et al. (2019). Comparing the qualitative performances of handheld NIR and Raman spectrophotometers for the detection of falsified pharmaceutical products. *Talanta* 202: 469–478.

127 Vickers, S., Bernier, M., Zambrzycki, S. et al. (2018). Field detection devices for screening the quality of medicines: a systematic review. *BMJ Glob. Health* 3 (4): e000725.

128 Roth, L., Biggs, K.B., and Bempong, D.K. (2019). Substandard and falsified medicine screening technologies. *AAPS Open* 5 (1): 2.

129 Esbensen, K.H. and Wagner, C. (2015). Theory of sampling (TOS) – fundamental definitions and concepts. *Spectrosc. Eur.* 27 (1): 1–4.

5

MOEMS and MEMS - Technology, Benefits & Uses

Heinrich Grüger

Fraunhofer Institute for Photonic Microsystems, Dresden, Germany

5.1 Introduction

5.1.1 Definitions

The term MEMS – micro-electromechanical systems – has shifted in its meaning through the last decades. Today, relevant characteristics are considered to be "a microscopic device" featuring "electrical functionality" and "moving parts" [1]. The definition includes sensors where a movement or displacement is converted into an electric signal [2], actuators where electrical driving principles are used to initiate motion or rotation of a deflectable part [3], as well as combinations of both, e.g. in ultrasonic transducers [4], accelerometers [5] and gyroscopes [6].

Most commercial MEMS are produced on silicon wafers, but other materials have been used as well, e.g. "Diamond MEMS" [7]. Mechanical motion is actuated by means of electrical driving principles, which can be electrostatic, magnetic or piezoelectric. For proper operation, MEMS need matching drive electronics, often operated in a close loop mode [8]. Ultracompact outlines can be achieved by the combination of MEMS and complementary metal-oxide-semiconductors (CMOS) in chip-scale assembly, or in selected cases, MEMS on top of CMOS [9].

The combination of electrical and mechanical functions enables devices, like microphones [10, 11], loudspeakers [12, 13], ultrasonic transducers [4, 14], pressure sensors [2], acceleration sensors [5, 15–17], gyroscopes [6, 18, 19], and many more. MEMS devices can be found in consumer electronic devices, like mobile phones, tablets, digital or video cameras, toys, entertainment equipment, drones, but also in more critical applications. For instance, in automobiles, a steadily increasing number of MEMS devices takes care of safety and comfort; tire pressure sensors are the most obvious example.

Optical MEMS are a specific class of actuators featuring interaction of MEMS actuation and light. Examples are scanner mirrors in rotational[1] [20–24] or piston-type configurations [25, 26], spatial light modulators [27, 28], also tiltable gratings [29], more complex devices [30], and optical MEMS systems in a chip [31]. Bulk silicon micro-machined scanner mirrors have been used for applications, like laser light deflection, from bar code scanning to projection, microscopy and spectroscopy [32]. Future systems, like LIDAR (light detection and ranging) are being developed [33]. Surface micro-machined mirror arrays in digital light processors [34] are used in digital cinema projectors and similar applications.

Optical MEMS contribute significantly to the development of systems for portable spectroscopy. The term "MOEMS" – "micro-opto-electromechanical systems" – is used for combinations of MEMS and micro-optics (see preface in [35]), but does not define a class of special MEMS. In the following, all devices are simply named MEMS.

1 https://www.hamamatsu.com/eu/en/product/type/S12237-03P/index.html.

Portable Spectroscopy and Spectrometry 1: Technologies and Instrumentation, First Edition.
Edited by Richard A. Crocombe, Pauline E. Leary and Brooke W. Kammrath.

Some devices which are fabricated using typical MEMS technologies, but lacking selected features of the MEMS definition, have been called "MEMS" or "MEMS technology-based" as well. Phrases like "no moving parts" are in contradiction to the basic MEMS definition. Nonetheless, devices based on such technologies have been used for spectrometer systems successfully and will be discussed here as well.

5.1.2 Brief History of MEMS

Microelectronic devices fabricated in CMOS technologies have changed the world since the detection of the rectifier effect by Ferdinand Braun in 1874. Today, multibillion transistors can be integrated in a single chip, and structures have feature sizes down to seven nanometers.

The availability of large wafer substrates and reliable process technology initiated the invention of devices which are not mainly electronics, but featuring electrically driven micro-mechanics. The term "Micro-electromechanical Systems – MEMS" was used first in a proposal to US Defense Advanced Research Projects Agency (DARPA) in July 1986. Details of the invention and history have been summarized in literature [35]. Here some relevant milestones shall be mentioned briefly. A special kind of silicon wafer substrate provides a lateral structure of silicon and silicon oxide. SOI – silicon on oxide – wafers have a top layer in the range from ¼ to 150 μm active silicon on 1–2 μm silicon oxide and a few hundreds of microns of silicon as handle layer. The advantage of this structure is the application of deep etch processes [36] which are extremely selective between silicon and silicon oxide. Thus, the oxide layer can be used as etch stop from top and bottom. The anisotropic deep etching, either wet, using fluid acids, or gas phase based, e.g. plasma etch or reactive ion etch (RIE), became the key to MEMS production. Much effort has been put into the optimization of anisotropic etch steps. Most relevant was the invention of the deep reactive ion etch ("DRIE") in the 1990s, often named "Bosch process" due to the inventing company [37]. The Bosch process has become the standard process for MEMS production. High aspect ratios can be achieved, which are used to realize MEMS structures with deep and steep trenches, or even complex 3D structures.

The continuous shift to larger substrates in microelectronics came later for MEMS, but high-volume production has been established long ago (Figure 5.1).[2]

The invention of optical MEMS devices followed the general MEMS technology development. Optical surfaces, e.g. mirrors, were integrated on moving parts. The main increase in complexity was in the field of optical

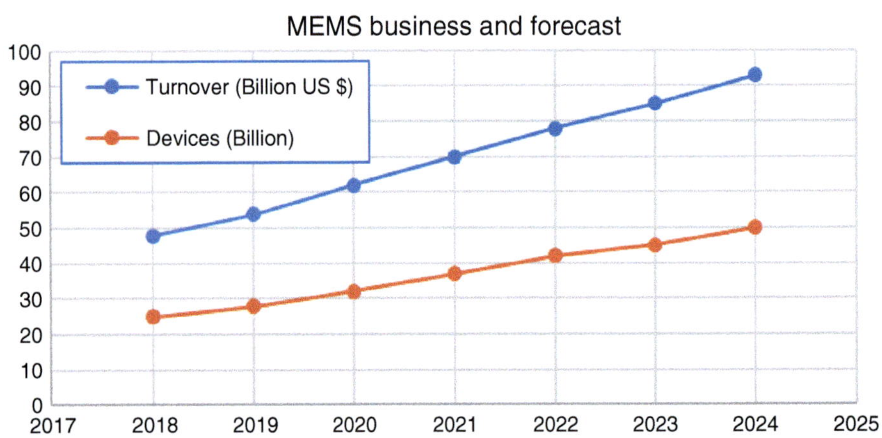

Figure 5.1 Development of the annual MEMS production volume. *Source*: Based on New market perspectives for the MEMS & Sensors. http://www.yole.fr/iso_upload/News/2017/ PR_STATUSMEMSINDUSTRY_NewMarketPerspectives_YOLE_June2017.pdf.

2 http://www.yole.fr/iso_upload/News/2017/PR_STATUSMEMSINDUSTRY_NewMarketPerspectives_YOLE_June2017.pdf.

MEMS assembly. The optical areas must be accessible. Thus, the assembly technology must be adapted to optical requirements like openings or windows.

A prominent example for optical MEMS-based systems were projectors for commercial and later private use. From cinema to meeting rooms and private entertainment, the use of a digital light processor, a MEMS device with a very high number of individually actuated mirrors [27], provided a significant increase in image quality compared to the earlier transmission panel based on liquid crystal displays (LCDs). The MEMS device was originally invented by Texas Instruments in 1987, and projectors based on digital light processing (DLP) entered the market in 1997.

Since the 1990s, monolithic MEMS scanner mirrors have been developed for applications, like barcode scanning and laser projection. The recent demand for LIDAR systems in automotive applications may lead to an increased demand for scanning mirrors from the automotive sector.

5.1.3 Basics of MEMS Technology

Pattern generation for MEMS devices on the wafer uses the same lithographic technology as used for CMOS components. A photoresist is applied, which is exposed to ultraviolet (UV) light through a mask bearing the structural information (e.g. chromium on glass, using 10× magnification). The structural information is transferred to the resist on the wafer before different deposition or etch steps is/are applied. These steps are repeated several times until the device in finished [38].

Typically, MEMS devices feature much wider structural feature sizes than CMOS devices. This is due to the fact that mechanical interaction, e.g. force sensing in acceleration or gyro sensors, as well as optical applications, require significant mass or optical active area to operate properly.

Characteristics for MEMS technologies are deep etch process steps with high aspect ratios. To realize MEMS structures, deep but narrow trenches have to be etched from the front side. For instance, the mirror plate of a bulk silicon microscanner mirror may be 150 μm thick, but the trenches should be only 2 μm wide. In turn, deep etch optimization reached aspect ratios up to 1 : 1000.

Freestanding structures formed by the front side trenches are released by first backside cavity etch, stopping on the buried oxide, and a final buried oxide removal. The main issue is the realization of deep trenches in silicon stopping on silicon oxide (Figure 5.2).

For 3D optical MEMS using the vertical areas of the etch structures or trenches, it is important to consider the so-called scallops originating from the repeated etch and deposition steps of the Bosch process. There are process options available to reduce the formation of these scallops.

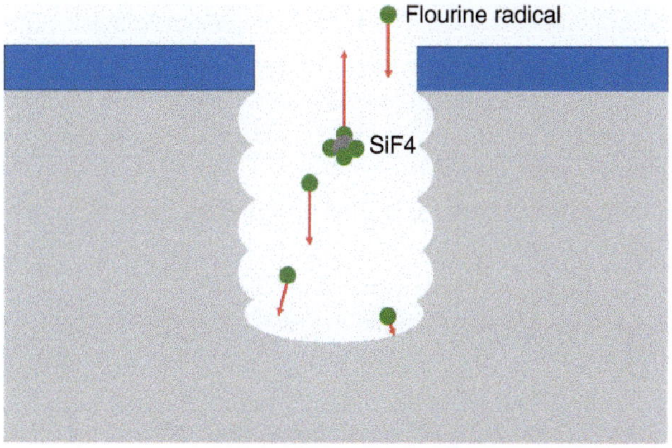

Figure 5.2 Anisotropic gas phase etch of silicon with photoresist.

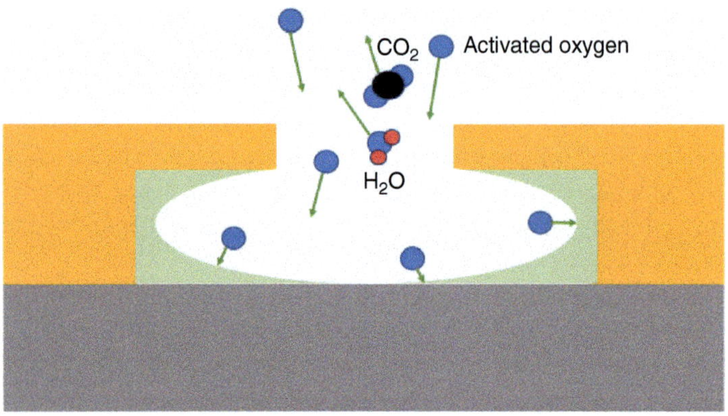

Figure 5.3 Removal of sacrificial layer by plasma etch.

Surface micro-machined MEMS differ in the etch process. Here a sacrificial layer is deposited and planarized [39]. Holes are etched into the sacrificial layer for the realization of hinges, before the surface MEMS material, typically a metal or alloy, is deposited. Applying another photolithography process step, the lateral MEMS structures and the etch holes are generated. Afterward, the sacrificial layer is removed by gas phase etching and the surface MEMS structures (Figure 5.3) are released.

MEMS devices are extremely reliable and robust: their small size means that they are very "stiff" with resonant frequencies much higher than found in ambient vibrations. Especially for bulk silicon micro-machined devices, the properties of the monocrystalline material leads to favorable system capabilities, while reproducibility ensures high standards for the systems based upon MEMS devices.

Once designed, the production of MEMS can easily be scaled up to high-volume production. The initial development requires an initial effort for setup, chip design, lithography mask manufacturing and process implementation. The process, consisting of hundreds of individual steps, may take multiple weeks to a month's time. Finally, production on the 200 or 300-mm wafers used today provides a low cost for the individual chip in volume production, which offers market opportunities for MEMS-based systems.

5.1.4 NIR Analysis

MEMS-based systems mostly enable small outline, high-volume and low-cost applications. For portable spectroscopy, this shifts the focus for MEMS-based systems toward the field of near-infrared (NIR) analysis, where the spectrometer and a light source will be the main components. The light source, e.g. a small halogen bulb, can be cost-efficient as well. No complex laser system is required as compared with Raman spectrometers. The potential for high-volume production addresses users of mobile devices. NIR spectroscopy, because both organic and inorganic matter can be analyzed, opens day-to-day applications in life, like food, dermatology, and medical. Furthermore, inorganic applications are possible in the geology, i.e. minerals.

For solid organic matter, the penetration depth of NIR radiation can be up to several millimeters, and, thus, more than the surface layers can be analyzed. The combination and overtone bands of organic molecules are used in both qualitative and quantitative analyses using chemoetrics-based mathematical models based on reference data.

Requirements for different spectrometers or spectral analyzers are summarized in Table 5.1.

5.2 Grating-Based Spectrometers

Historically, prism-based spectroscopes, invented in 1814 by Joseph von Fraunhofer, were the first spectral analyzers used for scientific applications – the evaluation of the sun's spectrum. Dispersive gratings were invented

Table 5.1 Requirements for multipurpose spectral analyzers.

	Portable desktop	Handheld	Mobile phone add-on	Mobile phone integrated
Size	$<100 \times 100 \times 100\,mm^3$ $<4'' \times 4'' \times 4''$	$<75 \times 50 \times 25\,mm^3$ $<3'' \times 2'' \times 1''$	$<50 \times 25 \times 15\,mm^3$ $<2'' \times 1'' \times 0.6''$	$<10 \times 10 \times 4.5\,mm^3$
Range	950–1900 nm/ 1250–2500 nm	950–1900 nm	1000–1850 nm	1000–1850 nm
Resolution	<10 nm	10 nm	12 nm	15–18 nm
Stability	0.1–0.5 nm	0.5 nm	0.5 nm	0.5–1 nm
SNR	1 : 4096	1 : 1024	1 : 1024	1 : 256 (512)

Devices with defined uses may differ mainly in their spectral ranges.

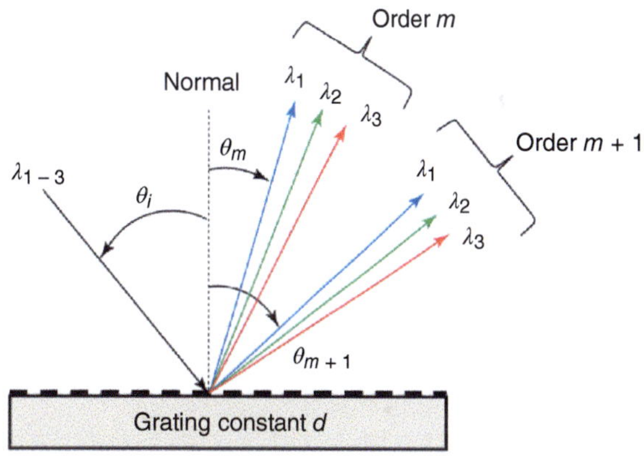

Figure 5.4 Reflection grating.

by David Rittenhouse in 1785 [40, 41]. Compared to prism-based systems, they have a higher dispersive power and, if used in reflection, there is no absorption through the material. Important spectrometer designs were presented by Otto von Littrow in 1862, H. Ebert [42] in 1889, followed by the work of Monk and Gillieson [43, 44], Czerny and Turner [45, 46], and the improved Ebert–Fastie [47, 48]. Other less prominent configurations have been published [49].

Grating-based spectrographs became the dominant technology in the NIR as soon as gratings with sufficient quality became available. The reproduction technologies to realize numerous gratings from a single master device were key to their commercial success. The diffraction pattern of a grating is based on the geometric path difference (Figure 5.4).

The grating equation (Eq. (5.1)) describes the mathematics criteria for constructive interference. Intensity maxima for a given wavelength will occur if the path difference, d_n, matches an integer multiple of the wavelength, including higher positive ($n > 1$) and negative ($n \leq -1$) diffraction orders.

$$d_n = n\,\lambda = g\,(\sin\,\alpha - \sin\,\beta) \tag{5.1}$$

The resolving power of a grating $\lambda/\Delta\lambda$ is depending on the number of elements N and the diffraction order n (Eq. (5.2)).

$$\lambda/\Delta\lambda = n\,N \tag{5.2}$$

The characteristic setup of a Czerny–Turner-type grating spectrometer is shown in Figure 5.5. The five major components of such a spectrometer are (i) entrance aperture – usually a slit, but could be a pinhole or the end

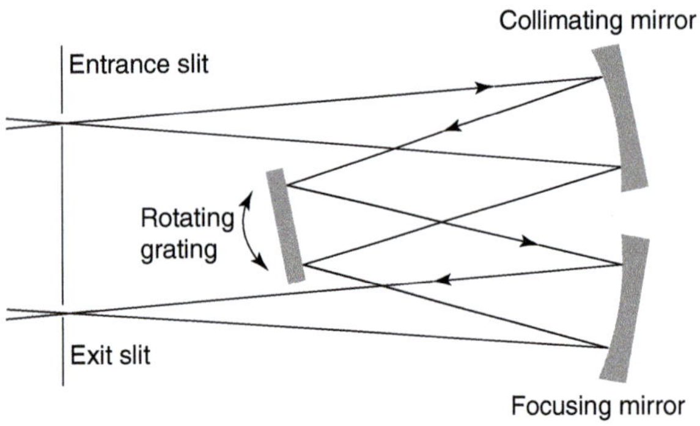

Figure 5.5 Czerny–Turner setup for a NIR spectrometer.

of an optical fiber as well, (ii) collimating mirror, (iii) grating, (iv) refocusing mirror, (v) exit aperture – slit and screen or detector behind it.

Today, scientific instruments are available either in the classic design with rotating gratings and flexible mounts for selected detectors, or small-sized systems using fixed gratings and detector arrays. Over the past few decades, significant work was put into miniaturization and efficient production of these spectrometers. Similar to digital photography, this may lead to their application in numerous fields, many of which are yet to be developed. The main reason is that spectroscopy is a very capable technique for the analysis of organic matter; its capabilities will be advantageous in an increasing field of applications.

Diffraction at a grating only occurs if the d-spacing is wider than half of the wavelength, which limits the resolution of grating-based spectrometers. The NIR bands of organic molecules in the condensed phase are typically wide, and, thus, the resolution required is not challenging. So, there is a growing market for NIR analysis.

5.2.1 Fixed Grating-Based Spectrometer

The first miniature spectrometer entered the market in 1992 from Ocean Optics, founded in 1989, now known as Ocean Insight. Based on a detector array, and applying a crossed Czerny–Turner configuration, the trend to miniaturization was initiated. Silicon detectors enable low-cost systems, but are limited to a long wavelength cutoff around 1100 nm. Similar systems have been developed for NIR applications using InGaAs detector arrays, which are still quite expensive. The slogan "no moving parts" has been used to characterize these systems. Thus, they do not exactly meet the MEMS definition.

Fixed grating spectrometers have been realized by applying MEMS technologies, i.e. reactive silicon deep etching of the slits, and the same chip housing the detector array. Hamamatsu has presented a "thumb size" spectrometer,[3] using a chip assembled with a concave grating, which was fabricated on a quartz body (Figure 5.6).

The first design with an outline of $27.6 \times 16.8 \times 13\ \text{mm}^3$ offered either a spectral range from 340 to 750 nm with 14 nm resolution or 640–1050 nm with 20 nm resolution. The second generation with a reduced outline of $20.1 \times 12.5 \times 10.1\ \text{mm}^3$ featured a detector with 256 elements each 12.5 μm wide and addressed a spectral range from 340 to 780 nm with 15 nm resolution.

New in 2019[4] was an ultracompact design optimized for the use in mobile phones with $11.5 \times 4.0 \times 3.5\ \text{mm}^3$ announced to be the world's smallest spectrometer for NIR (640–1050 nm, 17–20 nm resolution).

In the end, the miniaturization of detector-array-based spectrometers faces limitations. The number of detector elements cannot be reduced without affecting either range or resolution. In the NIR region, the element size strongly correlates with the detector noise. Ultracompact systems, i.e. below 10 mm size, require the reduction of the array outline to 256 elements, which may be only 25 μm wide. In turn, to cover a spectral range of 1000 nm,

3 https://www.hamamatsu.com/us/en/product/type/C12666MA/index.html.
4 https://www.hamamatsu.com/us/en/product/type/C14384MA-01/index.html.

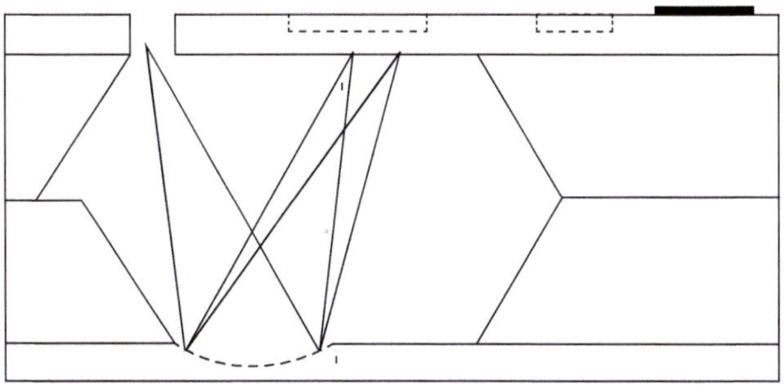

Figure 5.6 Chip for miniaturized spectrometer with slit and detector array.

each element has to cover 4 nm. This affects already the theoretical limits for resolution, and the stability of the wavelength scale as well.

5.2.2 MEMS Scanning Grating

Standard scanning monochromators use a mechanical drive to rotate the grating and thereby control which wavelength falls on the exit slit. Considering this, it became clear that a grating structure on a deflectable MEMS scanner device (Figure 5.7) could be used to realize a similar monochromator setup [29, 50].

Researchers at Fraunhofer IPMS initiated the development of MEMS scanning-grating devices based upon an existing 1D scanner mirror device with a resonant drive. This mirror is based on bulk silicon micro-machining, with typical device silicon thicknesses ranging between 30 and 90 μm. The grating is fabricated on the front side of the vibrating silicon plate by a wet anisotropic etch, using an additional mask for the grating structure, and is generated as grooves along the 111 crystal faces. Thus, the grating facets are atomically smooth, but the angle is given by the crystal orientation of the wafer used. Regular 100-oriented wafers result in facets along the 111 direction in an angle of 54.7°. Specific blazes are possible applying BSOI wafers with tilted crystal orientations for

Figure 5.7 Ultracompact outline spectrometer (Hamamatsu). *Source*: Hamamatsu Photonics K.K.

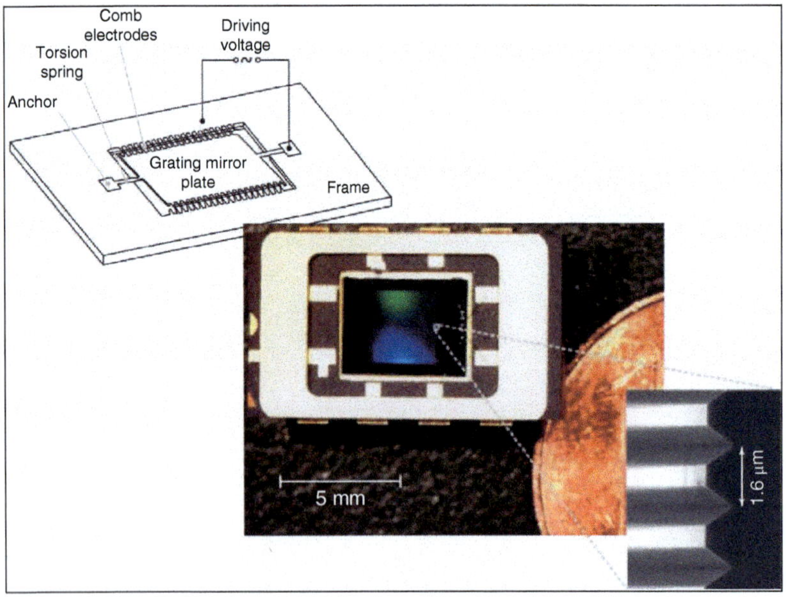

Figure 5.8 1D MEMS scanner device with grating. *Source*: Fraunhofer IPMS.

the active layer, but the huge effort is not justified by the results. For NIR applications, sufficient efficiency in a broad range is more relevant than highest efficiency at a single small band.

Different grating structures have been investigated (Figure 5.8), including refill process steps, leading to sinusoidal structures. Each structure has its specific advantages and disadvantages, but, in the end, the v-grooves have been selected for the system implementation [51].

Based on the MEMS scanning grating device, a spectrometer for NIR was developed (Figure 5.9). It features a Czerny–Turner configuration with an SubMiniature version A (SMA) coupler to connect standard optical fibers. An entrance slit can be added, and then an ultraprecision off-axis mirror is used for collimation. The grating is mounted on an adjustable stage, including a position readout sensor. The diffracted light is collected by a refocusing second mirror, and then captured by a standard single-element indium gallium arsenide (InGaAs) photodiode detector behind an exit slit. The signals are amplified by a three-stage transimpedance amplifier. A digital control unit computes the spectra from the intensities captured, and the grating position information. The sinusoidal movement, or an improved motion model, are used for highly accurate wavelength information and stability. A resonant MEMS device opens the option for long-term MEMS motion detection, or even active control in a close loop configuration, resulting in very accurate wavelength stability (Figure 5.10).

This device may be suitable for initial portable applications, but for real mobile use, a more miniaturized design would be advantageous. This, and a demand for high-volume capabilities, was the motivation for further research work. Size reduction is limited by the need for component alignment, but a promising option is the integration of more functions into the MEMS chip. Besides the grating, both slits can be integrated into the MEMS chip design, with the alignment determined by the lithographic accuracy of the MEMS process. In classic Czerny–Turner designs, the grating must be tilted out of the plane of grating and slits. However, this is difficult in a MEMS chip, and requires a more complex design [52]. A solution is the use of the first negative diffraction order. The entrance and exit slits were placed on one side of the grating, and the so-called *W*-configuration shifted to a folded double-V. For the realization, a new MEMS chip was fabricated. The weak point of this design is the need for complex off-axis mirrors; both mirrors were fabricated by ultraprecision micro-machining (UPM) in one aluminum substrate. The spectrometer is the size of a sugar cube (Figure 5.11), and was integrated by stacking the mirror substrate, a space

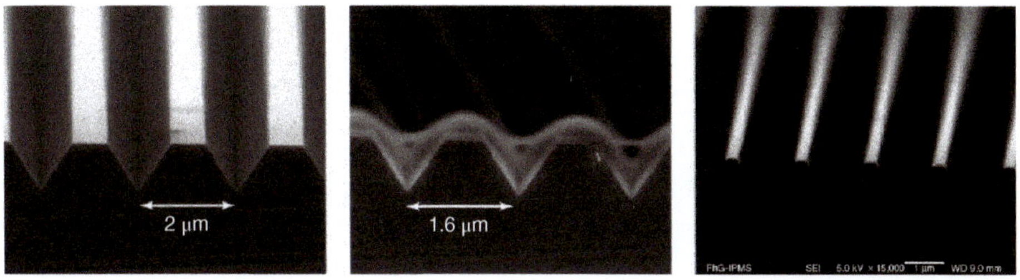

(a) (b) (c)

Figure 5.9 SEM images of different grating structures etch into silicon. (a) V-grooves etched into silicon. (b) Glass refill resulting in sinusoidal surface structures. (c) Blaze gratings realized using wafers with 4° … 11° angle misfit of the 111 orientation. *Source*: Fraunhofer IPMS.

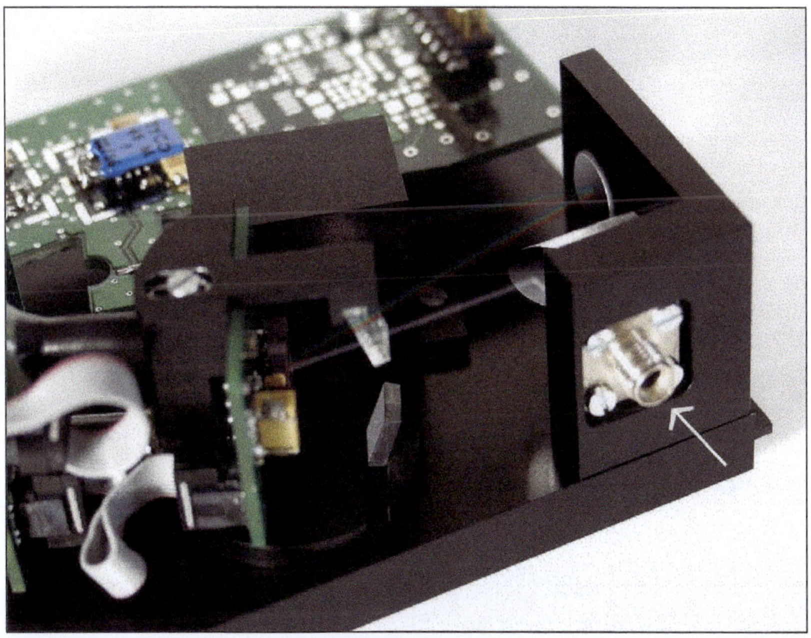

Figure 5.10 SGS 1900 MEMS-based scanning grating spectrometer. *Source*: Fraunhofer IPMS.

including structures for stray light suppression, and the MEMS board with a bar die detector chip, placed in a cavity in the MEMS chip [53].

The sugar cube-sized spectrometer achieved performance close to the existing SGS 1900 system. Besides this, some major disadvantages came up. The mirrors are complex and no technology was available for the production in volume and on large substrates. Stacking a huge number of spectrometers from a few large substrates and dicing afterward turned out to be complex too, as no reliable technology for dicing this stack could be found. Thus, the further development of a more miniaturized MEMS-based spectrometer, which would fit into mobile phones, had to be based on a different approach. The availability of a high throughput assembly tool that could put optical elements into selected places of complex 3D bodies was key. This next crossed Czerny–Turner design was further miniaturized. Spherical mirrors with rotation symmetric surfaces would be sufficient, if the astigmatism and coma could be controlled by the optic design. In the end, longer slits and a larger detector turned out to be advantageous.

Figure 5.11 MEMS scanning grating spectrometer in sugar cube size. *Source*: Fraunhofer IPMS.

Figure 5.12 Ultracompact MEMS scanning grating spectrometer. *Source*: Fraunhofer IPMS.

Based on the optic design, a spectrometer body was fabricated by means of 3D printing. Both mirrors and the entrance slit were assembled to the body, which was then placed on the MEMS board containing the MEMS scanning grating device, exit slit and detector (Figure 5.12).

5.2.3 DLP-Based Spectrometers

A different kind of deflectable mirror device can be realized using surface micro-machining. A thin material – most often a metal alloy – is deposited on a sacrificial layer which is removed afterward to generate free-standing structures. Due to the thickness limit of the sacrificial layer, the size of these types of MEMS mirrors must be

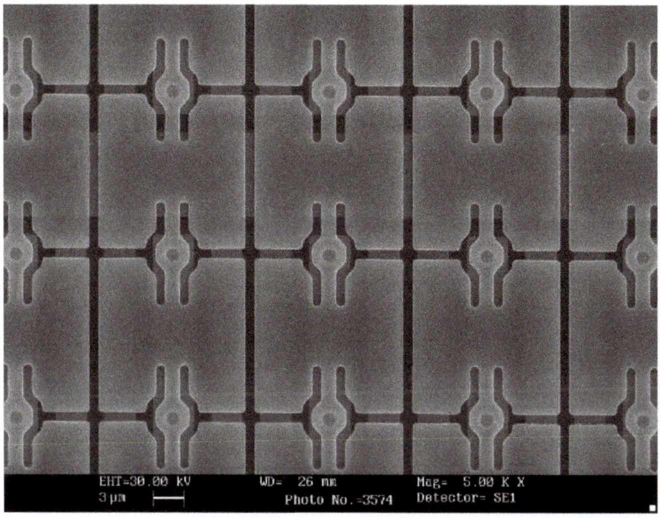

Figure 5.13 SEM image of a surface micro-machined spatial light modulator. *Source*: Fraunhofer IPMS.

Figure 5.14 Design of a fixed grating and DLP spectrometer (InnoSpectra). *Source*: InnoSpectra Corporation.

small, typically in the tens-of-micrometers range. High numbers of individual mirrors can be realized on one device. The freestanding structures can be deflected by means of electrostatic drive (Figures 5.13 and 5.14).

DLPs enable different spectrometer concepts and designs [54].[5] The most relevant types are a fixed grating spectrometer with DLPs for the selection of a pass wavelength to a single detector element and Hadamard spectrometers.

In fixed grating spectrometers, the detector array used in the image plane can simply be replaced by a tilting element light processor which deflects the light of one selected wavelength onto a single detector device, and the rest into an optical stop. This is advantageous if the detectors array is more expensive than the DLP and a single element detector.[6]

A weak spot of the grating spectrometers discussed previously is the very low signal level and the strong influence of noise in the electronic readout system. Here, Hadamard systems offer an alternative approach with similar components in the setup. The basic idea of a Hadamard transform spectrometer [55, 56] is to put half of the wavelengths of light on the single detector and then scan the change of the individual wavelengths within that half, instead of scanning small wavelength intervals. Afterward, the spectrum is calculated. The advantage is much higher flux on the detector, and, thus, less influence of detector noise. As compared with conventional array readout, a higher resolution analog-to-digital conversion is necessary. Based on this basic idea, much more complex approaches have been realized, which can also be used for more sophisticated spectral imaging systems.

5 http://www.ti.com/lit/wp/dlpa048a/dlpa048a.pdf
6 http://www.inno-spectra.com/index_en

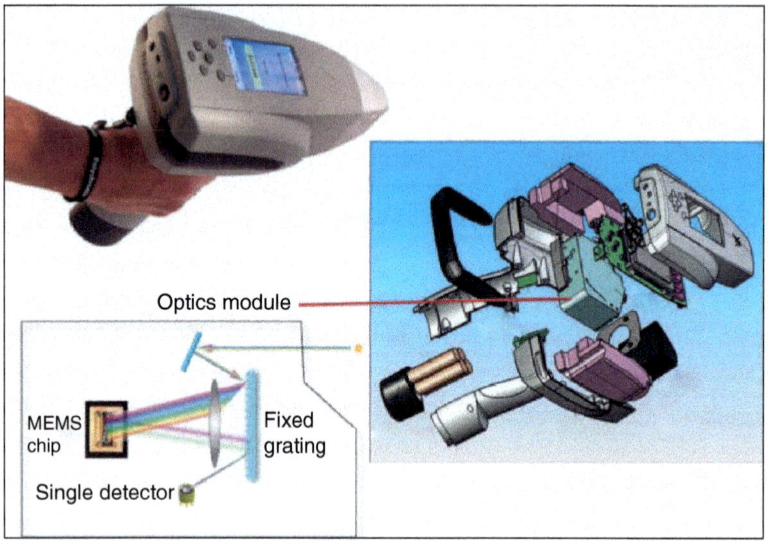

Figure 5.15 PAHZIR. *Source*: Encyclopedia of Analytical Chemistry in 2013 by John Wiley & Sons, Ltd. DOI: 10.1002/9780470027318.a9376

One of the first systems for portable spectroscopy was the "PHAZIR," developed by Polychromix which became part of Thermo Fisher Scientific later. A MEMS chip using surface micro-machining was realized by MEMS provider Boston Micromachines. It was operated as a programmable mask in a Hadamard-type configuration. A handheld system (Figure 5.15) was applied in handheld field applications, like recycling of carpet, which requires discrimination of carpet fiber materials, typically wool, nylon, polyethylene terephthalate, etc. the database and matching algorithm was stored in the system's memory.

5.2.4 Scanner Mirror-Based Micro-Spectrometer

Technically, there is no need to realize a grating structure on a MEMS scanner device to build a monochromator or spectrometer. Applying a simple scanning mirror device, and illuminating a grating mounted in an appropriate direction, also enables a monochromatic optical setup (Figure 5.16). The scanner essentially changes the incident angle of the collimated beam onto the grating. This design needs the grating as an additional component, but it offers advantages as well. The technology for the scanner is simplified, and the deflection required is only half of a scanning grating device which addresses the same spectral range. This contributes to possible cost reduction on the electronic side for the driver, but potentially the MEMS itself is less demanding too.

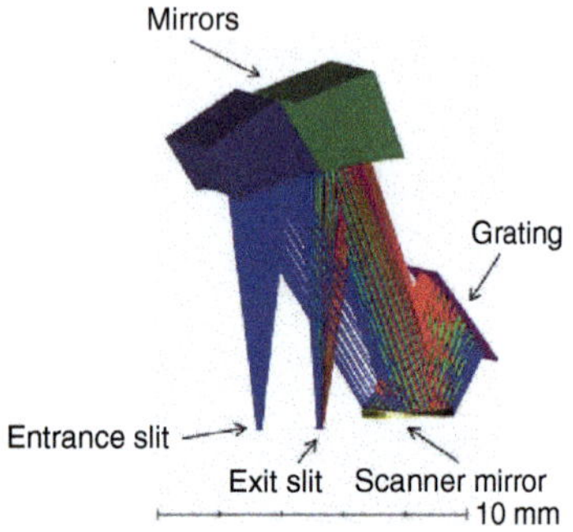

Figure 5.16 Optic design for a scanner mirror micro-spectrometer. *Source*: Grüger et al. [57]. © 2019, Society of Photo-Optical Instrumentation Engineers.

There are some more advantages of this concept. The grating can be simply changed, which opens flexibility for different *d*-spacing and blaze, and modular approaches may be possible. There are also no limitations from wet etching in silicon. Thus, the same basic system concept can be modified to fit the requirements of different applications.

Fraunhofer ENAS presented an early version of this kind of spectrometer. A system was realized that operates in the 1100–2100 nm range with a resolution of 12 nm [58].

A highly integrated spectrometer, based on resonant 1D scanner mirrors, was presented in 2019 [57]. The system is aimed at a 950–1900 nm wavelength range, addressable with uncooled extended InGaAs detectors. An ultracompact setup, in $25 \times 15 \times 15\,mm^3$ outline, includes the optical bench, MEMS, slits and detector on a printed board. The MEMS deflection can be reduced to $\pm 5°$ instead of $\pm 9.6°$, used for the previous scanning grating approach, without decrease in the spectral range. The resolution is still slightly below 10 nm.

5.3 Fourier Transform Spectrometer

Fourier transform (FT) or FT infrared (FTIR) spectrometers use interference in an interferometer instead of diffraction at a dispersive element. While scanning the interferometer, the signal is captured by a detector in the image plane, and then converted to a spectrum via a Fourier transformation.

FTIR spectrometers are the most accurate general-purpose devices for infrared spectroscopy [59]. Besides the use in laboratories, extremely miniaturized systems are now available for the mobile device market.

The reason for the success of FTIR spectrometers are the well-known advantages of the approach:

"Fellgett advantage": Multiplex operation, i.e. all wavelengths are captured simultaneously;

"Jacquinot advantage": High optical throughput as no slits limit the aperture, resulting in a good signal-to-noise ratio (SNR); "Connes advantage": Linearity of the wavelength scale can easily be calibrated, e.g. using a reference laser.

FTIR spectrometers can address a broad spectral range limited by their detector's sensitivity only. No higher diffraction orders, or related complications, appear. For proper operation, FTIR spectrometer-based systems need a highly stable light source because fluctuations of the illumination will generate noise across all the wavelengths. Sufficient computation power is required for the Fourier transform itself, but today this is no longer a real concern.

The most common option to realize an FT spectrometer is the use of a Michelson interferometer. The central part is the beam splitter, which divides the incident radiation into two beams. The reference beam is reflected by a fixed mirror, and the second beam by a moveable mirror (Figure 5.17). After recombination of both beams, the interference can be observed at the image plane.

To acquire spectroscopic information, the interferometer is scanned by translating the movable mirror. A detector in the image plane captures the signal. The mirror path L is the only parameter which limits the spectral resolution (Eq. (5.3)) [59].

$$\Delta\nu = 1/L \tag{5.3}$$

However, in a practical FTIR spectrometer, the beam angles cannot be limited to infinite. Thus, diameter of source and detector as well as the focusing optics will reduce the resolving power beyond the theoretical limit [59].

The interferogram pattern is captured as a function of the mirror position, digitized at regular optical retardation intervals. Afterward, it is converted to spectrum (linear in frequency – usually units of reciprocal centimeters) by a Fourier transformation (Eq. (5.4)).

$$F(w) = \frac{1}{\sqrt{2\pi}} \int_{-\infty}^{\infty} f(t) e^{iwt} dt \tag{5.4}$$

The fast Fourier transform (FFT) algorithm computes frequency domain information from time domain information. Historically, the basic algorithm was already used by Carl Friedrich Gauss in 1805. For implementation on a digital computer, one of the most prominent contributions was given by James Cooley and John W. Tukey in 1965 [60].

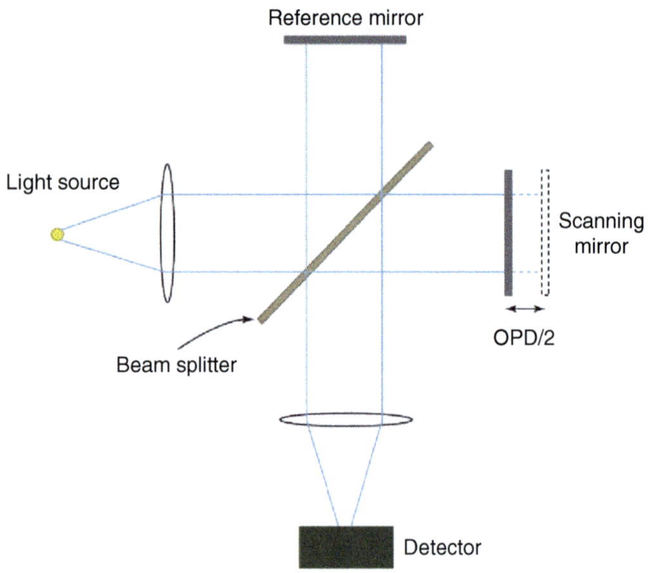

Figure 5.17 Michelson interferometer for an FT spectrometer setup.

5.3.1 MEMS Piston Mirror-Based FTIR Spectrometer

Different kinds of piston-type MEMS scanner mirror devices have been fabricated. The main difference is the deflection of the mirror planes. These can either be actuated in the chip plane or perpendicular out of the chip plane.

In an early stage, the University of Neuchatel group [61, 62] fabricated mirrors actuated in plane. The mirror plate had a size of 75 × 500 μm, and a Michelson interferometer was realized based upon this device. The displacement reached 80 μm by applying a driving voltage of ±10 V. This was a demonstration that the MEMS FTIR spectrometer concept had been realized. For commercial use, this MEMS device is too large for low cost, and the mirror plane is too small for adequate performance.

Piston-type MEMS mirrors, moving in the out-of-plane direction, have been developed and fabricated by Fraunhofer IPMS. The first trials were performed with bending springs [25] before the development of pantograph bearings (Figure 5.18) took place [26]. The mirror diameters reached up to 5 mm.

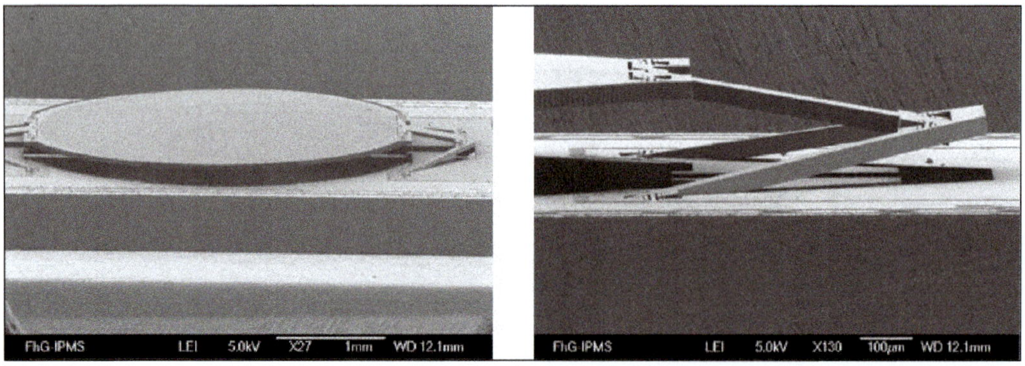

Figure 5.18 MEMS piston scanner mirror with pantograph bearings. *Source*: Fraunhofer IPMS.

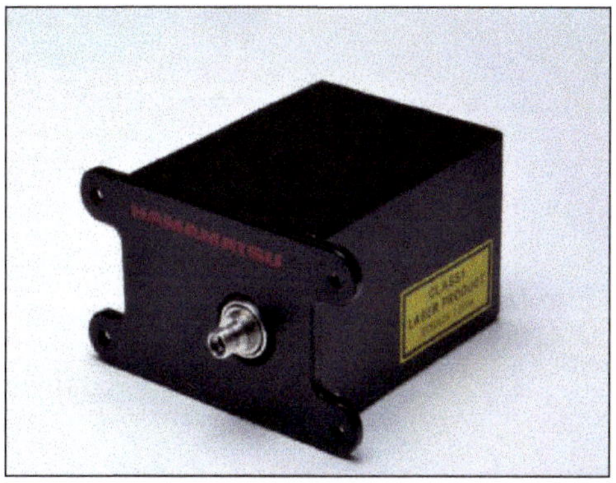

Figure 5.19 Hamamatsu FTIR spectrometer. *Source*: Hamamatsu Photonics K.K.

Air damping (i.e. air resistance to motion of the mirror) is a serious issue for these devices. Applying reduced air pressure (10 Pa), the driving voltage can be as low as 10 V, while under normal atmospheric pressure, more than 100 V are necessary. Deflections can reach up to ±600 μm.

Using the mirrors with pantograph design, FTIR spectrometers have been developed [63–66]. The systems were operated in a spectral range between 2.5 and 16 μm and a resolution of 8 cm^{-1} has been achieved. The commercialization of the FTIR spectrometer build around this MEMS device faced drawbacks however; no product is available yet.

Recently, new and promising work was presented by Hamamatsu.[7,8] Based on a resonant piston-type MEMS scanner mirror, with a mirror plate diameter of 3 mm, vibrating in a frequency range of 275 ± 50 Hz with an amplitude of 160–180 μm, a compact FTIR spectrometer has been realized (Figure 5.19). Applying a drive voltage of up to 80 V, a spectral range of 1100–2500 nm is covered, using an extended range InGaAs detector. The spectral resolution reached 5.7 nm, together with a favorable SNR of 7 500–55 000. The system offers an SMA fiber coupler with an incident focal distance of 6.24 mm, and a numerical aperture (NA) of 0.4.

5.3.2 Monolithically Integrated Interferometers

Extremely small outlines can be achieved when the entire systems is integrated monolithically. MEMS chip-integrated FTIR spectrometers, based on the Michelson interferometer, have been realized [67]. The optical path inside the chip plane requires the mirrors to be etched vertically into the active silicon layer of the MEMS substrate. A limiting factor is the quality of the reflecting surfaces, and for a long time, scallops inhibited sufficient planarity. With the availability of improved etch steps, the Egypt-based company Si-Ware realized monolithically integrated MEMS-based interferometers successfully (Figure 5.20).

The beam splitter, fixed mirror and moveable mirror have been fabricated in the same chip. Simple external mirrors or reflecting prisms can be adapted for optical coupling into the cavity, and to the detector.

The system based on this chip was called NeoSpectra Micro.[9] The outline is 32 × 32 × 22 mm^3, and the weight is 17 g. It addresses a wavelength range from 1350 to 2500 nm, and offers a resolution of 16 nm; wavelength accuracy is stated to be ±1.9 nm, and repeatability ±0.35 nm, respectively. The SNR is in a 1 : 1500–2000 range. Besides the ultrasmall interferometer chip, the detector requires thermoelectric cooling which affects the detector size. In the end, the system is available as a handheld with approximately 1 kg weight. It is battery operated for 1000 s of measurements per battery charge [68].

---

7 https://www.hamamatsu.com/resources/pdf/ssd/ftir_engine_kacc9012e.pdf.
8 https://www.hamamatsu.com/resources/pdf/ssd/c15511-01_kacc1294e.pdf.
9 https://www.neospectra.com/our-offerings/neospectra-micro/neospectra-micro-development-kit

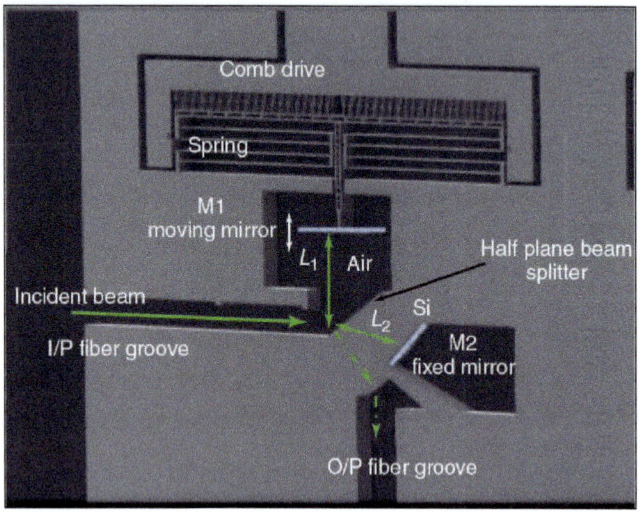

Figure 5.20 Monolithically integrated Michelson interferometer.

5.4 Tunable Fabry–Perot Interferometer

The Fabry–Perot Interferometer (FPI), as realized first in 1897 by Charles Fabry and Alfred Pérot, is an optical bandpass filter made through a cavity formed between two reflecting surfaces [69, 70]. Inside the interferometer, optical waves can propagate, and those matching the resonance criteria are enhanced, the others interfere destructively and vanish. Technically, two partially transparent mirrors are adjusted with a well-defined air gap. If the width of the gap can be varied, a tunable filter is realized. Deflectable MEMS mirrors can be employed, which is a very promising option to construct spectroscopic equipment. An advantage of the FPI approach is that no beamsplitter or reference system is necessary for operation, which makes the setup simple, thus cost efficient and compact.

The incident radiation is reflected multiple times between the two mirrors (Figure 5.21). To simplify the considerations, it shall be assumed that the reflectivity of both mirrors is identical, which is reasonable for most cases.

Depending on the geometry (l, α) a path difference δ is generated (Eq. (5.5)).

$$\delta = (2\pi/\lambda)\, 2n\, l \cos \alpha \tag{5.5}$$

The transmitted intensity T_e can be calculated if the reflectance R is known.

$$T_e = (1 - R)^2/(1 - 2R \cos \delta + R^2) = 1/(1 + F \sin^2[\delta/2]) \text{ considering } F = 4R/(1 - R)^2 \tag{5.6}$$

T_e is shown graphically in Figure 5.22. It becomes 1 for $2\, n\, l \cos(\alpha) = m\, \lambda$

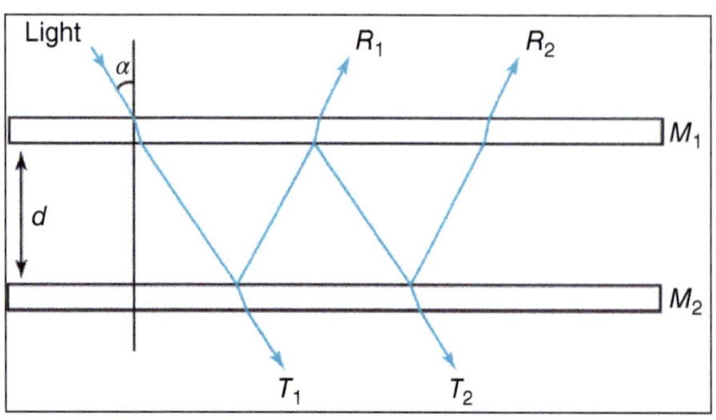

Figure 5.21 Multiple reflection between two mirrors.

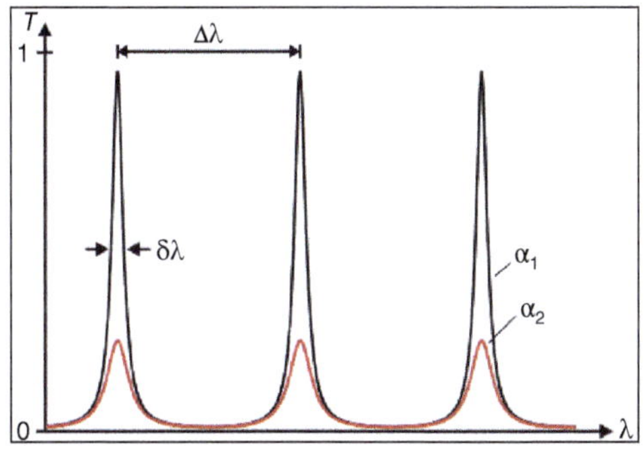

Figure 5.22 Transmission T_e of an FPI.

The resolution strongly depends on the number of passes inside the interferometer. Close to infinite passes, the filter function becomes very sharp. Depending on the reflectivity of the mirrors, the number of reflections in a real FPI can be very high, up to $\approx 10^5$ and the transmittance peaks become very narrow. There is a simple relationship between the free spectral range ($\Delta\lambda$) and the resolution ($\delta\lambda$), expressed as the Finesse ($\mathcal{F}$). The simplification is justified for $R > 0.5$.

$$\mathcal{F} = \Delta\lambda/\delta\lambda = \pi/(2\ \arcsin(1/\text{sqrt}(F))) \approx \pi\ \text{sqrt}(F)/2 = \pi\ \text{sqrt}(R)/(1-R) \tag{5.7}$$

MEMS-based system designs are very promising for a tunable FPI. The setup is simple, and does not require complex optical alignments, as needed for FTIR or grating spectrometers. Nonetheless, there are some general limitations. The free spectral range is limited by principle, and additional bandpass filters are required.

For proper operation, some requirements must be fulfilled, and the stability of the actuated mirror plate during movement is the most relevant one. No tilting of the actuated mirror plane is tolerable: the mirror plate must remain planar with a $\lambda/50$–$\lambda/100$ planarity requirement. The alignment of both mirrors must be accurate in this range as well.

Three examples described here show different implements of FPIs: bulk micro-machined silicon, surface micro-machined silicon and piezo drive-based systems.

A bulk silicon micro-machining device, similar to piston-type scanner mirrors, can be used to design an FPI. In particular, Fraunhofer ENAS [71] has presented a specific design in bulk micro-machined silicon, featuring double bearings for more stability. The MEMS chip has a size of $5 \times 5\ \text{mm}^2$ with an optical aperture measuring $1.8 \times 1.8\ \text{mm}^2$. The FPI spectrometer is realized in TO 39 outline. Applying driving voltages in a 15–60 V range, a bandwidth of 50–200 nm was achieved. The spectral ranges in the 3–11 μm field can be selected 3–5, 5–8 or 8–11 μm. The transmittance of the filter itself is higher than 70%. A consortium of InfraTec and Jenoptic Optical Systems has commercialized this technology.

A surface micro-machined MEMS device, using sacrificial layer technology, has been fabricated by Hamamatsu.[10] The device is mounted to a TO 5 carrier. The weight of the system is only 1 g. An InGaAs detector with 0.3 mm diameter is used, with an integrated bandpass filter. Driving voltages between 10 and 50 V are necessary. The FPI transmittance is around 60%. The system can address short-wave infrared (SWIR) ranges 1350–1650 nm with 18 nm resolution, 1550–1850 nm with 20 nm resolution, or 1750–2150 nm with 22 nm resolution.

The piezoelectric-driven FPI systems have been launched by VTT for SWIR systems and imagers [72, 73]. InGaAs detectors enable a spectral range from 1100 to 1600 nm. The piezo-driven FPI is using three actuators, and both mirrors use dielectrical layers for enhanced reflectivity. An example for an imager features 256×320 pixels. The field of view is 30°, focal length $f = 12$ mm, the opening offers $F = 3.2$, and the spectral resolution is 15 nm.

A spectral sensor was developed. The product NIRONE was launched through the VTT spin-out "Spectral Engines" (now part of Nynomic). The system has an outline of $25 \times 25 \times 17.5\ \text{mm}^3$, and a weight of 15 g. Using a

10 https://www.hamamatsu.com/eu/en/product/optical-sensors/spectrometers/mems-fpi-spectrum-sensor/index.html.

single-element InGaAs detector, an SNR between 1 500 and 15 000 has been achieved. Different spectral ranges can be addressed with slightly varying resolution: 1.1–1.35 μm @ 12–16 nm/1.35–1.65 μm @ 13–17 nm/1.55–1.95 μm @ 15–21 nm/1.75–2.15 μm @ 16–22 nm/2.0–2.45 μm @ 18–28 nm.

5.5 Integration Strategies for MEMS-/MOEMS-Based Spectrometers

Any spectroscopic device has high requirements for precision. The accuracy of the alignment of the optical components inside must be adjusted during production, and has to remain constant throughout its entire lifetime. Deviations affect the spectral range, the dispersion, the resolution or the optical throughput. Two different kinds of deviation may occur: some may be compensated by calibration, for example, if the wavelength range shifts. This is solvable, but requires undesired effort. Other deviations affect the system performance, and must be controlled by the accuracy of the assembly process.

In the past, benchtop spectrometers typically used alignment structures as micrometer screws and elongated holes, which enabled a sub-micrometer precision of the alignment of the individual components, albeit in a time-consuming procedure. The alignment accuracy reaches in the same range as the thermal expansion of a macroscopic system for just a few Kelvin temperature change.

Miniaturization reduces the options for manual alignment, and, thus, increases the complexity of the procedure. MEMS-based systems have specific requirements: the MEMS device must be assembled to a carrier or directly on a printed circuit board and electrical contacts must be connected. The aim is always an ultracompact system. There are some options for the integration of such optical systems. The design choices depend on the number of systems which will be produced. High-volume integration typically requires individually designed components, which may result in high initial effort, but low cost per piece.

5.5.1 Classic System Assembly for MEMS-Based Systems

The basic integration procedure for spectroscopic equipment can be adopted for MEMS-based systems as well. The core element is a chassis or baseplate, mounted in either a floor frame or a case, where bearings or assembly structures are designed at the right positions. All relevant components can be manually mounted and systematically adjusted in all six degrees of freedom. This manual process could be very accurate, but it is time-consuming, and the miniaturization options are limited. Furthermore, the production volume may be limited as well by the efficiency of the integration procedure, i.e. the manual alignment time required.

An example for this integration approach is the scanning grating spectrometer SGS 1900. The MEMS component and the detector are assembled in carriers. The MEMS in package grants access to the grating as optical active area from the front side, and an optical position readout from the backside. Wire bonds onto pads on the chip front side realize electrical contacts. Both mirrors are mounted on adjustable stages. The slits are placed at the fiber optical entrance stage and in front of the detector.

It is clear that here limitations for size reduction arise. Even if optical components are directly integrated into the housing, e.g. by LIGA (lithography, electroplating and molding) technology for mirrors and gratings, the alignment is critical, due to the complex assembly, and limits any reduction in size before optical limitations arise.

5.5.2 Stacking

Stacking components is a very efficient and well-established way to integrate axial optical systems. Lenses for cameras are a prominent example, where even multiple elements are integrated with high accuracy. MEMS-based Fabry–Perot interferometer represent axial optical systems for spectroscopic applications. Here stacking works fine.

For the realization of complex off-axis systems, like spectrometers or Michelson interferometers, stacking becomes quite complex. In principle, components could be stacked, but the alignment in 3D will make the situation complicated and the alignment tricky.

An idea for the efficient integration of miniaturized systems was the use of large substrates which bear components for a high number of systems, stack those and dice them afterward into the individual systems. Thus, a single alignment step could enable many devices.

This works especially fine if a lot of functionality can be integrated in each of the substrates. For a MEMS-based scanning-grating spectrometer, it is possible to integrate both slits into the same chip using the accuracy of the lithographic technologies. As the grating is not deflected in the rest position, changes to the normal Czerny–Turner have to be applied. Instead of a tilt of the grating to adjust the wavelength [53] of the W-configuration of the spectrometer, here the first negative diffraction order has been used to shift the configuration toward a folded double-V. An appropriate design is possible without predeflection of the grating. A small drawback is the complexity of the mirrors, which have to be off-axis and should have bi-conic areas. These mirrors could be realized by ultraprecision surface micromaching, and it is possible to realize both in a single substrate. Thus, the alignment task is reduced to two components only.

Grating spectrometer requires a focal length to operate properly; thus, between the MEMS and the mirror substrate, a space has to be implemented. The thickness is in a 6–8 mm range: shorter focal distances would result in a strongly reduced resolution for the NIR, which cannot be compensated through higher line densities, due to the $d < \lambda/2$ limitation. Direct wafer stacking using a wafer bond process, thus, is not an option. In-plane integration of a MEMS-based optical path is not possible, as no technology is available today that can realize appropriate mirrors, or grating structure, in vertical integration in silicon.

5.5.3 Place and Bend

A new integration concept was invented recently [74]. A need for high production volume of complex off-axis optics arises if a spectrometer is integrated into a mobile phone. This cannot be satisfied efficiently by existing integration technologies. Either new tools for high-volume 3D integration will have to be designed, or new concepts enabling integration with existing equipment, i.e. highly efficient and accurate 2D assembly tools, need to be developed.

This concept is based on the Japanese art "Origami," where a complex 3D body is folded from a planar "substrate" – a simple sheet of paper. The idea of "Place and Bend" assembly employs a substrate that features bending structures along designated lines. The substrate can be made from a plastic material by means of additive manufacturing (3D printing), casting, injection molding or other technologies. The electronic and optical components are placed in 2D assembly technology ("place"), afterward the walls are folded ("bend"), and the body with the optical path inside forms automatically if the accuracy is high enough. The concept was proven [75] and first spectrometers (Figure 5.23) have been realized using this "place and bend" assembly [57]. Further work on materials [76], process technology optimization and system design is planned. Fabrication of an FTIR spectrometer might be possible as well, and there are ideas how to design an interferometer efficiently.

5.5.4 Chip-Scale Integration

The highest level of MEMS integration can be reached by putting the entire optic design directly into the MEMS chip. A lot of limitations have to be faced, but if successful, the miniaturization will reach the highest level possible and production will be efficient and scalable up to any number. Up to now, this was shown successfully for FTIR spectrometer using Michelson interferometers by the scientists of Si-Ware Systems [67, 68, 76].

Figure 5.23 NIR spectrometer realized by "Place and Bend Assembly." *Source*: Fraunhofer IPMS.

5.6 Use of MEMS-Based NIR Spectrometers

The actual interest in miniaturized spectrometers, especially for NIR applications, is based on the possibilities for material analysis. In combination with the availability of access to data by mobile Internet services, the evaluation of measurements in the field in real time becomes possible. In principle, as a mobile phone add-on, or integrated to such a device, everybody would be able to perform composition analysis of solid objects, especially organic matter like food, cloth, and skin. Fluids in containers could be treated in a similar way if the container is sufficiently transparent. Liquid analysis requires different sample preparation, for example, cuvettes, and measurements in transmission instead of diffuse reflection.

The potential of mobile spectral analysis has attracted the attention far beyond the spectroscopic system providers. Mobile phone manufacturers, service providers and also user groups are seeking solutions suitable for use in the field.

NIR offers a good balance of effort of measurement and quality of the results. As noted above, the penetration depth gives an insight into an objects bulk composition, not just the surface. The broad structure of partially overlapping overtone and combination bands in the NIR require mathematical methods for the evaluation (chemometrics), and access to a database service with reference data that has been obtained using standard laboratory methods. Today, mobile devices offer sufficient computation power, and both access to online data ("the cloud") or storage on the device itself.

The general setup of a mobile spectral analyzer consists in a light source with illumination optics, optics for signal acquisition, the spectroscopic device, processing unit and data storage, and optionally online data access components (Figure 5.24). Besides this, a user interface (screen, indicator lamps, keyboard, etc.) is necessary. Furthermore, measurements in reflection require that a white standard be measured for reference from time to time. Results may also be communicated through a data interface.

The evaluation is based on chemometrics, which uses reference data from previous measurements of known samples, or samples which have been analyzed by reference methods. In the future, this may become simplified by the use of artificial intelligence methods and the possibility to include feedback and statistics from many measurements in the field. There will be a huge volume of information available if everybody is analyzing everything with an individual mobile spectral analyzer, and this may open options for the future of reference data.

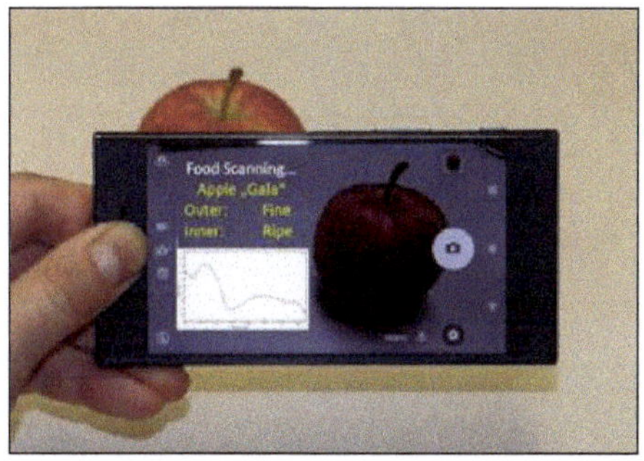

Figure 5.24 Sketch of a spectral analyzer. *Source*: Fraunhofer IPMS.

Systems for either a selected single purpose or for multipurpose measurements will be designed. They may differ in the spectroscopic setup, especially range and resolution. The user could be a professional, or even an expert, but also nonscientific people may use those systems. In the second case, guidance will be mandatory to achieve valuable results. Software "Apps" will lead the user through the measurement procedure. A combination of a spectral analyzer, and a camera integrated in one device, looking along the same optical axis, opens the option to guide the user by using the display, which may serve to initiate the measurement at the right spot and in the correct distance by showing crosshairs and color indicators for distance [77]. Also, image evaluation may provide collection of additional information about the object under test, e.g. specify kind ("apple") or sort ("Elstar apple"), and select an optimized reference database.

In the end, the results of the material analysis, solids and fluids, especially organic matter, can be evaluated and relevant information extracted. The user is interested in this compositional information, or data correlated to the composition, and the displayed result could be just one bit of information like "pass" and "fail" or "buy" – "not buy."

Acronyms and Abbreviations

CMOS	Complementary metal-oxide-semiconductor
BSOI	Bonded silicon on insulator
DARPA	Defense Advanced Research Projects Agency
DLP	Digital light processing
DRIE	Deep reactive-ion etching
InGaAs	Indium gallium arsenide
LCD	Liquid crystal display
LIGA	German acronym for "**Li**thographie, **G**alvanoformung, **A**bformung" (lithography, electroplating and molding)
MEMS	Micro-electromechanical systems
MOEMS	Micro-opto-electromechanical systems
NIR	Near infrared
FPI	Fabry–Pérot interferometer
FT	Fourier transform
FTIR	Fourier transform infrared
RIE	Reactive-ion etching
SMA	SubMiniature version A
SNR	Signal-to-noise ratio
SOI	Silicon on insulator

TO	Transistor outline
UPM	Ultraprecision machining/ultraprecision micromachining
UV	Ultraviolet

References

1 Rai-Choudhury, P. (ed.) (2000). *MEMS and MOEMS Technology and Applications*, SPIE Press Monograph, vol. PM85, preface, IX, ISBN 9780819437167. SPIE.

2 Clark, S.K. and Wise, K.D. (1979). Pressure sensitivity in anisotropically etched thin-diaphragm pressure sensors. *IEEE Transactions on Electron Devices* 26 (12): 1887–1896.

3 Schenk, H., Wolter, A., Dauderstädt, U. et al. (2005). Micro-opto-electro-mechanical systems technology and its impact on photonic applications. *Journal of Microlithography, Microfabrication and Microsystems* 4 (4): 041501–041511.

4 Ladabaum, I., Jin, X.C., Soh, H.T. et al. (1998). Surface micromachined capacitive ultrasonic transducers. *IEEE Transactions on Ultrasonics, Ferroelectrics and Frequency Control* 45 (3): 678–690.

5 Roylance, L. and Angell, J. (1979). A batch-fabricated silicon accelerometer. *IEEE Transactions on Electron Devices* 26: 1911–1917. https://doi.org/10.1109/T-ED.1979.19795.

6 Bernstein, J., Cho, S., King, A. et al. (1993). A micromachined comb-drive tuning fork rate gyroscope. *Proceedings of IEEE Micro Electro Mechanical Systems Workshop (MEMS '93)*, Fort Lauderdale, FL, pp. 143–148.

7 Kohn, E., Gluche, P., and Adamschik, M. (1999). Diamond MEMS – a new emerging technology. *ELSEVIER Diamond and Related Materials* 8 (2–5): 934–940, ISSN 0925-9635, https://doi.org/10.1016/S0925-9635(98)00294-5.

8 Tortschanoff, A., Holzmann, D., Lenzhofer, M., and Sandner, T. (2013). Closed-loop control for quasi-static MOEMS mirrors. *Proceedings of SPIE, vol. 8252, MOEMS and Miniaturized Systems XI, 82520T*. https://doi.org/10.1117/12.907825.

9 Fischer, A.C., Forsberg, F., Lapisa, M. et al. (2015). Integrating MEMS and ICs. *Microsystems & Nanoengineering* 1: 15005. https://doi.org/10.1038/micronano.2015.5.

10 Scheeper, P.R., van der Donk, A.G.H., Olthuis, W., and Bergveld, P. (1994). A review of silicon microphones. *Sensors and Actuators A: Physical* 44 (1): 1–11. https://doi.org/10.1016/0924-4247(94)00790-X.

11 Lee, S.S. and Lee, W.S. (2008). Piezoelectric microphone built on circular diaphragm. *Sensors and Actuators A* 144 (2): 367–373. https://doi.org/10.1016/j.sna.2008.02.001.

12 Cheng, M.C., Huang, W.S., and Huang, S.R.S. (2004). A silicon microspeaker for hearing. *Journal of Micromechanics and Microengineering* 14: 859–866. https://doi.org/10.1088/0960-1317/14/7/004.

13 Kaiser, B., Langa, S., Ehrig, L. et al. (2019). Concept and proof for an all-silicon MEMS micro speaker utilizing air chambers. *Microsystems & Nanoengineering* 5: 43. https://doi.org/10.1038/s41378-019-0095-9.

14 Amelung, J., Klemm, M., Elsäßer, L. et al. (2016). Back end of line integration technology of capacitive micromachined ultrasonic transducers (CMUT). *Proceeding of the Smart Systems Integration 2016*, Munich (9–10 March 2016).

15 Drabe, C., Schenk, H., Roscher, K. U. et al. (2004). Accelerometer by means of a resonant micro actuator. *Proceedings of SPIE, vol. 5344, MEMS/MOEMS Components and their Applications*. doi: https://doi.org/10.1117/12.524130.

16 Lee, I., Yoon, G.H., Park, J. et al. (2005). Development and analysis of the vertical capacitive accelerometer. *Sensors and Actuators A* 119: 8–18. https://10.1016/j.sna.2004.06.033.

17 Amarasinghe, R., Dao, D.V., Toriyama, T., and Sugiyama, S. (2007). Development of miniaturized 6-axis accelerometer utilizing piezoresistive sensing elements. *Sensors and Actuators A: Physical* 134: 310–320. https://doi.org/10.1016/j.sna.2006.05.044.

18 Palaniapan, M., Howe, R.T., and Yasaitis, J. (2002). Integrated surface-micromachined z-axis frame microgyroscope. In: *International Electron Devices Meeting 2002*, 203–206. IEEE https://doi.org/10.1109/IEDM.2002.1175813.

19 Park, S., Horowitz, R., and Tan, C.W. (2008). Dynamics and control of a MEMS angle measuring gyroscope. *Sensors and Actuators A: Physical* 144 (1): 56–63. https://10.1016/j.sna.2007.12.033.

20 Schenk, H., Sandner, T., Drabe, C. et al. (2009). Single crystal silicon micro mirrors. *Physica Status Solidi C* 6 (3): 728–735. https://doi.org/10.1002/pssc.200880714.

21 Schenk, H., Wagner, M., Grahmann, J., and Merten, A. (2018). Advances in MOEMS technologies for high quality imaging systems. *Proceedings of SPIE 10587, Optical Microlithography XXXI*, 1058703. doi: https://doi.org/10.1117/12.2297399.

22 Kurth, S., Kaufmann, C., Hahn, R. et al. (2005). A novel 24-kHz resonant scanner for high-resolution laser display. *Proceedings of SPIE* 5721: 23–33.

23 Senger, F., Hofmann, U., Wantoch, T. et al. (2015). Centimeter-scale MEMS scanning mirrors for high power laser application. *Proceedings of SPIE* 9375: 9375_7. https://doi.org/10.1117/12.2079600.

24 Kasturi, A., Milanovic, V., Hu, F. et al. (2019). MEMS mirror module for programmable light system. *Proceedings of SPIE 10931, MOEMS and Miniaturized Systems XVIII*, 109310L. https://doi.org/10.1117/12.2511074.

25 Kenda, A., Drabe, C., Schenk, H. et al. (2006). Application of a micromachined translatory actuator to an optical FTIR spectrometer. *Proceediongs of SPIE 6186, MEMS, MOEMS, and Micromachining II,* 618609, pp. 1–11.

26 Sandner, T., Grasshoff, T., Schenk, H., and Kenda, A. (2011). Out-of-plane translatory MEMS actuator with extraordinary large stroke for optical path length modulation. *Proceeding sof SPIE 7930, MOEMS and Miniaturized Systems X,* 79300I-1-9. doi: https://doi.org/10.1117/12.879069.

27 Oden, P.I. (2008). Texas instruments' DLP products massively paralleled MOEMS arrays for display applications: a distant second to mother nature. *Proceedings of the SPIE 6959, Micro (MEMS) and Nanotechnologies for Space, Defense, and Security II*, 69590D. doi: https://doi.org/10.1117/12.778499.

28 Gehner, A., Durr, P., Kunze, D. et al. (2018). Novel 512 × 320 Tip-Tilt Micro Mirror Array in a CMOS-Integrated, Scalable Process Technology. *IEEE Proceedings of the International Conference on Optical MEMS and Nanophotonics,* Lausanne Switzerland (29 July–2 August 2018). doi: https://doi.org/10.1109/OMN.2018.8454624.

29 Schenk, H. and Grüger, H. (2003). Spektrometer. Patent WO 002003069289.

30 Wang, Y., Kanamori, Y., Sasaki, T., and Hane, K. (2009). Design and fabrication of freestanding pitch-variable blazed gratings on a silicon-on-insulator wafer. *Journal of Micromechanics and Microengineering* 19: 025019. https://doi.org/10.1088/0960-1317/19/2/025019.

31 Saadany, B., Omran, H., Medhat, M. et al. (2009). MEMS tunable Michelson interferometer with robust beam splitting architecture. *IEEE Proceedings of the International Conference on Optical MEMS and Nanophotonics,* Cleanwater, FL, USA (17–20 August 2009). doi: https://doi.org/10.1109/OMEMS.2009.5338601.

32 Zimmer, F., Grüger, H., Heberer, A. et al. (2005). Scanning Micro-Mirrors: From bar-code-scanning to spectroscopy. *Proceedings of SPIE, vol. 5873, Optical Scanning, 5873*, pp. 84–94. doi: https://doi.org/10.1117/12.614895.

33 Sandner, T., Grasshoff, T., Wildenhain, M., and Schenk, H. (2010). Synchronized microscanner array for large aperture receiver optics of LIDAR systems. *Proceedings of the SPIE, vol. 7594, MOEMS and Miniaturized Systems IX, 75940C*. doi: https://doi.org/10.1117/12.844923.

34 Hornbeck, L. J. (1996). Digital light processing and MEMS: reflecting the digital display needs of the networked society. *Proceedings of SPIE, vol. 2783, Micro-Optical Technologies for Measurement, Sensors, and Microsystems.* doi: https://doi.org/10.1117/12.248477.

35 Motamedi, M.E. (2005). *MOEMS Micro-Opto-Electro-Mechanical Systems*. Bellingham Washington, USA: SPIE Press.

36 Liebermann, M.A. and Lichtenberg, A.J. (2003). *Principles of Plasma Discharges and Material Processing*, chapter 5, 2ee. Wiley-VCH https://doi.org/10.1002/0471724254. ISBN: 0-471-72001-1.

37 Inventors Franz Laermer and Andrea Schlip (1996). Method of anisotropically etching silicon. US Patent 5,501,893, filed 5 August 1994, published 26 March 1996.

38 Hilleringmann, U. (2014). *Silizium-Halbleitertechnologie*, 6ee. Wiesbaden: Springer Vieweg Verlag. ISBN: 978-3-8348-1335-0.

39 Krishnan, M., Nalaskowsk, J.W., and Cook, L.M. (2010). Chemical mechanical planarization: slurry chemistry, materials, and mechanisms. *Chemical Review* 110: 178–204. https://doi.org/10.1021/cr900170z.

40 Loewen, E.G. and Popov, E. (1997). *Diffraction Gratings and Applications*. New York: Marcel Dekker.

41 Palmer, C. (2014). *Diffraction Grating Handbook*, 7ee. Rochester, NY: Richardson Gratings.

42 Ebert, H. (1889). Zwei Formen von Spectrographen. *Annalen der Physik* 274: 489–493.

43 Monk, G.S. (1928). A mounting for the plane grating. *Journal of the Optical Society of America* 17: 358–362.

44 Gillieson, A.H.C.P. (1949). A new spectrographic diffraction grating monochromator. *Journal of Scientific Instruments* 26: 334–339.

45 Czerny, M. and Turner, A.F. (1930). Über den Astigmatimus bei Spiegelspektrometern. *Zeitschrift für Physik* 61: 792–797.

46 Czerny, M. and Pletti, V. (1930). Über den Astigmatimus bei Spiegelspektrometern II. *Zeitschrift für Physik* 63: 590–595.

47 Fastie, W.G. (1952). A small plane grating monochromator. *Journal of the Optical Society of America A* 42: 641–647.

48 Fastie, W.G. (1952). Image forming properties of the Ebert monochromator. *Journal of the Optical Society of America A* 42: 647–650.

49 Neumann, W. (2014). *Fundamentals of Dispersive Optical Spectroscopy Systems*. SPIE Press Book. ISBN: 9780819498243.

50 Grüger, H., Wolter, A., Schuster, T. et al. (2003). Realization of a spectrometer with micromachined scanning grating. *Proceeding of Photonics West 2003*, San Francisco CA, USA.

51 Zimmer, F., Grüger, H., Heberer, A. et al. (2006). Development of high-efficient NIR-scanning gratings for spectroscopic applications. *Proceedings of SPIE, vol. 6114, MOEMS Display, Imaging, and Miniaturized Microsystems IV*, 611407. doi: https://doi.org/10.1117/12.644481.

52 Grüger, H., Knobbe, J., and Pügner, T. (2019). Optical arrangement for a spectral analysis system, method for its production and spectral analysis system. US Patent 10,247,607 granted 2 April.

53 Pügner, T., Knobbe, J., and Grüger, H. (2016, 2016). Near-infrared grating spectrometer for Mobile phone applications. *Applied Spectroscopy* 70 (5): 734–745.

54 Pruett, E.. (2015). Latest developments in Texas Instruments DLP near-infrared spectrometers enable the next generation of embedded compact, portable systems. *Proceedings of SPIE Vol. 9482, Next-Generation Spectroscopic Technologies VIII*, 94820C. doi: https://doi.org/10.1117/12.2177430A.

55 Harwit, M. and Sloane, N.J.A. (1979). *Hadamard Transform Optics*. New York: Academic Press.

56 Treado, P.J. and Morris, M.D. (1989). A thousand points of light: the Hadamard transform in chemical analysis and instrumentation. *Analytical Chemistry* 61: 723A–734A.

57 Grüger, H., Knobbe, J., and Pügner, T. (2019). MEMS based NIR spectrometer with extended spectral range. *Proceeding of Photonics West*, San Francisco, CA, USA.

58 Otto, T., Saupe, R., Weiss, A. et al. (2006). Principle and applications of a new MOEMS Spectrometer. *Proceedings of SPIE, vol. 6114, MOEMS Display, Imaging, and Miniaturized Microsystems IV*.

59 Griffiths, P.R. and De Haseth, J.A. (2007). *Fourier Transform Infrared Spectrometry*, 2ee. Hoboken, NJ: Wiley.

60 Cooley, J. and Tukey, J. (1965). An algorithm for the machine calculation of complex Fourier series. *Mathematics of Computation* 19: 297–301.

61 Manzardo, O., Herzig, H.P., Marxer, C.R., and de Roonij, N.F. (1999). Miniaturized time-scanning Fourier transform spectrometer based on silicon technology. *Optics Letters* 24 (23): 1705–1707. https://doi.org/10.1364/OL.24.001705.

62 Montgomery, P.C., Montaner, D., Manzardo, O. et al. (2004). The metrology of a miniature FT spectrometer MOEMS device using white light scanning interference microscopy. *Thin Solid Films* 450: 79–83. https://doi.org/10.1016/j.tsf.2003.10.055.

63 Sandner, T., Kenda, A., Drabe, C. et al. (2007). Miniaturized FTIR-spectrometer based on optical MEMS translatory actuator. *Proceedings of SPIE, vol. 6466, MOEMS and Miniaturized Systems VI*, 646602.

64 Sandner, T., Kenda, A., Scherf, W. et al. (2008). Translatory MEMS actuators for optical path length modulation in miniaturized Fourier-transform infrared spectrometers. *Journal of Micro/Nanolithography MEMS and MOEMS* 7 (2).

65 Kenda, A., Kraft, M., Tortschanoff, A. et al. (2014). Development, characterization and application of compact spectrometers based on MEMS with in-plane capacitive drives. *Proceedings of SPIE, vol. 9101 Next-Generation Spectroscopic Technologies VII*, 910102. doi: https://doi.org/10.1117/12.2053347.

66 Sandner, T., Gaumont, E., Grasshoff, T. et al. (2018). Translatory MEMS actuator with wafer level vacuum package for miniaturized NIR Fourier transform spectrometers. *Proceedings of SPIE 10545, MOEMS and Miniaturized Systems XVII*, 105450W. doi: https://doi.org/10.1117/12.2290588.

67 Ghoname, A. O., Sabry, Y. M., Anwar, M. et al. (2019). Ultra wide band MIR MEMS FTIR spectrometer. *Proceedings of SPIE 10931, MOEMS and Miniaturized Systems XVIII*, 109310Z. doi: https://doi.org/10.1117/12.2509378.

68 Fathy, A., Sabry, Y. M., Amr, M. et al. (2019). MEMS FTIR optical spectrometer enables detection of volatile organic compounds (VOCs) in part-per-billion (ppb) range for air quality monitoring. *Proceedings of SPIE 10931, MOEMS and Miniaturized Systems XVIII*, 1093109. doi: https://doi.org/10.1117/12.2508239.

69 Hernandez, G. (1986). *Fabry-Perot Interferometers*. Cambridge, UK: Cambridge University Press.

70 Vaughn, J.M. (1989). *The Fabry-Perot Interferometer: History, Theory Practice and Applications*. Bristol, UK: Adam Hilger.

71 Ebermann, M., Neumann, N., Hiller, K. et al. (2016). Tunable MEMS Fabry-Pérot filters for infrared microspectrometers: a review. *Proceedings of SPIE 9760, MOEMS and Miniaturized Systems XV*, 97600H.

72 Mannila, R., Holmlund, C., Ojanen, H. J. et al. (2014). Short-wave infrared (SWIR) spectral imager based on Fabry-Perot interferometer for remote sensing. *Proceedings of SPIE 9241, Sensors, Systems, and Next-Generation Satellites XVIII*. doi: https://doi.org/10.1117/12.2067206.

73 Rissanen, A., Guo, B., Saari, H. et al. (2014). VTT's Fabry-Perot interferometer technologies for hyperspectral imaging and mobile sensing applications. *Proceedings of SPIE 10116, MOEMS and Miniaturized Systems XVI*, 101160I, doi: https://doi.org/10.1117/12.2255950.

74 Grüger, H. Knobbe, J., and Pügner, T. Bendable substrate with a device. Patent application US 2019/0250032.

75 Grüger, H., Knobbe, J., Pügner, T. et al. (2017). Concept for a new approach to realize complex optical systems in high volume. *Proceeding of Photonics West*, San Francisco, CA, USA.

76 Grüger, H., Knobbe, J., and Sabiha, M. (2019). Investigation of mechanical and optical properties of 3D printed ma-terials serving as substrate for place and bend as-sembly. *Proceeding of Photonics West*, San Francisco, CA, USA.

77 Grüger, H. Knobbe, J., and Pügner, T. System und entsprechendes Verfahren zur Vermessung eines Objekts. Patent DE102017204740.2.

6

Portable Raman Spectroscopy: Instrumentation and Technology

Cicely Rathmell, Dieter Bingemann, Mark Zieg and David Creasey

Wasatch Photonics, Inc., Logan, UT, USA

6.1 Introduction

Nearly one century ago, C.V. Raman discovered that light scattered from molecules contained a "feeble fluorescence" that carried information about the vibrational states of those molecules [1]. Upon further study, the source was discovered to be a new form of inelastic scattering of light that is rich in chemical information. The field of Raman spectroscopy was born.

The initial experiments performed by Raman and a parallel team in Russia required exposure times on the order of hours for crystals, and up to 180 hours for gases [2]. Thanks to decades of technological advancements, Raman spectra of many solids and liquids can now be collected using a handheld device in just seconds. This chapter explores the current state of that technology, and the many competing factors and tradeoffs that must be considered when designing a portable Raman instrument for applied use. First, we will look at the benefits of Raman spectroscopy as an analytical technique, and review the theory behind the Raman effect so as to better understand the demands on the instrumentation required, then delve deeper into the pertinent aspects of portable system design.

6.2 The Case for Raman: Capabilities and Scope

Raman spectroscopy's greatest strength is its ability to rapidly provide a fingerprint representative of the chemical structure of a material. This fingerprint can be used for identification with high chemical specificity, and may be applied to solids, liquids, and gases. Modern-day portable Raman instruments are noncontact, easy to use, require no special sample preparation, and often employ straightforward point-and-shoot operation. They are available with flexible sample interfaces, and can be configured for through-barrier measurements to enable detection of substances in bags, vials, and even colored or opaque bottles.

In contrast to infrared spectroscopy, Raman spectroscopy is insensitive to water content, eliminating a major source of interference, and allowing direct analysis without sample preparation. When configured with robust analysis software and appropriate sampling accessories, a portable Raman instrument need not be operated by an expert, vastly increasing the applications in which this powerful technology can be used. Simply put, portable Raman has the ability to provide answers at the point of use.

These benefits have propelled the use of portable Raman instruments into several fields spanning quality, safety, and health. In the pharmaceutical industry, the technology allows for easy identification of materials on site, from incoming inspection to outgoing quality control. Here, a portable Raman instrument allows the user to bring the

Portable Spectroscopy and Spectrometry 1: Technologies and Instrumentation, First Edition.
Edited by Richard A. Crocombe, Pauline E. Leary and Brooke W. Kammrath.

instrument to the sample when the sample cannot easily or should not be moved, and provides instant answers without disrupting the flow of inventory and operations. In security and forensics, portable Raman instruments offer rapid substance identification in the field, providing the instantly actionable information needed by first responders and security personnel.

In the medical field, portable Raman facilitates use at point-of-care, allowing immediate patient diagnostics and monitoring throughout the course of treatment to improve clinical outcomes. It expands the reach of quality diagnostics to remote and under-resourced settings, as well as facilitating in-home use for medical conditions which require frequent monitoring.

It is important to note that the use of Raman is not limited only to the classification of bulk materials or Raman-active compounds. Raman-active markers or tags extend the use of the technique for identification or validation in authentication or process-monitoring applications. Raman tags may also be functionalized to bind to specific chemicals or biological compounds and agents for use in assays, expanding the potential of Raman in point-of-care settings. Specific applications are discussed in the chapter in Volume 2 of this book by Hargreaves.

While the benefits of Raman are many, the technique comes also with limitations. Raman spectroscopy can struggle to measure low concentrations, samples prone to fluorescence (particularly complex organic molecules and geological samples containing trace amounts of transition metal ions), and dark-colored, easily ignited, or delicate materials. Reliance on laser excitation requires attention to eye safety through sampling accessories or protective gear. Special techniques and operating conditions have been developed to mitigate each of these circumstances, and will be discussed in a "Special Cases" section near the close of this chapter.

None of these capabilities or applications could have progressed beyond the laboratory without significant evolution of the supporting technology. Before Raman spectroscopy could become more widely used as an analytical technique, it needed to become smaller, lighter, and easier to use. The past 25 years have seen advances in the hardware to achieve greater sensitivity in a smaller footprint, allowing Raman to escape the lab. It has also seen advances in science as regards our understanding of ***how*** to use Raman spectroscopy to answer key problems. As a result of this progress, many new application spaces and use cases have emerged in which portable Raman can be applied.

6.3 The Theory of Raman Spectra

Raman spectroscopy is an analytical technique based upon the Raman effect, in which light is scattered inelastically from a molecule, yielding information about its vibrational states. As the theory has been treated exhaustively by others [3–5], we will provide only a brief overview, focusing on those aspects most pertinent to applied use.

In classical terms, when light interacts with molecules, the electron cloud resonates with the incident light. This light will almost always be scattered unaltered, but there is a probability of one part in a million that the photon will lose energy to (or even more rarely, gain energy from) a vibration that couples strongly to the polarizability of the molecule. This causes light of a different wavelength to be emitted, an effect known as Raman scattering.

Raman spectroscopy is complementary to infrared spectroscopy, in which light couples to the molecular dipole moment (light couples to the polarizability of the molecule in Raman spectroscopy). Both techniques provide information about vibrational states, but differ in their sensitivity to specific molecular vibrations. Water, for example, with its strong dipole moment and low polarizability, has a strong infrared absorbance, but a very small Raman resonance.

In theory, Raman scattering can be achieved with any wavelength of light, provided it is of sufficient intensity. As described previously, the vast majority of the incident photons will be elastically scattered – a phenomenon known as Rayleigh scattering. Of the inelastically scattered photons, many will have lost energy, yielding longer wavelength photons (Stokes radiation), and a smaller fraction will have seen an increase in energy, yielding shorter wavelength photons (anti-Stokes radiation), as shown in Figure 6.1.

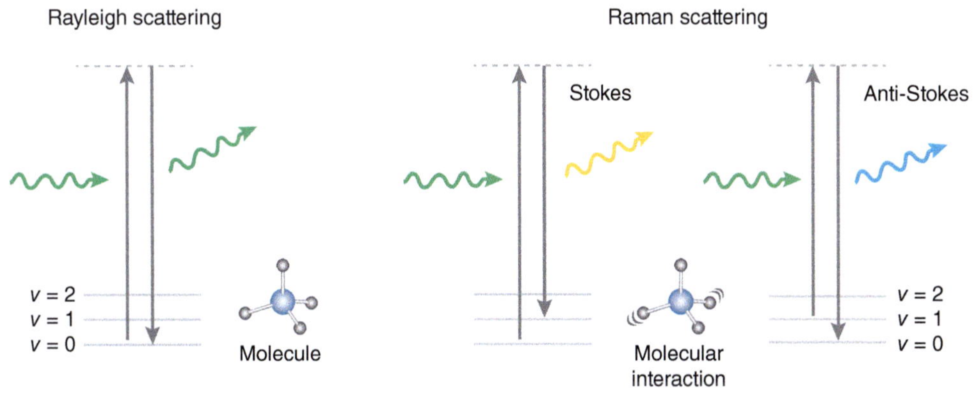

Figure 6.1 In Rayleigh scattering, the molecule returns to the same state, resulting in no change in energy for the scattered photon. In Raman scattering, the molecule returns to a different state, either losing energy (resulting in a Stokes shift, to longer wavelength) or gaining energy (resulting in an anti-Stokes shift, to shorter wavelength). *Source*: Courtesy of Wasatch Photonics.

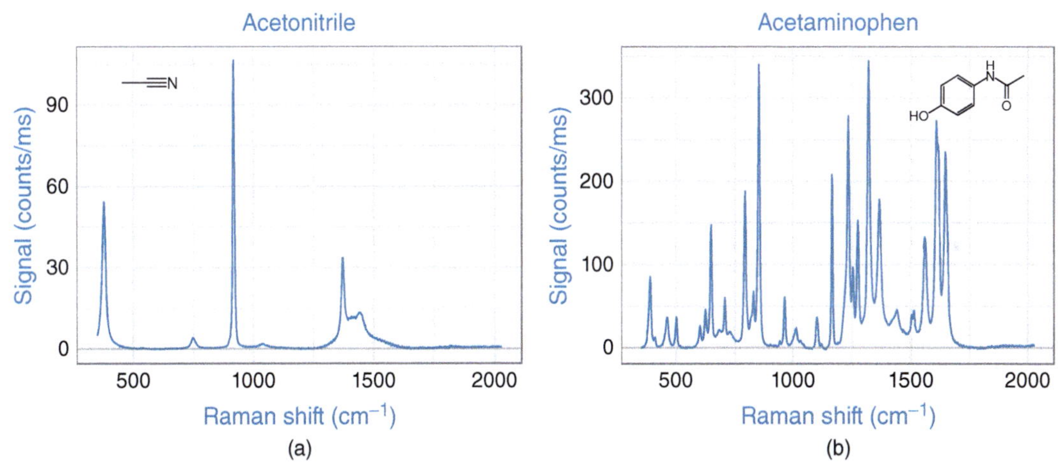

Figure 6.2 (a) Acetonitrile and (b) Acetaminophen as measured using 785 nm Raman. *Source*: Courtesy of Wasatch Photonics.

Information is contained in the frequency shifts of the emitted photons relative to the input light; each frequency shift or peak observed maps to one of the vibrational frequencies in the molecule. A typical Raman spectrum consists of several to many peaks. For simple molecules, like acetonitrile, the Raman spectrum may feature narrow and well-separated peaks, while for complex organics, like acetaminophen, the Raman peaks are often more frequent and closely spaced and/or overlapping, as shown in Figure 6.2.

The pattern of peaks composing a Raman spectrum is very specific to the material being interrogated, providing a unique fingerprint by which it can be identified. The lower frequency range up to 1500 cm^{-1} Raman shift is often sufficient to confirm identity, and, for this reason, is called the "fingerprint" region. Valuable information is also found in the "functional" region, out to 3600 cm^{-1} Raman shift, as shown in Table 6.1. As the name indicates, the frequencies in this extended range primarily indicate the presence of specific functional groups in a molecule; they represent local molecular information, rather than information about the overall molecular structure.

Table 6.1 Raman frequencies of common chemical functional groups.

Functional group	Position (cm^{-1})	Remarks
>S–S<	500–550	
C–C	~1060 and 1127	Polyethylene
C–C	700–1260	Highly mixed in complex molecule
Aromatic ring	~1000	Monosubstituted
Aromatic ring	~1000	1,3 disubstituted
Aromatic ring	~1000	1,3,5 trisubstituted
Aromatic ring	~860	1,4 disubstituted
CH_3 umbrella mode	~1375	
CH_3 and CH_2 deformations	1410–1460	
>C≡C<	~1650	
>C≡C<	~1623	Ethylene
>C≡O mixed with NH deformation	1620–1690	Amide I
>C≡O	1710–1745	Changes for ketone, aldehyde, and ester
C≡C	2100–2300	
SH	2540–2600	
>CH_2	2896 and 2954	Ethane
>CH_2	2845 and 2880	Polyethylene
CH_3	2870 and 2905	Polypropylene
$R_1R_2R_3CH$	2880–2890	Methine CH
CH	~2900	Cellulose
CH	~3015	Olefinic CH
CH	~3065	Aromatic CH
CH	3280–3340	Acetylenic CH
NH	3150–3340	Broadened and shifted by H-bonding
OH	3000–3600	Broadened and shifted by H-bonding

Source: Adar [6]. © 2016, UBM.

A Raman spectrum displays the intensity of the Raman emission as a function of the frequency shift relative to the incoming light frequency. Raman spectra are reported in terms of the "Raman shift," the frequency difference relative to the excitation source. This is reported in wavenumbers (cm^{-1}), a frequency space convention adopted in IR spectroscopy, and one which allows the two complementary techniques (infrared and Raman) to speak a common language for the identification and study of molecules. Raman shift may be calculated using the following equation, where ν_{Raman} signifies the Raman shift, and λ signifies wavelengths of the excitation laser and the Raman emission, as indicated by the subscript:

$$\upsilon_{Raman} = \frac{1}{\lambda_{Excitation}} - \frac{1}{\lambda_{Raman}}$$

By working in frequency space, Raman spectra measured with different excitation sources may be directly compared; each peak marks the frequency of molecular vibration, independent of the excitation wavelength.

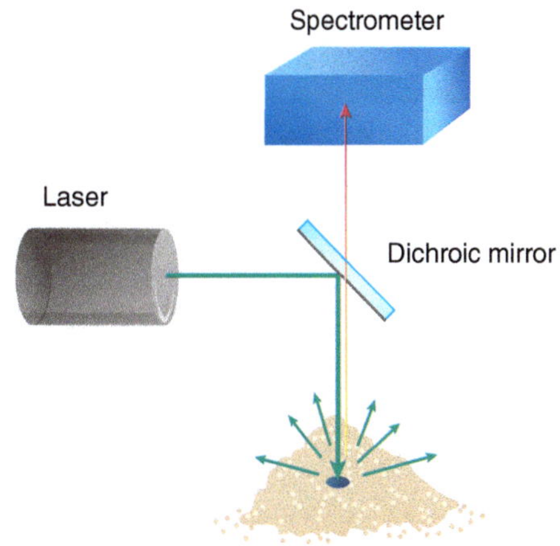

Figure 6.3 Epi-illumination optical configuration, typical in handheld Raman systems. *Source*: Courtesy of Wasatch Photonics.

6.4 Basics of a Raman System

A portable Raman system consists of three main functional units: a laser, a sample interface, and a spectrometer. For now, these three main components will be described only briefly to provide an overview of the whole system; each element will be discussed in more detail in the sections to follow.

The laser acts as the excitation source, providing the high intensities needed to balance the low probability of Raman scattering and yield sufficient Raman signal for detection. The laser beam is focused onto the sample to achieve better imaging of the resulting Raman emission into the spectrometer.

The sample interface provides delivery of the laser light to the sample, and collection of the Raman emission. In portable Raman instruments, the emission is most often detected as backscattered light in an "epi-illumination" arrangement (also termed 180° collection), as shown in Figure 6.3, with added sampling optics to both focus the laser and collect the Raman emission. A longpass dichroic mirror typically reflects the shorter wavelength laser onto the main beam path, then transmits the Raman-scattered light for routing to the spectrometer (note: some portable systems may use a different optical configuration). An additional longpass filter integrated into the sampling optics or spectrometer performs further rejection of the laser light and passes the Raman signal. The sample optics may be contained within the body of a fiber-coupled probe for increased flexibility, but more often are integrated with the laser and spectrometer as a single unit.

The spectrometer performs detection of the Raman signal; an array detector is used to capture all wavelengths simultaneously. As only one in 10^6 of the incident photons will be returned as Raman scatter, an extremely sensitive spectrometer is needed; high throughput is mandatory. The spectrometer reports a spectrum, which is translated in software to a Raman shift relative to the laser frequency.

These three components may be designed and integrated differently depending on system size and operational constraints, performance requirements, application, and sampling optic needs. Even within the subset of compact Raman instrumentation, a variety of form factors exist.

6.5 "Portable" Versus "Handheld" Versus "Mini"

When considering "portable" Raman systems, the first form factor that comes to mind is a handheld system, epitomized by early designs like the Ahura First Defender (acquired by ThermoFisher Scientific) [7] and Snowy Range CBEx [8], acquired by Metrohm Raman [9]. Handheld Raman instruments are roughly the size of a digital multimeter, with an integrated processor, user interface, and display screen, as shown in Figure 6.4. They run on

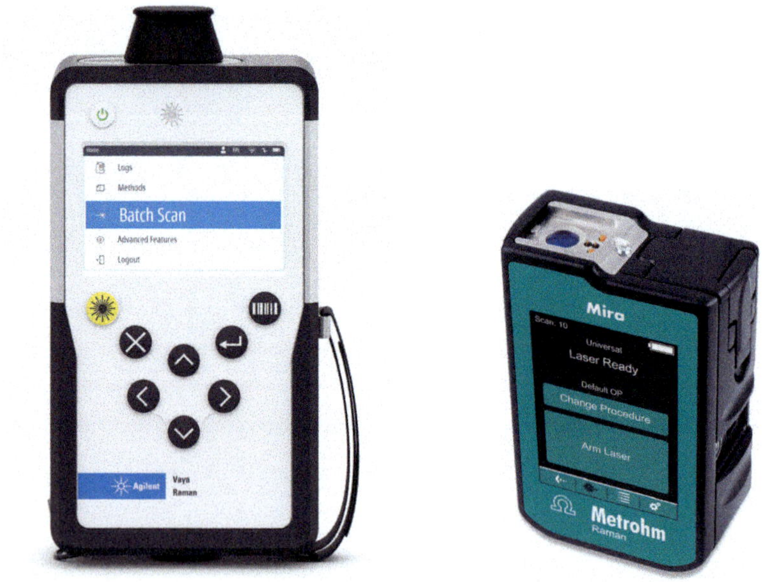

(a) (b)

Figure 6.4 Handheld Raman systems: (a) Vaya Raman Raw Material Identity Verification System. *Source*: © Agilent Technologies, Inc. Reproduced with permission of Agilent Technologies, Inc. (b) Mira DS Flex. *Source*: Metrohm AG. Reproduced with permission of Metrohm AG.

battery power, are operated in "point and shoot" mode, and provide answers in minutes or less using onboard or cloud-based library matching software. More than a dozen vendors on the market provide solutions in this form factor for material identification in pharma, hazmat, and authentication [7–26].

At a slightly larger size are "portable" Raman systems – too large to be operated handheld, but still small, lightweight, and robust enough to be easily moved from one location to another, as shown in Figure 6.5. These have footprints ranging in size from roughly half to a full sheet of letterhead or A4 paper, and may be packaged for transport in a hard, waterproof case containing specialized sample chambers, or have a fiber-coupled sample interface. They may be operated as field analyzers, or deployed for use in the clinic or the plant. Power may be provided by a battery, mains power (i.e. corded), or a combination of both. A laptop or tablet may be used for data processing and analysis, or all functions may be integrated and accessible via a user interface and display screen. A number of the handheld Raman system vendors also offer systems in this form factor, as do many application-specific vendors and compact spectrometer providers [27–33].

At the other extreme would be "mini" or pocket-sized Raman systems small enough to fit in the palm of your hand. While several companies offer mini-spectrometers [34–36], work in this area has not yet progressed to a fully integrated "mini" Raman system design capable of displacing the current generation of handheld Raman instruments. Industry focus has instead been on application optimization and sample interfaces, software graphical user interfaces (GUIs), and library matching/analysis, but will surely come back to miniaturization in the very near future.

6.6 Performance Needs in Portable Raman Instruments

Having developed a general understanding of the theory, basic system design, and possible form factors pertinent to portable Raman spectroscopy, we can now address its performance needs. Some are dictated by the technique

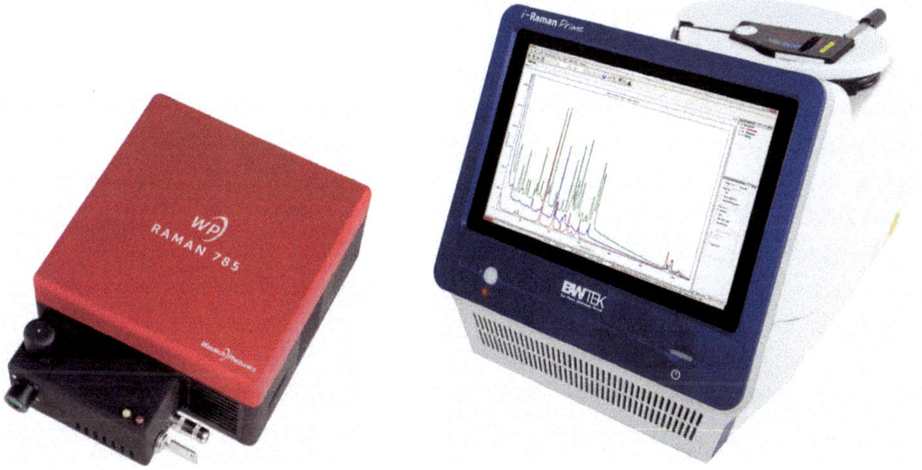

(a) (b)

Figure 6.5 Portable Raman systems: (a) WP 785L, an integrated Raman system designed for use with external data processing. *Source*: Courtesy of Wasatch Photonics. (b) i-Raman Prime 785H, an integrated Raman system including onboard user interface screen. *Source*: Metrohm AG. Reproduced with permission of Metrohm AG.

itself, while others are imposed by the required form factor, use case, or operating environment. All of these factors compete when designing a portable Raman instrument, and a general design principle of "good enough" for the application must be ruthlessly applied. In portable Raman systems, the goal is to generate an accurate answer at the required confidence level, in the smallest footprint and measurement time, and at the lowest cost and power draw possible. In doing so, one must also consider whether the answer must be qualitative (identification) or quantitative (how much), as a portable Raman instrument is most often used for identification, but is also capable of identifying components in mixtures and assessing relative concentrations.

As a technique, Raman spectroscopy requires extremely high sensitivity, and depends on highly blocking filters to reject Rayleigh scatter from the excitation laser. Even still, there is the potential for fluorescence to interfere – from the sample itself, and from optical components in the system; this must be minimized through a choice of excitation wavelength or other special configurations, which we will touch on near the close of this chapter.

The system as a whole must offer sufficient spectral range and optical resolution to acquire the peaks needed for analysis, and good signal strength across the full spectrum. Most portable Raman instruments cover 200–2400 cm^{-1}, with optical resolution from 6 to 15 cm^{-1} FWHM (full-width at half-maximum), depending on the application space. The spectra generated must be stable over time and temperature to allow for consistent spectral substance identification against a general library; even more so when quantitative answers are needed. Low unit-to-unit variation is desirable to facilitate matching against a master library, with as little correction as possible.

Portable operation introduces its own requirements, both in ruggedization and reduction of SWaP (*S*ize, *W*eight, *a*nd *P*ower). Operating environments may span indoor and outdoor operation, and the mechanical design must be robust against shock and vibration. Some degree of self-calibration is needed to validate the operation and minimize downtime, and spectral processing and analysis must be integrated to yield immediate answers.

Handheld operation takes these needs a step further, as the instrument must be smaller and lighter, with low power draw to allow for battery operation. The instrument must be even more sturdy, possibly water-resistant, and temperature-insensitive to handle field operation by a nonexpert. A specific IP (ingress protection) rating may be required to guard against intrusion from dirt, dust, other foreign bodies, and moisture [37]. Measurement times must be a few seconds or less when the user is holding the instrument against the sample, both for operator convenience and to minimize power draw. Lower power lasers are attractive not only to extend battery life, but also to avoid burning of high value, fragile, or explosive samples; the consequently reduced signal generated by lower laser powers underscores the need for highly sensitive detectors. Miniaturized, phone-sized screens typically require simplified software interfaces, providing conclusive answers in a minimum number of steps.

With these general performance needs in mind, we can now look at each component in the system individually to see how they contribute, consider the relevant design factors and options, and explore the current state of the art in portable Raman instrument design.

6.7 Excitation Laser

While any laser, in theory, can be used to generate a Raman spectrum, the choice of wavelength is the greatest factor in determining the strength and quality of that spectrum. The intensity of the Raman signal scales proportionally with the intensity of excitation light, and inversely with its wavelength to the fourth power:

$$I_{\text{Raman}} \propto \frac{I_{\text{laser}}}{\lambda^4}$$

Shorter laser wavelengths thus yield far more Raman signal, but also generate more autofluorescence background in organic and biological samples due to the excitation of electronic transitions. This fluorescence appears as a broad background, but can be significantly reduced by working at longer excitation wavelengths, particularly in the near infrared (NIR), as shown in Figure 6.6. Working in the NIR comes at the expense of Raman signal, however, and results in higher cost, lower sensitivity, and increased size, weight, and power draw due to the need for cooled indium gallium arsenide (InGaAs) detectors.

There is also a window in the ultraviolet (UV) prior to the onset of fluorescence yet within the sensitivity range of silicon detectors, shown in Figure 6.7. This is accessible with some limitations: UV laser sources offer lower power, have limited lifetimes, and are inconvenient from a size and power draw standpoint. UV Raman spectroscopy also places more stringent requirements on longpass filter performance and the spectrometer (including the grating), to achieve equivalent optical resolution and starting wavenumber. In addition, glass becomes opaque in the UV range, and portable UV Raman instruments are, thus, reserved for specialized applications, such as standoff spectroscopy.

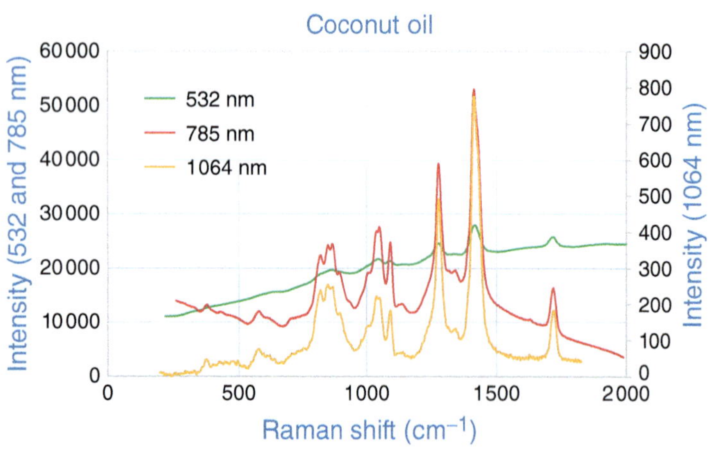

Figure 6.6 Choice of excitation wavelength can help to mitigate fluorescence background, as observed when comparing Raman spectra for coconut oil using 532, 785, and 1064 nm excitation. *Source:* Courtesy of Wasatch Photonics.

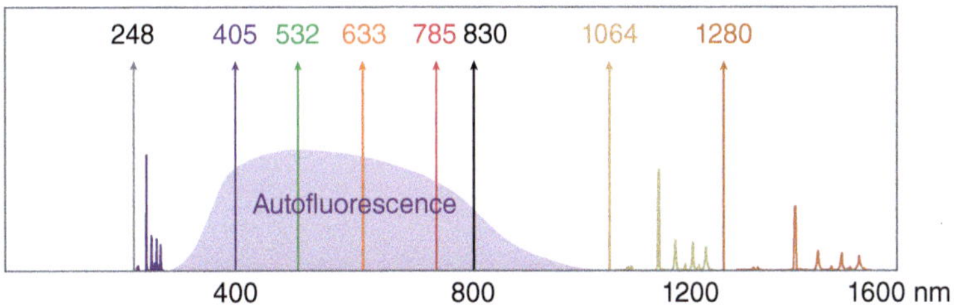

Figure 6.7 Plotting some typical Raman excitation wavelengths together with the autofluorescence band generated by intense excitation light illustrates the benefits of working at both UV and IR wavelengths for Raman spectroscopy. *Source*: Courtesy of Wasatch Photonics.

The development of stable, compact laser sources for nonspectroscopy applications has been a key enabling technology in the evolution of portable Raman instruments [38, 39], and has determined the specific wavelengths across the spectrum that have become easily accessible using commercial off-the-shelf (COTS) components. Blu-ray disc players gave us 405 nm diode lasers, solid state Yttrium Aluminum Garnet (YAG) lasers used in materials processing yielded 1064 nm and its doubled 532 nm wavelength, and 633 nm diode lasers have replaced the ubiquitous helium-neon (He-Ne) laser used in research and industry. Optical storage applications drove the development of 785 nm lasers, while 830 nm is a common diode pump laser wavelength.

Of these, the most popular wavelength used for handheld Raman instruments is 785 nm, as it yields the best trade-off between Raman signal strength and fluorescence background for the widest range of samples, with minimal sample heating. It allows measurement of Raman-active materials with Raman shifts to 3500 cm^{-1} using a silicon detector, reducing cost and power draw while yielding good signal strength.

As the calculation of Raman shift is predicated on accurate knowledge of the excitation wavelength, laser wavelength purity, stability, and repeatability are key requirements. The laser linewidth is typically <3 cm^{-1} to avoid line-broadening of Raman peaks, and a wavelength-selective element may be integrated into the laser or at its output to ensure consistent center wavelength and suppression of sidebands; a volume Bragg grating and/or dichroic laser line filter may be used. Minimal thermal drift, excellent mechanical ruggedness, and short warmup times help support the generation of consistent spectra in portable Raman instruments, where the laser may be switched on and off for each measurement and encounter a wide variety of operating environments.

Portable Raman instruments generally offer 10–499 mW of output laser power (Class 3B), focused to a spot size 50–200 μm in diameter. Laser safety must be considered in instrument design, whether in the form of a mechanical interlock at the sample interface or physical shielding, particularly for longer operating focal lengths. Handheld Raman systems generally have fixed laser power, but portable systems may offer adjustable output laser power for better control in avoiding sample damage or ignition of explosives. The laser may also be rastered across the sample, linearly or in an orbital pattern, to minimize heating of any single location. This has the added advantage of providing a more representative view of the sample, particularly for crystalline or inhomogeneous samples and mixtures.

The intensity stability of the laser's output is important for quantitative measurements, and calibration is needed when comparing answers from multiple units. A good beam profile, or spatial mode quality (as measured by the M^2 factor, or beam quality factor) is also very important to ensure the quality of focus on the sample at the desired operating distance.

Raman instruments use either single-mode (SM) or multimode (MM) lasers, where "mode" refers to the spatial mode or beam profile, as shown in Figure 6.8. SM lasers are lower in output power, typically up to 100 mW at 785 nm and <300 mW at 1064 nm; however, they offer the ability to focus the laser spot to a tighter spot due to their Gaussian beam profile. Typical SM lasers have an M^2 <1.1 and a spectral bandwidth <100 kHz.

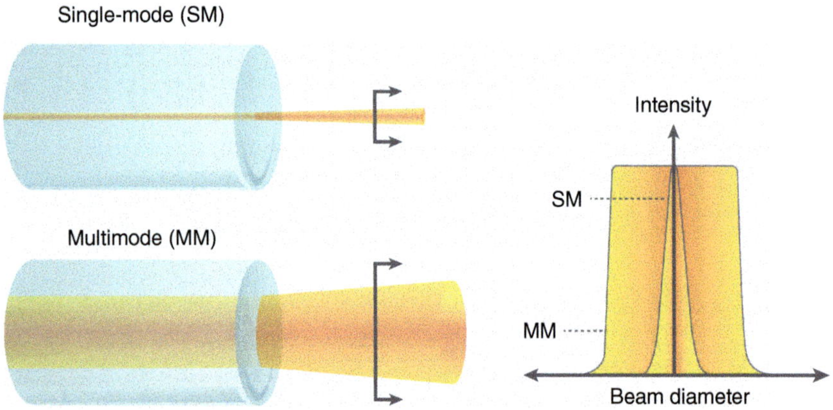

Figure 6.8 Single-mode and multimode lasers differ in both beam size and spatial intensity distribution profile. The radial distribution of a Gaussian beam can be more tightly focused to generate greater power density at the sample. *Source*: Courtesy of Wasatch Photonics.

MM lasers offer the user more output power, typically greater than or equal to 500 mW at 785 nm and up to 800 mW at 1064 nm, and are less susceptible to mode hopping. However, MM lasers cannot be focused as tightly as SM lasers due to lower native beam quality. Also, the use of the highest MM laser power levels may be limited in portable Raman systems by regulatory compliance, operator safety, and sample preservation.

6.7.1 Laser Packaging

Raman systems based on SM lasers traditionally use a small, monolithic TO-56 can in open beam format or the larger 14-pin butterfly package as shown in Figure 6.9, in either a fiber-coupled or open beam configuration. The TO-56 packaged product delivers a collimated output beam in a very compact package, making it ideal for integration into lightweight instrumentation. The SM laser 14-pin butterfly package is available at wavelengths ranging from 405 to 1064 nm for Raman spectroscopy. In this form factor, the laser is wavelength stabilized and delivers output power with ultranarrow spectral bandwidth and a diffraction-limited output beam. TO-56 SM lasers are less thermally stable than equivalent butterfly packages, and can suffer greater intensity fluctuations.

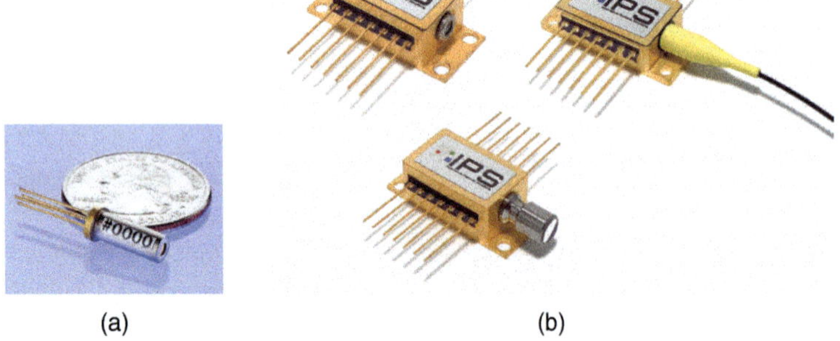

Figure 6.9 Laser packages used for Raman spectroscopy: (a) TO-56 can and (b) 14-pin butterfly packages with different coupling options (going clockwise, fiber-coupled, beam shaping and homogenizing STUB output, and open beam output). *Source*: Innovative Photonic Solutions. Reproduced with permission of Innovative Photonic Solutions.

Raman instrumentation using MM lasers also uses a 14-pin butterfly package (open-beam or fiber-coupled). These multimode spectrum-stabilized lasers offer high wavelength stability over time, temperature and vibration. MM lasers are available from 532 to 1064 nm and feature side mode suppression techniques to ensure the laser's stabilized central peak wavelength remains "locked" regardless of case temperature (15–45 °C). Other telecom industry diode lasers have been used in research applications, notably at 1280 and 1550 nm.

In general, MM lasers have the additional benefit of higher mean time before failure (MTBF) and typical lifetimes exceeding 10 000 hours, with exact lifetime and operating parameters being dependent upon the manufacturer, wavelength, model, and output power.

6.8 Optical Filters and Sampling Optics

The next elements in the optical path of a portable Raman instrument are the optical filters and sampling optics sometimes referred to together as the "probe." These probe elements may be housed separately via fiber coupling for some larger portable systems, but are always fully integrated into the unit in handheld Raman spectrometers. Together, they route clean laser light to the sample, and isolate Raman signal from Rayleigh scatter for delivery to the spectrometer (see Figure 6.10).

The first of these elements is a laser line cleanup filter. It can serve two purposes, both with the same goal of achieving the cleanest laser line possible, with no extraneous wavelength emission to cloud the Raman spectrum. A laser line cleanup filter is a narrow bandpass filter with a FWHM of 1.5–4 nm, with OD 5–OD 6 of blocking at shorter and longer wavelengths (OD = optical density). It may be used at the output of a multimode diode laser to block side modes and clean up laser output, removing the broadband amplified spontaneous emission (ASE). When used in a fiber-coupled Raman system, it also targets Raman scattering from the delivery fiber itself – background signal generated due to propagation of the intense laser light. By applying a laser line filter to the output of the laser routing fiber inside the probe head, this background can be stripped away [40].

The second element is the dichroic mirror used for laser and Raman signal routing. It is typically oriented at 45° to the laser, reflecting laser light to route it to the sample, and passing the longer wavelength Stokes emission (Raman signal) for routing to the spectrometer. While it is also possible to instead transmit the laser and reflect the Raman signal in a shortpass configuration, it is much easier to manufacture longpass dichroic mirrors than shortpass dichroic mirrors, particularly when broad bandwidths are desired.

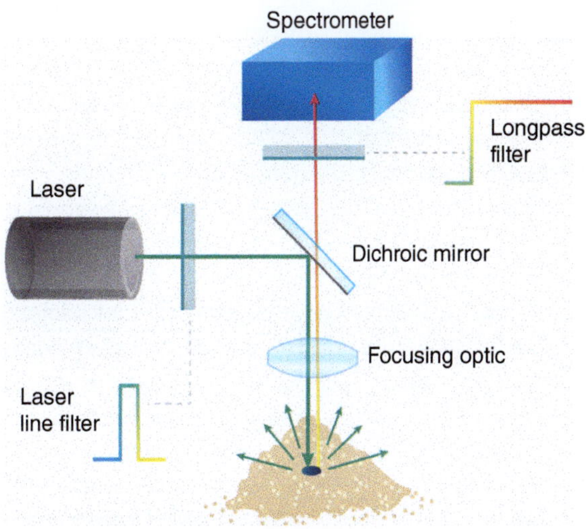

Figure 6.10 Epi-illumination schematic typical in a portable Raman system, with all the optical filters shown. *Source*: Courtesy of Wasatch Photonics.

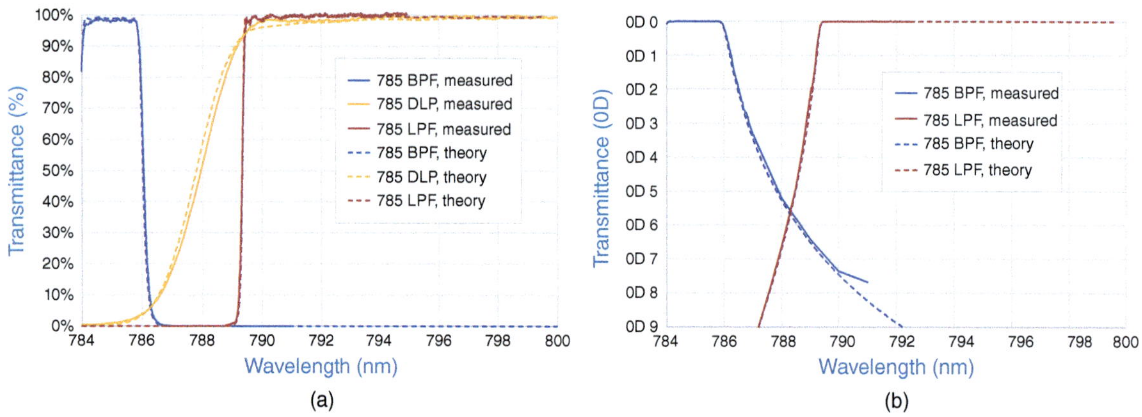

Figure 6.11 A matched 785 Raman filter set composed of a bandpass filter for laser line cleanup (BPF), a dichroic longpass mirror (DLP), and longpass filter (LPF), showing both theoretical and measured curves. (a) Transmittance curves plotted in percentage show the steep edges used to isolate the 785 nm laser line and Raman signal, while (b) transmittance curves plotted on an OD scale illustrate the deep blocking required to obtain good SNR. *Source*: 785 Raman filter. © 2020, Iridian Spectral Technologies Ltd.

The third element is the focusing optic, which both focuses the laser light onto the sample and collects the resulting Raman emission. This will be addressed at a later point when discussing the sample interface, once the groundwork for all considerations in its design has been laid. At this point, it is sufficient to say that these optical elements should be made of low fluorescence glass to avoid the addition of background.

The final element in the filtering and sampling optics is the longpass or "edge" filter used to reject the otherwise overwhelming Rayleigh scatter. A minimum of OD 6 absolute blocking at the laser wavelength is required (with OD 5 average blocking over a broader band to address any other background). Optical densities greater than this are difficult to manufacture, even more difficult to measure accurately, and of no benefit in the "good enough" philosophy of portable Raman.

Longpass edge filters come in a variety of grades, commensurate with the edge steepness, aka "transition width" or "cutoff." This may be quantified in percentage relative to the laser wavelength, or in cm^{-1} for more direct comparison between wavelengths. Edge steepness is determined by the application, though very few key Raman peaks lie close to the laser in typical portable applications.

Achieving a given edge steepness in cm^{-1} is more difficult at shorter excitation wavelengths, and can lead to excessively thick filter coatings in the NIR. These considerations favor the use of shortwave infrared longpass filters, as the most common 250–400 cm^{-1} edge transitions can easily be achieved to match 785 or 830 nm lasers, and are sufficient for the needs of handheld Raman.

It is also important that all three filters in use work together to achieve the needed OD and passband. Off-the-shelf Raman filters often come matched to one another in terms of edge location, steepness, and combined laser and autofluorescence blocking, as shown in Figure 6.11. These three are often designed together for original equipment manufacturer (OEM) applications to achieve peak performance, particularly if uncollimated light is used anywhere in the system.

Dichroic filters exhibit a blue shift with increasing angle of incidence, which is compounded with polarization splitting for angles above ~10°, and, thus, a steep edge designed for collimated light can quickly become washed out when used with a range of input angles. This is one of the most common sources of laser bleedthrough, which manifests itself as a trailing background in the first ~200 cm^{-1} of a Raman spectrum.

6.9 Spectrometer Design

All the optical elements described to this point have concentrated on the generation and collection of clean, stable Raman-scattered light. A Raman spectrometer is then required to capture the spectrum with sufficient range, optical resolution, signal strength, and consistency to allow analysis and the generation of robust answers. We call out a Raman spectrometer specifically, as the consequences of tradeoffs between range, optical resolution, and optical throughput are unique to this technique; it is not sufficient to repurpose an off-the-shelf broadband spectrometer when designing for the needs of portable Raman instruments. First, we shall look at the performance requirements of the Raman spectrometer as a whole, then at the options for bench design and optical elements within, as well as how each impacts performance.

While the specific application will dictate the requirements for range, optical resolution, and optical throughput, handheld Raman spectrometers typically measure the fingerprint region and beyond to 2400 cm^{-1} Raman shift, with <15 cm^{-1} FWHM optical resolution; they are designed with the aim of acquiring data in under five seconds. Depending on the materials to be identified, better optical resolution or an extended range to capture functional-group information may be required. Spectral bandwidth, optical resolution, and throughput are determined by the focal length of the spectrometer, its dispersion, slit width, and the numerical aperture (NA), combined with the array detector's pixel count and pitch. These are the dials to be turned, and their interdependencies must be managed in designing for "good enough" performance with the aim of robust library matching and/or quantification [41].

Additional constraints on the spectrometer design may also be imposed by the choice of laser wavelength, as mentioned previously. Excitation laser wavelengths longer than ~850 nm require the use of an InGaAs detector, while shorter excitation wavelengths require better optical resolution (in nm space) to achieve the same optical resolution in cm^{-1}. For example, 10 cm^{-1} resolution when using 830 nm excitation equates to a spectrometer resolution of 0.74 nm, while 0.29 nm spectrometer resolution is needed to maintain the same 10 cm^{-1} resolution using 532 nm excitation (taking a peak located, for example, at 400 cm^{-1} from the excitation laser). Since we are working in frequency space, instrument optical resolution also varies across a Raman spectrum, improving toward the long-wavelength end. Some manufacturers specify an average resolution, while others quote guaranteed resolution over the full operating bandwidth.

When manufacturing portable Raman systems, it is preferable to minimize unit-to-unit variability, though the spectral-response function of each unit must still be corrected for prior to library matching. The spectral-response function is the product of the efficiency of each individual optical component within the full system, and is heavily dependent on the spectral response of the spectrometer.

Finally, thermal and mechanical stabilities of the spectrometer are of utmost importance for portable or handheld Raman instruments, both as regards wavelength calibration (key for identification applications) and spectral sensitivity (essential for quantitative work).

As we consider each aspect in the optical design of a Raman spectrometer, we will refer back to the abovementioned requirements to inform our discussion of the available design options and constraints.

6.9.1 Optical Bench Design

The general design for any Raman optical bench includes an input slit, collimation optics, a dispersive element, and focusing optics for imaging onto a detector. The fundamental question in optical bench design is then whether to use reflection or transmission optics, illustrated in Figure 6.12. This choice has a number of impacts, played out through (and resulting from) the optical elements employed in each case.

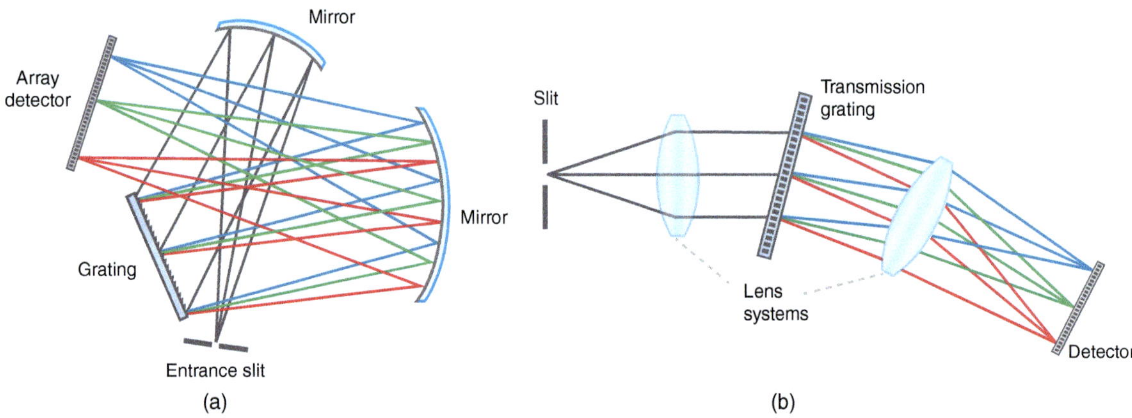

Figure 6.12 Two typical optical bench designs used in portable Raman spectrometers: (a) Reflective grating and optics in a crossed Czerny–Turner configuration, and (b) Transmissive grating and optics in a Littrow configuration. *Source*: Courtesy of Wasatch Photonics.

Reflective designs often use a Czerny–Turner optical bench geometry, with a reflective grating and concave focusing mirrors forming a W-shaped optical path. If the beam path is folded (as shown in Figure 6.12a), a compact optical bench is created that can be manufactured using readily available, low-cost optics. This approach epitomizes the "good enough" philosophy, yielding adequate signal in a compact footprint at low cost. It does, however, come with trade-offs. The folded design adds astigmatism, and any angle shift in the optics due to misalignment leads to a wavelength shift, impacting calibration.

Transmissive designs most often employ a Littrow configuration, with lens sets for collimation and focusing, and dispersion provided by a transmission grating. This on-axis approach minimizes astigmatism and other aberrations, the balance of which can be compensated through careful optical design. Transmissive designs are far less angle sensitive than reflective designs, and, thus, are much more robust against vibrations or temperature changes. This same property makes them easier to align, and spectral-response functions are more consistent from unit to unit. Edge filters may also be integrated directly into the design in the collimated space between the first lens set and grating (a design option which is far more challenging to implement in a reflective design).

While these factors are important, the primary design criterion for a portable Raman spectrometer must be sensitivity. Overall light collection power is dominated by the NA, the range of angles over which the system can accept the light. Put simply, the more light collected by the spectrometer, the more light can be detected, allowing for shorter measurement times. The throughput of a spectrometer (or étendue, G) scales with slit area (A_{slit}) and NA^2:

$$G = A_{\text{slit}} \times \text{NA}^2 \ \ [\text{mm}^2 \times \text{sr}]$$

The light collection power of a spectrometer may also be expressed in terms of f-number ($f/\#$). A high NA equates to a low f-number, and, thus, more throughput.

$$f/\# \approx \frac{1}{2\text{NA}}$$

Coming back to our two geometries, lenses allow for a higher NA than mirrors of the same size, favoring a transmissive design when greater collection efficiency is needed. They also win on the preservation of light within the optical bench as lenses with antireflection (AR) coatings offer higher efficiency than metallic mirrors. Though this comes with some increase in cost, it is often merited when overall performance to price ratio is considered.

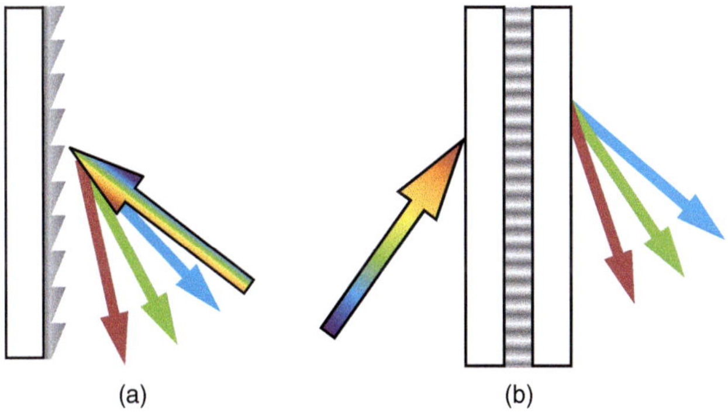

Figure 6.13 (a) Diffraction from a surface relief grating in reflection, (b) diffraction from a volume phase holographic grating in transmission. *Source*: Courtesy of Wasatch Photonics.

6.9.2 Diffraction Grating

The diffraction grating serves to provide the needed dispersion for the spectrometer using interference from a periodic structure. Diffraction gratings may operate in reflection or transmission as shown in Figure 6.13, and, thus, the choice may be dictated by preferred optical bench geometry. Understanding the performance of each can influence the choice of geometry, however.

Reflection gratings are manufactured either by ruling parallel grooves in a metallic coating or are replicated (pressed) from a master. This allows them to be mass produced at low cost, but the use of a mechanical process results in lower reproducibility and eventual deterioration of the master. This impacts unit-to-unit variability, with further risk to spectral response consistency in volume if the master must be replaced.

The efficiency of a ruled reflection grating tends to peak at the blaze wavelength, then fall off to either side and may vary widely over the operating bandwidth. The diffracted light is also split across multiple orders, and, thus, typical efficiency curves for first-order diffracted light range from 25 to 60% over the bandwidth [42]. This may be further decreased by dust, fingerprints, or damage to the grating surface; while the reflection grating will remain operational, it cannot be cleaned, and efficiency is permanently reduced. The presence of ghost images and scattered light can add noise to spectra, which is undesirable in Raman spectroscopy, especially given the low signal levels which are typically measured.

Transmission gratings may be manufactured by etching a periodic structure into a transmissive substrate, with similar efficiency curve challenges as ruled reflection gratings. If manufactured holographically, however, the grating can be designed so as to divert the majority of transmitted light into the first order for greater efficiency. Holographic gratings are all originals, and as such, have high reproducibility, but are more expensive than ruled gratings.

Surface-relief holographic gratings are etched, while volume phase holographic (VPH) gratings are created by imaging an index of refraction variation into gelatin film capped with glass. VPH gratings offer the additional benefit of being easy to handle and clean and are AR-coated to further reduce loss. Holographic transmission gratings, in general, offer good efficiency over a broad range as shown in Figure 6.14 (>70–80% is typical), and display far less polarization dependence than ruled reflection gratings. When applied to Raman instruments, these benefits increase sensitivity and reduce signal dependence on the orientation of sample crystals, respectively.

6.9.3 Stray Light

Stray light is an excess of light of a given wavelength in regions on the detector intended to sense other wavelengths. It arises due to scattering from optical components or surfaces within the spectrometer, focusing limitations, and

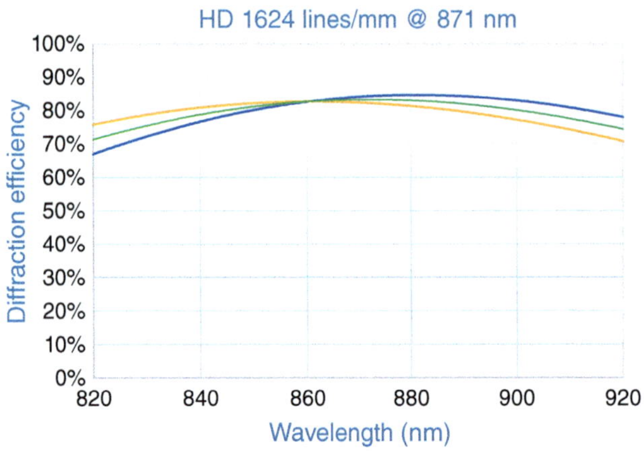

Figure 6.14 A volume phase holographic (VPH) transmission grating for Raman, 1624 lines/mm @ 871 nm. *Source*: Courtesy of Wasatch Photonics.

diffraction from higher orders in the grating. This often takes the form of spillover to wavelengths close by, but some light is scattered at larger angles, and, hence, over larger wavelength differences. In Raman instruments, stray light can lead to a slight broadening of a peak, formation of a pedestal on a peak, or a raised baseline. Use of low-scatter optics, careful correction for aberrations in focusing, and use of adequate, well-placed baffles within the optical design can help mitigate stray light.

6.9.4 Detectors

The detector in a portable Raman spectrometer converts photons into electrons, as measured over a period of time (the "integration time"). Portable Raman spectrometers use a linear array detector to read out the entire spectrum simultaneously using 512–4096 pixels. Each pixel generates an analog voltage output which is digitized to yield a raw intensity spectrum of "counts" versus pixels. This data can then be translated into a Raman spectrum using the wavelength calibration coefficients of the instrument and the Raman shift equation described earlier.

A detector used for portable Raman spectroscopy must be sensitive and have low inherent dark noise to yield the required signal-to-noise ratio (SNR). In practical terms, it must be relatively low-cost, consume less power than detectors used in larger instruments, and fit within a miniaturized bench. This combination of attributes could not have been imagined possible prior to the adoption of low-cost commercial array detectors by spectroscopy in the 1990s [43–45]. Used in devices like photocopiers, these commercial charge coupled device (CCD) detector arrays, based on silicon technology, offered reasonable sensitivity and SNR across the visible and short-wave infrared (SWIR) at a much lower price point than traditional scientific-grade detectors – the epitome of "good enough" technology. This, in turn, inspired detector manufacturers to expand the selection of lower cost array detectors for applied spectroscopy, optimizing factors like spectral-response curve, dark noise, speed, power consumption, size, cooling, and number of pixels to suit different applications and price points [46].

Raman spectrometers typically use either a CCD or complementary metal-oxide semiconductor (CMOS) linear array detector to collect and read out a spectrum, similar to the examples shown in Figure 6.15. Some detectors consist of a 2D array of pixels, designed to be summed vertically (known as "binning") to produce a final 1D array of higher sensitivity than would otherwise be possible. Most sensors deployed in portable Raman instruments are based on silicon technology; InGaAs technology is used for >1064 nm Raman signals. The exact sensor choice is based upon excitation wavelength, desired optical resolution, instrument sensitivity, application or use case, and cost [47].

(a) (b)

Figure 6.15 Examples of linear array detectors used in portable Raman: (a) Hamamatsu S10420 series back-thinned type CCD area image sensor, (b) Hamamatsu G9214 series InGaAs linear image sensor. *Source*: Hamamatsu Corporation. Reproduced with permission of Hamamatsu Corporation.

The sensitivity of a detector is gauged by its quantum efficiency (QE), the ratio of electrons produced versus incident photons. Most sensor manufacturers supply a graph of the relative sensitivity versus wavelength to help designers choose the correct detector for each specific application, as different fabrication methods can be used to optimize QE for specific wavelength ranges. Back-thinned detectors take this a step further, thinning the silicon wafer and reversing its orientation relative to the electrodes so as to increase interaction with light. This method greatly improves quantum efficiency, but can lead to etaloning, a sinusoidal modulation in the detector response that can be detrimental to low-signal applications, like Raman, especially when looking at highly fluorescent samples.

The most common CCD used for 633, 785, and 830 nm excitation has been the back-thinned NIR-enhanced S1151× series from Hamamatsu, offered in 1024 and 2048 pixel formats. The post-processing method used in the manufacture of this particular back-thinned sensor served to not only increase the native QE above 750 nm, but also reduced the fringing or etalon characteristics of the sensor, an ideal combination for 633–830 nm Raman. Due to recent litigation [48], however, these sensors are no longer commercially available, and, as such, instrument providers are evaluating new, alternate sensors with a view to minimizing the impact on Raman performance. Shorter wavelength excitation Raman, e.g. 532 nm, often utilizes non-NIR-enhanced sensors or front-illuminated CCD detectors.

Most 1064 nm-excitation-based instruments use either 512-pixel or 1024-pixel linear photodiode InGaAs arrays [49, 50]. These sensors are typically cooled to below −10 °C to offer optimal SNR. InGaAs sensors have a higher power draw due to the high demands of the thermoelectric coolers (TECs) used to lower the detector temperature and are considerably more expensive than silicon sensors. As such, they place additional constraints on portable Raman instrument design and performance. InGaAs linear detectors also have unique gain/offset correction requirements, as each pixel can be considered an independent photodiode. Due to the interleaved array design often used to control alternating pixels, multiple calibrations may be required to produce a smooth spectrum.

Detector technology continues to evolve to support the next generation of portable Raman instruments. Uncooled InGaAs sensor availability is on the rise, driven by technology advancements in other sectors, such as night vision and security. These offer some interesting avenues for 1064 nm Raman development, especially when used with high throughput optical benches. In the silicon region of the spectrum, new and novel sensors are also gaining in prevalence, largely through markets like consumer electronics. Here, CMOS-based sensors from Hamamatsu (S11639) and Sony (IMX series) show promise for sensors that could enable more compact and cost-effective Raman instrumentation in the coming years.

6.9.5 Electronics Design

Although scientific-grade detectors and digital camera sensors have improved immensely in recent years, they still require a significant amount of custom electronics to be used optimally for Raman spectroscopy. In addition, portable Raman spectrometers incorporate a variety of other components requiring electrical power and control, all of which must be integrated together in a low-power, miniaturized set of boards and interface cards.

Typically, the electrical design will start (or end, depending on your perspective) with a "daughtercard" holding the detector itself. This daughtercard will often have tight dimensional and mechanical requirements, as the detector will need to be mounted and aligned with great precision at the end of the optical collection path. Ambient light leakage at this terminus can be ruinous for acquisition, so various gaskets and optical baffles may be introduced. The detector itself will likely be thermally cooled, meaning a TEC, thermistor and heatsink will be competing for direct access to the detector's surface. Any heat or electrical noise generated by other electrical components can potentially degrade the raw signal generated by the sensor, and, thus, power-hungry components should be kept as far from the detector as possible.

All of these are reasons for the detector daughtercard to be as small and minimal as possible, typically containing only the basic analog front-end (AFE) needed to digitize the pixel voltage into digital intensity counts through a carefully gained and filtered analog-to-digital converter (ADC). The ADC bit-resolution will determine the detectable dynamic range (vertical extent in counts) of the system. For instance, a 12-bit ADC can only report 4 096 different levels (2^{12}) of intensity, regardless of the detector sensitivity, while a 16-bit ADC supports a much richer scale of 65 536 intensity levels (2^{16}). The daughtercard then passes the integral intensities back to a "mainboard" for processing and communications.

The daughtercard will also be responsible for the various clocks and control signals used to drive the detector acquisition. For deterministic high-speed timing, these clocks are typically generated by a field-programmable gate array (FPGA) which will often read out the digitized intensity values from the ADC. While FPGAs provide robust and reliable timers, they can be complicated to design and maintain, so once a spectrum of intensity values has been read out from the daughtercard, it may be passed straight to a standard microcontroller or microprocessor for disposition.

There are many types of microcontrollers and microprocessors in use for portable Raman spectrometers, but most are converging on the ARM family of processors [51] to leverage the broad commercial and open-source toolchains supporting that platform. The choice of a processor will largely depend on (or drive) the determination of the types of spectral processing which can be applied, and where those processes will occur. A low-power microcontroller may serve primarily as a communications gateway to offload the raw data as expediently as possible (indeed, such could be provided by the FPGA alone); or a more powerful microprocessor may elect to provide significant post-processing and even Raman matching internally.

Beyond the basic collection and routing of spectral data, there are many other functions which the spectrometer's electrical design needs to support. It will be responsible for distributing and level-shifting clean power on various "rails," often including 12, 5, 3.3 V and other voltages in a single system. It must also power and control TECs on detectors and/or lasers. It will control the lasers themselves, including safety interlock and status LEDs. Battery subsystems will require charging electronics, possibly during measurement operations, and provide "gas gauge" status indicators to monitor charge state. Communication subsystems can be extensive, depending on the variety of communication buses and protocols supported, from low-level I^2C or serial peripheral interface (SPI) hooks to power-hungry USB 3.0, Wi-Fi and Ethernet ports.

Medical Raman systems may include integrated photodiodes for laser power monitoring, and advanced instruments may include articulated motors to raster the laser beam or offset the optical path between measurements. Fully handheld devices will need electrical control of the user keypad and display screens. Consideration must be made toward external connectors if required for firmware or library updates, factory maintenance or cables for triggering or standoff operation. Nonvolatile storage must be provided via electrically erasable programmable read-only memory (EEPROM) or similar to store configuration and calibration data specific to the device.

All of the above electrical functions need to be integrated into as few boards as possible, occupying as little real estate as possible. It is clear, therefore, that careful consideration of the specific requirements for a given application is merited to ensure that a sufficient and complete, yet minimal, set of features are implemented in any particular design to optimize performance and operational runtime for the specified use case. Given the above, the electrical design of portable Raman systems will particularly benefit from attention to the following areas.

Low-noise electronics are desirable for the highest SNR, to yield the best analytical sensitivity. In most cases, electronic noise is significantly lower than other noise contributions, and, thus, should only be noticeable when looking at very small signals on a very low dark-current background.

The primary driving factor in electronics design for portable Raman systems is the need for low power consumption to maximize battery life. This also enables the use of a smaller battery, which results in a smaller form factor overall. To reduce the proportion of space taken up in the instrument by the battery, and to prolong the time between charges, the microprocessor firmware can be programmed with sleep options to conserve power. The measurement cycle can also be designed to preserve power by turning off the laser diode and TECs when not in use, as these represent major power draws within the instrument.

Communication options may include wireless protocols, such as Bluetooth, Wi-Fi, or similar to parlay with a phone app or the cloud [26]; low-power Bluetooth (BLE) is attractive as regards power consumption, but can be slow for spectral acquisition. If a laptop is used for analysis, USB or Ethernet connection is typical.

6.9.6 Mechanical Design

The mechanical demands on Raman instrumentation have never been greater, with a wider range of environmental operational requirements than ever before. A cursory glance at manufacturer datasheets shows the importance of shock, drop, and vibrational specifications tested to international and military standards along with IP ratings, like IP67 [37], which describes ingress protection against dust and water. Size and weight are minimized, as the lower the size and mass, the more user-friendly and more likely the acceptance in certain operational environments. The specifications also show the wide-ranging temperature and humidity conditions expected from Raman instrumentation.

All of this means that a Raman instrument needs to be designed with mechanical and thermal stability in mind from the beginning, selecting key components, like excitation laser, optical bench design, sample interface, power management, and display so as to withstand these demands individually as well as at the system level. Achieving this level of mechanical stability means unit-to-unit reproducibility is maximized, and material identification via library matching is less arduous due to improved spectroscopic data integrity on all platforms in the same product family.

There are various mechanical features and accessories of particular use in portable Raman instruments, though the convenience and value of each must be weighed against all critical cost and SWaP goals for inclusion in any particular system. Integrated mechanical shutters allow "dark" or "laser-off" measurements to be taken at any point (and automated by software) to help normalize spectra and improve library matching; however, holding shutter arms in fixed positions (independent of orientation and gravity) represents a minor but constant power draw, and incorporation of a rarely articulated component represents opportunity for breakage under stress.

High-availability applications may require a replaceable battery to minimize downtime when the charge runs low, adding requirements for a housing door or removable lid to accommodate hot-swapping of batteries. Various external connectors may require IP ratings, often leading to bulkier and more expensive ports. Even nominally handheld devices may include provisions for mounting on tripods or unmanned platforms. Field calibration standards can be built into lens caps, and standoff aides, like removable nosecone guides, can simplify measurement uniformity against different samples. Minor amenities like lanyard rings may be included to ease operator workflow.

A comprehensive thermal model of the system using computational fluid dynamics (CFD) may be required to balance the various thermal sources within the housing (e.g. processors and TECs) and ensure the heat they produce flows consistently and predictably away from thermally sensitive electro-optical components (detector and laser). A systematic thermal design will account for the thermal design power (TDP) of electrical components under load across a range of battery states and ambient environmental conditions and may involve Monte Carlo simulation to cover all the corner cases. Simulation results can advise the placement of fans, vents and heatsinks to direct stray heat and minimize overall system thermal noise.

6.10 Sample Interface and Accessories

Our earlier discussion of the excitation laser, sampling optics, and spectrometer took us as far as the dichroic for routing laser light to the sample and Raman signal to the spectrometer, with only brief mention of the optics to focus light onto the sample and collect the resulting Raman emission. This is an extremely important aspect of design for portable Raman, as the sample interface and accessories must be tailored to the specific application and/or sample type so as to produce a high level of sample signal, with the lowest level of background signal arising from the container or ambient light. For quantitative work, it is important to achieve highly reproducible alignment of optics to the sample.

For portable applications, it is strongly preferred to measure the sample in its native container, without transfer to a sample vial. Noncontact examples include white powder in plastic bags, chemicals inside a polyethylene container or an amber glass bottle (through-barrier), or identification of medication in a blister pack. For a bulk liquid or powdered sample, a dip probe in direct contact may be used. In either case, ambient light should be minimized – the sample interface and coupling optics should be designed with shielding, backed by the common practice of recording an ambient light signal under the same ambient conditions to account for both dark current and ambient light levels.

In the typical epi-illumination setup discussed under sampling optics, a lens is used to focus the collimated laser beam onto the sample, and to collect the Raman emission into a collimated beam. The beam is focused with a second lens onto the spectrometer slit (in a fully integrated system configuration) or a fiber (in case of an external probe). This second lens needs to be matched in NA to the input of the spectrometer (or spectrometer-coupling fiber, when an external probe is used). In the case of clear liquids, a higher NA sample lens is ideal to collect a larger solid angle of Raman emission, but from a shorter focal depth. In the case of solid samples, the imaged focal volume is shortened due to scatter, but the imaged area increases through backscatter. These counteracting effects mostly cancel, which opens up some flexibility for adjusting the working distance to the sample without loss of signal.

When deciding between a fiber-coupled probe and direct coupling via "integrated probe" (free-space optics), the direct coupling approach is chosen most often. It is more integrated, more compact, offers greater efficiency by using less optics, and can image a taller emission area from the sample onto the full slit. Fiber-coupled sample optics provide flexibility for instruments that cannot be held against a sample, as shown in Figure 6.16. The disadvantages of this flexible sample interface are the potential for fiber breakage, additional losses from fiber coupling, and the sacrifice of signal when the circular fiber is imaged onto a rectangular slit.

Working through a barrier for analysis of powders or liquids as described above is possible because most barriers allow a surprising degree of transmission of laser light into the container, and of Raman signal back to the system, particularly at NIR wavelengths beyond the range of our vision. Also, as the container and sample working distances are different, a long-working-distance lens can be made to focus deep inside the container to increase the proportion of Raman signal from the target material relative to the container emission. See the discussion on spatially offset Raman spectroscopy (SORS) below, and also in the chapter in Volume 2 by Hargreaves.

Handheld Raman instruments often come with useful accessories, as illustrated in Figure 6.17, to facilitate measurement of different sample types. These may include lenses of different focal length, a vial holder, or reader for slides used in surface-enhanced Raman scattering (SERS), to be discussed later. Portable Raman instruments may be configured with various probe options, including dip probes for process applications.

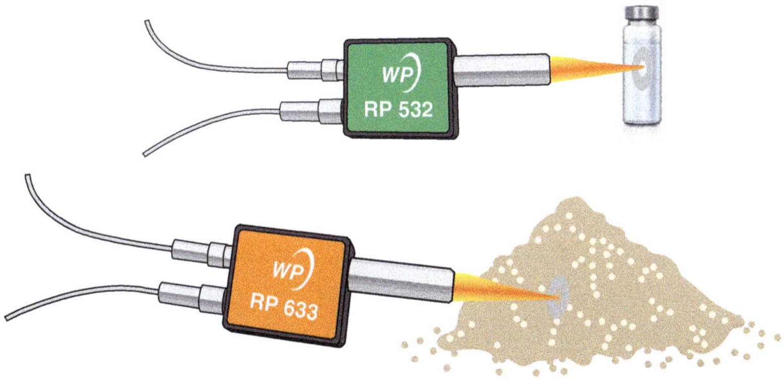

Figure 6.16 Probe-based portable Raman systems can work through-barrier to probe materials in containers, or sample powders directly. *Source*: Courtesy of Wasatch Photonics.

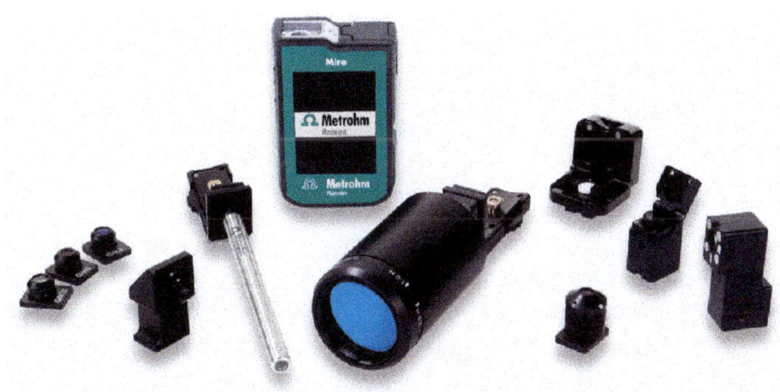

Figure 6.17 Some handheld Raman systems like the Mira DS are offered with a variety of attachments for measurement of powders, liquids, solids, and SERS substrates. *Source*: Metrohm AG. Reproduced with permission of Metrohm AG.

Sampling accessories are also important to ensure compliance with pharmacopeia standards. Wavelength calibration is typically performed using ASTM method E1840, which uses a standard compound with Raman signal across the full spectral range, such as a mixture of acetonitrile and toluene [52]. Raman shift accuracy may then be validated via a recommended reference standard, such as polystyrene, in compliance with United States Pharmacopeia (USP) chapter <1120> or European Pharmacopeia (EP) chapter <2.2.48> [53, 54].

The final point of consideration in the discussion of sample interface is the need for laser safety to be integrated into the sample interface, as most portable Raman systems utilize a Class 3B laser. It should only be possible to fire the laser if the laser is in a closed beam path when working distances are long. A short-focal-length lens may also be deemed adequate to ensure safe laser operation. In all cases, laser safety warning labels must be clearly visible on the instrument, and guidelines for safe operation and use cases included in the manual.

6.11 Spectral Processing and Analysis

Beyond the acquisition of a Raman spectrum in a portable Raman instrument lies the computer hardware, software, spectral processing, and libraries needed to generate answers from the data. In this section, we will discuss each, focusing on the aspects most pertinent to portable Raman spectroscopy.

6.11.1 Processing Hardware

Converting the many peaks of a Raman spectrum into an actionable answer requires more powerful hardware than is convenient to package inside a miniaturized optical bench. Commercial Raman systems have, therefore, converged around two basic configurations: either integrating an embedded computer alongside the spectrometer within a shared housing or performing processing and user interface on a physically separate phone or laptop which communicates with the spectrometer through wired or wireless protocols. In the ubiquitous "Internet of Things" (IoT), either option can be backed by cloud-based services if communication options allow.

Full integration of computer hardware onto the instrument adds size, weight, power consumption, and, of course, cost. Matching libraries must be stored or at least cached on the instrument, requiring additional storage and periodic updates, and causing security concerns for some applications. These are far outweighed, however, by the benefits of having all elements integrated into one compact package with push-button control; this is the standard for the many handheld Raman instruments deployed in the field.

Use of separate processing on a phone, tablet, or laptop allows COTS hardware to be used, is more easily updated, and allows additional flexibility and options in the user interface versus buttons or touchscreens. Communication between the instrument and remote processor can utilize a variety of consumer and industrial standards, like USB, Bluetooth, or Wi-Fi; data may be stored (or even processed) remotely in the cloud. This multiple-device approach, however, complicates measurement operation and makes the system susceptible to all the information technology (IT) vulnerabilities that come with nightly updates, operating system upgrades, and the like. The necessity to transmit spectral data between devices can also present a security issue for some applications.

6.11.2 Spectral Processing

A key step in the analysis of the acquired spectrum is post-processing to prepare the data and remove instrument variability. Library matching and quantitative analysis are both sensitive to spectral response mismatches between the portable instrument used in the field and the benchtop equipment used to establish the sample library (or determine calibration). Third-party or legacy Raman libraries may have been collected at a different laser wavelength, a different spectral resolution and range, or with different optics, which impacts the spectral shape and consequently matching accuracy. It is, thus, necessary to correct for as many of these differences as possible after acquisition of sample measurements in the field, and before comparing spectra via library matching [55].

An important step in most spectroscopic measurements is the subtraction of an ambient signal (laser-off) measurement, which must be taken at the same integration time and ambient conditions as the sample measurement. Also, many spectra may be averaged together over time to improve SNR and reduce spurious noise peaks. Linearity correction may be needed to account for varying detector sensitivity as pixels near saturation. Gain and offset corrections may also be needed in post-processing if not automated within the spectrometer.

As detectors can develop "bad pixels" over time (similar to LCD screens), the software may need to average or interpolate across gaps in the input data. High-frequency noise may be removed through simple convolutions, such as "boxcar averaging" of adjacent pixels. The sequencing of these steps is important and worthy of careful consideration, but will vary depending on what features are integrated into spectrometer hardware, versus what remains to be done in software.

Beyond standard spectral cleanup, there are other operations which can improve matching results against a particular library or sample type, depending on the particulars of an application. Sample or library spectra may need to be interpolated against a common x-axis, or normalized in intensity. Intensity correction can be applied with a calibration generated against a Raman intensity standard to account for overall system spectral response of the field instrument (as per ASTM E2911-13 [56]).

Baseline correction can be used to eliminate any remaining sample fluorescence, for which a number of methods have been developed [57, 58]. Finally, a Raman shift calibration of the field instrument against a Raman standard sample (see, for example, ASTM E1840-96 [52]) is typical to account for environmental effects and ensure proper calibration in cm^{-1} space [55, 59].

Raman signal quality may degrade at either end of the Raman shift detection range, due to laser blooming, detector response roll-off, filter cutoff edges, and general vignetting by the optics. Therefore, matching results may further improve through careful trimming of the spectrum to an optimal detector ROI (region of interest).

With all corrections complete, the Raman spectrum can then be presented to the source library for matching. In-house library development is a significant investment for Raman providers, particularly as they must be both comprehensive and targeted in order to meet the needs of a specific industry. Specialist companies who provide Raman libraries are on the increase, helping to bridge the gap between spectrum and answer for OEMs and researchers alike [60, 61]. Two possible use cases for libraries are described below.

6.11.3 Analysis: Identification Using Libraries

The primary goal of many handheld Raman instruments is to determine the identity of unknown substances and/or validate their identities. This is done via matching of the spectral profile against a library of known substances. This requires access to an existing, validated library, or the ability for a user to create her, his, or their own library. The option of creating a dedicated user-defined library is popular, as samples under test may be very specialized, and this ensures the direct comparison of known substances against unknowns for the particular use case. This also allows the user to ensure the library spectra were collected under similar conditions and equipment to what will be used in the field; this flexibility is balanced against the cost in time and materials to generate such measurements.

Library matching may be performed using a number of possible methods. (See the chapter in Volume 2 by Zhang, Lee and Shreyer for a discussion of algorithms, and the chapter in Volume 2 by Schreyer for a discussion of libraries.) One simple method is correlation analysis, in which the two spectra are compared wavenumber for wavenumber on a common (usually interpolated) scale to determine the degree of correlation between the two spectra. Another method uses peak identification, looking at position and relative intensity to generate and compare peak lists for the unknown versus a library.

Mixture analysis is somewhat more complex, but is a valuable tool given that it can identify adulteration or improperly prepared materials, as well as aid identification of samples in a container or deliberately concealed materials by removing the container signal [62]. Mixture analysis is an iterative process: in simple terms, the closest first match is subtracted from the spectrum and the remaining difference spectrum is used to search the library for a second match. The spectrum is then reconstructed as the best linear combination of these first two matches and the routine repeated for additional matches using proprietary algorithms [63].

6.11.4 Analysis: Quantification Using Chemometrics

Some handheld Raman instruments, and many application-specific portable instruments, seek to answer the question "how much?" for evaluation of purity, process analytics, or diagnostics. This adds a need for access to spectral response calibration parameters for the instrument to produce an intensity-corrected spectrum. As the Raman peak intensities will be proportional to the analyte's concentration, it is possible to develop a predictive model for concentration based on the evaluation of a series of standards across the concentration range of interest.

One approach to this is to look at the integrated peak intensity for one or more characteristic Raman peaks. As the relationship between signal and concentration is known, the user can rely on a limited calibration set and interpolate between calibration points. The downside of this method is that it is sensitive to cross talk from substances with overlapping Raman spectral peaks. This risk can be reduced by analyzing multiple characteristic peaks.

An alternative approach uses multivariate analysis (chemometrics) to look at the spectrum as a whole [64]. By using the entire spectrum rather than just individual peaks, it is possible to automatically account for cross-sensitivities and overlapping peaks. Development and execution of the model, however, are less transparent and requires much more extensive calibration effort to guard against interferences.

6.11.5 User Software

Above the spectral processing and identification or quantitative analysis algorithms lies the user interface – the software employed by the user to yield an answer. User software in portable Raman is almost always application specific to reduce the need for user training, knowledge, and support. On the simplest level, it must cover auto-settings, like integration time adjustment, background subtraction, and prompts for calibration against standards. It must also have built-in safeguards to identify when poor-quality data have been acquired, or when the instrument requires service.

For portable devices, the user software must make operation as simple as possible, from spectrum to identification or quantification with the push of a button. Results are displayed with the greatest clarity possible: yes/no, pass/fail, identity and confidence level, identities and concentrations, etc. The operating software may interface to a cloud analysis routine or feed into a global database for record-keeping. As libraries may evolve over time, either with added compounds or improved algorithms, mobile software should be easily updatable by end users.

Applications supporting identification or authentication will often include privacy or security aspects which must be carefully considered in software design, especially with regard to data storage and transmission. As part of application-specific support, requirements may be placed on data acquisition and storage to comply with Food and Drug Administration (FDA) regulations for electronic record-keeping (FDA 21 CFR Part 11) [65]. This is common for medical and pharmaceutical applications to ensure data integrity.

Given the inherent potential for injury with any laser-based system, Raman system software must be designed to minimize the opportunity for accidental exposure to laser radiation (FDA CFR 21.I.J.1040.10-11) [66]. While a comprehensive laser-safety protocol cannot be completely implemented in software, a well-designed user interface and rigorous use of watchdogs, timeouts, and conservative defaults will reduce chances for accidental laser exposure.

The needs for user software in handheld and portable Raman can vary widely depending on the application and operating environment, but all share a common baseline of requirements: ease of use, clear communication, and robust answers.

6.12 Special Cases

6.12.1 Fluorescence Mitigation

Although fluorescence background can be reduced through the use of excitation at wavelengths shorter than (<250 nm) or longer than (1064 nm) the autofluorescence region, it is not always practical. As such, alternative techniques have been developed to post-process data to reduce the broadband fluorescence signal and allow discrete Raman lines to be viewed [67, 68]. John Cooper et al. demonstrated in their work at Old Dominion University how multiple excitation wavelengths can be used to this effect [69]. This approach has been taken up commercially by Bruker in their Bravo instrument [21]. Recently, IPS/Metrohm presented a dual-wavelength excitation and deployment of their own mathematical algorithm to this effect also [70]. Further developments are underway at other Raman instrument vendors, and more are expected in this area in the near future.

Figure 6.18 Representation of an SORS geometry showing "zero offset" versus "offset" laser excitation relative to the detector, with a separation distance ΔS.

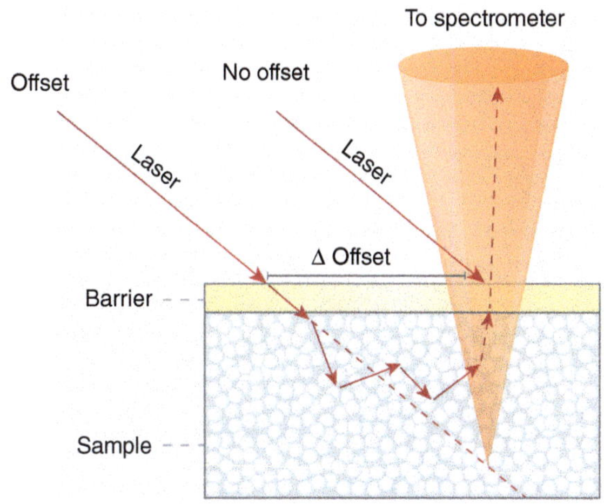

6.12.2 Through Barrier

Sample preparation is time consuming and can sometimes even be dangerous (as in hazmat) or costly (as in pharma), so it is highly advantageous to be able to provide reliable material identification through barriers to avoid the time or risk associated with removing a sample from its container. Matousek et al. first developed the SORS approach [71] subsequently implemented by Cobalt Light Systems [72] and acquired by Agilent, who have deployed the technology in several of their instruments [73]. By comparing measurements in which a single excitation laser at 830 nm is spatially offset to a measurement with no offset as shown in Figure 6.18, it is possible to differentiate the Raman signal of the material under investigation from the barrier or barriers in the near field. Similar approaches involving manually or dynamically adjusting laser spot size and focus distance with respect to the barrier and sample are deployed by other vendors.

In the medical field, Raman measurements through skin or tissue are gaining prevalence, with imaging geometry and excitation wavelength being tailored to the application. One such example of this is RSP Systems' GlucoBeam technology, whereby they are able to reproducibly and accurately determine glucose concentrations through all skin tones [74].

6.12.3 Standoff and Proximity

Traditional Raman instrumentation are point detectors; however, for safety purposes, it may be a requirement to sample at a longer, safer working distance. With this in mind, Pendar has developed a handheld Raman instrument which provides sensing up to 1 m [75], reputed to offer limited fluorescence interference and lower laser energy than many other commercially available options. Metrohm also offers the Mira DS with an attachment designed for remote Raman identification up to 1.5 m [9]. With the increase in unmanned aerial vehicle (UAV) technology and the potential reduction in the payload of miniature Raman instrumentation, future advancements in standoff Raman are possible.

6.12.4 SERS

The intensity of Raman scattering is prohibitively low for a number of desired applications, including the detection of low concentrations in liquid, and of trace materials on surfaces. This limitation may be overcome using SERS, a technique which uses noble metal nanostructures to enhance the Raman signal by a factor of 10^6–10^8 or more [76]. It requires molecules to be adsorbed on a metal surface or "captured" between coalescing nanoparticles in a solution, but has the benefit of minimizing acquisition time and reducing the amount of sample needed [77]. Most portable Raman instruments offer a substrate accessory or can be quite easily adapted for the measurement of SERS-based materials.

One of the challenges in deploying SERS commercially is reproducibility, as the strength of the SERS signal depends strongly on the geometric characteristics of the nanostructures. This necessitates very tight control over fabrication, and the resulting properties, in order to ensure consistent answers and reliable calibration curves for quantitative measurements [78]. SERS materials must be cost-effective and easy to fabricate, or at least reusable so as to keep the cost of assay low. Commercial SERS materials must also be highly stable over time, temperature, and humidity to be a viable solution for mission-critical applications in medicine and defense.

SERS can be performed using solid substrates or colloidal nanoparticles (NP) in solution. Solid substrates are fabricated using lithography, chemical etching, or plasma deposition, and may be composed of very well-defined geometric structures or deposited nanoparticles [79–83]. These substrates offer a general enhancement in signal (the magnitude of which may vary), but are generally not fabricated to look for a specific analyte.

Colloidal NPs offer benefits in reproducibility, tunability of optical properties via particle shape and size, and the benefit of optional surface functionalization to allow use in different environments and for specific proteins, nucleic acids, bacteria, and other biological markers. The ability to conjugate NPs to be specific to a biological target also facilitates multiplexing, as has been demonstrated for bacterial pathogens using a portable Raman system [84]. They can be fabricated cost-effectively and with high reproducibility, showing promise for applications in biomedical, environmental, and food quality applications. As yet, however, few commercial suppliers of colloidal SERS sensors have emerged, though we expect significant growth in this area in the coming years [85].

6.13 Conclusion

Raman spectroscopy has come a long way since C.V. Raman's early experiments hinted at its existence. Seeing how far the technique has come and what it is now able to accomplish, it seems quite incredible that Raman was able to see this "feeble fluorescence" at all.

Bit by bit, technological advances have broken down the barriers to the technique's widespread application, so that the evolution of portable Raman technology reads much like an anthology of the history of applied photonics itself. Each borrowed technology has helped make Raman smaller, more powerful, and more accessible, from lasers developed for Blu-ray players to ion beam-sputtered (IBS) filter technology perfected for telecom, gratings manufactured holographically, and detectors evolved from photocopiers.

In its current incarnation, Raman spectroscopy is capable of yielding definitive answers about identity and even concentration in critical aspects of security, pharma, and health – and the possible applications are only increasing. There will always be tradeoffs between size, cost, sensitivity, and power consumption, but each new appropriation of technology and optimization of design erodes those barriers further, allowing Raman to find its way into new use cases and new fields.

This is important, as the greatest potential for the expanded use of Raman lies in its increased use by nonexperts at the point of need or the point of care. To fully harness the potential of the technique, it is important that we continue to lift the current limitations on sensitivity, fluorescence mitigation, and power levels, while simultaneously driving for even smaller point-and-shoot devices geared to the specific needs of each new application and user base.

What will this look like? It may be a smartphone with integrated Raman, or a device the size of a deck of cards interfaced with cloud computing. One thing is for certain – it will take advantage of the latest technology to deliver "good enough" performance with better-than-ever fit, form, and function.

Acronyms and Abbreviations

ADC	Analog-to-Digital Converter
AFE	Analog Front End
ASE	Amplified Spontaneous Emission
BLE	Bluetooth Low Energy
CCD	Charge Coupled Device
CFD	Computational Fluid Dynamics
CMOS	Complementary Metal Oxide Semiconductor
COTS	Commercial Off-The-Shelf
EEPROM	Electrically Erasable Programmable Read-Only Memory
FDA	US Food and Drug Administration
FPGA	Field Programmable Gate Array
FWHM	Full Width Half Maximum
GUI	Graphical User Interface
IBS	Ion Beam Sputtering
I^2C	Inter-Integrated Circuit protocol for microprocessors
IoT	Internet of Things
IP	Ingress Protection
LCD	Liquid Crystal Display
LED	Light Emitting Diode
M^2 factor	Beam quality factor
MM	Multimode (fiber)
MTBF	Mean Time Between Failure
NA	Numerical Aperture
NP	Nanoparticles
OD	Optical Density
OEM	Original Equipment Manufacturer
QE	Quantum Efficiency
SERS	Surface-Enhanced Raman Spectroscopy
SM	Single-Mode (fiber)
SNR	Signal-to-Noise Ratio
SORS	Spatially Offset Raman Spectroscopy
SPI	Serial Peripheral Interface
SWaP	Size, Weight and Power
TDP	Thermal Design Power
TEC	Thermoelectric Cooler
USB	Universal Serial Bus

References

1 Raman, C.V. (1927). A new radiation. *Journal of Physics* 2: 387.

2 Singh, R. (2002). CV Raman and the discovery of the Raman effect. *Physics in Perspective* 4 (4): 399–420.

3 Koningstein, J.A. (2012). *Introduction to the Theory of the Raman Effect*. Dordrecht: Springer Science & Business Media.

4 Long, D.A. (2002). *The Raman Effect: A Unified Treatment of the Theory of Raman Scattering by Molecules.* Chichester: Wiley.
5 Ferraro, J.R., Nakamoto, K., and Brown, C.W. (2003). *Introductory Raman Spectroscopy.* USA: Elsevier Science.
6 Adar, F. (2016). Introduction to interpretation of Raman spectra using database searching and functional group detection and identification. *Spectroscopy* 31: 16–23.
7 ThermoFisher Scientific. *FirstDefender™ RMX Handheld Chemical Identification* [online]. https://www.thermofisher.com/order/catalog/product/FIRSTDEFENDERRMX#/FIRSTDEFENDERRMX (accessed 13 March 2020).
8 CBEx. *Snowy Range Instruments* [online]. http://www.wysri.com/cbex (accessed 13 March 2020).
9 Metrohm AG. *Mira Handheld Raman Analyzers* [online]. https://www.metrohm.com/en-us/products-overview/spectroscopy/mira-handheld-raman-spectrometers (accessed 13 March 2020)
10 ThermoFisher Scientific. *TruScan™ RM Handheld Raman Analyzer* [online]. https://www.thermofisher.com/order/catalog/product/TSRMTRUTOOLS#/TSRMTRUTOOLS (accessed 13 March 2020).
11 Chemring Group PLC. *PGR-1064® Handheld Raman Spectrometer* [online]. http://www.chemringds.com/what-we-do/sensors-and-information/chemical-and-biological-detection/chemical-detection (accessed 13 March 2020).
12 Field Forensics. *HandyRam II™ Handheld Raman Spectrometer* [online]. https://www.fieldforensics.com/introducing-handyram-ii (accessed 13 March 2020).
13 Smiths Detection. *ACE-ID Hand-Held Non-contact Chemical Identifier* [online]. https://www.smithsdetection.com/products/ace-id (accessed 13 March 2020).
14 Serstech AB. *Serstech 100 Indicator* [online]. https://serstech.com/our-offer/serstech-100-indicator (accessed 13 March 2020).
15 Coda Devices. *CDI 2 Handheld Raman Analyzer* [online]. https://codadevices.com/cdi2 (accessed 13 March 2020).
16 All Safe Industries Inc *ChemPro 100 Indicator Handheld Raman Spectrometer* [online]. https://www.allsafeindustries.com/chempro-100i-handheld-chemical-detector.aspx (accessed 13 March 2020).
17 Rigaku. *Advanced 1064 nm Handheld Raman Spectrometer for Raw Material Identification* [online]. https://www.rigaku.com/products/raman/progeny (accessed 13 March 2020).
18 B&W Tek. *Handheld Raman Spectrometers* [online]. https://bwtek.com/technology/handheld-raman (accessed 13 March 2020).
19 Agiltron. *PinPointer™ Handheld Raman Spectrometer* [online]. https://agiltron.com/product/handheld-raman-spectrometer (accessed 13 March 2020).
20 Anton Paar. *Handheld Raman Spectrometer: Cora 100* [online]. https://www.anton-paar.com/us-en/products/details/handheld-raman-spectrometer-cora-100 (accessed 13 March 2020).
21 Bruker. *BRAVO Handheld Raman Spectrometer* [online].https://www.bruker.com/products/infrared-near-infrared-and-raman-spectroscopy/raman/bravo.html (accessed 13 March 2020).
22 Agilent. *Resolve Handheld Through-Barrier Identification System* [online]. https://www.agilent.com/en/products/raman-spectroscopy/raman-spectroscopy-systems/handheld-chemical-identification/resolve (accessed 13 March 2020).
23 Agilent. *RapID Raw Material ID Verification System* [online]. https://www.agilent.com/en/products/raman-spectroscopy/raman-spectroscopy-systems/pharmaceutical-analysis/rapid (accessed 13 March 2020).
24 SciAps. *CHEM-200 Raman Analyzer* [online]. https://www.sciaps.com/raman-spectrometers/chem-200 (accessed 13 March 2020).
25 Optosky. *What is Handheld Raman?* [online]. https://optosky.com/handheld-raman.html (accessed 13 March 2020).
26 CloudMinds. *Smart Raman XI* [online]. http://airaman.com (accessed 13 March 2020).

27 Wasatch Photonics. *High Performance Raman* [online]. https://wasatchphotonics.com/product-category/spectrometers/raman (accessed 13 March 2020).

28 B&W Tek. *Portable Raman Spectrometers* [online]. https://bwtek.com/raman-technology/portable (accessed 13 March 2020).

29 Horiba Scientific. *AnywhereRaman* [online]. https://www.horiba.com/us/en/scientific/products/raman-spectroscopy/raman-spectrometers/benchtop-and-portable-raman/details/anywhereraman-tm-33805 (accessed 13 March 2020).

30 Agiltron. *Benchtop Raman Spectrometers* [online]. https://agiltron.com/category/raman-chemical-detection-spectrometers/benchtop (accessed 13 March 2020).

31 StellarNet Inc. *Raman Spectrometers Lasers and Probes* [online]. https://www.stellarnet.us/systems/raman-spectrometers-lasers-and-probes (accessed 13 March 2020).

32 Tornado Spectral Systems. *HyperFlux™ PRO Plus Raman Spectroscopy System* [online]. https://tornado-spectral.com/solutions/hyperflux/https-tornado-spectral-com-solutions-hyperflux-hyperflux-pro-plus (accessed 13 March 2020).

33 Ocean Insight. *QE Pro Raman Series* [online]. https://www.oceaninsight.com/products/spectrometers/raman/qepro-raman-series (accessed 13 March 2020).

34 Hamamatsu. *Mini-spectrometers* [online]. https://www.hamamatsu.com/us/en/product/photometry-systems/mini-spectrometer/index.html (accessed 13 March 2020).

35 Ibsen Photonics. *FREEDOM Raman* [online]. https://ibsen.com/products/oem-spectrometers/freedom-spectrometers/freedom-raman (accessed 13 March 2020).

36 Imec. *Handheld Raman Spectrometer Technology* [online]. https://www.imec-int.com/en/imaging-vision-systems/handheld-raman-spectrometer-technology (accessed 13 March 2020).

37 IECEE (1989). *Degrees of Protection Provided by Enclosures (IP Code).* https://www.iecee.org/certification/overview/ (accessed 22 September 2020).

38 Angel, S.M., Carrabba, M., and Cooney, T.F. (1995). The utilization of diode lasers for Raman spectroscopy. *Spectrochimica Acta Part A: Molecular and Biomolecular Spectroscopy* 51 (11): 1779–1799.

39 Pan, M.-W., Benner, R.E., and Smith, L.M. (2006). Continuous lasers for Raman spectrometry. In: *Handbook of Vibrational Spectroscopy*. Chichester: Wiley.

40 Carrabba, M.M. and Rauh, R.D. (1992). Apparatus for measuring Raman spectra over optical fibers. United States Patent 5,112,127, 12 May.

41 Scheeline, A. (2017). How to design a spectrometer. *Applied Spectroscopy* 71 (10): 2237–2252.

42 Newport Corporation. *Richardson Gratings™: Plane Ruled Reflection Gratings* [online]. https://www.gratinglab.com/Products/Product_Tables/T2.aspx (accessed 13 March 2020).

43 Kruschwitz, J.D. (2006). From small fish to oceans of opportunity: the story of Ocean Optics Inc. *Optics & Photonics News* 17 (2): 10–11.

44 Williamson, J.M., Bowling, R.J., and McCreery, R.L. (1989). Near-infrared Raman spectroscopy with a 783-nm diode laser and CCD array detector. *Applied Spectroscopy* 43 (3): 372–375.

45 Wang, Y. and McCreery, R.L. (1989). Evaluation of a diode laser/charge coupled device spectrometer for near-infrared Raman spectroscopy. *Analytical Chemistry* 61 (23): 2467–2651.

46 Pommier, C.J., Walton, L.K., Ridder, T.D. et al. (2006). *Array Detectors for Raman Spectroscopy*. Chichester: Wiley.

47 Heintz, R. (2019). *Back Illuminated vs. Front Illuminated CCD-Based Imaging Sensors and How it Impacts Raman Spectra, White Paper WP53197*. ThermoFisher Scientific.

48 Saylor, J. (2019). *SiOnyx, LLC v. Hamamatsu Photonics K.K.*, No. CV 15-13488-FDS (D. Mass. 25 July 2019). https://www.jdsupra.com/legalnews/massachusetts-patent-litigation-wrap-up-48739.

49 Hamamatsu. *InGaAs Linear Image Sensors* [online]. https://www.hamamatsu.com/us/en/product/optical-sensors/image-sensor/ingaas-image-sensor/ingaas-linear-image-sensor/index.html (accessed 13 March 2020).

50 Sensors Unlimited. *Linear Photodiode Arrays* [online]. http://www.sensorsinc.com/products/focal-plane-arrays (accessed 13 March 2020).

51 Arm Limited. *Arm Processors for the Widest Range of Devices – from Sensors to Servers* [online]. https://www.arm.com/products/silicon-ip-cpu (accessed 2 April 2020).

52 ASTM E1840-96 (2014). *Standard Guide for Raman Shift Standards for Spectrometer Calibration*. West Conshohocken, PA: ASTM International.

53 The United States Pharmacopeia Convention (2020). *USP 43-NF 38 General Chapter <1120>*.

54 European Directorate for the Quality of Medicines & Healthcare (2016). *European Pharmacopeia 8.7 Edition* 2016, *Chapter 2.2.48*.

55 Rodriguez, J.D., Westenberger, B.J., Buhse, L.F. et al. (2011). Standardization of Raman spectra for transfer of spectral libraries across different instruments. *Analyst* 136 (20): 4232–4240.

56 ASTM E2911-13 (2013). *Standard Guide for Relative Intensity Correction of Raman Spectrometers*. West Conshohocken, PA: ASTM International.

57 Cadusch, P.J., Hlaing, M.M., Wade, S.A. et al. (2013). Improved methods for fluorescence background subtraction from Raman spectra. *Journal of Raman Spectroscopy* 44: 1587–1595.

58 Chi, M., Han, X., Xu, Y. et al. (2019). An improved background-correction algorithm for Raman spectroscopy based on the wavelet transform. *Applied Spectroscopy* 73 (1): 78–87.

59 McCreery, R.L. (2006). Photometric standards for Raman spectroscopy. In: *Handbook of Vibrational Spectroscopy*. Chichester: Wiley Online Library. https://onlinelibrary.wiley.com/doi/abs/10.1002/0470027320.s0706.

60 S.T. Japan USA LLC. *Spectra Databases* [online]. http://www.stjapan-usa.com/spectradb.html (accessed 13 March 2020).

61 Wiley Science Solutions. Know It All Spectroscopy Edition Software [online]. https://sciencesolutions.wiley.com/knowitall-spectroscopy-software/ (accessed December 2020).

62 Wilcox, P.G. and Guicheteau, J.A. (2018). Comparison of handheld Raman sensors through opaque containers. *Proceedings of SPIE 10629 Chemical Biological Radiological Nuclear and Explosives (CBRNE) Sensing XIX*, Orlando, FL (16 May 2018).

63 Yaghoobi, M., Wu, D., Clewes, R.J. et al. (2016). Fast sparse Raman spectral unmixing for chemical fingerprinting and quantification. *Optics and Photonics for Counterterrorism, Crime Fighting, and Defence XII, International Society for Optics and Photonics*, vol. 9995.

64 Varmuza, K. and Filzmoser, P. (2016). *Introduction to Multivariate Statistical Analysis in Chemometrics*. Boca Raton, FL: CRC press.

65 U.S. FDA (2003). CFR part 11: electronic records. *Electronic Signatures* 21: 1–12.

66 U.S. FDA (2019). 21CFR1040.10, performance standards for light emitting products. *Code of Federal Regulations* 8.

67 Wei, D., Chen, S., and Liu, Q. (2015). Review of fluorescence suppression techniques in Raman spectroscopy. *Applied Spectroscopy Reviews* 50 (5): 387–406.

68 Winfield, G. (2019). *Over the Spectral Rainbow: Christina Baxter, CEO at Emergency Response TIPS, Talks to Gwyn Winfield about the Road to Raman Riches*. CBRNe World. https://cbrneworld.com/index.php?option=com_content&view=article&id=1335%3Aover-the-spectral-rainbow&catid=22%3Ajuly-2019 (accessed 13 March 2020).

69 Cooper, J.B., Abdelkader, M., and Wise, K.L. (2013). Sequentially shifted excitation Raman spectroscopy: novel algorithm and instrumentation for fluorescence-free Raman spectroscopy in spectral space. *Applied Spectroscopy* 67 (8): 973–984.

70 Chimenti, R. (2017). *Dual Wavelength Applications in Portable Raman Spectroscopy*. Reno, NV: SciX.

71 Matousek, P., Clark, I.P., Draper, E.R. et al. (2005). Subsurface probing in diffusely scattering media using spatially offset Raman spectroscopy. *Applied Spectroscopy* 59 (4): 393–400.

72 Parker, W. (2018). Scanner for spatially offset Raman spectroscopy. US Patent 9,880,099.

73 Agilent. *Spatially Offset Raman Spectroscopy (SORS)* [online]. https://www.agilent.com/en/technology/spatially-offset-raman-spectroscopy (accessed 13 March 2020).

74 Lundsgaard-Nielsen, S.M., Pors, S.M., and Banke, S.O. (2018). Critical-depth Raman spectroscopy enables home-use non-invasive glucose monitoring. *PLoS One* 13 (5): 1–11.

75 Pendar Technologies. *Pendar X10* [online]. https://www.pendar.com/products/pendar-x10 (accessed 13 March 2020).

76 Haynes, C.L., McFarland, A.D., and Van Duyne, R.P. (2005). Surface-enhanced Raman spectroscopy. *Analytical Chemistry* 77 (17): 338A–346A.

77 Langer, J., Jimenez de Aberasturi, D., Aizpurua, J. et al. (2020). Present and future of surface-enhanced Raman scattering. *ACS Nano* 14 (1): 28–117.

78 Fan, M., Andrade, G.F., and Brolo, A.G. (2019). A review on recent advances in the applications of surface-enhanced Raman scattering in analytical chemistry. *Analytica Chimica Acta* 1097: 1–29.

79 Silmeco. *SERS Substrates – SERStrate* [online]. https://www.silmeco.com/products/sers-substrate-serstrate (accessed 13 March 2020).

80 Hamamatsu. *SERS Substrate J12853* [online]. https://www.hamamatsu.com/us/en/product/type/J12853/index.html (accessed 13 March 2020).

81 Integrated Optics UAB. *SERS Substrates* [online]. https://integratedoptics.com/products/sers-substrates (accessed 13 March 2020).

82 SERSitive *SERSitive Substrates* [online]. https://www.sersitive.eu (accessed 13 March 2020).

83 Ocean Insight. *SERS Substrates* [online]. https://www.oceaninsight.com/products/sampling-accessories/raman/sers (accessed 13 March 2020).

84 Kearns, H., Jamieson, L., Graham, D. et al. (2019). Rapid portable pathogen detection with multiplexed SERS-based nanosensors. *Spectroscopy Raman Technology* 34: 20–31.

85 Real-Time Analyzers. *SERS Products* [online]. http://www.rta.biz/products/sers-products (accessed 13 March 2020).

7

Optical Filters – Technology and Applications

Oliver Pust

Delta Optical Thin Film A/S, Hørsholm, Denmark

7.1 Overview on the Use of Optical Filters in Spectroscopy

As the title of this chapter suggests, we look at the use of optical filters in spectroscopy. The use of optical filters includes spectrometers that detect electromagnetic radiation from the ultraviolet (UV) to the mid-infrared (MIR) part of the spectrum. It excludes gamma-ray spectrometers, X-ray spectrometers, and microwave spectrometers but also mass spectrometers. Sometimes, the term *optical* is used as a synonym for light that is visible to the human eye (380 to 780 nm)[1] . However, this definition seems too narrow, and we want to consider the above-mentioned larger part of the electromagnetic spectrum. Another method that we do not consider here is polarisation spectroscopy. Although such spectrometers might contain filters, they are not elemental to the method.

7.1.1 Types of Spectrometers

Spectrometers can be differentiated using different parameters. One of these parameters is size or portability. The use of optical filters in large or small spectrometers can be very different. For this reason, we divide spectrometers into three categories: bench-top, portable, and hand-held spectrometers. As the book focuses on portable spectroscopy, the use of optical filters in bench-top spectrometers will be discussed only briefly. Portable spectrometers inherently comprise hand-held spectrometers as well. However, the type and purpose of optical filters in these two types of spectrometers are typically very different. For this reason, they will be discussed in different sections.

7.1.1.1 Bench-top Spectrometers

Bench-top spectrometers are large, complex instruments that are designed for high performance and high flexibility. Typically, they can perform both excitation and emission scanning. For an emission scan, the excitation or illumination wavelength is kept fixed and the emission wavelength of the sample is recorded over a wider spectrum of interest either by a rotating grating or other filtering devices. For an excitation scan, this procedure is repeated for many different excitation wavelengths that are either originating from a number of discrete laser lines or light-emitting diodes (LEDs), or by filtering a broadband light source such as halogen or deuterium lamps or more recently supercontinuum lasers or laser-ignited plasma sources.

1 There are many species such as birds, bees, or mantis shrimps that can see UV radiation invisible to the human eye.

Portable Spectroscopy and Spectrometry 1: Technologies and Instrumentation, First Edition.
Edited by Richard A. Crocombe, Pauline E. Leary and Brooke W. Kammrath.

7.1.1.2 Portable Spectrometers

Portable or miniature spectrometers are certainly the most used spectrometer types and are produced in quantities of many ten thousands per year worldwide (Tematsys reports 40,000 U sold in 2016 and projects 170,000 U in 2021, [16]). They are most often fibre-coupled, use a grating as dispersive element and a charge-coupled device (CCD) or complementary metal–oxide–semiconductor (CMOS) line scan array as detector. Miniature spectrometers can be used as stand-alone systems connected to a personal computer (PC) or laptop or as integrated part of a larger spectrometer system. The key in both cases is that there are no moving parts and that the spectrum is measured instantaneously without scanning.

7.1.1.3 Hand-held Spectrometers

Hand-held or micro-spectrometers are the latest step in miniaturisation. Their volume is only of the order of a few cubic centimetres or even smaller. These days they find their way into smartphones and wearables for consumer applications. Possible use cases are checking the ripeness of fruit, the freshness and unaltered condition of meat or the blood oxygen concentration.

7.1.2 Types of Optical Filters

Optical filters are often overlooked but nevertheless important or even crucial key components in every optical instrument – especially in spectrometers. Out of personal experience working in an optical filter company, the author can confirm that there is a widespread and increasing tendency to consider optical filters at late stages during the development process of a new optical instrument, only to find out that filters with the desired requirements cannot or can only be designed or produced at prohibitive cost.

A good optical measurement almost always starts with a high-performance optical filter.[2] Therefore, it is necessary to understand optical filters and consider them in the design process from the very beginning. It is also a good idea to involve the filter manufacturer of choice early on. They might have valuable insight and tips to prevent blind roads and unnecessary cost.

Not unexpectedly, there are several types of optical filters. And each type has its own features, benefits, and disadvantages. In the following, we will consider coloured glass or absorptive filters, liquid crystal tunable filters (LCTFs), acousto-optic tunable filters (AOTFs), and thin-film or interference filters. Our focus will lie on thin-film filters because they are by far the most used filters.

7.1.2.1 Coloured Glass Filters

The oldest type of filters are coloured glass filters. These consist of glass that is on purpose doped with metallic salts. These absorb certain wavelengths, which makes the glass appear in the complementary colour when a white light source such as the sun is viewed through the glass. It is not known who invented (coloured) glass, but it is known that production of glass including colouring was known in ancient Egypt and Phoenicia. There is evidence, which suggests that glass was already in use as far back as the fourteenth century BC [6].

The major advantage of coloured glass filters is their insensitivity to the angle of incidence (AOI) of incoming light. The spectral characteristic does not change with AOI or opening angle (OA). This makes them ideal for applications with large angles or a large field of view (FOV). Another advantage is that undesired radiation is absorbed in the filter and not reflected within the instrument, where it could propagate uncontrolled and decrease the quality of the measurement signal. At the same time, the fact that undesired light is absorbed limits the use of coloured glass filters to low light intensities. They cannot be used as filters for strong light sources because they might heat up and be destroyed quickly.

Another drawback is their low resistance to temperature and humidity. Colored glass filters degrade over time and compromise the instrument's performance. Even protective coatings cannot fully prevent this process. As

2 Just consider the path of light or radiation into and through an optical instrument.

degradation is a slow and creeping process, it is typically only discovered when the adverse effects become strong. As opposed to instruments equipped with modern hard-coated thin-film interference filters, instruments with coloured glass filters require regular check-ups and periodic exchange of the filters, which incurs increased cost of ownership and potentially negative user experience.

While coloured glass filters can have deep blocking depending on their thickness, it is not possible to freely design their spectral performance. Deeper blocking goes hand in hand with reduced transmission. Their cut-on or centre wavelengths are determined by the metal salts, and the selection is limited. Last but not least, some materials of coloured glass filters are covered by RoHS[3] (Restriction of Hazardous Substances) regulations, and although there are exemptions for now, this might not be the case in the future. Many manufacturers of optical instruments have preemptively begun to exchange coloured glass filters with thin-film filters. This is not always possible without at least a partial redesign of the instrument due to the different characteristics of the two filter types.

7.1.2.2 Liquid Crystal Tunable Filters

LCTFs use electrically controlled liquid crystal elements to select a specific wavelength of light for transmission through the filter while excluding others based on wavelength-dependent polarisation created in birefringent crystals. This type of filter is ideal for use with imaging devices, because it offers a large clear aperture, excellent imaging quality, and a simple linear optical path. A typical LCTF is constructed from a stack of fixed filters consisting of interchanging birefringent crystal/liquid crystal combinations and linear polarisers. The spectral region passed by LCTFs depends on the choice of polarisers, optical coatings, and the liquid crystal characteristics.

7.1.2.3 Acousto-optic Tunable Filters

An AOTF uses an acousto-optical crystal that consists of an optically anisotropic crystal of tellurium dioxide, lithium niobate, or quartz that is coupled to a piezoelectric crystal that acts as an acoustic transducer. When a radio frequency (RF) signal in the range of 150–300 MHz is applied to the piezoelectric crystal, an ultrasonic sound wave is generated that travels through the optical crystal and changes its refractive index. The change in refractive index changes the angles in which light rays are diffracted. Different wavelengths can be selected at different angles at the exit side of the crystal.

7.1.2.4 MEMS-based and Piezo-Actuated Fabry–Perot Tunable Filters

VTT has developed tunable filters that are based on the Fabry–Perot interferometer principle [2]. The length of a cavity (here air gap) between two mirrors determines the wavelength that is transmitted through the Fabry–Perot interferometer. When the cavity length is changed, the transmitted wavelength changes as well and a tunable filter is achieved. The micro-electro-mechanical systems (MEMS)-based Fabry–Perot tunable filter can be mass-produced and is applicable to small sensors such as point detectors or imaging sensors in smart phones, while the piezo-actuated Fabry–Perot tunable filter can be produced with larger clear apertures. A general limitation of the Fabry–Perot approach is the small tuning range of the filters. In the first case, it is limited by the free spectral range and in the latter case to approximately $\pm 10\%$ around the centre wavelength.

7.1.2.5 Thin-Film Filters

While the phenomenon of constructive and destructive interference of light waves at the surfaces of thin films has been known for several centuries, it was only in 1939 that Walter H. Geffcken created the first interference filters using dielectric coatings [8].

3 Restriction of the Use of certain Hazardous Substances in Electrical and Electronic Equipment (RoHS-Directive 2011/65/EU).

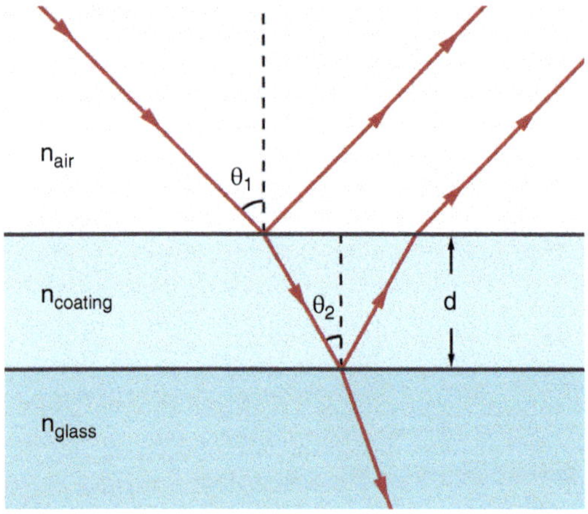

Figure 7.1 A wave of light reflecting off the upper and lower boundaries of an anti-reflection coating on glass. The reflected waves interfere with one another.

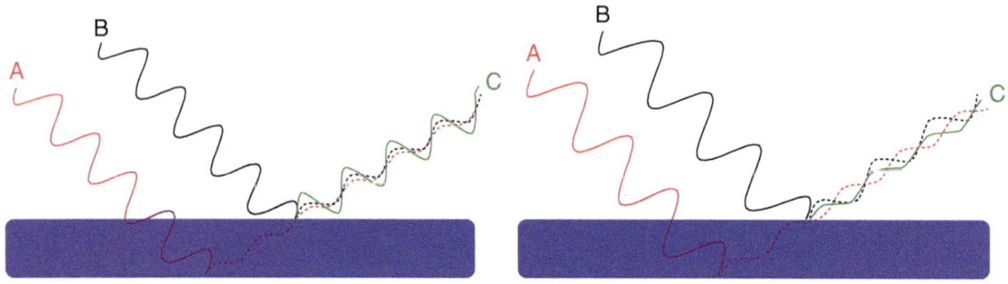

Figure 7.2 Constructive and destructive interference of two light waves that are partially reflected at the interfaces of a thin film. (Jhbdel [BY-SA https://creativecommons.org/licenses/by-sa/3.0]).

7.1.2.5.1 Principle of Thin-Film Interference

Thin-film interference occurs when light waves reflected off boundaries between thin films of materials with different refractive indices interfere with one another (see Figure 7.1). In this process, the intensity of the resulting light wave can either be enhanced (constructive interference) or attenuated (destructive interference).

When the film thickness is an odd multiple of one-quarter wavelength of the light wave, the reflected waves from both surfaces interfere destructively (see Figure 7.2 right). As the wave can no longer be reflected, it is completely transmitted through the thin film. When the film thickness is a multiple of a half-wavelength of the light, the two reflected waves interfere constructively, enhancing the reflection and attenuating the transmission (see Figure 7.2 left).

An odd-numbered sequence of layers of quarter-wave thickness with alternating high and low refractive index is called a quarter-wave stack and is the basic building block of any filter type. The structure reflects strongly over a region around that wavelength for which the layers are quarter waves, usually known as the reference wavelength. By combining quarter-wave stacks with different reference wavelengths and modifying them by modern methods such as the needle optimisation technique [17], it becomes possible to design filters with almost any desired spectral performance – manufacturability, especially economical, is a separate topic though and the subject of ongoing research and technical development. For an in-depth discussion of thin-film filter technology, refer to [12].

7.1.2.5.2 Angular Sensitivity of Thin-film Filters

A well-known but nevertheless often neglected inherent property of thin-film filters is their angular sensitivity. The wavelengths, at which constructive and destructive interference occurs, depend on the optical path lengths through the media or the phase thicknesses in a multilayer coating. It is obvious that the phase thickness Φ changes with the AOI:

$$\Phi = \frac{2\pi}{\lambda} n(\lambda) d \cos\theta \tag{7.1}$$

where

λ is the wavelength of light,
$n(\lambda)$ is the refractive index of the layer,
d is the physical thickness of the layer,
θ is the angle of light propagating in the layer.

The angle of light in the layer is linked to the AOI in air γ by Snell's law:

$$n \sin\theta = \sin\gamma \tag{7.2}$$

Hence,

$$\cos\theta = \sqrt{1 - \left(\frac{\sin\gamma}{n}\right)^2} \tag{7.3}$$

The coating mostly consists of a high number of layers, alternating between two materials with a high refractive index n_H and low refractive index n_L. However, it is acceptable to introduce an effective refractive index, representing the whole multilayer structure:

$$n_{eff} \approx \sqrt{n_H \times n_L} \tag{7.4}$$

The coating treated in Figure 7.3 is made of Ta_2O_5 and SiO_2. At a wavelength of 720 nm, those materials have refractive indices of approximately 2.1085 and 1.4771, giving an effective refractive index of 1.7648.

Assuming a wavelength of 721.5 nm at normal incidence (the black curve at the T = 50 % point), we may calculate the following values for the rest of the angles shown in Figure 7.3 with the formula:

$$\lambda_\theta = \lambda_0 \frac{\sqrt{n^2 - \sin^2\theta}}{n} \tag{7.5}$$

$\lambda(\theta = 5°) = 720.6$ nm, $\lambda(\theta = 10°) = 718.0$ nm,
$\lambda(\theta = 15°) = 713.7$ nm, $\lambda(\theta = 20°) = 707.8$ nm,
$\lambda(\theta = 25°) = 700.5$ nm, $\lambda(\theta = 30°) = 691.9$ nm.

The values fit the curves quite well. However, polarisation effects clearly cause a severe distortion of the edge at angles of incidence larger than 20°. It is recommended to keep angles of incidence lower than 15°. The curves shown in Figure 7.3 tell why.

A similar effect is caused by cone half-angle (CHA) or OA, where OA = 2 CHA. The effects of a non-collimated beam passing through an interference filter can be considered as averaging the effects of angles of incidence across a certain angular range. Figure 7.4 shows the passband of a nominally 17-nm-wide (full width at half maximum (FWHM)) bandpass filter. With increasing OA, the centre wavelength shifts toward shorter wavelengths, the edges of the passband become less steep, and as a consequence peak transmission drops because the edges cross each other earlier. It is noteworthy though that FWHM increases only slightly from 17 to 20 nm. It is not recommendable to use an OA larger than ±16° in most cases, as the effects of increasing OAs grow progressively. The same is true for the effects of AOIs.

The blue-shift with increasing AOI can be exploited to achieve wavelength tunability in an optical system. By tilting the filter in the beam path, a bandpass filter can be tuned to a different centre wavelength.[4] However, as

4 This effect is also used to compensate for the production tolerances within a batch of extremely narrow bandpass or Raman long-wave pass filters.

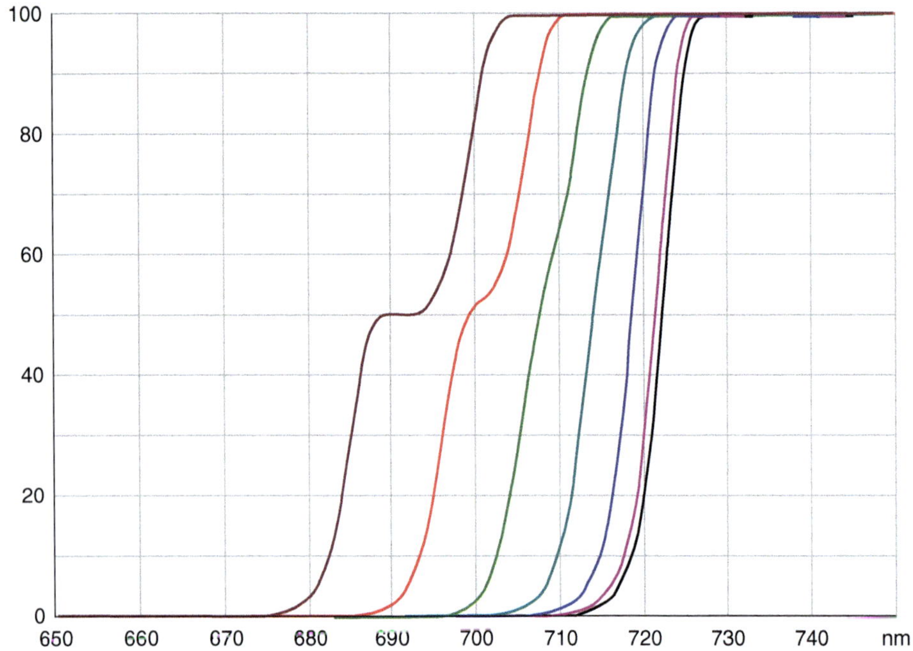

Figure 7.3 Angular dependency of a long-wave pass filter in steps of 5° from AOI = 0° (black curve) to AOI = 30° (brown curve). Source: Courtesy of Delta Optical Thin Film A/S.

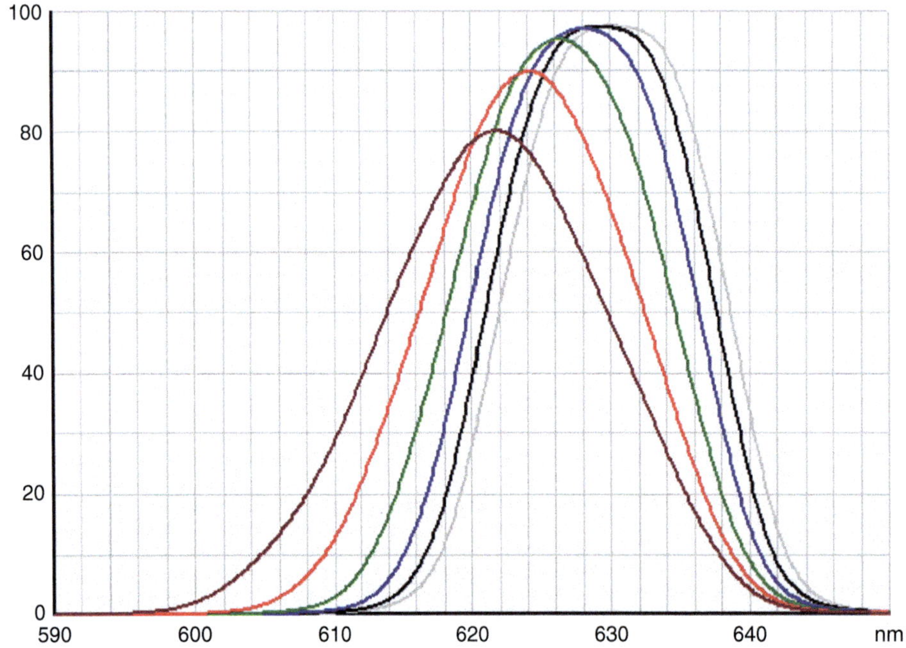

Figure 7.4 Predicted bands formed for opening angles (OA) of ±0° (light-grey), ±8° (black curve) ±12° (blue), ±16° (green), ±20° (red), ±24° (brown). The centre wavelength of the band shifts toward shorter wavelengths as the OA increases, and the band becomes less box-shaped. Source: Courtesy of Delta Optical Thin Film A/S.

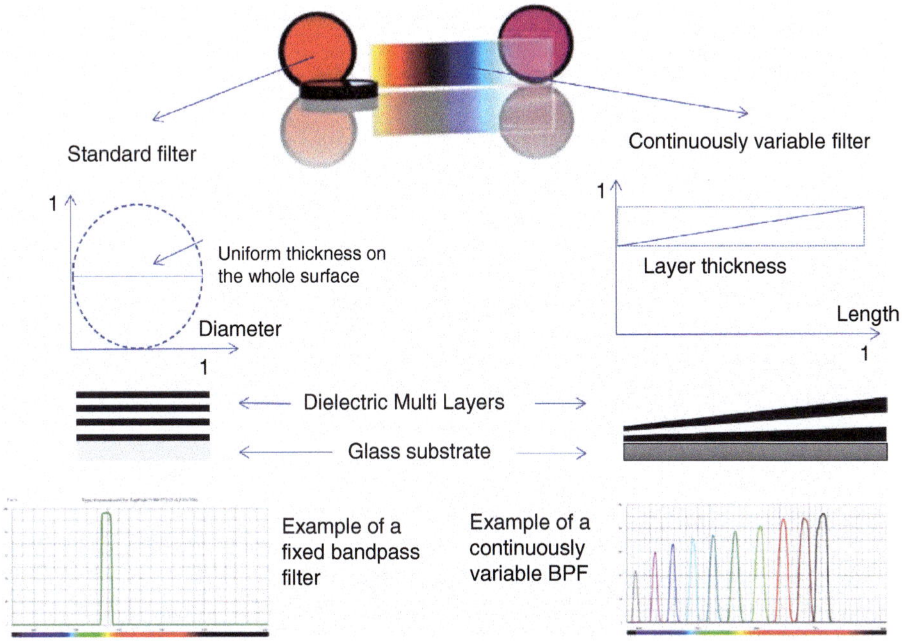

Figure 7.5 Construction principles of standard fixed and continuously variable filter.

shown above, the tuning range is limited to a few tens of nanometres before the adverse effects become too large. An elegant and purposeful way of realising tunable filters is described in the next paragraphs.

7.1.2.5.3 Continuously Variable Filters

For many years, interference filters were produced as fixed wavelength, that is spatially homogeneous filters. Irrespective of the specific type of filter (short-wave pass, long-wave pass, or bandpass filter), the goal of filter production was and is to yield filters that have a uniform spectral characteristic across their clear aperture and a high percentage of identical filters within a batch.

The centre wavelength of a bandpass filters depends on the optical thickness of the individual thin-film layers within the coating stack. As such, it is an obvious idea to create a variable filter by changing the total thickness of the stack continuously in one spatial dimension of a filter (Figure 7.5).

The thickness and wavelength gradients can be established along a circle segment (circular variable filter) or a straight line (often called linear variable filter). In both cases, the spectral characteristics change *continuously* along one spatial dimension of the filter. These filters are not segmented (butcher block filters) but are truly continuous. The word *linear* in linear variable filter must not be confused with the property of wavelength dispersion. The dependency between wavelength and position on the filter is not necessarily linear but can, for example, also on purpose be exponential (see Figure 7.6).

Like traditional filters with fixed wavelength characteristics, continuously variable filters (CVFs) can be produced as edge or bandpass filters and as dichroic beamsplitters as well. Although CVFs have been known and manufactured for more than a couple of decades, their spectral performance reached the necessary level for spectroscopic applications and durability for use with strong light sources only in recent years. Early CVFs were often produced with coloured glass or included thin metal layers (induced transmission filters) to achieve out-of-band blocking. This limited transmission, lifetime, and laser-induced damage threshold (LIDT).

Advances in design and production technology made it possible to produce CVFs that reach the same spectral performance as traditional fixed-wavelength filters. Modern CVFs reach transmission and blocking levels, and

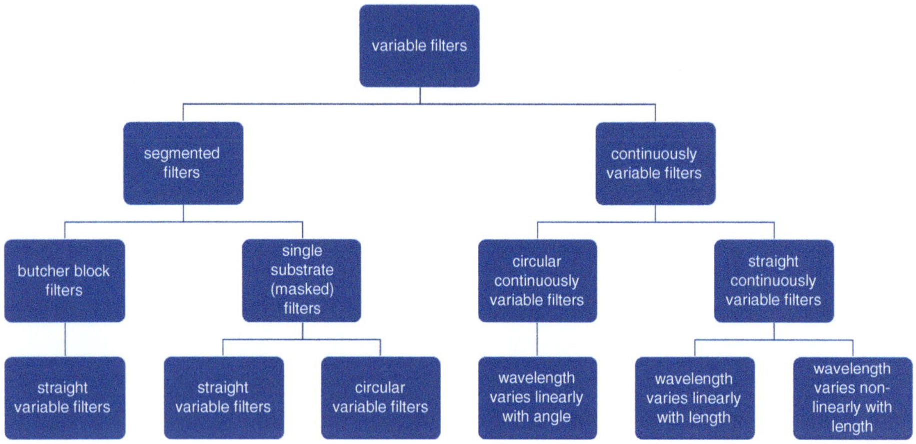

Figure 7.6 Classification and nomenclature of variable filters (author's suggestion).

edge steepness that are sufficient for demanding fluorescence measurements. Figures 7.7 and 7.8 show the performance of continuously variable long-wave pass filters (CVLWPs) and continuously variable short-wave pass filters (CVSWPs). Each of the filters can be used separately. Combining CVLWPs and CVSWPs enables the construction of bandpass filters that can be tuned continuously with centre wavelengths across several hundred nanometres, with the added benefit of tunable bandwidth. As CVF monochromators, these filters are for example used in fluorescence microplate readers or laser scanning microscopes.

The filters offer blocking better than OD4[5] over the complete reflection range of several hundred nanometres. It is possible to increase the blocking to beyond OD6 by placing two identical CVF in series. The filters are coated on single fused silica substrates for minimal autofluorescence and high LIDT, which makes it possible to use them as wavelength selector for high-power broadband light sources such as supercontinuum lasers. Examples of the use of CVFs will be described in Sections 7.2.3, 7.3, 7.4.3, 7.4.4.

7.2 Optical Filters as Auxiliary Filters

Optical filters have two main uses in spectroscopy. They can either be the primary wavelength-selective elements or serve as auxiliary filters for different purposes. In the latter use case, they do not replace gratings but still perform important additional functions in a spectrometer.

7.2.1 Neutral Density Filters

Neutral density (ND) filters are used to control the light's intensity before it reaches the detector. Sometimes, the light intensity is too high and will saturate the sensor. To prevent saturation, an ND filter is inserted into the light path to decrease the light intensity to a level close to the saturation level. This is achieved by selecting the ND filter that has the optimal extinction ratio from a series of different filters, or using a variable ND filter. The word neutral refers to the ideal property of an ND filter to offer a constant extinction ratio across a certain wavelength range. However, in practice, there will be some variation of the extinction ratio with wavelength that should be taken into account.

5 OD stands for optical density and is a quantitative measure expressed as a logarithmic ratio between the radiation falling upon a material I_0 and the radiation transmitted through a material I_1, more specifically $\mathrm{OD} = -\log_{10}(I_1/I_0)$.

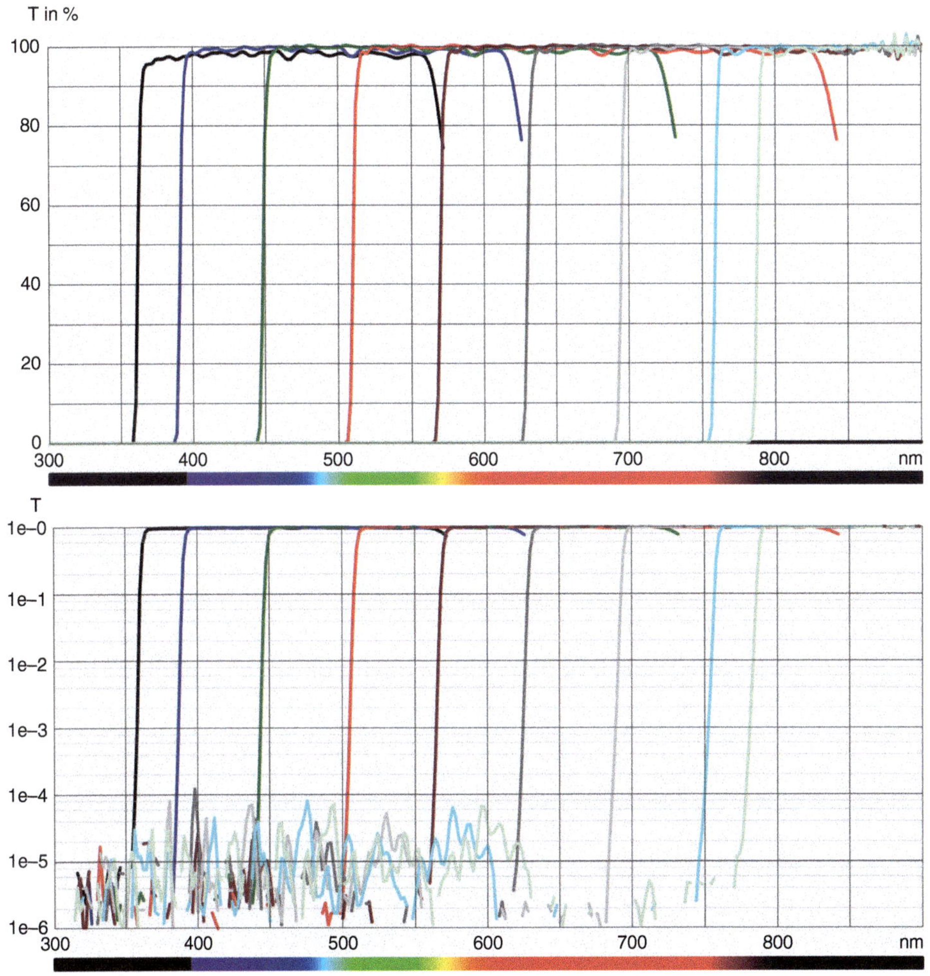

Figure 7.7 Measured transmission and blocking performance of modern continuously variable long-wave pass filter. The spectra were measured with a PerkinElmer 900 spectrophotometer using a 200 μm narrow slit at nine discrete locations to demonstrate the spectral properties. Given the continuous nature of the filter, any other cut-on wavelength can be found at intermediate positions. Source: Courtesy of Delta Optical Thin Film A/S.

7.2.2 Optical Filters as CCD/CMOS Equalizer Filters

Silicon-based photo detectors detect radiation between approximately 200 and 1200 nm with their maximum sensitivity between 500 and 600 nm. The spectral sensitivity curve can be optimised for high UV or near-infrared (NIR) sensitivity at the expense of the overall sensitivity or sensitivity in other wavelength ranges. For many applications, it is desired to work with a spectral sensitivity curve that spectrally is as flat as possible. This can be achieved by equaliser filters, also called gain flattening filters.

Figure 7.9 shows a typical curve of such a filter. In wavelength regions where the detector has low sensitivity (UV and NIR), the filter has high transmission. As the sensitivity of the detector increases from UV and NIR toward its maximum, the transmission level of the filter gradually decreases, so that the product of sensor sensitivity and filter transmission becomes flat.

Ideally, the filter curve would follow the inverse of the sensor sensitivity curve (see Figure 7.10, orange curve). The result would be a combined response curve that is flat in the whole specified wavelength range (see Figure 7.10,

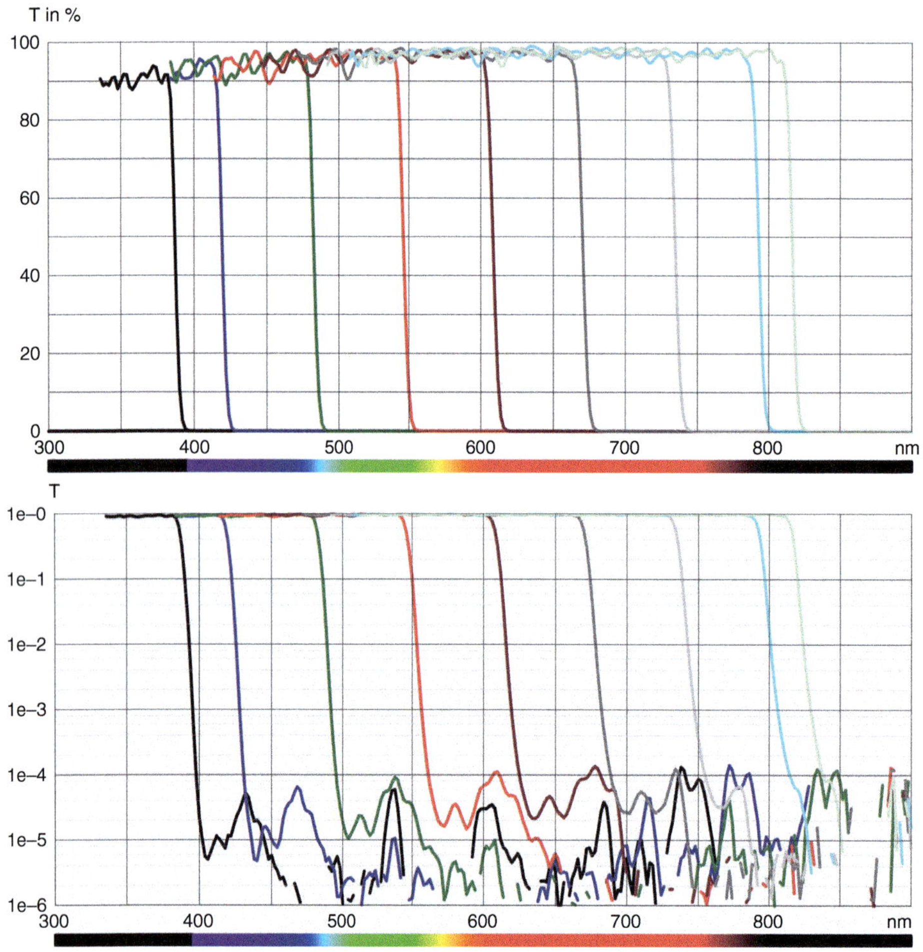

Figure 7.8 Measured transmission and blocking performance of modern continuously variable short-wave pass filter. The spectra were measured with a PerkinElmer 900 spectrophotometer using a 200 μm narrow slit at nine discrete locations to demonstrate the spectral properties. Given the continuous nature of the filter, any other cutoff wavelength can be found at intermediate positions. Source: Courtesy of Delta Optical Thin Film A/S.

green curve). An undesired consequence would be that the filter transmission would drop below 10 % where the sensor sensitivity is high to compensate for the low sensitivity in UV and NIR.

In practice, a compromise is made between overall flatness and combined response level (see Figure 7.11). The filter transmission remains high where the sensor sensitivity has its highest gradients. In the example shown here, the filter transmission is allowed to drop to 30 %, which in turn keeps the combined response flat between 240 and 950 nm. Other compromises are possible with different filter designs. It is noteworthy that this approach requires that arbitrary filter transmission characteristics can be designed and produced. Both has become possible in recent years with new design techniques for synthesising filter performance, and coating technologies that allow very accurate control of the thickness of individual coating layers.

7.2.3 Optical Filters as Order-Sorting Filters

The wavelength dispersive element in most mini-spectrometers is a grating. Besides the primary wavelengths (λ_0) that are to be detected, gratings also generate wavelengths at higher orders n. These higher-order wavelengths are

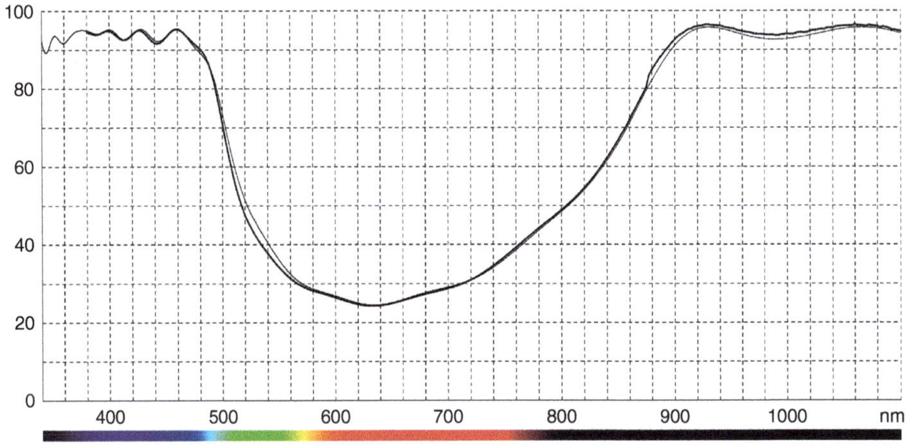

Figure 7.9 Spectral curve of a realised equaliser filter. Source: Courtesy of Delta Optical Thin Film A/S.

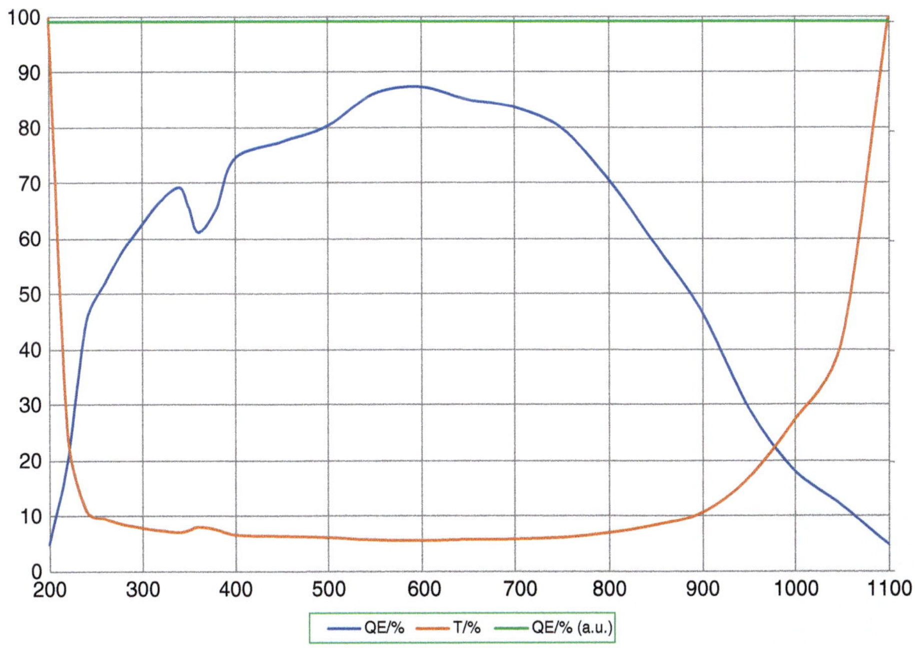

Figure 7.10 Ideal equaliser filter: quantum efficiency curve (blue), filter curve (orange), combined response (green).

fractions of the primary wavelength ($\lambda_n = \lambda_0/n$ with $n = [2, 3, 4, \ldots]$). These higher orders are spatially separated by angle, if the spectrometer covers less than one octave ($\lambda_2/\lambda_1 < 2$). If λ_2/λ_1 exceeds a factor of 2, the orders begin to overlap (see Figure 7.12) and need to be removed or *sorted out* before they reach the detector. For spectrometers that cover less than one octave, this can easily be done by a simple long-wave pass filter with a fixed cut-on wavelength. In this case, the cut-on wavelength must just be a little smaller than the lower wavelength of the spectrometer.

For spectrometers that cover more than one octave, a more advanced order-sorting filter is needed. Traditionally, segmented order-sorting filters were used. These could either be different segments of glass with long-wave pass filters of different cut-on wavelengths that were glued together, or a single substrate that was coated with areas

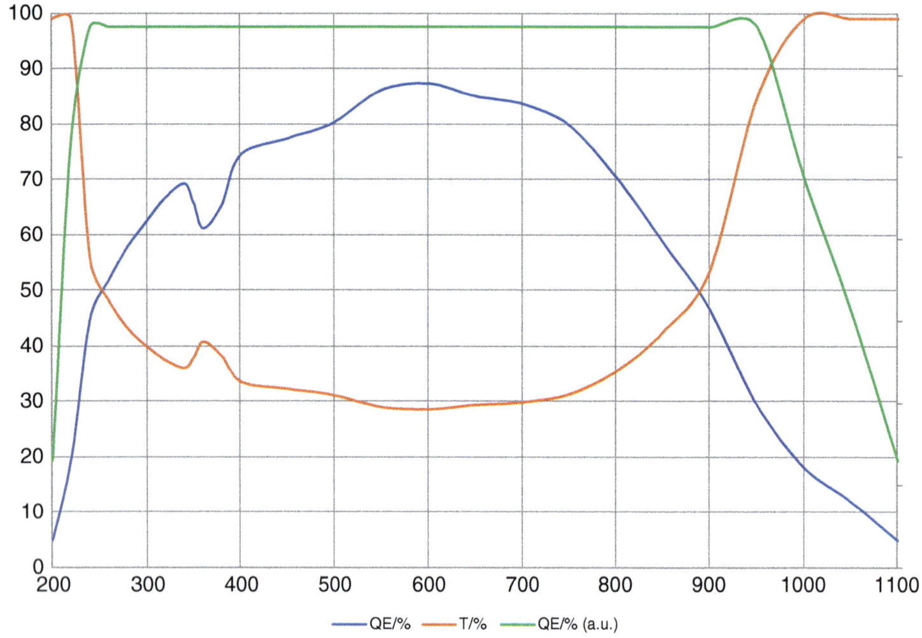

Figure 7.11 Real equaliser filter: quantum efficiency curve (blue), filter curve (orange), combined response (green).

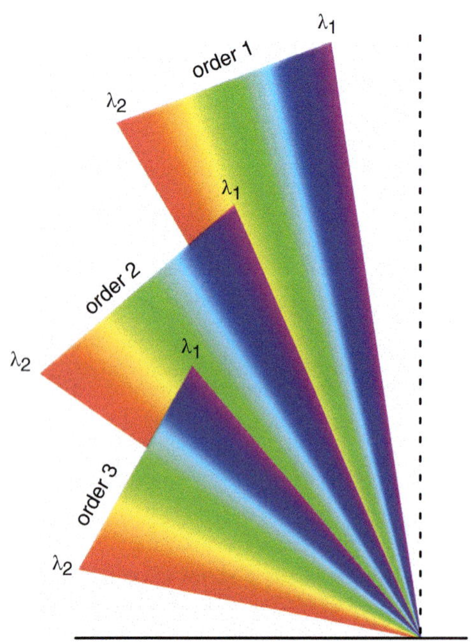

Figure 7.12 Overlapping orders in a grating spectrometer.

with long-wave pass filters of different cut-on wavelengths (see red curve in Figure 7.13). In either case, there are discontinuities where the glass segments or filter coatings meet. These discontinuities can also be seen in measured spectra, typically as dips. For a standard UV–NIR spectrometer that makes use of the sensitive wavelength range of silicon-based detectors (200–1100 nm), three segments are needed.

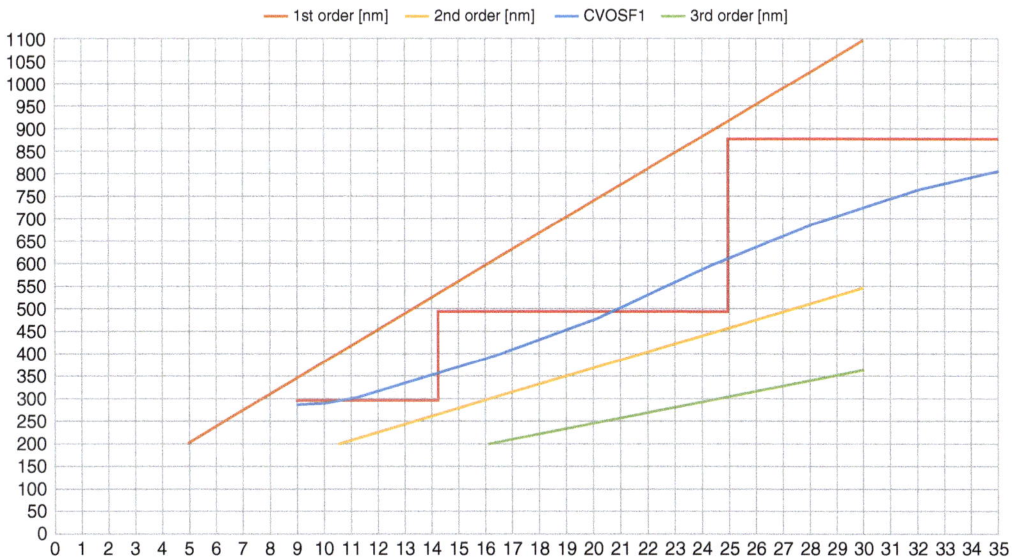

Figure 7.13 Comparison of segmented and continuously variable order-sorting filters (horizontal axis: detector length in mm, vertical axis: wavelength in nm): first order (orange curve), second order (yellow curve), third order (green curve), cut-on wavelength of segmented order-sorting filter (red curve), cut-on wavelength of continuously variable order-sorting filter (blue curve).

A more advanced and elegant way are continuously variable order-sorting filters (CVOSFs) because their coating is continuous, and the long-wave pass cut-on wavelength changes continuously along the filter (see blue curve in Figure 7.13). Furthermore, a CVOSF suppresses more stray light because the cut-on wavelength moves together with the dispersion of the grating, whereas with segmented filters the distance from the cut-on wavelength to the first-order one wants to transmit gradually increases and allows more stray light to pass through the filter.

Another advantage of CVOSFs is their insensitivity to misalignment with respect to their intended position in front of the detector. In Figure 7.13, misalignment can be seen as a horizontal shift of the filter curve. It is obvious that the red curve will quickly intersect with either the first-order curve or the second-order curve. A segmented order-sorting filter that is less sensitive to misalignment eventually requires a fourth segment. The CVOSF on the other hand can be misaligned by several millimetres before the curves intersect. However, for both filter types, an additional safety margin needs to be kept because of the finite steepness of the filter edge. If the nominal $\lambda_{50\%}$ wavelength[6] is positioned too close to the first order, the filter transmission has not yet reached its maximum and signal strength is lost. If the nominal $\lambda_{50\%}$ wavelength is positioned too close to the second order, the blocking has not yet reached its maximum and higher-order radiation will reach the detector.

7.3 Optical Filters as Complementary Filters

Gratings are today the standard component in spectrometers to analyse light by its spectral content. They can be used in a fixed configuration with linear arrays in mini-spectrometers (as described in the previous subsection) or as rotating elements with collimated light sources (excitation monochromator) and with point detectors (emission monochromators) in larger spectrometers.

6 There are two definitions of $\lambda_{50\%}$: (a) $\lambda_{50\%}$ is the wavelength at which transmittance has reached 50% or (b) where the filter achieves 50% of the maximum transmission. For the discussion of CVOSF, this distinction is not critical because CVOSFs have approximately 95% transmission and because the safety margin of CVOSF is large.

In principal, the quality of spectral measurements is affected by two factors: the occurrence of second- and higher-order reflections from the grating, and scattered light reaching the detector. Inaccuracies in a variety of measurements between 1 and 10% are typical without proper calibration and suppression of stray radiation. This scattered light can arise from any optical surface, from light leaks, and insufficient baffles, but the dominant source of scattered light is the diffraction grating itself.

Optical bandpass filters that are centred at the selected wavelengths of the excitation or emission gratings can help suppress most of the scattered light. During a wavelength scan, a series of bandpass filters needs to be moved sequentially into the beam path – either on a slider or filter wheel. The more the scattered light needs to be suppressed, the narrower the filters need to be, the more filters are needed to cover a given wavelength range. In practice, a compromise is chosen to keep the spectrometer feasible.

Another way of reducing scattered light is a second monochromator stage. For applications requiring the highest levels of performance, a double grating is used whereby the light from the first grating is passed through a second unit, reducing scatter and giving even greater resolution. However, each optical unit which the incident light impinges on results in a loss of energy. Hence, there is a trade-off between wavelength resolution and the intensity of the emitted light. In addition, the second grating makes the spectrometer bulky, expensive, prone to misalignment, and reduces its sensitivity because of the loss in signal strength.

Edinburgh Biosciences has developed an alternative approach using a pair of continuously variable edge filters to form a fully tunable bandpass filter – both in centre wavelength and bandwidth. Although the described solution is not applied in a portable instrument, it is mentioned here to demonstrate the usefulness of CVF in spectroscopy.

Figure 7.14 shows the typical setup of a single-monochromator spectrometer that is complemented with additional tunable bandpass filters. The tunable bandpass filters consist of a pair of a CVLWP filter and a corresponding CVSWP filter with spectral properties as shown in Figures 7.7 and 7.8. The filters are mounted on computer-controlled linear drives and placed in series in the beam path of the spectrometer. The rising and the falling edges of the edge filters form a bandpass filter whose centre wavelength and bandwidth can continuously *and* independently be tuned by either moving the filters in the same direction or in opposite directions. The centre wavelength can be adjusted across several hundred nanometres, and the bandwidth can be as narrow as a few nanometres or as wide as several dozens of nanometres. A full wavelength scan can automatically be performed by automatically tuning the centre wavelength to those of the grating and selecting a bandwidth that is appropriate for the measurement, and scattered light can be suppressed while the signal can be maximised.

The advantage of this approach over both a single- and a double-monochromator is clearly demonstrated at the example of the cyclohexane Raman spectrum that was measured with the EI FLS920 single monochromator fluorescence spectrometer (Figure 7.14). The measurement of high-resolution Raman scattering requires very high sensitivity and a clean background level as the probability for a Raman scattering event is only in the order of 10^{-6}. Cyclohexane exhibits several Raman peaks between 509.8 and 586.1 nm (Raman shift between 384.1 and 2938.3 cm^{-1}) when excited at 500 nm.

A high level of background is observable in the spectrum measured without filters (Figure 7.14, red/upper curve), and noise renders the weak peaks almost unobservable. The introduction of the synchronously tuned variable bandpass filters shows a significant reduction of the background noise (Figure 7.14, green/middle curve) by one order of magnitude (because of the deep broadband blocking of the CVFs) without compromising the signal strength (because of the high transmittance in the passband of the CVFs). The resulting reduction in photonic noise makes the weak peaks observable (see e. g. the weak peak at 576.8 nm). The signal-to-noise ratio (SNR) improves by a factor of 3.4 for the strongest peak. Obviously, even better ratios are achieved for the weaker peaks. For comparison, the same experiment was repeated on a double-monochromator system (Figure 7.14, blue/lower curve). The use of two additional diffraction gratings in the beam path severely reduces the signal strength which – despite the reduction in background levels – renders the weaker peaks unobservable (for more details refer to [14]) (Figure 7.15).

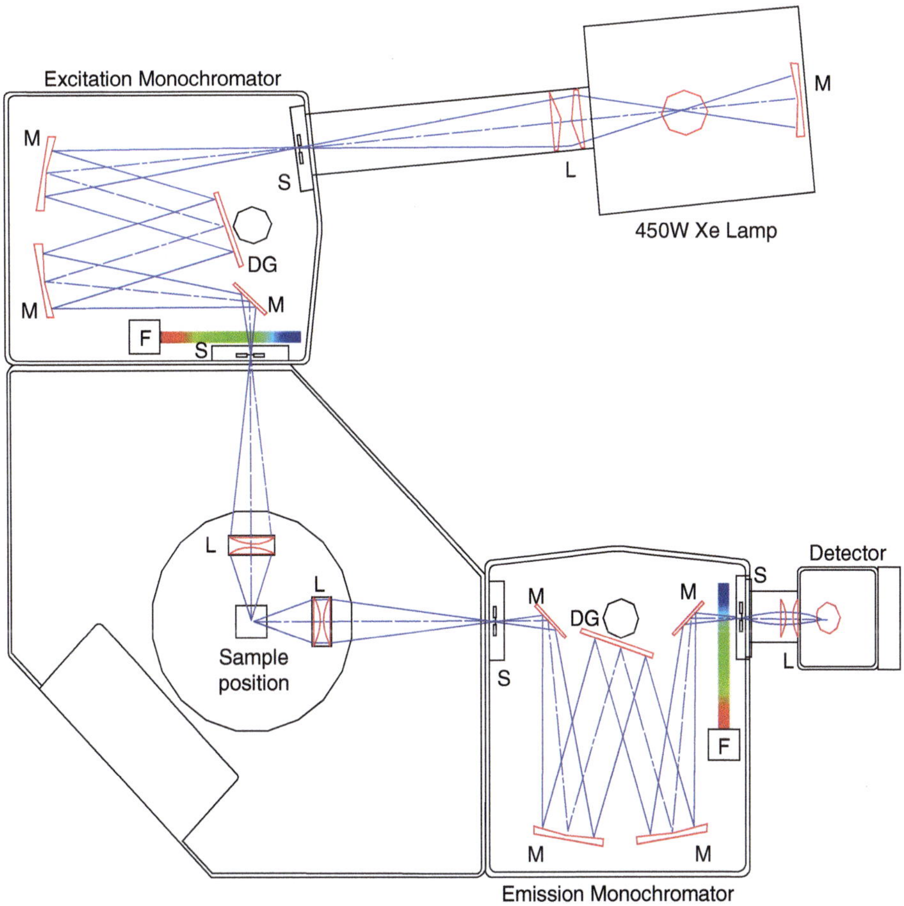

Figure 7.14 Typical setup of a spectrometer with single excitation and emission monochromators complemented with additional tunable bandpass filters. M: Mirror, L: lens, S: slit, DG: diffraction grating, F: new continuously variable filter stages. Source: Courtesy of Edinburgh Biosciences.

7.4 Optical Filters as Wavelength Selective Element

Optical filters are often found in bench-top spectroscopic instruments, such as micro-plate readers, as wavelength-selective elements. They are mounted on sliders or in filter wheels to facilitate wavelength selection in the excitation or emission channel. Even though filters have advantages over gratings because of their potentially much higher light grasp (for definition and details see [12] pp. 659), only a limited number of fixed filters can be installed in an instrument at a time. This limits the flexibility of the instruments and makes signal optimisation very difficult and time-consuming. BMG Labtech has developed a new generation of multi-mode micro-plate readers that use CVF and offer the high light grasp and thus sensitivity of filter-based micro-plate readers and at the same time full tunability of a grating-based instrument with the added benefit of narrow and wider passbands, which allows collection of a stronger signal. In addition to the tunable bandpass filters described in the previous section, the reader also includes a continuously variable dichroic (CVD) mirror that separates the excitation from the emission light in an epifluorescence setup (see Figure 7.16).

In hand-held spectrometers on the other hand, there is not enough space for a filter solution with moving parts. For this reason, hand-held spectrometers were until recently only based on gratings. Recent developments in the

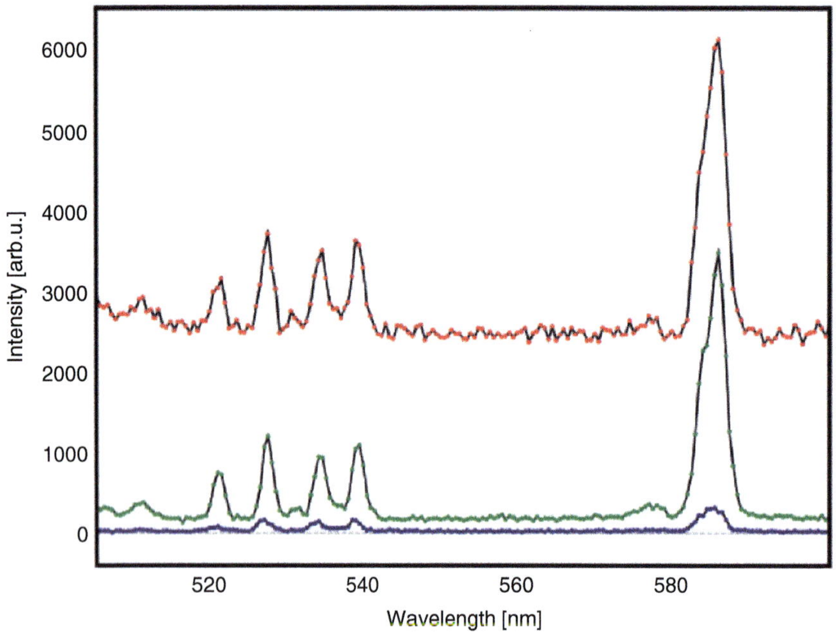

Figure 7.15 Raman spectra of cyclohexane measured at an excitation wavelength of 500 nm. The red curve (upper) denotes a spectrum taken at 1 nm spectral resolution without the use of variable filters. The green spectrum (middle) is taken with the variable filters in place. For comparison, the same measurement was repeated on a double-monochromator spectrometer (blue curve – lower). The signal is severely reduced. Source: Courtesy of Edinburgh Biosciences.

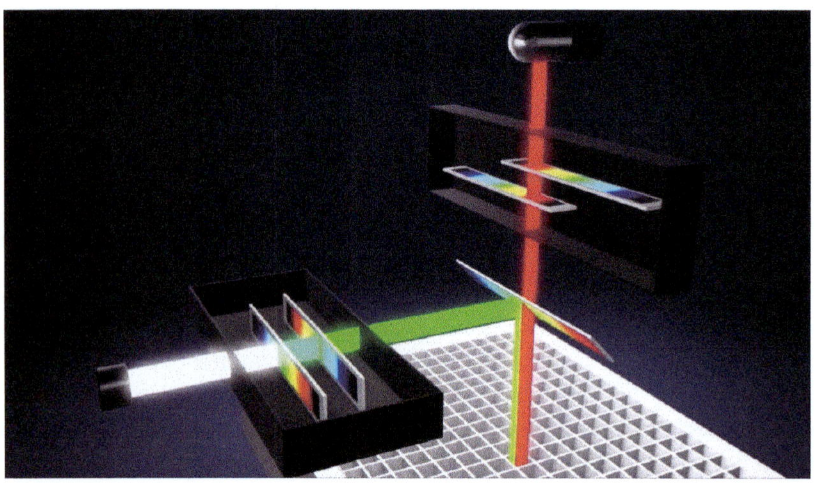

Figure 7.16 Optical engine of a CVF-based multi-mode micro-plate reader. To the left, the light from a Xenon flash lamp is filtered by a CVF monochromator to the optimal excitation wavelength range of the fluorophore in the micro-plate well. The filtered beam is reflected by the CVD into the well. The red-shifted fluorescence emission passes the CVD and is filtered by a second CVF monochromator to the optimal emission wavelength range of the fluorophore before it reaches the detector at the top of the image. Source: Courtesy of BMG Labtech GmbH.

field of continuously variable bandpass filters (CVBPF) enabled completely new spectrometer designs. Previous CVBPF – for example SCHOTT'S Veril filter[7] – were not designed for general purpose spectroscopic use and are typically too big for hand-held instruments. Another limiting factor is the fact that the blocking range does not cover the complete wavelength range in which silicon-based detectors are sensitive to radiation (typically from 200 to 1200 nm).

Newer generations of CVBPF overcome these limitations. Figures 7.17 and 7.21 show two CVBPF for the NIR, respectively, Visible/NIR (VIS/NIR) wavelength range. Common to these and similar filters is that they provide OD4 blocking or even better in the sensitive wavelength range of silicon-based detectors (CCD or CMOS).[8] This provides a high signal to background ratio without the need of any additional filters and allows the use of the detector in sunlight or other uncontrolled ambient light situations.

Besides their blocking capabilities, modern CVBPFs feature passbands that are narrow and steep enough for many spectroscopic applications (for example, fluorescence or colour measurements) and high transmission levels. Their size has shrunken so much that they now can be combined with typical line or array sensors. The combination with a line sensor enables point measurements and is described in the next section. Combining a CVBPF with an array sensor opens up for novel designs of multi- or hyperspectral cameras and is described in Section 7.4.4.

An important parameter of any spectrometer is its wavelength resolution. In case of diffractive spectrometers, it is mainly defined by the size of the slit and can achieve sub-nanometre resolution in exchange of optical throughput. For spectrometers that are based on CVBPFs, the spectral resolution is a combination of several factors. The dominating factor is the optical design bandwidth of the CVBPF. It is a theoretical value that is defined for collimated light with an infinitely small spot size on the filter, and which is the result of the bandpass layer structure of the filter. Typically, a multi-cavity design is used that naturally exhibits narrow bands with steep edges. The bandwidth is directly proportional to the centre wavelength in the first approximation (see for example Figure 7.21; the measured curves shown there are very close to the design performance).

When a CVBPF is coupled to a pixel-based sensor, the spectral performance is no longer defined by the slit size in the spectrometer but by the pixel size of the sensor, which is typically a few micrometres. For all practical purposes, this means that a single pixel receives light of a spectral width that corresponds to the design bandwidth of the CVBPF, which is typically several nanometres wide. The spectral resolution then is a combination of the filter's bandwidth, the filter's wavelength gradient, and the sensor's pixel pitch. As an example let us look again at the filter from Figure 7.21: its bandwidth is approximately 2% of its centre wavelength, i. e. 14 nm at 700 nm. Its wavelength gradient is (880–450 nm)/35 mm = 12.3 nm/mm. With a pixel pitch of say 10 μm, the change of peak wavelength between neighbouring pixels becomes 12.3 nm/mm × 10 μm = 0.123 nm.

It is obvious that in this example, the spectra are oversampled by a factor of 100. This allows for pixel binning to enhance signal and reduce noise at the same time. It also makes it possible to apply post-processing methods that are known from classical spectroscopy – such as spectral deconvolution – to enhance apparent spectral resolution beyond that of the filter's optical bandwidth. However, typically, it will not be possible to reach sub-nanometre spectral resolution as with spectrometers that employ a slit. On the other hand, the light throughput or light grasp can be considerably higher, and the spectral resolution is still high enough for applications like fluorescence or colour measurement. As such both methods can be considered complementary and have their optimal applications in their own right. In a real optical instrument, more factors such as AOI and OA that were discussed in Section 7.1.2.5 and distance of filter to sensor surface need to be considered to assess final spectral resolution.

7.4.1 Point-of-Care Instruments

Point-of-care (PoC) instruments are the subject of research and development worldwide. The goal is to bring medical diagnostic and treatment methodologies from large central laboratories and operation theatres to local

7 https://www.schott.com/advanced_optics/english/products/optical-components/optical-filters/interference-filters/bandpassfilter .html.

8 It is noteworthy that all side bands, that narrow bandpass filter designs typically exhibit, are suppressed.

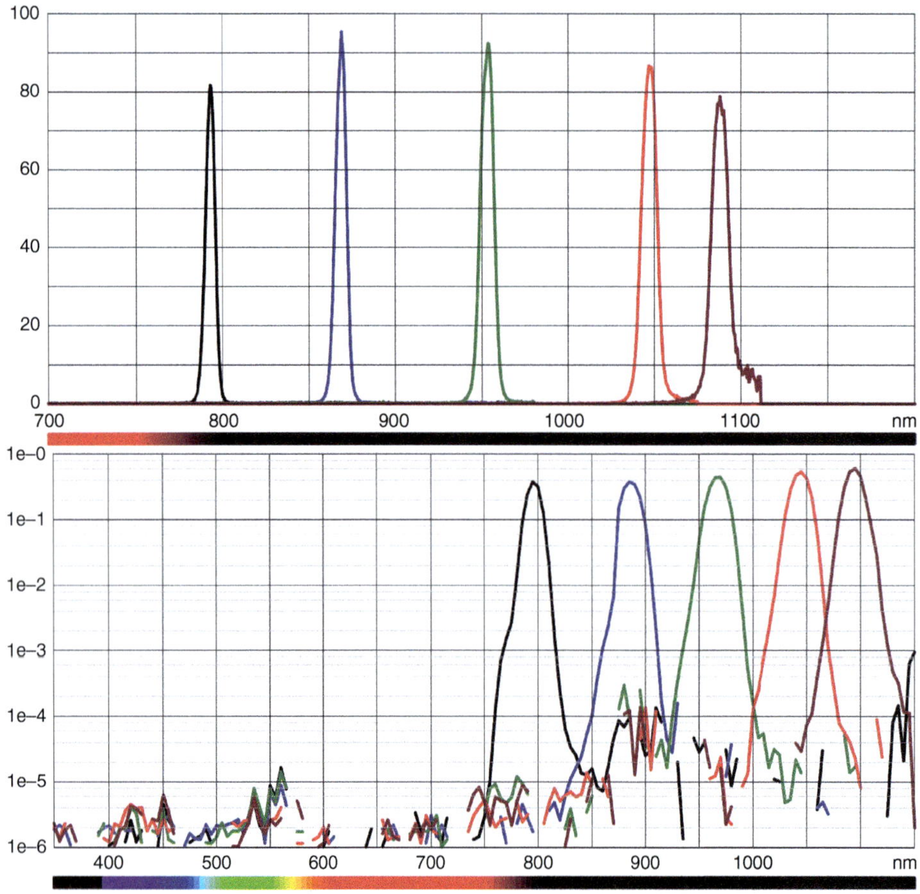

Figure 7.17 Measured transmission and blocking characteristics of continuously variable NIR bandpass filter covering 800–1100 nm along 18.5 mm. The spectra were measured with a PerkinElmer 900 spectrophotometer using a 200 μm narrow slit at five discrete locations to demonstrate the optical bandwidth. Given the continuous nature of the filter, any other peak wavelength can be found at intermediate positions. Source: Courtesy of Delta Optical Thin Film A/S.

physicians and to patients' beds or homes. Nowadays blood or other body liquids are sampled in the doctor's office and sent to a central laboratory. This causes a time delay of several days that can make the difference between life and death, but at least puts psychological stress on the patient. PoC also has the prospect to make advanced diagnostics available for patients in remote locations or developing countries.

PoC medical technologies require portable, small, and robust instrumentation for practical implementation. Most of the potential applications of PoC technology require a detection sensitivity at least equivalent to existing methods. This means that the optics in this system need to be at least as good as in larger equipment. The challenge is to reduce the size of all optics without compromising the optical output. Hand in hand with the size reduction of the instrument, often OAs increase to collect a sufficient signal. This has potentially adverse effects on filter performance as discussed in Section 7.1.2.5. These need to be managed by modern design and production technology. Moreover, small PoC instruments that are based on fluorescence detection typically do not have enough space to accommodate a dichroic beamsplitter (that helps suppress the excitation light in the emission channel) to separate the excitation from the emission channel. This puts higher blocking requirements on the bandpass filters in the instrument.

A good example for successful miniaturisation of a diagnostic method previously only available in large desktop instruments is Cepheid's GeneXpert Omni. It provides portable clinical molecular diagnostic polymerase chain reaction (PCR) testing using cartridge technology. The cartridge's microfluidics regulate all aspects of the testing process from sample preparation, nucleic acid extraction, amplification, and fluorescence-based detection. The instrument itself contains a large number of custom-designed bandpass filters that enable more tests beyond the initially planned human immunodeficiency virus (HIV) viral load test.[9]

7.4.2 Cataract Diagnosis and Treatment

Cataracts are a large and growing problem with a globally ageing population. The demand for cataract operations creates major stress on medical services worldwide (20 million surgeries per year). Future requirements of 44 million surgeries per year are projected for 2030.

The cloudiness of a cataractous eye lens limits visual diagnostic decisions on the extent of a cataract. Edinburgh Biosciences has shown that fluorescence emission from the eye can solve this problem quantitatively [7]. The fluorescence spectrum from a healthy lens changes significantly during cataract formation. This allows early diagnosis and permits more quantitative measurement of the cataract development.

Currently, the replacement of the lens is a plastic implant in the course of a major surgery. This is highly successful but demands full operating theatre resources and highly trained surgeons. The capsule retaining the lens is necessarily opened to allow insertion of the plastic lens. Such treatments can cause opacity of the capsule that reaches around 30% after three years and 50% after five years. It is treated by capsulotomy creating an aperture with a high-powered laser. Fragility of the capsule sometimes indicates a non-recommendation for this procedure.

Edinburgh Biosciences has developed a photobleaching treatment method of the cataract that retains the natural lens. This also allows the accommodation of the human lens to be preserved. The photobleaching process needs to be closely monitored. The process returns the lens spectrum to that of a healthy lens. Therefore, a PoC instrument was developed that contains both the tryptophan fluorescence-based detection and monitoring, and the photobleaching treatment module (see Figure 7.18).

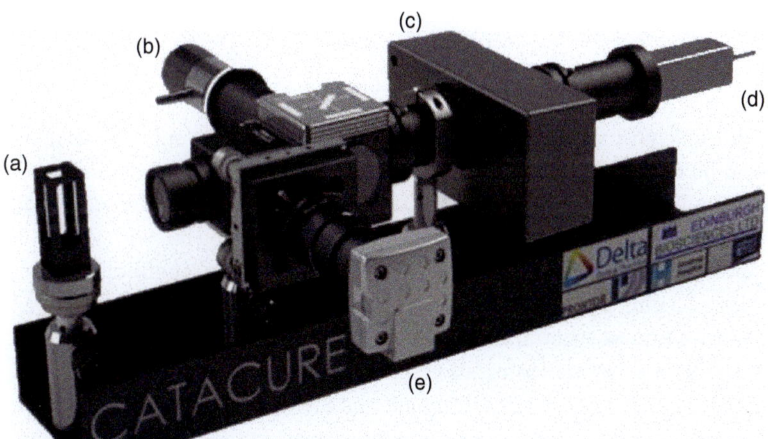

Figure 7.18 Functional prototype of a combined cataract diagnostic, monitoring, and treatment of PoC instrument with (a) sample (eye) location, (b) LEDs for excitation of fluorescence and photobleaching, (c) spectrometer based on continuously variable bandpass filter, (d) single photon-counting photomultiplier tube, (e) sample (eye) imaging system. Source: Courtesy of Edinburgh Biosciences.

9 https://www.cepheid.com/en/component/phocadownload/category/6-press-releases?download=132:2016-september-8-press-release.

Figure 7.19 Hand-held NIR spectrometer for quality control of grain. Source: Courtesy of GrainSense Oy.

At the heart of the monitoring module is a CVBPF on a linear drive. The spectrum is collected with a single photon-counting photomultiplier. The compact optical system which together with LED excitation, treatment sources, photomultiplier, and phase-sensitive electronics is 95% smaller in volume than today's instruments while retaining high light grasp and operation in ambient lighting conditions.

7.4.3 Point Measurements – Hand-Held NIR Spectrometer for Quality Control of Grain

The device shown in Figure 7.19 is a portable NIR spectrometer with integrated integration sphere that analyses protein, moisture, carbohydrates, and oil contents – the key constituents of crop value – in grain and other crops. Only a few kernels are needed, and results are obtained in a few seconds. The device applies a new method of NIR spectroscopy. As mentioned above, wavelength dispersion is not achieved with a grating but with a CVBPF that was specifically designed for this application (see Figure 7.17). The filter has an active length of 18.5 mm (in the wavelength direction) and a height of 3 mm. It is mounted flush with a custom photodiode array of 64 rectangular pixels that discretises 800–1100 nm into 64 spectral channels.

As the kernels are placed in the mid-cross-section of a small integration sphere, their orientation does not matter. Another advantage of this design is the improvement of both signal strength and SNR. In traditional NIR absorption spectrometers, the light passes the sample only once. By illuminating the kernels in the integration sphere, the light bounces back and forth and passes through the kernels several times before it reaches the detector. This also reduces the measurement time.

The new device allows farmers, cereal seed producers, and plant breeders for the first time to control the quality of their products in the field before and during harvest without needing to rely on large laboratories or the quality and price assessment by the buyers, because this new spectrometer is robust and easy to use. The device also helps meat producers monitor and adjust the protein content of feed in real time during the growth of their livestock. This has a major impact on the profitability of meat production.

7.4.4 Imaging Applications – Continuously Variable Bandpass Filters for Hyperspectral Imaging

Hyperspectral imaging (HSI) has been used for a couple of decades in applications such as satellite imaging, air reconnaissance, and other not overly price-sensitive markets. Still, there is no clear definition of the term *hyperspectral* imaging. Sometimes techniques that produce 2D images with more than the typical three RGB colours (or spectral channels) – for example by inclusion of a NIR channel – are already called hyperspectral. Mostly though, this is not considered sufficient. Typically, even 10 spectral channels are still to be called multispectral rather than hyperspectral. In the following, we require that certain criteria are fulfilled for an imaging technique to be called hyperspectral:

- For every pixel in the image, we measure the spectrum of the incident light or radiation.
- The measured spectrum is continuous and not discretised to a limited number of channels or bands.
- The spectrum covers more than one sub-wavelength range, for example, UVA, visible, and NIR, or NIR and SWIR.

The advent of alternative approaches makes HSI attractive for volume markets or even consumer products, for example, cancer detection, precision farming with unmanned aerial vehicles (UAVs) or directly at the plant, food testing in supermarkets, and many more. Alternative approaches comprise sensors that are coated at wafer level with fixed wavelength bandpass filters (e. g. PIXELTEQ or imec). Common are also thin-film coatings on glass substrates that can be patterned during deposition (in situ), or using a photolithographic process over the coating to block the addition or subtraction of materials deposited on the substrate surface (e. g. Materion). These micro-patterning techniques allow (a) filters that have a staircase of different centre wavelengths in one direction (also called stepped filters, suited for the so-called push-broom technique) or (b) 2D mosaics (suited for the so-called snapshot technique).

However, according to the above criteria, strictly speaking these solutions do not provide hyperspectral capability but are inherently multispectral due to their discrete changes in centre wavelength. Typically, between 10 and 100 different wavelengths or channels are offered. A truly hyperspectral sensor offers a continuous change in centre wavelength and as such a virtually unlimited number of channels.

A different approach to filters for HSI is followed with CVBPFs. They are fabricated for mid-size and full-frame CCD or CMOS sensors (e. g. 25 mm × 25 mm or 24 mm × 36 mm). These filters offer very high transmission levels and are fully blocked in the light-sensitive wavelength range of silicon-based detectors (200–1150 nm or higher). The combination of CVBPFs with silicon detectors allows the design of very compact, robust, and affordable HSI detectors (see Figure 7.20) that offer several advantages and benefits over conventional approaches:

- Large aperture compared with grating and prism,
- higher transmission than grating and prism,
- short measurement time,
- high suppression of stray light and
- excellent signal to background ratio.

These thin-film filters are coated with silicon dioxide and metal oxides on a single fused silica substrate without the use of glue, coloured glasses, or thin metal layers. The resulting filters are very robust against environmental conditions like temperature and humidity, and spectrally and mechanically stable. The filters are ideally suited for long-term use in airborne or space applications without any degradation. They are produced in coating chambers that are specifically modified to create a thickness gradient of the deposited coating material along one direction of the filter substrates. As the centre wavelength of a bandpass filter depends on the optical thickness of the interference layers, the thickness gradient creates a filter, whose centre wavelength changes continuously along the filter length.

Figure 7.20 Continuously variable filter based hyperspectral imaging detector.

Figure 7.21 shows the transmission characteristics of a CVBPF that covers a centre wavelength range from 450 to 880 nm with a bandwidth of approximately 2% of its centre wavelength. In a wide wavelength range, the transmission is higher than 90%. The transmission below 550 nm is underestimated due to the limited ability of the spectrophotometer to resolve narrow bands. But even more important than the peak transmission is that all undesired light from 200 to 1150 nm is suppressed better than OD4.

7.4.4.1 Comparison with Grating and Prism-Based Systems

Due to the diffractive nature of gratings and prisms, their use requires a certain large distance between sensor and diffractive element. This results in a large instrument that is prone to misalignment due to mechanical influences. Furthermore, a slit is needed to obtain high spectral resolution (see Figure 7.22). The slit limits the light throughput. The SNR level is typically not better than 200 : 1 [3].

CVBPFs, on the other hand, can be mounted directly on top of or close to the sensor. Options include gluing onto the sensor surface, replacing the cover glass with the filter or a mechanical holder. The resulting detector is very compact and robust at the same time. The optical design does not require the use of a slit. Light is collected through the full aperture of the lens. Together with the high transmission and the optical bandwidth of the filter, the HSI camera becomes very light efficient. The deep broadband blocking of the filter ensures a high SNR and minimises spectral crosstalk.

Without a slit, every acquired image shows the complete scene. This makes it possible to use the so-called windowing technique. It allows to arbitrarily image the scene from different positions without the need for precise synchronisation of lateral movement and image acquisition like with the push-broom technique. With the windowing technique, it is possible to construct the hyperspectral data cube using image pattern recognition techniques (see Figure 7.23). Another advantage is that 3D height information is readily available with standard stereoscopic reconstruction techniques because the objects in space are imaged from different angles throughout the image sequence (see Section 7.4.4.3).

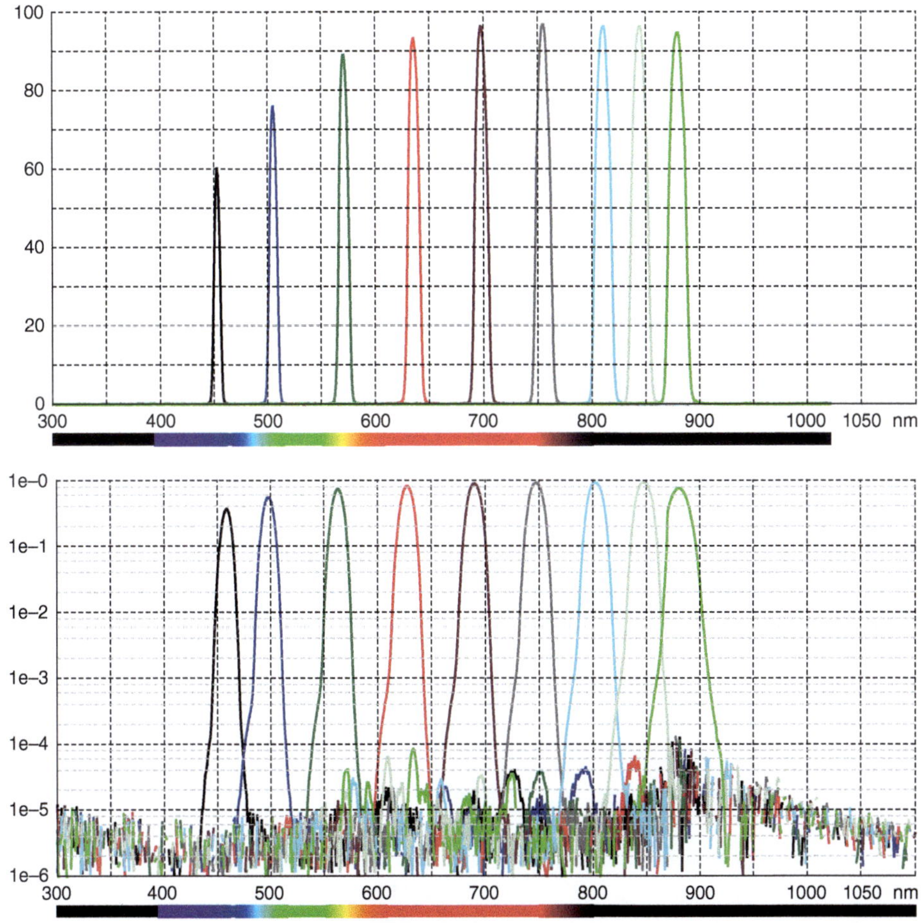

Figure 7.21 Measured transmission and blocking characteristics of a continuously variable bandpass filter covering 450–880 nm along 35 mm. The spectra were measured with a PerkinElmer 900 spectrophotometer using a 200 μm narrow slit at nine discrete locations to demonstrate the optical bandwidth. Given the continuous nature of the filter, any other peak wavelength can be found at intermediate positions. Source: Courtesy of Delta Optical Thin Film A/S.

7.4.4.2 Comparison with Wafer-level-Coated Detectors

What makes wafer-level-coated detectors (also known as filter-on-chip) attractive is the possibility to coat filters in arbitrary patterns also on small imaging sensors. Small size can be a crucial factor in certain applications. The same is true for snapshot capability (acquisition of the hyperspectral data cube with only one image). In this case, the sensor is coated with a 2D pattern of bandpass filters with different but constant centre wavelengths. The snapshot capability of course comes at the cost of reduced spatial resolution. Another typical trade-off of coating at wafer level is the limited spectral complexity of the filters. Imec, for example, offers simple Fabry–Perot filters that do not create sharp bands and have strong side bands that limit the overall free spectral range and the obtainable SNR.

CVBPFs are suited for larger sensors. To maintain their high performance and large wavelength range, the filters cannot be made much smaller than 20 mm in the variable direction with current production technology. An approach to enable snapshot HSI with CVBPFs was demonstrated by Fraunhofer IOF and recently commercialised by Cubert GmbH. It makes use of 2D lens arrays (plenoptics cameras) that are, for example, used in light field cameras [18]. A more detailed discussion takes place in Section 7.4.4.4 below.

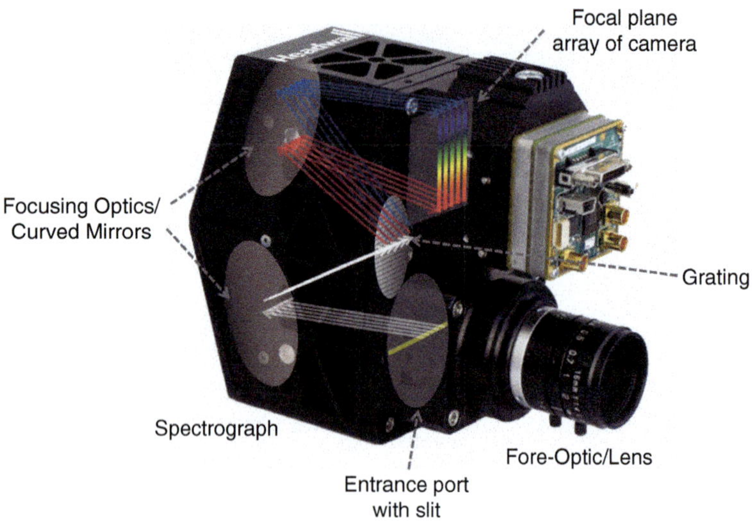

Figure 7.22 Design of a classical hyperspectral imaging camera with grating. Source: Courtesy of Headwall Photonics.

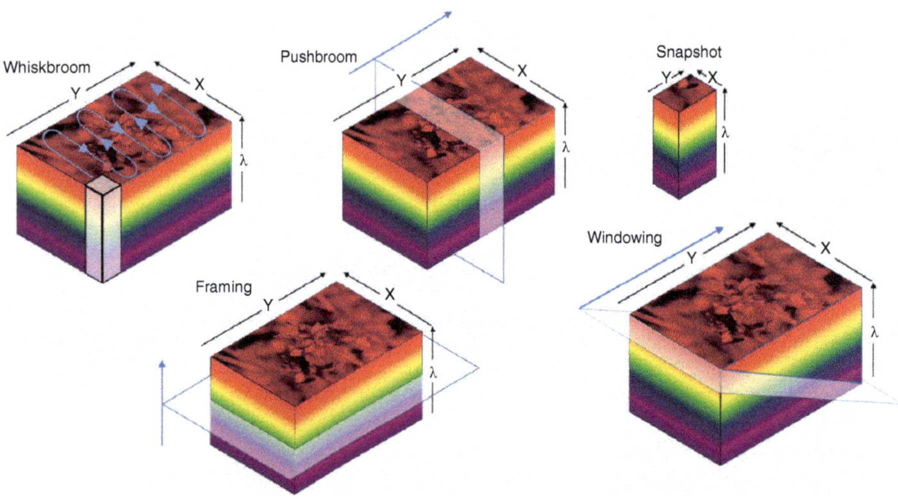

Figure 7.23 Different modes of acquiring the hyperspectral data cube. Source: Courtesy of Matt Gunn, Aberystwyth University.

7.4.4.3 3D Hyperspectral Imaging

For some applications, for example, precision farming or quality control of manufactured components, it is advantageous to combine hyperspectral data with 3D surface information [1]. In precision farming, this provides information on both nutrition and health of crop as well as its height and growth. However, it is not possible to extract the height information from traditional hyperspectral cameras that use slits and therefore only image a narrow slice of the object, especially under imaging conditions in which the relative movement between camera and object cannot be precisely controlled like with UAVs. Even with snapshot cameras, this is difficult because of their limited spatial resolution that typically is not sufficient for 3D reconstruction.

On the other hand, hyperspectral cameras using a CVBPF have superior spatial resolution that allows 3D reconstruction of the imaged scene by standard stereoscopic image processing [1]. This makes use of the fact that each object in the scene is viewed from different angles while passing by.

7.4.4.4 Snapshot Hyperspectral Imaging

The acquisition of the whole hyperspectral data cube with just one instantaneous image acquisition or measurement is not possible with classical push-broom or framing hyperspectral cameras. One approach to snapshot HSI was presented and discussed earlier with filters coated directly at wafer level in a mosaic pattern. Another approach to enable snapshot HSI combines 2D lens arrays with CVBPFs. Several groups have demonstrated prototypes ([9, 13, 15]). In the following, we discuss Fraunhofer IOF's solution [9] that was recently commercialized by CubertGmbH [11].

In order to overcome the restrictions of using scanning techniques or wafer-level-coated detectors, Fraunhofer IOF Jena proposes a multispectral imaging concept based on a multi-aperture system approach using a customised microlens array (MLA) combined with a CVBPF and a silicon-based image sensor (see Figure 7.24). In addition, a tailored baffle array is utilised for preventing optical crosstalk between adjacent optical channels. A customised multispectral analysis tool features the capabilities for advanced object classification.

The main advantage of choosing a micro-optical imaging system in combination with a CVBPF is the simultaneous capturing of spectral and spatial information in a single shot due to distinct spectrally coded channels. The multi-aperture principle allows one degree of freedom between the spectral and spatial sampling according to the

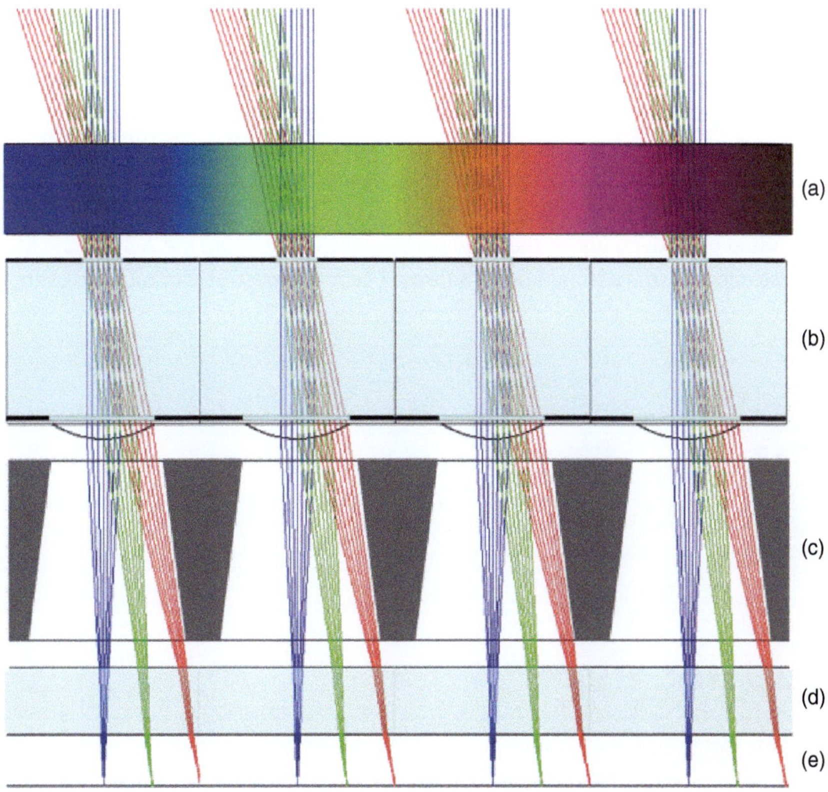

Figure 7.24 Cross section through Fraunhofer IOF's camera. The main elements are (a) continuously variable bandpass filter, (b) microlens array (MLA), (c) baffle array, (d) cover glass, and (e) image sensor. The light ray angles are color-coded. The colours do not represent wavelengths. Source: Courtesy of Fraunhofer IOF.

Table 7.1 Overview of system parameters of the snapshot multispectral camera.

System parameter	Value
Channels	11 × 6 (on Cartesian grid)
Optical path length	7.2 mm
f-Number	(*F*/#) 7
Field of view (FOV)	68° (diagonal)
Image resolution	400× 400 pixels (per channel)
Spectral range	450–850 nm
Spectral sampling	6 nm (linear)

constraints of the CVBPF and the size of the image sensor. The number of spectral channels is therefore equal to the number of microlenses. As a proof of concept, the optical design of the MLA employs a single microlens surface. The optimisation yields the system parameters that can be seen in Table 7.1. The CVBPF is slightly rotated around the optical axis with respect to the MLA in order to achieve a linear spectral sampling over the extended spectral range (see Figure 7.25). If the filter was not tilted, all lenses in one specific column would be exposed to the same wavelength.

The ultra-compact micro-optical system comprises microlenses in an array within diameters and sag heights in the range of hundreds of micrometres. Hence, the fabrication of the MLA was performed by the state-of-the-art wafer-level optics technologies. A spherical microlens master is fabricated by UV lithography and reflow of photoresist. A replication tool is created and used for moulding of the final lens elements. The CVBPF, the MLA, and the Baffle array were mounted in a mechanical holder, actively aligned to the image sensor and fixed to the housing. The snapshot multispectral camera demonstrator has an overall size of only $60 \times 60 \times 28\ \text{mm}^3$.

The spectral response of each individual pixel depends on the bandwidth of the CVBPF and the AOI on the filter. A spectral calibration corrects these effects using a tunable light source. The multispectral camera enables the

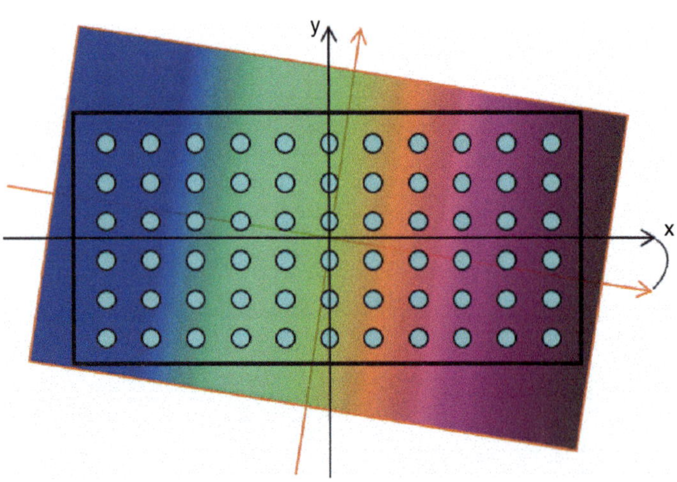

Figure 7.25 By tilting the filter axis with respect to the lens array axis, a simultaneous acquisition of 66 spectrally equidistant channels is possible.

Figure 7.26 Raw image of an extended scene in the lab (right, the red square shows an enlarged subimage) and in comparison, an image captured with a standard RGB camera (left). Source: Courtesy of Fraunhofer IOF.

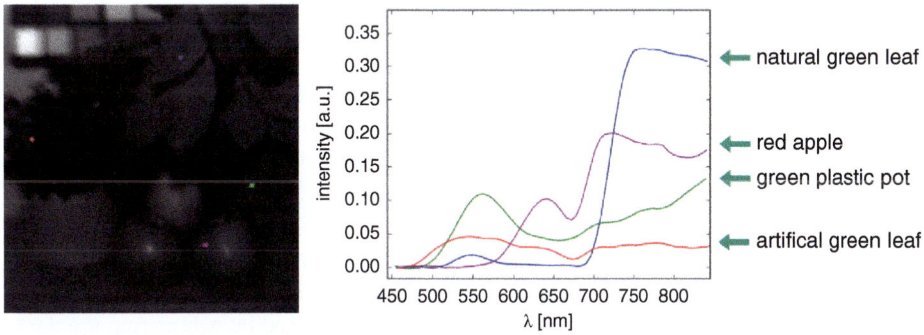

Figure 7.27 Test scene in the lab (left) and spectra extracted from the four marked objects (right). Source: Courtesy of Fraunhofer IOF.

snapshot acquisition of 66 spectral channels with a linear spectral sampling of approximately 6 nm over a wavelength range of 450–850 nm with a spectral resolution between 10 and 16 nm. In addition, an object distance dependent spatial calibration of the channels provides an accurate overlay of the individual sub-images in the data cube.

Figure 7.26 depicts the raw image of an extended scene in the lab (right) and in comparison, an image captured with a standard RGB camera (left). Due to the short focal length of the microlenses, the imaging module exhibits a large depth of field and therefore every object in the subimage is in focus (red square). The custom-developed software tool allows for a comprehensive analysis of spectra of objects in the scene as seen in Figure 7.27. For example, four processed and corrected spectra provide detailed object information, which constitutes the basis for an advanced object classification.

The camera demonstrator combines state-of-the-art micro-optical manufacturing methods and a multi-aperture imaging principle with a commercial CVBPF. It enables the realisation of highly compact and cost-efficient devices, capable of capturing spectrally resolved, extended object fields in a single shot with high resolution. Moreover, the proposed system concept provides a high flexibility with respect to spatial and spectral resolution by tailoring the number of spatial and spectral channels. The prospective fields of application of the developed system include environmental and agriculture monitoring, industrial surveillance and sorting, as well as biomedical imaging.

7.4.4.5 Spectral Quality of the Different Approaches

Cubert GmbH compared their three generations of hyperspectral video cameras side by side in an identical setup. The FireflEYE Q285 was, at the time of its launch, the first hyperspectral camera that could acquire a complete hyperspectral data cube (x, y, λ) with one single image shot. The prism-based sensor technology offered light efficiency of up to 70%. With 125 spectral channels (450–950 nm) and an image size of $50 \times 50\ px^2$, 2500 spectra were recorded simultaneously. With just $50 \times 50\ px^2$, it is difficult to resolve spatial structures in the image. Therefore, a second image sensor with only one band (panchromatic) was integrated into the camera, imaging the same view with $1000 \times 1000\ px^2$. With this additional information, the users can perform pan-sharpening of the spectral data that transform the final data cubes to a maximum size of $1000 \times 1000 \times 125$.

The multispectral ButterflEYE camera is based on the filter-on-chip technology. The number of spectral channels decreased to 16 or 25 depending on sensor type. The combined model ButterflEYE X2 provides an image resolution of $512 \times 272\ px^2$.

The Ultris camera was developed based on the light field technology described above. The camera reaches a native image resolution of $400 \times 400\ px^2$ with 100 spectral channels that consecutively cover the wavelength range from 450 to 850 nm. This means that the previously unheard number of 160 000 spectra is acquired simultaneously. The 12-bit sensor of the camera makes it possible to detect minute intensity differences in the spectral content while keeping the noise level very low. The dual GigE camera interface guarantees an image frame rate of up to 6 Hz.

Looking at the distribution of the spectral channels across their respective wavelength ranges between the three cameras, the improvements become obvious. Due to its prism-based optics, the Q285 has a nonlinear[10] but steady distribution of the channels. The camera based on a filter-on-chip sensor suffers additionally from the fact that the distance and position of the spectral channels cannot accurately be determined because of the complicated production process. The channels of the Ultris on the other hand are spaced equidistantly with a bandwidth (FWHM) of 2% of the centre wavelength.

The three cameras (ButterflEYE X2, FireflEYE Q285, and Ultris Q20) were installed in the same setup for a comparison. The exposure time was optimised with the help of a white reference to derive the maximum dynamics. As illumination, a stabilised tungsten light source (50 W) was used with exposure times of 10 ms (Q285), 16 ms (Ultris), and 120 ms (x2). The reflectance properties of the test samples were calculated by subtracting the dark current image from the measurement image and subsequently dividing it by the image of a calibrated 95% white reference (Zenith Lite). The dark current measurement and the white reference were averaged 20 times to achieve a good noise reduction.

The measurement, however, was acquired without any averaging and post-processing in order to faithfully represent the inherent noise. The data are presented in their original form without averaging, sharpening, or smoothing to show the true spectral quality of each sensor. Figure 7.28 (right side) shows quantities derived from the hyperspectral data cubes. The first line shows typical RGB representations (true colour), the two bottom lines show typical indices that are used for vegetation analysis. Each pixel of the different images represents one spectral curve of the respective sensors.

In the case of the FireflEYE Q285, the low spatial resolution is obvious. The colour representation on the other hand is very clear, which is also confirmed by the noise-free images of the vegetation indices. The filter-on-chip camera X2 has a higher spatial resolution but exhibits the typical high noise level of this type of sensor. Especially, the vegetation indices can only be used after intensive post-processing. The new Ultris combines high spatial resolution with low noise levels. Both the image quality and the spectral quality are excellent. The image noise is on a comparable level with the FireflEYE Q285.

Figure 7.28 (left side) shows the spectral signature of the sensors using three differently coloured paper samples. The spectra of all pixels in a predefined area with homogeneous colour were averaged. Because the standard deviation represents the noise equivalent of the sensors, it is used as error bars for each channel. The result shows clearly that the new Ultris can easily level with the spectral quality of the FireflEYE even though the Ultris has a tremendously higher spatial resolution.

---

10 The refractive index of the prism material increases with decreasing wavelength.

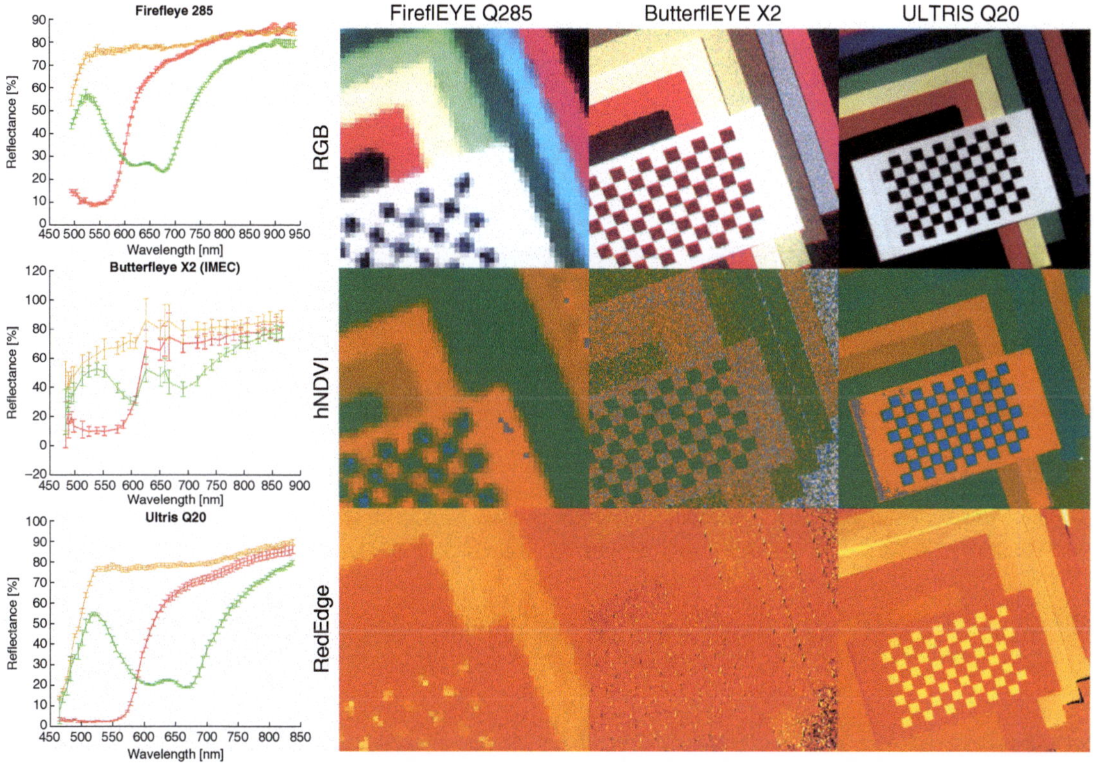

Figure 7.28 Comparison of the spectral quality of the three hyperspectral video cameras. For each camera, the spectra of the red, green, and yellow samples are shown along with the respective noise indicating standard deviation (left). Image comparison of the three hyperspectral video cameras FireflEYE Q285, Buttefleye X2 und Ultris Q20. For each camera, an RGB image (true colour) and two vegetation indices (hNDVI and RedEdge) are shown (right). Source: Courtesy of Cubert GmbH.

7.5 Conclusion and Outlook

All three approaches have their specific strengths. Classical hyperspectral cameras with gratings and prisms achieve the highest spectral resolution and are well suited for demanding applications in research. Sensors that are coated at pixel level allow the most compact cameras, offer total flexibility with respect to filter pattern, and are readily suited for snapshot imaging. Cameras based on CVBPFs combine high light efficiency, high SNR, and high spectral resolution with compactness and robustness. An in-depth discussion is made available by Renhorn et al. [4].

Although some of the filters that are discussed above are usually called linear variable filters, they are typically not perfectly linear with respect to centre wavelength versus position on the filter. This can easily be compensated for by a calibration curve or a one-time calibration of the assembled detector. However, there are good reasons for designing filters that deliberately are nonlinear.

As can be seen from the transmission curves in Figure 7.21, the bandwidth is proportional to the centre wavelength. This is a natural property of the multi-cavity design that is used in the filters. Consequently – if we for example compare light with 450 and 900 nm – there are twice as many neighbouring pixels on the sensor that see light of 900 nm as there are pixels that see light of 450 nm. This can be compensated for with an exponential relation between centre wavelength and its spatial position on the filter. Other design targets are possible. Desirable are filters that cover a larger wavelength range over a shorter length. Recently, filter designs were realised that, for example, cover 450–950 nm over 35 mm with an exponential dispersion function (see Figure 7.29).

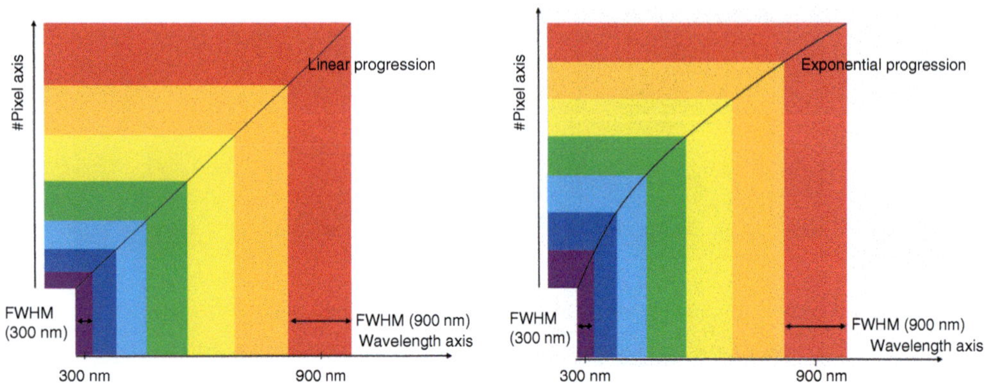

Figure 7.29 Comparison of CVBPF with linear and exponential dispersion.

Although the emphasis of this section lies on HSI, it should be mentioned that the same technology can be used to replace gratings in compact spectrometers – making them even more compact. Produced with ultra-hard coating technology, the filters can be diced to very narrow stripes that can be mounted in front of a line scan detector. This concept is, for example, used in wearable devices to measure the oxygen content of blood. Another promising application is the fusion of fluorescence microscopy and HSI as hyperspectral fluorescence microscopy. The technology is applicable to both laser scanning microscopy and wide-field microscopy [5, 10].

References

1 Ahlberg, J., Renhorn, I.G., Chevalier, T.R., Rydell, J., and Bergström, D. (2017). Three-dimensional hyperspectral imaging technique. *SPIE Proceedings Volume 10198, Algorithms and Technologies for Multispectral, Hyperspectral, and Ultraspectral Imagery XXIII*, San Francisco (5 May 2017), Bd. 1019805. doi:10.1117/12.2262456.

2 Antila, J., Miranto, A., Mäkynen, J. et al. (2010). MEMS and piezo actuator based Fabry–Perot interferometer technologies and applications at VTT.*Proceedings of the SPIE Defense, Security, and Sensing, 2010*, Orlando, FL, USA (27 April 2010). 7680, 04, S. 76800U. doi:10.1117/12.850164.

3 Bachmann, C.M., Eon, R.S., Lapszynski, C.S. et al. (2019). Foote: a low-rate video approach to hyperspectral imaging of dynamic scenes. *J. Imag.* 5 (6): 119. doi:10.3390/jimaging5010006.

4 Renhorn, E.I., Bergström, D., Hedborg, J., Letalick, D., and Möller, S. (2016). High spatial resolution hyperspectral camera based on a linear variable filter. *Opt. Eng.* 55 (11). doi:10.1117/1.OE.55.11.114105.

5 Favreau, P.F.U.A. (2014). Excitation-scanning hyperspectral imaging microscope. *J. Biomed. Opt.* 19(4). doi:10.1117/1.JBO.19.4.046010.

6 Fleming, S.J. (1999). *Roman Glass: Reflections on Cultural Change*. Philadelphia, PA: University of Pennsylvania Museum. ISBN 9780924171727.

7 Gakamsky, D.M., Dhillon, B., Babraj, J., Shelton, M., and Smith, S.D. (2011). Exploring the possibility of early cataract diagnostics based on tryptophan fluorescence. *J. R. Soc. Interf.* 8. doi:10.1098/rsif.2010.0608.

8 Geffcken, W. (1942). *Interferenzlichtfilter*. German Patent 742 463, filed 15 July 1942.

9 Hubold, M., Berlich, R., Gassner, C., Brüning, R., and Brunner, R. (2018). Ultra-compact micro-optical system for multispectral imaging. In: *MOEMS and Miniaturized Systems XVII* Bd. 10545 (ed. P., Wibool, P., Yong-Hwa, and Z., Hans), 206–213. International Society for Optics and Photonics, SPIE, Hrsg.

10 Gao, L. and Smith, R.T. (2015). Optical hyperspectral imaging in microscopy and spectroscopy – a review of data acquisition. *J. Biophoton.* 8 (6): 441–456. doi:10.1002/jbio.201400051.

11 Locherer, M. (2019). Lichtfeld für HSI. https://www.invision-news.de/download/ausgabe-3-2019-september/.

12 Macleod, H.A. (2010). *Thin-Film Optical Filters*. Boca Raton: CRC Press. ISBN 978–1–4200–73027.

13 Mu, T., Han, F., Bao, D., Zhang, C., and Liang, R. (2019). Compact snapshot optically replicating and remapping imaging spectrometer (ORRIS) using a focal plane continuous variable filter. *Opt. Lett.* 44 (5): 1281–1284. doi:10.1364/OL.44.001281.

14 Smith, S.D., Kerr, A., Shelton, M., and Fenske, R. (2012). Upgrade of diffraction grating spectrometers for multiple purposes. edinburghbiosciences.com/wp-content/uploads/2013/04/Upgrade-of-diffraction-grating-spectrometers.pdf.

15 Stanley, R.P., Chebira, A., Ghasemi, A., and Dunbar, A.L. (2017). Hyperspectral imaging using a commercial light-field camera. In:*Proceedings Volume 10110, Photonic Instrumentation Engineering IV*; 1011013, SPIE OPTO, San Francisco, CA, USA (28 April 2017). doi:10.1117/12.2253599.

16 Tematys (2016). Miniature and micro spectrometers: end-user needs, markets and trends. https://tematys.fr/Publications/en/spectroscopy/42-miniature-and-micro-spectrometers-end-user-needs-markets-and-trends-.html. Version: 2016.

17 Tikhonravov, Alexander V., Trubetskov, M.K., DeBell, G.W. (1996). Application of the needle optimization technique to the design of optical coatings. *Appl. Opt.* 35 (28): 5493–5508. doi:10.1364/AO.35.005493.

18 Zhou, Z., Yuan, Y., and Xiangli, B. (2010). Light filed imaging spectrometer: conceptual design and simulated performance. *Frontiers in Optics 2010/Laser Science XXVI OSA Technical Digest*. Rochester, NY, USA (2428 October 2010). doi:10.1364/FIO.2010. FThM3.

8

Portable UV–Visible Spectroscopy – Instrumentation, Technology, and Applications

Anshuman Das

Labby Inc., Cambridge, MA, USA

8.1 Introduction

Spectroscopy instrumentation is transforming, thanks to developments in advanced manufacturing techniques in the integrated circuits (IC) and optical components (Crocombe 2013, 2018). With better lithographic methods, such as nanoimprint lithography (Chou et al. 2014; Mohamed 2019) and extreme ultraviolet lithography (Ogletree 2016; Radamson et al. 2018), it is now possible to pack millions of electronic components within a single chip and even embed optical functionality. This has led to the realization of faster and powerful on-chip computing, for instance, in our smartphones. High demand for these computing devices has driven down costs and paved the way for field-level applications in virtually every industry. Whether it is in the form of distributed sensing networks or *Internet of Things* (IoT) (Karanassios and Fitzgerald 2018)-based approaches (Gubbi et al. 2013; Zanella et al. 2014), field-level instrumentation has dramatically changed over the last decade.

Although spectrometers are routinely used in material analysis today, they have been more suited to controlled laboratory environments (Nelson 2019). Several factors are responsible for this trend including the cost of the instrument, need for an expert for interpretation, need for sample preparation, noise from ambient light sources, and sensitivity to ambient conditions, among others. Among all the different spectral ranges for spectrometers, from ultraviolet-visible (UV-Vis), near infrared (NIR), shortwave infrared (SWIR), mid-infrared (MIR) and far infrared (FIR), the most successful field deployments have been possible for the UV–Vis instruments. By far, instrumentation costs have limited the use of other optical spectrometers. Silicon (Si)-based sensors and devices are low-cost, while those based on Germanium (Ge) and Indium–Gallium–Arsenide (InGaAs) are expensive, and do not scale as well as Si. As a result, linear or image (2D) sensors based on these detectors are not field-compatible due to much higher costs. Hence, innovative single-detector approaches have been the focus of recent developments (Gelabert et al. 2016), although the principles behind these techniques (e.g. Hadamard transform) have been known for many years. Due to these constraints, UV–Vis spectrometers have been more successfully demonstrated in portable configurations, although for a limited range of applications, such as water quality analysis (Langergraber et al. 2004).

In this chapter, the basic instrumentation configurations for portable UV–Vis spectrometers will be introduced. This will be followed by an in-depth review of various approaches for achieving spectral acquisition from digital micromirror devices (DMD) to do-it-yourself (DIY) approaches for teaching or hobby to smartphone spectrometers. Finally, some applications and challenges of portable UV–Vis spectrometers will be discussed, followed by challenges and future prospects.

Portable Spectroscopy and Spectrometry 1: Technologies and Instrumentation, First Edition.
Edited by Richard A. Crocombe, Pauline E. Leary and Brooke W. Kammrath.

8.2 Typical Instrumentation of a Portable UV–Vis Spectrometer

Although technically the UV–Vis spectrum spans from around 190–800 nm, most spectrometers will offer a higher wavelength limit up to 1100 nm. This is because Si has an appreciable responsivity into the NIR region (Pavesi et al. 2013). However, most common spectrometer configurations will not operate over this entire wavelength range. There is generally a tradeoff between spectral bandwidth and wavelength resolution. So, manufacturers will offer spectral ranges that are a subset of 190–1100 nm. Additionally, there are no light sources that have such a broad emission spectrum. Generally, a combination of two or more light sources is needed to achieve coverage from 190 to 1100 nm. Finally, most applications typically do not require their entire bandwidth along with a high resolution. Due to these factors, spectrometers operate in limited spectral bandwidths within the entire UV–Vis–NIR spectrum.

8.2.1 Basic Configuration of UV–Vis Spectrometers

A UV–Vis spectrometer, like most other spectrometer technologies (James 2007), comprises a broadband light source, a dispersing element like a prism or grating, slits and a detector (Ball 2006; Milton et al. 2009), as shown in Figure 8.1. Depending on the arrangement, the detector could be a point detector or a linear array. Incident light is either focused or directed through a narrow slit on to the dispersing element, which spatially separates the spectral components. These components are either directly collected at a detector or pass through the test sample before being received at the detector. Depending on the instrumentation, the spectrum could be captured sequentially by rotating the dispersing element or using a linear detector array.

8.2.2 Light Sources

Traditionally, Deuterium lamps are used as the excitation source in the UV spectrum with an emission in the range of 190–400 nm. Xenon (Xe) lamps are also used in the 200–1000 nm range, sometimes in flash mode, as opposed to continuous mode, but they have intense peaks in the NIR region. More recently, a range of UV light-emitting diodes (LEDs) have also been used although with lower bandwidths. Tungsten lamps are by far the most common Vis–NIR sources with emissions in the 350–2500 nm. Again, broadband Vis LEDs are also available with various phosphors to extend emission bandwidths. Although Tungsten and Xe lamps require cooling in order to dissipate excess heat, they can be made quite compact and designed for portable spectrometers. A comparison of emission spectra of various sources is shown in Figure 8.2.

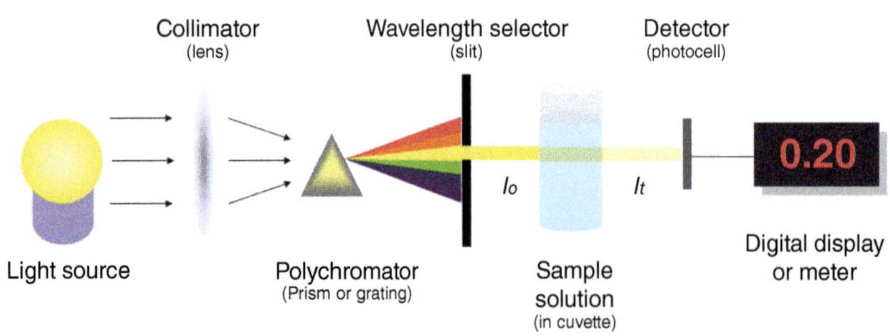

Figure 8.1 Typical spectrometer instrumentation.

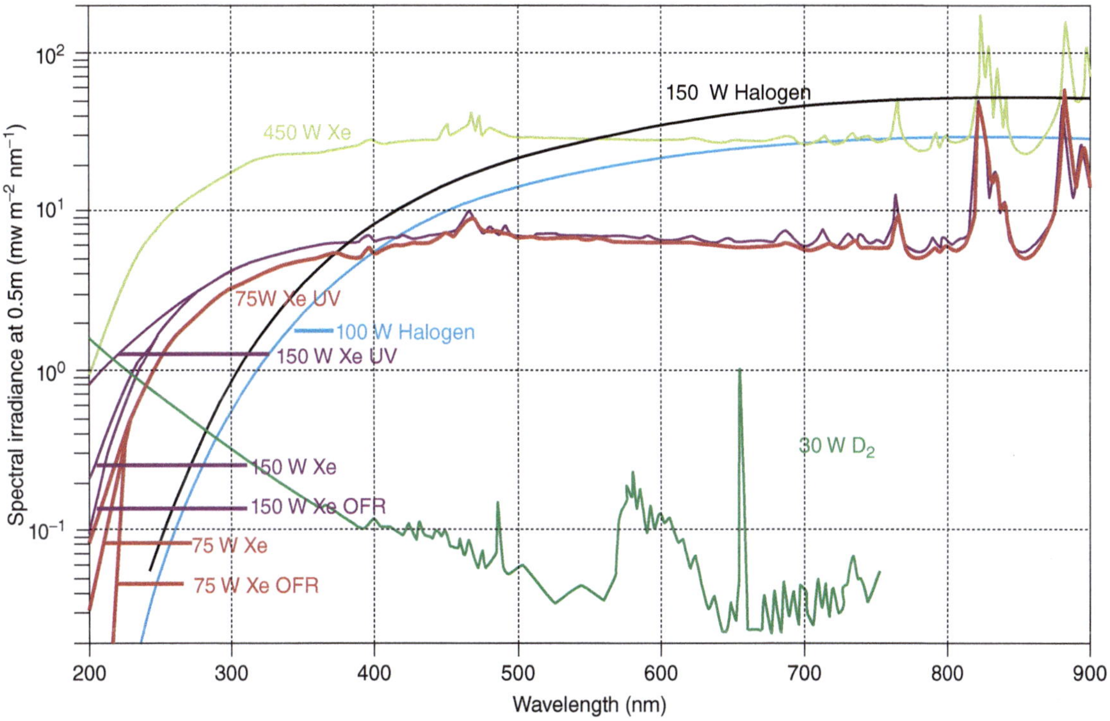

Figure 8.2 Spectra of common UV–Vis light sources. *Source*: Spectroscopy Instruments Catalog, Zolix. © Zolix Instruments Co.,Ltd.

8.2.3 Dispersing Elements

Although early designs of spectrometers involved prisms, most modern configurations have gratings (James 2007; Loewen and Popov 2018). Compared to prisms, gratings are lighter, more compact and offer flexibility in terms of resolution, size and bandwidth. Typically, ruled gratings are used in spectrometers with high diffraction efficiency, customizable coatings and a choice of blaze parameters. A typical diffraction grating is shown in Figure 8.3.

Figure 8.3 Gratings as dispersing elements. *Source*: Photo courtesy of Newport Corporation.

Recently, there have been demonstrations of spectrometers without a traditional dispersing element. Configurations involving a miniature 2D sensor with suitable filters over each pixel have shown limited applicability due to low spectral resolution (Oliver et al. 2013). However, they offer a simpler approach to achieve wavelength separation and may find niche applications.

In many benchtop configurations, these gratings are mounted on precision rotation stages for wavelength selection. Although this approach offers very high spectral resolution, it is complex to implement and sensitive to vibrations. Therefore, they are not generally suited for portable spectrometers. There are two most commonly used grating mount configurations. One is a Czerny–Turner configuration, where a plane grating is used with two concave mirrors. Light is incident from an entrance slit on to one of the concave mirrors that collimates the beam on to the grating. The diffracted light from the grating is focused by the second concave mirror on to the detector. The other is an aberration-compensated Paschen mount configuration, where the grating is etched on a concave surface that acts as a dispersing and focusing element, as shown in Figure 8.15. The Paschen configuration has fewer optical components and can be monolithically integrated on a chip. However, such gratings are challenging to fabricate and less flexible to employ.

8.2.4 Detectors

Almost all spectrometers that operate in the UV–Vis region have Si detectors (Acree 2006). Si has an excellent responsivity over the entire UV–Vis–NIR range; and it is readily available and customizable into regular photodetectors, avalanche detector or photomultiplier tubes, depending on the sensitivity required. Most portable UV–Vis spectrometers today use diode arrays in the complementary metal oxide semiconductor (CMOS) or charge coupled device (CCD) configurations and can have several thousands of pixels for high-resolution applications, as shown in Figure 8.4. These arrays are compact with fast readout times and can have inbuilt high-resolution analog-to-digital (A/D) converters.

8.2.5 Interfaces, Displays, and Data Storage

Depending on the application, benchtop and portable spectrometers can have their own display. Although several general-purpose spectrometers may require a desktop computer, there are scenarios where some amount of computing is done on the spectrometer itself and results are displayed on a built-in display unit. There is a range of portable spectrometers that require universal serial bus (USB) interfaces and work with laptops or desktop computers, while others have wireless capabilities to transmit data to a *Bluetooth*- or *Wi-Fi*-enabled device. Hence, a

Figure 8.4 Photodiode arrays used in portable UV–Vis spectrometers. *Source*: Photo courtesy of Hamamatsu Photonics.

(a) (b)

Figure 8.5 (a) Standalone spectrometer with display. *Source*: Permission from Sekonic. (b) Typical USB-type spectrometer. *Source*: USB4000 Fiber Optic Spectrometer, Ocean Optics. © 2006 Ocean Optics, Inc.

distinction can be made that portability always does not imply field compatibility. Several popular spectrometers like the one shown in Figure 8.5, do offer portability but their interface is USB-based, which means they require an external laptop or desktop computer to visualize and process data. This does not support standalone operation and is generally suited to lab-based studies. Standalone spectrometers with inbuilt displays have local data storage and can have secure digital (SD) cards or USB storage capability for portability of data.

8.2.6 Other Instrumentations

In addition to the basic instrumentation, portable spectrometers have circuitry to convert incident analog signals into digital signals for storage and postprocessing. A/D converters are usually present on the spectrometer board itself along with microcontrollers, buffers, and other control circuits. If some computing is done locally on the spectrometer, additional circuitry involving field-programmable gate arrays (FPGAs) could be present as well. Finally, data transfer circuitry in the form of USB, local area network (LAN), or Bayonet Neill–Concelman (BNC) ports may be found on spectrometers. In wireless devices, transmitting and receiving antennae will be present.

8.3 Measurement Configurations

A spectrometer is always used in a specific configuration depending on the sample state (solid, liquid, or gas), material interaction, and type of assay among other factors. For instance, liquid samples are typically examined using a transmission or absorption arrangement where the illuminating light source and detector are on either side of the sample. A few common illumination-collection geometries are presented in this section.

8.3.1 Reflection

In a reflection arrangement, the light source and detector are on the same side of the sample that is being examined. Upon illumination, the sample reflects certain wavelengths, which are collected by the detector. Depending

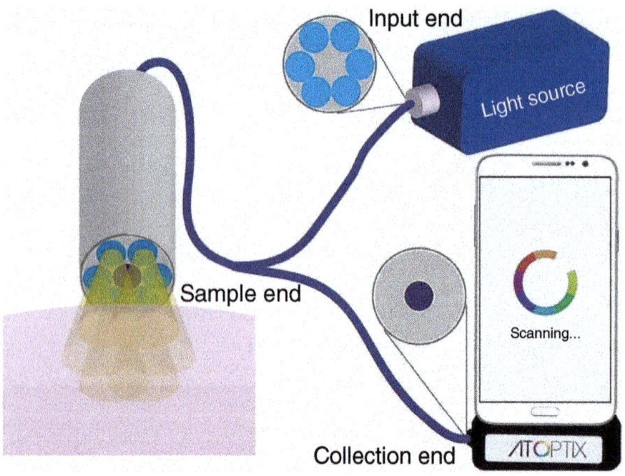

Figure 8.6 Typical setup for reflection spectroscopy using optical fibers. Creative Commons license. *Source*: Edwards et al. (2017). Licensed under CC BY 4.0.

on the sample surface, there can be diffuse and/or specular components. In some cases, an integrating sphere may be required to capture both the diffuse and specular components simultaneously. Figure 8.6 shows a typical reflection arrangement using optical fibers. A fiber bundle with six illuminations and one centrally placed collection fiber is used as the probe. The sample is illuminated by the six fibers on the periphery and the reflected light is guided to the spectrometer using a splitter arrangement. This is a common reflection configuration in portable spectrometers and the use of optical fibers minimizes any alignment issues. A similar setup can be realized in a free-space arrangement without optical fibers.

This configuration is most commonly used for color-sensing applications, where the reflection spectrum of a sample needs to be determined. Color measurements are used across several industries to characterize and standardize paints, pigments, and displays among others. Section 8.5 will discuss specific examples in color sensing.

8.3.2 Absorption

In an absorption measurement setup, the light source and detector are on either side of the sample, as shown in Figure 8.7. Light first passes through a reference, which is typically a nonabsorbing material like distilled water. The reference transmission spectrum is captured and saved. Next, the sample is introduced and another transmission spectrum is recorded. By assuming that there are no other interactions like scattering or reflection, the absorption spectrum can be deduced from the transmission spectrum by applying Beer–Lambert's law. Benchtop spectrometers generally have a double beam configuration for measuring absorption, wherein the reference and sample measurements can be done simultaneously. However, in portable spectrometers, those are not feasible, and measurements for the reference and sample need to be done sequentially. In such cases, fluctuations in the light source limit the sensitivity of detection. An example of a portable absorption spectrometer is shown in Figure 8.7b.

8.3.3 Fluorescence

Another common configuration that is especially seen in biomedical applications involves detecting fluorescence (Lakowicz 2006). Depending on the use case and sample state, there are several geometries for detecting fluorescence from thin films, solutions, droplets to powders among others. Typically, a narrowband source is used for exciting the fluorophore and the wavelength of the source chosen such that it coincides with an absorption maximum of the fluorophore. In most cases, laser sources are used to obtain suitable signal strength, but high-intensity LEDs have also been used in recent implementations. On the detector side, longpass filters are generally used to reject the excitation wavelength and only transmit the fluorescence signal comprising longer wavelengths. This is a versatile configuration where the source and detector can be arranged in absorption or reflection-like geometries.

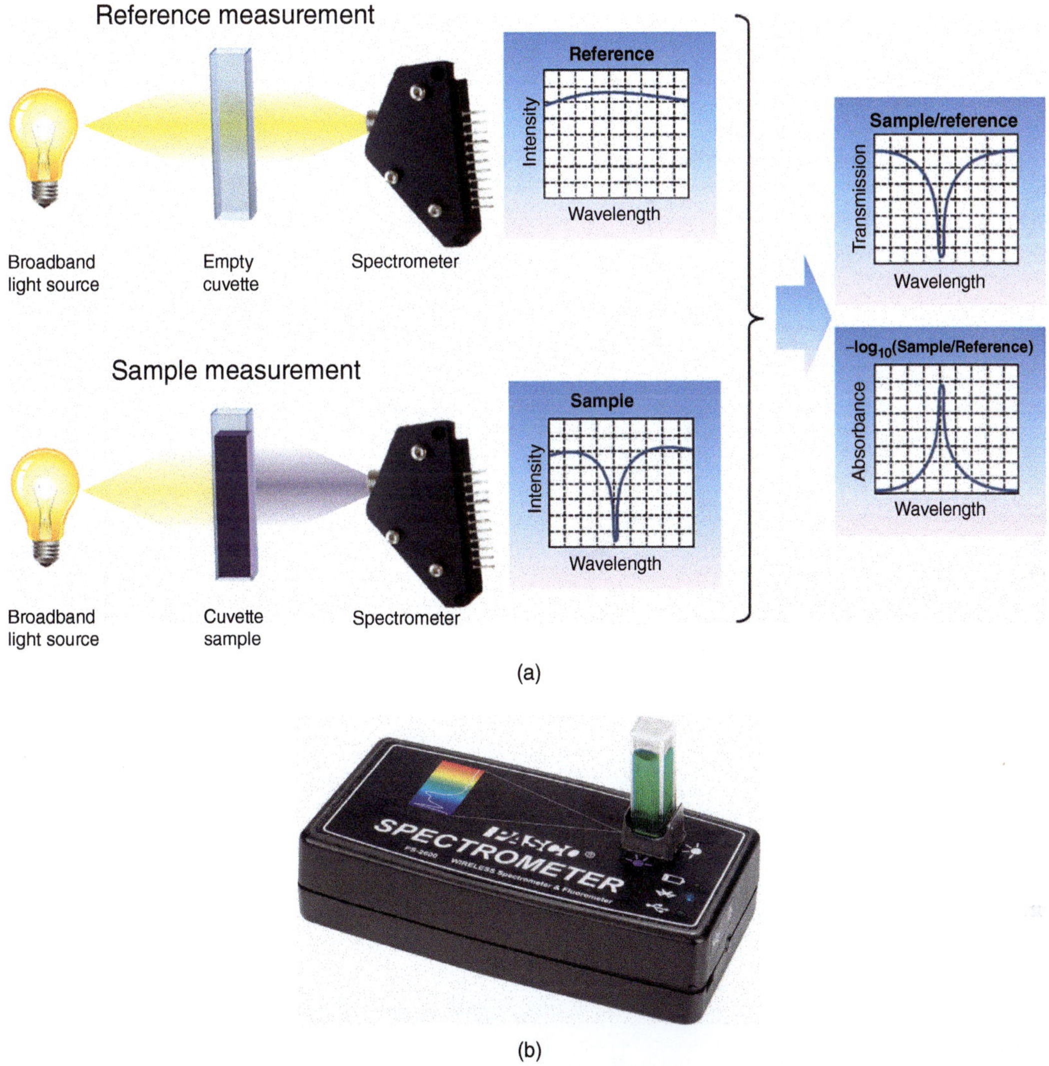

Figure 8.7 (a) Absorption spectroscopy setup. *Source*: Permission from Ibsen Photonics. (b) Example of a portable absorption spectrometer. *Source*: Permission from PASCO.

In applications where the fluorophore has a weak fluorescence or exists in low concentrations, additional focusing optics are employed to increase the signal collected by the spectrometer. This geometry is routinely used in confocal microscopy where a spectrometer is coupled to the microscopy objective and fluorescence spectra are captured at each point as a raster scan is carried out. However, in portable fluorescence measurement configurations, optical fibers are typically used to minimize alignment errors. In some cases, a custom sample holder has been designed to house the filters and other optical components, as shown in Figure 8.8. Yu et al. demonstrated a smartphone-based spectrometer, where the fluorescence signal was incident on a grating placed at the camera of the smartphone that played the role of a detector (Yu et al. 2014). The image captured by the camera could be calibrated with known sources and a spectrum could be extracted based on that.

Although a spectrum can be obtained using the geometry shown in Figure 8.8, that is only the first step toward a usable device. Several questions need to be addressed regarding the stability of the wavelength calibration, the

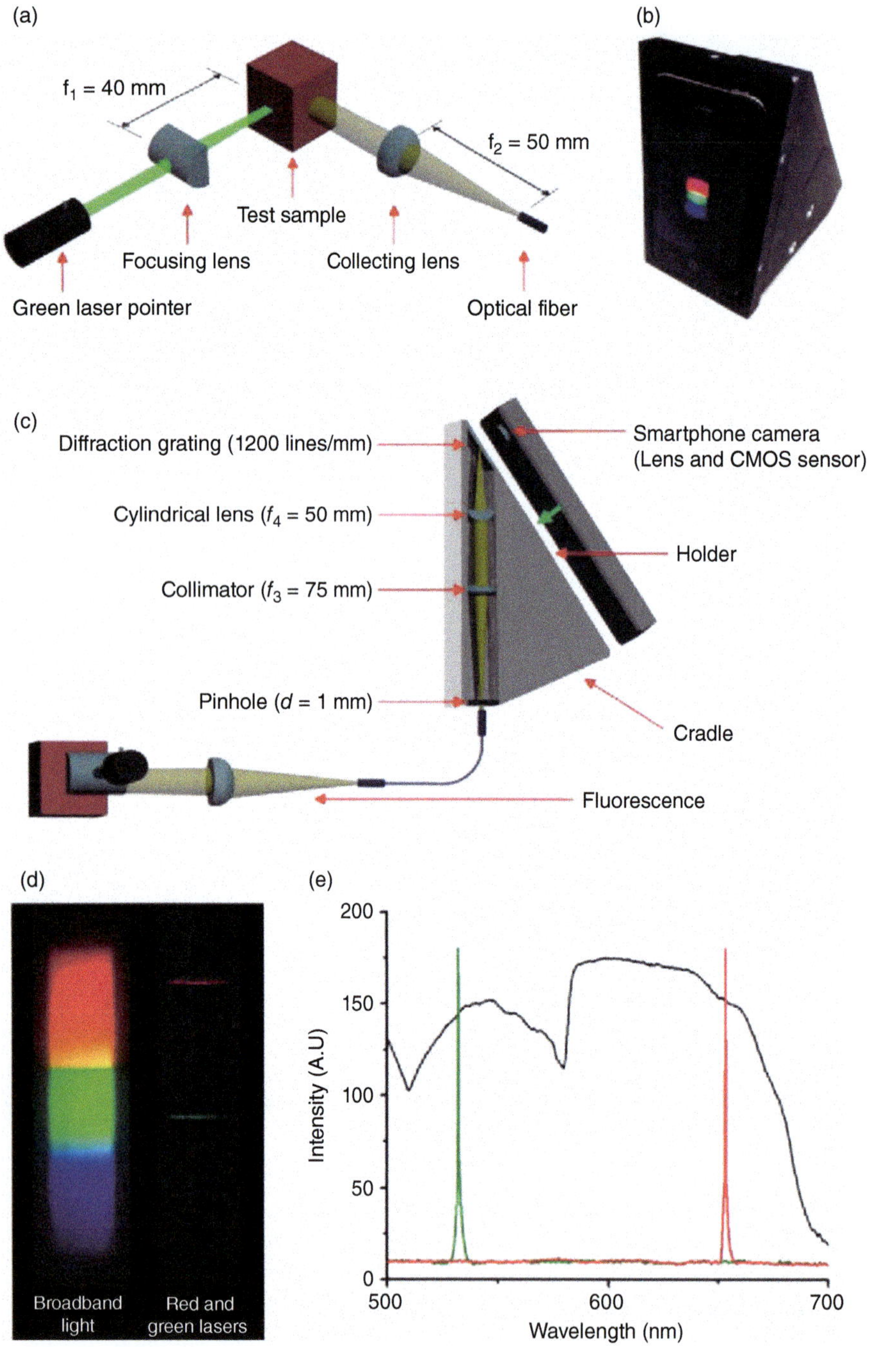

Figure 8.8 Smartphone-based portable fluorescence spectrometer arrangement. *Source*: Yu et al. (2014). © 2014 American Chemical Society.

linearity of the response, effect of background noise, and summing of adjacent pixels. Without a complete understanding of the device performance and limitations, developing applications, particularly in the biomedical field can lead to low sensitivity and specificity of tests. Section 8.6 discusses some of the parameters to keep in mind when developing applications based on such spectrometer configurations.

8.4 Types of Instrumentation Used in UV–Vis Spectroscopy

Due to the need for robust field operation and cost constraints, several innovative designs have been demonstrated in portable UV–Vis spectrometers. Some of these are unique to portable spectrometers with marked differences compared to traditional spectrometer instrumentation. In general, the bulk of the instrumentation is focused on accurately converting an incident light signal into its spectral components. Light sources, guiding optics, A/D conversion and other peripheral electronic components are straightforward to implement. The process of reliably dispersing the incident light signal, capturing it with low noise and then converting it to a calibrated optical spectrum is a challenging aspect of spectrometer instrumentation. The accuracy of this process determines the eventual possible applications. For instance, a robust implementation allows for use of such devices in highly sensitive biomedical applications, whereas a basic lower quality version would be good for teaching purposes.

In this section, various approaches for achieving light dispersion and subsequent conversion to the light spectrum will be introduced.

8.4.1 Grating with Linear Detector Array

This approach can be considered as a compressed version of a traditional benchtop spectrometer. Almost all the instrumentation is identical except for a reduction in component size and distances involved. A diffraction grating is used to spatially disperse the incident light signal and a suitably placed linear detector array simultaneously captures its constituent spectral components, as shown in Figure 8.9a. Typically, the linear array is comprised of 2^n ($n = 8,9,10,11,12\ldots$) CCD or CMOS pixels. The grating density, size, and the number of pixels determine the wavelength resolution of the spectrometer. Additionally, the input aperture diameter, optical magnification, and aberrations can affect the final spectral resolution. These spectrometers are also called array or grating spectrometers due to the instrumentation type, as shown earlier in Figure 8.5.

8.4.2 Digital Micromirror Devices

A major drawback of linear array-based spectrometers is the need for a sensor with multiple pixels or detectors. Typically, to achieve a wavelength resolution of around 1 nm or lower, 4096 pixels may be required. This is not a problem for Si-based detectors as image sensors in consumer cameras generally have a comparable number of pixels. However, if the detector is made from other materials like InGaAs or Ge in the case of NIR spectrometers, the cost of the device increases significantly. In such cases, single detector arrangements are desirable. An alternate approach involves using Fourier transform (FT)-based techniques. Although FT spectrometers are quite popular, they have some drawbacks. For instance, they do not directly measure spectra, but obtain interferograms, which need to be transformed into spectra by postprocessing. This enforces a higher requirement for computing power. In 2003, Spudich et al. came up with an innovative approach that eliminates linear arrays and addresses the drawbacks of FT-based systems (Spudich et al. 2003). They introduce a digital micromirror array where the orientation of the mirror can be digitally controlled. After a grating splits the incident light into the spectral components, each micromirror sequentially reflects light on to a single detector, almost behaving like a rotating grating. A spectrum can be generated after all the micromirrors have reflected their respective spectral components. Since these mirrors have a size of 10 μm and a pitch of about 5.4 μm, they can be operated at microsecond-switching speeds.

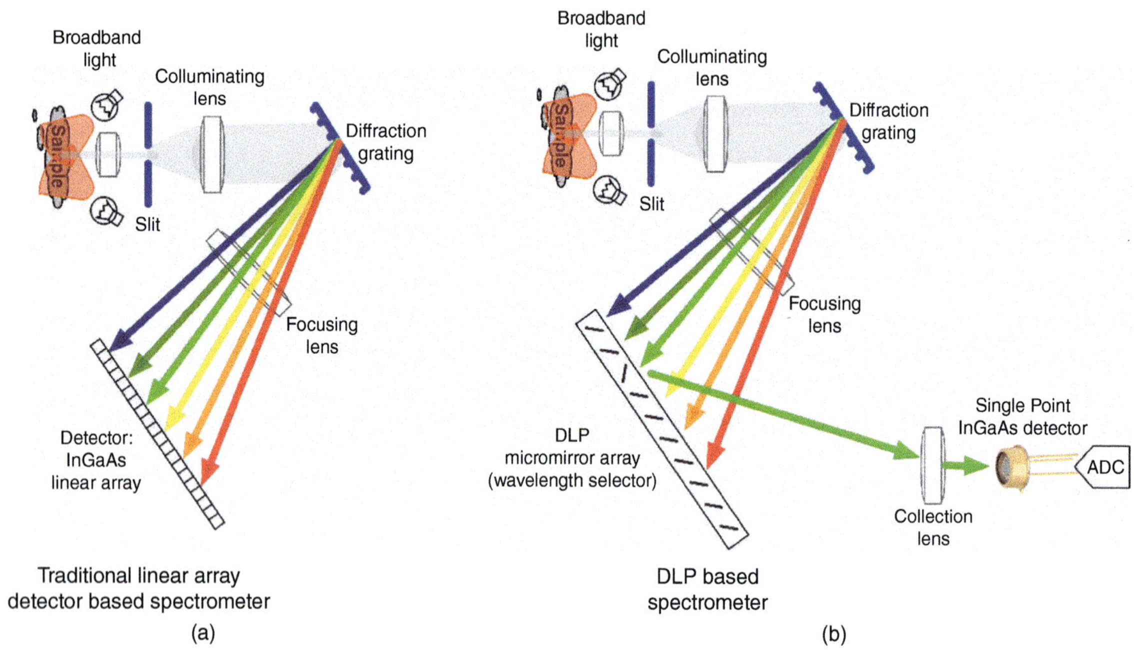

Figure 8.9 (a) Traditional array-based spectrometer. (b) Micromirror-based spectrometer. *Source*: © Texas Instruments.

This allows for fast acquisition of spectra with minimal postprocessing requirements, based on a Hadamard transform. Recently, Texas Instruments (TI) commercialized micromirror-based spectrometers using their digital light projector (DLP) technology (Gelabert et al. 2016), as shown in Figure 8.9b. Figure 8.10 shows a micromirror-based spectrometer using the DLP2010NIR module, which can operate from 700 to 2000 nm.

8.4.3 2D Filter Array-Based Devices

For grating-type spectrometers to be effective, there has to be a gap introduced between the grating and the linear detector array. A larger gap allows for better spatial separation of wavelengths, which results in higher resolution. Due to this inherent requirement, conventional grating-type spectrometers cannot be made into miniature chip-like devices unless monolithic fabrication is carried out. As a result, these portable grating-based spectrometers often have dimensions in the tens of centimeters. There are applications that require ultracompact spectrometers, where resolution can be traded for size. For instance, in space applications, there is a need for lightweight and compact sensors. In such scenarios, spectrometers fabricated with grating-free elements, such as filter arrays are more suitable (Oliver et al. 2013). In this approach, a filter array with varying transmission spectra is placed over a CCD image sensor, as shown in Figure 8.11a.

In this arrangement, the filters have a selective transmission spectrum that exclusively allows a narrow wavelength band. By designing several filters that cover a given spectral region, a spectrum can be reconstructed by analyzing the intensities obtained at individual pixels of the CCD. Wang et al. demonstrated a 128-channel filter array-based spectrometer with dimensions of 12 mm × 12 mm × 5 mm, giving rise to a volume of 1 cm^3 (Wang et al. 2007). Although this technique yields extremely compact spectrometers, fabricating a 128-channel filter array with each pixel having a different transmission pattern is quite tedious.

On similar lines, quantum dots (QDs) have also been used as filters due to ease of manufacturing. Chemical approaches can be used to alter the transmission properties of QDs and create the desired filter array. Bao and Bawendi demonstrated a QD spectrometer (Bao and Bawendi 2015) comprising a filter with 195 unique QDs, as shown in Figure 8.12.

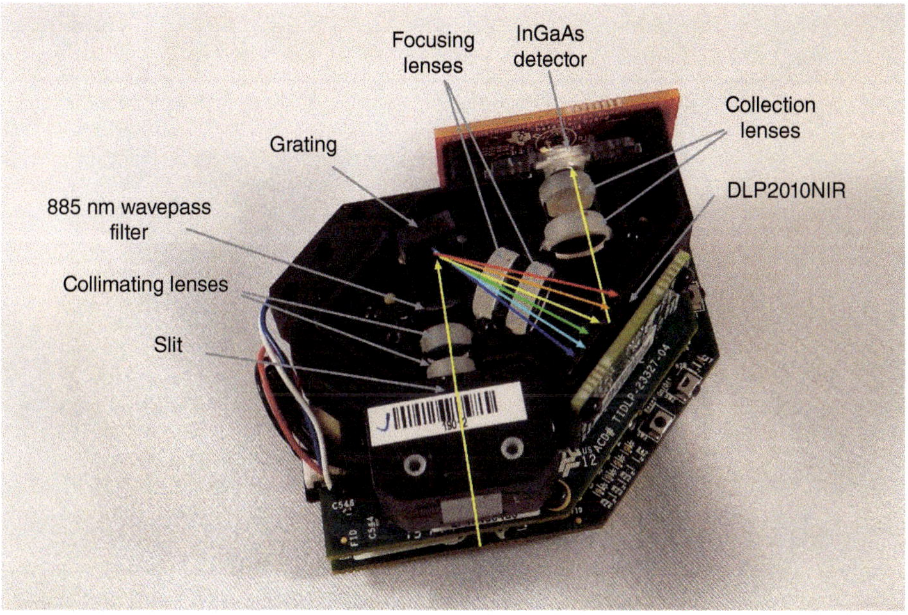

Figure 8.10 Micromirror-based spectrometer developed by Texas Instruments. *Source*: Texas Instruments Incorporated.

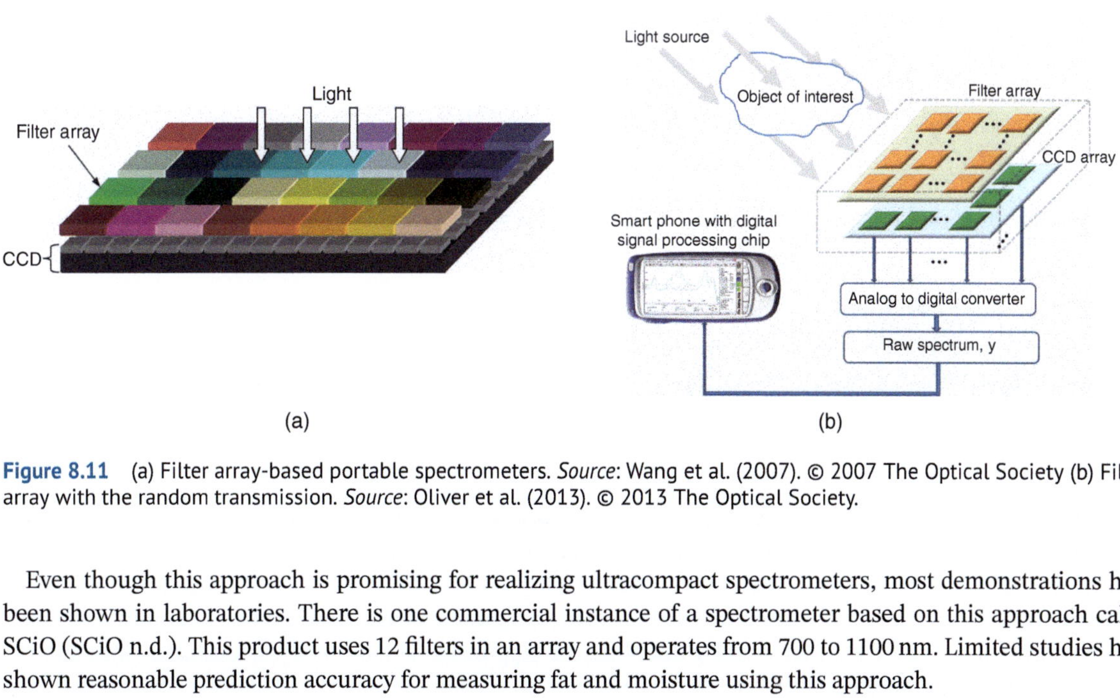

Figure 8.11 (a) Filter array-based portable spectrometers. *Source*: Wang et al. (2007). © 2007 The Optical Society (b) Filter array with the random transmission. *Source*: Oliver et al. (2013). © 2013 The Optical Society.

Even though this approach is promising for realizing ultracompact spectrometers, most demonstrations have been shown in laboratories. There is one commercial instance of a spectrometer based on this approach called SCiO (SCiO n.d.). This product uses 12 filters in an array and operates from 700 to 1100 nm. Limited studies have shown reasonable prediction accuracy for measuring fat and moisture using this approach.

8.4.4 Smartphone Camera-Based Approaches

In recent years, due to the ease of accessibility, high computational power and high-resolution cameras of smartphones, several groups have utilized smartphones as the central detection and computation module. This approach

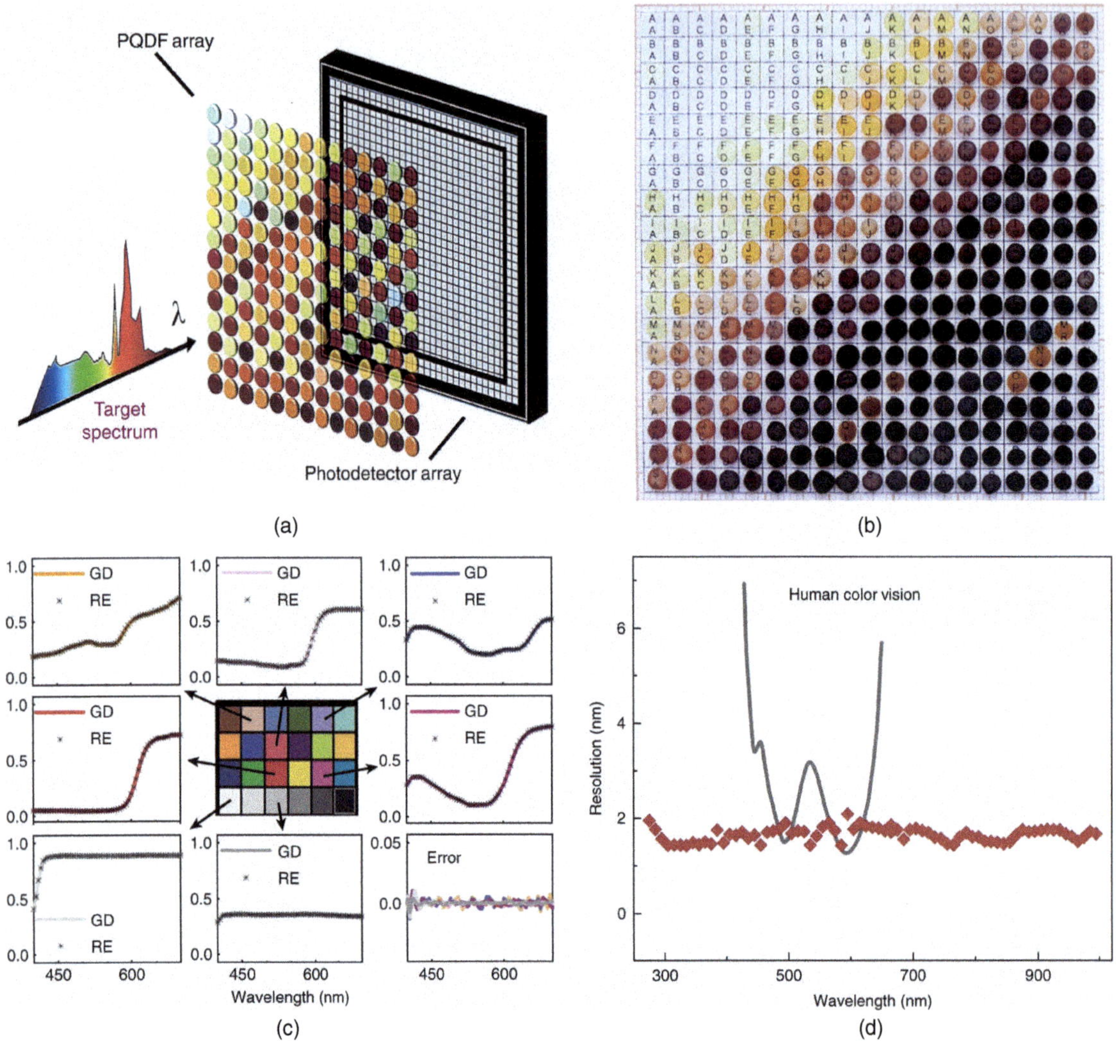

Figure 8.12 Quantum dot spectrometer. (a) Schematic and (b) detailed configuration. *Source*: Bao and Bawendi (2015). © 2015 Springer Nature.

eliminates the need for components, like displays, A/D converters, and other peripheral communication circuitry (Long et al. 2014; McGonigle et al. 2018). An external battery-powered light source can be used with a grating directly in line with the smartphone camera, as shown in Figure 8.13. A test sample is introduced in this path and the spatially dispersed light is captured on the smartphone camera (de Oliveira et al. 2017). By accessing the raw optical data, the effect of Bayer color filters can be suppressed. Finally, using light sources with known wavelengths, a calibration can be set up.

This approach offers a cost-effective way to realize spectrometers without designing displays and other electronic components. However, not many commercial spectrometers based on this approach have reached the market, except for some do-it-yourself (DIY) kits that will be introduced in the next subsection. The technical details of smartphone-based spectrometers are described in the chapter in this Volume by Scheeline, and the field clinical applications in the chapter by Algar in Volume 2.

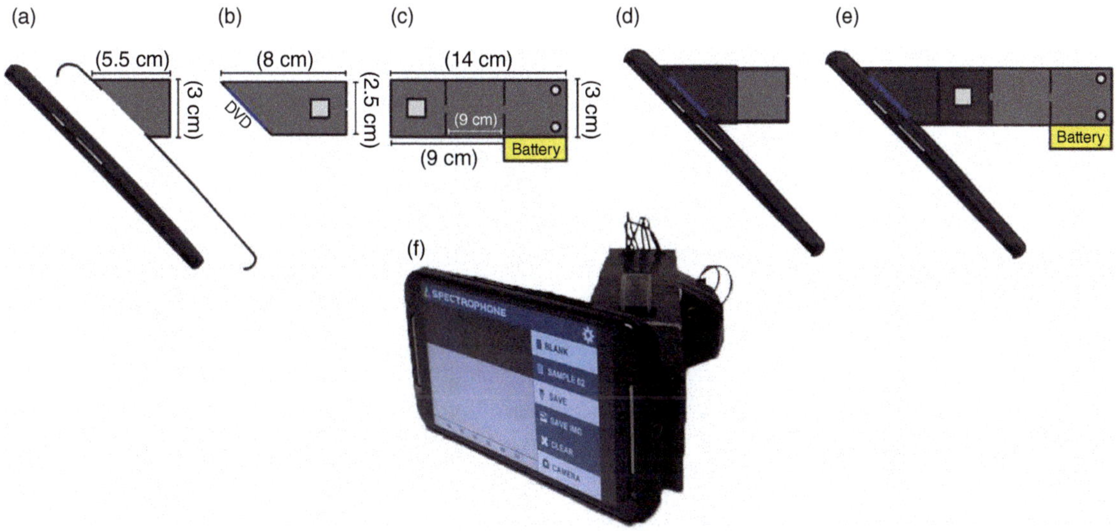

Figure 8.13 Spectrometer design utilizing the camera of a smartphone as the detector. *Source*: Permission from Oliviera et al. (2017). © 2017 Elsevier.

8.4.5 "DIY" Approaches

The emergence of rapid prototyping using 3D printers, laser cutters, and computer numerical control (CNC) machines has also ushered in DIY approaches for building spectrometers from readily available and inexpensive components (Hopkins 2014; Scheeline and Kelley 2013). For instance, Public Lab offers spectrometer kits for under $10 consisting of a laser-cut cardboard attachment housing a DVD fragment that serves as the grating, as shown in Figure 8.14a. This attachment can be aligned with the camera of a smartphone to observe spectra using any camera app. This approach offers a great way to introduce spectroscopy into a teaching curriculum without the need for expensive spectrometers. Since these devices are suitable for education and hobby activities, they have

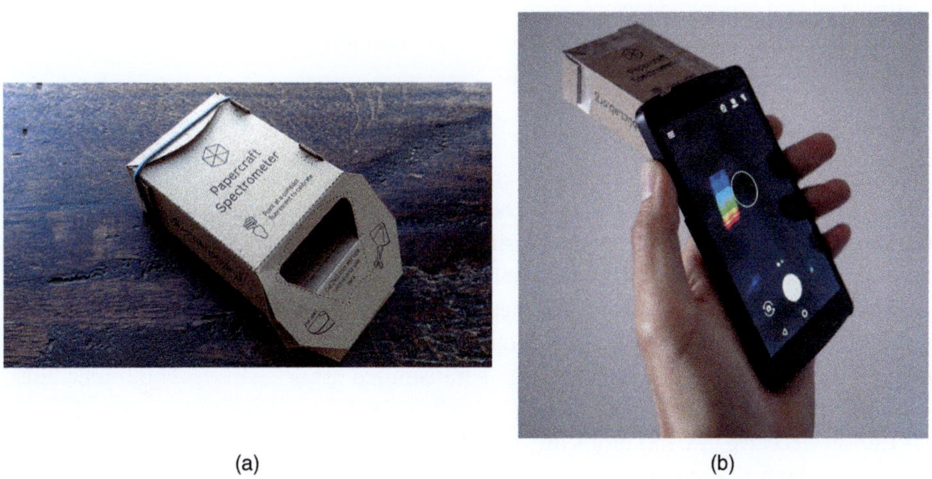

Figure 8.14 DIY Spectrometer Kits. (a) Attachment for a smartphone. (b) Complete assembly. *Source*: Permission from Jeffrey Yoo Warren and Public Lab.

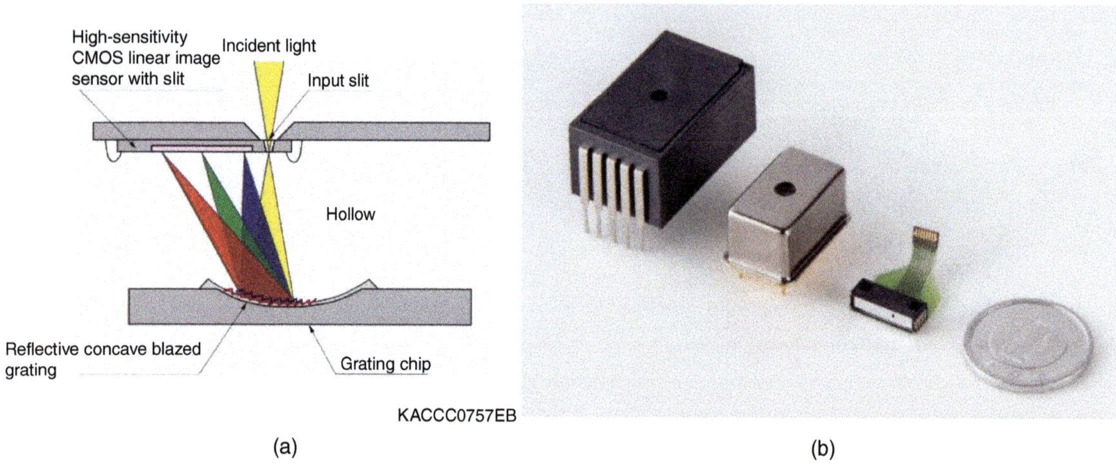

Figure 8.15 (a) Architecture of MEMS-based spectrometer. *Source*: Structure/Technology | Mini-spectrometers, Hamamatsu. © Hamamatsu Photonics K.K. (b) Line of micro-spectrometers from Hamamatsu Photonics. *Source*: Photo courtesy of Hamamatsu Photonics.

limited applicability in scenarios that require high sensitivity and robust operation. In one instance, a webcam was modified with fiber optic cables to perform the action of an *n*-channel spectrometer (Sumriddetchkajorn and Intaravanne 2012).

8.4.6 Compact Chip-Based Spectrometers

With the advent of advanced lithographic techniques for fabricating nanostructures, it is becoming possible to integrate optical assemblies at a chip level. For instance, Hamamatsu Photonics (Shibayama and Suzuki 2008) introduced a microspectrometer chip that had a blazed grating and CMOS linear sensor integrated into a single chip, as shown in Figure 8.15a. The spectrometer chip was fabricated using techniques commonly used in micro-electromechanical systems (MEMS) manufacturing. These chips only weigh about 5 g, have a wavelength resolution around 10–15 nm, and can operate from 340 to 840 nm. Recently, an SMD spectrometer chip (Figure 8.15b) was introduced that is designed to operate in the 640–1050 nm range and has a weight of 0.3 g, thereby making it suitable for applications where lightweight sensors are desirable.

8.4.7 Standalone Smartphone Spectrometers

Smartphone-based spectrometers could be defined as devices where a significant process of the spectral conversion and postprocessing is carried out in a smartphone. This could be either using the phone's camera as an image sensor or capturing wireless spectral data followed by postprocessing on the phone through an app. Most examples introduced in previous sections used the camera of a smartphone and postprocessing to obtain a spectrum. In this subsection, some examples of standalone smartphone spectrometers are presented. Here, the spectrometer does not use the smartphone camera but has an inbuilt detector along with dispersing elements and other circuitry (Das et al. 2016). Most commonly, data captured from the sensor is transmitted using a wireless communication framework.

Bluetooth antenna and transmit/receive (Tx/Rx) protocols have been used in most such devices. An app on the smartphone is able to control the operation of the spectrometer, trigger data capture and Rx/Tx operations. Das et al. (2016) demonstrated a high-performance smartphone spectrometer design that used the Hamamatsu microspectrometer chip, as shown in Figure 8.16. They designed and fabricated all the wireless and control

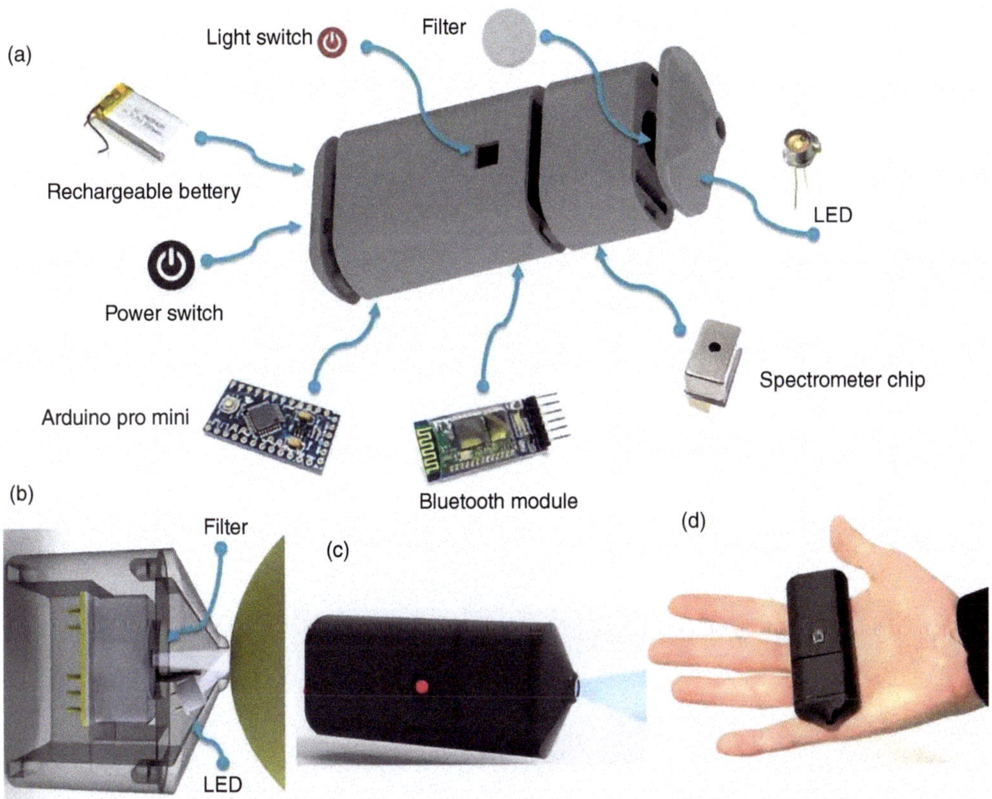

Figure 8.16 Components involved in a standalone smartphone spectrometer developed by Labby Inc. *Source*: Das et al. (2016). Licensed under CC BY 4.0.

circuitry along with a customized smartphone app that could calibrate and process the data on the phone itself. This configuration is advantageous for remote operation where the smartphone needs to be at a distance from the spectrometer. Bluetooth antenna can communicate with the smartphone at a distance of up to 9–12 m for classic chips and beyond 100 m for low-energy chips. Additionally, it can work with any smartphone that has the app installed, without having to deal with specific camera configurations. However, they are more expensive mainly due to the chip.

8.5 Applications

In this section, various applications of portable UV–Vis spectrometers will be discussed.

8.5.1 Color Sensing

By far, the most common use of portable UV–Vis spectrometers is in color sensing (Giusti and Wrolstad 2001). The ability to quantifying color characteristics in terms of RGB, Lab, XYZ, ΔE (color difference), and other metrics is central to several industries. Color matching is an important process for selecting paints in virtually any application that uses them. Portable color sensors are routinely used for color selection and matching in dentistry (Brewer et al. 2004; Sproull 1973), automotive (Streitberger and Dössel 2008), displays (Streitberger and Dössel 2008), food and

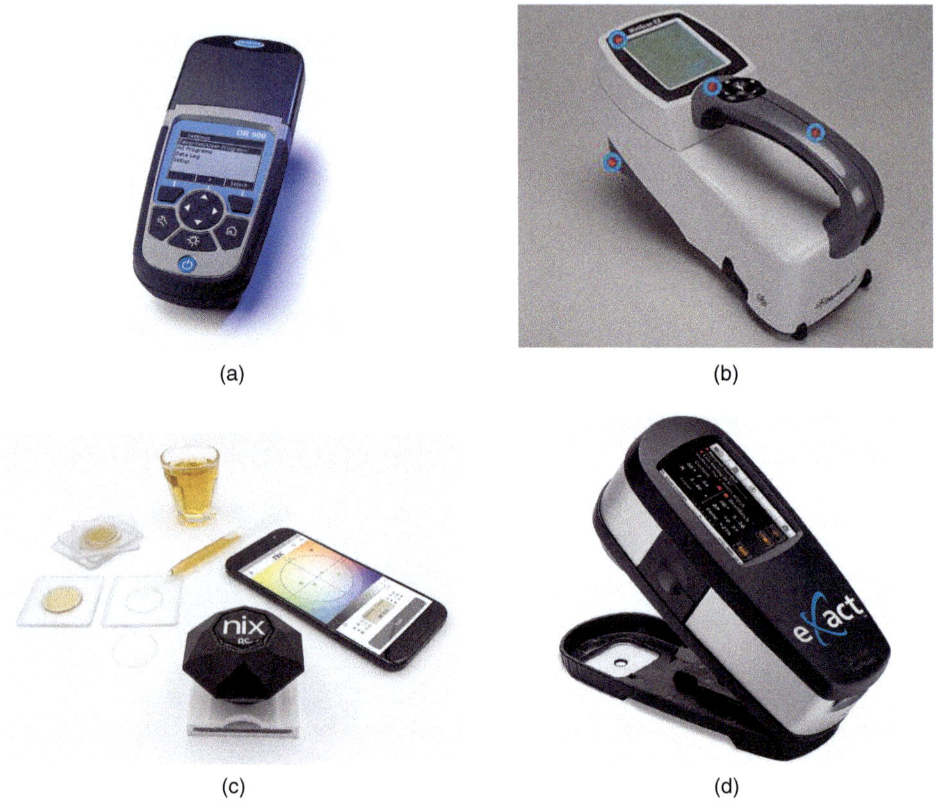

Figure 8.17 Some portable color-sensing spectrometers. (a) Hach DR900. *Source*: Courtesy of Hach. (b) HunterLab Miniscan EZ 4500L. *Source*: Hunter Associates Laboratory, Inc. (c) Nix QC. *Source*: NIX SENSOR, LTD. (d) X-Rite eXact. *Source*: X-Rite, Incorporated.

beverage (Giusti and Wrolstad 2001), anthocyanins (Giusti and Wrolstad 2001), and textile (Berns 1992; Zhang and Li 2008) industries among many others. Spectrometers are used to obtain a calibrated reflection spectrum which is the first step in color measurement. Depending on the illuminant and International Commission on Illumination (CIE) color space, a reflection spectrum can be transformed to RGB, Lab, cyan, magenta, yellow, and key (CMYK), and other systems. Notable examples of portable color sensors are X-Rite, HunterLab, Nix, and Hach, as shown in Figure 8.17. Some color sensors, based on discrete filter sensors, are the size of a lipstick and can cost as low as $59.

In addition to color systems like RGB and Lab, industrial applications have their own standards for quality control of oils, lubricants, beer, wine, etc. For instance, ASTM D1500D and D6045 are color-based methods for quantifying oil quality. ASBC 10 (standard reference method, SRM) is a standard for measuring beer color, whereas OIV-MA-AS2-07B is a standard for wine color (tone and intensity) using spectrophotometric methods (Streitberger and Dössel 2008).

8.5.2 Life Science Assays

Portable UV–Vis spectrometers have shown great potential in point-of-care (POC)-based assays, especially in enzyme-linked immunosorbent assay (ELISA) (Thiha and Ibrahim 2015). Most implementations of ELISA use either fluorescence or absorption spectroscopy in the UV–Vis spectrum, which makes UV–Vis spectrometers ideal for such applications (Schmid 2001). Long et al. demonstrated a smartphone-phone spectrometer-based ELISA for detecting peanut allergens (Long et al. 2014). They used the camera of the smartphone as the detector, earlier

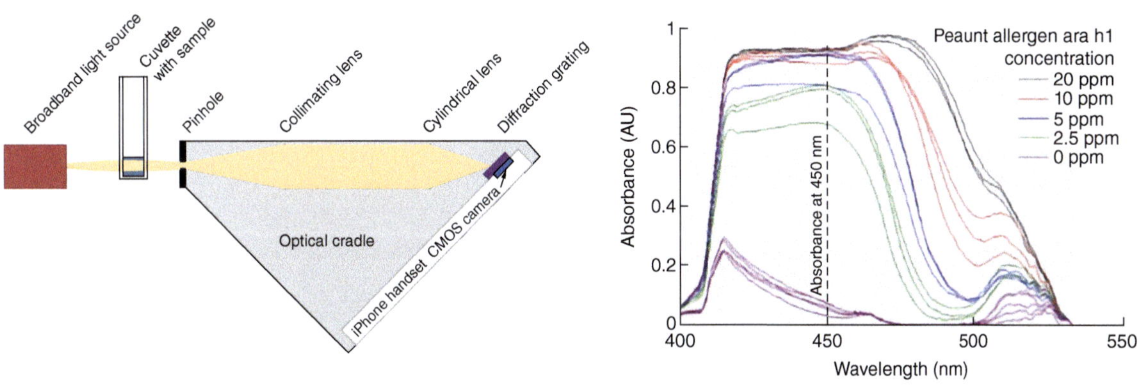

Figure 8.18 (a) Smartphone spectrometer arrangement for ELISA. (b) Absorption spectra correlated to peanut allergen. *Source*: Long et al. (2014). © 2014 Optical Society of America.

described in Section 8.4.4. A schematic of their setup is shown in Figure 8.18a along with the absorption spectra in Figure 8.18b.

Nucleic acids like deoxyribonucleic acid (DNA) and ribonucleic acid (RNA) have a characteristic absorption peak at around 260 nm, which can be used to detect and quantify their concentration (Desjardins and Conklin 2010). Additionally, proteins have an absorption centered at around 280 nm. Portable UV spectrometers are well-suited for these applications. For instance, the NanoDrop™ is a portable spectrophotometer that only intakes 1–2 μl of a sample and can analyze absorption or fluorescence spectra to detect concentrations of nucleic acids and proteins. Figure 8.19a shows the NanoDrop One/OneC spectrophotometer manufactured by Thermo Fisher. Another small-volume spectrometer is the UV5Nano from Mettler Toledo, as shown in Figure 8.19b.

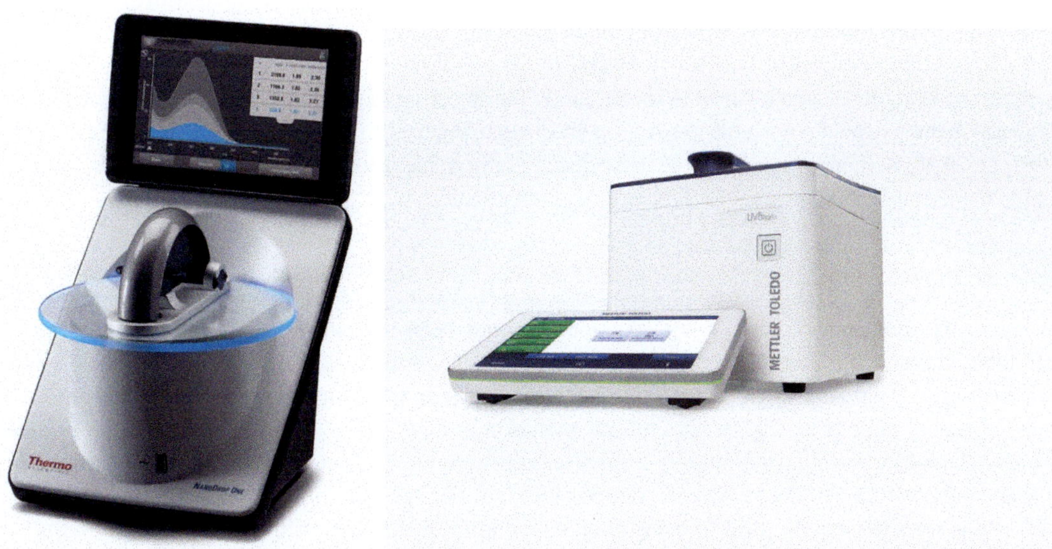

Figure 8.19 (a) NanoDrop One/OneC spectrophotometer. *Source*: Permission from Thermo Fisher. (b) UV5Nano small volume spectrometer. *Source*: METTLER TOLEDO.

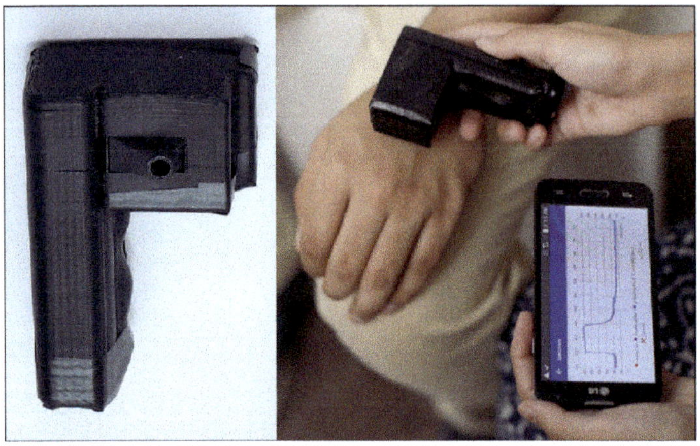

Figure 8.20 Smartphone-based fluorescence spectrometer for the screening of skin disease. *Source*: Poojary et al. (2019).

8.5.3 Biomedical Applications

Several molecules that constitute blood, urine, and skin have absorption or fluorescence in the UV–Vis spectrum. As a result, spectrometers operating in this spectral region can be used for not only detecting concentrations of chromophores like hemoglobin (Edwards et al. 2017), but also screen for any disease or abnormalities in a POC setting (Das et al. 2015). Wahi et al. (2018) demonstrated a smartphone spectrometer to excite fluorophores in the skin, such as collagen and NADH (nicotinamide adenine dinucleotide (NAD) plus hydrogen (H)) with 385 nm UV-A light for diagnostic applications. Poojary et al. (2019) used a portable smartphone spectrometer to screen for skin diseases like vitiligo by observing UV fluorescence, as shown in Figure 8.20. Their device was effective in detecting hyperpigmentation and hypopigmentation in the skin with the advantage of tracking variations over time. This could be useful in monitoring treatments of patients who are receiving different medications and determining efficacies (Figure 8.21).

Edwards et al. demonstrated a Grating-Fresnel (G-Fresnel) spectrometer design that could measure hemoglobin concentration from diffuse reflectance spectra from tissue phantoms (Edwards et al. 2017). They obtained errors under 10% for hemoglobin concentration, which establishes the efficacy of these devices for POC settings. In another study, Maity et al. developed a portable spectrometer for detecting malaria in blood samples (Maity et al. 2019). Thiha et al. demonstrated a portable absorption spectroscopy-based ELISA reader for diagnosing dengue fever (Maity et al. 2019). Although only a few of these devices have made it to the market, these studies point to the potential of the portable spectrometer for biomedical applications.

8.5.4 Water Testing

An established application of portable UV–Vis spectrometers has been in water quality testing, with several products available today in the market. Water quality analysis plays an important role in the individual- and community-level health. There is always a threat for water contamination from various sources, such as industrial waste, sewage (Langergraber et al. 2004), agricultural pollution among others. In such scenarios, a portable device can prove crucial for testing water quality from reservoirs and other sources. For instance, measuring the absorption spectrum of water can detect several colored materials like dyes (Antonov et al. 1999; Özdemir et al. 2017). Additionally, reagents can be used to bring about some kind of contrast to improve the detectability of trace or nonabsorbing contaminants. The Hach DR3900 is an excellent example of a portable water quality tester that has been used for detecting a host of contaminants, such as chromium, ammonium, and phosphorus, among others, as shown in Figure 8.22.

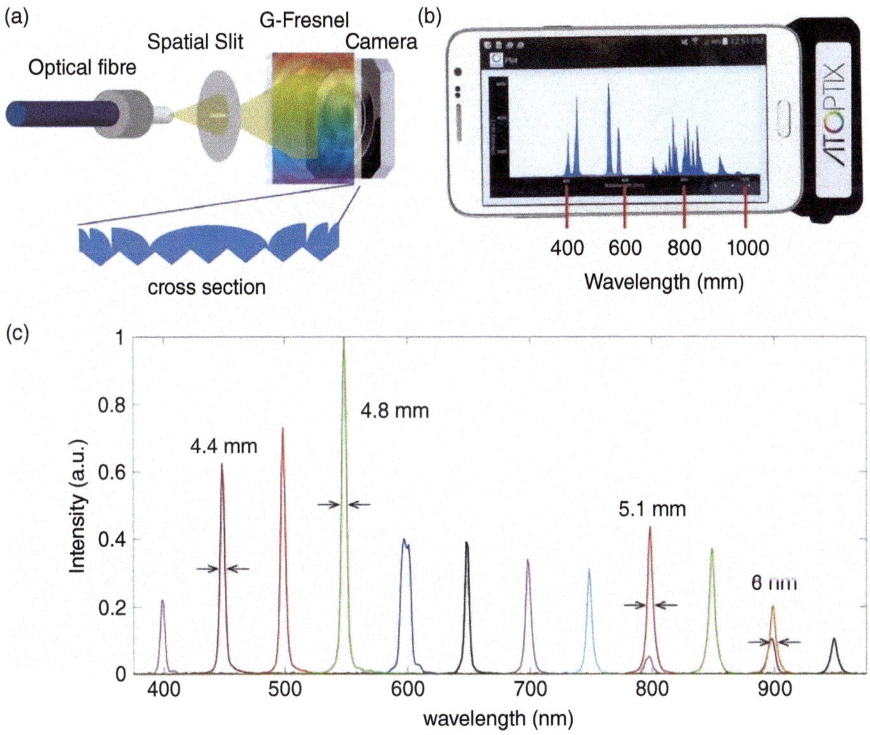

Figure 8.21 G-Fresnel spectrometer for detecting hemoglobin concentration. *Source*: Edwards et al. (2017). Licensed under CC BY 4.0.

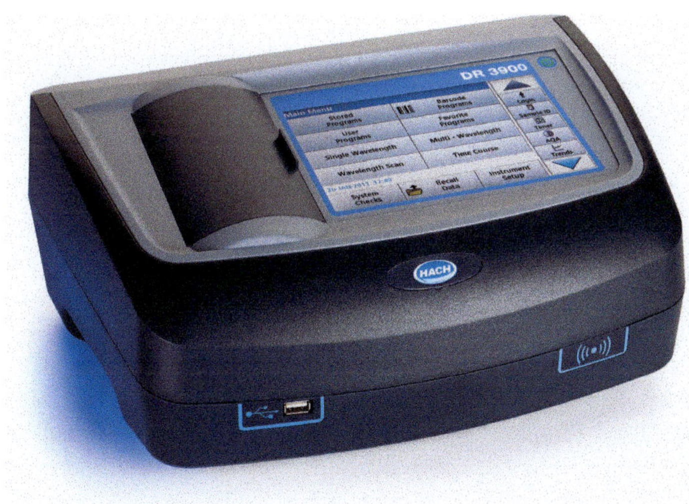

Figure 8.22 Hach DR3900 spectrophotometer for water quality analysis. *Source*: Courtesy of Hach.

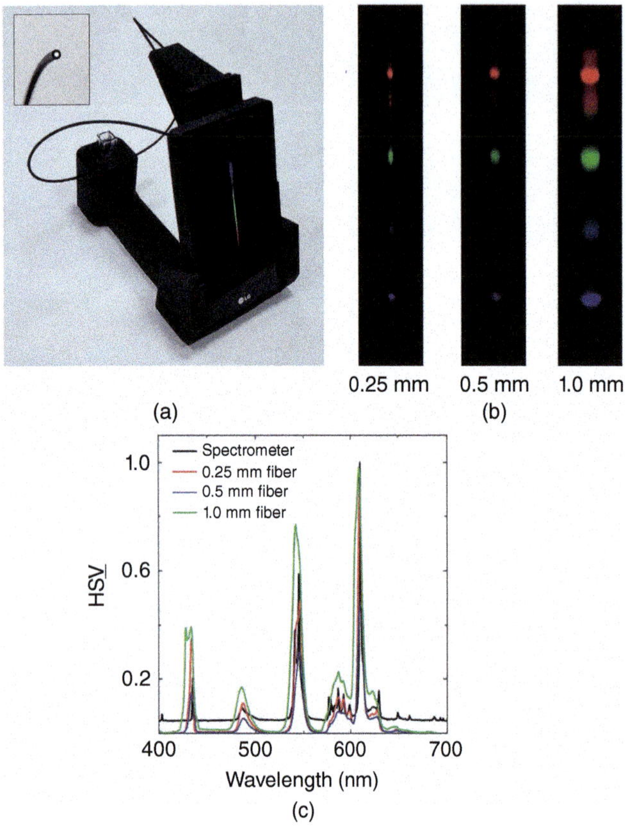

Figure 8.23 Smartphone spectrometer for detecting methylene blue in the water. (a) Smartphone spectrometer prototype with an optical fiber coupled to the camera flash. (b) Images of light output from the fibers. (c) Spectra corresponding to the fibers compared with a standard spectrometer. *Source*: Özdemir et al. (2017). © 2017 Royal Society of Chemistry

In a recent study, Özdemir et al. (2017) demonstrated a smartphone spectrometer for detecting dyes like methylene blue by adsorbing dye molecules on polyacrylonitrile (PAN)/zeolite nanofibers, as shown in Figure 8.23. In another example, Wang et al. used a DVD grating-based smartphone spectrometer to detect neurotoxins in water (Wang et al. 2016).

8.5.5 Food and Beverage Applications

By far the most commonly researched application of portable spectrometers is in the space of food and beverage quality analysis. They have been used for a range of applications, such as detecting ripeness (Das et al. 2016), composition, chlorophyll content (Maxwell and Johnson 2000; Porra et al. 1989), anthocyanins (Boulton 2001; Giusti and Wrolstad 2001; Gupta et al. 2018), phenols (Acevedo et al. 2007; Martelo-Vidal and Vázquez 2014), and meat freshness (Boulton 2001; Giusti and Wrolstad 2001; Gupta et al. 2018), among others. This is attributed to the fact that fruits, vegetables, cheese, meats, and seafood all contain components that absorb, reflect, or fluoresce under UV–Vis light. Many of these components also change as foods age, and, hence, can be good indicators of ripeness or freshness. For instance, Das et al. (2016) developed a portable smartphone spectrometer that utilized a longitudinal variation of chlorophyll fluorescence to determine the ripeness of a variety of apples, as shown in Figure 8.24.

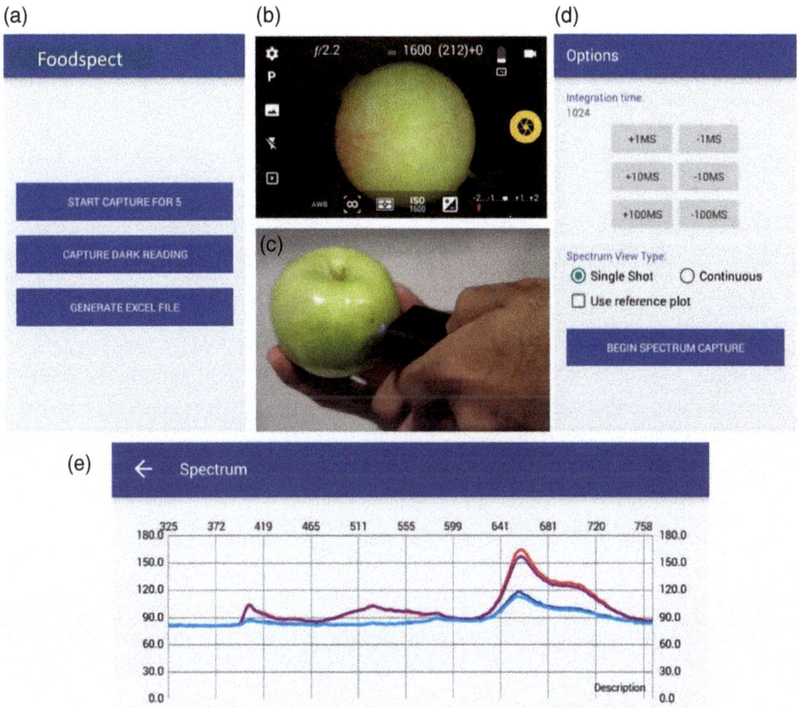

Figure 8.24 Smartphone spectrometer for detecting ripeness of apples using chlorophyll fluorescence. *Source*: Permission from Das et al. (2016).

Figure 8.25 Linksquare smartphone spectrometer being used in fruit freshness analysis. *Source*: Permission from Linksquare and You et al. (2017).

In this application space, there are a few commercial portable Vis spectrometers out in the market, such as Linksquare, which have shown some success in classifying food and detecting freshness, as shown in Figure 8.25a. You et al. (2017) used the Linksquare spectrometer to classify identical-looking food powders, like sugar, salt, rice, etc., using reflection spectroscopy, as shown in Figure 8.25b. They used machine learning algorithms for classification and achieved near 100% accuracy for eight different types of food powders.

In another report, Cai et al. (2017) demonstrated pencil-like spectrometer using a prism-grating configuration with an image sensor, as shown in Figure 8.26a. The spectrometer operated in the 400–675 nm region with a

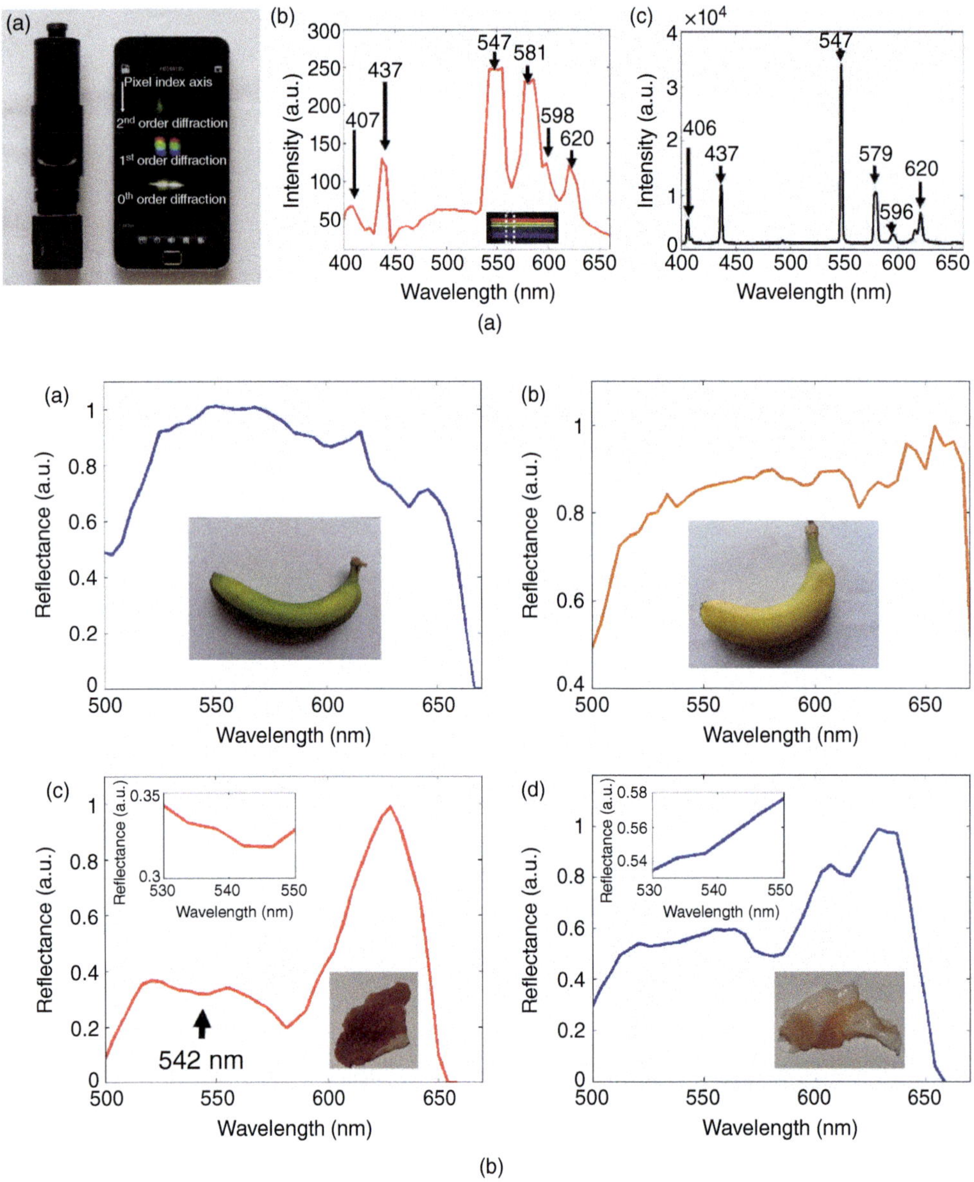

Figure 8.26 Top. Panel A, Pencil-like smartphone spectrometer. Panel B, Mercury lamp spectrum captured by the pencil spectrometer. Panel C, Mercury lamp spectrum obtained using a commercial spectrometer. Bottom. Panel A, Reflectance spectrum of unripe banana. Panel B, Reflectance spectrum of ripe banana. Panel C, Reflection spectrum of pork muscle. Panel D, Reflection spectrum of pork fat. *Source*: Cai et al. (2017). © 2017 Optical Society of America.

resolution of 17 nm. The spectrometer was used for several applications including detecting ripeness of bananas and freshness of pork, as shown in Figure 8.26b.

8.5.6 Geosensing Applications

Spectroscopy has been a powerful technique for atmospheric and environmental remote sensing. In recent years, portable spectrometers have been increasingly used in volcanology, aerosol sensing, and detection of atmospheric

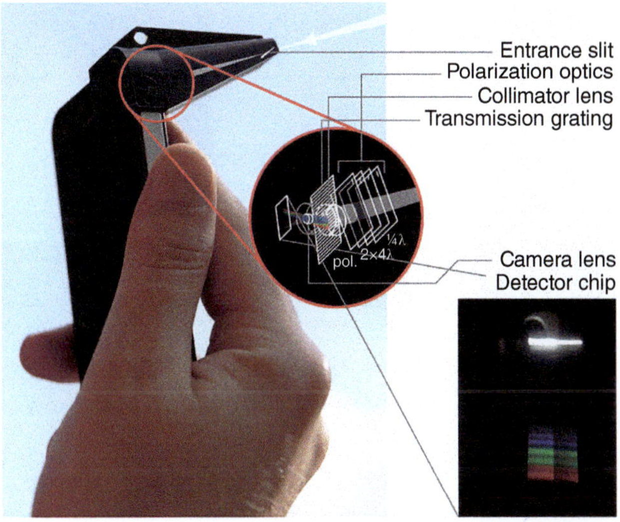

Figure 8.27 Smartphone spectropolarimeter for sensing atmospheric aerosols. *Source*: Snik et al. (2014). © 2014 John Wiley & Sons

gases, like SO_2, among other applications. For instance, Snik et al. (2014) demonstrated a smartphone spectropolarimeter for aerosol mapping, which measured the degree of linear polarization as a function of wavelength, as shown in Figure 8.27. They designed a low-cost add-on comprising polarizers and a transmission grating, which was attached to the camera of the smartphone. Using this miniature arrangement, multiple (over 3000) measurements could be conveniently made by "citizen spectroscopists." Reasonable agreement between ground-based spectrometers was observed, which establishes the potential of portable spectrometers for aerosol sensing.

Galle et al. (2003) developed a portable UV spectrometer for sensing SO_2 emissions for applications in volcanic surveillance. The spectrometer had a Czerny–Turner configuration and operated in the 245–380 nm spectral range for achieving differential optical absorption. A telescope was used to capture the scattered UV light from aerosols, which was then coupled to the spectrometer using an optical fiber cable, as shown in Figure 8.28a. The telescope-spectrometer arrangement could be mounted on a car, a helicopter, or a tripod, such that measurements

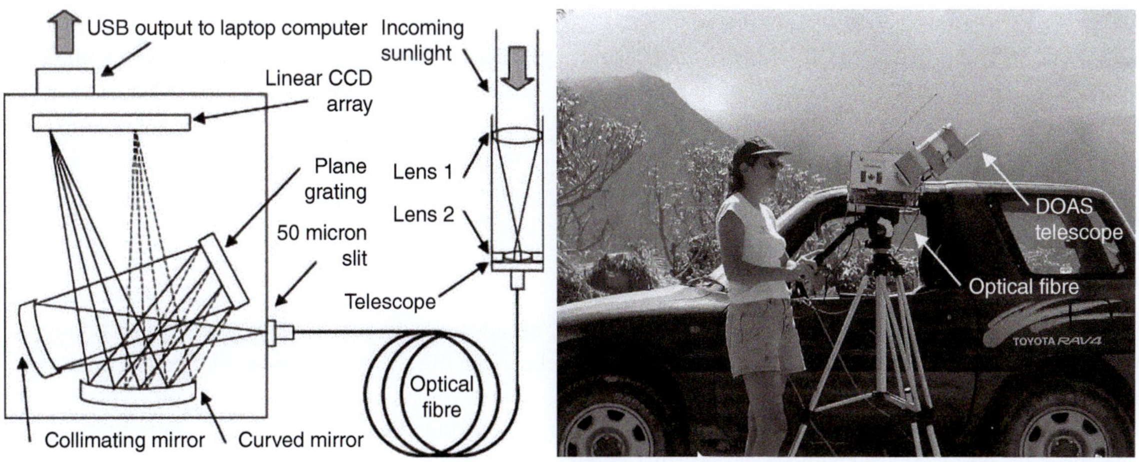

Figure 8.28 Portable UV spectrometer for SO_2 detecting during volcanic eruptions. (a) Spectrometer design. (b) Field implementation. *Source*: Permission from Galle et al.

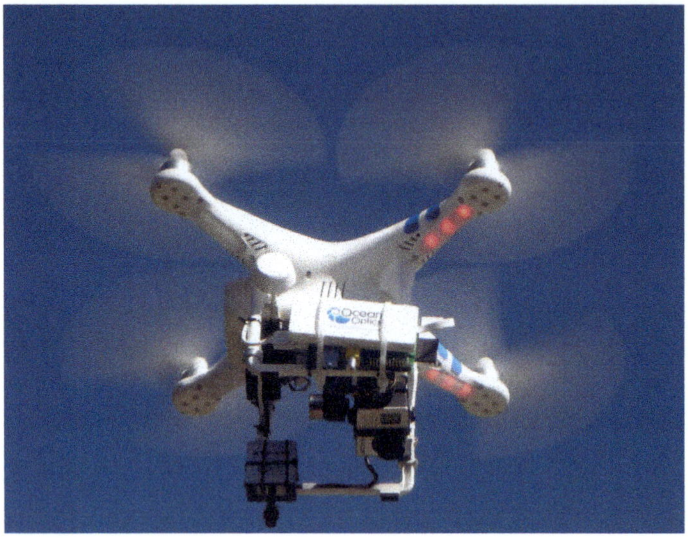

Figure 8.29 Ocean Optics STS spectrometer mounted on a drone. *Source*: Ocean Optics Inc. and VOXearch Inc.

could be made close to the site of the volcanic eruption, as shown in Figure 8.28b. Compared to the existing SO_2 detectors, this approach was attractive as the spectrometer could capture the entire UV spectrum, thereby providing greater flexibility to identify errors and apply better statistical methods. Furthermore, a good agreement was observed with a standard correlation spectrometer, which established the accuracy of their prototype.

8.5.7 Spectrometers on Drones and Aerial Platforms

Due to the recent availability of compact microcontrollers, like Raspberry Pi and Arduino, and the ease of interfacing with a host of peripheral components, it has become increasingly possible to integrate spectrometers on to unmanned aerial vehicles (UAVs) or drones and other aerial platforms (Dwight et al. 2018; Hakala et al. 2018). This allows for surveillance of vegetation and crops, forests, and other natural resources on a large scale. For instance, Ocean Optics has developed a spectrometer kit designed for drone applications, as shown in Figure 8.29. This spectrometer operates in three ranges – UV (190–650 nm), Vis (350–800 nm), and NIR (650–1100 nm). It has a 2.4 GHz Wi-Fi link that operates in a range of up to 150 m. Furthermore, the kit works on a Raspberry Pi microcontroller, which makes it straightforward to control and capture data. Due to this feature, it can work with most imaging and navigation systems. The drone-spectrometer has already been used by research groups for monitoring for pesticides and fertilizers in farms.

8.6 Challenges for Portable Spectrometers

8.6.1 Tools Versus Toys

Portable UV–Vis spectrometers have been in the market for decades now and have shown promise in a range of applications. However, with the emergence of new approaches, like filter arrays, chip-based sensors, micromirrors, and image-based spectrometers, there has been an explosion in researchers demonstrating proof-of-concept demonstrations. While this opens up new opportunities for applications, there are instances where there is not much focus on performance characterization. Just being able to detect a spectrum is not a sufficient criterion for the method to be applicable in a realistic application. Spectroscopic measurements go beyond mere spectral acquisition and involve several steps from sample information/preparation, dark and reference acquisition, sample acquisition, spectral data conditioning, postprocessing, and, finally, interpretation. Without careful assessments at each stage, errors can propagate and result in less-effective measurements. A distinction should be made whether the spectrometer is a "toy" or a "tool." A toy version would be great for educational and hobby use, whereas a tool version would be imposed with stringent performance requirements. The issues involved in the construction of appropriate libraries and calibrations are described in the chapter by Schreyer.

Although there are several approaches and reports of new spectrometer designs, only a few of them have made it to the market. Within them, a smaller subset has been commercially successful. By looking at established spectrometers in the market, it is possible to impose certain constraints, such as ease of manufacturing, scalability, robustness, among others, that will determine the successful deployment of newer portable spectrometer technologies. This section highlights a few aspects that will aid in the development of robust spectrometers.

8.6.2 Need for Robust Protocols

Establishing a protocol is critical for obtaining effective test results. A robust protocol starts with proper attention to the sample. Is there sample preparation needed? Is the sample contaminated? Is the sample homogenous? Are there relevant spectral changes in the region being measured? These are a few questions to ask before actually going ahead with tests. Unfortunately, there have been instances of commercial demonstrations where a portable spectrometer is shown to operate with random sample geometries, and accurate results are claimed. Materials like animal feed and hay are very inhomogenous, and a forage testing lab would make pellets out of them for accurate testing. Yet, we can see some portable spectrometers testing piles of hay in the field and claiming some kind of utility from that. This issue is explored in the context of NIR spectroscopy in the chapters in Volume 2 by Siesler and colleagues.

Additionally, if the same spectrometer is used for analyzing multiple materials, like produce, cheese, or meats, then there needs to be different protocols as the raw signal from them is different. The signal will have different noise characteristics, the texture of various materials can have different specular and diffuse reflection features, and the homogeneity may be different. In such cases, there is a need for customizing the protocol to suit each fruit, vegetable, meat type, etc. In general, without a robust testing protocol, it will not be feasible to obtain meaningful test information.

8.6.3 Noise Suppression

Since portable spectrometers are often operated in the field, there is a chance of noise entering the system from a range of sources. Unlike tests done in a controlled laboratory setting, noise plays a significant role in uncontrolled environments. Specifically, in the case of UV–Vis spectrometers, ambient light can be a source of the noise. Hence, care must be taken to isolate the sample from ambient noise, i.e. eliminate stray light. In reflection mode, the distance between the sample and spectrometer can increase the chance of stray light and hinder repeatability. The design and user protocol need to ensure that the distance remains fixed. Finally, noise can also originate from electronics. Dark noise subtractions must be performed routinely before testing samples. Additionally, light source fluctuations should be studied and their effect needs to be quantified on the accuracy of testing.

8.6.4 Calibration

Wavelength calibration is a critical process in spectroscopy and plays an important role in determining the accuracy of measurements. All factors that can affect calibration need to be considered, such that their effect can be quantified and minimized (Burggraaff et al. 2019). For instance, in camera-based spectrometers, internal Bayer filters in the image sensor can disrupt the calibration. The use of RAW format images has shown to minimize the effect of these filters. The positioning of the grating must be robust and not change, as that can affect the calibration. In the case of filter array-based spectrometers, the spectral overlap of the filters needs to be considered.

8.6.5 Device-to-Device Repeatability

As the fabrication method becomes more complex, device-to-device performance repeatability can become a challenge. For instance, filter array-based spectrometers require several hundreds of optical filters to achieve wavelength separation. Small changes in the spectral characteristics of these filters will change the calibration. In the case of chip-based spectrometers, small tolerances during fabrication can change the performance. Protocols for calibrating every unit are needed, such that their performance is identical.

8.7 Outlook

The notion of a "lab" is taking a new meaning thanks to new sensors, the increased computing power of smartphones and the emergence of cloud functionality. On the one hand, today's smartphones boast of 6 GB random access memory (RAMs), 2 GHz processors, and a range of wireless communication approaches with the availability of programmable open source components. On the other hand, miniaturized spectrometers have been realized due to advanced fabrication approaches. The merger of these technologies offers promising opportunities for a range of applications that were not feasible earlier. Today, portable spectrometers can be seamlessly integrated into smartphones via apps and be made more effective through cloud computing and machine learning. It will be possible to generate large spectroscopic datasets for a range of food, beverage, and agricultural products that will pave the way for intelligent sensors that "learn" over time and enable new applications like counterfeit detection, predict harvest times, and improve the distribution of produce among other uses: so-called "smart agriculture."

References

Acevedo, F.J., Jiménez, J., Maldonado, S. et al. (2007). Classification of wines produced in specific regions by UV–visible spectroscopy combined with support vector machines. *Journal of Agricultural and Food Chemistry* 55 (17): 6842–6849. https://doi.org/10.1021/jf070634q.

Acree, W.E. (2006). Detectors, absorption and luminescence. In: *Encyclopedia of Analytical Chemistry*. John Wiley & Sons, Ltd https://doi.org/10.1002/9780470027318.a5405.

Antonov, L., Gergov, G., Petrov, V. et al. (1999). UV-Vis spectroscopic and chemometric study on the aggregation of ionic dyes in water. *Talanta* 49 (1): 99–106. https://doi.org/10.1016/S0039-9140(98)00348-8.

Ball, D.W, Society of Photo-optical Instrumentation Engineers (2006). *Field guide to Spectroscopy*. SPIE.

Bao, J. and Bawendi, M.G. (2015). A colloidal quantum dot spectrometer. *Nature* 523 (7558): 67–70. https://doi.org/10.1038/nature14576.

Berns, R.S. (1992). Instrumental colour measurements and computer aided colour matching for textiles, by H. S. Shah and R. S. Gandhi, Forwand, by Fred W. Billmeyer, Jr., Mahajan book distributors, Ahmedabad, India, 1990, 370 pp., 12 color plates, hardbound, $60.00. *Color Research & Application* 17 (1): 62–62. https://doi.org/10.1002/col.5080170111.

Boulton, R. (2001). The copigmentation of Anthocyanins and its role in the color of red wine: a critical review. *American Journal of Enology and Viticulture* 52 (2): 67–87.

Brewer, J. D., Wee, A., & Seghi, R. (2004). Advances in color matching. *Dental Clinics of North America*, 48(2), v, 341–358. https://doi.org/10.1016/j.cden.2004.01.004

Burggraaff, O., Schmidt, N., Zamorano, J. et al. (2019). Standardized spectral and radiometric calibration of consumer cameras. *Optics Express* 27 (14): 19075. https://doi.org/10.1364/oe.27.019075.

Cai, F., Wang, D., Zhu, M., and He, S. (2017). Pencil-like imaging spectrometer for bio-samples sensing. *Biomedical Optics Express* 8 (12): 5427. https://doi.org/10.1364/boe.8.005427.

Chou, S.Y., Krauss, P.R., Renstrom, P.J. et al. (2014). Nanoimprint lithography. *Nanoimprint Lithography* 4129 (1996): 9–14. https://doi.org/10.1116/1.588605.

Crocombe, R.A. (2013). *Handheld Spectrometers: The State of the Art* (eds. M.A. Druy and R.A. Crocombe). SPIE https://doi.org/10.1117/12.2017892.

Crocombe, R.A. (2018). Portable spectroscopy. *Applied Spectroscopy* 72 (12): 1701–1751. https://doi.org/10.1177/0003702818809719.

Das, A., Swedish, T., Wahi, A. et al. (2015). *Mobile Phone Based Mini-Spectrometer for Rapid Screening of Skin Cancer* (eds. M.A. Druy, R.A. Crocombe and D.P. Bannon). SPIE https://doi.org/10.1117/12.2182191.

Das, A.J., Wahi, A., Kothari, I., and Raskar, R. (2016). Ultra-portable, wireless smartphone spectrometer for rapid, non-destructive testing of fruit ripeness. *Scientific Reports* 6 https://doi.org/10.1038/srep32504.

Desjardins, P. and Conklin, D. (2010). NanoDrop microvolume quantitation of nucleic acids. *Journal of Visualized Experiments* 45 https://doi.org/10.3791/2565.

Dwight, J.G., Tkaczyk, T.S., Alexander, D. et al. (2018). Compact snapshot image mapping spectrometer for unmanned aerial vehicle hyperspectral imaging. *Journal of Applied Remote Sensing* 12 (04): 1. https://doi.org/10.1117/1.JRS.12.044004.

Edwards, P., Zhang, C., Zhang, B. et al. (2017). Smartphone based optical spectrometer for diffusive reflectance spectroscopic measurement of hemoglobin. *Scientific Reports* 7 (1): 12224. https://doi.org/10.1038/s41598-017-12482-5.

Galle, B., Oppenheimer., C. et al. (2003). A miniaturised ultraviolet spectrometer for remote sensing of SO_2 fluxes: A new tool for volcano surveillance. *Journal of Volcanology and Geothermal Research* 119, 1–4, 1.

Gelabert, P., Pruett, E., Perrella, G. et al. (2016). *DLP NIRscan Nano: An Ultra-Mobile DLP-Based near-Infrared Bluetooth Spectrometer* (eds. M.R. Douglass, P.S. King and B.L. Lee). SPIE https://doi.org/10.1117/12.2231054.

Giusti, M.M. and Wrolstad, R.E. (2001). Characterization and measurement of Anthocyanins by UV-visible spectroscopy. *Current Protocols in Food Analytical Chemistry* 00 (1): F1.2.1–F1.2.13. https://doi.org/10.1002/0471142913.faf0102s00.

Gubbi, J., Buyya, R., Marusic, S., and Palaniswami, M. (2013). Internet of things (IoT): a vision, architectural elements, and future directions. *Future Generation Computer Systems* 29 (7): 1645–1660. https://doi.org/10.1016/j.future.2013.01.010.

Gupta, O., Das, A.J., Hellerstein, J., and Raskar, R. (2018). Machine learning approaches for large scale classification of produce. *Scientific Reports* 8 (1) https://doi.org/10.1038/s41598-018-23394-3.

Hakala, T., Markelin, L., Honkavaara, E. et al. (2018). Direct reflectance measurements from drones: sensor absolute radiometric calibration and system tests for forest reflectance characterization. *Sensors (Switzerland)* 18 (5) https://doi.org/10.3390/s18051417.

Hopkins, J.L. (2014). DIY Spectroscopy, Chapter 4. In: *Using Commercial Amateur Astronomical Spectrographs*, 101–123. Springer https://doi.org/10.1007/978-3-319-01442-5_4.

James, J.F. (2007). Spectrograph design fundamentals. In: *Spectrograph Design Fundamentals*, vol. 9780521864. Cambridge University Press https://doi.org/10.1017/CBO9780511534799.

Karanassios, V. and Fitzgerald, R. (2018). The Internet of Things (IoT) for a smartphone-enabled optical spectrometer and their use on-site and (potentially) for Industry 4.0. In: *Next-Generation Spectroscopic Technologies XI* (eds. M.A. Druy, R.A. Crocombe, S.M. Barnett, et al.), 4. https://doi.org/10.1117/12.2305466.

Lakowicz, J.R. (2006). Principles of fluorescence spectroscopy. In: *Principles of Fluorescence Spectroscopy*. Springer https://doi.org/10.1007/978-0-387-46312-4.

Langergraber, G., Fleischmann, N., Hofstaedter, F., and Weingartner, A. (2004). Monitoring of a paper mill wastewater treatment plant using UV/VIS spectroscopy. *Water Science and Technology* 49 (1): 9–14. https://doi.org/10.2166/wst.2004.0004.

Loewen, E.G. and Popov, E. (2018). Diffraction gratings and applications. In: *Diffraction Gratings and Applications*. Taylor and Francis https://doi.org/10.1201/9781315214849.

Long, K.D., Yu, H., and Cunningham, B.T. (2014). Smartphone instrument for portable enzyme- linked immunosorbent assays. *Biomedical Optics Express* 5 (11): 3792. https://doi.org/10.1364/BOE.5.003792.

Maity, M., Gantait, K., Mukherjee, A., & Chatterjee, J. (2019). Visible Spectrum-based Classification of Malaria Blood Samples on Handheld Spectrometer. *2019 IEEE International Instrumentation and Measurement Technology Conference (I2MTC)*, pp. 1–5. https://doi.org/10.1109/I2MTC.2019.8826860.

Martelo-Vidal, M.J. and Vázquez, M. (2014). Determination of polyphenolic compounds of red wines by UV-VIS-NIR spectroscopy and chemometrics tools. *Food Chemistry* 158: 28–34. https://doi.org/10.1016/j.foodchem.2014.02.080.

Maxwell, K. and Johnson, G.N. (2000). Chlorophyll fluorescence – a practical guide. *Journal of Experimental Botany* 51 (345): 659–668. https://doi.org/10.1093/jexbot/51.345.659.

McGonigle, A., Wilkes, T., Pering, T. et al. (2018). Smartphone spectrometers. *Sensors* 18 (2): 223. https://doi.org/10.3390/s18010223.

Milton, E.J., Anderson, K., Kneubühler, M., and Fox, N. (2009). Progress in field spectroscopy. *Remote Sensing of Environment* 113: S92–S109. https://doi.org/10.1016/J.RSE.2007.08.001.

Mohamed, K. (2019). Nanoimprint lithography for Nanomanufacturing. In: *Comprehensive Nanoscience and Nanotechnology*, 357–386. Elsevier https://doi.org/10.1016/b978-0-12-803581-8.10508-9.

Nelson, D.L. (2019). Introduction to spectroscopy. In: *Spectroscopic Methods in Food Analysis*, 3–34. Taylor and Francis https://doi.org/10.1201/9781315152769-1.

Ogletree, D. F. (2016). Molecular excitation and relaxation of extreme ultraviolet lithography photoresists. In *Frontiers of Nanoscience* (vol. 11, pp. 91–113). Elsevier. https://doi.org/10.1016/B978-0-08-100354-1.00002-8

de Oliveira, H.J.S., de Almeida, P.L., Sampaio, B.A. et al. (2017). A handheld smartphone-controlled spectrophotometer based on hue to wavelength conversion for molecular absorption and emission measurements. *Sensors and Actuators B: Chemical* 238: 1084–1091. https://doi.org/10.1016/J.SNB.2016.07.149.

Oliver, J., Lee, W.-B., and Lee, H.-N. (2013). Filters with random transmittance for improving resolution in filter-array-based spectrometers. *Optics Express* 21 (4): 3969–3989. https://doi.org/10.1364/OE.21.003969.

Özdemir, G.K., Bayram, A., Kılıç, V. et al. (2017). Smartphone-based detection of dyes in water for environmental sustainability. *Analytical Methods* 9 (4): 579–585. https://doi.org/10.1039/C6AY03073D.

Pavesi, L.V., Lorenzo, L., Vivien, L. et al. (2013). *Handbook of Silicon Photonics*. Taylor and Francis http://books.google.com/books?id=lypGfgJqSM8C&pgis=1.

Poojary, S., Jaiswal, S., Wahi, A. et al. (2019). A portable fluorescence spectrometer as a noninvasive diagnostic tool in dermatology: a cross-sectional observational study. *Indian Journal of Dermatology, Venereology and Leprology* 85 (6): 641. https://doi.org/10.4103/ijdvl.IJDVL_440_18.

Porra, R.J., Thompson, W.A., and Kriedemann, P.E. (1989). Determination of accurate extinction coefficients and simultaneous equations for assaying chlorophylls a and b extracted with four different solvents: verification of the concentration of chlorophyll standards by atomic absorption spectroscopy. *BBA – Bioenergetics* 975 (3): 384–394. https://doi.org/10.1016/S0005-2728(89)80347-0.

Radamson, H.H., Simeon, E., Luo, J., and Wang, G. (2018). Scaling and evolution of device architecture. In: *CMOS Past, Present and Future*, 19–40. Elsevier https://doi.org/10.1016/b978-0-08-102139-2.00002-1.

Scheeline, A., & Kelley, K. (2013). *Cell Phone Spectrometer*. http://www.asdlib.org/onlineArticles/elabware/Scheeline_Kelly_Spectrophotometer/index.html.

Schmid, F.-X. (2001). Biological macromolecules: UV-visible spectrophotometry. In: *Encyclopedia of Life Sciences*. John Wiley and Sons https://doi.org/10.1038/npg.els.0003142.

SCIO (n.d.). SCiO – The World's First Pocket Sized Molecular Sensor. https://www.consumerphysics.com (accessed 2 December 2019).

Shibayama, K. and Suzuki, T. (2008). Patent No. US8027034B2. https://patents.google.com/patent/US8027034.

Snik, F., Rietjens, J.H.H., Apituley, A. et al. (2014). Mapping atmospheric aerosols with a citizen science network of smartphone spectropolarimeters. *Geophysical Research Letters* 41 (20): 7351–7358. https://doi.org/10.1002/2014GL061462.

Sproull, R.C. (1973). Color matching in dentistry. Part II. Practical applications of the organization of color. *The Journal of Prosthetic Dentistry* 29 (5): 556–566. https://doi.org/10.1016/0022-3913(73)90036-X.

Spudich, T.M., Utz, C.K., Kuntz, J.M. et al. (2003). Potential for using a digital micromirror device as a signal multiplexer in visible spectroscopy. *Applied Spectroscopy* 57 (7): 733–736. https://doi.org/10.1366/000370203322102799.

Streitberger, H.J. and Dössel, K.F. (2008). Automotive paints and coatings. In: *Automotive Paints and Coatings*, 2e. John Wiley and Sons https://doi.org/10.1002/9783527622375.

Sumriddetchkajorn, S. and Intaravanne, Y. (2012). Home-made N-channel fiber-optic spectrometer from a web camera. *Applied Spectroscopy* 66 (10): 1156–1162. https://doi.org/10.1366/11-06522.

Thiha, A. and Ibrahim, F. (2015). A colorimetric enzyme-linked Immunosorbent assay (ELISA) detection platform for a point-of-care dengue detection system on a lab-on-compact-disc. *Sensors* 15 (5): 11431–11441. https://doi.org/10.3390/s150511431.

Wahi, A., Jaiswal, S., Sahoo, A. et al. (2018). Smartphone-based fluorescence spectroscopy device aiding in preliminary skin screening (erratum). In: *Optics and Biophotonics in Low-Resource Settings IV*, vol. 10485 (eds. D. Levitz, A. Ozcan and D. Erickson), 21. SPIE https://doi.org/10.1117/12.2289857.

Wang, S.-W., Xia, C., Chen, X. et al. (2007). Concept of a high-resolution miniature spectrometer using an integrated filter array. *Opt. Lett.* 32: 632–634.

Wang, L.-J., Chang, Y.-C., Ge, X. et al. (2016). Smartphone optosensing platform using a DVD grating to detect neurotoxins. *ACS Sensors* 1 (4): 366–373. https://doi.org/10.1021/acssensors.5b00204.

You, H., Kim, Y., Lee, J.H. et al. (2017). Food powder classification using a portable visible-near-infrared spectrometer. *Journal of Electromagnetic Engineering and Science* 17 (4): 186–190. https://doi.org/10.26866/jees.2017.17.4.186.

Yu, H., Tan, Y., and Cunningham, B.T. (2014). Smartphone fluorescence spectroscopy. *Analytical Chemistry* 86 (17): 8805–8813. https://doi.org/10.1021/ac502080t.

Zanella, A., Bui, N., Castellani, A. et al. (2014). Internet of things for smart cities. *IEEE Internet of Things Journal* 1 (1): 22–32. https://doi.org/10.1109/JIOT.2014.2306328.

Zhang, B., & Li, H. (2008). Research on application for color matching in textile dyeing based on numerical analysis. *Proceedings – International Conference on Computer Science and Software Engineering, CSSE 2008*, 1, pp. 357–360. https://doi.org/10.1109/CSSE.2008.609

9

Smartphone Technology – Instrumentation and Applications

Alexander Scheeline[1,2]

[1] *Department of Chemistry, University of Illinois at Urbana-Champaign, Urbana, IL, USA*
[2] *SpectroClick Inc., Savoy, IL, USA*

9.1 Introduction and Context

Array detectors have been a spectrometric tool for approximately half a century. In 1988, I obtained my first charge coupled device array (CCD), a Thomson CSF liquid nitrogen-cooled 586 by 384 pixels, 26 μm^2-pixel device with a controller the size of a PC-AT, connected to a PC-XT that, in turn, could communicate with the rest of the world over a 1200-baud modem [1, 2]. This CCD allowed more measurements to be made in one hour than the technology it replaced, photographic plates, could make in a week, with greater convenience, linearity, and dynamic range. Dark current (when the CCD was below −110 °C) was undetectable. Linear diode array spectrometers in the ultraviolet (UV) and visible were already common. What more could a spectroscopist desire?

In due time, the answer became clear: portability, lower cost, and less need for an expert user were all desirable. A variety of highly portable instruments have become available, some as small as a pack of playing cards. Still, spectrometry is typically carried out by skilled personnel. However, the populace at large has many measurement needs: medical tests, water quality tests, characterization of food, clothing, surface contamination, and a myriad of determinations that would be more timely and useful if made at a point of need rather than in a laboratory.

Thus, when cell phones appeared, it was a short extrapolation to think that cell phone spectrometry would soon become common. Everything that was part of a 1988 array spectrometry instrument, the detector, controller, computer, storage, and communications device, was an all-in-one package. In the Netherlands, an inexpensive spectropolarimeter attachment for iPhones was used to measure air quality in a large citizen science activity in 2013 [3, 4]. While numerous literature articles suggest such instruments are feasible, none are commonly available in the United States as of fall, 2019, and only two commercial cell phone spectrometers have come to the author's attention, one by Huawei in China [5] and one by AlphaNov in France [6]. One suspects they are unavailable in the United States for two reasons: (i) in the case of the Huawei device, a web search reveals it is only for sale in Burma, suggesting that the hopes of its designers were not met by the manufactured product and (ii) both devices would violate a foundational United States patent [7]. Samsung has filed a patent for a reflectance spectrometer built into a phone [8–10], though similarities to the SCIO device [11] and the Huawei offering are apparent. In addition to visible measurements, they suggest vibrational spectra will be obtained using the Si-Ware NeoSpectra infrared spectral engine[9].

One has two choices in describing cell phone spectrometry. One may report research to date for noncommercial systems, speculate on systems specifically designed to do spectrometry and small enough to fit into a cell phone along with the rest of the normal components, or discuss the challenges of working with devices engineered for taking selfies and landscapes when desiring output to have scientific merit. This chapter focuses on the

Portable Spectroscopy and Spectrometry 1: Technologies and Instrumentation, First Edition.
Edited by Richard A. Crocombe, Pauline E. Leary and Brooke W. Kammrath.

latter. Data processing takes place on the phone; this differs from an early approach by Whitesides and coworkers, where the camera transmitted a Joint Photographic Experts Group (JPG) to a central processing station for image correction and interpretation [12]. Such transmission avoided the need for on-site computation and was utterly dependent on web access. While 5G communications may someday make the web ubiquitous, there are still many places where cellular or Wi-Fi reception is slow or nonexistent.

Other recent reviews include [13–16], while the role of smartphone (or, equivalently, cell phone) spectrometry in contrast to other portable instruments is reviewed in [17]. Some authors focus on developing chemistry that is robust for use by nonprofessional analysts, but assume that high-quality spectrometry is readily achieved [18–20], a viewpoint rejected by the current author. Use of filters to turn cell phones into filter photometers or hyperspectral imaging cameras (with a series of filters) are not discussed here.

Additionally, this chapter does not focus on the use of a smartphone as a computer connected to an external, portable instrument. A handheld spectrometer connected to a laptop or smartphone will give identical results, absent software bugs. The design of portable standalone instruments, while having much in common with smartphone spectrometry, has the extra design freedom to employ a detector optimized for the measurement rather than for inclusion in a thin, lightweight device, can have an independent source of electrical power, and need not share detecting duties with a camera optimized for snapshots.

Furthermore, the emphasis is on quantitative measurement of the trace to minor components of specimens. Identification of major components is often most easily done with near-infrared spectrometry. Minor or trace components either do not have a spectrum of sufficient magnitude to sense against a fluctuating background (differences in lighting and major components), or require amplification of their optical properties to allow detection (colorimetric reactions, fluorescence excitation). In some cases, external reactant packages allow for semi-quantitative testing using cellcams without additional wavelength resolution other than the red, green, blue (RGB) pixel combination common in color cameras. For example, qualitative detection of psychoactive drugs can be performed with a "swipe card" and picture test [21]. Qualitative testing can be thought of as a binary test of whether the concentration of a sought-for-substance (or interferents) is above a threshold. This places negligible demands on a detector for linearity. Quantitative work does require a linear response, or nonlinear response sufficiently reproducible and smooth to allow linearization. Furthermore, because the resolution of most small, portable systems is 1 nm or coarser, atomic spectrometry is ignored.

We also note that smartphone spectrometry is a subset of smartphone chemical analysis. Separations, electrochemistry, and even mass spectrometry may someday be included in a cellular device, but that is beyond current technology and the scope of this chapter.

9.2 Challenges of Smartphone Spectrometry

Any spectrometer must produce a result that can be verified. This requires both precision and accuracy. Calibration is typically carried out under controlled conditions, yet a portable, citizen-science instrument must operate under conditions of temperature, humidity, dust, vibration, battery or power status, and human operation that often is not conducive to stability, reproducibility, and quality control. Let us contrast how a laboratory instrument and a putative cell phone spectrometer are designed, built, calibrated, maintained, and used. See Table 9.1.

To understand the engineering challenges of cell phone spectrometry, let us expand the information in Table 9.1. Power is an obvious problem. If wall socket power is available, it makes little difference whether an instrument requires milliwatts or hundreds of watts. Running from a battery, power consumption limits instrument and controller lifetime. If a cell phone battery stores 100 Ah of charge and can run the phone for 24 hours, then an instrument drain of 3 A (the upper limit for universal serial bus (USB) 2.0 [22] using microconnectors) could reduce battery lifetime by a factor of 2 (USB 3.0 supports higher currents, and, thus, is potentially even a greater strain on batteries). Since USB power is nominally at 5 V, maximum sustained power is limited to 15 W, with

Table 9.1 Summary comparison of laboratory and cell phone spectrometer characteristics.

Characteristic	Laboratory instrument	Cell phone spectrometer
Power consumption	Minor consideration	Minimize to avoid battery load
Temperature stability	Lab temperature roughly constant; materials chosen to minimize performance change with temperature	Wide and unpredictable temperature range requires either extraordinary measures for stability or temperature-monitoring and real-time compensation for changes
Humidity	20–80% in air-conditioned spaces	0% – supersaturated
Optical train	Optimized for performance	Simplicity and light weight essential
Sample/spectrometer alignment	Rigid	Varies from rigid to arbitrary to random
Detector choice	Optimized for performance	Built-in camera or dedicated, microfabricated spectral engine
Wavelength range	Optimized for problem-solving	Limited by low-power light sources, built-in detector constraints
Intensity dynamic range	Limited by detector and stray light	Limited by detector, stray light, and access to raw detector data
Calibration: wavelength	At factory; reconfirmed intermittently	Must be confirmed at the time of use; alignment highly variable
Calibration: linearity	Detector and electronics ensure typically excellent	Must be characterized; typically quite nonlinear
Calibration: slope response	At factory; pattern noise removed in firmware or software	Must be characterized by an end user
Likely user expertise	Trained professional	Unknowable
Dominant source of variance	Typically sample	Ambiguous

lower power desirable. Such power levels preclude the use of many, if not most, arc lamps and blackbody radiators. Light-emitting diodes (LEDs) light sources, including use of the LED flash built into many, if not most, cell phones, are common. Note that this power limitation thus restricts wavelength range for luminescence and absorbance measurement, independent of detector characteristics. Even sunlight has been employed, though the variability of cloud cover and the earth's rotation present obvious challenges [23]. The authors of [23] confuse dispersion with resolution, and while they obtain linear working curves, slope and precision do not always match the performance of laboratory instruments for reasons discussed throughout the current chapter.

Any portable instrument must deal with temperature variations. Maintaining instrument stability even in laboratories is challenging, as materials change size with temperature, semiconductors change conductivity, light sources change steady state and noise behavior, and detectors shift dark current and shot noise levels. There are two ways to deal with temperature effects. One is to carefully engineer thermal expansivity and temperature response of components to minimize drift with temperature. The other is to allow the system to drift but to calibrate its behavior at the temperature, location, and time of employment. Both approaches are challenging, though the latter often allows a lighter instrument to be devised. Similarly, the humidity not only can lead to fogged optics (high humidity) and static damage to electronics (low humidity), but also varies the refractive index of air, subtly influencing wavelength calibration. Any material porosity can add hysteresis to responses to humidity changes, further complicating instrument stabilization.

Laboratory instruments can employ mirrors, prisms, gratings, lenses, and fiber optics in a wide range of geometries. Portable instruments typically use thin polymer optics (Fresnel lenses, transmission gratings, thin film polarizers) to minimize mass. Such optics may limit image quality and, consequently, throughput, resolution, and channel crosstalk.

Spectrometers may have built-in sample holders or may be connected to samples with fiber optics or other relaying optics. Either of these is more stable than a handheld cell phone which is typically not equipped with a sample holder or a fiber optic connector [24]. It is common to place the phone in a frame [25–28] or to mount a sample holder and grating in front of the phone's camera [6, 29–31]. While such arrangements give adequate precision in short-term demonstrations, the author is not aware of any long-term stability characterization nor of interuser or interinstrument transferability studies.

Most daunting is detector choice. Laboratory instruments can be engineered with any detector that optimizes wavelength coverage, resolution, dynamic range, noise, and so on for a particular measurement. In cell phones, the most obvious detector is one of the built-in cameras, whose choice is forced on the spectroscopist and user by engineers optimizing for cost, the field of view, color balance, and "selfie" appeal. Often, the default behavior is to preprocess images for white balance, brightness, and hue, thus destroying any information on intensity. The wavelength range is limited by interposing a near-infrared filter to block light in the wavelength regions to which silicon most sensitively responds (750–950 nm).[1] Fortunately, both Android and iOS operating systems have system calls allowing control of exposure and gain and allow access to raw detector signal. One should never use JPG images for doing quantitative spectrometry; JPG is a lossy compression scheme that is guaranteed to distort working curves [33]. Whether the cost/performance benefits of adding dedicated spectrometric engines to phone detectors will prove to be commercially viable is unclear.

Phone cameras are naturally optimized for detection between 400 and 700 nm, though filters frequently reduce throughput past 650 nm to prevent excessive redness in images (silicon response increases in the near infrared while human response fades). Plastic lenses and other components cut off UV light below ~315 nm, and silicon has poor response below ~400 nm. A Nikon D50 single lens reflex (SLR) camera observing a high-pressure sodium vapor lamp through a diffraction grating easily sees the visible sodium emission and absorption lines (see Section 9.3.1.6 in this chapter) and the continuum between 420 and 685 nm. Observation of a mercury vapor lamp with an OmniVision OV5640 camera (a common 5 MPix cell phone camera) showed only the visible multiplets (404, 435, 546, 579 nm), not the 256 nm line. The wide availability of silicon detectors coated to allow UV detection contrasts the wavelength range restriction of cell phone cameras.

The aforementioned OV5640 chip has 10-bit on-chip digitization but typically reports 8-bit data to the observer. Digitization noise thus limits single-pixel precision to 1 part in 256 or 1 part in 1024 [34, 35]. Realistically, one cannot use the detector's entire dynamic range, as linearity suffers near pixel full well exposure [36]. As most CCD arrays have deep wells (holding over 10^5 electrons) and digitizers with 14-bit or 16-bit precision and some linear diode arrays have 20-bit dynamic range, the range available on a cell phone instrument will be challenging. A number of reported instruments appear to use large numbers of pixels, together with careful choice of exposure time, to allow dynamic range approaching three orders of magnitude [26–28, 33, 37, 38]. The author has championed the use of gratings that generate a large number of orders with varying throughput to broaden the dynamic range of an instrument beyond the dynamic range of the detector [30, 39–41]. An argument for the use of CCD or complementary metal oxide silicon (CMOS) arrays with narrowband filters overlaying the array as a separate spectral analyzer added to phone is that dynamic range would be higher [5, 42].

Calibration for wavelength requires a standard. If a mercury lamp or other line source is available, such calibration is straightforward and can use long-standing algorithms [43, 44]. Unfortunately, mercury lamps require high voltage, generate UV light, are fragile, and in many cases, are physically large, complicating alignment. Alternatively, neon bulbs or fluorescent lamps may be employed, as further discussed in Section 9.3.1.6. White LEDs show promise as continuum sources with known peak emission wavelength. However, peak wavelength varies

1 Silicon wavelength response varies based on doping and thickness. See e.g. figure 1–11, p. 6 in [32] for examples of a range of responses. Since absent calibration, one rarely has knowledge of the wavelength response details of the sensor in a smartphone, generalizations above response beyond "visible region" are difficult.

with temperature [45] and even nominally identical LEDs have a range of peak wavelengths [46] that shift as the LED ages [47, 48].

Thanks to the Central Limit Theorem [49], massively parallel measurements have a higher precision in measuring a mean signal than individual detector measurements. When using a cellular camera as a detector, some subsets of the pixels are employed, so precision depends on camera size, pixel size, pixel-to-pixel variation, light flux, temperature (which influences dark current), and the optical train's uniformity in illuminating pixels. An excellent study of the relative importance of global and individual pixel response has recently been published, together with algorithms for calibrating spectroradiometric response of individual cellcams and classes of cellcams [33].

Not to be overlooked is that the typical cell phone user is not a trained spectroscopist. Regardless of the precision and accuracy of the cell-based spectrometer, the human user must prepare a specimen in a way that allows the instrument to function optimally. How can one ensure that a determination performed with a cellular spectrometer has validity if the performance of the user is unknown and unknowable? How many users will be interested in undergoing any level of training? If the operation is only feasible where user skill is minimal (as is currently the case for home pregnancy tests or the use of test strips for pH, blood glucose, and such), is there anything to be gained by adding spectrometric capability to the cell cam?

Finally, for a measurement to be useful, the dominant source of signal variance has to be the property that the user is trying to study. If an instrument's variance from all causes except the analyte is one order of magnitude lower than the variance due to the analyte, the instrument is good enough for semiquantitative work. If the instrument contributes variance as large as that of the analyte, no useful information can be obtained on a statistically sound basis. Every factor in Table 9.1 has to be understood, quantified, and validated for cell phone spectrometry to have a commercial future.

One thus sees that making a viable cell phone spectrometer is a daunting task. Many of the engineering choices one would like to have are unavailable. Essentially everyone on earth owns or knows someone who owns a smartphone [50]. While no statistics are available, personal experience strongly suggests that a much lower percentage of humanity is trained in chemical analysis. There is, thus, a danger in making cell phone spectrometers without adding a training and quality control component to the operating software. Without adequate guidance, the untrained user could easily generate meaningless data even from the best engineered device, but a cost-effective device will surely have performance trade-offs that will demand care in specific areas of sample preparation from the user.

9.3 Progress to Date

This chapter emphasizes device engineering, not applications, which are described in the chapter in Volume 2 by Peveler and Algar. Rather, we discuss aspects of instrument engineering which influence analytical performance, but which are not necessarily emphasized in some prior work.

9.3.1 Dispersion and Resolution

9.3.1.1 Conventional Transmission Gratings

Undispersed, zero-order light transits transmission gratings undeviated, regardless of grating rotation. If one wishes 400–700 nm to display across the detector, the resolution is more than twice as high if zero order is not visible on the detector, so that the wavelength range observed is 300 nm, not 700 (or more) nm. Figure 9.1 shows the problem and a common solution.

In inset (a), the grating is interposed on incoming light just in front of the camera lens. When focused, zero order is approximately in the center of the scene. Plus and minus first order are dispersed across the camera field

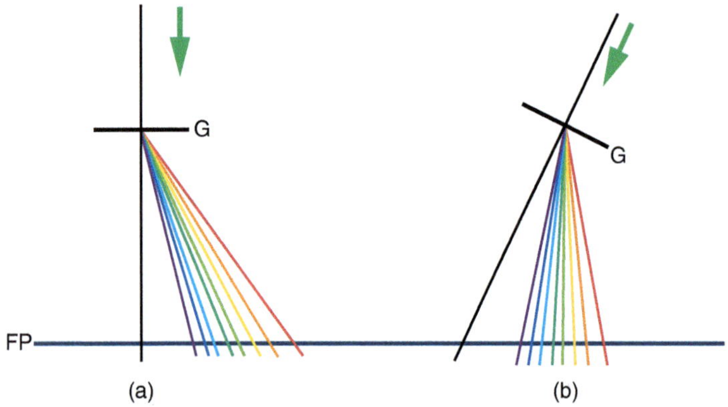

Figure 9.1 Diffraction through transmission gratings. FP = focal plane, G = diffraction grating. Arrow indicates the direction of incoming light. (a) Grating normal to the entering light beam and parallel to the focal plane. (b) Focal plane normal to diffracted beam at some selected wavelength.

of view. Groove spacing on the grating is chosen so that the desired wavelength range appears between the edge of the visible field and the center of the camera, with the corresponding negative order on the other side of the grating normal (not shown) [51]. Geometry (a) is simple to align, but the resolution is typically poor. The range between 0 and 400 nm is typically devoid of information; for illustrative purposes, assume that red cutoff is at 700 nm and dispersion is linear. If the diffracted light covers 2100 pixels, then 450 are illuminated by −1 order, 450 by +1 order, and the remaining 1200 pixels are illuminated by zero order and the UV region to which native silicon is unresponsive (and low-power light sources provide negligible illumination). Dispersion is 0.66 nm per pixel, but optical aberrations (coma, spherical, chromatic, field curvature) typically limit resolution to significantly poorer values, especially if the incident light is incompletely collimated and the instrument entrance aperture is larger than one pixel. Geometry (b) disperses light only in a single order. While wavelength calibration is more difficult since zero order is not visible, the dispersion may place a 300 nm range across the same 2100 pixels, giving dispersion of 0.143 nm/pixel, thus improving dispersion, and potentially resolution, by about a factor of 4. A subtle additional limiting factor is that the maximum relative resolution possible from first-order diffraction, $\lambda/\Delta\lambda = N$, the number of grating grooves illuminated. Cell phone cameras typically have entrance apertures of the order of 2–3 mm, and coarse gratings suitable for low-resolution spectrometers have 200–600 grooves/mm. Thus, a 200-groove grating coupled to a 2 mm entrance aperture limits resolution to ~400 (or 1 nm at 400 nm), while a 3 mm aperture and 600 groove grating could potentially have resolution ~1800. The imposition of a Bayer pattern (red/green pixels in even camera rows, green/blue pixels in odd rows) [52] further degrades resolution and data recovery (see also Section 9.3.3). When reviewing manuscripts, the author has seen numerous cases of others equating dispersion with the resolution, ignoring finite pixel size, aberrations, and so on. An example of geometry (b) can be seen in figure 1 of reference [27], with several more examples in figure 2 of reference [13].

9.3.1.2 Stacked, Mutually Rotated Gratings

Several of the problems of coupling linear gratings to areal detectors may be overcome, in principle, by averaging data over multiple orders. If orders are projected onto nonoverlapping regions of a multimegapixel array, then each of the orders may be independently examined, and the dynamic range of the instrument as a whole is the product of the dynamic range of the camera and the variable throughput of the multiplicity of orders. This is the basis of the author's developments in portable spectrometry [30, 40, 53]. Two or more gratings are positioned parallel to the focal plane, slightly separated along the light propagation direction, with the collimated illuminating beam centered on the camera, as in Figure 9.1a. The dispersion direction of one grating is rotated about the optical

Figure 9.2 An example of the many orders visible when stacked, double-axis diffraction gratings are illuminated with collimated light from a white LED. Zero order appears in the approximate center of the image.

axis so that orders are dispersed at a multiplicity of radii and angles. Particularly useful patterns appear when double-axis gratings [54] are employed.[2] An example appears in Figure 9.2.

The chief theoretical advantages of this grating arrangement are (i) the peak emission of the LED occurs at a known wavelength (slightly influenced by device temperature and LED fabrication precision [45]), allowing real-time wavelength calibration. The peak wavelength response of the camera may be different than peak emission output of the LED, so the LED must be taken as a secondary standard and the system initially calibrated using a source where wavelength information is not easily skewed by differences in camera sensitivity with wavelength. Atomic emission lines serve this purpose well, but are unlikely to be available in a safe, portable unit. Fluorescent lamps that emit mercury neutral atom lines are widely available and are discussed in Section 9.3.1.6. (ii) Different orders have strikingly different throughput, so, once calibrated, can be used to expand the dynamic range of the instrument beyond the dynamic range of the camera. The author has used an 8-bit camera to get 12 bits of dynamic

2 Ordinary plane gratings are ruled with equally spaced, parallel grooves. Double axis gratings consist of a square (or, plausibly, rectangular) grid of bumps so that dispersion is in both x and y Cartesian directions [54].

range, as the various orders gave throughput differences of up to a factor of 20. Using a 2 mW 633 nm laser projected through a double axis grating and measuring throughput with a photodiode, observed power ranged from 869 μW (zero order) to 200 μW (first order) to 300 nW (third order), meaning that a single grating inherently has nearly three orders of magnitude dynamic range. Since, approximately, the throughput of a beam diffracted by two gratings is the product of the efficiency of each grating, it is clear that stray light rather than camera performance will limit the dynamic range of instruments with stacked, mutually rotated gratings. (iii) Resolution degrades gracefully as the amount of light available for a measurement decreases. At 100% transmittance, second-order spectra may be readily observed, while first-order spectra are saturated (as is the case of Figure 9.2). At lower flux, weak orders disappear into the background and strong orders come out of saturation. Lower orders have higher throughput and lower dispersion, so throughput/resolution tradeoff is readily evident. (iv) Signal averaging (to overcome shot noise and digitization granularity) and resolution are accommodated on a single detector, in contrast to using single axis gratings that give a single wavelength range and resolution. It is unclear whether processing data from the hundreds of thousands of usefully illuminated pixels will provide sufficient precision in contrast to employing the entire field of view (with a single axis grating) in which field curvature complicates both wavelength resolution and calibration.

The position of each order depends on the rotation of the gratings about their common normal, the rotation of the camera with respect to the grating rulings, the order of diffraction in x and y from each grating, and any off-axis illumination angle. For normal incidence, the relevant equations are:

$$n_{\text{eff}}\lambda = d_0 \sin\beta \tag{9.1}$$

$$n_{\text{eff}} = (n_{x1}^2 + n_{y1}^2 + n_{x0}^2 + n_{y0}^2 + 2((n_{x1}n_{x0} + n_{y0}n_{y1})\cos\theta_g + (n_{y0}n_{x1} - n_{x0}n_{y1})\sin\theta_g))^{1/2} \tag{9.2}$$

$$\theta_{n_{\text{eff}}} = \arctan\frac{n_{y1}\cos\theta_g + n_{x1}\sin\theta_g + n_{y0}}{n_{x1}\cos\theta_g - n_{y1}\sin\theta_g + n_{x0}} \tag{9.3}$$

where

λ = Wavelength of interest (nm)
d_0 = Grating spacing (nm). If the spacing is different for x and y (i.e. if the diffracting regions on the double axis gratings form a rectangle rather than a square), one may readily revise Eqs. (9.2) and (9.3).
n_{eff} = Effective order number. Counterintuitively, n_{eff} is not necessarily an integer. For example, for $n_{x0} = n_{y0} = 1$, $n_{x1} = n_{y1} = 0$, $n_{\text{eff}} = \sqrt{2}$
n_{wm} = Diffraction order with respect to a cardinal axis in a grating plane w is either x or y, Cartesian directions, and m is an integer order in the Cartesian direction specified.
θ_g = Angle between the x axes for the two double-axis gratings.
β = Diffraction angle with respect to the grating normal, regardless of rotation from the reference x orientation
$\theta_{n_{\text{eff}}}$ = Rotation of the order from the x-axis, taken as the orientation of order (1,0,0,0).

Commercial implementations of this design are not yet widely used, in part because the software to implement all the calibrations required has proven difficult to write. If such calibration becomes facile, the match to typical cell phone cameras (3–15 MPix, 8–10 digitization bits per pixel) is excellent.

9.3.1.3 Polarization Sorting

Because light scattering is polarization dependent, skylight and scattering from aerosols are polarized. Combining a diffraction grating, quarter-waveplate, "2×4λ" phase retarder, and a thin film polarizer allows a cell phone camera to do spectropolarimetry in a single image [4]. The mass-produced instrument is shown in Figure 9.3.

As explained in [55], the relative intensity of x and y polarized light can be obtained by using wavelength-independent waveplates. Fresnel rhombs using total internal reflection for polarization come close to this ideal but

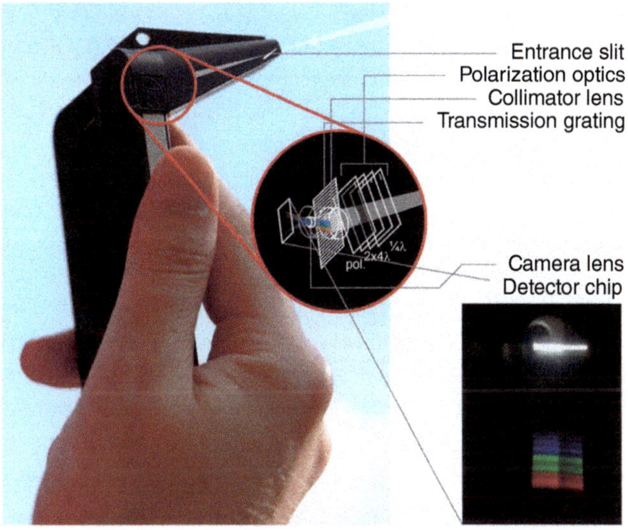

Figure 9.3 Cell phone spectropolarimeter. *Source*: Sink et al. [4].

are ungainly for portable use. Thin film polarizers and waveplates are hardly achromatic, and temperature changes significantly influence retardance. However, for citizen science experiments, accuracy is good enough and nonidealities can either be compensated (by modeling) or averaged (central limit theorem). Thus, in Figure 9.3, intensity modulation is not cleanly periodic in either wavelength or wavenumber, but relative *p*- and *s*-polarization can be extracted throughout the visible. Based on an understanding of polarized light, Figure 9.3 appears to be slightly inaccurate. Instead of two "4λ" waveplates, there are likely two $\frac{1}{4}\lambda$ waveplates, giving a total of $\frac{1}{2}$ wave phase shift. Those waveplates occlude only one half of the entering beam. Thus, the part of the beam that traversed only the correctly labeled quarter-waveplate shows linear polarization in one plane, while the portion of the beam that traverses three quarter-waveplates generates data in the plane perpendicular to that of the singly occluded beam. The linear polarizer ensures that the grating is illuminated by only one polarization, avoiding throughput differences for orthogonal polarizations as is common for diffraction gratings. Because the alignment of the phone with respect to reference directions is imprecise, errors in polarization $\sim\cos\theta$ where θ is the tilt of the phone from some reference direction can be anticipated. In general, one must presume that a cell phone camera has a random orientation with respect to the outside world, and so calibration algorithms must allow for arbitrary tilts.

9.3.1.4 Fabry–Perot, Wedge Filtered, and Nanoarray Optical Sensors

Early cell phone cameras sensed radiation to ~1000 nm. Near-infrared light made strong near-infrared emitters look violet in color images. Most current color cameras include a near-infrared blocking filter so that, other than at 940 nm, employed by devices, such as video remotes to communicate, smartphones no longer respond to infrared light. Prior to the advent of near-infrared filters, cellcams allowed measurement of such atmospheric constituents as ozone or water vapor [56].

Omitting USB-equipped instruments that can communicate with any USB-equipped device including cell phones, one thus expects few cell phone interferometric instruments. A visible instrument based on a wedge interferometer (linear variable filter) as in Ocean Optics' (now Ocean Solutions') discontinued Spark line of miniature spectrometers [57, 58] would be a possibility. Another, pursued by Consumer Physics [11] in the near infrared, is to make a thin film [59] array color-selective array sensor small enough to fit in a phone. Such sensors with 8–16 spectral bands are also available in the visible [60–62] from vendors who have had less publicity. Greater number of spectral bands may be available in planar arrays in the future. An early paper described how to extract over 100 passbands with better than 1.5 nm resolution [63]. The Si-Ware NeoSpectra MEMS interferometer looks promising [9]. A Fabry–Perot interferometer [64–66] tuned to measure CO_2 is available in a size that could clamp onto a cell phone [67]. Precision is ±100 ppm for CO_2 levels between 100 and 1000 ppm.

All the interferometric approaches are technically viable, but their routine use depends on reaching a scale where unit cost is acceptable. Most require that a detector other than the cell phone camera be employed.

An ingenious approach to absorbance measurement has been described by Ding et al. [68]. Broadband light is polarized, partitioned by a microlens array, and transmitted through a wedge interferometer (alternatively, a linear variable optical filter), the output of which is refocused through the smartphone camera's lens and projected onto the detector. Because each lens sends light through a small portion of the wedge interferometer, each spatial region provides intensity information over a small band of the spectrum. By averaging throughput over many pixels, the effect of pixel-to-pixel response variation is minimized. Resolution is set by the size of the wavelength range transmitted by each illuminated region of the wedge. Polarization is necessary to avoid chromatic aberration from wedge birefringence and the consequent difference in wavelength transmission at each *xy* location for different polarizations. The long-term viability of the approach depends on the economics of fabricating wedge filters, the stability of suitably mounted filters as a function of temperature, humidity, and aging, and the adequacy of wavelength resolution with this approach.

Another recent development is the fabrication of arrays of semiconducting sensors with differing wavelength response along the array [69]. Because sensors are about the size of camera pixels, one expects that noise characteristics can be made comparable to current cameras. However, signal averaging with multiple pixels requires more complicated geometry than reported as of September 2019. Reproducibility of wavelength response, with reproducible changes as a function of temperature, will also be critical. If microfabrication can be made sufficiently reproducible, then spectrometric sensors even less expensive than the interferometers described in the previous paragraph may be practical, at which point use of cameras for spectral parsing may be unnecessary.

9.3.1.5 Why Not Prisms or Conventional Instruments?

Prism spectrometers, while less common than grating instruments, are not unusual [70]. The nonlinear dispersion due to changes in refractive index with wavelength complicates wavelength calibration and compresses the spectrum in the red, but can be readily handled by fitting to an appropriate spline or polynomial function. However, prisms weigh more than transmission gratings and can be more challenging to align. However, use of a multi-component cemented cuboid prism has recently been reported, in which the prism and camera are colinear and parameters are designed specifically to minimize response differences from one device to another [71]. Optimally, the detector should be large compared to the prism, which means that many cell phone cameras are suboptimal for use with this approach. A separate, large, embedded CCD would be preferable.

Because most cell phones have a USB port, almost any conventional instrument could be interfaced to a phone if the instrument also is USB-compatible. One typically does not call such arrangements a cell phone spectrometer because the phone is acting purely as a computer and communications device, not a data collector. Only if the camera or some other sensor built into the phone is used to transduce light to numbers is the instrument a cell phone spectrometer. This is consistent with usage in the defining patent for the field [7].

Simply shrinking a conventional Czerny–Turner or Rowland circle instrument is conceivable and has been patented [72]. It is unclear whether such a device has ever been built or sold, or what resolution or throughput may have been realized. Narrower applicability, but simpler construction, has been reported for cell phone colorimeters [37, 73, 74]. Here, either LEDs provide wavelength selection or the color selection among red-, green-, and blue-responding pixels defines the spectral range. Because the wavelength range is often wider than the absorption feature being monitored, working curve nonlinearity is expected.

9.3.1.6 Wavelength Calibration Sources

Convenient calibration sources must be stable, encapsulated, safe, and readily available. There are numerous sources for atomic emission sources which work well in the laboratory but are not available in general commerce.

Portable, inexpensive wavelength calibration standards may include selected LEDs, but should not be random LEDs that are becoming common light sources, as precise values for peak emission wavelengths may vary. Neon lamps are simple line emission sources [75], emitting a dense set of lines in the yellow and red part of the spectrum, but providing few lines in the blue. If one can trust that calibration will be stable for days at a time, mercury vapor street lamps and sodium vapor lamps (either high-pressure or low-pressure) are widely observable after dark (though LED lamps are beginning to replace such vapor lamps for environmental reasons). Indoor fluorescent lamps are also common, but present additional difficulties. Laser pointers have two, likely insurmountable, problems. They are too intense for direct use, and, if based on diode lasers, have poor wavelength precision.

Examples of possible calibration sources are shown in Figures 9.4 and 9.5 (fluorescent light), Figure 9.6 (high-pressure sodium vapor lamp), and Figure 9.7 (simulation of a neon lamp). While the examples have been chosen to emphasize specific problems, in fact, all sources must be evaluated for such limitations.

Because fluorescent lights are common, mercury atomic emission is almost always present, and the phosphor composition often includes terbium- and europium-doped yttrium oxide, these sources might be considered universal transfer standards. If a fiber optic collects light from the lamp and sends the light to an entrance aperture of a spectrometer, this approach is plausible (although lamp aging may alter the relative intensity at various wavelengths). However, given the grating/phone camera geometry commonly employed as discussed in conjunction

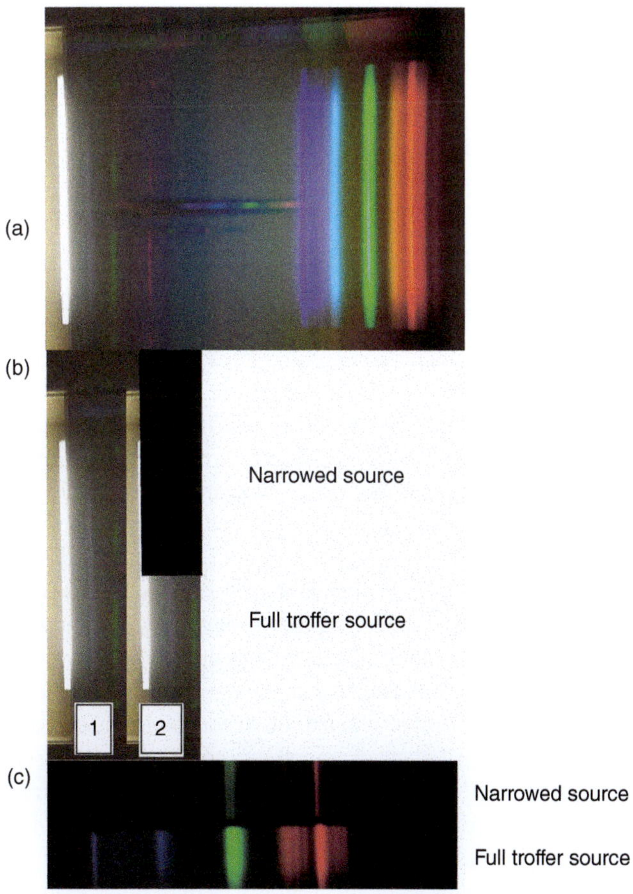

Figure 9.4 Linear fluorescent bulb used as a reference. (a) Unmasked fluorescent troffer and full field spectrum. (b) Masking of lamp; inset 1 unmasked, inset 2 heavily masked. (c) Comparison spectra for masked and unmasked lamp.

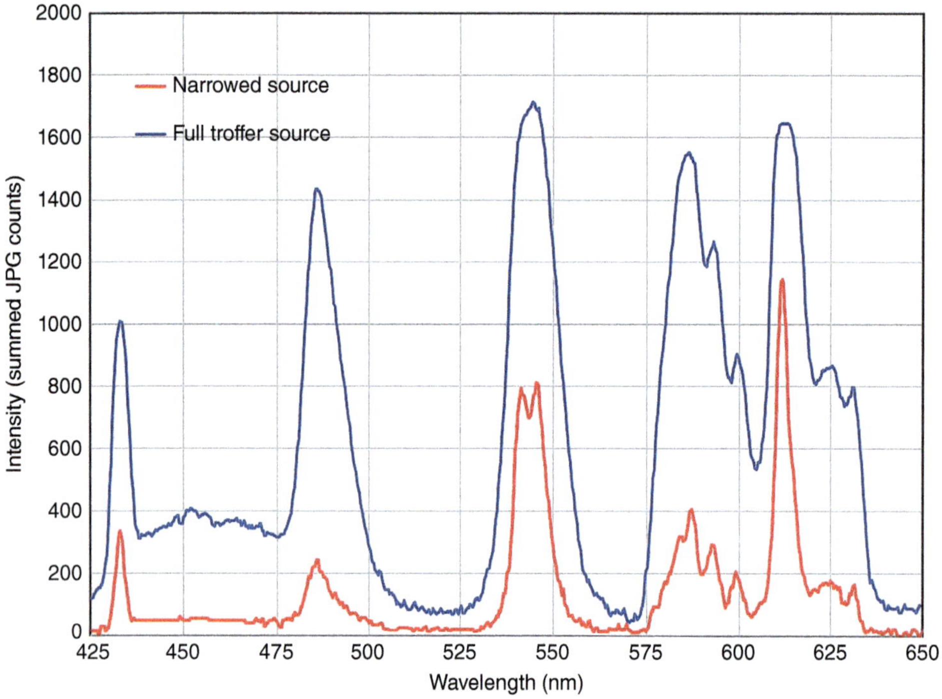

Figure 9.5 Comparison of full height troffer spectrum with masked lamp spectrum. For line and band identification, see [76]. *Source*: Alexander Scheeline.

with Figure 9.1, one is tempted to point the phone at a fluorescent lamp through the diffraction grating and use the resulting image for calibration.

In Figure 9.4a, the 5 MPix camera in an LG 4S cell phone is used to observe a small fluorescent light in a compact troffer (rectangular, recessed light fixture) through a 600 line/mm transmission grating. The initial impression is that the fluorescent light could provide reference wavelength data. Inset (b) sketches how the troffer was masked for the remaining images in Figure 9.4, with inset (b).1 showing the unmasked troffer (in fact, the same image as in inset A), while inset (b).2 shows how the troffer was masked with black paper (a drawing of the masking, rather than a photo, is used for clarity). Spectra of the partially masked troffer are photographed in Figure 9.4c and plotted in Figure 9.5, the latter extracted using software available online [77, 78]. The extracted spectra sum the 8-bit signals in selected regions of the JPG image in Figure 9.4c over a height of 9 pixels. Lines and bands are identified in [76]. Because the center of the illuminated pixels in zero order (white band of the troffer image) is 29 pixels wide in the unmasked area and 13 pixels wide in the masked area, a shift of 8 pixels to align wavelengths was necessary for generating the blue trace in Figure 9.5 compared to the red trace. Such shifting due to the size of the source is a general feature of spectrometers without fixed entrance apertures and limits the wavelength calibration precision, whether the instrument is portable or fixed. In addition, barrel or pincushion distortion in the camera and the nonlinear relationship between angle and distance on the focal plane of any camera complicates fitting of wavelength to a position in the camera. Image curvature, common to any spectrometer where the illumination is off axis, may further compromise the uniformity of calibration as a function of height in the image. In Figure 9.4a, the green band top and bottom ends are 8 pixels to the right of the center of the band over an image height of 980 pixels. Since the dispersion is about 0.71 nm/pixel, this corresponds to an apparent wavelength range along a single column of the image of 5.6 nm. Of course, data processing can compensate for such curvature, provided it is noticed, characterized, and employed.

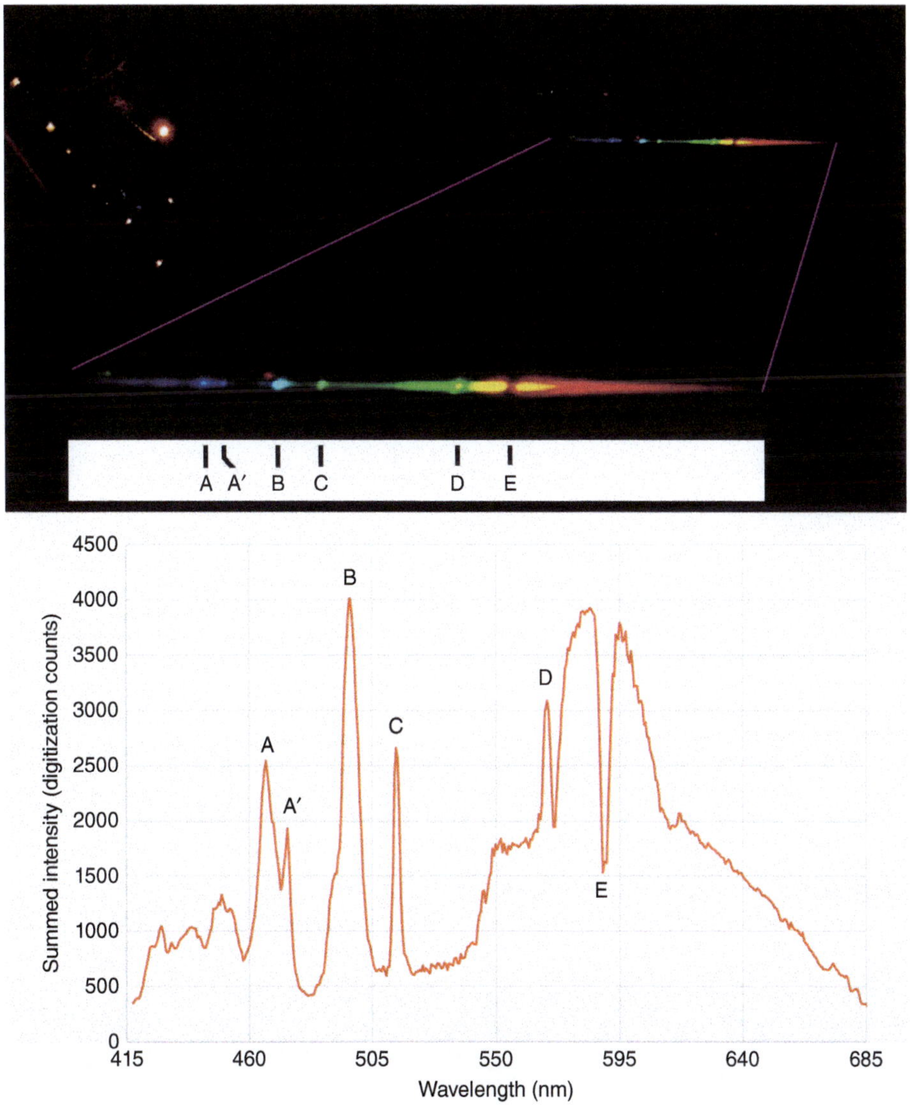

Figure 9.6 High-pressure sodium vapor lamp spectrum, annotated.

Figure 9.5 also shows how source size can obscure spectral structure and, thus, mislead one as to the wavelength of a particular line or band. When viewing the full troffer height, the band near 550 nm appears to be a single entity, most likely due to the 546 nm Hg I line. When the troffer is masked, it becomes evident that two transitions are present, one due to Hg I emission, and the other due to Te^{3+} phosphorescence or the cluster of Hg I lines emitting near 579 nm.

Figure 9.6 shows emission from a high-pressure sodium vapor lamp street light, taken with the same 600 line per mm grating as used for the fluorescent lamp spectrum and using the same Nikon D50 camera employed for the calibration image in Figure 9.4c. The top of the image shows the lamp at left and dispersed spectrum stretching to the right. An enlarged rendering of the spectrum is shown below, with violet guidelines superimposed on the figure. Atomic absorption by the ~589 nm sodium D lines is evident as feature E ($2p^2S_{1/2} - 2p^2P^{\mathrm{o}}_{1/2,3/2}$). Feature D is easy to mistake for the Hg I 546 nm line (plausibly from a NaHg amalgam), but is, in fact, a sodium line

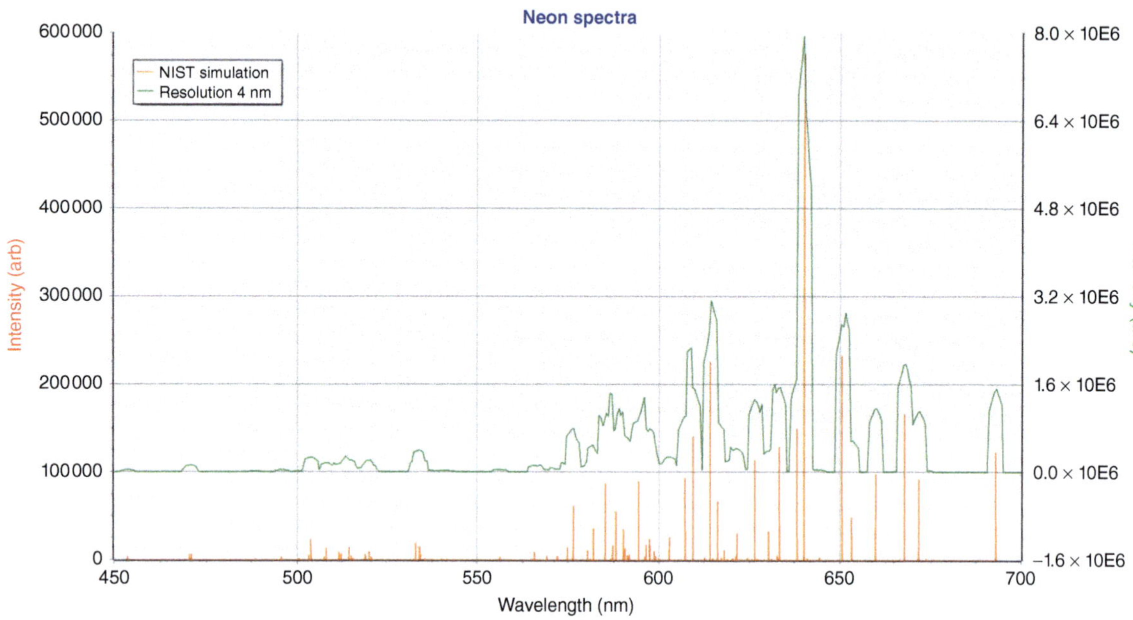

Figure 9.7 Simulated neon atomic emission and effect of low resolution on observations.

$(3p^2P^o_{1/2,3/2} - 4d^2D_{3/2})$ at 568 nm. Other features are: A $(3p^2P^o_{1/2,3/2} - 6d^2D_{3/2,5/2})$ (466 nm), A′ $(3p^2P^o_{1/2,3/2} - 7s^2S_{1/2})$ (475 nm), B $(3p^2P^o_{1/2,3/2} - 5d^2D_{3/2})$ (498 nm), and C $(3p^2P^o_{1/2,3/2} - 6s^2S_{1/2})$ [79–81].

Last, in Figure 9.7, we show the spectrum expected from a neon bulb. Data were simulated using National Institute of Standards and Technology (NIST's) atomic spectral database, assuming a temperature of 2 eV, with Doppler broadening, and electron density of 10^{14} cm^{-3} [81]. The orange trace shows what a high-resolution spectrometer might observe – a spectrum dense in red and yellow lines, but weak in the blue and green, thus forcing the use of the frequently poor assumption of linear dispersion across the wavelength range. The green trace simulates the appearance with a spectrometer with 4 nm resolution, using pixels spaced every 0.5 nm. Offsets from expected bright-line wavelengths and blurring of adjacent lines are readily seen.

In short, calibration with any portable light source or handheld instrument is difficult, of limited precision, and a likely source of measurement error.

9.3.2 Focus

Most smartphone cameras have lenses that are focused for distant objects. Field of view ranges from a few degrees in zoom or telephoto mode to quite wide angles (the author has seen up to 60°) for group selfies. Accurate wavelength calibration requires constant zoom, and optimum resolution requires tight, consistent focus. Fortunately, both iOS and Android operating systems include commands to set focus and zoom. In neither case, however, can one be confident of the stability or reproducibility of the zoom and focus functions. In the author's experience, autofocus is the least controllable parameter of the camera modules commonly found in phones.

One common camera, available from OmniVision since at least 2012, is the OV5640, 2592 × 1944 pixels, approximately 4 mm focal length, and ±22.5° field of view in the x (wide) direction. For an ideal planar detector, distance from scene center, r, and off-axis angle of the object viewed, β, are related by

$$r = r_0 \tan\beta \tag{9.4}$$

where r_0 is a scaling constant. For this particular detector, when $r = 1296$ pixels (half the width of the detector) and $\beta = 22.5°$, $r_0 = 3129$ pixels. If the small-angle approximation applied, each pixel would subtend $22.5°/3129 = 0.0072° = 26''$, but, in fact, differentiation of Eq. (9.4) shows

$$\frac{dr}{d\beta} = r_0 \sec^2\beta \tag{9.5}$$

Since pixel size is constant ($dr = 1$), the angle subtended by each pixel scales as

$$d\beta = \frac{\cos^2\beta}{r_0} \tag{9.6}$$

Thus, in the limit that pixels are large compared to the resolution set by the rest of the instrument, wavelength range per pixel appears to improve as $\cos^2 \beta$, or by a factor of 0.85 at 22.5°. In the real world, pixel-limited resolution is uncommon. Almost always, pixels are small compared to the optical blur generated by the rest of the optics [82]. Under these circumstances, a fixed $d\beta$ is stretched across a region scaling as $\sec \beta$. Given the commonly derived dispersion of a spectrograph,

$$\frac{d\lambda}{dx} = \frac{d \cos \beta}{nf} \tag{9.7}$$

where n is the grating diffraction order, f the spectrometer camera mirror or lens focal length, and d (other than when meant as differential operator) the grating spacing, an image of width w at the center of the detector blurs to width $w \sec \beta$. The off-axis blurring exactly compensates for the increase in dispersion with angle, so that resolution, $\lambda/\Delta\lambda$, scales as

$$\frac{\lambda}{\Delta\lambda} = \frac{\frac{d \sin\beta}{n}}{\frac{\Delta x_{\text{eff}} d}{nf}} = \frac{f \sin\beta}{\Delta x_{\text{eff}}} = k\lambda \tag{9.8}$$

Independent of diffraction order, resolution increases with wavelength, with a scaling constant dependent on camera focal length and the effective size of the entrance aperture image in zero order. Since focus adjustments change f and Δx_{eff}, image blur influences resolution as well as throughput, but the dependence is hidden from the casual user.

9.3.3 Pixel-to-Pixel Uniformity

Even the most uniform detectors have some response variation from pixel to pixel. As almost all cell phone cameras are color cameras, the variation in color filter transmission and blocking from pixel to pixel multiplicatively increases pixel response variance. Because human eyes compensate for minor variations in image quality, a good selfie camera may display significant nonuniformity, untroubling to humans, but entirely inadequate for spectrometry. Quantum efficiency and slope response (change in signal per unit light flux) vary across the detector not only because of changes in dopant levels in the semiconductor but also because of vignetting (variation in light transmission efficiency across a scene) by the lens. Such variation was recently characterized [33] for two common cellular phones as shown in Figure 9.8.

Two characteristics are evident. First, the distributions in detector response are different for the two cameras; the iPhone camera has a nearly Gaussian distribution with gain ~1.75 ± 0.25 digitization units per electron (with slight bimodality for the green channel), while the Samsung Galaxy response is non-Gaussian and averages 1 count per electron. Human vision responds approximately logarithmically to light intensity, so the camera variations are irrelevant for their intended use. However, for spectrometry, the errors may be catastrophic. If I_0 in Beer's Law is measured by a different pixel than I, the transmittance is meaningless unless pixel response has been calibrated. This problem has rarely been noted in enthusiastic papers on the possibilities of cell phone spectrometry, as averaging over thousands of pixels has concealed the problem. The problem of transferring a method from one phone to

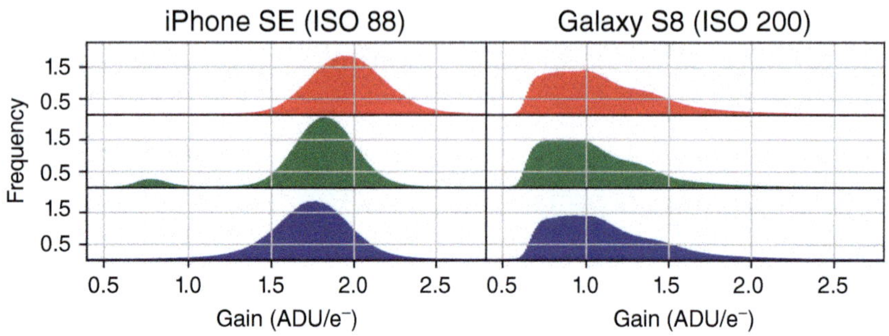

Figure 9.8 Burggraaff's original caption: "Histogram of gain values in the R (top), G and G_2 (middle), and B pixels (bottom) in the iPhone SE (left; ISO speed 88) and Galaxy S8 (right; ISO speed 200) sensors. The vertical axes were normalized to account for the different amounts of pixels." *Source*: Burggraaff et al. [33]. Licensed under CC BY 4.0.

another, given the variation in camera response, is obvious. While standardizing color correction between Bayer pattern color cameras has received significant attention, e.g. [83], such standardization is most useful for optimizing human perception of pictures, not for standardizing wavelength-dependent slope response, as the relative sensitivity within a particular color range (e.g. red, green, or blue) is not addressed when balancing perception color response.

Further complicating use of cell phone cameras is pixel response nonlinearity, shown, for example, in [84] or, unrecognized in turbidimetric measurement, in [85]. At some level of exposure, usually above 2/3 of pixel full well capacity, charge accumulation is sublinear, leading to saturation (no increase in signal for increasing exposure). The cited reference shows how to linearize the data. Excessive exposure leads to blooming, where adjacent pixels register signal from light not incident on that pixel. Such blurring is common to many array detectors; for low dynamic range cameras, the problem is exacerbated. To some degree, a nonlinear response can be compensated through the use of appropriate chemometrics, for example, classification and cluster analysis of RGB signals rather than fully dispersed spectra [74].

Alternatives to Bayer pattern sensing have been developed. Direct detection of red, green, and blue in a single pixel occurs in Foveon's chips [86]. Postprocessed to a JPG, bitmap (BMP), or tagged image file format (TIFF) image, a human might not, at first glance, see the difference from photos taken using a Bayer pattern. Firmware in most cameras depixelates images, thus destroying intensity data while improving perceived snapshot quality. The last thing a spectroscopist wants is any processing that destroys absolute intensity information or blurs information from adjacent pixels of different wavelength response. Getting useful information in the presence of pattern noise, shot noise, and varying responsivity is hard enough. "Helpfully" synthesizing composite colors by blurring data compounds calibration difficulties and increases the likelihood of artifact generation that is analyte dependent.

The nature of the pixelation problem is illustrated in Figure 9.9. We posit an atomic emission line and spectrometer such that the line spreads over 3 pixels in a Bayer pattern. Perhaps it is the mercury 546.1 nm line which is sensed with relative response 0.01 for blue, 1 for green, and 0.05 for red (not the values seen in measurement the author made on one particular camera, but similar). The figure shows where the image falls with respect to the pixels (ignoring curvature as discussed in conjunction with Figure 9.4) and assumes that any charge in excess of 255 counts spills into adjacent pixels. The model ignores nonlinearity short of saturation, and so more precisely shows where saturation occurs than is seen in real cameras. Precision is shot noise limited with noise just short of saturation of ± 8 counts ($\pm 3\sigma$).

Despite such simple geometry and spectroscopy, the plot is rich with measurement artifacts. When the green pixels become saturated, adjacent pixels begin showing the signal from the overflow. The solid lines in the plot show data that would be obtained if exposures were taken once per arbitrary exposure time unit. However, most calibration schemes sample only a few exposure times, then fit the data to a regression line. The dashed lines are

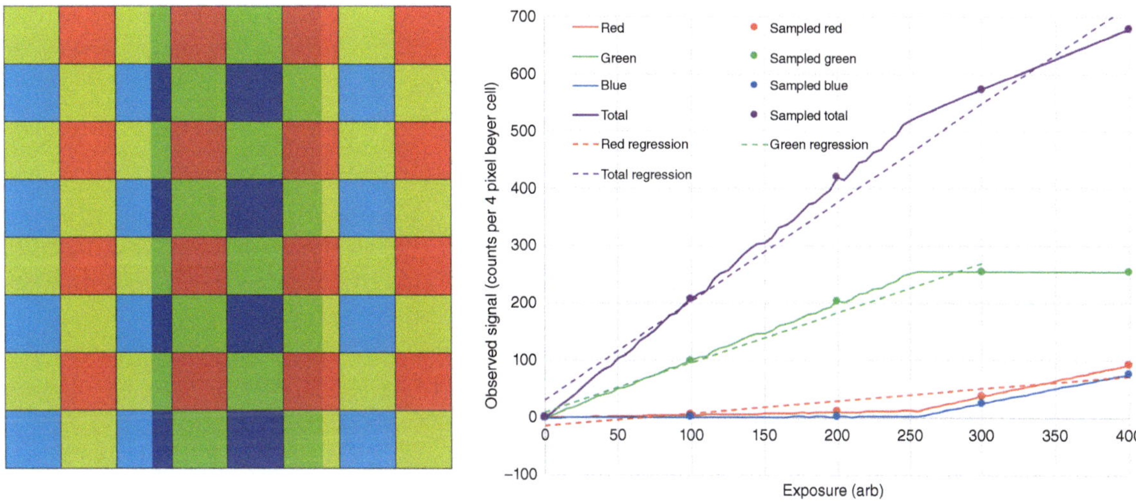

Figure 9.9 Bayer pattern, showing illumination by a green atomic emission line spanning three columns. Related growth curves as a function of exposure shown in the plot. See text for explanation.

examples of regression fits (no blue data regression was performed to avoid clutter in the figure). "Obviously," the saturation of the green data would be recognized by any competent measurement scientist, so regression would halt at the point where saturation is observed. Perhaps one would interpret the red (or blue) data as having a small slope that only rises above baseline at high exposure, not recognizing that blooming from overflowing green pixels is the cause. Simply summing across all pixel colors, ignoring the masking by the Bayer pattern has its own pathology. Unless one recognizes that the noise level is much lower than the fitting precision, there is little reason not to accept the regressed calibration line. Even if one knows the noise is small compared to the fitting error, it is unclear how to interpret the summed data. Add to this the pixel-to-pixel variations, focus limitations, and the impossibility of measuring the transmission profiles of the Bayer filters outside the camera, and clearly, the use of consumer-grade cameras presents severe problems for performing scientific measurements.

9.3.4 Coupling Cellular Cameras to Sample Holders

Already cited have been cell phone spectrometers that observe ambient light. Any instrument that has an entrance aperture can be coupled to a fiber optic, the distal end of which may observe any specimen as if the cell phone spectrometer were any benchtop or portable instrument. Thus, within this section, what is of interest is specimen holders other than fiber optic couplers that place the sample and cell phone spectrometer in close proximity or in contact so that the design is specific to the constraints of portability, size, cost, and repeatability of such instruments.

The most common sample holders are 1-cm cuvettes and 96-well plates. In addition, colorimetric test strips may be employed. We discuss each in turn.

9.3.4.1 1-cm Cuvettes

For path length through a cuvette to be uniform, collimated light must impinge thereupon. Collimated light is also optimal for grating illumination. Benchtop instruments separately collimate light for the cuvette and grating, avoiding distortion from refraction and scattering/light diffusion in the cuvette. Scattering not only offsets the absorbance baseline but degrades wavefront planarity essential for grating performance. Using single collimation thus risks spectral blurring and dramatic increases in the background, but simplifies the optical system when scattering is only minor. Such an approach has been used in the handheld instruments the author has employed, where the camera is the OmniVision unit previously cited. Light is collimated with an off-axis parabolic mirror, passed through the cuvette and grating stack, then observed with the camera module. The slight wedging of cuvette walls (draft) necessary to allow cuvettes to be removed from their forming molds means that the spectrum seen through the cuvette is translated on the camera from that seen for an air path spectrometer. The prior discussion of pixel-to-pixel variation indicates why this is problematic: different pixels measure I and I_0 [30, 39–41, 53].

A 3D-printed dual-beam sample and grating holder that clips onto a phone were recently reported [31]. Two raw spectra are dispersed across the cell camera focal plane. While such geometry compensates for variance in light source output, it exacerbates problems in wavelength calibration, field curvature, and pixel-to-pixel response variance. The authors discuss none of the precision or accuracy problems of this geometry.

The sunlight-source spectrometer of reference [23] uses a 1-cm cuvette. Light is partially collimated and refocused with biconvex lenses. While light traversing the cuvette is close to collimated, spherical aberration makes the collimation imperfect. In contrast, an adapter for a cell phone that uses the cell phone's flash LED as the light source illuminates a cuvette with uncollimated light [87]. Here, too, working curve slopes are less than those from a multiwell plate reader, and working curve nonlinearity is evident. A spectrometer appropriate for classrooms, where light is dispersed with a fragment of a DVD and the predispersed, uncollimated light passed through a cuvette and into the cell camera gives linear working curves over narrow concentration ranges, but also displays background absorbance, at least at blue wavelengths [88]. Seemingly better performance with a 1-cm cuvette is obtained if an LED is used as light source, narrowing the spectral range but simplifying optics, wavelength selection, and signal averaging [73].

9.3.4.2 96-Well Plates

Running samples in parallel allows for greater throughput but complicates instrument design. Should a single wavelength be passed through the wells, and then perform assays by averaging signal over those pixels observing a particular well? Should each well have white light passed through it, after which the signal is dispersed? Or should a dispersed spectrum illuminate each well, after which an image of the dispersed illumination provides a spectrum, spatially resolved within the image of each well? Postdispersion is the only one of these options which does not appear in the literature, and, probably, for good reason. Adequate precision would be difficult to obtain.

The previously cited chemometric data reduction has been performed on multiwell plates. The only wavelength resolution is in the RGB color separation of the camera pixels [74]. The wavelength precision is, thus, set by the camera manufacturer and the temperature sensitivity of the individual pixel bandpasses.

A well-developed approach to the use of multiwell plates comes from Ozcan's group [89]. Rather than trying to image the entire plate, a fiber optic bundle routes light from each well to a compact header readily photographed by the cellcam. A bank of LEDs illuminate all 96 wells at a putatively identical wavelength. One can imagine interchangeable LED banks to allow different wavelengths to be used, or the use of multicolor LEDs with appropriate diodes turned on for each determination. The use of white LEDs for tri-channel measurement is reported. This colorimeter does not permit resolution better than the LED's bandwidth, nor does it solve the limited digitization precision problem except by allowing signal averaging. LED wavelength precision is adequate for screening.

Glucose monitoring with accuracy comparable to an electrochemical blood glucose meter uses immobilized horseradish peroxidase (HRP) and colorimetric reactants in a 96-well format [90]. The light source is a tablet computer with a uniformly illuminated display, while the detector is a cellcam. The software selects for portions of the field of view not biased by line-of-sight obscuration. Neither the influence of off-axis viewing nor characterization of a reference light intensity is discussed; the authors use $I_0 = 255$, regardless of the actual backlight level. As long as the backlight intensity is constant and a reference sample is available for each experiment, this should not pose a problem. Instead of $A = -\log_{10}(I/I_0)$, one is essentially saying $A = k - \log_{10}(I)$, with the value of k implicit in the intercept of the working curve. It appears that the authors were focused on the chemistry of the glucose-selective reaction more than the quality and complexity of the spectrometry.

9.3.4.3 Test Strips

Test strips have been among the earliest analytical approaches employed with cellular telephone cameras. Whitesides developed an elaborate, low-cost test system using wax-impregnated paper and colorimetric reactions for Third World chemical and biomedical testing [12, 91–94]. Because calibration was so challenging, images were sent to a central computer for processing, rendering of diagnoses, and return to the user. A decade later, were the work to be repeated, it is likely that all processing would occur on the smartphone. Imaging has been augmented with electrochemical measurements [95, 96]. A fully optical pH meter, using only images of dye-embedded test strips, is now available [97]. Whitesides has opined (and the author concurs) that what most limits the development of inexpensive chemical and biomedical tests is that the areas that have the greatest need for such capabilities are least able to afford them, so that financing, rather than science or technology, impedes the broad dissemination of autonomous phone/test strip technology. One can imagine making tests that cost pennies apiece, but commercial investors are more interested in tests with healthy profit margins. Development of such tests will require support from non-governmental organizations (NGOs), governments, and foundations.

Suslick has championed the use of colorimetric arrays for "smell seeing," sensing airborne species by quantifying patterns in color changes of porphyrins and other organometallics upon exposure [98, 99]. His group has used flat-bed scanners, cell cams, digital single-lens reflex cameras, and dedicated spectrometers for quantifying the patterns. Because signal averaging is most easily achieved and digitization bit depth greatest for the dedicated spectrometer, precision is highest. While "color depth" is 12 bits, this amounts to 4 bits each of red, green, and blue. As shown in figure 3 of [100], the authors distinguish the effect of averaging over multiple pixels within a color spot from the precision of a single pixel but do not characterize photometric precision of any individual pixel. For an 8-pixel diameter image, an iPhone 5S gave precision of 18 counts out of 4095, while a linear CCD, with optics optimized to match the test strip geometry, gave precision of 2 counts. While such practical contrasts are useful for comparing specific devices or measurement conditions, a more fundamental view of precision is needed for adequately designing and contrasting cell phone spectrometers in general.

9.3.5 Precision and Error Propagation

Without adequate precision, linearity, and accuracy, a cell phone spectrometer is worse than useless – it may be misleading. The formalism for explicating spectrophotometric precision has been developed for laboratory instruments. Selected overviews can be found in [34, 35, 101–108], and authors who have significantly contributed to such thinking include Winefordner, Crouch, Ingle, and Tellinghuisen. We will not derive quantitative expressions here but rather will highlight areas that are of significant concern, frequently with higher variance for a cell phone instrument than for common laboratory instruments. Some noise sources may also be related to deterministic as well as random errors, and some errors are cross-correlated.

i. Digitization quantization noise: The charge stored in each camera pixel is converted to an integer before any signal processing can be done. For an N-bit converter, the minimum relative uncertainty for each digitization is 1 part in 2^N or 2^{-N}.
ii. Shot noise in signal and dark current: While there are a few situations where shot noise can be suppressed, typical incoherent light has Poisson noise and so for M detected photons has uncertainty $M^{1/2}$ for a signal-to-noise ratio of $M^{1/2}$. Bigger pixels can store higher charges and, thus, show lesser shot noise. For a typical cell cam with $\sim 1.4\ \mu m^2$ pixels, full well capacity is $\sim 2 \times 10^4$ electrons. Maximum precision at the full well is, thus, ~ 140, which explains why 8-bit precision, as is common in JPG files, fits well with camera performance. Some cameras have 10-bit digitizers, and so generate 1 count for every 20 electrons at full scale. However, the last 2 bits fluctuate due to shot noise. The example in Figure 9.8 suggests one count approximately equals one electron. In such cameras, one requires 15-bit digitization to be able to quantify full well charge. Alternatively, for 11-bit digitization, a full well would hold only 2000 electrons, for a maximum

signal-to-noise ratio of 45. The author has not yet figured out how to resolve the discrepancies among these calculations, although one suspects that the digitization ratio (counts/electron) is some multiple of that shown in Figure 9.8.

iii. Flicker noise: Even when power or current to a light source is precisely controlled, the output often fluctuates at low frequency (thus having a noise power spectrum $\sim 1/f^{\alpha}, f =$ frequency in Hertz, $0 < \alpha < 2$). Flicker noise is commonly minimized by using dual-beam measurements. Few cell phone instruments are capable of instantaneous double beam measurement, so flicker noise is high in such systems. The flash LED is used as a light source for some spectrometers, and that LED is controlled only sufficiently to provide pleasing snapshots, not for high precision photometry. Electronic flash, either white LEDs or xenon flashlamps, are also possible. Without careful characterization, their precision is suspect, again requiring dual-beam operation.

iv. Sample cell positioning noise: This is perhaps the most difficult noise source to control in portable instruments loosely attached to cell phones. Any change in position or angle of the sample-holder changes throughput and path length of light from the source to sample and sample to camera/sensor. If the flash LED in the cell phone is used as the light source, then the movement of a sample holder changes not only sample/camera alignment but also light source/sample alignment. Cuvettes and multiwell plates have strains, dust, and other imperfections whose position challenges precision measurement even in the lab. In portable settings, there is no way to identify, much less stabilize, their influence. To make meaningful measurements, sample cell positioning error must be small compared to the variance caused by the analyte. When such problems are ameliorated using a rigid, precisely alignable stand, portability is often compromised. Any claim of detection limits or quantification limits should indicate how much of the degradation from such limits in the laboratory is due to imprecise alignment in field instruments.

v. Thermal noise (electronics): Random motion of electrons in resistors (including wires) is called Johnson noise. The root mean square noise voltage, v_N, increases with temperature T, resistance R, and bandwidth Δf:

$$v_N = \sqrt{4kTR\Delta f} \tag{9.9}$$

Thus, noise depends on the environment and the temperature of individual components. If a camera is not properly heat sunk, then heavy use will gradually increase noise magnitude. If Wi-Fi coverage is poor so that transmitter amplitude must be high, or a battery is heavily loaded and releasing waste heat, phone temperature will increase. Without monitoring component temperature, noise amplitude becomes a significant, unquantified, drifting parameter. A 1 kΩ resistor at 300 K and 1 kHz bandwidth has an root mean square (RMS) noise amplitude of 129 nV or, equivalently, 129 pA. Bringing a pixel with 20 kiloelectron capacity to saturation in 1/30 second is a current of 0.1 pA. Clearly, noise suppression is critical to measurement quality, and more noise can be tolerated in selfie images than in scientific measurements.

vi. Thermal noise (mechanical structure): Solid objects can appear to have the rigidity of gelatin as temperature changes. Consider the coefficient of linear expansion of common optical and instrument materials (Table 9.2): Clearly, materials commonly used in laboratory instruments are less subject to temperature-induced dimensional changes than are materials employed in cell phones, injection-molded accessories, and plastic optics. A grating made of Mylar®, as are many inexpensive transmission gratings, will have, e.g. 600 lines/mm at 25 °C, 600.9 lines/mm at 0 °C, and 599.3 lines/mm at 45 °C. Mylar stretched flat at 0 °C with an unsupported span of 2 cm will expand to a length of 2.003 cm at 25 °C, allowing a departure from planarity of ~0.67 mm, corresponding to a departure from the desired mechanical plane of approximately 5° at the mounting points (assuming continued planarity of the supporting frame, with maximum displacement at the optical axis). For molecular absorbance or molecular fluorescence spectrometry, the change in grating d spacing is small, leading to less than a 2 nm shift in calibration at any pixel if dispersion is 1 nm/pixel, but the change in the angle of incidence at the edges of the grating may be significant. A 1 cm plastic cuvette will change path length by 0.2% between 0 and 25 °C (or any comparable temperature change). Refractive indices are also temperature

Table 9.2 Common coefficients of thermal expansion [105] except as noted.

Material	Coefficient of thermal expansion ($\Delta L/(L\Delta T)$, K^{-1}) × 10^6
Invar	1.5
Polystyrene	70
Polycarbonate	65–70
Polyolefins	32–108
Mylar	59.4
Aluminum	21–24
Titanium	8.5–9
Pyrex	4
Fused quartz	0.55
BK-7	7.1 [117]

Source: Based on Rieke [105].

dependent. Assembly of optical components using materials of different coefficients of thermal expansion may lead to alignment shifts. All these thermally driven distortions decrease precision.

vii. Thermal noise (wavelength calibration): As already noted, gratings or prisms change dispersion with temperature. If peak emission of an LED is used as a calibration reference, that too changes with temperature. Atomic emission lines are reliable for setting calibration wavelengths, but portable, low-power atomic line sources are rare. It is unfortunate that plasma panels (http://edenpark.com/products.html) are not optimized for use in the visible, neon lights have few emission wavelengths in the blue or green, mercury lamps require high-voltage transformers that are heavy, and flame emission poses assorted hazards in addition to being inconvenient to align.

viii. Thermal noise (chemistry): The position of chemical equilibrium and the rate of the chemical reaction are both temperature dependent. Thus, the slope of analytical working curves is temperature dependent, even for an instrument with thermostated optics. Further, if the specimen undergoes chemical reaction during measurement, the reaction may heat or cool the sample holder/cuvette, in turn, changing refractive index, optical path, reflection losses at interfaces, and (if illuminated off-normal) beam position.

ix. $1/f$ noise: light source aging: The color temperature of white LEDs shifts with age, generating a gradual drift in the wavelength dependence of I_0 for absorbance measurements. While LEDs drift more slowly and more monotonically than hot filament incandescent bulbs or arc lamps, they are not perfectly stable and, thus, also have $1/f$ noise. Gradual aging of camera response due to humidity, the decay of detector Bayer filters, and diffusion of semiconductor dopants contributes additional $1/f$ drifts. Camera quantum efficiency, thus, has a $1/f$ component. One may view the substitution of one camera model for another in a particular series of cell phones as a low-frequency but a high-amplitude event.

9.3.6 User Training and Method Validity

Even experts display limited precision in carrying out analytical procedures, and nonexpert users are unlikely to have instincts for detecting sample handling or preparation actions that could lead to invalid or inaccurate results. If cell phone spectrometry is to be useful for anyone but plaintiff's attorneys, the spectrometers, methods, and operating procedures must be coordinated using computer intelligence and real-time feedback. Measuring the contributions of instrument, environment, and user to measurement error will require at least a Gage Repeatability and Reproducibility study [109–111] and potentially error propagation modelling as well.

It is likely that most users will not want to see raw data, will not wish to perform calibration activities, and will be intimidated if they are informed they made an error [112, 113]. Thus, at each step of sample handling and measurement, data must be critically examined by software and the sequence of operations tuned to what users actually do as opposed to assuming they have proceeded in some desirable way. The Goedel Theorem [114, 115] ensures that no computer program can anticipate all the missteps of which analysts are capable, but ensuring that many anticipatable errors can be trapped and handled will be just as important as devising well-calibrated hardware if cell phone chemical analysis is to succeed. The structure of such variable path programming has much in common with programmed learning system units and their interconnection [116].

9.4 Conclusion and Prospective

Technical, business, and psychological barriers must be overcome if cell phone spectrometry is to be more than a chimera. The difference between qualitative and quantitative measurements, between instruments that report to a smartphone and instruments employing the phone as a measurement device needs to be clear to producers and consumers alike. Applications that seek to support human perception (color-matching) and those that seek to make scientific measurements (quantitative analysis) must be carefully distinguished, as the former must mimic the nonlinearities and tastes/biases of the user, while the latter must seek to be as linear and objective as possible. For quantitative measurements, as emphasized here, the environmental variability in which the systems will be used means that either instruments must be unrealistically stable, or they must be rapidly calibrated either autonomously or by unskilled personnel at nearly the moment of use. The technical sophistication required must be produced at an extremely low cost so that the potential mass market can be served, and distribution channels established that allow for reasonable profit margins. It is likely that niche markets requiring large numbers of determinations for a few analytes will seed this market; lead in drinking water, coliform bacteria in water, and nitrate or phosphate in agricultural runoff are likely possibilities. User psychology is also important. The potential user must not be intimidated, must be guided to perform operations adequately, and must get a satisfactory, actionable result without excessive effort or time investment. While such outcomes may be possible for assay of major components of materials by, e.g. Raman, infrared, or near-infrared spectrometers mated to cellular telephones, it will be considerably more challenging to devise instruments, reactant kits, and procedures for minor and trace components.

References

1 Mork, B.J. and Scheeline, A. (1988). Wavelength-resolved single-spark emission images using a charge coupled device detector. *Appl. Spectrosc.* 42: 1332–1335.

2 Mork, B.J. and Scheeline, A. (1989). Observations of high-voltage atmospheric-pressure spark discharges in argon using a charge-coupled device detector. *Spectrochim. Acta A* 44B: 1297–1323.

3 Leiden University (2013). Measuring particulates with your smartphone. http://www.research.leiden.edu/news/measuring-particulates-with-your-smartphone.html (accessed 1 September 2019).

4 Snik, F., Rietjens, J.H.H., Apituley, A. et al. (2014). Mapping atmospheric aerosols with a citizen science network of smartphone spectropolarimeters. *Geophys. Res. Lett.* 41: 7351–7358.

5 Oswald, E. (2017). With a built-in molecular spectrometer, this phone can identify any material. *Digit. Trends.* http://www.research.leiden.edu/news/measuring-particulates-with-your-smartphone.html (accessed 6 September 2019).

6 GoSpectro Spectrometer for Smart Phone (2019). https://www.alphanov.com/en/news/gospectro-power-spectroscopy-your-fingertips (accessed 16 September 2019).

7 Wang, S.X. and Zhou, X.J. (2008). Spectroscopic sensor on mobile phone. U.S. Patent 7,420,663, issued 2008.

8 Levenson, J. (2019). Samsung Galaxy S11 has a secret weapon to take on the iPhone 12. *T3 Smart Living*. https://www.t3.com/news/samsung-galaxy-s11-spectrometer (accessed 7 October 2019).

9 NeoSpectra (2019). MEMS + FT-IR. https://www.neospectra.com/our-offerings/neospectra-micro (accessed 16 September 2020).

10 Kang, S., Son, D., and Park, J. (2019). Electronic device comprising plurality of light sources. US Patent Application 16/359,189, published 2019.

11 Consumer Physics (2019). SCIO by consumer physics. https://www.consumerphysics.com (accessed 1 September 2019).

12 Martinez, A.W., Phillips, S.T., Carrilho, E. et al. (2008). Simple telemedicine for developing regions: camera phones and paper-based microfluidic devices for real-time, off-site diagnosis. *Anal. Chem.* 80 (10): 3699–3707.

13 McGonigle, A.J.S., Wilkes, T.C., Pering, T.D. et al. (2018). Smartphone spectrometers. *Sensors* 18 (223): 1–15.

14 Dutta, S. (2019). Point of care sensing and biosensing using ambient light sensor of smartphone: critical review. *TrAC* 110: 393–400.

15 Rezazadeh, M., Seidi, S., Lid, M. et al. (2019). The modern role of smartphones in analytical chemistry. *TrAC* 118: 548–555.

16 Saini, S.S., Sridhar, A., and Ahluwalia, K. (2019). Smartphone optical sensors. *OPN* 30 (2): 34–41.

17 Crocombe, R.A. (2018). Portable spectroscopy. *Appl. Spectrosc.* 72 (12): 1701–1751.

18 Ross, G.M.S., Bremer, M.G.E.G., and Nielsen, M.W. (2018). Consumer-friendly food allergen detection: moving towards smartphone-based immunoassays. *Anal. Bioanal. Chem.* 410: 5353–5371.

19 Nelis, J., Elliott, C., and Campbell, K. (2018). "The smartphone's guide to the galaxy": in situ analysis in space. *Biosensors* 8 (4): 96–126.

20 Alawsi, T. and Al-Bawi, Z. (2019). A review of smartphone point-of-care adapter design. *Eng. Rep.* 1 (2): e12039.

21 DetectaChem. MobileDetect Pouches. https://detectachem.com/products/mobiledetect-pouches (accessed 22 November 2019).

22 USB Implementers Forum, Inc (2018). USB Power Delivery. https://www.usb.org/document-library/usb-power-delivery (accessed 16 September 2020).

23 Jian, D., Wang, B., Huang, H. et al. (2019). Sunlight based handheld smartphone spectrometer. *Biosens. Bioelectron.* 143: 111632.

24 Hossain, M.A., Canning, J., Cook, K., and Jamalipour, A. (2016). Optical fiber smartphone spectrometer. *Opt. Lett.* 41 (10): 2237–2240.

25 Hong, J.I. and Chang, B.-Y. (2014). Development of the smartphone-based colorimetry for multi-analyte sensing arrays. *Lab Chip* 14 (10): 1725–1732.

26 Yu, H., Tan, Y., and Cunningham, B.T. (2014). Smartphone fluorescence spectroscopy. *Anal. Chem.* 86 (17): 8805–8813.

27 Long, K.D., Yu, H., and Cunningham, B.T. (2014). Smartphone instrument for portable enzyme-linked immunosorbent assays. *Biomed. Opt. Express* 5 (11): 3792–3806.

28 Long, K.D., Yu, H., and Cunningham, B.T. (2015). Smartphone spectroscopy: three unique modalities for point-of-care testing. *Next-Gen. Spectrosc. Technol.* VIII: 94820J-1-94820J–8.

29 Mudanyali, O., Dimitrov, S., Sikora, U. et al. (2012). Integrated rapid-diagnostic-test reader platform on a cellphone. *Lab Chip* 12 (15): 2678–2686.

30 Scheeline, A. and Bui, T.A. (2014). Energy dispersion device. U.S. Patent 8,885,161, issued 2014.

31 Bogucki, R., Greggila, M., Mallory, P. et al. (2019). A 3D-printable dual beam spectrophotometer with multi-platform smartphone adaptor. *J. Chem. Ed.* 96 (7): 1527–1531.

32 Hamamatsu Corp (2019). Silicon photodiodes. https://www.hamamatsu.com/resources/pdf/ssd/e02_handbook_si_photodiode.pdf (accessed 22 November 2019).

33 Burggraaff, O., Schmidt, N., Zamorano, J. et al. (2019). Standardized spectral and radiometric calibration of consumer cameras. *Opt. Express* 27 (14): 19075–19101.

34 Rothman, L.D., Crouch, S.R., and Ingle, J.D. (1975). Theoretical and experimental investigation of factors affecting precision in molecular absorption spectrophotometry. *Anal. Chem.* 47: 1226–1233.

35 Ingle, J.D. and Crouch, S.R. (1988). *Spectrochemical Analysis*. Englewood Cliffs, NJ: Prentice Hall.

36 Holst, G.C. and Lomheim, T.S. (2007). *CMOS/CCD Sensors and Camera Systems*. Winter Park, FL: JCD Publishing.

37 Hussain, I., Ahamad, K.U., and Nath, P. (2017). Low-cost, robust, and field portable smartphone platform photometric sensor for fluoride level detection in drinking water. *Anal. Chem.* 89 (1): 767–775.

38 Das, A.J., Wahi, A., Kothari, I., and Raskar, R. (2016). Ultra-portable, wireless smartphone spectrometer for rapid, non-destructive testing of fruit ripeness. *Sci. Rep.* 6: 32504.

39 Scheeline, A. (2018). Progress towards low resolution visible spectrometry with COTS components. *Next-Gen. Spectrosc. Technol.* X: 10657-06-1–10657-06-8.

40 Scheeline, A. and Bui, T.A. (2016). Stacked, mutually-rotated diffraction gratings as enablers of portable visible spectrometry. *Appl. Spectrosc.* 70 (5): 766–777.

41 Scheeline, A. (2015). Is "good enough" good enough for portable visible and near-visible spectrometry? *Next-Gen. Spectrosc. Technol.* VIII: 94820H-1-94820H–9.

42 Zhu, W., Yi, S., Chen, A. et al. (2019). Single-shot on-chip spectral sensors based on photonic crystal slabs. *Nat. Commun.* 10 (1): 1020–1025.

43 Sadler, D.A., Littlejohn, D., and Perkins, C.V. (1995). Automatic wavelength calibration procedure for use with an optical spectrometer and array detector. *J. Anal. Atomos. Spectrosc.* 10 (3): 253–257.

44 Miller, D.L. and Scheeline, A. (1993). A computer program for the collection, reduction, and analysis of Echelle spectra. *Spectrochim. Acta A* 48B: E1053–E1062.

45 Cao, X.A., LeBoeuf, S.F., and Stecher, T.E. (2006). Temperature-dependent electroluminescence of AlGaN-based UV LEDs. *IEEE Electron. Dev. Lett.* 27 (5): 329–331.

46 OSRAM (2012). ChipLED 0603 datasheet version 1.0, 1–23.

47 Xie, R.-J. and Hiosaki, N. (2007). Silcon-based oxynitride and nitride phosphors for white LEDs – a review. *Sci. Technol. Adv. Mater.* 8: 588–600.

48 Wagner, M., Herzog, A., Ganev, H., and Khanh, T.Q. (2015). LED aging acceleration – an analysis from measuring and aging data of 14,000 hours LED degradation. *12th China International Forum Solid State Lighting*, Shenzhen, China (2–4 November 2015), pp. 75–78.

49 Routledge, R. (2019). Central limit theorem. https://www.britannica.com/science/central-limit-theorem (accessed 19 September 2019).

50 Taylor, K. and Silver, L. (2019). Smartphone ownership is growing rapidly around the world, but not always equally. https://www.pewresearch.org/global/2019/02/05/smartphone-ownership-is-growing-rapidly-around-the-world-but-not-always-equally (accessed 1 September 2019).

51 Hernandez, T.C.M., Gonzalez-Valencia, E., Torres, P., and Ramirez, D.L.A. (2017). Low-cost spectrometer for educational applications using mobile devices. *Opt. Pura Apl.* 50 (3): 221–228.

52 Bayer Filter (2019). https://en.wikipedia.org/wiki/Bayer_filter (accessed 3 October 2019).

53 Scheeline, A. (2019). Energy dispersion cuvette. U.S. Patent 10,324,023, issued 2019.

54 Plymouth Grating Laboratory (2019). Two-dimensional (2D) diffraction gratings. https://www.plymouthgrating.com/product/two-dimensional-2d-diffraction-gratings (accessed 22 November 2019).

55 Snik, F., Karalidi, T., and Keller, C.U. (2009). Spectral modulation for full linear polarimetry. *Appl. Optics* 48 (7): 1337–1346.

56 Raju, L. (2012). Development of a low cost infrared spectrophotometer and a Matlab program to detect terrestrial and extraterrestrial water vapor. *Cent. Alabama Reg. Sci. Eng. Fair* 2013.

57 Coates, J. (2010). New miniaturized spectral measurement platforms covering the visible to the mid-IR. PDF of PowerPoint on linear variable filter spectra (7 October 2013). http://depts.washington.edu/cpac/Activities/Meetings/Fall/2010/documents/NeSSI_Workshop_November_2010_Coates.pdf (accessed 21 October 2019).

58 Coates, J. (2011). Integrated sensing module for handheld spectral measurements. U.S. Patent 7,907,282, issued 2011.

59 Macleod, H.A. (2017). *Thin Film Optical Filters*. Boca Raton, FL: CRC Press.

60 PixelSensor™ Multispectral Sensors (2019). https://pixelteq.com/pixel-sensor (accessed 1 September 2019).

61 11 Channel Spectral Color Sensor (2019). https://ams.com/as7341 (accessed 1 September 2019).

62 LightShift (2019). https://surfaceoptics.com/products/hyperspectral-imaging/lightshift (accessed 1 September 2019).

63 Wang, S.-W., Xia, C., Chen, X. et al. (2007). Concept of a high-resolution miniature spectrometer using an integrated filter array. *Opt. Lett.* 32 (6): 632–634.

64 Hercher, M. (1968). The spherical mirror Fabry-Perot interferometer. *Appl. Optics* 7: 951–966.

65 Wikipedia (2019). The Fabry Pérot Interferometer. https://en.wikipedia.org/wiki/Fabry–Pérot_interferometer (accessed 21 November 2019).

66 Vaughan, M. (1989). *The Fabry-Perot Interferometer: History, Theory, Practice and Applications*. Boca Raton, FL: CRC Press.

67 Mannila, R., Hyypiö, R., Korkalainen, M. et al. (2015). Gas detection with microelectromechanical Fabry-Perot interferometer technology in cell phone. *Next-Gen. Spectrosc. Technol.* VIII: 94820P.

68 Ding, H., Chen, C., Zhao, H. et al. (2019). Smartphone based multispectral imager and its potential for point-of-care testing. *Analyst* 144 (14): 4380–4385.

69 Yang, Z., Albrow-Owen, T., Cui, H. et al. (2019). Single-nanowire spectrometers. *Science* 365 (6457): 1017 LP–1020.

70 Learner, J. (2019). PARISS® imaging spectrograph. https://lightforminc.com/products/imaging-spectrograph (accessed 4 October 2019).

71 Choi, S., Kim, Y., Lee, J.-Y. et al. (2019). Minimizing device-to-device variation in the spectral response of portable spectrometers. *J. Sensors* 2019 (8392583): 7.

72 Zhang, J. (2013). Spectrometer miniaturized for working with cellular phones and other portable electronic devices. U.S. Patent 8,537,343, issued 2013.

73 Hussain, I., Bora, A.J., Sarma, D. et al. (2018). Design of a smartphone platform compact optical system operational both in visible and near infrared spectral regime. *IEEE Sensors* 18 (12): 4933–4939.

74 Coleman, B., Coarsey, C., and Asghar, W. (2019). Cell phone based colorimetric analysis for point-of-care settings. *Analyst* 144 (6): 1935–1947.

75 International Light Technologies, Inc.(2019). Neon Lamps: Application and Technical Notes. https://www.intl-lighttech.com/specialty-light-sources/neon-lamps (accessed 7 October 2019).

76 Wikimedia Commons (2009). File:Fluorescent Lighting Spectrum Peaks Labelled.png. https://commons.wikimedia.org/wiki/File:Fluorescent_lighting_spectrum_peaks_labelled.png (accessed 8 October 2019).

77 Kelley, K.D. and Scheeline, A. (2009). Cell phone spectrophotometer. *J. Anal. Sci. Digit. Lib.*, entry 10059. https://www.asdlib.org/onlineArticles/elabware/Scheeline_Kelly_Spectrophotometer/index.html (accessed 16 September 2020).

78 Scheeline, A. and Kelley, K. (2014). CellPhoneSpecIntEd. http://scheeline.scs.illinois.edu/~asweb/CPS/CellPhoneSpecIntEd.exe (accessed 8 October 2019).

79 PhysicsOpenLab (2017). Spectral Lines Broadening. http://physicsopenlab.org/2017/09/07/spectral-lines-broadening (accessed 2 October 2019).

80 Miokovic, Z., Balkovic, D., and Veza, D. (2010). Shift and broadening of sodium nS-3P and nD-3P transitions in high pressure NaCd and NaHg discharges. *Arxiv*. https://arxiv.org/ftp/physics/papers/0410/0410146.pdf (accessed 1 October 2019).

81 Kramida, A., Ralchenko, Y., and Reader, J. (2018). NIST atomic spectra database (Ver 5.6.1) [Online]. https://physics.nist.gov/cgi-bin/ASD/lines1.pl?spectra=Na+I&limits_type=0&low_w=400&upp_w=600&unit=1&submit=Retrieve+Data&de=0&format=0&line_out=0&en_unit=0&output=0&bibrefs=1&page_size=15&show_obs_wl=1&show_calc_wl=1&unc_out=1&order_out=0&max_low_enrg=&s (accessed 7 October 2019).

82 Scheeline, A. (2017). Focal point: how to design a spectrometer. *Appl. Spectrosc.* 71 (10): 2237–2252.

83 Zhao, Y., Elliott, C., Zhou, H., and Rafferty, K. (2019). Spectral illumination correction: achieving relative color constancy under the spectral domain. *IEEE Int. Symp. Signal Process. Inf. Technol.* 2018: 690–695.

84 Ding, H., Chen, C., Qi, S. et al. (2018). Smartphone-based spectrometer with high spectral accuracy for mHealth application. *Sens. Actuat. A Phys.* 274: 94–100.

85 Bayram, A., Yalcin, E., Demic, S. et al. (2018). Development and application of a low-cost smartphone-based turbidimeter using scattered light. *Appl. Optics* 57 (21): 5935–5940.

86 Fovean Corp (2010). X3 technology direct image sensors. http://www.foveon.com/article.php?a=67 (accessed 7 October 2019).

87 Wang, Y., Liu, X., Chen, P. et al. (2016). Smartphone spectrometer for colorimetric biosensing. *Analyst* 141 (11): 3233–3238.

88 Hosker, B.S. (2018). Demonstrating principles of spectrophotometry by constructing a simple, low-cost, functional spectrophotometer utilizing the light sensor on a smartphone. *J. Chem. Ed.* 95 (1): 178–181.

89 Berg, B., Cortazar, B., Tseng, D. et al. (2015). Cellphone-based hand-held microplate reader for point-of-care testing of enzyme-linked immunosorbent assays. *ACSNano* 9 (8): 7857–7866.

90 Pla-Tolós, J., Moliner-Martínez, Y., Molins-Legua, C., and Campíns-Falcó, P. (2018). Solid glucose biosensor integrated in a multi-well microplate coupled to a camera-based detector: application to the multiple analysis of human serum samples. *Sens. Actuat. B Chem.* 258: 331–341.

91 Carrilho, E., Phillips, S.T., Vella, S.J. et al. (2009). Paper microzone plates. *Anal. Chem.* 81 (15): 5990–5998.

92 Ellerbee, A.K., Phillips, S.T., Siegel, A.C. et al. (2009). Quantifying colorimetric assays in paper-based microfluidic devcies by measuring the transmission of light through paper. *Anal. Chem.* 81 (20): 8447–8452.

93 Whitesides, G.M. (2009). Paper diagnostics – using first world science in developing economies. *Pittsburgh Conf. Anal. Chem. Appl. Spectrosc*: 12.

94 Martinez, A.W., Phillips, S.T., Whitesides, G.M., and Carrilho, E. (2010). Diagnostics for the developing world: microfluidic paper-based analytical devices. *Anal. Chem.* 82 (1): 3–10.

95 Christodouleas, D.C., Nemiroski, A., Kumar, A.A., and Whitesides, G.M. (2015). Broadly available imaging devices enable high-quality low-cost photometry. *Anal. Chem. (Washington, DC, United States)* 87 (18): 9170–9178.

96 Ainla, A., Mousavi, M.P.S., Tsaloglou, M.-N. et al. (2018). Open-source potentiostat for wireless electrochemical detection with smartphones. *Anal. Chem.* 90 (10): 6240–6246.

97 Gotor, R., Ashokkumar, P., Hecht, M. et al. (2017). Optical pH sensor covering the range from pH 0–14 compatible with mobile-device readout and based on a set of rationally designed Indicator dyes. *Anal. Chem.* 89 (16): 8437–8444.

98 Rakow, N.A. and Suslick, K.S. (2000). A colorimetric sensor array for odor visualization. *Nature* 406 (6797): 710–713.

99 Suslick, K.S. and Rakow, N.A. (2001). A colorimetric nose: "smell – seeing". *Proc. Electrochem. Soc.* 2001–15 (A): 8–13.

100 Askim, J.R. and Suslick, K.S. (2015). Hand-held reader for colorimetric sensor arrays. *Anal. Chem.* 87 (15): 7810–7816.

101 Ingle, J.D. and Crouch, S.R. (1972). Evaluation of precision of quantitative molecular absorption spectrometric measurements. *Anal. Chem.* 44: 1375–1386.

102 Galbán, J., de Marcos, S., Sanz, I. et al. (2007). Uncertainty in modern spectrophotometers. *Anal. Chem.* 79: 4763–4767.

103 Winefordner, J.D., Svoboda, V., Cline, L.J., and Fassel, V.F. (1970). A critical review of atomic emission, atomic absorption, and atomic fluorescence flame spectroscopy. *Crit. Rev. Anal. Chem.* 1 (1): 233–272.

104 Tellinghuisen, J. (2000). Statistical error calibration in UV-visible spectrophotometry. *Appl. Spectrosc.* 54 (3): 431–437.

105 Rieke, G. (2003). *Detection of Light: From the Ultraviolet to the Submillimeter*. Cambridge: Cambridge University Press.

106 Sweedler, J.V., Ratzlaff, K., and Denton, M.B. (1994). *Charge Transfer Devices in Spectroscopy*. New York: VCH.

107 McCreery, R.L. (2000). *Raman Spectroscopy for Chemical Analysis*. New York: Wiley-Interscience.

108 Tellinghuisen, J. (2019). Calibration: detection, quantification, and confidence limits are (almost) exact when the data variance function is known. *Anal. Chem.* 91 (14): 8715–8722.

109 Quality-One International (2019). Gage repeatability & reproducibility (Gage R&R). https://quality-one.com/grr (accessed 22 November 2019).

110 Peralta, M. (2012). *Measurement System Error Analysis: Analyzing and Reducing Measurement Errors in Test Systems*. CreateSpace Independent Publishing Platform.

111 Gupta, S.V. (2012). *Measurement Uncertainties: Physical Parameters and Calibration of Instruments*. Springer GmbH.

112 Scheeline, A. (2011). Democratization of Detection: Every Citizen a Sensor? *An Integr. Water Secur. Summit Dedic. to Defense-in-Depth*, University of the Pacific.

113 Scheeline, A. and Bui, T.A. (2014). *Millions of Shallow CMOS Pixels and the Art of Spectroscopy*. Pittcon.

114 Raatikainen, P. (2018). Gödel's incompleteness theorems. *Stanford Encycl. Philos.* (Fall 2018 Ed.) (accessed 21 October 2019). https://plato.stanford.edu/archives/fall2018/entries/goedel-incompleteness (accessed 16 September 2020).

115 Hofstadter, D. (1979). *Gödel, Escher, Bach: An Eternal Golden Braid*. New York: Basic Books.

116 Sherwood, B.A. (1977). *The Tutor Language*. Minneapolis, MN: The Control Data Education Co.

117 Glass Dynamics LLC. (2015). BK-7 optical glass. http://www.glassdynamicsllc.com/bk7.html (accessed 5 October 2019).

10

Portable Standoff Optical Spectroscopy for Safety and Security

Matthew P. Nelson and Nathaniel R. Gomer

ChemImage Corporation, Pittsburgh, PA, USA

10.1 Introduction

10.1.1 Motivation

In today's world, biological, explosive, chemical, and narcotic threats are a constant worry, with the number of incidents on the rise. Military, law enforcement, first responders, public health officials, and civilians alike are facing this new reality that has forever changed the way we live our lives. Post-9/11 biological threats emerged in the form of anthrax being sent through mail, while improvised explosive devices (IEDs) became the weapon of choice for Al-Qaeda and Islamic State of Iraq and Syria (ISIS) terrorist groups during the post-9/11 wars in Iraq and Afghanistan, and continue to be prevalent even today. In 2018, chemical attacks were reported to have taken place in Syria, resulting in the deaths of 40–50 people and potentially injuring more than a hundred (Anon 2020a). Abuse of opioids has rapidly become epidemic, being responsible for 68% of drug overdose deaths nationally in 2017 (Scholl et al. 2019). Even more recently, the COVID-19 pandemic has sickened millions and claimed the lives of hundreds of thousands of people around the globe (Anon 2020b).

As threat prevalence grows, so does the need for technological advancements that counter these dangers to society, whether through intelligence gathering, threat detection and identification, or obliteration of the hazard. In recent years, optical spectroscopy has demonstrated that it plays a role in addressing the problem since it has the potential to address a number of unmet needs associated with practical concepts of operation (CONOPS).

An ideal portable, standoff, optical spectroscopic sensor has several distinct characteristics. Most CONOPs demand that sensors be capable of real-time, or near real-time, measurements so as not to interrupt the flow of established procedures. Sensors need to be standoff in nature to increase the safety of operators/bystanders and equipment. Specific performance requirements are generally CONOPs-driven, but all demand high sensitivity and high specificity. Threat *identification* is often necessary in addition to *detection* alone. Given the broad range of end users, sensors need to be highly automated, and with easy-to-use user interfaces. Since the location of a threat in an area of interest is often unknown, sensors also need to be capable of wide-area scanning. Portability to enable ease in fielding, operating and supporting is often important, as well as the sensor being affordable, in terms of both acquisition and maintenance costs.

10.1.2 Definitions

In order to set the stage for this chapter, we first need to define a few key terms in the context of portable standoff optical spectroscopic sensors for safety and security.

Portable Spectroscopy and Spectrometry 1: Technologies and Instrumentation, First Edition.
Edited by Richard A. Crocombe, Pauline E. Leary and Brooke W. Kammrath.

Table 10.1 Optical radiation wavelength regions and applicable detector technologies.

Optical radiation region	Wavelength range (μm)	Common detector technologies
Ultraviolet (UV)	0.010–0.35	PMT, vacuum phototube, UV-CCD, Si photodiode, UV-CMOS
Visible (VIS)	0.35–0.770	PMT, vacuum phototube, CCD, Si photodiode, CMOS
Near-infrared (NIR)	0.77–1	PMT, vacuum phototube, CCD, Si photodiode, CMOS
Shortwave-infrared (SWIR)	1–3	InGaAs, HgCdTe, InSb
Midwave-infrared (MWIR)	3–8	HgCdTe, InSb, PbSe
Longwave-infrared (LWIR)	8–15	HgCdTe, Microbolometers
Far-infrared (FIR)	15–1000	Doped silicon

The first term we will define is "portable" or "portability." In the context of this chapter, we consider sensors that may be moved from one location to another by vehicle, or by man, to be portable. There are varying degrees of portability. Portability, as we define it here, will span vehicle-/robot-integrated, to man-portable (1 or 2 man-carry), to handheld (1–2 hand-carry), to wearable (heads-up display, cell phone-integrated, etc.).

Another term we will define is "standoff." Although the majority of sensing technologies in this chapter fall in the meter-to-kilometer standoff range, our simplest definition of standoff is noncontact.

In the following sections, we will frequently refer to specific optical radiation regions. There is little agreement in the literature (Byrnes 2009; Henderson 2007; Ingle and Crouch 1988; IPAC 2012) on specific starting and ending points of each of these spectral regions. Therefore, it is prudent we define the spectral regions, associated wavelength ranges, and applicable detector technologies (see Table 10.1).

Related to the optical radiation region definitions is the associated detector terminology. When using the term "camera," we will be referring to a device that acquires a digital record in the visible portion of the spectrum. "Charge coupled device (CCD)" and "complementary metal-oxide-semiconductor (CMOS)" detectors will generally be associated with the ultraviolet (UV), visible (Vis) and near-infrared (NIR) spectral regions. The use of the term "focal plane array (FPA)" detector will commonly be used to describe how the imaging detector technology is associated with the infrared (IR) spectral regions.

The final sets of terms we will define are single point vs. multipoint spectroscopy (multispectral, hyperspectral). Single-point spectroscopy refers to single spectral measurements acquired from a target of interest. Multipoint spectroscopy (also known as spectral imaging) describes multiple spectral measurements acquired from a target of interest, often resulting in an image where each pixel has an associated spectrum. There are two general classes of multipoint spectroscopy, namely multispectral and hyperspectral. We define multispectral as spectral information containing 2–10 bands, and images captured within discrete wavelengths and/or wavelength ranges of the electromagnetic spectrum. Multispectral measurements take advantage of the fact that many material characterization applications need only a small number of spectral bands to provide the desired analytical information for the analyte(s) of interest. A reduced number of spectral bands enables faster speeds (near real-time), and less data, and, therefore, computational burden. Finally, "hyperspectral" describes spectral information containing many 10s to 100s or even 1000s of bands. In hyperspectral imaging (HSI), a plurality of images are captured at discrete wavelengths, and/or wavelength ranges of the electromagnetic spectrum. Alternatively, hyperspectral images are obtained by the acquisition of full sets of spectra. These methods take advantage of full spectral library searching algorithms, but at a cost of sacrificing speed of acquisition and analysis. These full hypercube methods can also be computationally burdensome, due to large amounts of data.

10.1.3 Chapter Scope

The scope of this chapter is portable standoff optical spectroscopy for safety and security applications. We will focus primarily on optical absorbance-based sensors ([NIR], shortwave-infrared [SWIR] and midwave-infrared [MWIR]), and Raman sensors. There are many other technologies (UV/Vis absorbance, millimeter wave, terahertz, fluorescence, photoacoustic, and laser-induced breakdown spectroscopy [LIBS]) that also play an important role in standoff sensing capabilities, but will not be covered extensively in this chapter.

10.1.4 Portable Standoff Optical Spectroscopy History

10.1.4.1 Standoff Optical Sensing Origins

In the broadest sense, the first standoff optical spectroscopic sensors were biological in origin: the eyes. The first proto-eyes evolved among animals 600 million years ago (Jones 2014). Like most standoff optical spectroscopic sensors, the eyes contain a light-gathering subsystem (i.e. lens and iris), a wavelength-selection device (i.e. the cone cells which distinguish color, along with rod cells which provide low-light contrast) and a detection subsystem (i.e. the retina which converts the collected light into a visual signal which is sent to the brain through the optic nerve for processing.

10.1.4.2 Standoff Optical Spectroscopic Instrumentation Origins

Astronomy was the first field to utilize and develop standoff sensing instrumentation that enhanced what the human eye was able to sense. The most important development being the telescope, which was first credited to Hans Lippershey, an eyeglass-maker from the Netherlands in 1608. A year later, Galileo refined the design and used it for astronomy, which he used to discover four satellites of Jupiter and to discover sunspots. Since then, numerous improvements to the design have improved its use and performance (Editors 2008; King 2011).

The advancement of telescopes led to the first standoff spectroscopic measurements made from celestial objects, including planets and stars in order to determine properties such as chemical composition, mass, or temperature. In the mid- to late 1600s, Isaac Newton used a prism to look at the refractive properties of light. In the early 1800s, Joseph von Fraunhofer, using Newton's work and theories of wavelength dispersion, replaced the prism with a diffraction grating to improve the spectral resolution, and allow for the more accurate accounting of the dispersed wavelengths. The distinct black lines in emission spectra of the sun became known as Fraunhofer lines. Fraunhofer used this approach to collect spectral information from Venus, the Moon, and Mars (Hearnshaw 1986). Today, our knowledge of the composition of the sun and planetary atmosphere is based on standoff optical spectroscopy (Hanel et al. 1980).

Long-range optics, inclusive of the contemporary telescope, along with advancements in light sources, wavelength selection devices and detector technologies, have been key enablers that have led to modern standoff optical spectroscopic instrumentation. Standoff Raman and IR sensors are two categories of such sensor technologies that have benefited immensely from these developments.

10.1.4.3 Standoff Raman Spectroscopic Instrumentation

Raman spectroscopy has long been a valuable technique for standoff and remote applications due to its ability to provide in situ analysis and a near unambiguous chemical fingerprint for a target. The first standoff Raman experimentation was performed by Tomas Hirschfeld in the 1970s, where long-range atmospheric measurements were made (Hirschfield 1974). Hirschfeld paved the way for the development of smaller standoff-capable systems in the late 1900s and early 2000s for a wide range of applications (Angel et al. 1992), including explosive detection (Carter et al. 2005), mineral and organic compound analysis (Sharma et al. 2006), and planetary exploration (Angel et al. 2012).

10.1.4.4 Standoff Infrared Spectroscopic Instrumentation

Development of standoff Fourier transform (FT)-based IR spectroscopic instrumentation for condensed phase samples has been encumbered by atmospheric interference and the necessity for moving parts and sample contact. The more recent development of quantum cascade lasers (QCLs) and supercontinuum laser sources (Dupont et al. 2012) has advanced the use of MWIR and longwave-infrared (LWIR) sensors for standoff sensing capabilities. QCLs have been widely used for gas-sensing applications (Pecharromán-Gallego 2017). Experimentation done by Thériault et al. (2004) showed the potential for passive IR to detect chemical warfare agents (CWAs) on surfaces.

NIR- and SWIR-based standoff sensors have seen faster implementation into standoff configurations, compared to MWIR and LWIR, due, in part, to their more robust performance under varied atmospheric conditions and cost-effective components and system design. Solar illumination and/or inexpensive broadband incandescent lighting may be used for source illumination and detectors can be uncooled.

10.1.4.5 Portable Standoff Optical Spectroscopy Instrumentation

Historically, spectroscopic sensing instruments have been large, costly, and difficult to implement in field-based applications. These impediments to portability are largely due to the maturity of the required components, typically the illumination source (i.e. laser) or spectral analysis system (i.e. spectrometer and detector). Recent technological advances have reduced the size, weight, and power consumption requirements for system components, which has allowed a tremendous rise in the implementation and production of portable sensors.

The development of smaller, more durable IR and Raman handheld spectrometers came out of the post-9/11 war effort, as they were used in Iraq and Afghanistan to detect IEDs and other hazardous threats or materials. The first modern, practical NIR handheld was introduced in 2005 by a company called Polychromix (now Thermo Fisher Scientific). The first MWIR sensor was developed by SensIR (now Smiths Detection), which was FT-based and the size of a suitcase. The first fully capable portable Raman system was developed by Ahura Scientific (now Thermo Fisher Scientific) (Crocombe 2018).

10.2 Portable Standoff Optical Instrument Types

Portable standoff optical spectroscopic sensors fall into three main categories based on the method of operation. The following section will introduce each method and associated subcategories of sensors.

10.2.1 Point Measurement Sensors

The first category of sensors are the point measurement sensors. These instruments focus the illumination source onto the target surface, followed by the collection of a full spectrum using a spectrometer or interferometer. This approach is the simplest overall approach, and provides a full spectrum that may be searched against a spectral library. The downside of this approach, as it relates to portable standoff sensors, is that it requires knowledge of where to scan, often requiring the aid of a supplemental targeting or cueing sensor.

10.2.2 Spatial Scanning (Mapping) Sensors

The second general category of sensors are the spatial scanning (also known as mapping) sensors. Two main subcategories of spatial scanning sensors include point-by-point (Boogh et al. 1992) and line-mapping (Bowden et al. 1990; Jestel et al. 1998, "push-broom") sensors.

Point-by-point sensors produce chemical "maps" by first focusing a light source (when using active illumination) onto a surface, followed by the collection of one or more spectra at each desired spatial location, using a spectrometer or interferometer. Spatially accurate chemical (spectral) images may be generated by acquiring spectra with regular x, y-coordinated collection of spectra. These sensors provide a full image and full spectra, but at the cost of lengthy acquisition times where the point measurement time, T_P is proportional to the number of measurements, "n."

Line-scanning sensors are the second subcategory of spatial scanning sensors. Line scanning is an extension of point-by-point mapping. "Maps" of the sample surface are generated by focusing an elongated illumination beam (when using active illumination) in one dimension (using a cylindrical lens or a scanning mechanism such as a moving mirror) with the target oriented parallel to the entrance slit of a dispersive spectrograph equipped with a two-dimensional FPA detector. Each row of the detector provides a means to generate a full spectrum at each spatial location along the length of the illumination source. Line-scanning approaches have the advantage of being full image- and full spectrum-capable. Compared to point-by-point scanning, line-scanning acquisition times, T_L, are reduced as defined by the following:

$$T_L = (1/\text{sqrt}(n))^* T_P \text{ (single point)}$$

However, the spatial resolution differs for x- and y-axes where the parallel direction is determined by optical properties of the collection optics, while the perpendicular direction is determined by the precision of the sampling mechanism.

10.2.3 Wide-field Imaging (Staring) Sensors

The third category of sensors is the wide-field or global imaging sensors. When applications demand rapid, accurate and more comprehensive characterization of target morphology and composition over a large area, wide-field imaging approaches are typically sought out. Advancements in the past 25 years have made wide-area, real-time spectroscopic imaging a reality.

Wide-field imaging sensors operate by illuminating the entire field-of-view, while images are captured at discrete wavelengths, and scanning portions of the spectrum (discrete waveband imagers), or by capturing all spatial and spectral information simultaneously (snapshot imagers). Discrete waveband imagers use imaging spectrometers (i.e. acousto-optical tunable filters (AOTFs) (Goldstein et al. 1996; Schaeberle et al., 1995, 1996; Treado et al. 1992), rotating dielectric filters (Batchelder et al. 1991), Fabry–Perot interferometers (Cristensen et al. 1995; Vaughn 1989) and liquid crystal tunable filters (LCTFs) (Morris et al. 1994, 1996; Turner and Treado 1997)) to scan through discrete wavelengths interleaved with image captures at each step. Alternatively, discrete waveband imagers can utilize source tuning (i.e. QCLs (Faist et al. 1994; Schlossberg and Kelley 1981) or light-emitting diodes (LEDs) (Schubert 2003) to illuminate the field-of-view one waveband at a time, while capturing images at each step. Snapshot imagers, on the other hand, operate using wavelengths/waveforms designed with a particular end use in mind ((i.e. snapshot (West et al. 2018), multivariate optical elements (MOEs) (Nelson et al. 1998b) or conformal filters (CFs) (Nelson et al. 2019, 2018, 2017, 2016)) or by using devices that enable multipoint full spectral image captures (i.e. fiber array spectral translators (FAST) – also known as dimension reduction arrays and fiber image compression devices) (Nelson et al. 1996, 1998a; Nelson and Myrick 1999a, b; Ma and Ben-Amotz 1997; Wentworth et al. 2007).

Advantages associated with wide-field imaging approaches include unparalleled speeds [near real-time (<10 s) and real-time (>10 detection frames/s)]. The disadvantage of wide-field imaging is that active source architectures require high source illumination power to achieve comparable illumination power densities, and, hence, signal levels, compared to point or line approaches.

For discrete wavelength imagers, the experiment time (T_{DW}) is proportional to the number of spectral channels "m" instead of the number of measurements, "n":

$$T_{DW} = (m/n)T_P \text{ or } T_W = (m/\text{sqrt}(n))^* T_L$$

The experiment time for snapshot imagers (T_{SS}) is proportional to the time to acquire a single point measurement, since there is not scanning of spectral channels (i.e. $m = 1$), and all spatial measurements are being acquired simultaneously (i.e. $n = 1$):

$$T_{SS} = (T_P/n)^*\text{ROI};\ T_{SS} = (T_L/\text{sqrt}(n))^*\text{ROI};\ T_{SS} = (T_{DW}/m)^*\text{ROI}$$

10.3 Portable Standoff Optical Instrument Technologies

This section is not meant to be a comprehensive treatment of technology approaches. We discuss the most common architectures in use by end users for each of the three standoff portable instrument types discussed in the previous section.

10.3.1 Point and Spatial Scanning

Point and spatial scanning instruments typically use either a dispersive-based optical spectrometer (Ingle and Crouch 1988) or an interferometer. Dispersive-based optical spectrometers are devices that measure light as a function of wavelength. Today, most devices are based on diffracted light produced through the use of a diffraction grating, in combination with a linear array or two-dimensional array detector. Point and spatial scanning devices use light-gathering optics to capture a portion of the photons emitted, scattered or reflected from a target surface, and direct it through the entrance slit of the spectrometer. A detector captures the dispersed light diffracted by the grating, and converts the photons to electric charge, which is read out by an analog-to-digital converter and computer, ultimately producing a single spectrum or multiple spectra of the target.

Two common interferometer-based spectrometers are the continuous-wave Michelson or FT interferometer (Ingle and Crouch 1988) and the spatial heterodyne spectrometer (SHS) (Gomer et al. 2011a, b, 2019).

Michelson or FT interferometers operate by splitting the light from a source into two beams by a half-silvered mirror to one fixed return mirror and one movable return mirror (introducing a time delay). The split light beams interfere producing a temporal-based interferogram. Subjecting the interferogram to an FT produces a frequency-based spectrum of the light.

SHS interferometers are fundamentally a Michelson interferometer with return mirrors replaced by fixed diffraction gratings oriented at particular angles. The return wavefronts produce wavenumber-dependent crossing angles resulting in a superposition of Fizeau fringes, containing spatial frequencies. A two-dimensional FPA records the fringe pattern. Applying an FT again produces a frequency-based spectrum of the light.

10.3.2 Wide-field Imaging

Wide-field imaging architectures are more diverse than the point- and line-scanning approaches. The following section discusses four main classes of wide-field architectures, namely fixed bandpass filter-based, tunable bandpass filter-based, optical fiber-based and source tuning-based.

10.3.2.1 Fixed Bandpass Filter-Based

Fixed bandpass filter-based architectures include dielectric interference filter wheels, rotating dielectric filters, snapshot imagers, and MOEs. Optical filter technology and its applications are described in detail in the chapter in this Volume by Pust.

Dielectric interference filter (Baillard et al. 2006) sensors typically use a wheel or linear translation stage incorporating optical filters that reflect one or more spectral bands and transmit others. Interference filters consist of multiple thin layers composed of dielectric material having different refractive indices. Placing a series of filters in a motorized filter wheel, in front of a two-dimensional detector array, and sequentially capturing images as the filters are mechanically moved in front of the imaging detector, enables a hyperspectral data cube to be collected. This approach is simple, and low cost, with good spatial resolution, and is high in optical throughput. Disadvantages of this approach include image shift from mechanical motion of filters, limitations around the number of filters used (i.e. wavelengths), slow speed and low spectral resolution.

Rotating dielectrics involve the mechanical rotation of dielectric filters (Batchelder et al. 1991), enabling tuning of the center bandpass. Like dielectric filter wheels, rotating dielectric designs are generally simple, of low cost and high spatial resolution and provide high optical throughput. Drawbacks of this approach include image shift from mechanical motion of filters and inhomogeneity of the bandpass across the field of view.

Snapshot architectures enable a method for capturing hyperspectral images during a single integration time of the camera (Figure 10.1). Most notable is the Bayer pattern approach (Bayer 1976) in which filters are fabricated onto a grid of detector pixels using common lithographic techniques (Buchsbaum and Morris 2000; Kutteruf et al. 2014). The grid is either applied to wafer-level glass and bonded to the sensor, or is deposited directly onto the semiconductor substrate of the sensor. Filter designs come in a variety of patterns including area (Bayer-like), row (stripe), per-pixel (Mosaic), wedge (line scan), or custom. Advantages of the snapshot design include high spatial resolution, real-time compatible, high throughput, no moving parts, the potential to be made to be very small, and environmental robustness. However, snapshot imagers lack reconfigurability. In other words, the snapshot technology cannot be field-upgraded to incorporate new threats, or to improve classification and false alarm performance. Since the filter array is bonded directly onto the FPA, upgrading the system not only likely requires new software and a new algorithm, but it also involves designing a new FPA if new or different wavelengths are necessary for detection. In addition, the fabrication time and cost of the snapshot's pixel-level filter array may be prohibitive at low production volumes. Other limitations include having a limited number of wavelength

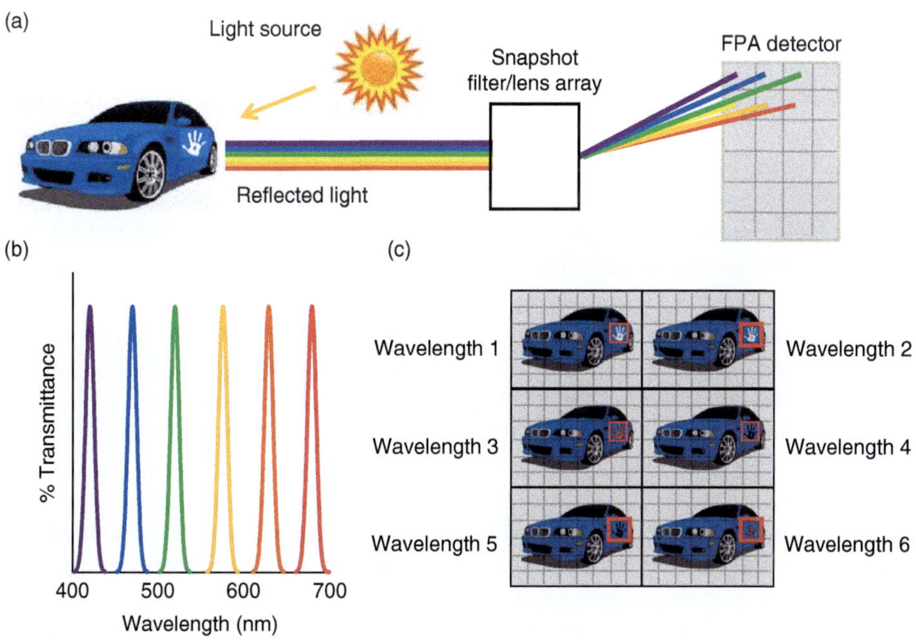

Figure 10.1 Snapshot sensor concept illustration. (a) Light captured from a targeted scene is filtered at the pixel level on the focal plane array. (b) Example transmittance spectral profiles associated with coated pixels. (c) Reconstructed wavelength-resolved images.

bands, susceptibility to image artifacts (i.e. edge effects, etc. due to spatially offset pixel patterns), reduced spatial resolution and manufacturing difficulties depending upon filter design patterns and transmittance properties. With snapshot imagers, a trade between spatial resolution and spectral resolution is often a consideration.

MOEs (Nelson et al. 1998b) are the key elements of a multivariate optical computer. Traditional methods use multivariate or chemometric methods, such as multivariate calibration, classification, etc. to extract analytical information through generation of a regression vector from data collected at many wavelengths spanning the analytical measurement of interest (i.e. concentration). Multivariate optical computing encodes the regression vector optically into the transmission function of the optical element, i.e. the MOE. Light from a sample contains the spectral information of the samples. The normalized light detected after passing through the MOE is proportional to the dot product of the regression vector with that spectrum, and from this, the analytical measurement of interest may be obtained. MOE benefits include high spatial resolution, real-time compatibility, high optical throughput, no moving parts, a low size and weight. Shortcomings of MOEs include lack of reconfigurability, large upfront development effort, and susceptibility to changing conditions that have the potential to compromise measurement robustness.

10.3.2.2 Tunable Bandpass Filter-Based

Tunable bandpass filter-based architectures include LCTFs, AOTFs, and Fabry–Perot interferometers.

LCTFs are composed of a series of stages consisting of fixed retardance birefringent elements, bonded to a nematic liquid crystal waveplate, and sandwiched between linear polarizers (Morris et al. 1994, 1996; Turner and Treado 1997). Narrow bandpasses of light are achieved by controlling the voltages applied across the liquid crystal waveplate, resulting in constructive and destructive interference. Tuning is achieved via varying the applied voltages in a controlled way with a computer. Common LCTF types include the Lyot, Solc, Evans Split-Element, Multiconjugate Filter (MCF), and CF designs (Lyot 1944).

The Lyot design exhibits a bandpass and free spectral range determined by thickest and thinnest birefringent elements. Overall transmittance is a sinc function. The design suffers from poor optical throughput due to large numbers of highly absorbing polarizers.

The Solc (1965) filter has a fanout and folded filter design that improves optical throughput using only two polarizers, at a cost of poorer out-of-band rejection efficiency. The Evans split-element design (Evans 1949; Evans and Solc 1958) provides a better balance of throughput and sidelobe characteristics, and also reduces the number of polarizers to approximately half of the Lyot design.

The MCF design (Wang et al. 2006; Figure 10.2) addresses the need for improved optical throughput and thermal stability by combining the high transmittance of Solc and the high out-of-band rejection efficiency of Lyot. Its finesse (number of passband widths in a spectral period) is 1.5X of Evans and 2.0X of Lyot, allowing manufacturing with fewer stages.

The CF is a multivariate HSI architecture that uses a subset of elements of commercial-grade MCFs (Lyot 1944). CF is engineered to transmit multiband waveforms as opposed to single. Selection of passbands is CONOPs-specific and is optimized for maximum discrimination between target and background/interferents. CF-imaging exploits concepts of compressive sensing and multivariate optical computing in that it allows collection of imagery approximating detection response of conventional HSI with only two discrete measurements (i.e. filter-tuning states), rather than hundreds. Real-time (>10 detection frames per second, fps) detection imagery is achieved using an approach called dual-polarization conformal filter (DP-CF) imaging by the simultaneous collection of two tuning states. DP-CF sensors divide light captured by the sensor's light-gathering optics into two optical paths using a polarizing beam splitter, such that each path respectively preserves the parallel and perpendicular polarizations of light. Light from each path is then optically filtered by two discrete CFs tuned to target/background discrimination optimized voltage states. Filtered images are then captured onto one or two cameras. Images are ultimately cropped, co-registered, and mathematically manipulated to produce a score image that is further processed to provide streaming detection imagery.

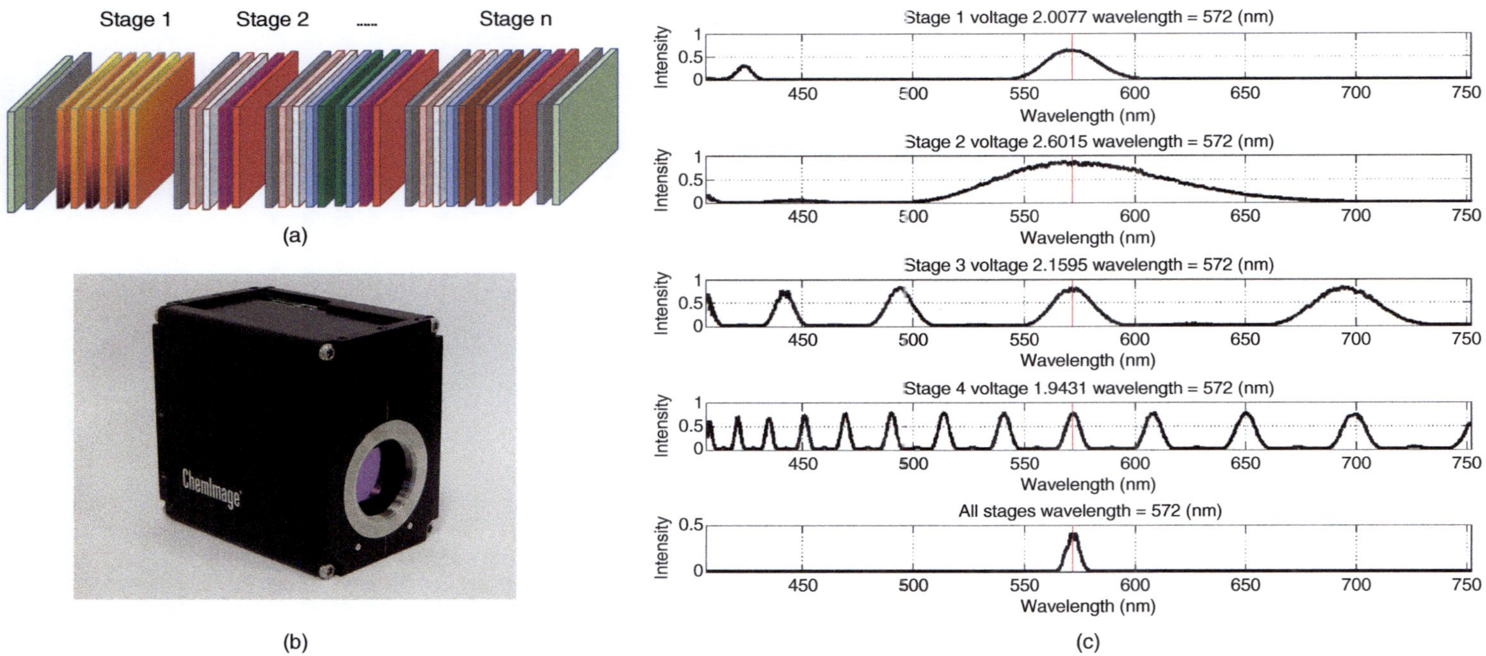

Figure 10.2 Multiconjugate filter (MCF) design. (a) Series of stages consisting of polarizers, quartz retarders, and liquid crystal waveplates. (b) Photograph of assembled filter. (c) Transmittance spectral profiles of individual stages (top four) and combined (bottom). *Source*: Picture and data courtesy of ChemImage Corporation.

Strengths of the LCTF approaches include real-time compatibility (DP-CF), reconfigurability/adaptability, near-diffraction-limited spatial resolution, high out-of-band rejection efficiency and no moving parts. The main weaknesses of the LCTF architectures are inefficient full spectral data collection (except CFs), and low-to-moderate optical throughput depending on the design and restrictive storage temperature limits (i.e. approximately −15 to +55 °C).

AOTFs are solid-state, no-moving-parts tunable filter devices, based on the interaction of light with a traveling acoustic sound wave in an anisotropic crystal medium (Treado et al. 1992; Goldstein et al. 1996; Schaeberle et al. 1995, 1996). The devices are based on changes in the refractive index of a medium when sound waves are present, producing a refractive index grating that is "seen" by the light propagating through it. A narrow bandpass of light is created by applying a radio frequency (RF) signal to the AOTF. Tuning is achieved by changing the applied RF signal frequency and power.

AOTF advantages include relatively high optical throughput compared to LCTFs (~40% for unpolarized light), broad spectral coverage (UV to MWIR) and rapid tuning speeds (~100 μs). Disadvantages include inefficient full spectral data collection, incompatibility with real-time imaging, broad spectral bandpass, a spatial resolution that is 5× worse than diffraction limits, and nonuniform spectral bandpass across the field of view.

Fabry–Perot interferometers (Cristensen et al. 1995; Vaughn 1989) (also known as etalons) incorporate an optical cavity made from two parallel reflecting surfaces with an air gap. Multiple instances of light interferences occur in the cavity. Light passes through the optical cavity when in resonance with the optical cavity, i.e. multiples of $\lambda/2$ fit exactly within the cavity. Tuning is achieved by changing the distance or refractive index between the two surfaces. Fabry–Perot interferometers boast high image quality and high finesse. The devices are limited in that they provide moderate spectral resolution, low out-of-band rejection, limited free spectral range, small acceptance angle, and require extensive thermal management to reduce passband drift.

10.3.2.3 Fiber Optic-Based

Fiber optic-based approaches utilize the flexibility that optical fibers provide to physically reduce the dimensionality of data via FAST, enabling simultaneous capture of spectral and spatial data. FAST may be coupled to a traditional dispersive spectrometer, or an SHS.

FAST-dispersive devices (Figure 10.3) operate by global or wide-field illumination of a target area (Nelson and Myrick, 1999a, 1999b; Nelson et al., 1996, 1998a, 1998b; Wentworth et al. 2007). A portion of the light is gathered from the target area and projected in the form of an image onto the two-dimensional proximal end of the fiber bundle. Fibers are reordered in a specific pattern to the one-dimensional end, which is inserted in parallel into the entrance slit of a dispersive spectrometer equipped with an imaging format FPA detector. Spatial information is preserved in the y-axis while spectral information is in the x-axis. A single snapshot contains two spatial and one spectral dimension of data. The system software reconstructs the hypercube from the focal plane image and the known configuration of fibers at the 2D end of the bundle, providing position-specific spectra and wavelength-specific images. FAST-dispersive device's main advantage is the speed for applications that are not light-limited. FAST provides a good balance between image fidelity and spectral information for certain applications where high image fidelity is not necessary. FAST also provides full spectral coverage. FAST disadvantages include low image fidelity (limited by the magnification of fibers and number of FPA rows).

FAST may also be coupled with an SHS (Figure 10.4) to enable wide-field HSI (Gomer et al. 2011a, b, 2019). Similar to FAST-dispersive, the target area is illuminated in a wide-field manner – i.e. laser for Raman measurements. The scattered light is collected, filtered with an appropriate collection optic and Rayleigh-rejection filter(s) and focused onto a FAST bundle. The output of FAST is collimated and directed into the SHS, and the output of each fiber is spatially dispersed onto a two-dimensional CCD. A single CCD image contains the spatially resolved interferograms which are transformed to the associated Raman spectra via an FT. SHS requires no moving parts and may be made to be very small. Unlike traditional dispersive spectrometers that require long focal lengths to achieve high spectral resolution, SHS is slitless, making it well-suited for wide-field illumination and functions

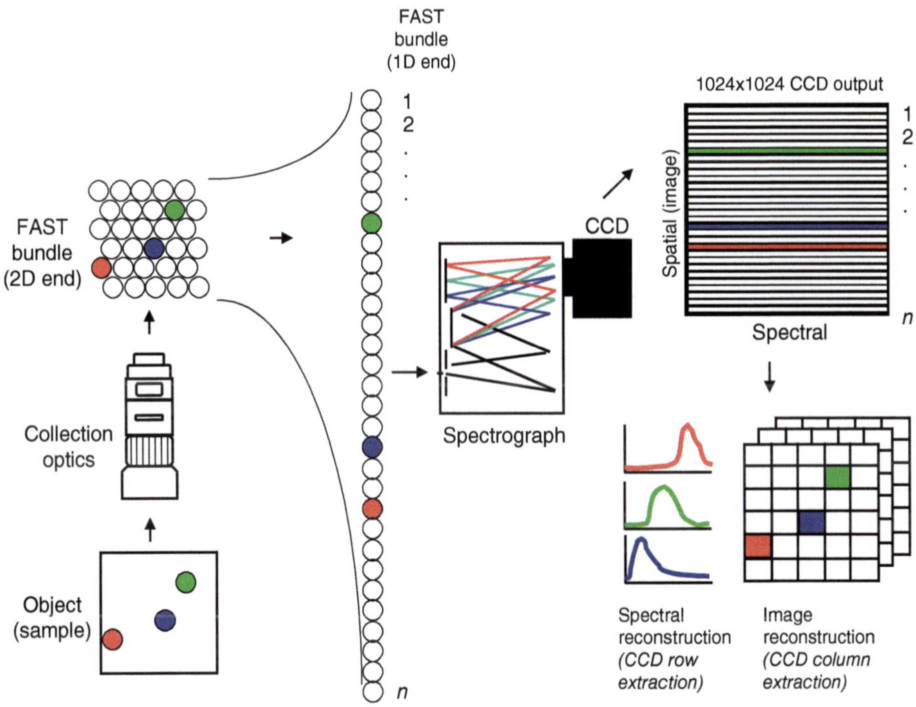

Figure 10.3 Fiber Array Spectral Translator (FAST) concept illustration. Light from target is captured and projected onto the 2D end of fiber array, which is reordered distally into a linear array at the entrance slit of a dispersive spectrograph equipped with a 2D FPA detector. A single snapshot image captures two spatial and one spectral dimension of data simultaneously.

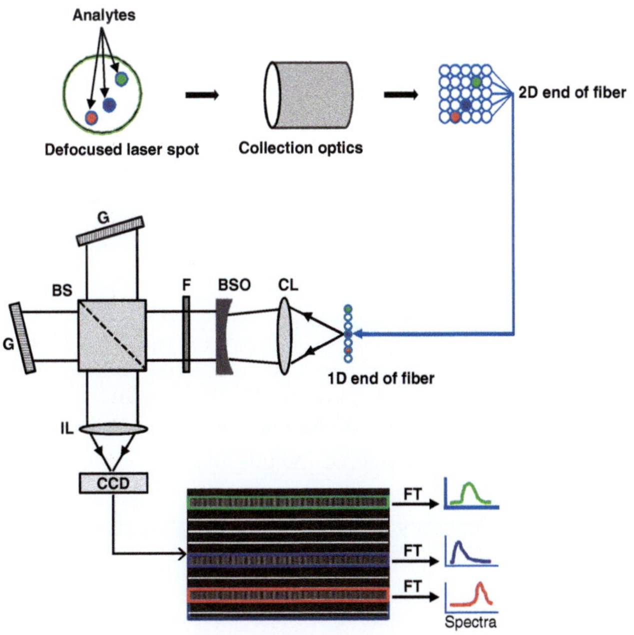

Figure 10.4 Fiber Array Spectral Translator-Spatial Heterodyne Spectrometer (FAST-SHS) concept illustration. Light captured and propagated through the FAST bundle is imaged by the SHS producing an interferogram on detector rows that can be mapped back to specific spatial locations on the target. A Fourier transformation and image reconstruction produces spatially resolved spectra and spectrally resolved images.
CL, Collimating lens; BSO, beam shaping optics; F, optical filter; BS, beamsplitter; G, grating; IL, imaging lens; CCD, charge couple device; FT, Fourier transform.

like an interferometer – capable of very high resolving powers. SHS is relatively inexpensive and exhibits a large étendue. FAST-SHS requires tight control of alignment and is susceptible to vibrations and manufacturing difficulties. However, recent advances in monolithic SHS designs have largely addressed these limitations (Waldron et al. 2020).

10.3.2.4 Source Tuning-Based

Up to this point, wide-field HSI techniques have been operating on the detection side. Alternatively, source tuning provides a means to achieving a similar result. Two popular source-tuning architectures are tunable QCLs and LEDs.

Tunable QCLs consist of semiconductor lasers emitting in the mid- to far-IR or terahertz region (Faist et al. 1994; Schlossberg and Kelley 1981). Laser emission is achieved through the use of inter-sub-band transitions in a repeated stack of semiconductor multiple quantum well heterostructures. QCLs operate in a fundamentally different mode than conventional laser diodes. They are comprised of dozens of alternating layers of semiconductor materials, which form quantum energy wells that confine the electrons to particular energy states. As electrons traverse the lasing medium, they transition from one quantum well to the next driven by the applied voltage. Photons are emitted as the electrons transition one valence band energy to a lower one. Output wavelengths are determined by the structure of the layers rather than the lasing material. Tuning may be achieved by varying temperature or through rotation of an external grating (external cavity design). Benefits of tunable QCLs include wide tuning range, fast response time, and compatible with operation in the spectral range associated with fundamental IR bands, where specificity and sensitivity are very good. Drawbacks include standoff range limits due to output power bounds as well as spectral range limitations in some cases.

LEDs are semiconductor light sources that emit light when current flows through them (Schubert 2003). Electrons in the semiconductor medium recombine with electron holes, releasing photons. The color (i.e. wavelength) emitted is determined by energy required to cross the bandgap of the semiconductor. Tunability is achieved by using a collection of LEDs with varying spectral outputs. LED advantages include low power consumption, long lifetime, small and fast switching speeds. Unfortunately, output power restricts standoff distance to just a few meters. LED availability outside the visible range is currently somewhat limited. Furthermore, output spectral profiles are relatively broad, and applications have the practical limitation of reliance on approaches to remove ambient light when used in outdoor, lighted conditions. This approach has been used in cultural heritage studies, where high-intensity, broadband, illumination of the sample is undesirable (Christens-Barry et al. 2009).

Table 10.2 summarizes the portable standoff optical spectroscopy technologies covered in this section. In the following section, we provide some direction on how to determine which technology is best suited to meet your application needs.

10.4 Portable Standoff Optical Spectroscopy Sensor Selection

With so many sensor and underlying technology options, selecting the most appropriate portable standoff optical spectroscopy sensor can be a daunting task. The answer to which option most often resides in the requirements governing the intended use scenarios. Primary factors that must be considered include the appropriateness of the underlying phenomenology and the spatial, spectral, and temporal measurement needs. Other important factors include environmental needs, size, weight, power and cost concerns and whether or not there is a need for reconfigurability.

10.4.1 Phenomenology Considerations

Having a sound understanding of the underlying spectral properties of the targets and backgrounds relevant to your application will go a long way in helping to select the most appropriate sensor. Knowing which spectroscopic

Table 10.2 Portable, standoff, optical instrument technologies summary.

Portable instrument category	Portable instrument type	Wavelength selection device	Principle of operation	Capabilities	Challenges
Point and spatial scanning	Dispersive-based	Diffraction grating	Spectra generated by diffraction grating spatially separating incoming light into component frequencies onto a 1D or 2D detector	• High spectral resolution • Full spectral coverage	• Inefficient image acquisition • Low area search rate • Poor spatial resolution
	Interferometer-based	Fourier transform (FT) interferometer	Spectra obtained by measuring the coherence of a light source, using time-domain or space-domain measurements	• High spectral resolution • Full spectral coverage	• Inefficient image acquisition • Low area search rate • Poor spatial resolution
		Spatial heterodyne spectrometer	Similar to FT interferometer, but with no moving parts. For each wavenumber in the wavefront entering the interferometer, two wavefronts exit with a wavenumber-dependent crossing angle that produces an interference pattern that is mathematically deconvolved to a spectrum	• High spectral resolution • Full spectral coverage	• Inefficient image acquisition • Low area search rate • Poor spatial resolution
Wide-field imaging	Fixed bandpass filter-based	Dielectric interference filter wheels	Hyperspectral images are generated by capturing images sequentially for a series of discrete wavelengths defined by the dielectric interference filters used	• Simple design • Low cost • High spatial resolution • High optical throughput	• Image shift artifacts • Limited number of wavelengths • Poor spectral resolution • Slow speed
		Rotating dielectric filters	Hyperspectral images are acquired sequentially as a dielectric filter is mechanically rotated	• Simple design • Low cost • High spatial resolution • High optical throughput	• Image shift artifacts • Limited number of wavelengths • Poor spectral resolution • Slow speed
		Snapshot imagers	Filters are fabricated onto a grid of detector pixels using common lithographic techniques. A single image captures all wavelengths simultaneously	• Real-time compatible • Moderate cost in high volumes • High optical throughput • No moving parts	• Image shift artifacts • Limited number of wavelengths • Moderate spatial resolution • Lack reconfigurability

(*Continued*)

Table 10.2 (Continued)

Portable instrument category	Portable instrument type	Wavelength selection device	Principle of operation	Capabilities	Challenges
		Multivariate optical elements	Fixed filter with regression vector associated with a particular chemical or physical measurement optically encoded into the transmission function	• Real-time compatible • Moderate cost in high volumes • High optical throughput • No moving parts • Small size, weight	• High upfront development cost and time • Lacks reconfigurability • Susceptibility to changing conditions
	Tunable filter-based	Single-passband liquid crystal tunable filter (LCTF)	Construct consists of fixed birefringent retarders, polarizers, and liquid crystal waveplates. Hyperspectral images obtained by applying prescribed voltages to LC waveplates in a sequential fashion	• Near real-time compatible • Reconfigurable • High spatial resolution • No moving parts	• Inefficient full spectral acquisition • Low-to-moderate optical throughput • Limited storage/operation temperature ranges
		Multipassband LCTF	Multipassband hyperspectral imaging architecture using a subset of elements of sequential LCTF. Two filters can be used in tandem to approximate a full hyperspectral image in only two measurements	• Real-time compatible • Reconfigurable • High spatial resolution • No moving parts	• Susceptibility to motion-induced edge artifacts • Limited storage/operation temperature ranges
		Acousto-optical tunable filter (AOTF)	Hyperspectral images achieved based on the interaction of light with a traveling acoustic sound wave in an anisotropic crystal medium. Tuning is achieved by changing the applied RF signal frequency and power	• High optical throughput • Broad spectral coverage • Rapid tuning speeds	• Inefficient full spectral acquisition • Broad spectral bandpasses • Moderate spatial resolution • Nonuniform spectral bandpass across FOV
		Fabry–Perot interferometer	Hyperspectral images obtained by sequentially collecting images as the distance or refractive index between two parallel reflecting surfaces are changed resulting in wavelength-dependent optical interference	• High image quality • High finesse	• Moderate spectral resolution • Low out-of-band rejection • Limited free spectral range • Small acceptance angle

Fiber-based	Fiber array spectral translator (FAST)	Captured light is projected onto the 2D array of optical fibers rearranged distally into a 1D array and inserted in parallel into the entrance slit of a dispersive spectrometer or SHS equipped with an imaging format FPA detector. Single image contains two spatial and one spectral dimensions of data from which software reconstructs hyperspectral imagery	• Near real-time compatible • Full spectral coverage	• Low image fidelity
Source-tuning based	Quantum cascade lasers	Source modulation achieved with tunable laser source comprised of dozens of alternating layers of semiconductor materials that are tuned by varying temperature or through the grating rotation	• Wide tuning range • Fast response time • Compatible with fundamental IR	• Limited standoff range • Limited number of wavelengths
	Light-emitting diodes (LEDs)	Source modulation achieved by sequentially applying current to an array of semiconductor light sources that emit light of unique wavelengths	• Low power consumption • Fast switching speeds	• Limited standoff range • Limited commercially available LEDs outside the visible range • Limited number of wavelengths

technique is compatible with, and will provide the greatest potential for detecting analytes of interest in relevant amounts, and distributions among applicable backgrounds, is extremely beneficial. Not all sensing technologies are compatible with every spectroscopic phenomenology. For instance, LCTFs are not compatible with MWIR measurements due to the highly absorptive nature of the polarizers beyond the SWIR spectral range. With any phenomenology, there will inevitably be tradeoffs between sensitivity and selectivity. Be sure to select a technique that will provide a suitable level of detection performance. If identification is necessary, a technique that has a high degree of specificity (i.e. Raman or MWIR) should be pursued.

10.4.2 Spatial, Spectral, and Temporal Considerations

It is important to know whether or not the application requires imaging. When sample heterogeneity is not a concern and the operator knows where the target of interest is located, a point-sensing strategy may suffice. Real-world targets and surfaces more often than not, however, are inherently heterogenous in their composition and distribution. Further, many applications demand looking for the proverbial "needle in a haystack." In other words, the sensor must locate a target of interest that is not readily visible to the naked eye. In these situations, scanning or imaging-based approaches may be most appropriate.

When imaging is required, knowing the spatial resolution, image fidelity, and area coverage requirements is important. Wide-field imaging approaches that use high-fidelity staring FPA detectors tend to provide the best spatial metrics performance. When the technique requires illumination with a laser source, one must choose point, scanning, or wide-field illumination approaches. This inevitably brings into play considerations, such as measurement time, eye safety, measurement uniformity across the field-of-view, and factors impacting data signal-to-noise ratios (SNR).

Laser sources contain etalons that give rise to temperature-dependent fringes in the output spectra, degrading data SNR. Speckle is another property exhibited by laser sources with which instrument platforms must contend. Approaches such as the use of rotating roughened silica for direct beam instruments, or integration of fiber agitators for fiber-delivery-based laser systems, have been used to mitigate speckle (Crocombe 2018). Eye safety and skin safety are other concerns that must be considered in the decision. Choice of laser wavelength, laser energy, or power density, and duration of exposure are all factors that go into computing key safety metrics, such as the maximum permissible exposure (MPE) values (Anon 2007; Schröder 2000).

If the measurement speed and the number of spectra are not large drivers in the decision, then a point or spatial scanning approach may be most appropriate since illumination may be confined to a small, localized area, limiting the opportunity for secondary scattering. However, if measurement speed and area search rate are critical drivers, then wide-field methods may be preferred.

Having knowledge of the spectral range and the number of spectral channels required to make the desired measurement is also important in sensor selection. The phenomenology often dictates the spectral range. To answer how many spectral channels are needed often requires subsetting high-resolution data, and modeling the outcome should a smaller number of wavelengths be used in the measurements. If a small number of spectral wavebands (i.e. <10) is sufficient, then fixed bandpass, tunable filter and source tuning-based instruments may be viable options. If the nature of the threat is likely to change spectrally, and sensor reconfigurability becomes necessary, then tunable filter-based or source-tuning-based approaches are good options to consider.

10.4.3 Size, Weight, Power, and Cost (SWaP-C) Considerations

Sensor size, weight, power and cost (SWaP-C) concerns are also important factors to evaluate when selecting a portable sensor. Often, performance is traded to some degree in order to meet SWaP-C requirements of the end user. For example, it is not practical for a soldier to carry a high-resolution, 25-pound spectrometer on his/her person for extended periods of time. Fortunately, the trend in all portable instrument types discussed in this chapter is toward lower SWaP-C options. Many of these technologies have or are undergoing miniaturization efforts in order to broaden market prospects and make portability and routine use more practical.

10.4.4 Environmental Considerations

A final key category of considerations is environmental. Portability implies field use which often requires survival through rough use and exposure to the elements. Not all sensors are equal in this respect, and becoming familiar with the environmental specifications (i.e. temperature, humidity, shock and vibration, drop, etc.) is advisable in the selection of a standoff sensor. All optical-based sensors are susceptible to vibration-induced misalignments and damage, and many are also susceptible to temperature-induced spectral drift. Be sure to inquire about built-in temperature compensation routines. Also, know the temperature limits for both storage and operation of the device and ensure that they meet the needs of the intended use.

10.5 Portable Standoff Optical Spectroscopy Sensors and Applications

10.5.1 Biothreat Detection and Identification

Rapid, reagentless detection and identification of biological threat organisms within complex backgrounds have long been an unmet need. Microbiology, immunoassays, genetic and molecular-based approaches for identification of bio-organisms require long analysis times for culturing, deoxyribonucleic acid (DNA) extraction and use of reagents. Reagentless optical methods, such as fluorescence or LIBS lack robust identification capability. Mass spectrometry is susceptible to false alarms due to background clutter and lack of signature robustness. Raman spectroscopy addresses a number of these deficiencies, but Raman alone is also susceptible to matrix effects. Bio-organism growth media and natural environmental contamination introduce signal-masking fluorescence, and nonagent Raman bands, into the measurement, resulting in low signal-to-background data. Raman imaging, however, provides a means to largely address the signal-to-background challenge by providing a means to efficiently generate "pure pixel" Raman spectra in the presence of clutter at a single-organism level.

Post-9/11, ChemImage Corporation led the development of an all-optical spectroscopic detection methodology and database of Raman signatures of biological threats, near-neighbors and confusants (Kalasinsky et al. 2007). The work was performed in collaboration with the former Armed Forces Institute of Pathology (AFIP) using a Falcon Raman Chemical Imaging Microscope platform equipped with red, green, blue (RGB) video, fluorescence chemical imaging, Raman chemical imaging and dispersive Raman spectroscopy. An extension of the initial signature work in collaboration with AFIP was the development of a portable FAST Raman and broadband fluorescence microscope platform called Eagle (Figure 10.5a).

10.5.1.1 Hardware

Eagle uses RGB video and broadband fluorescence for targeting suspected biological material. Fluorescence imaging provides a reasonably good discriminator between biological and nonbiological material. A mercury arc lamp provides illumination for both visible brightfield reflectance and UV-excited fluorescence-imaging modes. In UV-fluorescence mode, sample fluorescence is induced using a 365 nm filtered mercury arc lamp in combination with a dichroic beam splitter, in an epi-illumination configuration, to direct UV light to the sample. Sample fluorescence is recovered in a back-scattered configuration through the dichroic beam splitter and an emission filter. Brightfield reflectance images are obtained with the use of a 50/50 beam splitter in place of the dichroic. In FAST Raman mode, the mercury lamp light is blocked using mirrors, allowing the beam from a 532 nm laser (frequency-doubled, diode-pumped neodymium-doped yttrium aluminum garnet (Nd:YAG) laser) to engage the sample. The resulting Raman radiation is split between LCTF and dispersive Raman channels. Eagle provides 13 mW of power (3000 W/cm^2) over the field-of-view of the imaging optics. The FAST Raman

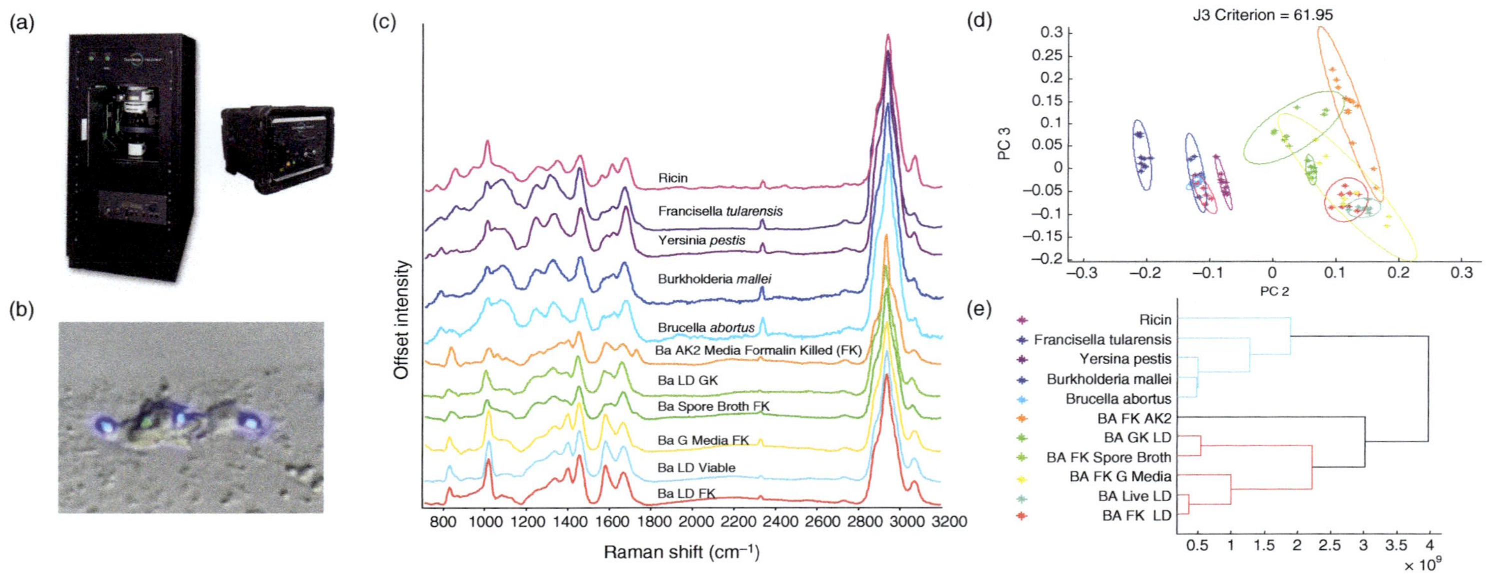

Figure 10.5 Raman-based biothreat detection and identification. (a) Falcon Raman Chemical Imaging Microscope (left) and Eagle Transportable Raman Microscope system (right). (b) Raman (green) and fluorescence (blue) overlays onto differential image contrast image revealing bacterial spores. (c) Representative Raman spectra of Centers for Disease Control and Prevention (CDC) category A/B pathogens. (d) Principle component analysis (PCA) score plots revealing underlying pathogen spectral groupings. (e) Dendrogram showing taxonomic relationships of pathogens. *Source*: Pictures and data courtesy of ChemImage Corporation.

dispersive channel is equipped with a dispersive spectrometer, equipped with a thermoelectrically cooled CCD FPA, providing a full spectrum signature, with 12 cm^{-1} spectral resolution.

10.5.1.2 Software and Algorithms

ChemImage XPert™ software provides acquisition control as well as data processing for the Falcon and Eagle instrument platforms. FAST Raman spectral data is typically corrected for instrument response using a National Institute of Standards and Technology (NIST) standard reference material and sample fluorescence. The spectra are then truncated to a usable Raman spectral range, followed by the application of a polynomial baseline correction. Common data analysis tools applied to the preprocessed data include Principal Component Analysis (PCA) for cluster analyses, Mahalanobis distance to evaluate the identity of agents, dendrograms to visualize hierarchical cluster analysis of ranked similarity of spectra with associated agent class, and confusion matrices to quantify the degree of specificity inherent in the analysis. LCTF Raman image processing includes cosmic ray removal, instrument response correction, Gaussian blurring (for high-frequency noise removal), application of a Euclidean Distance classifier trained against known organism spectral responses, and fusion of video, fluorescence and/or Raman image data.

10.5.1.3 Example Results

Figure 10.5c shows dispersive Raman spectra of pure Centers for Disease Control and Prevention (CDC) Category A/B pathogens along with an associated PCA scatter plot (Figure 10.5d), and dendrogram (Figure 10.5e). The Raman/fluorescence/brightfield differential interference contrast (DIC)-fused image (Figure 10.5b) shows the detection of bacillus globigii (Bg) spores in a complex background. A Bg suspension was intentionally added to outdoor ambient particulate matter collected into the aqueous collection fluid buffer by a Joint Biological Point Detection System (JBPDS) sensor. The sample was subsequently deposited in droplet form onto a microscope slide, and imaged using the Falcon equipped with a 100× magnification objective. Bg spores are evident with both fluorescence and Raman imaging mode. However, due to the excitation wavelength, the Raman image reveals spores that are embedded deeper into the background clutter.

Figure 10.6 shows a FAST Raman result from a virus-simulant MS2 bacteriophage sample mixed with cellulose. The color-coded underlying Raman spectra reveal the spatial locations of the MS2 bacteriophages, cellulose and an unknown contaminant.

10.5.2 Condensed Phase CWA Contamination Detection/Avoidance and Surveying/Mapping

In the event of a CWA attack, there is a need to be able to detect contamination on surfaces (ground, equipment, etc.) to avoid exposure to military personnel and civilians. A sensor addressing this need must operate in near to real time, and be capable of detecting CWAs, toxic industrial chemicals (TICs), nontraditional agents (NTAs), and

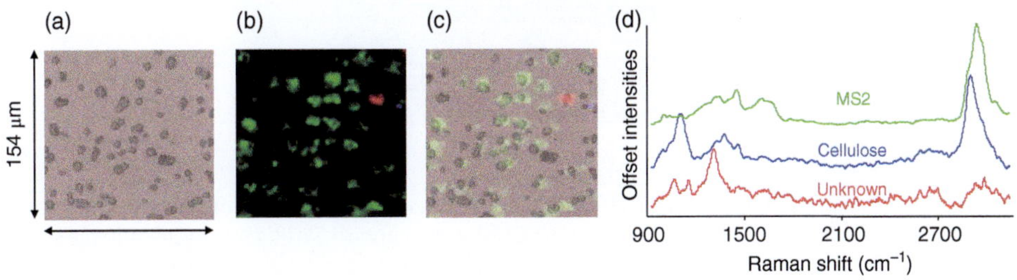

Figure 10.6 FAST Raman chemical imaging example for biothreat detection and identification. Brightfield reflectance image (a), color-coded Raman chemical image (b), Raman/brightfield overlay (c), and underlying Raman spectra (d) of MS2 and cellulose mixture. *Source*: Pictures and data courtesy of ChemImage Corporation.

other chemical agents in solid and liquid forms while on the move. Such a sensor would provide advanced notice to warn personnel of potential threats, as well as ensure that military equipment is free of contamination since the availability of water on the battlefield is often limited for decontamination purposes.

10.5.2.1 Sensor Type I: SWIR HSI LCTF

In 2014, ChemImage developed a SWIR HSI sensor called the VeroVision® Chemical Detector (VVCD). SWIR HSI requires no laser and can operate in an eye-safe manner, in a true standoff mode, using sunlight and/or onboard halogen lighting.

10.5.2.1.1 Hardware

VVCD incorporates an LCTF coupled to an InGaAs FPA detector. The filter is tuned through computer control and the operator can either collect a series of images at specific wavelengths or acquire sequential images over a preprogrammed wavelength range. These images form a hypercube of data, which provides a SWIR reflectance spectrum at every image pixel.

10.5.2.1.2 Software and Algorithms

The detection algorithms and software operate on the captured wavelength images to produce a detection image showing pseudocolored areas of contamination overlaid on a monochrome image of the detection scene.

10.5.2.1.3 Example Results

Figure 10.7 shows a photograph of the VVCD prototype and detections of droplets applied using a spray bottle of CWA-simulant dimethyl methylphosphonate (DMMP) sprayed onto a scene featuring paint cans and containers at a 1 m standoff distance.

10.5.2.2 Sensor Type II: SWIR HSI Snapshot

To mitigate storage temperature and motion sensitivity risks of the LCTF-based instrument, a SWIR snapshot imager called VVCD-S was also developed and evaluated. In order to implement this approach, ChemImage worked with Pixelteq, a manufacturer of micropatterned bandpass filter arrays.

10.5.2.2.1 Hardware

The snapshot imager incorporates a 3 × 2 filter-pixel pattern tessellated over a 640 × 512, 25 μm pixel pitch InGaAs FPA, resulting in an effective FPA format of 213 × 256 pixels.

(a) (b) (c)

Figure 10.7 Shortwave-infrared hyperspectral imaging (SWIR-HSI) detection of chemical warfare agent (CWA) simulant at 1 m standoff using a liquid crystal tunable filter (LCTF)-based sensor. (a) LCTF-based standoff SWIR sensor (ChemImage's VeroVision® Chemical Detector [VVCD]). (b) Digital photograph of a real-world scene with a complex 3D structure that has been sprayed with CWA simulant. (c) SWIR detection image of CWA simulant (red) overlaid with SWIR pass-through image. *Source*: Pictures and data courtesy of ChemImage Corporation. Used with the permission of the Joint Program Executive Office for Chemical, Biological, Radiological and Nuclear Defense (JPEO-CBRND) – Approved for Public Release, Distribution Unlimited.

10.5.2.2.2 Software and Algorithms

Spectral Kitchen software was used to acquire image frames. Factory corrections were applied to correct for bad pixels and flat fielding. Wavelength image frames were reconstructed from the raw snapshot frame by extracting the intensity information from the pixels corresponding to each respective wavelength. Math operations were applied to the imagery to generate score imagery followed by spectral, spatial and temporal filters similar to the LCTF variant. Further, VVCD used ANNs (artificial neural networks) as the basis of the detection algorithm.

10.5.2.2.3 Example

Figure 10.8 shows representative SWIR snapshot detection images at a 1 m standoff distance of scenes containing CWA-simulant diethyl phthalate (DEP) deposited onto various surface types. Figure 10.8a shows a digital photograph of the VVCD-S sensor. Figure 10.8b shows detections in the green of 1 μl depositions of DEP on linoleum (left), Formica® (middle), and drywall (right). Figure 10.8c shows a volume study of DEP-simulant on drywall (left), Formica® (middle) and soapstone (right) surface at concentrations between 0.2 and 1 μl drops (i.e. left to right on surface – 1.0, 0.8, 0.6, 0.4, 0.2 μl). While the snapshot imager readily detects the DEP droplets on drywall and Formica®, no detections were made on the soapstone due to the highly absorptive properties of the substrate.

10.5.2.3 Sensor Type III: LWIR QCL

LWIR standoff detection systems have also emerged in response to a need for detection and identification of CWA hazards. Both passive and active illumination architectures exist. Pendar Technologies developed a detector that addresses the challenges of laser packaging and optical design, integrated laser electronics, statistically robust decision algorithms and deep chemical knowledge for QCL-based LWIR sensing of CWAs.

10.5.2.3.1 Hardware

The sensor utilizes a quantum cascade laser array (QCLA) source that provides a near-continuous, fully monolithic tunable source, ideal for field and handheld applications requiring speed, ruggedness and stability. The current design incorporates four 32-element QCLAs, with each array covering a distinct LWIR band within the 6.5–10.5 μm region (1540–950 cm^{-1}). This spectral range is particularly useful for sensitive and specific detection of CWAs. The emitted light in these four bands is then spectrally beam-combined, using dielectric mirrors and polarization techniques. The entire sensor is in an 8 × 8 × 81.5 cm package.

10.5.2.3.2 Software and Algorithms

The software provides control of these lasers in pulsed mode, projecting a 2 mm focused spot on targets. Speckle is mitigated by dithering the combined beams pointing angle slightly, resulting in a 2 × 2 cm illumination area

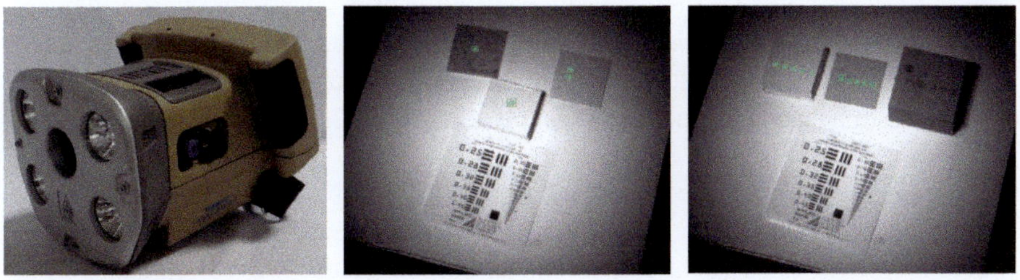
(a) (b) (c)

Figure 10.8 SWIR HSI detection of CWA simulant at 1 m standoff using a snapshot imager-based sensor. (a) Snapshot imager-based standoff SWIR sensor (ChemImage's VeroVision® Chemical Detector-Snapshot [VVCD-S]). (b) SWIR detection image of 1 μl droplets of diethyl phthalate (DEP) deposited on various substrate types (detection in green). (c) Limit-of-detection study showing varying volumes of DEP (i.e. left to right on surface – 1.0, 0.8, 0.6, 0.4, 0.2 μl) on various surface types. *Source*: Pictures and data courtesy of ChemImage Corporation. Used with the permission of JPEO-CBRND – Approved for Public Release, Distribution Unlimited.

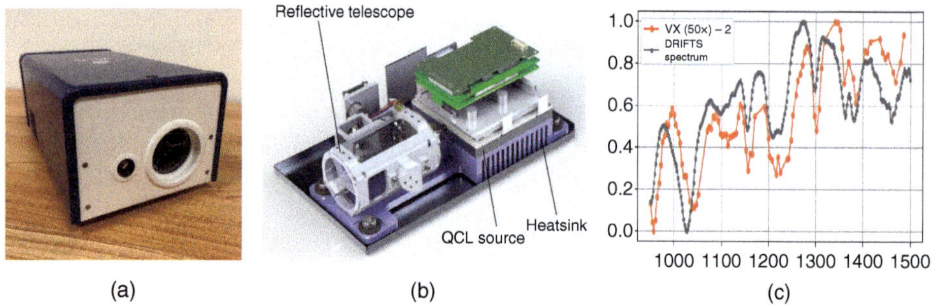

Figure 10.9 Quantum Cascade Laser Array (QCLA)-based longwave infrared (LWIR) detection of CWA. (a) Integrated Pendar Technologies QCLA sensor (2.25 l volume, 4.5 lbs). (b) Drawing of the instrument overall architecture, showing the QCLA source, Reflective telescope used to emit and receive the LWIR light, and system electronics. (c) Example of a spectrum measured at the Defence Science and Technology Laboratory (Dstl, UK) of VX deposited on sandblasted stainless steel at 500 μg/cm^2 loading, acquired at 1.3 m standoff distance. The spectrum is compared to a reference acquired using a Bruker LUMOS FTIR microscope. *Source*: Pictures and data courtesy of Pendar Techologies. Used with the permission of the Defence Science and Technology Laboratory (Dstl, UK). Licensed under OGL v3.0.

at ~1.3 m. Taking into account scan time, duty cycle, scan repetition and processing times, a total acquisition time of 3.3 seconds was achieved.

10.5.2.3.3 Example

Figure 10.9a shows a photograph of the integrated QCLA system. Figure 10.9b shows the system architecture. Figure 10.9c shows a measurement taken at the Defence Science and Technology Laboratory (Dstl, UK) of VX deposited on sandblasted stainless steel at 500 μg/cm^2 loading, acquired at 1.3 m standoff distance (Blanchard and Vakhshoori 2020). This measurement was obtained by scanning the IR beam across a deposited VX droplet (25 × 25 measurements) to acquire a hyperspectral image (~5 minutes acquisition time), and then segmenting the image between background vs. contaminated pixels. The spectrum shown in Figure 10.9c is obtained by dividing the mean signal across the assumed contaminated pixels by the mean signal across the assumed clean pixels.

10.5.3 Gas-Phase Remote Chemical Cloud, CWA, TICS, and Detection

Gas-phase chemical releases threaten life, whether a release from an industrial accident or intentional terroristic release. Detection and monitoring of organic and inorganic gases and vapors (i.e. chemical cloud) from a remote distance are critical in this situation in order to provide advanced notice to warn personnel. Standoff IR sensors – either active or passive – are well-suited for this application given the high sensitivity and specificity of the available measurement tools. LWIR is relevant because the most condensed-phase and gas-phase materials exhibit unique and intense spectral features in the LWIR due to the fundamental vibrational bands in this part of the spectrum. LWIR provides near real-time detection and identification of CWAs, TICs, NTAs and other chemical agents in gas form.

There are numerous vendors and sensors available with capabilities that range in degree of specificity (specific chemical vs. general chemical family), sensitivity (from ppb-meter to hundreds of ppm-meter), spectral region (1.5–14.5 μm), measurement type (filter-based vs. interferometer-based), weight (~2–65 kg), data output (single-pixel spectrum vs. single/multi/hyperspectral image), range (2 m–10 km) and price ($100–$600 K) (National Security Technologies, LLC 2016). We report here on one handheld unit from Block Engineering called PORTHOS™.

10.5.3.1 Hardware

Porthos is an FTIR-based LWIR sensor that operates in the 7.5–13 μm spectral band. Single-pixel spectra are produced with a selectable resolution between 4 and 8 cm^{-1}. Images are created by point scanning and overlaying on a visual image for analysis. The 7.7 kg sensor detects and alarms in a little at 2 seconds over a standoff range of 0.1–5 km. The sensor has two field-of-view options (i.e. 1.5° vs. 5°) and is capable of being powered on battery for up to 4 hours of operation under passive illumination mode.

10.5.3.1.1 Software and Algorithms

The Porthos can be operated from an external host computer using remote interface software to analyze spectral data with complete identification capability. The Porthos detects certain chemicals of interest and displays the chemical type, intensity, and records the information on an internal hard-drive. The supplied detection processing program can read the data, process the data, and create results.

The data output file of the Porthos instrument is read by proprietary software which shows the detection levels on the instrument display. The detection levels are overlaid at the bottom of the image collected by the monochrome video camera. The information collected and interpreted by the Porthos is performed in real time.

10.5.3.1.2 Example

Figure 10.10 shows the operation of Porthos portable standoff chemical detection system.

10.5.4 Explosives Detection and Identification

IEDs are a real threat to military personnel and civilians, and detection of such devices is critical. Explosive Ordinance Disposal (EOD) technicians are trained to detect and remove IEDs. There is a need for mobile, standoff sensors with high area search rate (near real-time) and high specificity/sensitivity to keep technicians at safe distance from explosive threats. Detecting explosive residues associated with the emplacement or manufacturing of IEDs is one approach to countering IEDs.

10.5.4.1 Sensor Type I: Deep-UV Raman Point Sensor

To address the need for standoff detection and identification of explosive residues associated with the emplacement of IEDs, Alakai Defense Systems developed a standoff deep-UV Raman sensor called the Portable Raman Improvised Explosive Detector (PRIED).

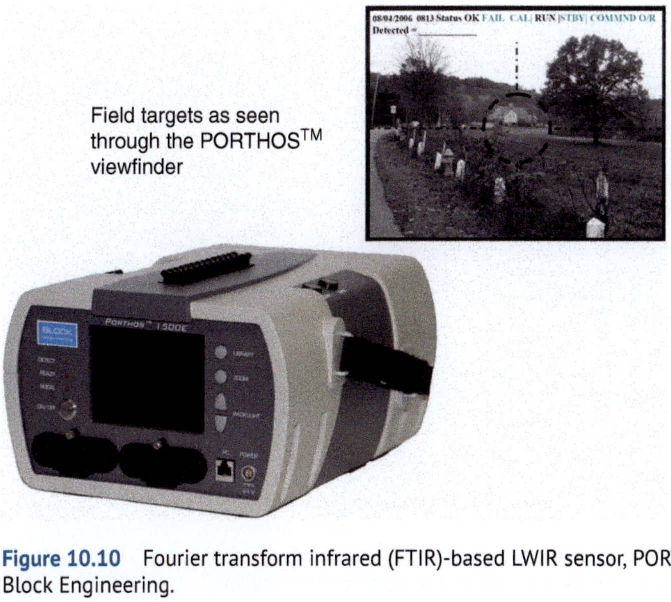

Figure 10.10 Fourier transform infrared (FTIR)-based LWIR sensor, PORTHOS in operation. *Source*: Pictures courtesy of Block Engineering.

10.5.4.1.1 Hardware

PRIED is a deep-UV (262 nm excitation) portable Raman sensor, capable of up to 50 m standoff for homemade explosives (HMEs), and 10 m for most other chemicals. PRIED delivers a 0.5 cm diameter laser spot to excite the sample. Material amounts as little as 10s of μg/cm^2 at 1 m, and higher amounts at 10 m, are detectable. It consists of a wand (5″ × 7″ × 15″) and backpack (17″ × 12″ × 7″), and, when combined, weighs 31 lbs (short mission mode), or 38 lbs (extended mission).

Eye-safe operation is achievable in two operating modes: "Eyesafe" and "Max Detect." The system powers up in eye-safe mode. In this mode, the overall system is rated as a Class II laser system. To accomplish this, the PRIED system utilizes Alakai's patented stimulated aversion technology (Pohl et al. 2012) that results in a Class II system even though the sensor contains a Class III UV laser. In many applications, users wish to either acquire results faster or to interrogate lower concentration samples. In these cases, assuming they have the proper laser training and can control the nearby area around them, they can turn up the laser power by entering Max Detect Mode. To enter this mode, the user enters a password and the laser power is increased and the PRIED sensor operates with faster integration times, at farther detection distances or at lower detection capabilities.

10.5.4.1.2 Operation

An operator pulls the wand trigger in half-way, points the visible targeting beam at the intended target, and then pulls the trigger all the way to start the Raman analysis. The operator holds the beam on target until the unit produces a red light (threat) or green light (no-threat) result.

10.5.4.1.3 Example

Figure 10.11 shows photographs of PRIED in operation (Figure 10.11a), subsystem components (Figure 10.11b) and 50 m range Raman data of a potassium chlorate target at various collection times (Figure 10.11c).

10.5.4.2 Sensor Type II: FTIR-Modulated Supercontinuum Laser

To exploit benefits of working with the fundamental infrared vibrational modes, Leidos developed a standoff FTIR-based instrument called the Supercontinuum Wide-band Infrared Fourier Transform (SWIFT) sensor.

10.5.4.2.1 Hardware

SWIFT uses supercontinuum laser pulses modulated by an FTIR spectrometer and captured with high-speed detectors. A standoff distance of up to 15 m has been demonstrated for trace material identification on surfaces and air. SWIFT produces an active IR beam in the 2–11 μm spectral region and captures data from a ~2° angular field-of-view on the target. Material amounts as little as 10s of μg/cm^2 for solids and liquids and 1 ppm for vapors have been demonstrated. SWIFT consists of a (24″ × 18″ × 18″) sensor head atop a tripod with a combined weight

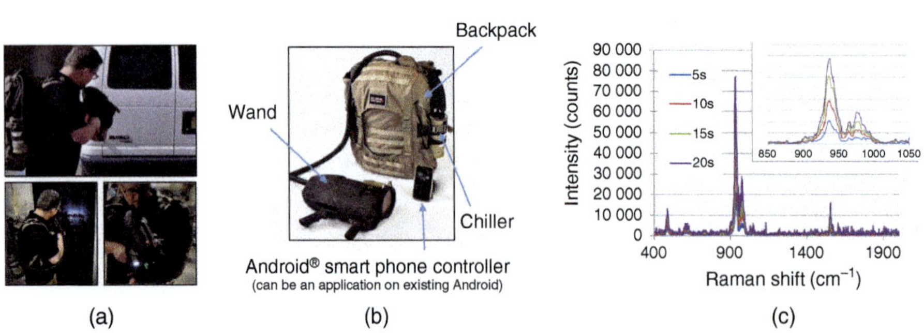

Figure 10.11 Raman-based standoff detection of explosive threats. (a) Alakai Defense System's Portable Raman Improvised Explosive Detector (PRIED) sensor in operational use case scenarios. (b) PRIED subsystem components – backpack with a chiller (top), wand (bottom, left), and Android® smartphone controller (right). (c) Raman spectra of potassium chlorate at variable detector integration times. *Source*: Pictures and data courtesy of Alakai Defense Systems, Inc.

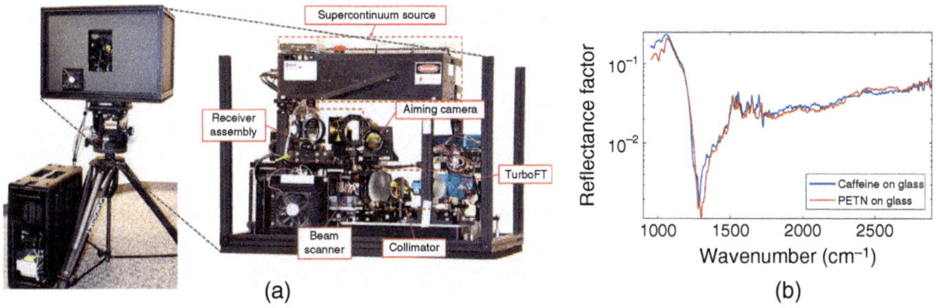

Figure 10.12 FTIR supercontinuum laser-based standoff detection of explosive threats. Leidos' SWIFT sensor (a) performs direct reflectance measurements (b) on surfaces up to 15 m away for automated detection and identification of trace material contamination. The 4 cm^{-1} spectral resolution enables discrimination of harmful materials from harmless ones. *Source*: Pictures and data courtesy of Leidos.

of 165 lbs. SWIFT currently is at a technology readiness level (TRL) of 5 for indoor applications and TRL 4 for outdoor applications. SWIFT is considered an eye-safe sensor having a Class 1 laser designation.

10.5.4.2.2 Operation

SWIFT uses a built-in aiming camera to establish the measurement location. A pushbutton operation follows that generates FTIR spectra of the reflected light from the target that may be searched against a spectral library.

10.5.4.2.3 Example

Figure 10.12a shows a photograph of the SWIFT sensor along with subsubsystem components (inset) and reflected light FTIR signatures of caffeine and pentaerythritol tetranitrate (PETN) on glass, respectively (Figure 10.12b).

10.5.4.3 Sensor Type III: SWIR HSI LCTF

SWIR HSI sensors provide wide-area imaging at near real-time speeds. They overcome two obstacles present in Raman detection sensors – slow area search rate (due to small laser spot sizes), and eye safety concerns. SWIR HSI sensors may be integrated into mobile, robotic and handheld variants for detection of chemical, biological, explosive (CBEs) and narcotics. To address the shortcomings of current generation systems, ChemImage developed the VeroVision™ Threat Detector (VVTD).

10.5.4.3.1 Hardware

VVTD is a portable, tripod-mounted sensor that utilizes LCTF-based SWIR technology. In operation, solar or supplemental broadband radiation illuminates the surface of interest. Photons are absorbed or reflected by materials depending on their composition. Threat detection and classification are facilitated through the multispectral acquisition of the target(s) at key wavelengths. VVTD is equipped with eight 35 W halogen light bulbs to provide close-range (<2.5 m) standoff detection capabilities. Greater distances are achieved with the use of at-target lighting and/or reliance on solar illumination. Light-gathering optics with continuous 2× optical zoom and focus capture a portion of the reflected light, which is, in turn, directed to an LCTF coupled to an InGaAs FPA detector. RGB video is also included to provide for enhanced situational awareness. A power control module provides power distribution to the sensor as well as provides a housing for a lithium-ion battery capable of 2–4 hours of life depending on duty cycle.

10.5.4.3.2 Software and Algorithms

VVTD's software is used to rapidly capture an abbreviated hyperspectral data cube, containing spatially resolved SWIR spectral signatures that may be compared to a SWIR-spectral library using pattern-matching algorithms to locate and identify threats in the scene. Once a threat is detected, higher resolution hyperspectral data may be acquired and autonomously processed to presumptively identify the threat component.

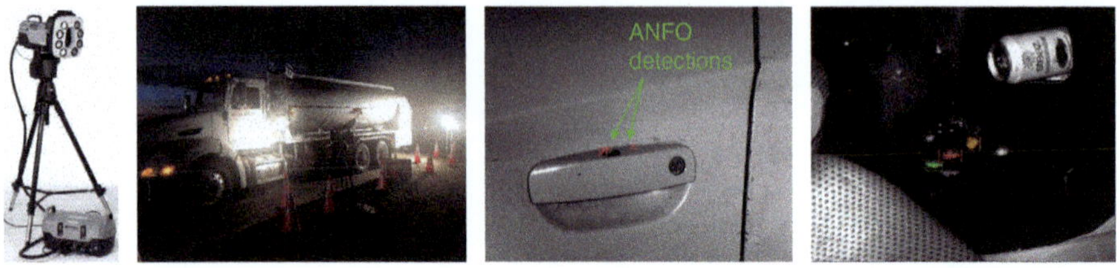

(a) (b) (c) (d)

Figure 10.13 SWIR-HSI-based standoff detection of explosive threats. (a) ChemImage VeroVision® Threat Detector (VVTD) in use screening vehicles for explosive residues at the 2019 US Open Championship (Pebble Beach, CA). (b) Ammonium nitrate fuel oil (ANFO) prill detection on a car door handle. (c) Narcotics detection in a vehicle console. *Source*: Pictures and data courtesy of ChemImage Corporation.

10.5.4.3.3 Example

Figure 10.13 shows a photograph of the VVTD sensor (a), VVTD in operation at the 2019 US Open in Pebble Beach, CA (b) and representative threat material detection results associated with an entry control point (ECP) screening CONOPs (c and d).

10.5.4.4 Sensor Type IV: Combined SWIR HSI LCTF/Raman FAST HSI Dispersive

In order to combat the threat of emplaced explosives (land mines, etc.), while ensuring the safety of military and civilian bystanders, ChemImage developed a multisensor, robot-mounted sensor capable of identification and confirmation of potential threats. The system, known as the Shortwave-infrared Targeted Agile Raman Robot (STARR), utilizes SWIR spectroscopy for the identification of potential threats, combined with a visible short-range standoff Raman FAST HSI system for material confirmation. The entire system is mounted onto a Talon UGV (Unmanned Ground Vehicle), giving the sensor an increased area search rate and reducing the risk of injury to the operator.

10.5.4.4.1 Hardware

STARR incorporates LCTF-based SWIR HSI, offering wide-area surveillance, for autonomous near real-time detection, and FAST Raman to provide rapid multipoint Raman spectral analysis as a confirmation tool. The robot may be operated remotely wirelessly at distances up to 70 m from the operator, and STARR is capable of a 1–3 m sensor to target standoff range operation. STARR is integrated on a TALON UGV from Foster-Miller Inc. (part of Qinetiq North America) for high mobility and remote, wireless control, enabled through an 802.11n link. TALON features two-stage gripper arm, four cameras for area surveillance, and is controlled by an operator control unit through a two-way RF or fiber optic protocol.

In operation, a video RGB camera provides real-time situational awareness, guiding operator control to suspicious locations. The SWIR subsystem consists of light-gathering optics, coupled to an LCTF operating from 900 to 1700 nm and an InGaAs FPA. Images are captured at discrete wavelength characteristics of explosives within the SWIR library. STARR's Raman subsystem incorporates a 532 nm pulsed laser (5 mJ/pulse, 20 Hz rep rate), a 100-fiber (10×10) FAST bundle coupled to a dispersive spectrometer equipped with a 1024×1024 intensified charge-coupled device (ICCD). Time gating of the ICCD allows reasonable Raman signals to be captured synchronously with the high-energy laser pulses in the presence of ambient light, while maintaining a moderate average power.

10.5.4.4.2 Software and Algorithms

STARR software provides acquisition controls for the RGB video, SWIR and Raman subsystems. The SWIR subsystem has two operating modes: near real-time (a small number of wavelengths), and full spectral mode for greater spectral information. SWIR wavelengths are mathematically processed to generate a score image, with pixel intensities corresponding to the degree of match with library signatures. Additional spectral, spatial, and temporal filters are applied to the score imagery, overlaid onto the passthrough SWIR image, to generate a detection image revealing the location of potential threats in the scene.

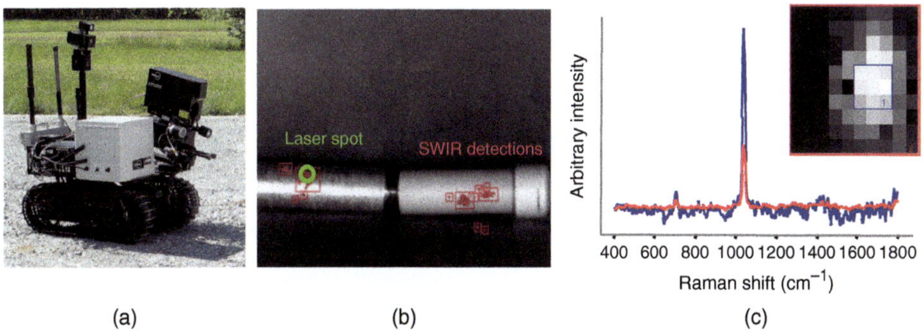

Figure 10.14 Hybrid Raman and SWIR standoff detection of explosive threats. (a) Robot-integrated ChemImage Shortwave-infrared Targeted Agile Raman Robot (STARR) sensor. (b) SWIR-HSI detection image revealing suspected explosive residues on a pipe bomb. (c) FAST Raman image and spectra provide confirmation of explosive composition for an area indicated by the "laser spot" in Figure 10.13b. *Source*: Pictures and data courtesy of ChemImage Corporation. Used with the permission of the Navy Explosive Ordnance Disposal Technology Division (EODTECHDIV) – Approved for Public Release, Distribution Unlimited.

If detections are observed in the SWIR mode, the operator may then "click" on potential threats enabling FAST Raman targeting. In Raman mode, 100 spatially resolved spectra collected simultaneously, reconstructed into a Raman FAST image and analyzed using partial least squares discriminant analysis (PLSDA) database of known Raman threat signatures. The operator is notified by visual and audible alarms for any confirmed threats.

10.5.4.4.3 Example

Figure 10.14 shows a photograph of the STARR system integrated on a TALON robot (Figure 10.14a), SWIR detection showing candidate threat locations (Figure 10.14b) and FAST-Raman confirmation of ammonium nitrate fuel oil (ANFO) residues on the exterior of simulated pipe bomb (Figure 10.14c).

10.5.5 Narcotics Detection and Identification

Today's intensifying opioid crisis presents challenges to law enforcement as well as public safety. Sensors having standoff capability are increasingly important, considering the increased presence of synthetic opioids, such as fentanyl, that are leading to an ever-increasing number of deaths. Federal, state and local law-enforcement agencies require standoff sensors for border patrol, hazmat, crime scene investigation, clandestine laboratory cleanup and venue security.

10.5.5.1 Sensor Type I: SWIR HSI LCTF

VVTD (described previously) offers additional capabilities for narcotics detection applications. Figure 10.15 shows standoff detection results form a simulated clandestine drug-manufacturing laboratory containing illicit narcotic simulants. The figure shows two images taken with VVTD of a scenario featuring illicit drug stimulants in plastic bags and glass containers, from a standoff distance. Figure 10.15a shows the scene acquired with the RGB camera, while Figure 10.15b shows the detection image generated by the SWIR HSI sensor and SWIR HSI software. The detection image highlights the potential threat detections using colored boxes, with each color unique to a specific class of compounds. The detections in the image are represented by red, green, and yellow boxes, indicating that three different classes of illicit narcotic stimulants are present. Since RGB imagery is collected simultaneously, it can provide additional information to confirm which containers have a potential threat and clarify any peculiarities in a SWIR region image by displaying the scene in the visible region.

10.5.5.2 Sensor Type II: SWIR CF Imaging – ChemImage VeroVision™ Moving Target (VVMT)

With few exceptions, all inbound international airmails are subject to inspection at International Mail Facilities (IMFs) by US Customs and Border Protection (CBP), part of the US Department of Homeland Security (DHS). International mail has been identified as a well-used distribution system for illicit opioids entering the United

(a) (b)

Figure 10.15 SWIR HIS-based standoff detection of narcotics in a simulated clandestine laboratory scene. (a) Digital photograph of the scene containing a variety of material types including narcotic simulants. (b) SWIR chemical image revealing the location of various narcotic simulant types (red, green, and yellow false colors). *Source*: Pictures and data courtesy of ChemImage Corporation.

States. The CDC has published reports showing that overdose deaths involving any opioid – prescription opioids (including methadone), synthetic opioids, and heroin rose to nearly 50 000 in 2017 (Scholl et al. 2019). However, many of these fatal overdoses are driven by the combination of traditional opioids with extremely small amounts of the extremely potent Fentanyl and Carfentanil. Consequently, large-scale drug trafficking, coinciding with this serious health epidemic in the United States population, can occur via very small packages sent in the mail. As stated by the DHS Office of Inspector General (OIG) in the Customs and Border Protection (CBP) Audit at JFK International Airport from 2018: "CBP actions to address identified deficiencies and more effectively inspect international mail will be essential to winning the war on opioids and countering their devastating impact on the nation's population."

In response to a need to address limitations of current sensors for high throughput mail screening, ChemImage is developing a sensor called VeroVision® Moving Target (VVMT) which is based on real-time, reconfigurable SWIR DP-CF technology.

10.5.5.2.1 Hardware

The VVMT prototype is a fully contained, portable (hand-carried – i.e. gripped and held with two hands) sensor. Illumination is provided by four (4) 35 W lamps for short standoff distance illumination. Light-gathering optics provide a 12° field-of-view (i.e. ~3.4 mm/pix at 10 m). An RGB video camera is incorporated for situational awareness. Other features include an integrated display with push-button, easy-to-use operational interface, a rechargeable battery and an optional tripod-mounting capability.

10.5.5.2.2 Software and Algorithms

Conformal Training Software (CFTS) was used for voltage selection optimization based on target vs. background discrimination, and Spectral Kitchen Software provides real-time autonomous detection imagery.

10.5.5.2.3 Example Results

Figure 10.16 shows a photograph of VVMTS in use (left) and representative feasibility results (right) for CWA surveying and mapping, high throughput mail screening and person-borne IED (PBIED) screening applications.

Table 10.3 provides a summary comparison of several portable standoff optical spectroscopy sensors that are in use for safety and security applications. Although far from being a comprehensive listing, the table provides the same perspective of key characteristics of a variety of instrument types.

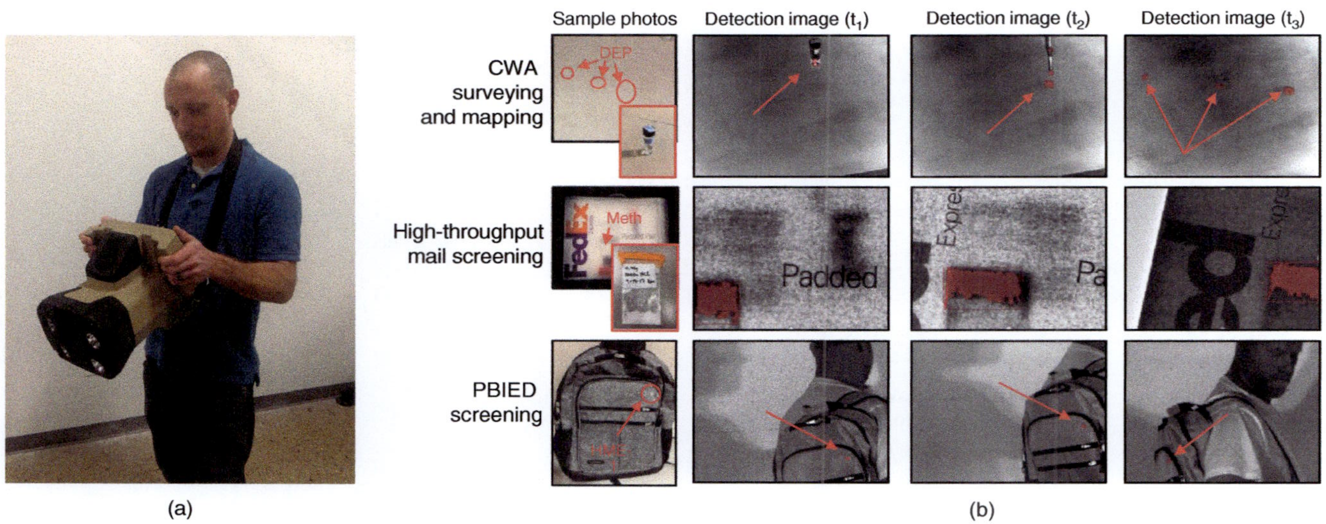

Figure 10.16 Real-time SWIR-HIS-based standoff detection of CWA, narcotic, and explosive threats. (a) The handheld, real-time, reconfigurable Conformal Imaging sensor (ChemImage VeroVision® Moving Target [VVMT]) sensor. (b) Digital photographs and representative time-sequenced SWIR detection images of a CWA threat simulant (top), narcotic threat (middle), and explosive threat (bottom). *Source*: Pictures and data courtesy of ChemImage Corporation. Used with the permission of the Defense Advanced Research Projects Agency (DARPA) – Approved for Public Release, Distribution Unlimited.

Table 10.3 Portable standoff optical spectroscopy instrument comparison.

No.	Parameter	Alakai defense systems PRIED	Block engineering LASERWARN™	Block engineering PORTHOS™	ChemImage VeroVision®	Leidos SWIFT	Photon systems STANDOFF 200
1	Name of product or technology	PRIED – portable Raman improvised explosive detection system	LASERWARN™	PORTHOS™	VeroVision® product family	Supercontinuum wide-band infrared Fourier transform (SWIFT)	STANDOFF 200
2	Model(s)	N/A	• Indoor • Outdoor models	N/A	• VeroVision® Chemical Detector (VVCD) • VeroVision® Chemical Detector Snapshot (VVCD-S) • VeroVision® Moving Target (VVMT) • VeroVision® Threat Detector (VVTD)	N/A	N/A
3	Manufacturer	Alakai defense systems	Block engineering	Block engineering	ChemImage Corporation	Leidos	Photon Systems, Inc.
4	Technology	Deep UV Raman	QCL and active IR absorption spectroscopy	Passive FTIR spectroscopy	SWIR-HSI	FTIR modulated supercontinuum laser	Deep UV Raman and fluorescence spectroscopy
5	Intended application	Standoff detection of threat chemicals	Fenceline, environmental, and leak detection gas monitoring	Military CWA and TIC gas detection	• VVCD, VVTD – near real-time standoff detection of threats • VVCD-S, VVMT – real-time standoff detection of threats	Trace material identification on surfaces and air	Real-time standoff CBE detection for first responders
6	Detected threat materials	CRNE/related materials, TICs/TIMs, narcotics/precursors	Most gases except symmetric and diatomic molecules	Military CWA and TIC gases	• VVCD, VVCD-S – priority CWAs • VVMT, VVTD – priority explosive, narcotic and CWA threats	Solid, liquid, and gas state threats	Broad range of CBE detection and classification
7	Identification capability	Yes	No	No	Presumptive ID capable	Yes	Yes

8	Time for detection (or identification)	3–30 s depending on range and concentration	2 s (typical), sub-second readings for rapid detection and immediate warning alerts	Seconds	• VVCD – <30 s • VVCD-S, VVMT – 0.1 s • VVTD – <10 s	15 s	<1 s to max about 10 s dependent on mode of detection, standoff, and concentration of target
9	Standoff distance range	Up to 50 m for HME's and 10 m for most other chemicals	250 m	0.1–5 km	• VVCD, VVCD-S – 1 m • VVMT, VVTD – 1–20+ m	15 m	0.6–5+ m
10	Optimum standoff distance	1–10 m	N/A	N/A	• VVCD, VVCD-S – 1 m for droplets • VVMT, VVTD – 1–3 m for residues	5–10 m	2 m
11	LOD at optical standoff distance	Trace concentrations at 1 m (10s μg/cm^2) and residues at 10 m (100s μg/cm^2)	Varies with gas and standoff distance (e.g. Sarin at 10 ppb over 500 m)	Not available	• VVCD, VVCD-S – Sub-microliter droplets • VVMT, VVTD: μg/cm^2 at short distances to bulk at long distance	10 μg/cm^2 for solids/liquids; 1 ppm for vapors	< 1 μg/cm^2 of C&E materials and < ~60 bacterial spores
12	Angular FOV	0.06° starting with a 0.5 cm diameter spot size	Low-cost retroreflectors and mirrors for large area coverage	0.5 and 1.5°	• VVCD – 17° • VVCD-S – 17° • VVMT – 12° • VVTD – 10–20° (continuous zoom)	~2°	Point detector
13	Size (external dimensions)	Wand: 5″ × 7″ × 15″, Backpack: 17″ × 12″ × 7″	18″ × 14″ × 6″ inches (indoor); 37″ × 14″ × 10″ inches (outdoor)	13.4″ × 10.7″ × 6.6″	• VVCD: 14.3″ × 9.9″ × 10.2″ • VVCD-S: 10.5″ × 9.4″ × 8.2″ • VVMT – 14.0″ × 9.8″ × 11.0″ • VVTD – 18.5″ × 7.9″ × 9.4″ (sensor head)	24″ × 18″ × 18″	7″ × 12″ × 17″

Table 10.3 (Continued)

No.	Parameter	Alakai defense systems PRIED	Block engineering LASERWARN™	Block engineering PORTHOS™	ChemImage VeroVision®	Leidos SWIFT	Photon systems STANDOFF 200
14	Weight	• Short mission mode: 31 lbs • Extended mode: 38 lbs	• 32 lbs (indoor) • 55 lbs (outdoor)	17 lbs	• VVCD, VVMT – 15 lbs • VVCD-S – 9.9 lbs • VVTD – 94 lbs (sensor, tripod, power box, computer)	165 lbs	10 lbs
15	Battery or line powered?	Both	Line	Battery	Both	Line	Onboard battery
16	Is the system eye safe?	Yes – two operating modes Eyesafe and Max detect	Yes	Yes	Yes	Yes	Yes
17	Degree of autonomy	None yet but capability could be added if desired	Fully automated operation	Fully automated operation	Can be fully autonomous	Pushbutton after targeting	Can be fully autonomous
18	Reconfigurable for new materials (targets)	Yes	Yes	Yes	Yes	Yes	Yes
	Form factor(s)	Handheld wand with portable backpack	Fixed or portable	Handheld, UAV, tripod	• VVCD, VVCD-S – handheld or tripod-mounted • VVMT – handheld, tripod-mounted, vehicle-mounted • VVTD – tripod-mounted	Fixed, tripod-mounted	Handheld and adaptable to robot-mounted

10.6 Conclusions and Future Direction

Portable standoff spectroscopy has advanced significantly since the first portable standoff optical spectroscopic sensors became commercially available in the early 2000s. Substantial improvements have been made in point scanning, spatial scanning and wide-field imaging platform designs. Instrumentation has branched out from breadboarded components on an optical table to robust, fieldable units that have moved the technology far beyond the research laboratory setting. Along with these developments has come the expansion of the technology, and associated applications, ranging from close standoff, submicron Raman imaging of biological spores, to multikilometer standoff IR imaging of CWA clouds. Significant advances in the underlying instrument technologies have now made real-time, autonomous, wide-area, standoff spectroscopic imaging a reality. Combining multiple spectroscopic imaging modalities has dramatically increased what is possible from a structural, elemental composition, and molecular composition materials characterization standpoint.

Future trends will inevitably continue to push the capabilities of instrumentation in areas, such as reduced SWAP-C, increased autonomy, and increased utilization of artificial intelligence and machine learning methods. Application and market demands will propel Raman instrumentation development to enable greater area coverage, increased data acquisition rates, through-barrier capabilities, improved data SNR and increased instrument stability and ease-of-use. Sensor and data fusion methods will become more widely used to improve overall sensor performance. Algorithm advancements including mixture analysis methods and fluorescence mitigation or avoidance will be necessary. With the realization of these continual enhancements in portable, standoff spectroscopic instrumentation, new applications and uses of these technologies will inevitably prove beneficial to mankind.

Acronyms and Abbreviations

AFIP	Armed Forces Institute of Pathology
ANFO	Ammonium Nitrate Fuel Oil
ANN	Artificial Neural Network
AOTF	Acousto-Optic Tunable Filter
CBE	Chemical, Biological, and Explosive
CBP	US Customs and Border Protection
CCD	Charge-Coupled Device
CDC	Centers for Disease Control and Prevention
CF	Conformal Filter
CMOS	Complementary Metal-Oxide-Semiconductor
CONOPS	Concept of Operations
CWA	Chemical Warfare Agent
DEP	Diethyl phthalate
DHS	Department of Homeland Security
DIC	Differential Interference Contrast
DMMP	Dimethyl methylphosphonate
DP-CF	Dual Polarization Conformal Filter
DRS	Diffuse Reflectance Spectroscopy
EOD	Explosive Ordnance Disposal
FAST	Fiber Array Spectral Translator
FPA	Focal Plane Array
fps	frames per second
FT	Fourier Transform

FTIR	Fourier Transform Infrared
HME	Homemade Explosive
HSI	Hyperspectral Imaging
ICCD	Intensified Charge-Coupled Device
IED	Improvised Explosive Device
JBPDS	Joint Biological Point Detection System
IMF	International Mail Facility
LCTF	Liquid Crystal Tunable Filter
LOD	Limit-of-Detection
LWIR	Longwave Infrared
MCF	Multiconjugate Filter
MOE	Multivariate Optical Element
MWIR	Midwave Infrared
NGCD	Next Generation Chemical Detector
NIR	Near-Infrared
NIST	National Institute of Standards and Technology
NTAs	Nontraditional Agents
OIG	Office of Inspector General
PCA	Principal Component Analysis
PETN	Pentaerythritol tetranitrate
PLSDA	Partial Least Squares Discriminant Analysis
PRIED	Portable Raman Improvised Explosive Detector
RGB	Red, Green, Blue
QCL	Quantum Cascade Laser
QCLA	Quantum Cascade Laser Array
SHS	Spatial Heterodyne Spectroscopy
SNR	Signal-to-Noise Ratio
SWaP-C	Size, Weight, Power and Cost
SWIR	Shortwave Infrared
TIC	Toxic Industrial Chemical
TRL	Technology Readiness Level
UAV	Unmanned Aerial Vehicle
UGV	Unmanned Ground Vehicle
UV	Ultraviolet
VIS	Visible
VVCD	VeroVision® Chemical Detector
VVMT	VeroVision® Moving Target
VVTD	VeroVision® Threat Detector

References

Angel, S.M., Kulp, T.J., and Vess, T.M. (1992). Remote-Raman spectroscopy at intermediate ranges using low-power CW lasers. *Appl. Spectros.* 46: 1085.

Angel, S.M., Gomer, N.R., Sharma, S.K., and McKay, A.C. (2012). Remote Raman spectroscopy for planetary exploration: a review. *Appl. Spectrosc.* 66: 137–150.

Anon (2007). *Safety of Laser Products – Part 1: Equipment Classification and Requirements*, 2ee. International Electrotechnical Commission.

Anon (2020a). *Wikipedia* [Online]. en.wikipedia.org/wiki/Douma_chemical_attack (accessed 6 April 2020).

Anon (2020b). *Worldometer* [Online]. www.worldometers.info/coronavirus (accessed 6 April 2020).

Baillard, X., Gauguet, A., Bize, S. et al. (2006). Interference-filter-stabilized external-cavity diode lasers. *Opt. Commun.* 266 (2): 609–613.

Batchelder, D., Cheng, C., Muller, W., and Smith, B. (1991). A compact Raman microprobe/microscope: Analysis of polydiacetylene Langmuir and Langmuir-Blodgett films. *Makromol. Chem. Macromol. Symp.* 46: 171.

Bayer, B.E. (1976). Color imaging array. US Patent No. 3,971,065A.

Blanchard, R. and Vakhshoori, D. (2020). Standoff detection of chemicals using infrared hyperspectral imaging and Raman spectroscopy. *Proc. SPIE Def. Sec. CBRNE Sensing XXI* 114160C: 114160C-1–114160C-10.

Boogh, L., Meier, R., and Kausch, H. (1992). A Raman microscopy study of stress transfer in high-performance epoxy composites reinforced with polyethylene fibers. *J. Polym. Sci. B* 30: 325–333.

Bowden, M., Gardiner, D.J., Rice, G., and Gerrard, D.L. (1990). Line-scanned micro Raman spectroscopy using a cooled CCD imaging detector. *Raman Spectros.* (21): 37–41.

Buchsbaum, P.E. and Morris, M.J. (2000). Method for making monolithic patterned dichroic filter arrays for spectroscopic imaging. US Patent No. US6638668B2.

Byrnes, J. (2009). *Unexploded Ordnance Detection and Mitigation*. Dordrecht: Springer.

Carter, J.C., Angel, S.M., Lawrence-Snyder, M. et al. (2005). Standoff detection of high explosive materials at 50 meters in ambient light conditions using a small Raman instrument. *Appl. Spectrosc* 59: 769–775.

Christens-Barry, W.A., Boydston, K., France, F.G. et al. (2009). Camera system for multispectral imaging of documents. *Electronic Imaging* 7249: 8–18.

Cristensen, K., Bradley, N., Morris, M., and Morrison, R. (1995). Raman imaging using a tunable dual-stage liquid crystal Fabry–Perot interferometer. *Appl. Spectrosc.* 49: 1120–1125.

Crocombe, R.A. (2018). Portable spectroscopy. *Appl. Spectrosc.* 72 (12): 1701–1751.

Dupont, S., Petersen, C., Thøgersen, J. et al. (2012). IR microscopy utilizing intense supercontinuum light source. *Opt. Express* 20 (5): 4887–4892.

Editors, E.B. (2008). *Cassegrain reflector astronomical instrument* [Online]. https://www.britannica.com/science/Cassegrain-reflector (accessed 6 April 2020).

Evans, J. (1949). The birefringent filter. *J. Opt. Soc. Am.* 39: 229–237.

Evans, J. and Solc, I. (1958). Birefringent filter. *J. Opt. Soc. Am.* 48: 142–143.

Faist, J., Capasso, F., Sivco, D. et al. (1994). Quantum cascade laser. *Science* 264 (5158): 553–556.

Goldstein, S.K., Kidder, L., Herne, T. et al. (1996). The design and implementation of a high-fidelity Raman imaging microscope. *J. Microsc.* 184: 35–45.

Gomer, N.R., Gordon, C.M., Lucey, P. et al. (2011a). Raman spectroscopy using a fixed-grating spatial heterodyne interferometer. *Spectroscopy* 26: 29.

Gomer, N.R., Gordon, C.M., Lucey, P. et al. (2011b). Raman spectroscopy using a spatial heterodyne spectrometer: proof of concept. *Appl. Spectrosc.* 65 (8): 849.

Gomer, N.R., Lamsal, N., Sun, H. et al. (2019). Explosive detection and identification using a wide-area hyperspectral raman imaging sensor. *Proc. SPIE Def. Sec. CBRNE Sensing XX* 110100H: 110100H-1–110100H-9.

Hanel, R., Crosby, D., Herath, L. et al. (1980). Infrared spectrometer for Voyager. *Appl. Opt.* 19: 1391–1400.

Hearnshaw, J. (1986). *The Analysis of Starlight*. Cambridge: Cambridge Univeristy Press.

Henderson, R. (2007). *Wavelength considerations* [Online]. web.archive.org/web/20071028072110/info.tuwien.ac.at/iflt/safety/section1/1_1_1.htm (accessed 6 April 2020).

Hirschfield (1974). Range independence of signal in variable focus remote Raman spectrometry. *Appl. Opt.* 13: 1435–1437.

Ingle, J. and Crouch, S. (1988). *Spectrochemical Analysis*. Prentice Hall.

IPAC, N. (2012). *Near, mid and far-infrared* [Online]. archive.is/20120529003352/http://www.ipac.caltech.edu/Outreach/Edu/Regions/irregions.html (accessed 6 April 2020).

Jestel, N.L., Shaver, J.M., and Morris, M.D. (1998). Hyperspectral Raman line imaging of an aluminosilicate glass. *Appl. Spectros.* 52: 64–69.

Jones, B.W. (2014). *Evolution of site in the animal kingdom* [Online]. https://webvision.med.utah.edu/2014/07/evolution-of-sight-in-the-animal-kingdom (accessed 6 April 2020).

Kalasinsky, K., Hadfield, T., Shea, A. et al. (2007). Raman chemical imaging spectroscopy reagentless detection and identification of pathogens: signature development and evaluation. *Anal. Chem.* 79: 2658–2673.

King, H.C. (2011). *The History of the Telescope*. New York: Courier Dover Publications.

Kutteruf, M.R., Yetzbacher, M.K., Deprenger, M.J. et al. (2014). 9-band swir multispectral sensor providing full-motion video. *Proc. SPIE Airborne Intelligence, Surveillance, Reconnaissance (ISR) Systems and Applications XI* 9076.

Lyot, B. (1944). The birefringent filter and its application in solar physics. *Astrophysics* 39: 229–242.

Ma, J. and Ben-Amotz, D. (1997). Rapid micro-Raman imaging using fiber-bundle image compression. *Appl. Spectrosc.* 51: 1845–1848.

Morris, H., Hoyt, C., and Treado, P. (1994). Imaging spectrometers for fluorescence and Raman microscopy: acousto-optic and liquid crystal tunable filters. *Appl. Spectrosc.* 48: 857–866.

Morris, H., Hoyt, C., Miller, P., and Treado, P. (1996). Liquid crystal tunable filter Raman chemical imaging. *Appl. Spectrosc.* 50: 805–811.

National Security Technologies, LLC (2016). *Passive Infrared Systems for Remote Chemical Detection Market Survey Report*. https://www.dhs.gov/sites/default/files/publications/PIS-RCD-MSR_0716-508.pdf.

Nelson, M. and Myrick, M. (1999a). Fabrication and evaluation of a dimension-reduction fiberoptic system for chemical imaging applications. *Rev. Sci. Instrum.* 70: 2836–2844.

Nelson, M. and Myrick, M. (1999b). Single-frame chemical imaging: dimension reduction fiber-optic array improvements and application to laser-induced breakdown spectroscopy. *Appl. Spectrosc.* 53: 751–759.

Nelson, M., McLestar, M., Aust, J., and Myrick, M., 1996. Distributed sensing of fiber optic arrays. The Pittsburgh Conference and Exposition on Analytical Chemistry and Applied Spectroscopy, Chicago, IL, March 1996.

Nelson, M., Bell, W., McLester, M., and Myrick, M. (1998a). Single-shot multiwavelength imaging of laser plumes. *Appl. Spectrosc.* 52: 179–186.

Nelson, M.P., Aust, J.F., Dobrowolski, D.J.A. et al. (1998b). Multivariate optical computation for predictive spectroscopy. *Anal. Chem.* 70: 73–82.

Nelson, M.P., Shi, L., Zbur, L. et al. (2016). Real-time short-wave infrared hyperspectral conformal imaging sensor for the detection of threat materials. *Proc. SPIE Def. Sec.* 9824 (42): 982416-1–982416-9.

Nelson, M.P., Tazik, S.K., Bangalore, A.S. et al. (2017). Performance evaluation and modeling of a Conformal Filter (CF) based real-time standoff hazardous material detection sensor. *Proc. SPIE Def. Sec. Next Gen. Spectros. Technol. X* 10210: 102100L-1–102100L-10.

Nelson, M.P., Tazik, S.K., Treado, P.J. et al. (2018). Real-time short-wave infrared hyperspectral conformal imaging sensor for the detection of threat materials. *Proc. SPIE Def. Comm. Sens. Next Gen. Spectros. Technol. XI* 106570U: 106570U-1–106570U-10.

Nelson, M.P., Tazik, S.K., and Treado, P.J. (2019). Real-time, reconfigurable, handheld molecular chemical imaging sensing for standoff detection of threats. *Proc. SPIE Def. Sec. CBRNE Sensing XX* 1101005: 1101005-1–1101005-10.

Pecharromán-Gallego (2017). An overview on quantum cascade lasers: origins and development, quantum cascade lasers. *IntechOpen* https://doi.org/10.5772/6500.

Pohl, K.R., Ford, A.R., Waterbury, R.D., Vunck, D., and Dottery, E.L. (2012). Optical hazard avoidance and method. US Patent No. US8724097B2.

Schaeberle, M., Karakatsanic, C., Lau, C., and Treado, P. (1995). Raman chemical imaging: noninvasive visualization of polymer blend architecture. *Anal. Chem.* 67: 4316–4321.

Schaeberle, M., Kalasinsky, V., Luke, J. et al. (1996). Raman chemical imaging: histopathology of inclusions in human breast tissue. *Anal. Chem.* 68: 1829–1833.

Schlossberg, H. and Kelley, P. (1981). Infrared spectroscopy using tunable lasers. In: *Spectrometric Techniques* (ed. G. Vanesse), 161–238. New York: Academic Press.

Scholl, L., Seth, P., Kariisa, M. et al. (2019). *Drug and opioid-involved overdose deaths – United States, 2013–2017. MMWR Morb. Mortal. Wkly Rep. 2019* 67: 1419–1427.

Schröder, K. (2000). *Handbook on Industrial Laser Safety*. Technical University of Vienna.

Schubert, E.F. (2003). *Light Emitting Diodes*. Cambridge, UK: Cambridge University Press.

Sharma, S.K., Misra, A.K., Lucey, P.G. et al. (2006). Remote pulsed raman spectroscopy of inorganic and organic materials to a radial distance of 100 meters. *Appl. Spectrosc.* 60: 871–876.

Solc, I. (1965). Birefringent chain filters. *J. Opt. Soc. Am.* 55: 621–625.

Thériault, J.-M., Puckrin, E., Hancock, J. et al. (2004). Passive standoff detection of chemical warfare agents on surfaces. *Appl. Opt.* 43: 5870–5885.

Treado, P., Levin, I., and Lewis, E. (1992). High-fidelity Raman imaging spectrometry: a rapid method using an acousto-optic tunable filter. *Appl. Spectrosc.* 46: 1211–1216.

Turner, J.I. and Treado, P. (1997). LCTF Raman chemical imaging in the near infrared. *Proc. SPIE – Infrared Technol. Appl.* XXIII (3061): 280–283.

Vaughn, J.M. (1989). *The Fabry–Perot Interferometer: History, Theory Practice and Applications*. Bristol, UK: Adam Hilger.

Waldron, A., Allen, A., Colón, A., Carter, J.C., and Angel, S.M. (2020). Monolithic spatial heterodyne raman spectrometer: prelizminary characterization results. *Appl. Spectrosc.*

Wang, X., Voigt, T., Bos, P. et al. (2006). *Evaluation of a High-Throughput Liquid Crystal Tunable Filter for Raman Chemical Imaging of Threat Materials*, 637808. Boston: SPIE – The International Society of Optical Engineering.

Wentworth, R., Neiss, J., Nelson, M., and Treado, P. (2007). Standoff Raman Hyperspectral Imaging Detection of Explosives. *2007 IEEE Antennas and Propagation Society International Symposium*, 4925–4928.

West, M., Grossman, J., and Galvan, C. (2018). *Commercial Snapshot Spectral Imaging: The Art of the Possible. Mitre Technical Report*. McLean, VA: Mitre Corporation.

11

Microplasmas for Portable Optical Emission Spectrometry

Vassili Karanassios

Department of Chemistry and Waterloo Institute for Nanotechnology, University of Waterloo, Waterloo, ON, Canada

11.1 Introduction

According to the Merriam Webster dictionary, a portable *object* is one "capable of being carried or moved about," likely because it is smaller or lighter or because it requires lower amounts of consumables for operation as compared to its original counterpart. In terms of small-sized plasmas (in the form of microplasmas), numerous examples have been described in the literature [1–12]. To provide a basis for comparison, an original plasma-based *object* relevant to this work is an inductively coupled plasma, or ICP for short. To further provide an idea of average size, an ICP is formed on top of a torch with typical dimensions of about 25 cm in length and 2 cm in diameter. The widely used atmospheric pressure plasma-torch is powered by a relatively heavy and bulky (and, thus, not a person-portable) radiofrequency (RF) generator, typically the size of an office laser printer. For chemical analysis, atmospheric pressure operation is preferred because there is no need to use power-consuming and heavyweight vacuum pumps, as these disable portability. Furthermore, during the operation, an ICP torch becomes hot, so much so that the torch requires water-cooling. Cooling or other thermal management adds weight and complexity, thus also disabling portability. In terms of consumables required for operation, an ICP of the type described above uses a total of about 20 l/min of expensive argon gas and requires about 1.5 kW of electrical power. In addition to adding to operating costs, the need to carry such large amounts of consumables clearly disables portability. One way of addressing plasma portability is to use small-sized, lightweight plasmas with low requirements for consumables in the form of a microplasma [2–12].

A microplasma has been arbitrarily defined as a quasineutral electrical discharge (i.e. an ensemble of ions and electrons) with one critical dimension (such as height, width, radius) of 1000 μm or less [2, 3]. But, as the dimensions are reduced and approach the micrometer range, in addition to becoming small-sized and lightweight (thus enabling portability), microplasmas also become nonequilibrium and nonthermal. By being nonthermal, they do not require cooling or other thermal management, thus enabling portability. In addition, due to their small volume, a power density per unit volume equal to (or larger than) that obtained by an ICP can be obtained using lower input power (e.g. as little as a few Watts). And, due to their small size, they also require small flow rates of inert gas, with some of them operating using freely available air.

To be useful in portable chemical analysis applications (i.e. for use out of a laboratory), microplasmas must be a part of an instrument that also includes a portable spectrometer [13, 14]. A block diagram of such a system is shown in Figure 11.1. For such applications, a portable sample processor (often including sample preparation, sample separation, and sample introduction) is required [15]. Low-power operation (e.g. requiring less than 100 W of continuous operation) is essential [16–18]. The somewhat arbitrary 100 W limit is set by current battery technology; for example, state-of-the-art laptops use anywhere between about 50 and 100 W. Actual electrical power

Portable Spectroscopy and Spectrometry 1: Technologies and Instrumentation, First Edition.
Edited by Richard A. Crocombe, Pauline E. Leary and Brooke W. Kammrath.

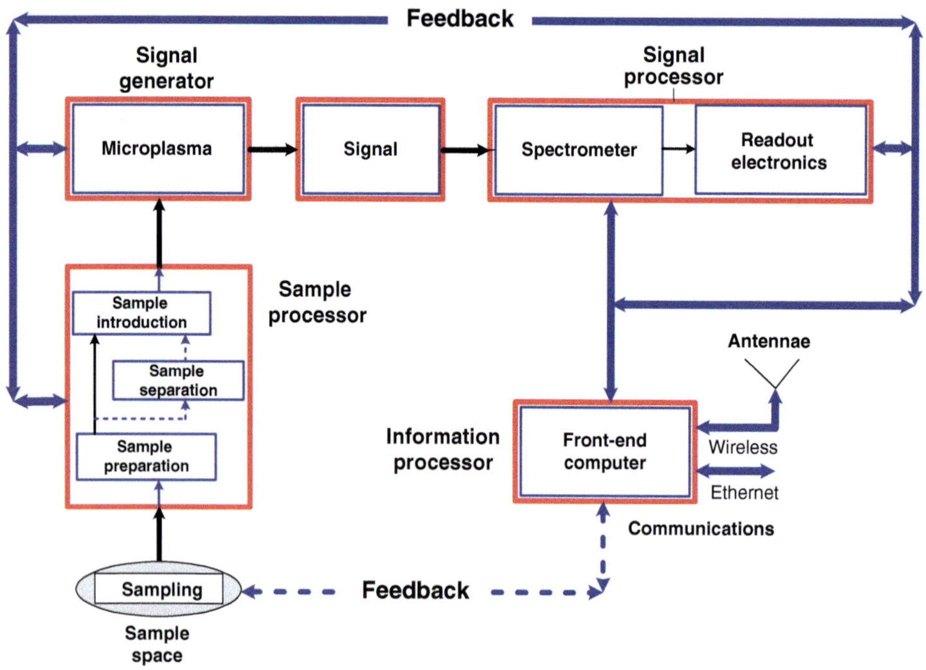

Figure 11.1 Block diagram illustrating a portable microinstrument that includes a microplasma as a "signal generator" and it has associated processors, such as a sample processor, a signal processor, and an information processor. By analogy to the definition used for a microplasma, a microinstrument has at least one critical component (or subsystem or processor) in the microregime. *Source*: Based on Karanassios [2].

usage depends on the number of software applications running concurrently. A portable signal processor is also essential, and it involves a spectrometer along with its associated readout electronics. Most attempts at implementing a microinstrument of the type shown in Figure 11.1 involve the inclusion of a portable optical spectrometer [13, 14]. The brief review included in this chapter primarily involves optical spectrometers because many types of such spectrometers are commercially available, while, at present, portable mass spectrometers are not widely commercially available. (See the chapters by Snyder, Blakemen and Miller, and Leary et al. in this Volume for discussions of portable mass spectrometry, high-pressure mass spectrometry, and gas chromatography–mass spectrometry systems, respectively.) However, in the literature, many nonportable, lab-based mass spectrometers have been evaluated using a microplasma as an ion source. Microplasma–mass spectrometry combinations are mentioned not because they are portable, but because they demonstrate that microplasmas can be useful ion sources for mass spectrometry. To convert data into useful information, an information processor is also required. To reduce complexity and weight, the front-end processor (Figure 11.1) can be replaced by a smartphone [19–21]. And, to address power requirements, a variety of energy-scavenging and energy-harvesting approaches have been described [16–18].

11.2 A Brief Review of the Portable Microplasma Literature

11.2.1 Micro ICPs (Micro Inductively Coupled Plasmas)

Hopwood and coworkers have pioneered the development of miniaturized ICPs or mICPs as they call them [22–26]. These were primarily developed for operation in vacuum. Initially, one of the first examples involved the

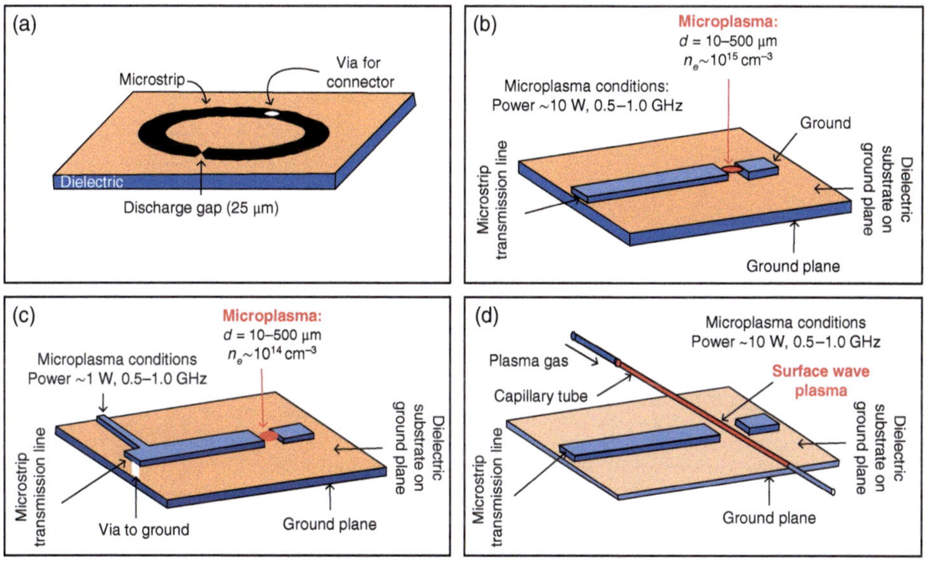

Figure 11.2 (a) Microwave-induced microplasma (with a 25 μm discharge gap); (b) a single transmission line microplasma; (c) a microwave plasma obtained using a microwave resonator; and (d) a surface wave microplasma [22, 23]. *Source:* Hopwood et al. [22]. Licensed under CC BY 3.0.

use of induction coils with diameters of 5, 10, and 15 mm. Following this initial implementation, the authors described various "plasmas-on-a-chip" designs operating using microwave frequencies. Using such frequencies, lightweight, inexpensive, and commercially available amplifiers could be used. Example designs are summarized in Figure 11.2. To put things in a scale-perspective, the traditional, widely used, lab-scale, ICPs operate at atmospheric pressure, they have a typical frequency of operation of 27 MHz (and sometimes 40 MHz or even 100 MHz), they require 1–2 kW of electrical power, and they use about 20 l/min of argon gas. Unlike the widely used lab-sized ICPs mentioned above, mICPs operate at low pressures (0.1–10 Torr), have frequencies ranging between 100 and 500 MHz (some more than that), and require less than 4 W of electrical power. A good analytical example of their application involves the determination of SO_2 using molecular emissions, using a 5 mm (diameter) mICP and an optical multichannel analyzer (OMA). Despite their lightweight and use of low electrical power, low-pressure operation limits their portability and use in the field. However, due to their operation at low pressure, they are very well-suited as diagnostic tools for the semiconductor industry, and, in fact, they are commercially available for this purpose. They have been included here because they have been commercialized for use in the vacuum chambers employed by the semiconductor industry [25, 26].

11.2.2 Electrolyte Cathode Discharge (ELCAD)

In electrolyte cathode discharges (ELCADs), as their name implies, one of the electrodes (the cathode) contains a liquid electrolyte, obtained by adjusting the pH anywhere between 1 and 2 [27–35]. Electrolyte cathode direct current (DC) glow discharges were primarily developed to address the introduction of liquid samples into a glow discharge. To provide a bit of a perspective here and regardless of plasma-size, the question of how to introduce an initially ambient-temperature liquid sample into a gas-phase plasma continues to be an active area of research. In ELCADs, a nebulizer (as required for use in conventional flames and plasmas) is not necessary because sample introduction is facilitated due to use of a sample mixed with a liquid electrolyte in the cathode (Figure 11.3).

Historically, ELCADs are considered an outgrowth of glow discharge electrolysis (GDE), originally described in 1887. An example of a GDE cell is shown in Figure 11.3a. In this case, the plasma is formed between an electrode

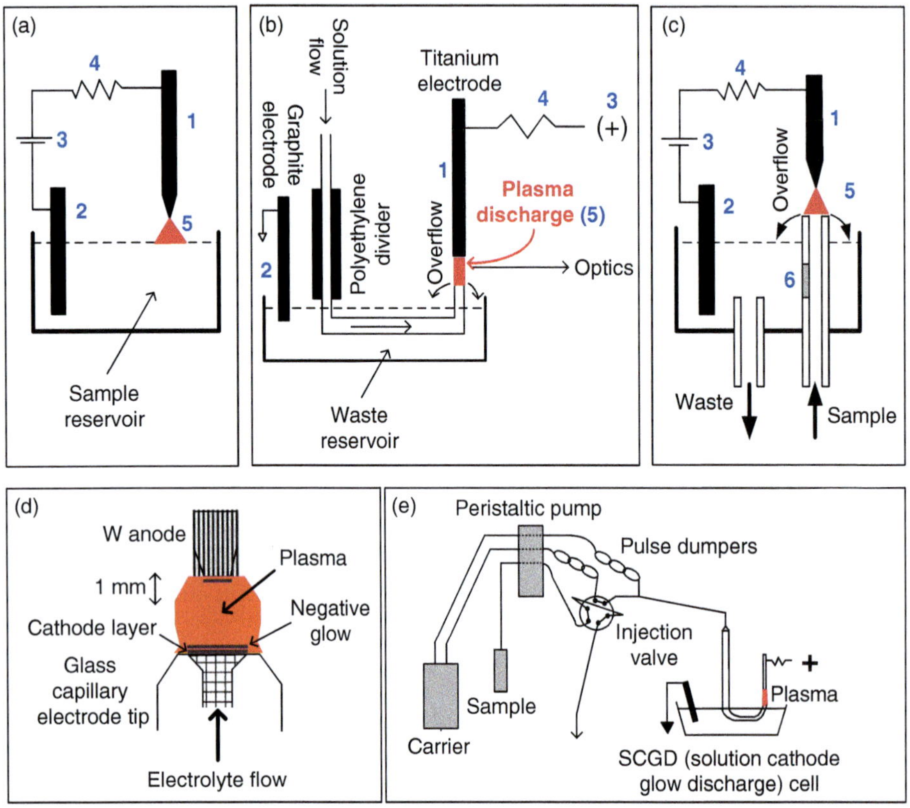

Figure 11.3 Simplified illustrations for (a) glow discharge electrolysis (GDE) cell; and (b) electrolyte cathode discharge (ELCAD) systems. In all cases, the solution level is marked using a dashed line. (c) There is a continuous flow of sample. *Source*: Webb and Hieftje [30]. © 2009 American Chemical Society. (d, e) The inner diameter of the sample tube can be less than 1 mm. (1) Metal electrode; (2) solid counter electrode; (3) high-voltage dc power supply; (4) current-limiting resistor; (5) plasma; (6) glass frit or cotton fiber (generally a diaphragm, but not always present). *Source*: Webb et al. [31]. © 2007 American Chemical Society.

(often made of titanium), and the surface of a sample (buffered using an appropriate electrolyte). The distance between the tip of the electrode and the surface of the liquid (labeled sample reservoir in Figure 11.3a) varies, but it typically is a few mm. This is true for all configurations shown in Figure 11.3a–e. A resistor is used to ballast the discharge (item 4 in Figure 11.3a). Typical operating values are: open circuit output voltage of about 1200 V (typically less), current that can be adjusted to between 30 and 100 mA, and the ballast resistor which is variable to between 1.5 and 6.0 kΩ, and a discharge gap between 1 and 4 mm. Although the maximum electrical power is 120 W, typical operating powers are about 70 W. A graphite electrode (item 2 in Figure 11.3a) is immersed into the sample reservoir to complete the circuit.

For ELCADs, there is typically a continuous flow of sample (Figure 11.3b and c). The inner diameter of the sample tube is a few mm, and flow rates range between 8 and 10 ml/min. In some of the most recent miniaturized ELCAD designs, the inner diameter of the sample tube (Figure 11.3d and e) was less than 1 mm (e.g. 380 μm), and the flow rate ranged between 2.5 and 3.5 ml/min. In addition to continuous flow, a multiport valve (Figure 11.3d) was used for transient sample introduction (25 μl injection). In such a configuration (Figure 11.3d), an ELCAD can be used for chromatographic applications. The detection limits were impressive for 25 μl volume samples, ranging from 5 pg (0.2 ppb) for lithium to 6 ng (270 ppb) for mercury.

Because the surface must be covered with the solution for stable operation, a large volume of sample is used, and, thus, a significant volume of waste is produced. For example, typical flow rates in the range of 8–10 ml/min have been reported for most of the early ELCAD designs. More recent designs, however, operate at much lower flow rates, for instance, between 2.5 and 3.5 ml/min. Despite the simplicity of the ELCAD designs, the discharge performs well. Detection limits for most metals tested, as reported in the literature, typically range in the 10s of ppb, and short-term precision was reported to be around 1%, obtained using 1 ppm single-element solutions.

11.2.3 Liquid Sampling-Atmospheric Pressure Glow Discharge (LS-APGD)

This atmospheric pressure discharge is similar to an ELCAD and, as its name implies, it also incorporates liquid sample introduction (Figure 11.4) [36–41]. Specifically, a conductive tube delivers the sample, typically 5 μl, to the discharge. The small amount of sample makes this discharge ideal as a detector for liquid chromatography. In addition, and unlike the ELCAD, a waste reservoir is unnecessary because the discharge is operated in total consumption mode. The typical 0.1–0.3 ml/min flow rate of the solution sample makes this parameter a key aspect of liquid sampling-atmospheric pressure glow discharge (LS-APGD), especially when used as a detector for liquid chromatography. Due to the low flow rate, a He-gas sheath (Figure 11.4) is necessary to reduce (or eliminate) signal fluctuations. Injection-to-injection, or sample-to-sample, precision of <10% RSD has been reported for concentrations in the range of hundreds of ppm. Relative unit-concentration detection limits have been reported to be in the 1–2 ppm range. Absolute unit (mass) detection limits of 5–10 ng have been reported for many elements across the periodic table. Remarkably, very little variation in detection limits has been reported for many of the elements tested.

11.2.4 Liquid Electrode Plasma (LEP)

A liquid electrode plasma (LEP)-optical emission spectrometry system (Figure 11.5) is described in a series of papers [42–46]. This system involves a combination of a quartz chip and sample flow system. In this system, an electroconductive solution sample is pumped into a microfluidic channel whose center is narrow (~100 μm wide). A pulsed, high voltage (about 1500 V) is applied at the opposite and wide ends of the channel, and a plasma is formed inside the narrow part of the channel. The sample that resides inside the narrow part of the channel

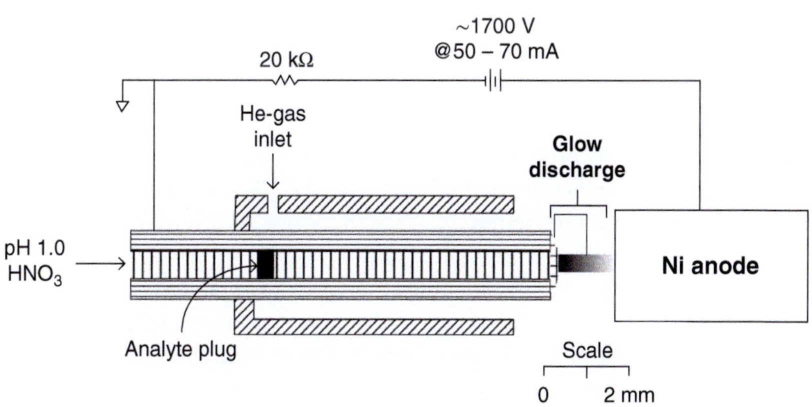

Figure 11.4 Illustration of a liquid sampling-atmospheric pressure glow discharge (LS-APGD). A high voltage is applied between a conductive capillary and an Ni anode. Typical voltages and currents are shown in the figure. A glow discharge is formed between the solution emerging from the capillary and the anode. A flow of He-gas, running concentric to the solution, is used to stabilize the solution-delivery, especially when low solution flow- rates are utilized. *Source*: Modifed from Marcus et al. [36].

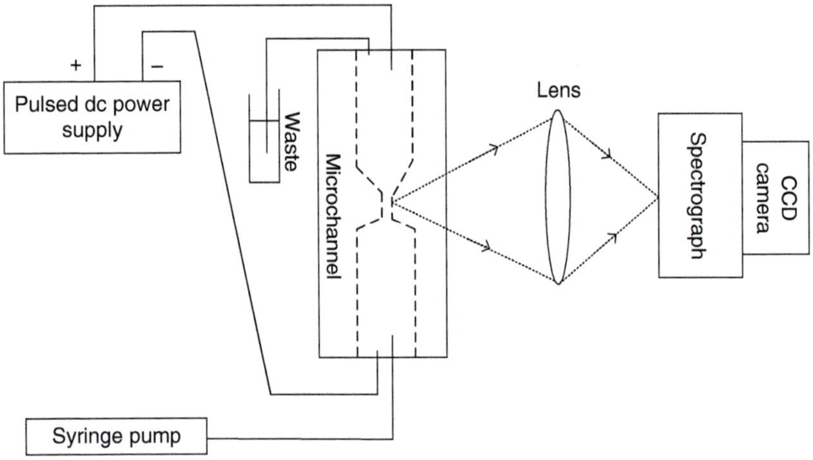

Figure 11.5 Instrumentation setup for measurement by a liquid electrode plasma (LEP). *Source*: Kohara et al. [42]. © 2015 Royal Society of Chemistry.

is vaporized, and its interaction with the microplasma generates atomic emission. Spectral lines so generated are used for the construction of calibration curves. The limits of detection were reported to be 0.52 μg/l for cadmium and 19.0 μg/l for lead. There is a commercially available system based on this design [46].

11.2.5 Postage Stamp Microplasmas on Polymeric Substrates

To reduce costs (e.g. those associated with ownership, operation, and fabrication), we developed and characterized many designs of microplasmas on polymeric substrates, and some examples are shown in Figure 11.6. For the microplasmas of the type shown in Figure 11.6, although their critical plasma dimension was in the μm regime, they were formed inside millifluidic channels (e.g. ~2 mm wide or less). This was done for rapid prototyping [47] and to avoid accidental contact of the microplasma with the channel walls, which is important during testing. Once prototypes were produced, the channel width was never revisited. Although polymeric substrates have high dielectric strength, to address poor transmission of polymers in the ultraviolet (UV), the channels were fitted with a quartz plate (Figure 11.6).

11.2.6 3D-Printed Microplasmas

To reduce fabrication costs, we used an inexpensive 3D printer to rapidly prototype 3D chips in a few hours or less [47–49]. This approach eliminated the need for cleanrooms and lithography. The 3D chips on the polymeric substrate were fitted with a quartz plate to obtain transparency in the UV. An example of a 3D-printed microplasma is shown in Figure 11.7.

To provide some operation details, a microsample (e.g. 3 μl) containing an analyte (i.e. the species of interest) was introduced into the microplasma using the sample introduction system shown in Figure 11.8. This microsample introduction system consists of a coil filament made of rhenium (Re). Briefly, the assembly carrying the recoil filament is retracted from the vaporization chamber (Figure 11.8), which is connected to an external power supply. Typically, 3 μl of a sample or of a diluted standard solution (e.g. 500 ppb of Zn) is deposited on the filament, and the coiled-filament assembly is reinserted into the vaporization chamber. Using the external power supply, low electrical power (e.g. 0.85 W) is applied to the coil to vaporize the solvent of the microsample that was deposited on

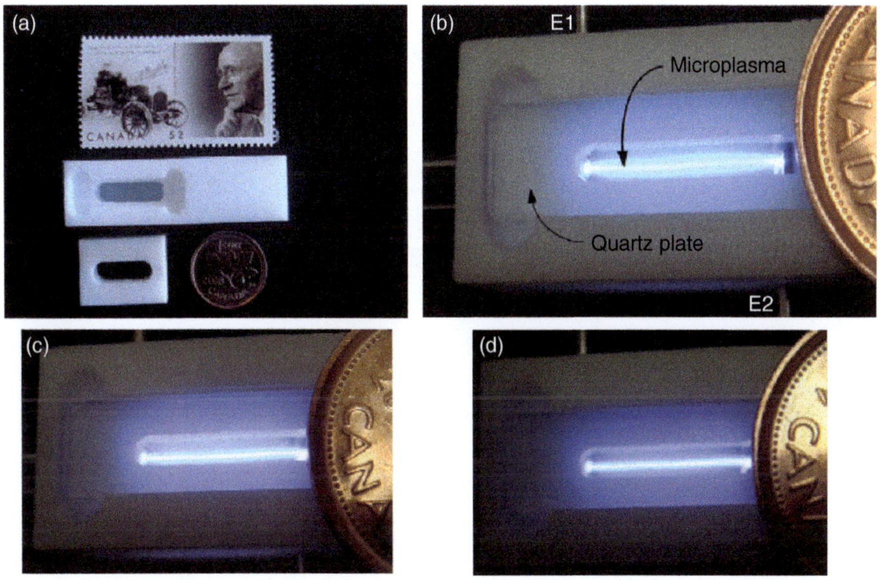

Figure 11.6 (a) Postage stamp-sized polymeric 3D-chips used for microplasma formation, and (b) microplasma formed between electrodes E1 and E2. The critical microplasma dimension depended on operating conditions. For example, microplasmas with diameters of (b) ~750 μm, (c) ~400 μm, and (d) ~200 μm were formed. In these photographs, a one-cent coin was included for size. It is worth noting that the microplasma fits inside the letter A of the coin.

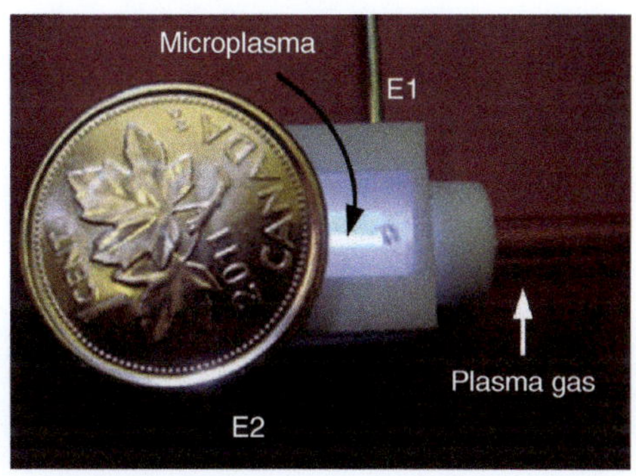

Figure 11.7 3D-printed microplasma fabricated on a hybrid 3D-chip. The microplasma is formed between electrodes E1 and E2. A one-cent coin has been included for size, the microplasma fits inside the letter A of the coin.

the coil. When the sample is dry, after about two minutes, the dried solution residue that remained on the coil was vaporized from the coil by applying higher electrical power to the coil (e.g. ~25 W). As shown in Figure 11.8, the outlet of the vaporization chamber of this sample introduction system is connected to the inlet of the microplasma. The sample vaporized from the coil was routed to the microplasma where it interacted with the microplasma, thus generating characteristic optical emission of the analyte present in the sample. Using the fiber-optic cable, optical emission was guided to the portable spectrometer, which could be controlled by a smartphone. An example of the signal so obtained is shown in Figure 11.9. The required signal processing, due to the transient nature of the signal generated by the recoil microsample introduction system, is briefly described in the caption of Figure 11.9.

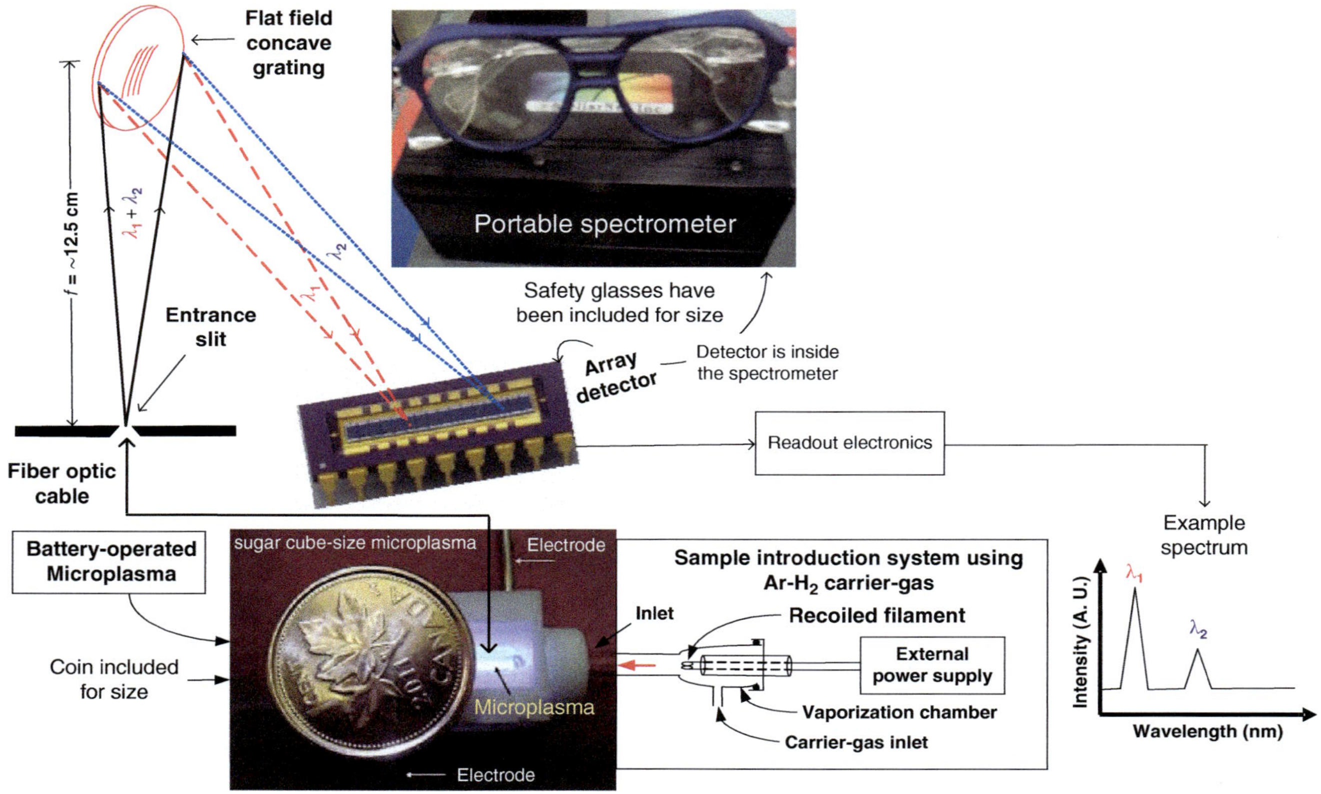

Figure 11.8 Example of the experimental setup showing the sample introduction system and one of the varieties of microplasma devices (MPDs) we designed, fabricated, and evaluated. The microsample introduction system we used is shown in the center frame of this figure.

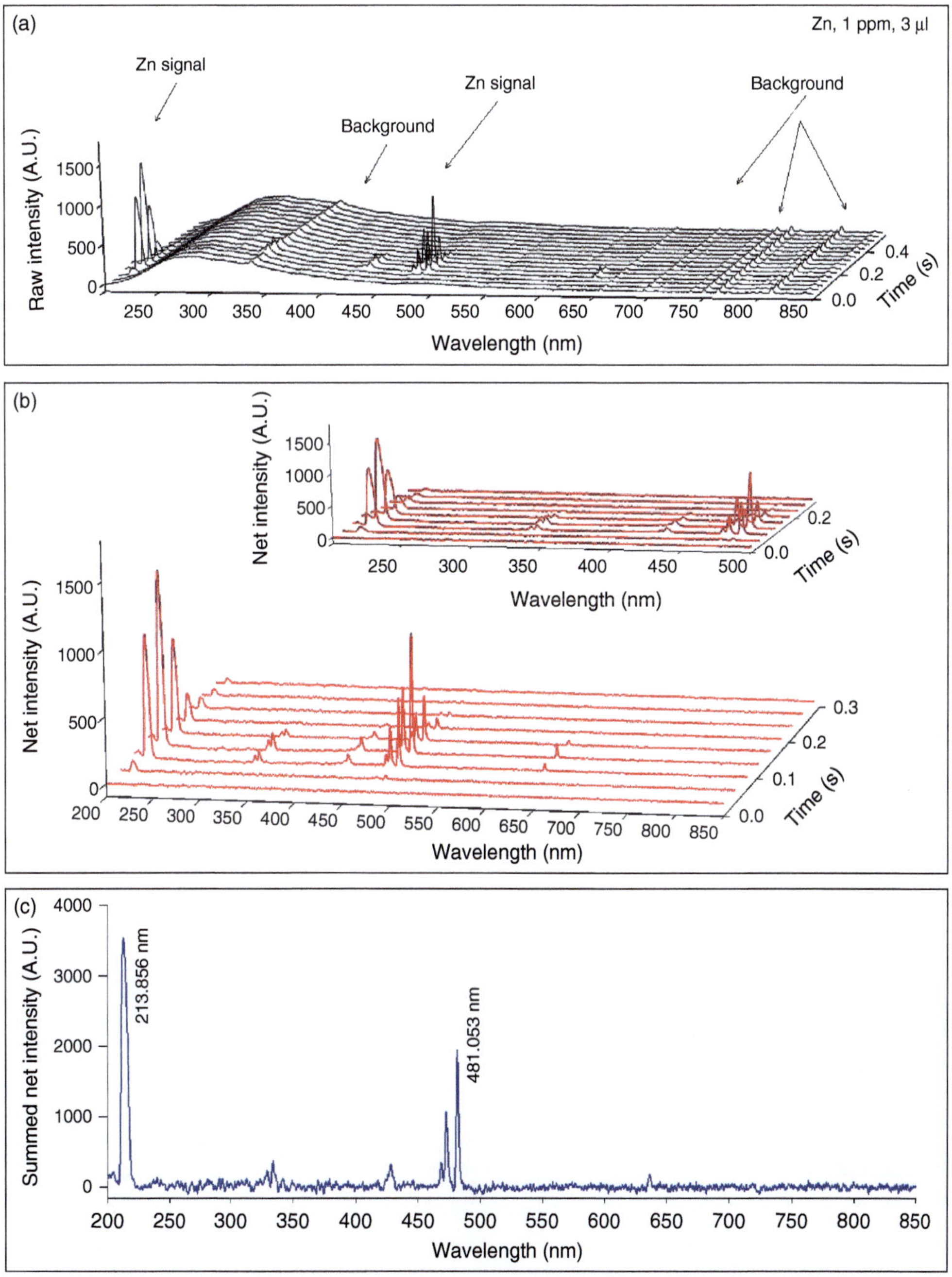

Figure 11.9 (a) Raw signal, (b) background-subtracted signal, and (c) 2D-signal obtained by summing only the spectra that contained a signal in (b).

11.3 Conclusion

Over the last several years, a variety of microplasmas have been described in the literature. For use outside of a laboratory, the low power consumption of microplasmas is one of their key characteristics. Many of the microplasmas described in the literature can also be used with portable, optical emission spectrometers, so that the entire, portable, microplasma-based spectrometer systems can be developed for use in chemical analysis *on-site* (i.e. in the field). In the future, it would be interesting to see further development in portable microplasma designs, develop their fundamental microplasma characteristics and apply them to a variety of anticipated applications.

Acronyms

3D	Three-dimensional
dc	Direct current
ELCAD	Electrolyte cathode discharges
GDE	Glow discharge electrolysis
ICP	Inductively coupled plasma
LS-APGD	Liquid sampling-atmospheric pressure glow discharge
mICP	Micro inductively coupled plasma
MPD	Microplasma devices
OMA	Optical multichannel analyzer
ppb	Parts per billion
ppm	Parts per million
RF	Radiofrequency
RSD	Relative standard deviation

Abbreviations

kΩ	kilo-ohm
kW	kilowatt
L/min	Liters/minute
MHz	Megahertz
mA	Milliamp
mL/min	Milliliters/minute
mm	Millimeter
ng	Nanogram
pg	Picogram
μg/l	Micrograms/liter
μl	Microliter
μm	Micrometer
V	Volt
W	Watt

Acknowledgments

Financial assistance from the Natural Sciences and Engineering Research Council of Canada (NSERC) and from Waterloo's Institute for Quantum Computing via a CFREF (Canada First Research Excellence Fund) is gratefully acknowledged.

References

1 Crocombe, R.A. (2018). Portable spectroscopy. *Applied Spectroscopy* 72 (12): 1701–1751. https://doi.org/10.1177/0003702818809719.

2 Karanassios, V. (2004). Microplasmas for chemical analysis: analytical tools or research toys? *Spectrochimica Acta Part B* 59 (7): 909–928. https://doi.org/10.1016/j.sab.2004.04.005.

3 Karanassios, V. (2018). Microfluidics and nanofluidics: science, fabrication technology (from cleanrooms to 3D printing) and their application to chemical analysis by battery-operated microplasmas-on-chips", Chapter 1. In: *Microfluidics and Nanofluidics* (ed. M.S. Kandelousi), 1–34. InTech Publishing https://doi.org/10.5772/intechopen.74426.

4 Becker, K.H., Schoenbach, K.H., and Eden, J.G. (2006). Microplasmas and applications. *Journal of Physics D: Applied Physics* 39 (3): R55. https://doi.org/10.1088/0022-3727/39/3/R01.

5 Franzke, J., Kunze, K., Miclea, M., and Niemax, K. (2003). Microplasmas for analytical spectrometry. *Journal of Analytical Atomic Spectrometry* 18: 802–807. https://doi.org/10.1039/B300193H.

6 Janasek, D., Franzke, J., and Manz, A. (2006). Scaling and the design of miniaturized chemical-analysis systems. *Nature* 442: 374–380. https://doi.org/10.1038/nature05059.

7 Bruggeman, P. and Brandenburg, R. (2013). Atmospheric pressure discharge filaments and microplasmas: physics, chemistry and diagnostics. *Journal of Physics D: Applied Physics* 46: 464001. https://doi.org/10.1088/0022-3727/46/46/464001.

8 Yuan, X., Tang, J., and Duan, Y. (2011). Microplasma technology and its applications in analytical chemistry. *Applied Spectroscopy Reviews* 46: 581–605. https://doi.org/10.1080/05704928.2011.604814.

9 Luo, D. and Duan, Y. (2012). Microplasmas for analytical applications of lab-on-a-chip. *Trends in Analytical Chemistry* 39: 254–266. https://doi.org/10.1016/j.trac.2012.07.004.

10 Tachibana, K. (2006). Current status of microplasma research. *IEEJ Transactions* 1: 145–155. https://doi.org/10.1002/tee.20031.

11 Becker, K., Kersten, H., Hopwood, J., and Lopez, J.L. (2010). Microplasmas: scientific challenges and technological opportunities. *European Physical Journal D: Atomic, Molecular, Optical and Plasma Physics* 60 (3): 437–439. https://doi.org/10.1140/epjd/e2010-00231-4.

12 Chiang, W.-H., Mariotti, D., Sankaran, R.M. et al. (2020). Microplasmas for advanced materials and devices. *Advanced Materials* 32: 1905508. https://doi.org/10.1002/adma.201905508.

13 Weagant, S. and Karanassios, V. (2009). Helium-hydrogen microplasma device (MPD) on postage-stamp-size plastic–quartz chips. *Analytical and Bioanalytical Chemistry* 395: 577–589. https://doi.org/10.1007/s00216-009-2942-2.

14 Weagant, S., Dulai, G., Li, L., and Karanassios, V. (2015). Characterization of rapidly-prototyped, battery-operated, argon-hydrogen microplasma on a chip for elemental analysis of microsamples by portable optical emission spectrometry. *Spectrochimica Acta Part B* 106: 75–80. https://doi.org/10.1016/j.sab.2015.01.009.

15 Badiei, H.R., Stubley, G., Fitzgerald, R. et al. (2018). Computational fluid dynamics (CFD) applied to a glass vaporization chamber for introduction of micro- or nano-size samples into lab-based ICPs and to a

CFD-derived (and rapidly prototyped via 3D printing) smaller-size chamber for portable microplasmas, Chapter 8. In: *Computational Fluid Dynamics: Basic Instruments and Applications in Science* (ed. A. Ionescu), 187–215. InTech Publishing https://doi.org/10.5772/intechopen.72650.

16 Trizcinski, P., Nathan, A., and Karanassios, V. (2017). Approaches to energy harvesting and energy scavenging for energy autonomous sensors and micro-instruments. *Proceedings of SPIE* 10194: 1019431-1. https://doi.org/10.1117/12.2262949.

17 Illar, J.A., Gao, Z., Sivoththaman, S., and Karanassios, V. (2019). From nanoenergy harvesting to self-powering of micro- or nano-sensors for measurements on-site or for IoT applications. *Proceedings of SPIE* 109831BA https://doi.org/10.1117/12.2519939.

18 Lim, N., Millar, J.A., and Karanassios, V. (2020). Triboeletric nanogenerators for powering fieldable sensors and systems. *Proceedings of SPIE 11390, Next-Generation Spectroscopic Technologies XIII* 113900T https://doi.org/10.1117/12.2559159.

19 Fitzgerald, R., Wang, E., and Karanassios, V. (2019). Smartphone-enabled data acquisition and digital signal processing: from current-output or voltage-output sensors for use on-site, to their use in IoT, in Industry 4.0 and (potentially) in Society 5.0. *Proceedings of SPIE* 1098304 https://doi.org/10.1117/12.2519941.

20 Fitzgerald, R. and Karanassios, V. (2018). The internet of things (IoT) for a smartphone-enabled optical spectrometer and their use on-site and (potentially) for industry 4.0. *Proceedings of SPIE* 10657: 1065705. https://doi.org/10.1117/12.2305466.

21 Fitzgerald, R., Wang, E., and Karanassios, V. (2018). Fast fourier transform of non-periodic signals generated from a microplasma: migrating from a desktop computer to an IoT-connected Smartphone. *Proceedings of SPIE 10657, Next-Generation Spectroscopic Technologies XI* 1065703: 7pp. doi: https://doi.org/10.1117/12.2305462.

22 Hopwood, J., Hoskinson, A.R., and Gregorio, J. (2014). Microplasmas ignited and sustained by microwaves. *Plasma Sources Science and Technology* 23: 064002. https://doi.org/10.1088/0963-0252/23/6/064002.

23 Minayeva, O.B. and Hopwood, J.A. (2003). Microfabricated inductively coupled plasma-on-a-chip for molecular SO2 detection: a comparison between global model and optical emission spectrometry. *Journal of Analytical Atomic Spectrometry* 18: 856–863. https://doi.org/10.1039/b303821a.

24 Hopwood, J. and Iza, F. (2004). Ultrahigh frequency microplasmas from 1 pascal to 1 atmosphere. *Journal of Analytical Atomic Spectrometry* 19: 1145–1150. https://doi.org/10.1039/B403425B.

25 Verionix commercialized microplasma optical emission technology as a process or residual gas analyzer. http://verionix.com (accessed 1 June 2020).

26 Inficon. https://jusmundi.com/en/document/decision/en-verionix-inc-n-k-a-k1tech-ventures-inc-v-inficon-inc-opinion-of-the-united-states-district-court-of-the-southern-district-of-new-york-thursday-21st-april-2016 (accessed 22 October 2020).

27 Webb, M.R., Andrade, F.J., Gamez, G. et al. (2005). Spectroscopic and electrical studies of a solution-cathode glow discharge. *Journal of Analytical Atomic Spectrometry* 20: 1218–1225. https://doi.org/10.1039/B503961D.

28 Webb, M.R., Andrade, F.J., and Hieftje, G.M. (2007). Use of electrolyte cathode glow discharge (ELCAD) for the analysis of complex mixtures. *Journal of Analytical Atomic Spectrometry* 22: 766–774. https://doi.org/10.1039/B616989A.

29 Cserfalvi, T. and Mezei, P. (2003). Sub-nanogram sensitive multi-metal detector with atmospheric electrolyte cathode glow discharge. *Journal of Analytical Atomic Spectrometry* 18: 596–602. https://doi.org/10.1039/b300544p.

30 Webb, M.R. and Hieftje, G.M. (2009). Spectrochemical analysis by using discharge devices with solution electrodes. *Analytical Chemistry* 81: 862–867. https://doi.org/10.1021/ac801561t.

31 Webb, M.R., Andrade, F.J., and Hieftje, G.M. (2007). High-throughput elemental analysis of small aqueous samples by emission spectrometry with a compact, atmospheric-pressure solution-cathode glow discharge. *Analytical Chemistry* 79: 7807–7812. https://doi.org/10.1021/ac0707885.

32 He, Q., Zhu, Z., and Hu, S. (2014). Flowing and non-flowing liquid electrode discharge microplasma for metal ion detection by optical emission spectrometry. *Applied Spectroscopy Reviews* 49: 249–269. https://doi.org/10.1080/05704928.2013.820195.

33 Wu, J., Yu, J., Li, J. et al. (2007). Detection of metal ions by atomic emission spectroscopy from liquid-electrode discharge plasma. *Spectrochimica Acta Part B* 62 (11): 1269–1272. https://doi.org/10.1016/j.sab.2007.10.026.

34 Manjusha, R., Shekhar, R., and Kumar, S. (2019). Direct determination of impurities in high purity chemicals by electrolyte cathode discharge atomic emission spectrometry (ELCAD-AES). *Microchemical Journal* 145: 301–307. https://doi.org/10.1016/j.microc.2018.10.052.

35 Zhao, M., Peng, X., Yang, B., and Wang, Z. (2020). Ultra-sensitive determination of antimony valence by solution cathode glow discharge optical emission spectrometry coupled with hydride generation. *Journal of Analytical Atomic Spectrometry* https://doi.org/10.1039/D0JA00009D.

36 Marcus, R.K., Paing, H.W., and Zhang, L.X. (2016). Conceptual demonstration of ambient desorption-optical emission spectroscopy using a liquid sampling-atmospheric pressure glow discharge microplasma source. *Analytical Chemistry* 88 (11): 5579–5584. https://doi.org/10.1021/acs.analchem.6b00751.

37 Venzie, J.L. and Marcus, R.K. (2006). Effects of easily ionizable elements on the liquid sampling atmospheric pressure glow discharge. *Spectrochimica Acta Part B* 61 (6): 715–721. https://doi.org/10.1016/j.sab.2006.02.005.

38 Marcus, R.K., Manard, B.T., and Quarles, C.D. Jr., (2017). Liquid sampling-atmospheric pressure glow discharge (LS-APGD) microplasmas for diverse spectrochemical analysis applications. *Journal of Analytical Atomic Spectrometry* 32: 704–716. https://doi.org/10.1039/C7JA00008A.

39 Quarles, C.D. Jr.,, Manard, B.T., Burdette, C.Q., and Marcus, R.K. (2012). Roles of electrode material and geometry in liquid sampling-atmospheric pressure glow discharge (LS-APGD) microplasma emission spectroscopy. *Microchemical Journal* 105: 48–55. https://doi.org/10.1016/j.microc.2012.01.012.

40 Paing, H.W. and Marcus, R.K. (2017). Parametric evaluation of ambient desorption optical emission spectroscopy utilizing a liquid sampling-atmospheric pressure glow discharge microplasma. *Journal of Analytical Atomic Spectrometry* 32: 931–941. https://doi.org/10.1039/C7JA00035A.

41 Lu, Q., Feng, F., Yu, J. et al. (2020). Determination of trace cadmium in zinc concentrate by liquid cathode glow discharge with a modified sampling system and addition of chemical modifiers for improved sensitivity. *Microchemical Journal* 152: 104308. https://doi.org/10.1016/j.microc.2019.1043082020.

42 Kohara, Y., Terui, Y., Ichikawa, M. et al. (2015). Atomic emission spectrometry in liquid electrode plasma using an hour glass microchannel. *Journal of Analytical Atomic Spectrometry* 30: 2125–2128. https://doi.org/10.1039/C5JA00059A.

43 Khoai, D.V., Kitano, A., Yamamoto, T. et al. (2013). Development of high sensitive liquid electrode plasma - atomic emission spectrometry (LEP-AES) integrated with solid phase pre-concentration. *Microelectronic Engineering* 111: 343–347. https://doi.org/10.1016/j.mee.2013.02.086.

44 Khoai, D.V., Yamamoto, T., Ukita, Y., and Takamura, Y. (2014). On-chip solid phase extraction–liquid electrode plasma atomic emission spectrometry for detection of trace lead. *Japanese Journal of Applied Physics* 53: 05FS01. https://doi.org/10.7567/JJAP.53.05FS01.

45 Kitano, A., Liduka, A., Yamamoto, T. et al. (2011). Highly sensitive elemental analysis for cd and Pb by liquid electrode plasma atomic emission spectrometry with quartz glass chip and sample flow. *Analytical Chemistry* 83: 9424–9430. https://doi.org/10.1021/ac2020646.

46 Commercialized by MICRO EMISSION (Japan). http://www.micro-emission.com/products/papers.html. (accessed 10 October 2020).

47 Weagent, S., Li, L., and Karanassios, V. (2011). *Rapid prototyping of hybrid, plastic-quartz 3D-chips for battery-operated microplasmas*, 209–226. InTech Publishing, Chapter 10. https://doi.org/10.5772/24994.

48 Shatford, R. and Karanassios, V. (2016). 3D printing in chemistry: past, present and future. *Proceedings of SPIE* 9855: 98550B–98560B. https://doi.org/10.1117/12.2224404.

49 Rue, W., Afzal, D., and Karanassios, V. (2020). An overview of 3D printing in chemistry. *Proceedings of SPIE* 11390: 1139016. https://doi.org/10.1117/12.2559157.

12

Portable Electro-Optical-Infrared Spectroscopic Sensors for Standoff Detection of Chemical Leaks and Threats

Hugo Lavoie[1], *Jean-Marc Thériault*[1], *Eldon Puckrin*[1], *Richard L. Lachance*[2], *Alexandre Thibeault*[2], *Yotam Ariel*[2] *and Jean Albert*[2]

[1] *Defence R&D Canada – Valcartier Research Center, Quebec, QC, Canada*
[2] *Bluefield Technologies Inc., San Francisco, CA, USA*

12.1 Introduction

For several decades, the advancement of spectroscopic technology has yielded systems more rugged and smaller than ever. This advancement has permitted the technology to leave the traditional laboratory setting and become usable in the field. Infrared (IR) spectroscopy and its corresponding technology that includes electric cryocoolers, rugged spectrometers or modulators and optical coatings contribute to the more prevalent use of the technology in the field. The development of detection, identification, and monitoring sensors has provided options for first responders, the military and environmental agencies to improve their threat-detection capabilities. Many materials of chemical interest (gas, liquid, or solid), such as greenhouse gases, air pollutants, or ground contaminants, possess good and discriminative signatures in the IR region. The present chapter will provide an overview of four IR spectroscopic techniques developed for the standoff detection and monitoring of various chemicals of interest. Three Fourier transform spectrometer (FTS)-based systems and one gas-filter correlation radiometric sensor will be discussed. The popular use of these technologies is on the near horizon.

12.2 A Differential FTIR Approach for Standoff Gas Detection

12.2.1 Background

Fourier transform infrared (FTIR) spectroscopy is a well-established technique for the passive remote monitoring of chemical vapors. There are several applications for which this technique is particularly well-suited, such as the remote monitoring of air pollutants and greenhouse gases, as well as the passive standoff detection and identification of toxic industrial compounds (TICs) and chemical warfare agents (CWAs). In this context, DRDC (Defence R&D Canada) Valcartier has been developing and optimizing a sensor technology for differential detection. This optimized configuration exploits the optical subtraction capabilities of a double-beam Michelson interferometer which provides an efficient means for the real-time suppression of the spectral background, which overlays the desired spectral signature, and the suppression of the internal IR emission contributions arising from the spectrometer itself. These subtraction attributes represent a definite advantage for portable spectroscopy. A review of the differential detection concept and its validation steps based on the double-beam Michelson FTIR spectrometer is given below.

Portable Spectroscopy and Spectrometry 1: Technologies and Instrumentation, First Edition.
Edited by Richard A. Crocombe, Pauline E. Leary and Brooke W. Kammrath.

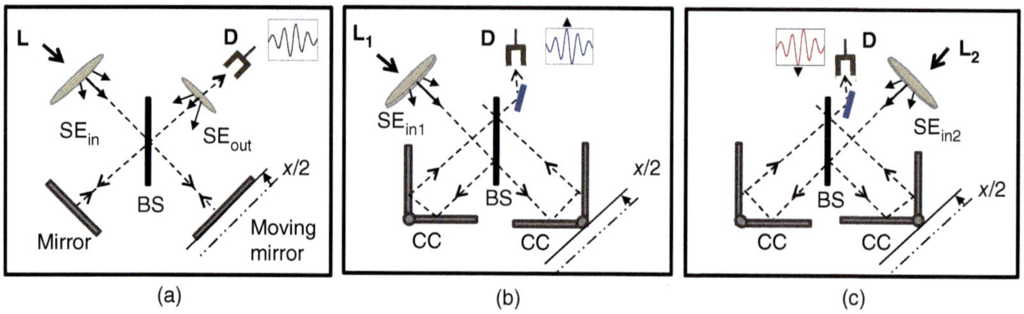

Figure 12.1 Schematic views of a single-input port FTIR Michelson interferometer (a) and a double-input port FTIR Michelson interferometer (b and c) with their associated self-emission contributions; where *cc*, *SE*, and *BS* stand for corner-cube reflector, self-emission, and beamsplitter, respectively.

12.2.2 FTS Background

The basic principle of an FTS is that a Michelson interferometer modulates the incident beam intensity according to the wavelength of the radiation. As shown in Figure 12.1a, the beam from the source is divided by a beamsplitter (BS) into two separate components which are recombined after traveling different optical path lengths. In an FTS, the Michelson interferometer is designed in such a way that one or both of its mirrors can move periodically around a mean position creating a periodic optical path difference between the two interfering beams. In this case, when a monochromatic source of unit amplitude is incident on the interferometer, the emergent amplitude $A(x)$ is given by

$$A(x) = (rt) + (tr)e^{i2\pi\nu x} \tag{12.1}$$

where ν is the wavenumber (inverse of the wavelength, $1/\lambda$), x is the optical path difference between the two interfering beams, and r and t represent the amplitude reflection and transmission coefficients of the beamsplitter, respectively. By definition, the emerging intensity $F(x)$ is found by the multiplication of the amplitude $A(x)$ with its complex conjugate $A(x)^*$ yielding

$$F(x) = A(x)A(x)^* = 2RT + RT(e^{i2\pi\nu x} + e^{-i2\pi\nu x}) \tag{12.2}$$

where the beamsplitter reflection and transmission in intensity, R and T are given by $(r \times r^*)$ and $(t \times t^*)$, respectively. The interferogram $I(x)$ is defined as the modulated component of the above intensity where

$$I(x) = RT(e^{i2\pi\nu x} + e^{-i2\pi\nu x}) \tag{12.3}$$

and RT acts as the optical responsivity for this simplified design. Equation (12.3) represents the interferogram obtained from a monochromatic source passing through an ideal interferometer. When a polychromatic beam of spectral intensity $S(\nu)$ (raw spectrum) propagates inside a Fourier transform (FT) interferometer, it generates an interferogram that can be represented by the continuous summation of individual modulations associated with each spectral element, i.e.

$$I(x) = \int_0^{\infty} S(\nu)(e^{i2\pi\nu x} + e^{-i2\pi\nu x})\, d\nu \tag{12.4}$$

Using a mathematical workaround (it does not affect the physical representation) which consists in extending the domain of integration to include the negative values of ν and assuming a symmetric spectrum such that $S(-\nu) = S(\nu)$, we find, after rearrangement, an interferogram expression more appropriate for processing. In this case and by application of the basic FT property, we have

$$I(x) = \int_{-\infty}^{\infty} S(\nu)e^{i2\pi\nu x}\, d\nu \tag{12.5}$$

$$S(\nu) = \int_{-\infty}^{\infty} I(x)\, e^{-i2\pi\nu x} d\nu \tag{12.6}$$

Equation (12.5) states that the interferogram generated by a Michelson interferometer is given by the inverse FT of the raw spectrum $S(\nu)$. Conversely, the raw spectrum $S(\nu)$ is given by the FT of the interferogram (Eq. 12.6), which represents the foundational result of Fourier spectrometry.

12.2.3 Standard FTIR Spectrometer

The schematic of the instrument shown in Figure 12.1a corresponds to a standard single-input beam Michelson FTIR interferometer. For a real instrument of this type, the output signal is actually composed of radiation from different origins. An important part of this radiation comes from the external source, L, under study. The remaining contributions are generated by the instrument itself. This parasite radiation is due to gray body emissions and stray light reflections from the interferometer components (lenses, mirrors, inner walls, and others). It is referred to as the self-emission of the instrument. Figure 12.1a identifies the various sources of radiation intervening in a single measurement: L is the spectral radiance of the external source (target source), $\mathbf{SE}_{\text{in}}$ and $\mathbf{SE}_{\text{out}}$ represent the self-emissions (raw spectra) generated by the input and the output optics, respectively, and $\mathbf{SE}_{\text{BS}}$ represents the beamsplitter self-emission (raw spectrum). In the IR, the self-emission terms mainly depend on the temperature of the intervening components. It is quite significant especially for an instrument operated at or near room temperature. Taken separately, each of these radiation sources generates a specific interferogram and, consequently, a corresponding raw spectrum. The sum of these individual raw spectra is given by

$$\mathbf{S} = (\mathbf{K}\, L) + \mathbf{SE}_{\text{in}} + \mathbf{SE}_{\text{out}} + \mathbf{SE}_{\text{BS}} \tag{12.7}$$

Equation (12.7) represents the usual way to link the complex raw spectrum $\mathbf{S}$ to the target spectrum L (the one to be evaluated) taking into account the instrument characteristics which are the complex spectral responsivity $\mathbf{K}$ and the complex spectral offset. The complex spectral offset of the instrument is a rather complicated expression involving the three different self-emission (**SE**) terms that are theoretically difficult to evaluate particularly for this type of instrument (single-beam). Usually, the self-emission terms are not stable in time due to thermal variations of instrument components. For this reason, the best way to acquire radiometrically accurate target spectra is by doing frequent calibration measurements. Ideally, this calibration is performed concurrently with each source measurement. The current method used for the accurate calibration (radiometric) of interferometric spectra is known as the two-temperature calibration method. In this method, two reference blackbodies of known radiances are required to solve the two unknowns, i.e. the spectral responsivity and the spectral offset of the instrument. Often for FTIR work in the long wave infrared (LWIR), the reference blackbodies are set to be close to the environmental temperature conditions; for instance, the hot reference blackbody is kept at a constant temperature near 60 °C, and the second one is at an ambient temperature near 20 °C.

The above discussion illustrates the usual attributes encountered with most standard FTIR instruments, i.e. the presence of variable and uncontrolled self-emissions, which necessitates three consecutive measurements (target and two references) for obtaining a single calibrated spectrum L. This represents a burden, especially for the real-time applications. In order to make the instrument more stable radiometrically and, consequently, more simple to calibrate, we have proposed to exploit and optimize certain attributes of the double-input beam interferometer design. As seen below, this type of interferometer is also suited for developing an FTIR spectrometer adapted for differential detection.

12.2.4 Double-Input Beam Interferometer

Among the few number of double-beam interferometers, the modified Michelson with corner-cube (CC) retroreflectors represents a relevant design from the point of view of its optimization for self-emission suppression.

Figure 12.1b and c give a schematic view of the double-beam Michelson interferometer. Note that both light sources L_1 and L_2 are actually installed in the same double-input-port interferometer. The use of two schematics (Figure 12.1b and c) is intended to simplify the ray tracing associated with each light source. The use of CC reflectors rather than standard flat plate mirrors has two main advantages for our application. First, the reflected beams from corner reflectors are offset laterally from the incident beams. This has the great advantage of decoupling input and output channels, as opposed to the standard single-beam interferometer where incident and reflected beams are superimposed. This attribute makes it possible to adjust independently the self-emission of each of the input ports without affecting in any way the output optics, which is not possible with the single-beam configuration. With this type of interferometer, the two beams from the two inputs are optically combined such that the two self-emission terms from each input port fully cancel out by optical subtraction, i.e. when the two self-emission terms are equal in intensity and 180° out of phase leading to a theoretically perfect cancellation.

The following equations serve to establish the expected attributes of the optimized instrument. A more complete analysis can be found in [1]. For an FTIR spectrometer based on a double-input port interferometer, the total signal **S** represents the sum of two individual complex raw spectra, $\mathbf{E}_1$ and $\mathbf{E}_2$, associated with each input. If input-1 and input-2 aim at target sources of radiance L_1 and L_2 (see Figure 12.1b and c), respectively, the resulting raw spectrum is given by

$$\mathbf{S} = \mathbf{E}_1 + \mathbf{E}_2 + \mathbf{SE}_{\mathrm{BS}} \tag{12.8}$$

with

$$\mathbf{E}_1 = \mathbf{K}_1(L_1 + \mathbf{SE}_{\mathrm{in1}}) \tag{12.9}$$

$$\mathbf{E}_2 = \mathbf{K}_2(L_2 + \mathbf{SE}_{\mathrm{in2}}) \tag{12.10}$$

where $\mathbf{K}_1$, $\mathbf{K}_2$ and $\mathbf{SE}_{\mathrm{in1}}$, $\mathbf{SE}_{\mathrm{in2}}$ are defined as the responsivities and the self-emissions associated with input-1 and input-2, respectively, and $\mathbf{SE}_{\mathrm{BS}}$ represents the self-emission from the beamsplitter.The total self-emission contributions can be minimized to zero by the fulfillment of the three following conditions:

$$\text{Balanced responsivities}: \mathbf{K}_1 = \mathbf{K}_2(\mathrm{e}^{i\pi}) = -\mathbf{K}_2 \tag{12.11}$$

$$\text{Balanced self} - \text{emissions}: \mathbf{SE}_{\mathrm{in1}} = \mathbf{SE}_{\mathrm{in2}} \tag{12.12}$$

$$\text{Beamsplitter transparency}: \mathbf{SE}_{\mathrm{BS}} = 0 \tag{12.13}$$

By application of these three conditions, the resulting raw spectrum derived from Eqs. (12.8)–(12.10) reduces to

$$\mathbf{S} = \mathbf{K}_1\,(L_1 - L_2) \tag{12.14}$$

And, consequently, the differential radiance, $L_1 - L_{2,}$ which represents the physical parameter of interest, is given by

$$\delta L = (L_1 - L_2) = \mathbf{S}/\mathbf{K}_1 \tag{12.15}$$

Equations (12.11)–(12.13) define in a mathematical sense the three conditions for an optimized FTIR spectrometer which does not contain any self-emission terms from the instrument itself. The resulting raw spectrum **S** is directly proportional to the difference between the two source radiances $L_1 - L_2$.

12.2.5 CATSI Prototype

DRDC Valcartier has developed a passive FTIR sensor for the differential detection and identification of chemical vapor clouds at long standoff ranges [2]. The sensor is based on a double-input beam FTIR interferometer optimized for self-emission cancellation. This proof-of-concept sensor developed between 1995 and 2008 is referred to as the Compact Atmospheric Sounding Interferometer (CATSI). With this configuration, the two probed scenes are optically combined onto a single detector resulting in a real-time optical subtraction. This yields a cloud vapor spectrum minimally perturbed by the background radiation. More precisely, the CATSI sensor consists of two identical 4-inch diameter Newtonian telescopes optically coupled to the dual-beam interferometer. Each telescope can be independently rotated to the selected scene. Figure 12.2a and b summarize the optical design and show the instrument mounted on a tripod.

An important design effort was done to optically balance the two input ports to achieve the optimization conditions (Eqs. 12.11 and 12.12). The CATSI system allows measurements of spectra according to the following specifications: scene fields of view (FOVs) from 4 to 11 mrad, spectral coverage from 3 to 18 μm, and spectral resolution of 1 cm^{-1} or greater. A double pendulum-scanning mechanism controls the periodic displacement of the two CC reflectors that generates the interferogram. The beamsplitter, which satisfies the third optimization condition (Eq. 12.13), consists of a thin air gap ($\lambda/4$ at 7 μm) squeezed between two identical zinc selenide (ZnSe) plates having identical antireflection coatings on their external faces [1]. In practice, the beamsplitter assembly is made of two ZnSe plates separated by a thin circular ring made of hard material and having a thickness of approximately 2 μm. For most of the work done in standoff chemical detection, the CATSI interferometer was optimized for the LWIR with a modulation efficiency (4RT) of 0.7 at 10 μm and an mercury cadmium telluride (MCT) detector having a peak D^* of approximately 4×10^{10} Jones at 10 μm. With this instrument, two scenes coming from adjacent FOVs are optically combined at the detector level yielding the spectral residual, δL, of the scenes. In its usual configuration of operation, the FOV of CATSI is set to 10 mrad and the angular tilt between the two FOVs is 2° (34.9 mrad). For instance, at a standoff range of 1 km, the distance between the two probed regions is 35 m and the diameter of the region is 10 m. Usually, this angular separation (2°) is sufficient to ensure that the gas cloud occupies partially only one FOV, the other being gas-free. However, in some field tests, we observed that the angular separation was too small to avoid cross contaminations, but the concentration variations induced by turbulence always yield a concentration differential that is easily detected by the differential detection with CATSI.

12.2.6 Target Gas Detection

In our approach, the passive remote monitoring of chemical vapors is based on the differential radiance equation (Eq. 12.15). Remote gas detection is performed from the optical subtraction of the radiance spectra received in each port (Figure 12.2c). The differential radiance is defined as the radiance of the scene containing the gas (L_{gas}) minus the radiance of the adjacent same scene without the gas (L_{clear}). With this condition, it can be verified [3] that for horizontal path scenarios and for a vapor cloud filling the entire FOV of the receiver, the differential radiance is simply given by

$$\delta L_{calc} = L_{gas} - L_{clear} = \Delta L_{clear} + (1 - \tau_{gas})(B_{air} - F_{far})\tau_{near} \tag{12.16}$$

Equation (12.16) gives the calculated differential radiance (δL_{calc}) of a vapor cloud with a provision for handling a possible background variation (ΔL_{clear}) between the two adjacent scenes. τ_{near} is the atmospheric transmittance between the gas and the sensor. The radiation source term ($B_{air} - F_{far}$) represents the radiative contrast between the Planck radiance of the gas assumed to be at the same temperature as the air (B_{air}) and the radiance of the background (F_{far}). The term ($1 - \tau_{gas}$) denotes the vapor cloud spectral emissivity where τ_{gas} is defined by

$$\tau_{gas} = \exp[-\alpha(\nu)\mathrm{CL}] \tag{12.17}$$

and $\alpha(\nu)$ is the spectrally dependent absorption coefficient (1/ppm-m) of the compound obtained from a spectral library. The column density (CL), the parameter to be estimated, is given here in ppm-m. A general procedure of radiometric monitoring adapted for differential detection with a double-input beam FTIR interferometer was developed and tested. To exploit the attributes of this differential detection method, a processing algorithm called

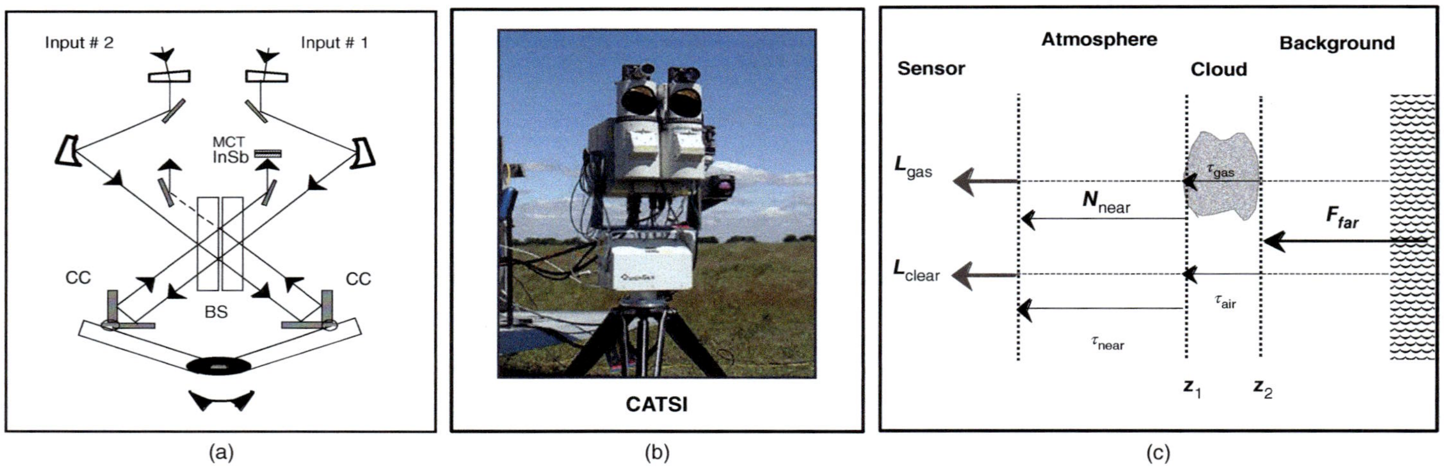

Figure 12.2 Schematic of CATSI optical design (a) and a picture of the instrument mounted on a tripod (b). (c) Schematic diagram and related terminology defining the three-layer geometry for differential detection.

GASEM (Gaseous Emission Monitoring) was developed. GASEM was built to control the interferometer acquisition and to perform the online fitting that yields in real time the CL of the probed chemical target vapor [3]. The greatest advantage of this differential detection approach is to provide, in the field, a spectrally relatively clean signature of the remote chemical plume which facilitates its processing in real time. Figure 12.3 gives typical examples of the standoff detection and monitoring with the CATSI-GASEM capability. Figure 12.3a shows the measured and the GASEM best-fit spectra for a gas mixture of DMMP-SF6 (dimethyl methylphosphonate–sulfur hexafluoride) released from a well-controlled stack situated at a distance of 1.5 km from the sensor. For this case, the best-fit spectrum was obtained using the CL reference estimates of 92 ppm-m for DMMP and 56 ppm-m for SF_6 as inputs for the GASEM model with T_{FIT} as the fitting parameter. In this case, the fitted plume temperature

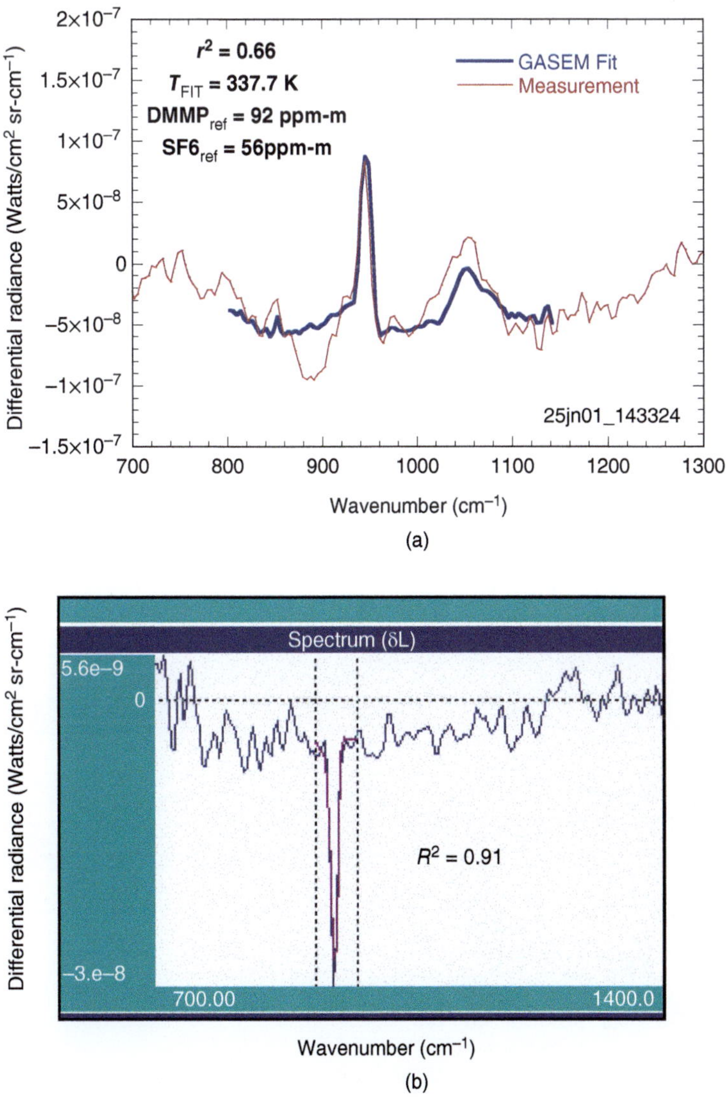

Figure 12.3 Examples of monitoring results: (a) GASEM output showing the result of the best-fit DMMP-SF6 spectrum (blue curve) with the measured differential radiance spectrum (red curve) at a standoff distance of 1.5 km; (b) GASEM result for the best-fit SF6 spectrum (red curve) at a standoff distance of 5.7 km.

was determined to be 337.7 K (more details in ref. [3]). Figure 12.3b shows results for the standoff detection of SF_6 vapor released at a distance of 5.7 km from the CATSI sensor. For this case, the SF_6 was released from a tank connected to a 3 m aluminum pipe having a series of 5 mm holes linearly distributed along the pipe. The pipe

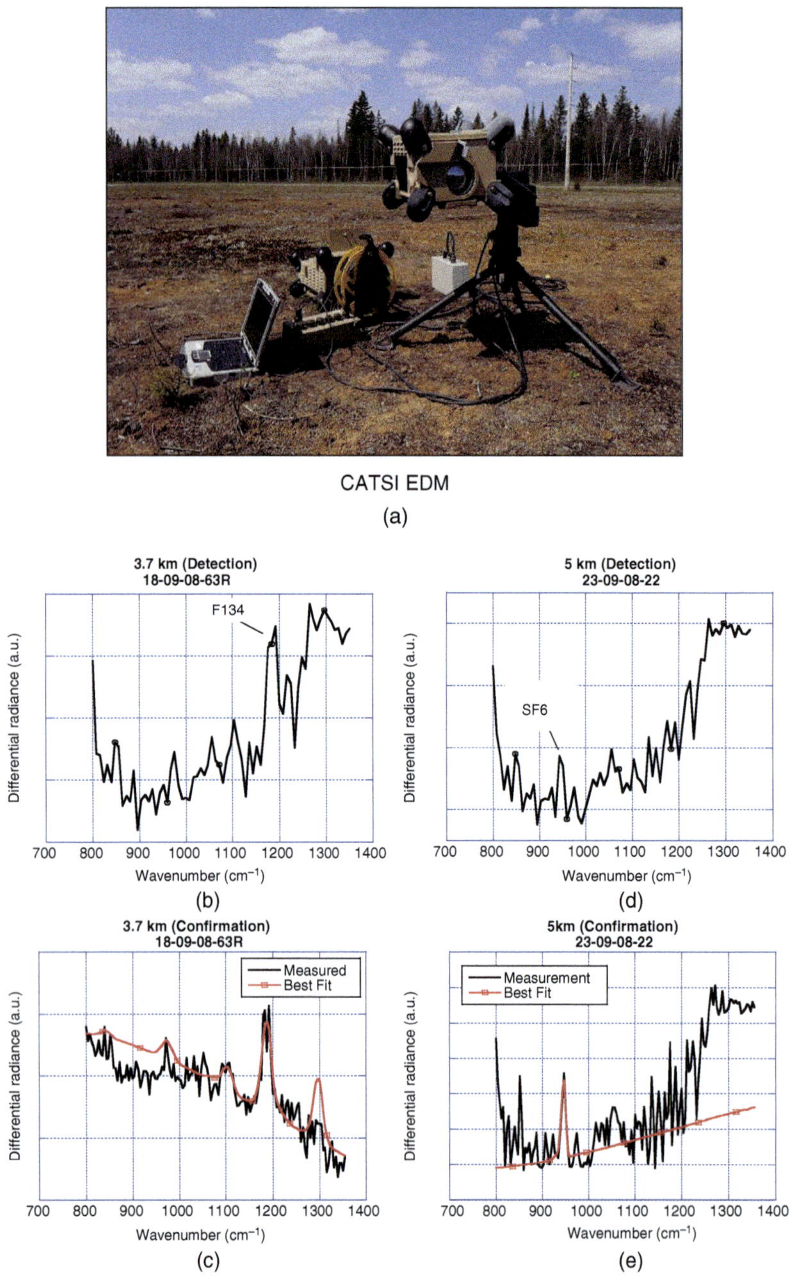

Figure 12.4 (a) Picture of the CATSI EDM with its hardware; (b) Example of spectral radiance measurement at 3.7 km for a release of F134 recorded in the detection mode, and (c) in the confirmation mode; (d) Example of spectral radiance measurement at 5 km for a release of SF6 recorded in the detection mode, and (e) in the confirmation mode.

was mounted vertically on the roof of the source-site building. The result of Figure 12.3b illustrates an example of the differential radiance measured (blue curve) with CATSI together with the GASEM best-fit calculation (red curve). In this case, the coefficient of determination (r^2) between the measured and the best-fit spectra is better than 0.9 indicating a very good agreement between the measured and modeled SF_6 absorption bands. More results are available elsewhere [4, 5].

From 2004 to 2010, DRDC Valcartier also completed the development and validation of a ruggedized sensor for the detection and identification of gaseous CWAs and TICs. It is referred to as CATSI EDM (Compact Atmospheric Sounding Interferometer Engineering Model). The CATSI EDM sensor uses a well-balanced passive IR double-beam FTS with two adjacent FOVs to optically suppress the instrument self-emission and background spectrum [6]. As opposed to the original CATSI, the EDM uses only one telescope (see Figure 12.4a) to perform the differential sensing.

In this case, the two adjacent FOVs are defined by two fixed circular pupils located in the focal plane of the telescope which generate two scenes having an angular separation of 2° and where each scene (beam) is sent to its dedicated input port of the double beam interferometer. When combined with the Chemical Agent Spectral SIgnature Detection and Identification (CASSIDI, [6]) algorithm, CATSI EDM provides autonomous real-time detection and identification of CWAs and TICs at distances up to 5 km. CASSIDI performs three distinct functions: detection, confirmation, and alarm. Figure 12.4 presents results for two typical cases recorded during the performance field tests for standoff detection at 3.7 km (b, c) and at 5 km (d, e). In both cases, CATSI EDM autonomously detects, identifies, and alarms the actual target gas. Figure 12.4b shows the differential spectral radiance (resolution of $16\,cm^{-1}$) recorded in the searching mode when the system detects the spectral signature of the target gas, F134 (1,1',2,2'-tetrafluoroethane). At this point, the gas location is recorded and the current searching pattern of the sensor is automatically changed to the confirmation mode. In the confirmation mode, the spectral resolution of the sensor is changed to $8\,cm^{-1}$ and the sensor concentrates its scanning and acquisitions to regions immediately surrounding the point of detection. This mode allows the collection of radiance spectra with sufficient signal-to-noise ratio (SNR) and spectral resolution to perform a spectral best fit that yields an estimate of the CL of the detected gas. This estimate is then used to make a decision to declare an alarm if the CL is above a reference threshold. Figure 12.4c presents the differential radiance measurement with the corresponding best fit obtained in the confirmation mode for the F134 gas. Figure 12.4d and e present the same type of results, but for the detection, confirmation, and alarm of SF6 recorded during the 5 km segment of the performance field tests. In summary, over the 203 valid runs performed during the two weeks of field test evaluations, the CATSI EDM sensor successfully performed at both distances. In both cases, it exceeded the test performance requirement of a probability of detection of 80% with a 95% confidence level.

12.3 iCATSI Sensor

12.3.1 Background

The iCATSI (Figure 12.5a) is an adaptation from the ABB MRi imaging spectroradiometer that includes a symmetrical dual input–dual output optical system and an MCT photodiode linear array (16 by 1 pixel) [7–9]. As in the case for the original CATSI, the optimized configuration of iCATSI also exploits the optical subtraction capabilities of a double-beam Michelson interferometer, but with some major design improvements. iCATSI is a completely modular and configurable instrument where the input optics (reflective or refractive telescope), and the two detector modules (one per output) can be easily changed. The iCATSI was developed over the period from 2008 to 2010 and is still used as an R&D sensor today.

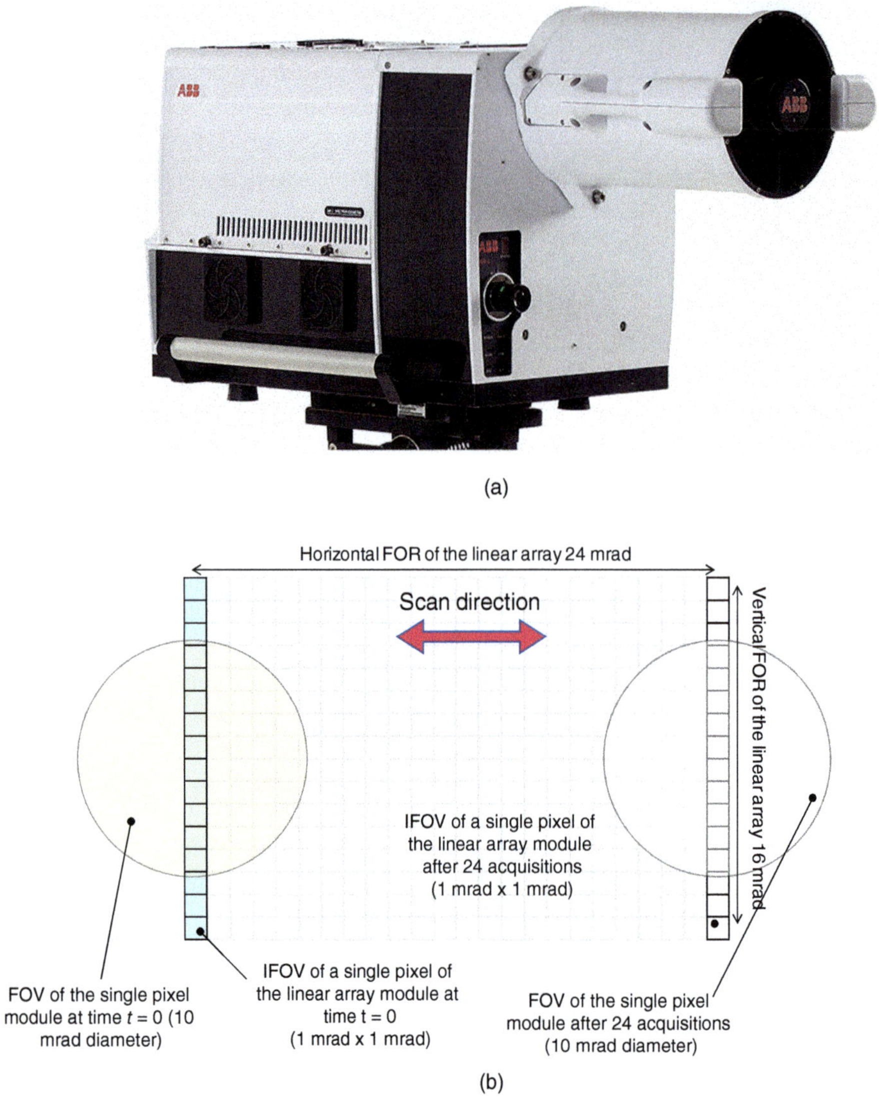

Figure 12.5 Picture of the iCATSI sensor with both mono- and multipixel (right) detector and the scanning telescope (a). Field of regard (FOR) and field of view (FOV) of iCATSI and the scanning strategy (b).

12.3.2 System Configuration and Design

The iCATSI interferometer output ports are both configured to focus the beam onto their respective detectors. Stirling cycle coolers cool both detectors. The signals from the detectors are amplified and digitalized by electronics placed close to the detectors. The digital signals are then sent through the control electronics to the remote computer. There is a single telescope common to both input ports. At the output of the telescope, a reflecting prism splits the FOV of the scene into two parts. Each part feeds one input port of the interferometer. The angular separation between both input ports is set to two degrees. The spectrometer of the iCATSI is built around a double pendulum Michelson interferometer with CC reflectors [7–9]. Both CCs are mounted on the two tips of a V-shaped scan arm which rotates back and forth around a ZnSe beamsplitter. The motion of the scan arm is monitored in real time by measuring the interference pattern of a helium-neon (He-Ne) laser that travels into the interferometer. The optical path difference (OPD) of the interferometer is approximately 1 cm. With this OPD, the highest spectral resolution achieved with a Boxcar-truncated spectral resolution (full width at half maximum [FWHM]) is approximately $0.8\,cm^{-1}$ at 8 μm and $1.0\,cm^{-1}$ at 3 μm.

The instrument is field deployable and can be considered as weatherproof. Thermal load, heat evacuation, and protection against the direct Sun and light rain are all implemented. The principal iCATSI configuration uses two MCT photodetectors, i.e. a monopixel detector with 10 mrad FOV and a 16-pixel linear array with 1 mrad/pixel FOV. The system generates a 16×24 mrad field of regard (FOR) image of the scene using a scanning mechanism in the telescope [7–9]. Figure 12.5b summarizes the scanning strategy of iCATSI.

12.3.3 Results with the iCATSI

Detection tests were performed with iCATSI at a 500 m range. Initial tests were performed with a Freon-type refrigerant (F-152a) released from a laboratory fume hood. Figure 12.6 shows a sequence of screenshots recorded with iCATSI for the real-time detection and identification of the gas plume release. The colored pixels represent the detection score processed in real time with the gas detection algorithm of iCATSI. Six images are shown at various times over a period of one minute in which 50 l of gas were released. Each screenshot shows the visible image on which are superimposed two IR images (mosaic) resulting from the differential mode FOR of iCATSI. The system provides the same algorithm output for both FORs. Even with a low spatial resolution (1 mrad/pixel), the evolution and the release origin of the plume are easily identified. The color code goes from light blue to red where red shows the higher detection ratio. The iCATSI also has been validated against several other threats like acetone, methanol, and other TICs.

12.4 Active FTIR for Ground Contamination Detection

12.4.1 The Active iCATSI

The second application discussed in this chapter is the standoff detection, identification, and monitoring of ground contamination. The first responder and environment protection agencies need detection systems to rapidly find and segregate leaks and contaminated terrain for avoidance and decontamination purposes. In this respect, a detection system was proposed and built around an active compact FT interferometer. The system uses a graphite Globar heated to approximately 1000 °C as an IR source to irradiate the interrogated surface. The system was optimized for the LWIR band (8–12 μm) because many ground contaminants of interest have significant and discriminative signatures in this band.

12.4.2 System Configuration and Design

The system (Figure 12.7) was built with two optional configurations, one for active and the other for passive sensing. These two configurations are interchangeable by swapping the Globar source with the detector. For both configurations, the optical design is based on a Michelson double pendulum single-input and single-output port configuration with an optical velocity of 1 cm/s. The modulator provides a spectral resolution between 1 and $32\,cm^{-1}$ (FWHM) over an optimized spectral range of 8–13 μm.

The source illumination is optimized to produce a uniform illumination spot size of 10 cm at 5 m (20 mrad). The collection port has a slightly smaller spot size than the illumination at 5 m to avoid the collection of parasite radiation from the background. To maximize the overlap between the illumination and the collection spots, two 650 nm Picotronic laser diodes were installed on top of the telescope mount to help manually align the two beams. Finally,

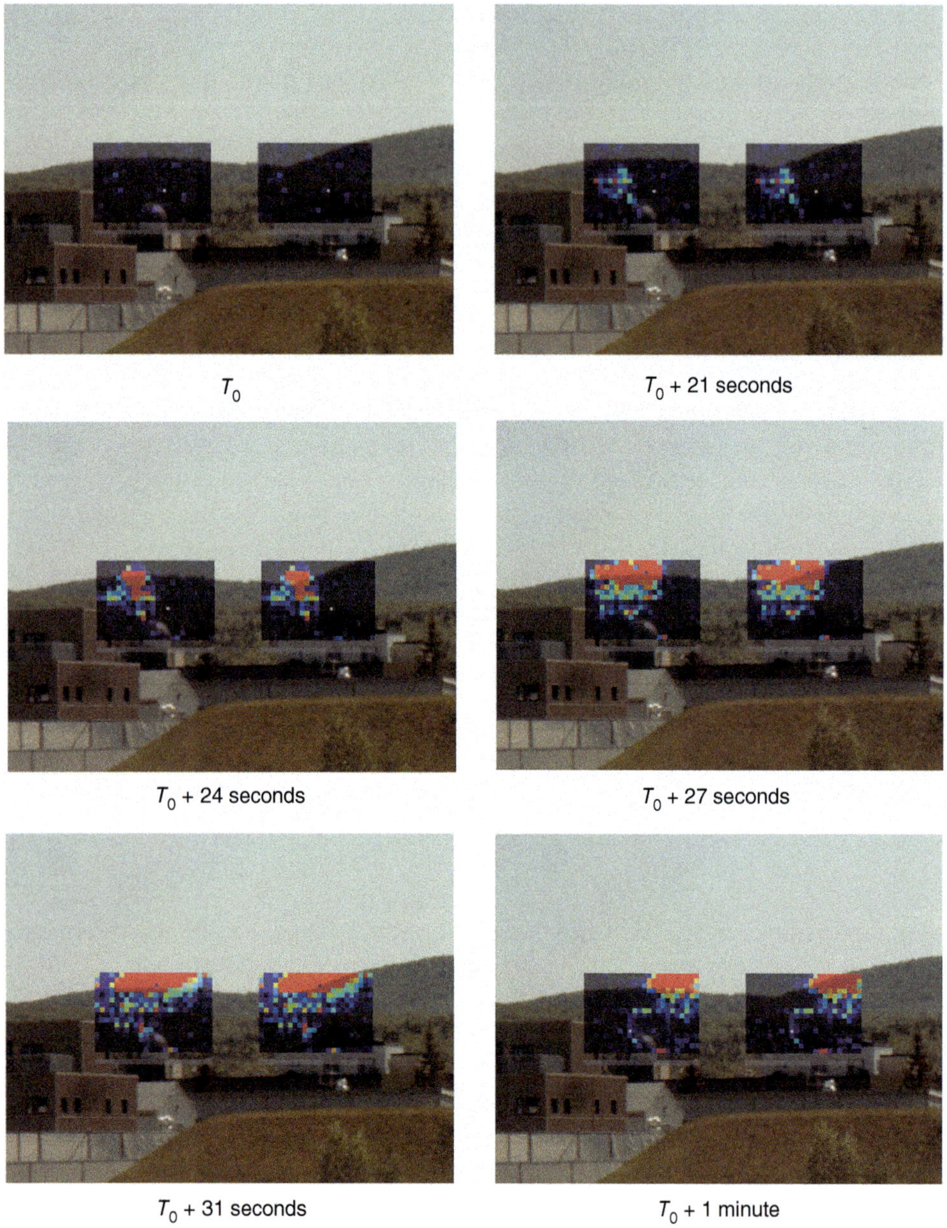

Figure 12.6 Evolution of a Freon gas plume over a period of one minute measured with iCATSI.

the system can be battery-powered with an autonomy of a few hours when used with a commercial-off-the-shelf (COTS) battery belt.

The innovation behind the system is the use of a modulated source to reduce and filter the effect of the background radiation. In this system, the FTS itself modulates the source prior to irradiating the surface. With this strategy, the collected background radiation is not modulated yielding a direct-current (DC) level at the detector that is rejected by the fast Fourier transform (FFT). This results in a real-time subtraction of the natural/environmental background emission radiance.

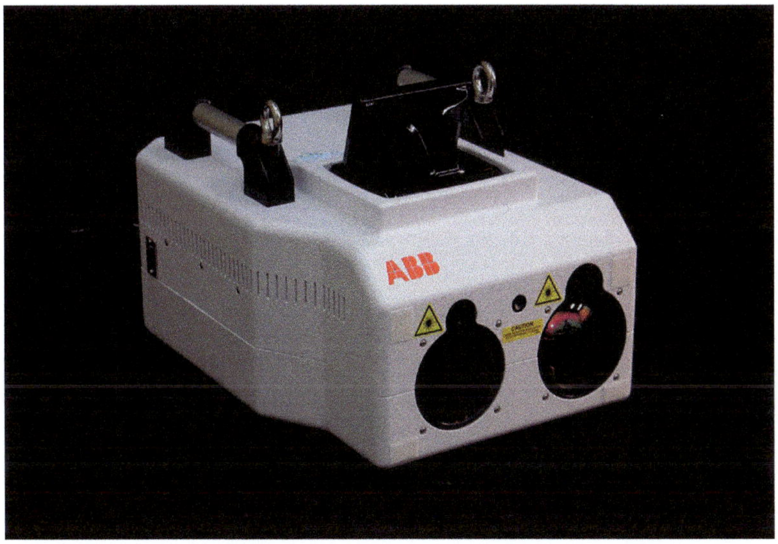

Figure 12.7 Picture of the active iCATSI from the front of the instrument.

12.4.3 Results with the Active iCATSI

Two main attributes can significantly affect the signal reflected from clean surfaces, i.e. the surface type and the surface reflection angle. As an example, Figures 12.8 and 12.9 show results for various surface types and the effect on the angle of incidence, respectively.

Comparison of the reflected signals from a Lambertian surface (Infragold), an aluminum mirror, a brushed aluminum panel, a piece of wood, and a piece of sandpaper demonstrate a large variability of reflected signal intensities. As expected, the brushed aluminum reflector and the aluminum mirror gave the highest reflected signal intensities, characterized by a fairly flat response between 800 and 1400 cm^{-1}. The Lambertian surface shows a reflected intensity with the same spectral characteristics, but two orders of magnitude lower than the flat mirror result. The wood and the sandpaper have a signal reflectance one order of magnitude lower than the Infragold with some spectral features. These results demonstrate the difficulties and constraints that the surface type will generate for the detection and identification of ground contaminants.

Moreover, the reflected signal will be affected by the angle of incidence and the bidirectional reflectance distribution function (BRDF) of the surface. Figure 12.9 shows the signal reflected from two clean surfaces, Infragold and aluminum mirror, as a function of the angle of incidence relative to the surface normal. These measurements show that the angle of incidence on the Infragold surface is less critical than on a pure reflective surface: the reflection on Infragold is reduced by an order of magnitude from 0 to 15° relative to the surface normal while the reflection on a flat mirror is dramatically reduced by four orders of magnitude for only a one-degree tilt from the surface normal.

The above results show how sensitive the sensor could be to misalignment and how the detection sensitivity can be affected by the surface type and the angle of incidence.

Figure 12.10 shows detection results for the sensor (active configuration) located 2 m away from a surface covered with SF96, a surrogate that acts as a liquid contaminant. The selected test surface is wood because it represents a difficult detection scenario due to its low reflectance. Figure 12.10 shows detection results of SF96 on wood at different concentrations (1 ml [0.0025 ml/cm^2], 5 ml [0.0125 ml/cm^2], and 10 ml [0.025 ml/cm^2]) and angles of incidence (0–15° relative to surface normal). The results of Figure 12.10a can be understood based on the

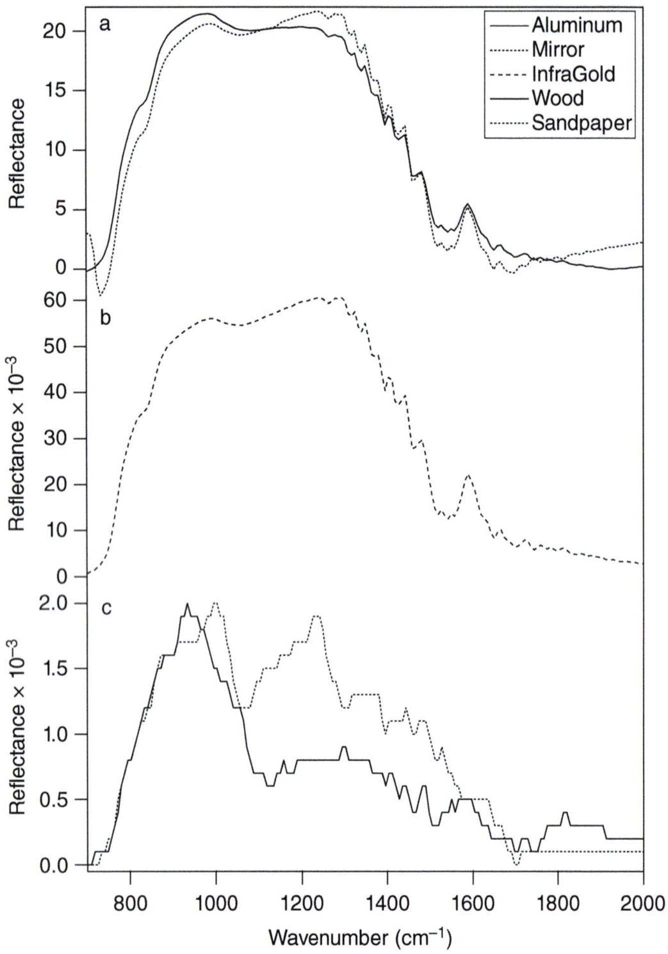

Figure 12.8 Reflective signal as a function of surface type, aluminum plate, and mirror (a), Infragold (b), wood and sandpaper (c).

reflection phenomenology associated with rough surfaces, such as clean or covered plywood. The reflection from a rough surface can be divided into two main components, i.e. the specular reflection component which is parallel to the surface normal and the diffuse component scattered off normal. A large portion of the reflected beam is lost by scattering because the collection angle of the sensor is too small and only the specular component is detected. Overall, the intensity of the specular reflection component increases with the SF96 coverage because the effective surface (a mixture of SF96 and plywood) appears smoother yielding a larger specular reflection component. The system shows detection sensitivities down to 1 ml spread on a $20 \times 20\,\text{cm}^2$ wood surface. Moreover, for Figure 12.10b, the system shows the detection of $0.025\,\text{ml/cm}^2$ of SF96 at an angle of incidence up to 15°. The SNR starts to become too low for detection around 15° and higher in this scenario.

In addition, DRDC Valcartier also developed a passive sensing technique for ground contamination using hyperspectral polarimetry. More information about this capability can be found in references [10, 11]. The system demonstrates a capability to detect and identify contamination on the ground and several surface types, such as metal, wood, and sandpaper at a standoff distance of up to 2 m. This technique could provide first responders with a new capability to identify and, hence, protect themselves against ground contamination.

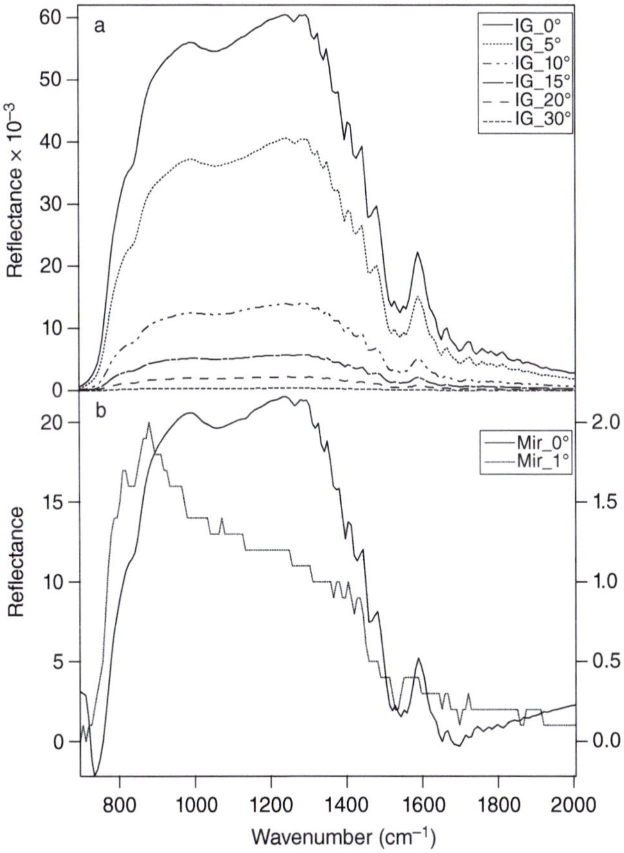

Figure 12.9 Reflective signal collected at different angles on two different surfaces: Infragold (a) and mirror (b).

12.5 Signature Collection: Broadband Portable Field Spectral Reflectometer

12.5.1 Background

Diffuse spectral reflectance measurements in the field are a common and important endeavor in the areas of mineral and geological analysis [12], crops and agriculture characterization [13], and ground-truthing for spectral remote sensing applications [14], which includes the development of spectral libraries used in remote sensing exploitation algorithms. To carry out these measurements, field-deployable reflectometers have been used in the solar-reflective region (visible, near-infrared [NIR] and shortwave infrared [SWIR]) for more than 20 years. However, relatively few field-portable sensors exist for measuring reflectance in the emissive part of the spectrum that includes the longwave and midwave infrared (MWIR) regions (3–14 μm). Even fewer, if any, sensors exist that are capable of measuring reflectance across the entire IR spectrum from the NIR to the LWIR (0.7–14 μm). One such technology, a full spectrum reflectometer (FSR), has recently been fabricated by ABB to address the requirement of providing highly sensitive measurements of diffuse reflectance from 0.7 to 13.5 μm with a lightweight portable instrument, which is battery-powered and simple to operate in the field.

12.5.2 Design Overview

The FSR is based on a FTIR spectrometer design in which an IR source is collimated and sent through a Michelson interferometer. The optical path difference between the arms of the interferometer varies in time so that the beam

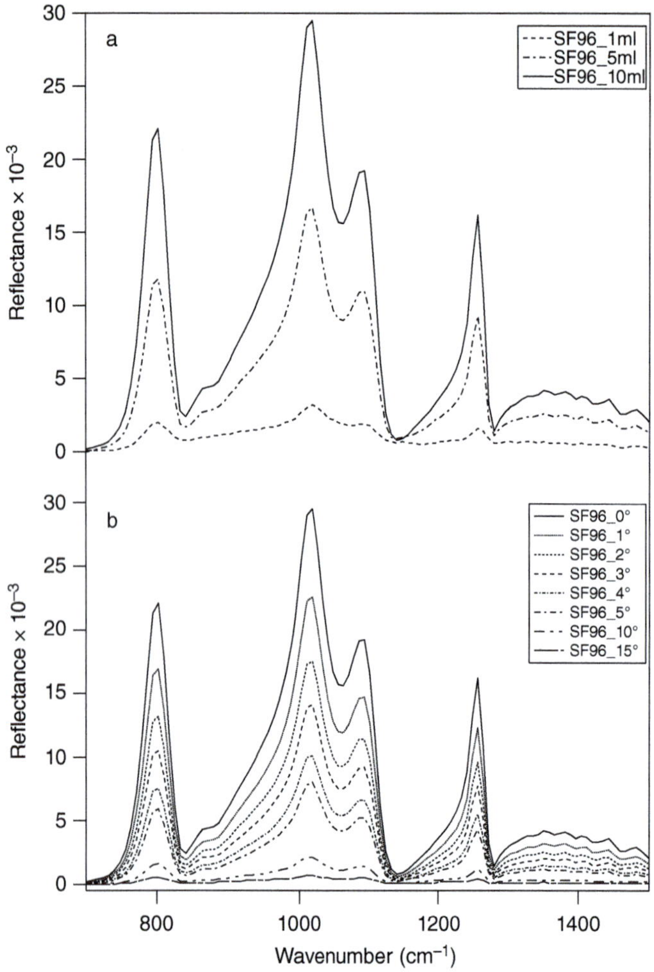

Figure 12.10 Reflective signal with contamination of SF96 on wood at different concentration (a) and at different angle of incidence (b).

at the output of the interferometer has an amplitude that is modulated in time, as described in detail in Section 12.2.2. The modulated output radiation, $I(x)$, is then focused on a target consisting of a sample or appropriate reference material. The radiation, $\rho I(x)$, reflected by the target is collected and focused on to the detector, which measures it as an interferogram of the spectral radiance reflected by the target, as described by Eq. (12.5). The interferogram is subsequently converted to a spectrum, $S(\nu)$, by a FFT, as given by Eq. (12.6). The IR radiation, L_{sample}, emitted by the target does not contribute to the interferogram since it is not modulated by the interferometer. This process is shown by the schematic diagram in Figure 12.11.

The diffuse spectral reflectance, $\rho(\nu)$, is simply given by the ratio of the spectrum measured for the sample and reference targets:

$$\rho(\nu) = R_{ref} S(\nu)_{sample} / S(\nu)_{reference} \tag{12.18}$$

where R_{ref} is the absolute reflectance of the reference target that accounts for its spectral features. If required, the absolute reflectance of the reference material can be determined in principle using an independent technique

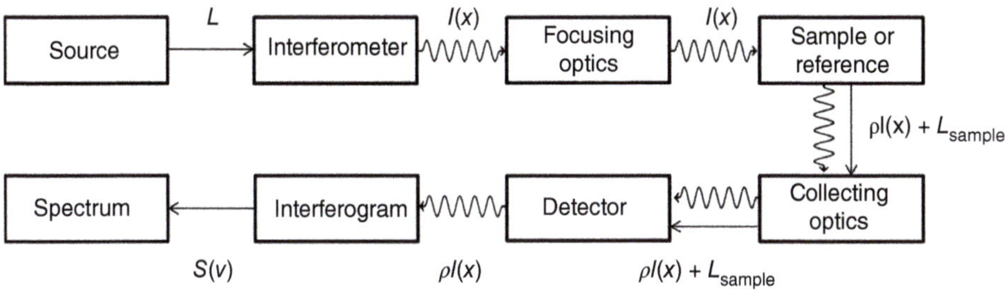

Figure 12.11 Schematic diagram of the radiation flow in a broadband spectral reflectometer.

such as that offered by the National Institute of Standards and Technology (NIST), or a NIST-traceable calibration facility.

In practice, Infragold and Spectralon reference targets are used for the 2–13.5 and 0.7–3.3 μm bands, respectively. Similarly, two IR sources are required to cover the large band; a silicon-carbide Globar for the 2–13.5 μm region, and a quartz–tungsten–halogen lamp for the 0.7–3.3 μm region. In addition, two detectors are required; a Stirling-cooled MCT detector for the 2–13.5 μm region, and a thermoelectrically cooled indium–arsenide detector for the 0.7–3.3 μm region, although there is a significant overlap of responsivity in the 2–3.3 μm region.

The Michelson-type interferometer module is rugged and compact, measuring 6 cm × 12 cm × 14 cm and weighing less than 900 g, and incorporates CC reflectors in each arm of the interferometer. The modulated beam can be focused on the reflectometer's input port for a measurement of the sample target, or switched by a movable mirror to measure the internal reference targets. The radiation reflected by the sample and reference targets is collected at an angle that minimizes the specular reflection component.

The reflectometer is portable, weighing less than 13 kg, and can function on standard alternating current (AC) (110 or 220 V) or a low-voltage (9–36 V) battery that can last for up to three hours.

12.5.3 Main Operating and Performance Parameters

The main operating and performance parameters of the reflectometer are listed in Table 12.1.

A measurement of the SNR for the two bands is shown in Figure 12.12. The measurement was obtained by averaging several individual spectra of the reference target (Infragold or Spectralon) and dividing the mean by the root-mean-square (RMS) noise of the measurements. An integration time of one minute was used at a spectral resolution of 16 cm^{-1}.

12.5.4 Example Measurements Obtained with the Broadband Portable Field Spectral Reflectometer

The FSR has been deployed in the field on several occasions since its development in 2010. An example includes reflectance measurements of soil samples of the former distant-early-warning (DEW) radar site at Cape Dyer, Nunavut. The DEW line radar sites were created in the Canadian Arctic during the 1950s to provide the first line of protection during the Cold War [15]. Since 1990, the 21 sites were no longer required and the task of cleaning up the sites became an important environmental issue. The possible types of contamination that may be present on these sites include metals, paints, fuel, batteries, and electrical equipment. Due to the difficulty in accessing these very remote sites, one effective way to monitor the contamination could be with an airborne remote-sensing platform. To verify if the contamination could be characterized through measurements of the spectral reflectance, the FSR was deployed to one of the former radar sites (Dye-Main) located on Cape Dyer on the eastern coast of Baffin Island.

Table 12.1 FSR operating parameters.

Characteristics	Values
Instrument type	Active reflectometer based on Michelson interferometer
Spectral resolution (cosine apodized)	4, 8, 16, and 32 cm^{-1}
Spectral range	Band 1 : 740–5 000 cm^{-1} (2–13.5 μm) Band 2 : 3 000–14 000 cm^{-1} (0.7–3.3 μm)
Spectral instability	Less than 1% over 15 minutes
Diameter of the sensed spot	4 mm at 30 cm from the interferometer
Signal-to-noise ratio (when looking at the built-in references, at 16 cm^{-1} resolution setting, for one minute of observation time)	Band 1: >10 000 (see Figure 12.12) Band 2: >500 (see Figure 12.12)
Mass	12.25 kg (excluding batteries)
Dimensions	30 cm × 35 cm × 43 cm
Operating temperature	Tested from 0 to 35 °C

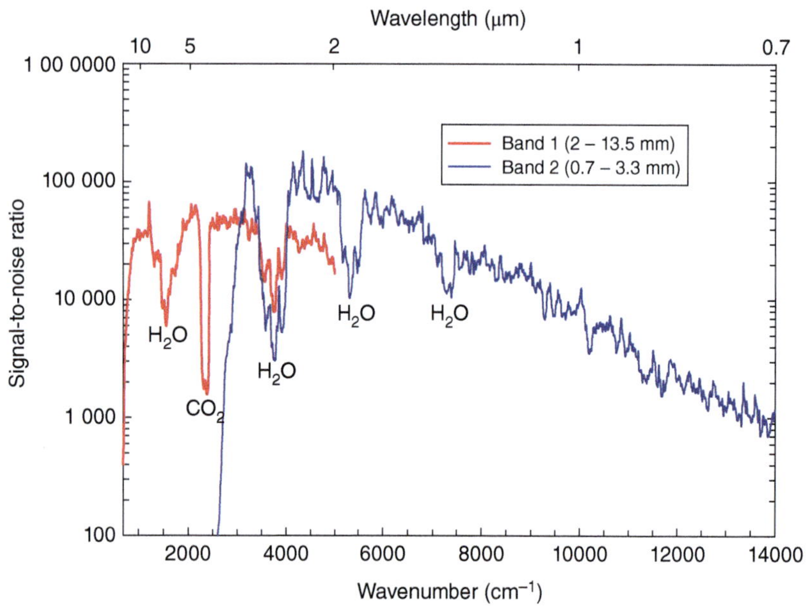

Figure 12.12 SNR of the FSR for the 2–13.5 and 0.7–3.3 μm bands. Absorption features associated with atmospheric water vapor and carbon dioxide within the instrument are present, as indicated in the graph (SNR measurements provided by ABB).

Some examples of reflectance measurements from areas contaminated with fuel and heavy metals are shown in Figures 12.13–12.15. Figure 12.13a shows a patch of ground, exposed for remediation, which smelled strongly of diesel fuel. The FSR was placed on the ground in various locations, as shown in Figure 12.13b, to obtain good-quality spectral reflectance measurements from 0.7–13.5 μm, as shown in Figure 12.13c. Soil samples sent to the Royal Military College in Ottawa for chemical analysis were also measured with the FSR. Figure 12.14 shows

(a)

(b)

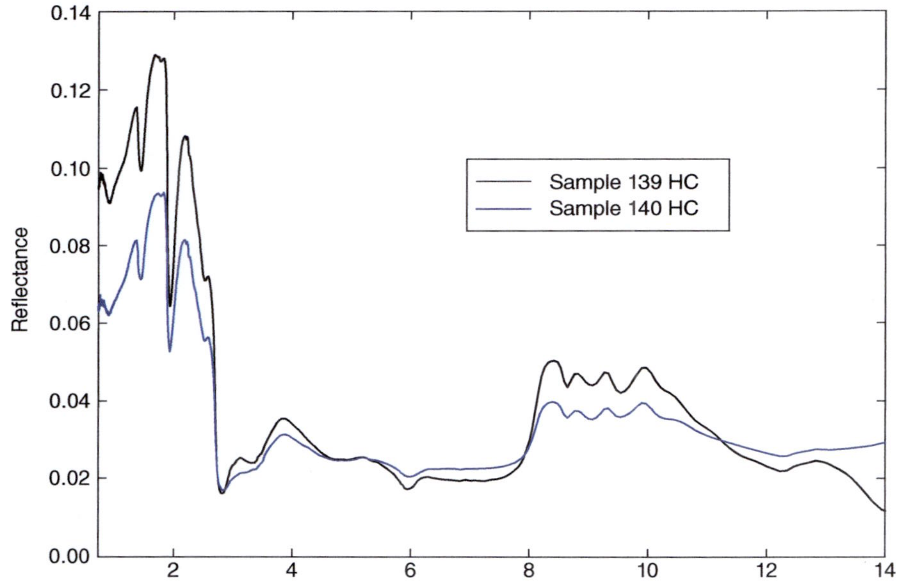

(c)

Figure 12.13 (a) Region of Dye-Main contaminated with fuel (hydrocarbons). (b) FSR positioned to measure soil reflectance. A computer is connected to make use of full software capability. (c) Spectral reflectance measurements made with the FSR.

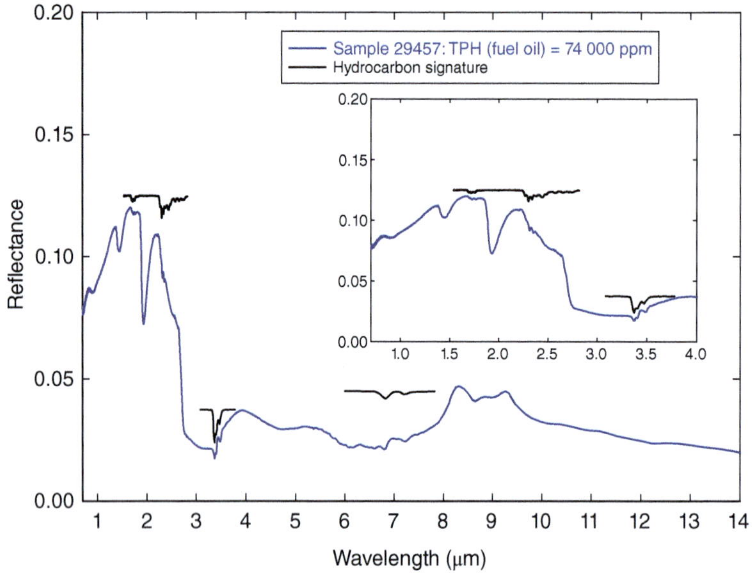

Figure 12.14 Reflectance of the soil sample from Dye-Main containing a high concentration of hydrocarbon. The spectral features associated with hydrocarbons are observed in the spectrum displayed in the inset view expanded over the 0.7–4 μm region. The reference hydrocarbon absorption features (black line) are derived from a reflectance spectrum of diesel fuel.

examples of the reflectance results from one sample containing 74 000 ppm of hydrocarbons. The spectral features associated with hydrocarbons are observed in the spectrum at 1.7, 2.4, 3.3, and 7 μm. Analysis of other soil samples showed that these features persisted for samples containing as little as 1000 ppm of hydrocarbons. Unfortunately, there were not any samples containing quantities of hydrocarbon in the 100 ppm range.

Figure 12.15a shows stains that were formed from the runoff from a pond that was contaminated with heavy metals from old equipment left on the site. Oxides of these metals formed from weathering processes may potentially be detected, thereby identifying the presence of the metal. The corresponding spectral measurements of these stains are shown in Figure 12.15b, which may be useful in identifying the contaminant.

12.6 Imaging Gas Filter Correlation Radiometry

12.6.1 Background

Gas Filter Correlation Radiometry (GFCR), a form of nondispersive IR spectrometry, has long been established as a simple, but efficient, means of detecting trace gas emissions, in terms of both sensitivity and selectivity levels. Taking advantage of the banded nature of gas-phase IR spectra, it is well-adapted to measuring target gas concentration in the section of the atmosphere located within the line of sight of an instrument focused on a given light source [16–18].

GFCR is an attractive technique for sensing applications. It combines the attributes of optical and mechanical simplicity, allowing the use of small, robust, low-powered and lightweight instruments with no moving parts, for improved durability and resistance to vibrations. GFCR detects target gases by using a bandpass filter to isolate measurements to a unique resonance band and comparing this with a gas cell that correlates to all of the absorption lines of a specific molecule within that band (Figure 12.16). The measured SNR is enhanced by the effect of the combined high spectral resolution (resolving power $\lambda/\Delta\lambda > 20\,000$) and increased sensitivity due to the large

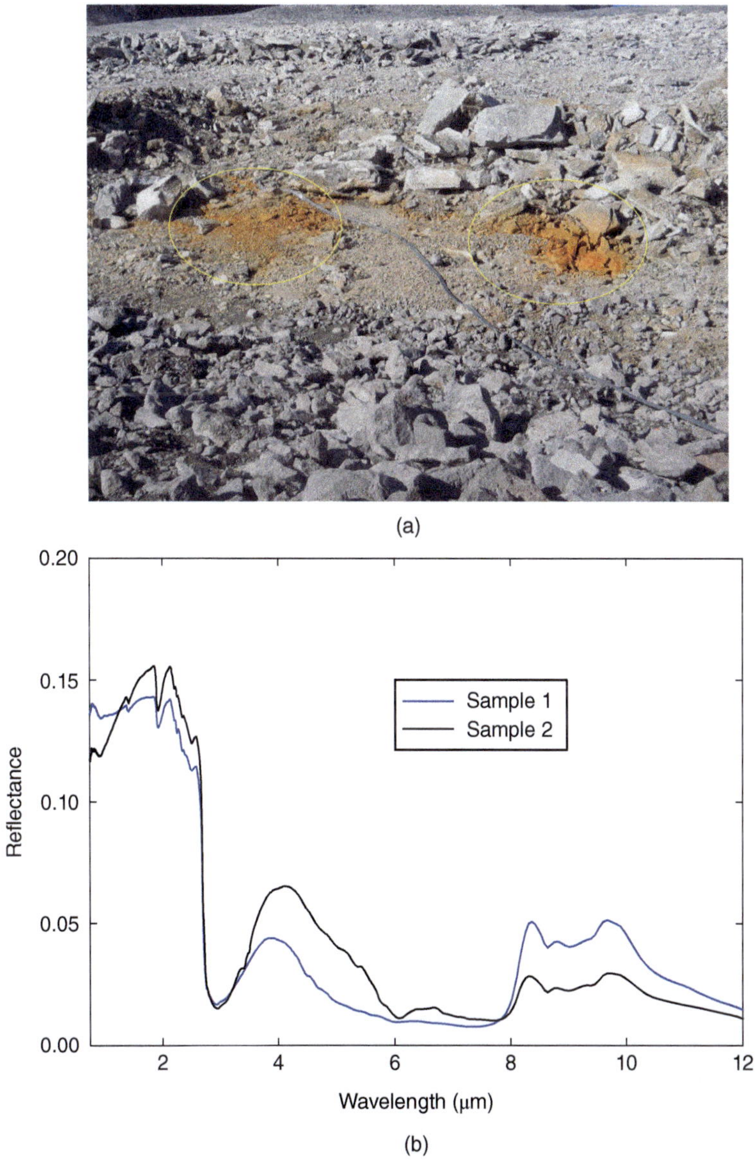

Figure 12.15 (a) Region of Dye-Main contaminated with heavy metals (stains encircled). (b) Spectral reflectance measurements of the stains made with the FSR may be useful in identifying the associated metal oxide.

optical throughput. The resulting enhancements in target gas selectivity and high rejection of contaminant gases and interferents compares favorably with conventional radiometers. Moreover, since recording, reconstruction and analysis of spectral information are done by optical correlation, the SNR is greatly improved and data analysis simplified; the electrical signal recorded at the focal plane arrays (FPAs) level is already proportional to the CL of the target gas in the FOV.

Figure 12.17 illustrates the gas correlation measurement principle, well-adapted to detect trace gas leaks from portable platforms.

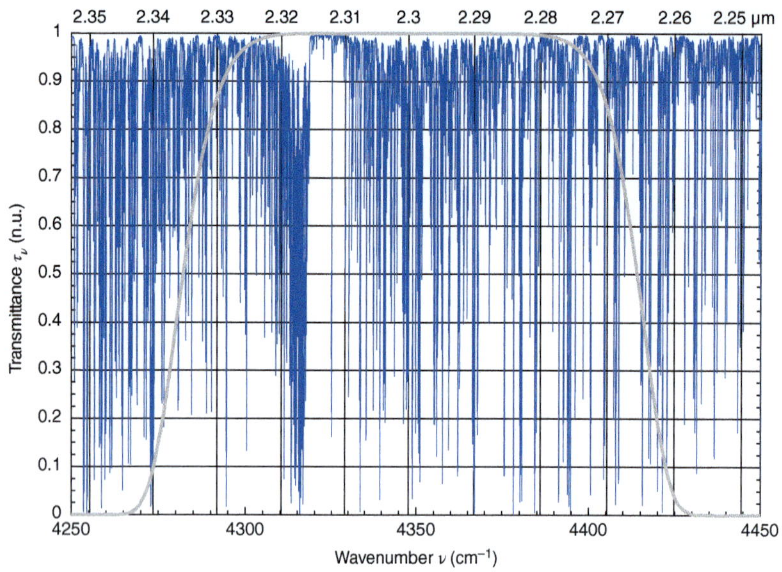

Figure 12.16 Methane lines (≈20 000) within the bandpass at 2.3 μm.

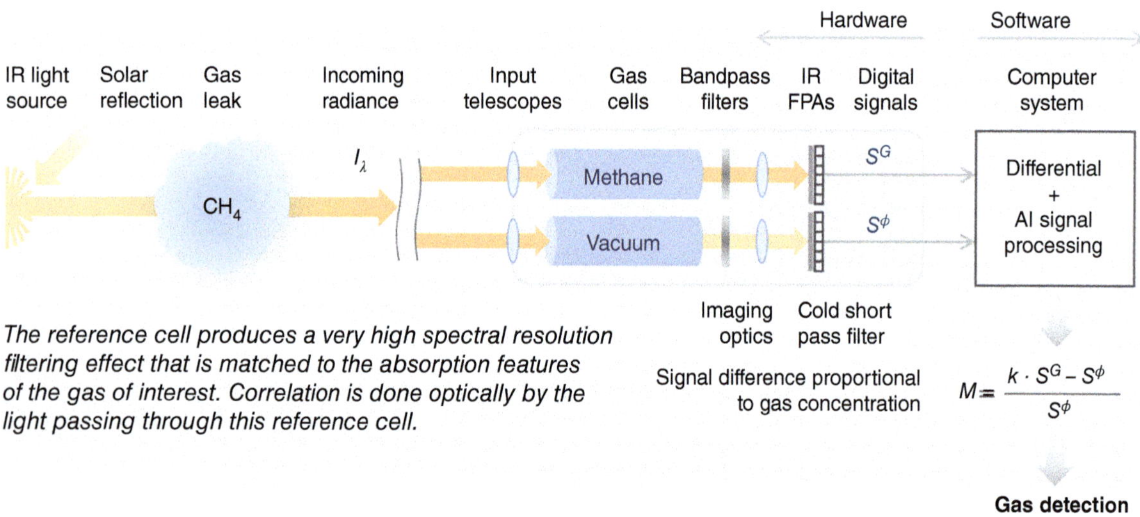

Figure 12.17 Gas Filter Correlation Radiometry (GFCR) general operation concept (imaging synchronized views' implementation).

Using an advanced GFCR-imaging version, Bluefield's sensor identifies gases, visualizes quantity distribution over terrain maps and estimates concentrations in two-dimensional images. Its synchronized-view imaging system can locate and quantify specific emitters to an unprecedented sensitivity level and spatial resolution.

The synchronized view implementation (where the two FPA image acquisitions are triggered simultaneously) allows the readings to be correlated temporally *and* spatially, where every snapshot contains the complete spectral content for every scene location, making the GFCR differential image proportional to the absorption of the target gas only, nulling correlated noise (to first order) in the two images. This eliminates problems induced by scanning or modulation techniques, minimizing influence by geospatial variations in the background surface albedo or any other variations in the scattering characteristics.

12.6.2 Advanced Image Analysis and Artificial Intelligence Techniques

To fully leverage the GFCR method and attain unprecedented sensitivity levels, Bluefield created and implemented innovative techniques. Original GFCR systems used the analog electrical difference at the electronic amplifiers level on single-pixel detectors. Only recently has the system been digitalized with straightforward computer image analysis of recorded 2D frames captured with a small number of pixels.

Bluefield brings this to the next level. Its cutting-edge technology employs massive oversampling and a full range of advanced image processing and signal analysis tools, complemented by machine learning (ML) and artificial intelligence-based data-mining techniques. This approach makes the most out of every recorded photon to achieve unprecedented detection levels. This results in the automated processing of large amounts of high frame rate recorded data from the platform (similar to the space sensor presented in Section 12.6.5).

12.6.3 Experimental Prototypes (Ground, Helicopter, and Balloon)

Bluefield's prototype sensor is a first-generation, advanced GFCR instrument, and field-portable, built using a broadband MWIR detector. This configuration was chosen to detect gases in the SWIR (CH_4 @ 2.3 μm) as well as in the MWIR regions (CH_4 @ 3.4 μm and SO_2 @ 4.0 μm), which was successfully demonstrated in the field. However, this broadband configuration increases detrimental self-emission effects. Further, if using a single-gas configuration with a fixed cold filter at the Dewar level optimizes performance, it also limits operation to a single gas (as chosen for a mission-specific detection shown in Section 12.6.5).

Figure 12.18 shows the compact prototype sensor configured with methane and nitrogen (acting as the empty channel) closed cells during an outdoor field campaign. This was built by Bluefield with support from NovaSyst/AIWorx (now Bentley Systems) and Opto-Mécanique de Précision (OMP).

As discussed with GFCR, correlation is directly done at the optical level. Even if the subsequent different calibrations and advanced gas-detection algorithms to be applied are complex, they are not computationally heavy compared with FFT analysis associated with classical FTIR spectrometers. Currently, the prototype sensor operates at a frame rate enabling 30–50 ms computation time per frame, allowing both real-time processing and gas plume display. Additional sensor details are presented in Table 12.2.

Figure 12.18 Bluefield's prototype sensor operating from a helicopter. *Source*: Bluefield Technologies Inc.

Table 12.2 Bluefield's prototype sensor configuration.

Parameter	Values	Comments
Wavelength range (@ 50%)	≈2.5–5.5 μm	Broadband integrated Dewar and cooler assemblies covering both SWIR and MWIR
Detector array size and pitch	640 × 512, 15 μm	MCT (HgCdTe), cryo-cooled to 90 K
Dark current density	1.5 nA/cm^2	Not a dominant noise source at short exposure time
Gas cells' length and pressure	10 cm, 760 Torr	Sealed CH_4 SWIR; refillable for other gases
Aperture diameter and FOV	35 mm, 11° × 8.8°	Stingray SR0955 broadband, 50 mm focal length
Instantaneous FOV (IFOV)	0.3 mrad	<Optical resolution
Spectral resolution	N/A ("*infinite*")	GFCR instrument uses all spectral lines in optical passband at native resolution (undigitized analog mode)
Temporal measurements per acquisition and frame rate	10–20 ms, ≤100 fps	Depends on scene illumination conditions (background reflectance, cloud cover, Sun angle, etc.)
Digital data interface	Camera link, 14 bits	Through a single USB-3 connection to the laptop computer
Operation modes	Solar reflected *and* thermal (absorption and transmission)	
Operational distances	15 m–1000+ m	FOV = 1.9 m × 1.5 m @ 10 m; FOV = 193 m × 154 m @ 1000 m

12.6.4 Laboratory Demonstration and Field Testing

Extensive experimental testing campaigns were performed at DRDC laboratories and outdoor test sites to study and demonstrate the prototype's ability to detect CH_4 and SO_2 under solar reflected and thermal modes with blackbodies in the background, as shown in Figure 12.19.

Controlled-release experiments were also performed during April 2019 at the Stoneham fire station, a secure facility located north of Québec City, Canada. Bluefield collaborated with an independent party, Flux Lab [19], to validate the results. Blind-test scenarios were used to evaluate the Bluefield technology detection capacity to transient methane plumes in realistic weather conditions and to determine the system's sensitivity when exposed to different methane flow rates. The prototype sensor threshold detection limit was evaluated at 0.1 kg/h (2 l/min) when measured horizontally on the ground at a 15 m distance, as shown in Figure 12.20. Blind tests from a helicopter showed a capacity to detect the presence of methane gas at release flow rates of 0.6 kg/h (14 l/min) at 100 m altitude, at wind speeds of about 16 km/h (10 mph) and varying amounts of sunlight and suboptimal cloud conditions. The sensor detected methane released at a flow rate of 12 kg/h from an altitude of more than 1.1 km (3000 ft) with similar wind speed levels.

To prepare for space flight and evaluate key components of the instrument (e.g. camera registration and gas cell behavior under low pressure and temperature environments), two high-altitude balloon tests were performed by Bluefield during summer 2019 to evaluate key components of the instrument (e.g. camera registration and gas cell behavior under low pressure and temperature environments). A "selfie" of the experiment can be seen in the bottom image of Figure 12.21.

12.6.5 Space Mission

Prior NASA missions, starting in the 1970s, validated the GFCR approach, including the Halogen Occultation Experiment (HALOE) [20] on the Upper Atmosphere Research Satellite (UARS) and the MOPITT instrument [21] on the *Terra* spacecraft. The development of microsatellites makes CH_4 monitoring from space affordable.

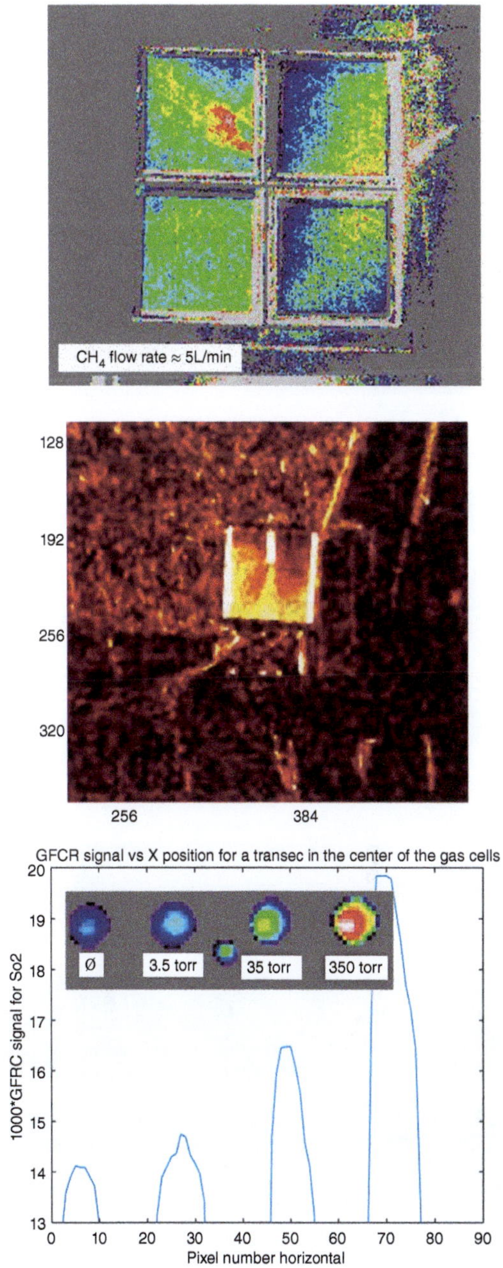

Figure 12.19 Thermal mode CH_4 plume detection (top, $L = 12$ m) and SO_2 plume flow visualization (middle and bottom, $L = 10$ m) (SO_2 sensitivity of 110 ppm-m from the 10 cm 0.5 Torr cell). *Source:* Bluefield Technologies Inc.

Bluefield is developing a space-grade version of its sensor to fit into compact 20 kg microsatellites for monitoring CH_4 from low Earth orbit at 530 km altitude (Figure 12.22). The orbit will be Sun-synchronous with a local nodal crossing time of noon ±90 minutes (optimized for Sun illumination and cold radiator orientation). Each satellite will carry and point sensors capable of detecting methane ground leaks at a fine spatial resolution of 20 m × 20 m (for more information see [22]). Satellites will measure nadir along the orbital track with a 13 km × 11 km FOV. An off-nadir pointing capacity of a ±25° angle with a 500 km FOR allows additional coverage and increased capacity for cloud avoidance.

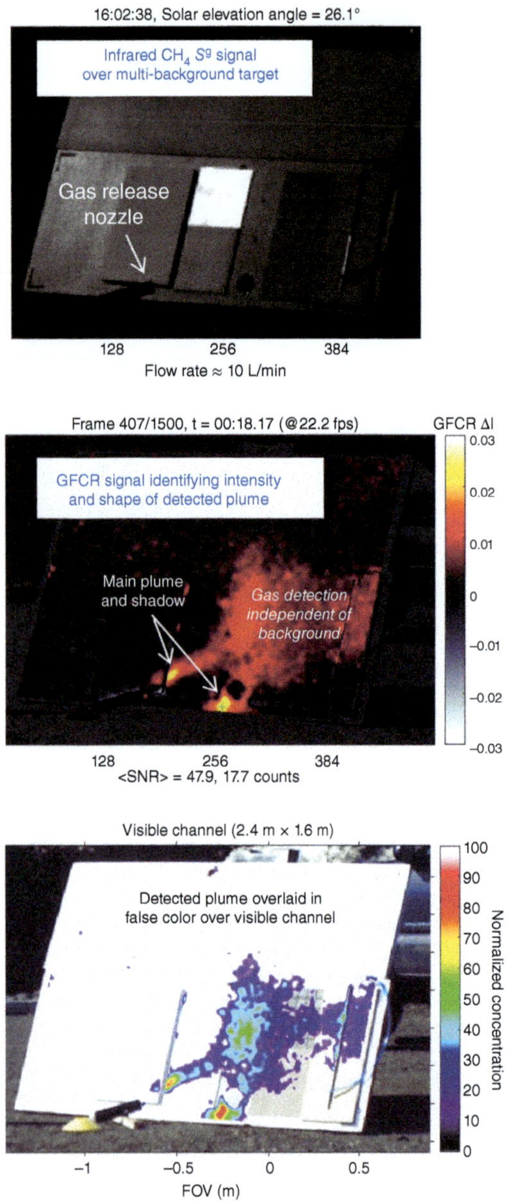

Figure 12.20 Sun-reflected mode example of CH_4 plume detection ($L = 15$ m) over a multialbedo background. *Source*: Bluefield Technologies Inc.

Maps of surface methane plumes (through column-averaged dry-air mixing ratios of CH_4) will be produced by measuring solar backscattered radiation in the SWIR over the whole globe.

The CH_4 passband at 2.3 μm was also chosen because it delivers excellent sensitivity to atmospheric methane (factor 3–5 stronger absorption compared with 1.6 μm), and, therefore, shorter cells can be used. Further, this wavelength is nearly immune to interference from water and N_2O contamination, as well as to changing Earth albedo, essential conditions for plume signature identification using GFCR back-reflected sunlight. Moreover, at 2.3 μm, the thermal emission components are at least three orders of magnitude less than the reflected solar signal, and the scattered downwelling radiation is relatively negligible, allowing quality illumination at low Sun angles and moderate cloud cover.

Figure 12.21 Bluefield's prototype sensor helicopter flight test campaigns (top) and Bluefield test instrument during high-altitude balloon ascent (bottom). *Source*: Bluefield Technologies Inc.

In addition, while imaging spectrometers traditionally use a "whiskbroom" (with a point detector) or a "push-broom" (with a linear detector array) observation method to generate images as they progressively move along, limiting photon throughput, Bluefield's sensor makes it straightforward to implement a spectral imaging method using FPAs, by operating in a "push-frame" observation mode, as shown in Figure 12.23 [23, 24]. In such a staring imaging mode, no FPA dimension is sacrificed to spectra creation (unlike standard diffractive or interferometric methods).

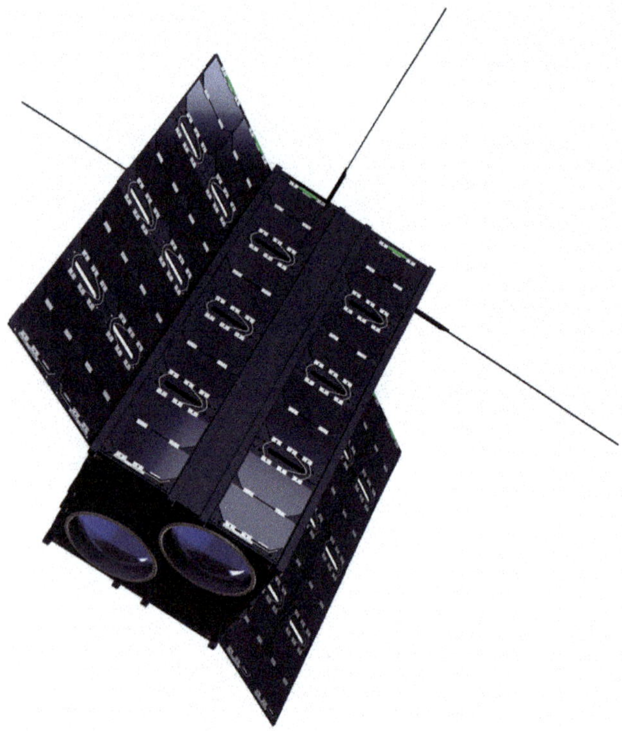

Figure 12.22 Bluefield microsat early design.

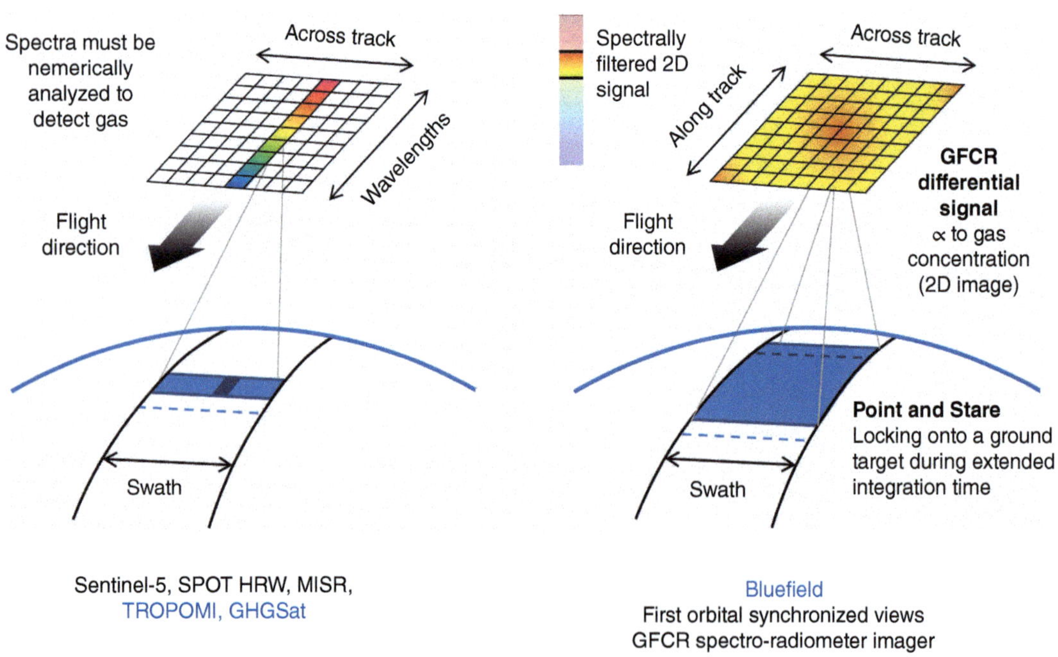

Figure 12.23 Push-broom (left) and push-frame (right) observation methods.

Finally, to achieve unprecedented levels of stability and accuracy and limit system complexity by the addition of a pointing mirror mechanism, an Attitude Determination and Control Subsystem (ADCS) is used to perform lock on-and-stare maneuvers for about 5–10 s on each ground target as it passes within their FOV.

Based on laboratory tests and data modeling, Bluefield satellites will detect methane at concentrations of <0.5% natural abundance (100 ppm-m integrated concentration threshold, or 65 kg/m^2), corresponding to leak flow rates below ≈75 kg/h (3σ). Such a detection level achieves main point emission sources detection under the 2016 EPA Greenhouse Gas Reporting Program which contribute to >92% of total US point source emissions.

The goal of this satellite deployment is to ensure the levels of accuracy and precision required to address GHG inventories. This is achieved through adequate spatial and temporal resolutions and the coverage needed to track weak anthropogenic individual point sources, which tend to be relatively small and spatially clustered (e.g. oil and natural gas operations, livestock operations, landfills, coal mine vents, etc.).

12.7 Conclusion

In the last 20 years, DRDC Valcartier has developed a standoff differential detection method that exploits the attributes of dual-input beam interferometer systems for background and instrument self-emission suppressions. The greatest advantage of this approach is to provide, in the field, a spectrally clean signature of the remote chemical plume, which facilitates its processing in real time. The original CATSI and its subsequent militarized version (CATSI EDM) have clearly demonstrated the capability of the differential radiometric approach for the detection and identification of chemical vapor clouds located at large standoff distances. More recently, the CATSI family technology has expanded to include imaging capabilities for gas mapping (iCATSI) and a compact portable version for active sensing (AC-iCATSI).

A broadband portable field spectral reflectometer has also recently been developed for making contact surface measurements using active sources. The instrument is sensitive over the 0.7–13.5 μm region and permits the acquisition of signatures in the field to properly consider the chemical mixing and morphology of the surfaces encountered.

GFCR is another remote-sensing technique well-adapted for measuring trace gases on compact and portable instruments, outperforming classic spectrometers that cannot reach the required size to be practically implemented in a small platform without substantial performance loss. In that case, the GFCR technique is particularly well-adapted to reveal tenuous amounts of a selected target gas.

References

1 Thériault, J.-M. (1999). Modeling the responsivity and self-emission of a double-beam Fourier-transform infrared interferometer. *Applied Optics* 38: 505.

2 Thériault, J.-M. (2001). Passive standoff detection of chemical vapors by differential FTIR radiometry. *DREV Technical Report, TR-2000-156.*

3 Thériault, J.-M., Puckrin, E., Bouffard, F., and Déry, B. (2004). Passive remote monitoring of chemical vapours by differential FTIR radiometry: results at a range of 1.5 km. *Applied Optics* 43: 1425–1434.

4 Lavoie, H., Puckrin, E., Thériault, J.-M., and Bouffard, F. (2005). Passive standoff detection of SF6 at a distance of 5.7 km by differential FTIR radiometry. *Applied Spectroscopy* 59 (10): 1189–1193.

5 Thériault, J.-M. and Puckrin, E. (2005). Remote sensing of chemical vapours by differential FTIR radiometry. *International Journal of Remote Sensing* 26: 981–995.

6 Thériault, J.-M., Lacasse, P., Lavoie, H. et al. (2010). CATSI EDM: a new sensor for the real-time passive stand-off detection and identification of chemicals. *Proceedings SPIE 7665, Chemical, Biological, Radiological, Nuclear, and Explosives (CBRNE) Sensing XI, 766513* (5 May 2010).

7 Moreau, L., Prel, F., Lavoie, H. et al. (2010). A novel multi-pixel imaging differential standoff chemical detection sensor. *Proceedings of SPIE* 7660: 76602C.

8 Moreau, L.M., Prel, F., Lavoie, H. et al. (2011). iCATSI: a multi-pixel imaging differential standoff chemical detection sensor. In: *SPIE Defense Security and Sensing* (27 April 2011). Orlando, USA: SPIE Defense.

9 Prel, F., Moreau, L., Lavoie, H. et al. (2011). iCATSI, Multi-pixel imaging differential spectroradiometer for standoff detection and quantification of chemical threats. In: *SPIE Defense, Security & Sensing* (20 September 2011). Prague: SPIE Defense. (12 pages, 28 Slides).

10 Thériault J.-M., Fortin, G., Bouffard, F. et al. (2013). Hyperspectral gas and polarization sensing in the lwir: recent results with moddifs. *Proceeding of 2013 5th Workshop on Hyperspectral Image and Signal Processing: Evolution in Remote Sensing (WHISPERS)*, Gainesville, FL, USA (26–28 June 2018). https://ieeexplore.ieee.org/document/8080742 (accessed September 2020).

11 Thériault, J.-M., Fortin, G., Lavoie, H. et al. (2011). A new imaging FTS for LWIR polarization sensing: principle and application. In: *Imaging and Applied Optics, OSA Technical Digest*. Optical Society of America, paper FTuD2.

12 Bonifazi, G., Picone, N., and Serranti, S. (2013). Ore minerals textural characterization by hyperspectral imaging. *Proceedings of SPIE 8655, Image Processing: Algorithms and Systems XI, 865510* (19 February 2013). doi: https://doi.org/10.1117/12.2003054.

13 Thenkabail, P.S., Smith, R.B., and De Pauw, E. (June 2002). Evaluation of narrowband and broadband vegetation Indices for determining optimal hyperspectral wavebands for agricultural crop characterization. *Photogrammetric Engineering & Remote Sensing* 68 (6): 607–621.

14 Kumar, A., Lee, W.S., Ehsani, R.J. et al. (2012). Citrus greening disease detection using aerial hyperspectral and multispectral imaging techniques. *Journal of Applied Remote Sensing* 6 (1) https://doi.org/10.1117/1.JRS.6.063542.

15 Lackenbauer, P.W., Farish, M.J., and Arthur-Lackenbauer, J. (2005). The distant early warning (DEW) line: a bibliography and documentary resource list. *The Arctic Institute of North America*, ISBN 1-894788-01-X.

16 Sandsten, J., Edner, H., and Svanberg, S. (1996). Gas imaging by infrared gas-correlation spectrometry. *Optics Letters* 21 (23): 1945–1947.

17 Chambers P. (2005). A study of a correlation spectroscopy gas detection method. Ph.D. Thesis. Faculty of Engineering, Science and Mathematics Optoelectronics Research Centre, University of Southampton, October 2005.

18 Mercier J.A., Smith M.W., Hunt J.P., and Ison A.M. (2012). *Modeling, sensor design, and performance predictions for gas filter correlation radiometers*. Sandia National Labs Report, SAND2012-7985, September 2012.

19 http://fluxlab.ca.

20 Russell, J.M. et al. (1993). The halogen occultation experiment. *Journal of Geophysical Research-Atmospheres* 98: 10 777–10 797.

21 Drummond, J.R. and Mand, G. (1996). The measurements of pollution in the troposphere (MOPITT) instrument: overall performance and calibration requirements. *Journal of Atmospheric and Oceanic Technology* 13: 314.

22 http://Bluefield.co.

23 L3Harris. Different types of scanning sensors. http://www.harrisgeospatial.com/Support/Self-Help-Tools/Help-Articles/Help-Articles-Detail/ArtMID/10220/ArticleID/16262/Push-Broom-and-Whisk-Broom-Sensors (accessed September 2020).

24 Fialka (2018). Meet the Satellites That Can Pinpoint Methane and Carbon Dioxide Leaks: European and Canadian orbiters can work together to catch wayward emissions. *Scientific American*. http://www.scientificamerican.com/article/meet-the-satellites-that-can-pinpoint-methane-and-carbon-dioxide-leaks (accessed September 2020).

13

Handheld Laser Induced Breakdown Spectroscopy (HHLIBS)

David Day

Sciaps Incorporated, Woburn, MA, USA

13.1 Introduction

The Laser Induced Breakdown Spectroscopy (LIBS) measurement technique has similarities to other Optical Emission Spectroscopy (OES) methods in that it excites electrons in atoms and small molecules within a sample and then monitors the decay of those electrons back to lower energy states (Cremers and Radziemski 2006; Miziolek et al. 2006; Musazzi and Perini 2014). By analyzing the emitted light during the decay process, both the elements present and their concentrations may be determined. The element is determined from the unique electron transition energy, which yields specific emission wavelengths, and concentration is derived from the intensity of that emission.

However, unique to LIBS is that a short-pulsed laser with energies ranging from microjoules to 100s of millijoules is used to heat a sample surface, vaporize a small amount of the material, and excite electrons all at the same time. As with other OES methods, the emitted light is collected and analyzed with one or more spectrometers. Due to the very short duration of the emission event, multichannel spectrometers are typically used as opposed to scanning monochromator methods. Figure 13.1 shows a simplified schematic of a typical LIBS instrument.

The LIBS technique began in earnest in the early 1980s with scientific publication increasing nearly exponentially since that time (Noll 2012, p. 4). After the year 2000, several groups worked on making "transportable" and "portable" LIBS instruments ranging from backpack-based units to small suitcase-sized units. These typically used an umbilical cable containing fiber optic cables for delivering signals back to a spectrometer and either laser pulse delivery optics or power cables to a remote laser in the measurement head. As of 2013, there were several such research and commercial instrument producers which included Applied Photonics, Berlin Technologies, Ivea-Solution, StellarNet (Rakovsky et al. 2014).

In the spring of 2013, RMG Technology introduced the first self-contained truly handheld LIBS instrument which utilized a microjoule (μJ) laser. In early 2014, TSI (Shoreview, MN, USA) and SciAps (Woburn, MA, USA) both introduced LIBS handhelds with millijoule (mJ) lasers. In subsequent years, several more handheld LIBS instruments appeared from various companies including Rigaku (Wilmington, MA, USA), B&W TEK (Newark, DE, USA), Hitachi (formerly Oxford, formerly RMG) (Westford, MA, USA), Thermo Fisher (Tewsbury, MA, USA), and Vela (Burlington, MA, USA) (Figure 13.2).

All these handhelds are self-contained with no umbilical cables. However, they have many feature and capability differences. The variety of the basic components required for handheld LIBS and the technologies behind them are discussed in this chapter.

Traditional laboratory OES and LIBS instrumentation enjoy the luxury of being used in a stationary location and controlled temperature environment. Handheld LIBS (HHLIBS), on the other hand, must be capable of operation under variable thermal conditions, in any orientation, and must also be able to withstand various

Portable Spectroscopy and Spectrometry 1: Technologies and Instrumentation, First Edition.
Edited by Richard A. Crocombe, Pauline E. Leary and Brooke W. Kammrath.

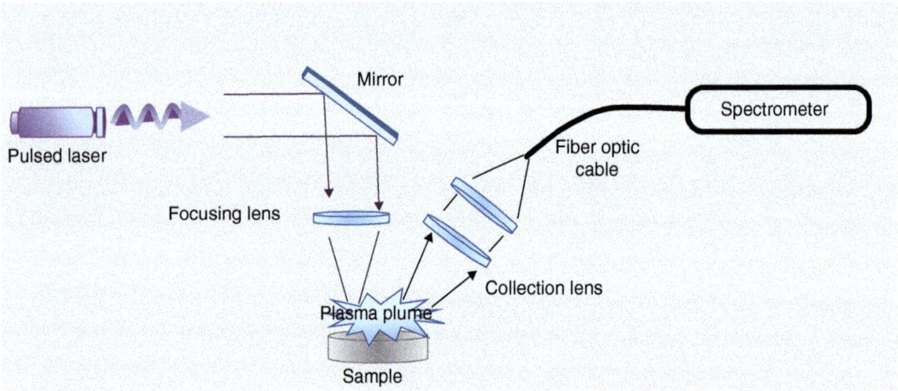

Figure 13.1 Schematic diagram of the laser-induced breakdown spectroscopy measurement.

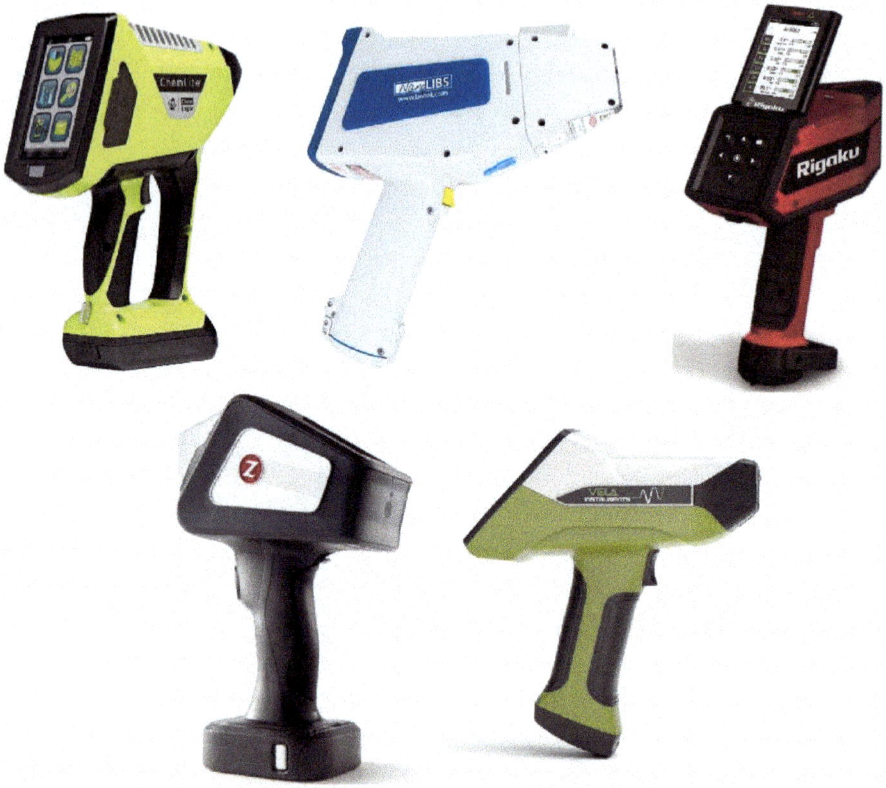

Figure 13.2 Examples of commercially available HHLIBS clockwise from the upper left: TSI-ChemLite (*Source*: Courtesy of TSI), BWTEK-NanoLIBS (*Source*: Courtesy of BWTEK), Rigaku-Katana (*Source*: Courtesy of Rigaku), SciAps-Z (*Source*: Courtesy of SciAps), and Vela-A1 (*Source*: Courtesy of Vela).

environmental situations including dust, rain, mild vibration and shock. It is very difficult to design lasers and spectrometers that are entirely temperature-stable and so designers must be careful to apply some form of thermal compensation or utilize thermoelectric devices to keep instrument components at a constant temperature. The drawback of precise temperature control is that it can consume a good deal of electrical power, lowering battery life. Additionally, any form of temperature control produces heat, especially for thermoelectric cooling which is very inefficient. The more heat that needs to be expelled from an instrument, the heavier and larger the outer cooling surfaces must be. A successful HHLIBS design becomes a sensitive balancing act between performance stability, power consumption, and instrument size and weight.

13.2 Handheld LIBS-Enabling Technologies

The technologies that enable handheld LIBS include:

13.2.1. Compact pulsed lasers
13.2.2. Compact spectrometers
13.2.3. Laser delivery optics, detection optics, and plasma containment
13.2.4. Beam rastering (in some instruments)
13.2.5. Gas purging (in some instruments)
13.2.6. Detector temporal gating (in some instruments)
13.2.7. Calibration (wavelength, intensity, and application)
13.2.8. Compact electronics and power supply

All of the above technologies have been evolving over the last several decades. However, it has really been during the last decade that miniaturization has progressed to the point of enabling all of them to be contained together in a fully standalone handheld instrument. In the following sections, these various components are reviewed.

13.2.1 Compact Pulsed Lasers

As of this writing, there are two types of pulsed lasers being utilized in commercially available fully self-contained HHLIBS instruments. These are microchip lasers, which are highly integrated with less than 1 mJ pulse energy, and lasers made from individual components with pulse energies from 100 μJ to 15 mJ. Both classes of lasers are Diode-Pumped Solid-State (DPSS) pulsed lasers with pulse energies limited to less than 15 mJ. "Transportable" LIBS systems with more powerful flashlamp-driven lasers in the 50 mJ range are available, such as Applied Photonics LIBSCAN 25 (www.appliedphotonics.co.uk/products_services/libscan_25.htm [accessed 1 November 2019]), but these systems are not fully self-contained handhelds and not covered in this chapter.

Microchip lasers, first reported in 1989 (Zayhowski and Mooradian 1989), are capable of pulse energies up to several hundred microjoules and pulse rates of up to several kHz. These lasers contain a resonant cavity consisting of an optical gain medium, a saturable absorber (SA or Q-Switch) and reflective coatings on each end (Aubert 2001). These components are fabricated as a single part with a total volume of only a few cubic millimeters. A pump diode is added to this which injects higher energy light into the gain medium through the back-mirror coating which is designed to be reflective at the laser output wavelength, but transparent at the pump diode wavelength. While there are various construction variations, Figure 13.3a shows a schematic diagram of a typical microchip laser. Further, output optics may be added either internally or externally to the laser package.

Microchip lasers are often associated with "micro-LIBS," which is roughly defined as a measurement regime utilizing laser pulse energies less than 1 mJ, typically in the 10s–100s of microjoules and with pulse frequencies up to several kHz. Microjoule lasers may be constructed with a microchip or individual optical components.

Higher power (>1 mJ), lower frequency lasers (<100 Hz) are used in some HHLIBS instrumentation (SciAps, Thermo Fisher, TSI) that are constructed from individual components which include pump diodes, crystal gain media, cavity mirrors, and either a saturable absorber (passive Q-switch) or an active electronically controlled Q-switch. As a result of the larger size of the gain medium, more energy can be stored and then quickly released with each pulse. Pulse frequencies must be held to lower rates due to the high amounts of power dissipated internally per pulse. Optical pumping (excitation) of the crystal gain medium can be carried out in a similar fashion to the microchip laser by "end pumping" through one of the cavity mirrors or by "side pumping" (Figure 13.3b).

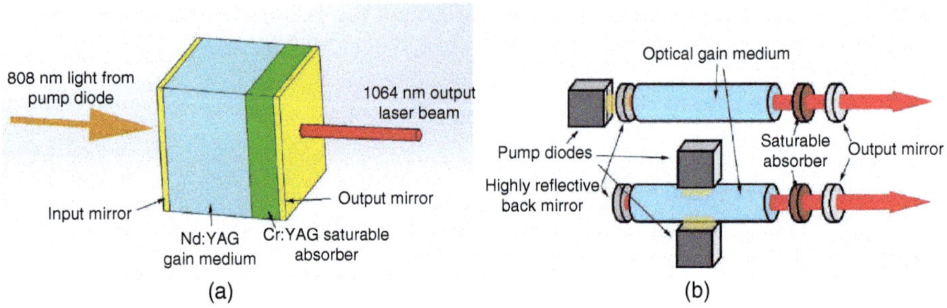

Figure 13.3 (a) Nd:YAG microchip laser diagram. Input mirror is transparent to pump light at 808 nm, but highly reflective to laser 1064 nm light. Output mirror is partially reflective to laser 1064 nm light. Total volume 0.03 cm^3 not including pump diode; (b) Diode-pumped solid-state laser. Top shows "end-pumped" configuration and bottom shows "side-pumped" configuration. Total volume 25–100 cm^3. Key components include pump diodes, gain media, mirrors, and the saturable absorber or "Q-switch." *Source*: Mukhopadhyay (2011). © 2011 Lambert Academic Publishing.

There are several issues that must be considered in designing and utilizing small compact lasers in HHLIBS instruments. Typical efficiencies are on the order of a few percent. The average optical power output of a laser firing at 20 Hz and 10 mJ is 200 mW. With a 2% input-to-output power efficiency, the power consumption in this case would be 10 W average during pulsing. If an instrument design utilizes thermoelectric cooling to stabilize the laser temperature, the power consumption increases from there significantly. Fortunately, the laser is not firing continuously; however, in some applications, such as metal analysis, the duty cycle can be quite high. It is quite difficult to design any handheld instrument to dissipate 10 W of average power without an external fan, and so the main options left for thermal management are total external surface area and thermal conductivity to that surface, both of which must be maximized for thermal dissipation, but minimized for reasonable overall size and weight. Thus, the laser alone has a very important influence on final product ergonomics, which must be carefully considered in the design.

The technology is still evolving on DPSS lasers. There are many lasers commercially available from companies, such as Quantel (Bozeman, MT, USA), RPMC (O'Fallon, MO, USA), Coherent (Salem, NH, USA), Kigre (Hilton Head, SC, USA), and Thorlabs (Newton, NJ, USA) for example. However, due to the fact that most of the commercial lasers are not quite suitable for HHLIBS due to either large size, large power consumption, poor output power/rep rate, or cost, most commercial HHLIBS vendors are currently using their own proprietary technologies with the components finely tuned for the LIBS application. For the HHLIBS application, the optical components are often being pushed to the limits just within damage thresholds for pump diodes, gain media, and optical coatings. Antireflective coatings and mirror coatings are still evolving and so it is expected improvement in this field will be an enabling technology for even higher power DPSS lasers in the future (Lyngnes et al. 2006 and https://www.photonics.com/Articles/High-Energy_Laser_Optics_Require_Coatings_in/p5/vo61/i482/a44258 [accessed 1 November 2019]).

The most common laser gain media used for HHLIBS microjoule and millijoule lasers are neodymium-doped yttrium aluminum garnet (Nd:YAG) or a neodymium-doped yttrium orthovanadate (Nd:YVO4) due to their relatively high efficiency and good thermal conductivity (Mukhopadhyay 2011, p. 70). A saturable absorber of chromium YAG (Cr:YAG) is typically used with 1064 nm emitting gain media. Other similar gain media materials producing output wavelengths from 800 to 1340 nm (Nd:YLF, Yb:YAG, Cr:LiSAF) have also been used to produce DPSS lasers (Noll 2012, p. 49).

At the power and repetition rate that HHLIBS 1064 nm lasers are operated, they all have a rating of Class 3B and are eye hazards. The HHLIBS from TSI and some first-generation SciAps instruments produce laser wavelengths in the range of 1500–1600 nm. TSI HHLIBS patents describe a laser that utilizes an Nd:YAG DPSS laser coupled with an Optical Parametric Oscillator (OPO) crystal which converts the 1064 nm light into two lower energy (longer wavelength) outputs, one of which is at 1574 nm (Quant et al. 2016). The first-generation SciAps instrument utilized a Kigre laser with an erbium-doped phosphate laser gain medium and output wavelength of 1535 nm (https://www.kigre.com/products/laser_glass.htm [accessed 1 November 2019]). Output wavelengths longer than 1500 nm are far safer because the light entering the human eye is highly absorbed by the intraocular fluid and reduces or prevents focused light from ever reaching the retina. This enables instrument laser safety classification of such long-wavelength instruments (>1500 nm) of Class 1 M (https://www.tsi.com/products/metal-analyzers/chemlite-laser-aluminum-analyzer-4235 [accessed 1 November 2019]) depending on laser power levels and beam divergence conditions. However, since all HHLIBS produce plasmas regardless of the laser wavelength, safety features for sample presence detection are typically incorporated into the instrument-operational features. Reported examples of sample presence detection schemes include both plasma chamber pressure and light/dark sensor (Thermofisher 2020) (https://www.thermofisher.com/order/catalog/product/NITONAPOLLO#/NITONAPOLLO [accessed 2 March 2020]), both video analysis and light-emitting diode (LED) spectrum sample reflection (Day 2018), and real-time "per shot" spectral analysis (Day and Sackett 2017). With such safety features and proper training, a laser safety officer may allow class 3B laser instruments to be used under class 1 M laser conditions.

13.2.2 Compact Spectrometers

Compact spectrometers have been around since the early 1990s and were first sold by Ocean Optics (https://oceanoptics.com/25-years-amazing-sensing-solutios [accessed 1 November 2019]). Since that time, several other companies, Avantes (Apeldoorn, Netherlands) and Stellar Net (Tampa, FL, USA) for example, offer similar and ever smaller products. These types of spectrometers have found their way into numerous handheld technologies including Raman, near infrared (NIR), ultraviolet-visible (UV-Vis), and LIBS (Crocombe 2018). HHLIBS vendors generally do not disclose many details about the spectrometers they have either designed or purchased; however, the general basics of compact spectrometer design are covered in this section.

13.2.2.1 Linear Detector Arrays

Key to the original Ocean Optics spectrometer's operation were front-illuminated linear CCD (charge coupled device) detector arrays, developed originally for bar code scanners and produced by companies, such as Sony, Hamamatsu, and Toshiba. In more recent times, back-thinned CCDs (BT CCDs) and complementary metal-oxide semiconductors (CMOS) photodetector arrays have also been introduced, each with their own advantages and disadvantages. BT CCDs have a higher sensitivity than standard CCDs but also have a higher cost, due to more difficult processing. They also can have a periodic wavelength response function, known as "etaloning," due to interference of internal surface layer reflections from the detector's thinned layers. Another advantage of BT CCDs is that they are sensitive into the deep UV (<200 nm), whereas traditional CCDs lose sensitivity below 400 nm due to reduced photon penetration depth into silicon at short wavelengths (Darmont 2009). Traditional CCDs and some CMOS devices must be coated with an appropriate photon down-converting phosphor, such as Lumogen (https://www.teledyne-e2v.com/content/uploads/2017/08/ccdtn103.pdf [accessed 1 November 2019]) or Metachrome (https://www.actonoptics.com/products/metachrome-coatings [accessed 1 November 2019]) for sensitivity into the deep UV.

The linear array detectors typically have between 1024 and 4096 individual detector elements (pixels), spanning approximately 30 mm, that all collect photons simultaneously and are then read out after a fixed amount of integration time (Figure 13.4). This type of operation is key for LIBS measurement because the signal is only present for 10s of microseconds or less. It is for this reason that scanning monochromator spectrometers are not appropriate for LIBS spectral measurement.

Larger and more expensive benchtop LIBS systems often utilize an intensified CCD (ICCD) which incorporates an image intensifier mounted to the front of the CCD. Not only does the ICCD intensify the signal, but it also allows very precise electronics shuttering or "time gating," i.e. the ability to start and stop collecting photons at

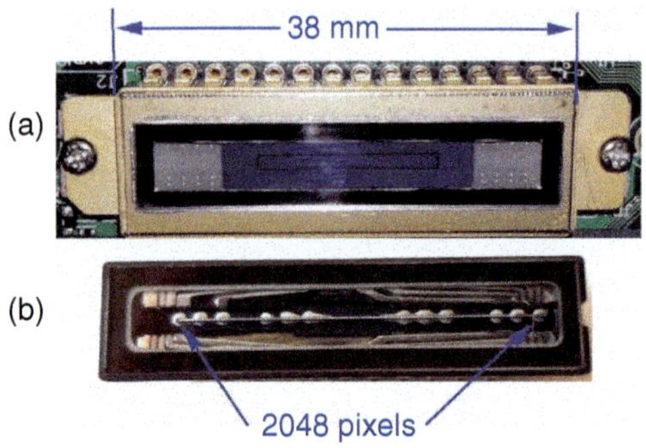

Figure 13.4 Examples of two linear detector arrays: (a) Hamamatsu S11511-1006; and (b) Sony ILX511B charge coupled devices. Both sensors are shown on the same scale.

very precise and very short microsecond time scales. For HHLIBS, ICCDs are impractical because of their cost and drive electronics. Fortunately, some standard linear detector arrays have various degrees of electronic shuttering capabilities. HHLIBS instruments employing higher power lasers (>1 mJ) benefit from electronic shutter control (temporal gating) so that the initial plasma continuum radiation can be rejected. For microjoule laser-based HHLIBS, electronic shuttering is not used or needed because continuum background is much reduced, and the laser repetition rate is much faster than the combined shutter speed and readout time of any available detector. For this reason, the detector specifications must be carefully reviewed and selected depending on how it will be used. A good example of some available detectors and specification can be viewed in the Avantes spectrometer catalog (ref. https://www.avantes.com/catalog/19/#zoom=z [accessed 1 November 2019]). See Section 13.2.6 for more details and examples of temporal gating with millijoule lasers.

13.2.2.2 Spectrometer Optics

Several current commercial HHLIBS utilize a spectrometer layout known as Czerny–Turner (CT) (Czerny and Turner 1930). The basic CT design incorporates in the order encountered along the optical path: an entrance slit, a collimating mirror, a diffraction grating, a refocusing mirror, and an exit slit. In the original CT design, the grating was slowly rotated so that the appropriate wavelength could be selected at the exit slit. For HHLIBS, the exit slit is replaced with the linear detector array and the different wavelengths impinge upon different pixels along the length of the array. In the classical CT design shown in Figure 13.5a, aberrations caused by the first off-axis collimating mirror 1 are partially compensated by the second refocusing mirror 2. A CT variant known as Crossed Czerny–Turner (Figure 13.5b) loses the aberration correcting feature, but gains an advantage in reduced overall size. Both types of spectrometers are currently used in commercial HHLIBS. Alternate forms of the CT design include the usage of transmission gratings (Quant et al. 2016), and lenses instead of mirrors. Figure 13.6 shows a commercial Avantes spectrometer that is used in some larger commercial LIBS instruments.

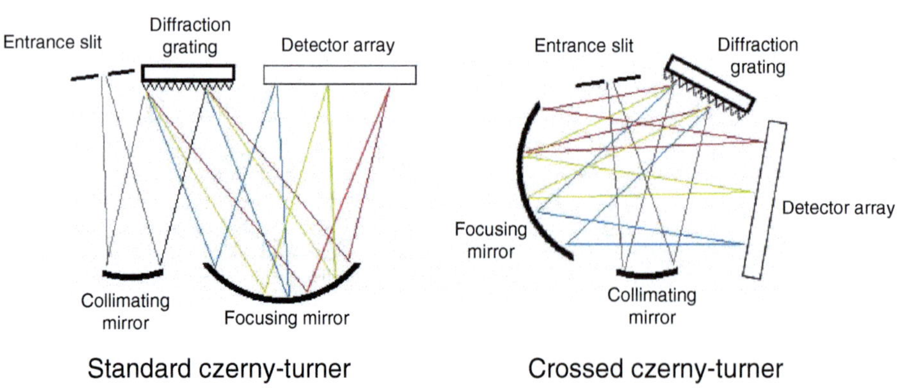

Figure 13.5 Classical Czerny–Turner (a) and crossed Czerny–Turner (b) spectrometer configurations, both with linear detector arrays replacing the exit slit. *Source*: Based on Czerny and Turner (1930).

Figure 13.6 Example of a small Avantes spectrometer that utilizes the standard CT configuration. 1, detector; 2, fiber optic connector; 3, diffraction grating; 4, input slit; 5, filter; 6, collimating mirror; 7, focusing mirror; 8 and 9, light traps. *Source*: Courtesy of Avantes.

Important to HHLIBS is the spectral resolution and spectral band needed to distinguish elemental emission lines. Factors that influence resolution and the spectral band are the diffraction grating and the mirror focal lengths. Higher diffraction grating groove densities and longer focal lengths both lead to a higher resolution. While detailed spectrometer design (Scheeline 2017) is beyond the scope of this chapter, suffice it to say that for spectrometer sizes that reasonably fit a handheld instrument and with current diffraction grating technology, the maximum resolution obtained in current commercial instrumentation is just below 0.1 nm. While this high resolution is not needed for most LIBS applications, it is important and utilized for certain applications such as carbon emission line quantification at 193.1 nm for better separation from neighboring lines (Section 13.4.2).

Another factor that must be considered is the spectral band or range captured by the linear detector. There is a trade-off between the desired resolution and range. HHLIBS designers carefully consider the resolution needed and where the elemental emission lines are for a given application and try to design a spectrometer that can meet both requirements. Several of the commercial HHLIBS used for metal alloy determination and sorting use a single spectrometer in the approximate range of 230–450 nm because most of the standard alloy base and alloying element emission lines can be found within that range. If we assume a detector with 2048 pixels, then the pixel resolution of the spectrometer would be (450–230 nm)/2048 pixels or 0.108 nm/pixel. However, for accurate peak location detection and quantification, the optical resolution must be about a factor of 2.5 lower (broader) than the pixel resolution, ensuring that the peak is spread over several pixels. It may seem counterintuitive, but by spreading the peak out over several pixels, the peak center can be located to better than a tenth of a pixel, whereas if the emission line were located entirely within one pixel, the location would only be known to the width of one pixel. Thus, the common resolution for these single-spectrometer HHLIBS units is on the order 0.2–0.3 nm. At least four commercial HHLIBS instruments use multiple spectrometers to get increased resolution and range (B&WTEK "NanoLIBS" utilizes two, Thermo Fisher "Apollo" utilizes two, TSI "ChemLite" utilizes two, and SciAps "Z" has various models with two, three, and four spectrometers) to open up the capability of more applications, such as carbon in steel, and a wider element suite including fluorine, lithium, sodium, potassium, and sulfur in geochemical analyses. Figure 13.7 shows an example of a spectrum from the SciAps Z300 which utilizes three spectrometers for a spectral range of 188–950 nm and resolution ranging from 0.15 to 0.35 nm.

13.2.3 Laser Delivery Optics, Detection Optics, and Plasma Containment

Laser delivery optics and plasma detection optics are implemented in different ways by the various HHLIBS manufacturers and not fully disclosed. Several of the basic methods are described.

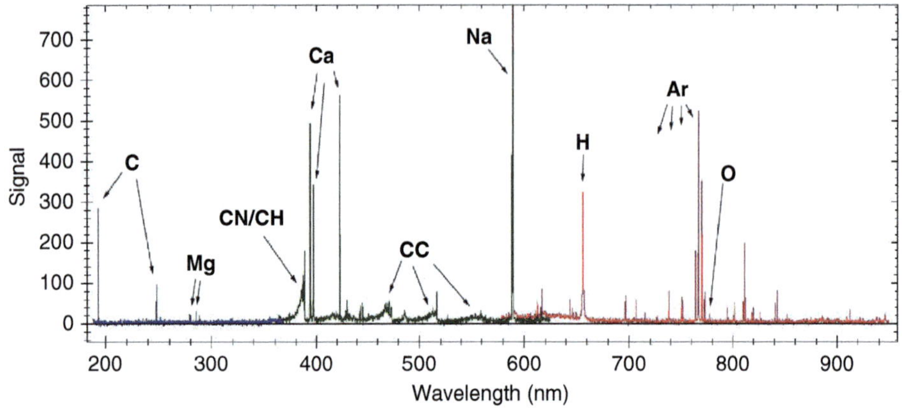

Figure 13.7 LIBS spectrum exported from SciAps Z300 wide spectral range instrument with three spectrometers. Sample: human fingernail under argon purge, 5 mJ laser, 650 ns gate delay.

13.2.3.1 Laser Delivery Optics

Most DPSS lasers are designed to emit collimated or near-collimated light. This is convenient as it enables the HHLIBS design to incorporate some distance between the laser exit window and the focusing optic. A focusing optic, either a lens or a concave mirror, is used to bring the collimated laser beam to a small spot on the sample surface. In order to induce plasma on aluminum, the beam must be concentrated to a power density of $\sim 10^7$–10^8 W/cm^2 during the laser pulse (Noll 2012, p. 20). Materials with lower absorption coefficients require more power, with gas plasmas requiring on the order of 10^{11} W/cm^2 (Cremers and Radziemski 2006, p. 40). Considering a laser with 100 μJ and a 2 ns pulse width as an example, there is 50 kW present during the pulse period. If that is focused on a 100 μm spot, it yields roughly 5×10^8 W/cm^2 and will result in plasma on aluminum and most other metals. However, it will often not be enough to create plasmas on lower absorbance materials. Conversely, a 10 mJ laser with a 2 ns pulse width will have two orders of magnitude more power density per pulse and, in some cases, using slightly tighter focusing to 50 μm, can create plasmas in air.

Optics must, therefore, be chosen to adequately focus to a spot small enough to reach sufficiently high-power densities during the pulse. The ultimate spot size is a function of beam diameter before the focusing optic, the focal length of the optic, and the laser beam quality. The interrelationships between these properties are beyond the scope of this chapter; however, the end result is that current HHLIBS instruments with the pulse energies described earlier utilize focusing optics with focal lengths on the order of 5–20 mm and attain power densities of 10^8 W/cm^2 or higher.

13.2.3.2 Plasma and Particulate Containment

The distance from the plasma to the focusing optic is critical not only for obtaining enough power density (shorter is better), but also for keeping optics clean (longer is better). LIBS plasmas often emit small molten particles of the material under test (Figure 13.8). The higher the power the laser has, the further these particles can travel while remaining hot. If a molten particle reaches the focusing optics, it can become welded to it and only be removed by polishing or replacement.

The Thermo Fisher Apollo HHLIBS has a cleaning/polishing kit and procedure that is recommended every 1000 measurements (https://players.brightcove.net/665001591001/dae1816b-a958-4ec8-b102-253bcd5c7d2a_default/index.html?videoId=5997407365001 [accessed 1 November 2019]). The SciAps Z models all have a disposable drop-in window placed between the focusing lens and plasma that is wiped off or replaced when necessary (Day et al. 2016). Similarly, a BWTek patent (Wang 2017) describes a protective window combined with an air-curtain to minimize buildup.

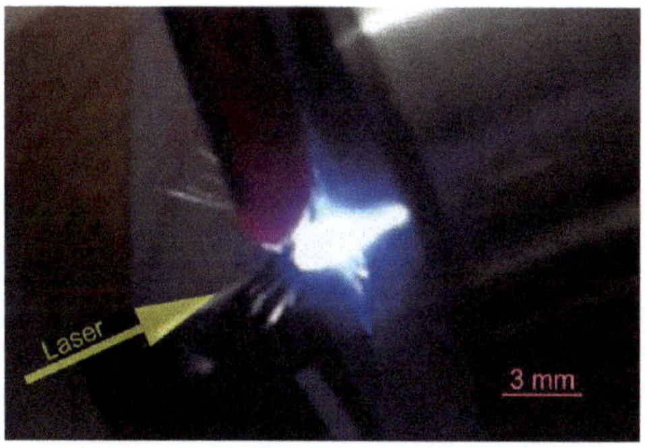

Figure 13.8 Plasma with molten incendiaries resulting from a 10 mJ laser pulse (1064 nm) on steel. The red object is an argon delivery tube.

13.2.3.3 Plasma Light Collection Optics

Light collection from the plasmas and delivery to the spectrometer(s) are accomplished by a variety of means. In the simplest case, an optical fiber tip can be located near the plasma and off to the side of the incoming laser light with the fiber then delivering light directly to the spectrometer(s). In other designs, a refocusing lens or mirror can be used to collect more light from the plasma and match it to numerical aperture input cone of the optical fiber. In another HHLIBS design, a single lens is used for both laser focusing and plasma light collection (Figure 13.9). Advantages of this configuration are the alignment of the laser light path to the plasma light collection path is more easily maintained and only one optic need be moved to raster the beam (see Section 13.2.4).

13.2.4 Beam Rastering

Several HHLIBS models (Bruker-EOS500, BWTEK-NanoLibs, Rigaku-Katana, Hitachi-Vulcan, SciAps-Z, Vela-A1) implement some form of beam rastering. Beam rastering has been classically used in benchtop systems for the purposes of better averaging of sample inhomogeneities and, in some cases, for elemental mapping. In the case of HHLIBS microjoule lasers, rastering is a necessity because stationary beams stop creating plasmas after a few shots and, thus, must be constantly scanned. This behavior has been attributed to a change in surface reflectivity after the first pulse (Tashuk et al. 2007, p. 187) and to the presence of melted material after the first few shots if the sample is not moved (Rakovsky et al. 2014, p. 280). Regardless, the moving laser spot serves to effectively average inhomogeneity concentration variation across a sample surface (Figure 13.10). Multiple passes over the same line

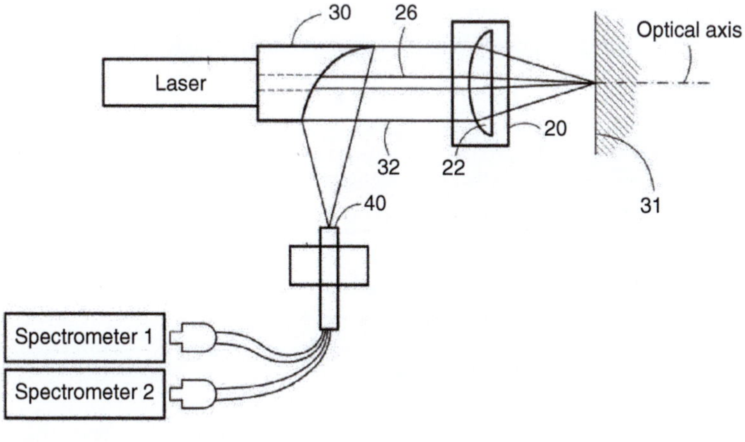

Figure 13.9 Multispectrometer HHLIBS optical configuration enabling beam rastering through the movement of a single lens (22). Collected plasma light passes through the laser-focusing lens (22), then reflected off a parabolic mirror (30), and refocused on to a fiber optic cable (40). *Source*: Day (2018).

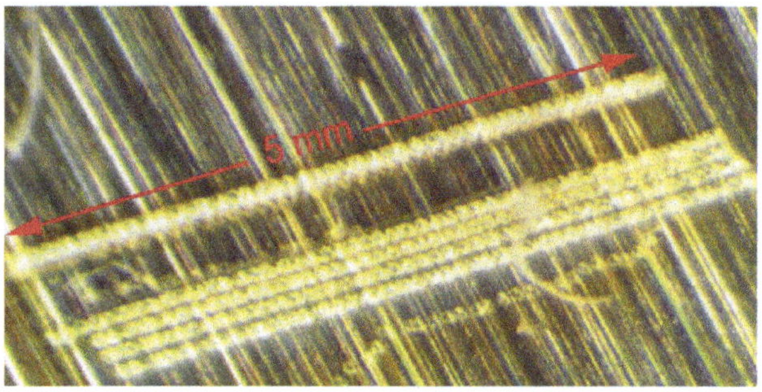

Figure 13.10 Microjoule laser one-dimensional surface scans.

are used by some microjoule instruments to preclean a location before valid data are collected for analysis (Noll et al. 2018, table 3) and are described as "drill down" technology in Rigaku Katana literature (https://www.rigaku.com/en/kt100s [accessed 1 November 2019]) or "preburn" in Hitachi Literature (https://hha.hitachi-hightech.com/assets/uploads/downloads/handheld-analysers/vulcan/Vulcan-LIBS-Analyser-For-QA-QC.pdf [accessed 1 November 2019]).

When higher power lasers are used (>1 mJ), LIBS plasmas can be consistently created from the same location without having to move the laser. This can be quite useful as initial laser shots can be used very effectively for "cleaning" the sample surface and burning through contamination and oxidation layers. TSI and Thermo Fisher units operate in this fashion using a single location. The SciAps HHLIBS not only uses cleaning shots, but also implements user-controlled two-dimensional rastering settings for the number of X- and Y-direction locations, the spacing between locations, number of cleaning shots per location, and number of data shots per location. While for alloy analysis, multiple locations are used for better averaging of inhomogeneities, in some geochemical applications, multiple sample locations are used to map out elemental distributions (Figure 13.11).

13.2.5 Gas Purging

Traditional benchtop LIBS systems incorporate the ability to purge samples with inert gases for the purposes of enhancing signal from the plasma. Gases with higher densities and lower thermal conductivity tend to keep the plasma localized, with higher plasma density and temperatures, which further results in more light emission (Noll 2012, p. 29). Argon is by far the most common purge gas used and has been shown to increase light output by more than a factor of 10 in some cases. Purge gas ionization energy can also play a role in transferring energy to the vaporized sample elements in certain cases. Fluorine, with an ionization potential of 17.4 eV, is a poor LIBS

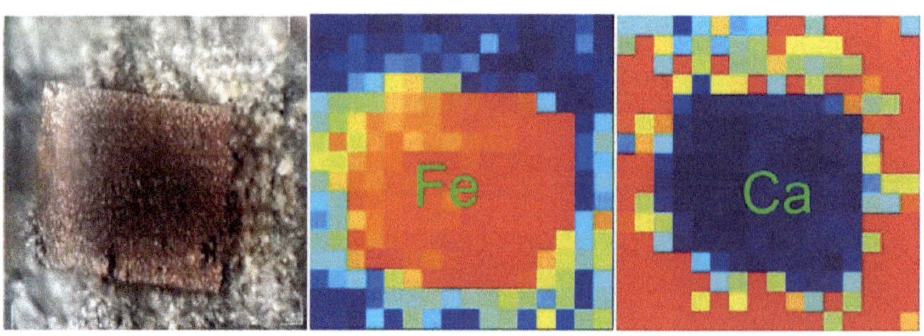

Figure 13.11 Left: photo of iron pyrite inclusion in rock sample; middle and right: 16 × 16 raster grid showing iron and calcium LIBS intensities or "heatmap" from the same sample. Grid-spacing 120 μm, full image width 1.8 mm. *Source*: Connors et al. (2016).

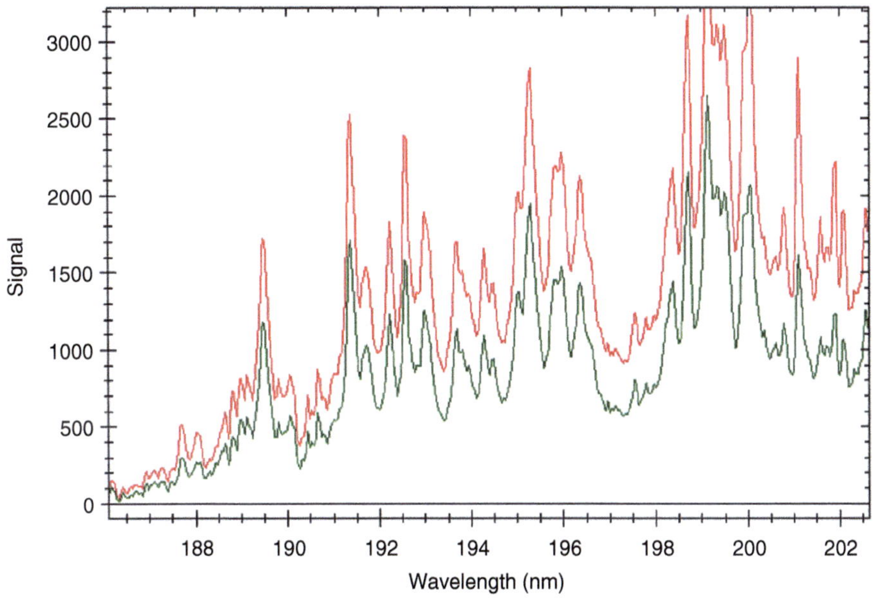

Figure 13.12 LIBS spectrum (1050 Carbon Steel) from 270 μJ laser, 1064 nm laser, 140 Hz, 300 mS data collection window, no gating (green = air, red = argon purge).

emitter; however, it can be enhanced with a helium purge gas due to its high ionization potential (24.6 eV) relative to that of argon (15.8 eV) resulting in lower background and lower plasma-shielding effects (Asimellis et al. 2005; Cremers and Radziemski 1983). The helium enhancement has been used in HHLIBS to detect relative amounts of fluorine in fast food packing (Connors and Day 2018).

While the argon enhancement is beneficial for laser pulse energies greater than 1 mJ, it has been shown that for microjoule lasers, there is very little effect from a purge gas. Figure 13.12 shows an example of the LIBS signal from a steel sample (SRM 1050) in an air and argon environment using a 270 μJ pulse energy. While the signal was slightly higher for argon, there was no significant signal-to-noise enhancement. The insensitivity of microjoule laser plasmas to the gas environment is not well-understood, but may be related to the very short plasma lifetimes and dominance of substrate thermal conductivity.

As with high laser power benchtop LIBS systems, higher power HHLIBS laser plasmas (>1 mJ laser pulse) exhibit significant signal enhancement in the presence of argon (Figure 13.13).

The SciAps "Z" and Thermo Fisher "Apollo" HHLIBS models (5 and 10 mJ, respectively) both utilize argon purge for signal enhancement, and have shown that it is absolutely necessary for low carbon concentration measurement in carbon steels and L-grade stainless steels. In both cases, argon cartridges are enclosed within the instrument and flow is carefully controlled to last for hundreds of measurements. While noncarbon alloy analysis can be carried without argon on >1 mJ laser HHLIBS, the use of argon leads to higher signal levels and better precision for all elements.

13.2.6 Detector Temporal Gating

In the typical higher energy plasma excitation process (>1 mJ), there is enough excitation to ionize a large percentage of the atoms (Noll 2012, pp. 146–149). Shortly after the laser pulse, which is typically a few nanoseconds in duration, the emitted light is observed over a large range of wavelengths with continuous broadband emission. This initial radiation is known as "continuum radiation" or "bremsstrahlung" (or braking light) and results from free electrons with random energies returning to vacant atomic energy levels. As a result of the free

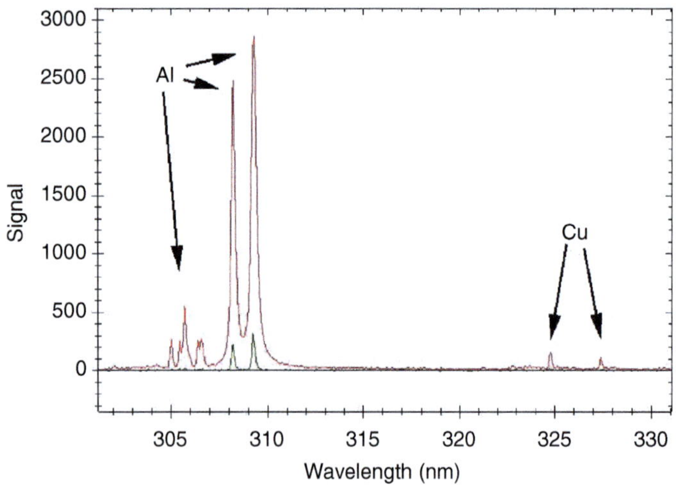

Figure 13.13 HHLIBS spectra of aluminum foil in AIR ambient (green) and ARGON ambient (red) under otherwise identical conditions: 5 mJ laser pulse, 650 nS gate delay.

electrons having a distribution of energies, this initial radiation has no specific emission wavelength and results in broadband white light.

The initial continuum radiation decays on a time scale of 100s of nanoseconds to a few microseconds depending on experimental conditions (laser power, purge gas, energy density). After this point, there still exists both excited neutral atoms and excited ions. While some of the excited states decay during the initial continuum radiation, many excited states still exist well beyond this period and the narrow emission lines can be observed as they decay. Figure 13.14a shows an example of observed temporal emission of a liquid lithium chloride solution at three wavelengths (Hydrogen 656 nm, Lithium 671 nm, and 665 nm where there are no specific element emission lines). The continuum radiation is observed at all three wavelengths and is nearly gone by 0.5 μs. The 665 nm emission falls to zero shortly thereafter. The 656 nm wavelength shows additional emission beyond 0.5 μs and this is due to the H-α decay of excited hydrogen electrons. The hydrogen emission is mostly gone under these conditions (5 mJ laser, 80 μm spot focus, 2 nS pulse duration, air ambient) by 1 μs. The 671 nm emission line, due to Lithium de-excitation, has a longer lifetime than hydrogen and persists out to 3 μs.

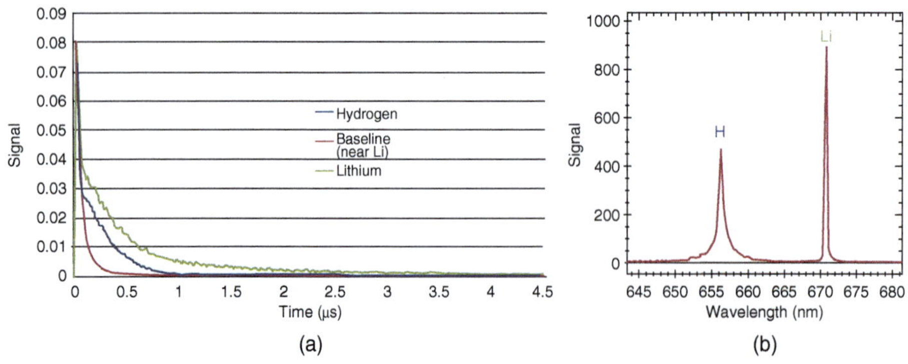

Figure 13.14 (a) Temporal emission from lithium chloride solution using silicon photomultiplier detectors located at three specific wavelengths: hydrogen at 656 nm, lithium at 671 nm, and baseline at 665 nm; (b) resulting HHLIBS spectrum using 250 ns gate delay under identical conditions: (5 mJ 2 ns 1064 nm laser pulse at solution surface, air ambient).

It is, thus, the nature of >1 mJ pulse energy LIBS that short lifetime states can be difficult to observe due to the initial continuum radiation masking them, and other longer lifetime states are easily observed. While it is not absolutely necessary, most benchtop and higher laser energy HHLIBS instruments incorporate "time-gated" spectrometer detectors which can block the initial continuum light through the use of high-speed electronic shutters. The advantages of doing so remove unwanted continuum background and, in many cases, yield narrower emission lines resulting from the lower plasma temperatures and electron densities at longer gate delay times. An example of this is shown in Figure 13.14b which utilized a 250 ns gate delay to eliminate continuum background; however, short lifetime species may be entirely lost by time-gating. This may be a disadvantage if their observation is desired, or it can be an advantage if there are nuisance short lifetime lines superimposed on other longer lifetime species. Figure 13.15 shows the influence of gate delay on the resulting spectrum.

In the lower energy pulse regime of microjoule lasers, the plasma behavior is somewhat different. The broadband continuum emission is much smaller than for the higher power lasers and the emission lifetimes are extremely short (Tashuk et al. 2007, p. 180). At low pulse energies below several 10s of microjoules, the plasma emission duration is similar to the length of the laser pulse itself (Gornushkin et al. 2004). In the several hundred microjoules range, plasma light emission durations are on the order of 10–100 ns, still about two orders of magnitude shorter than the higher energy pulsed lasers (Häkkänen and Korppi-Tommola 1995). At these short lifetimes, gating is simply not possible with standard CCD or CMOS detectors. However, as previously mentioned, the continuum radiation is very low with microjoule lasers and good spectra are obtained without gating. The pulse repetition rate of these lasers is often in the multiple kHz regime and so data are typically collected with the detector shutter left open over hundreds of pulses. This enables a significant amount of light to be accumulated within one detector integration lasting several hundred milliseconds and yields a good signal to noise ratio (SNR). Figure 13.16 shows an example of spectra collected from the same sample (86L20 Steel) and the same HHLIBS spectrometer using an ungated microjoule laser and a gated millijoule laser. In the region near 225 nm, which is dominated by ionic iron emission lines, the resulting spectra are very similar with some continuum and line broadening observed from the ungated microjoule laser.

In the region near 193 nm, where the critical carbon emission line is for steel analysis, there is a larger difference between the two techniques. The microjoule laser exhibits some line broadening and large emission peaks that do not appear in the gated millijoule laser spectrum. Carbon concentration in this steel is near 0.2%. The millijoule

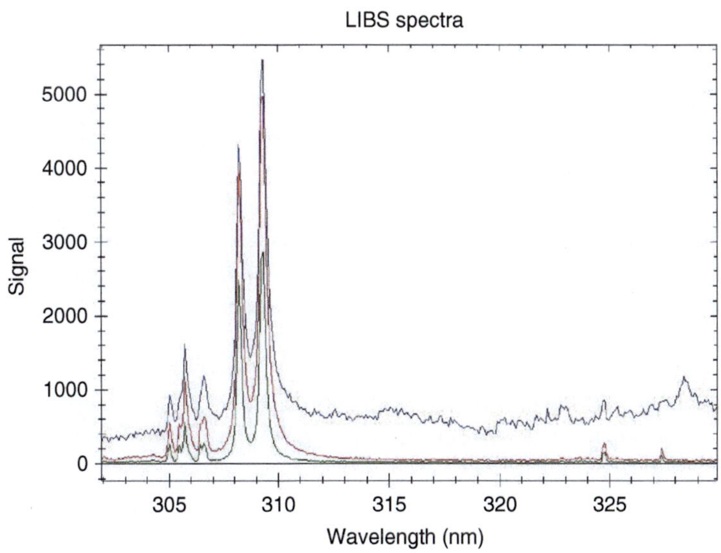

Figure 13.15 HHLIBS spectrum of aluminum foil with various gate delays (green = 650 nS, red = 250 nS, blue = 0 nS) (5 mJ pulse, 2 nS pulse, argon ambient).

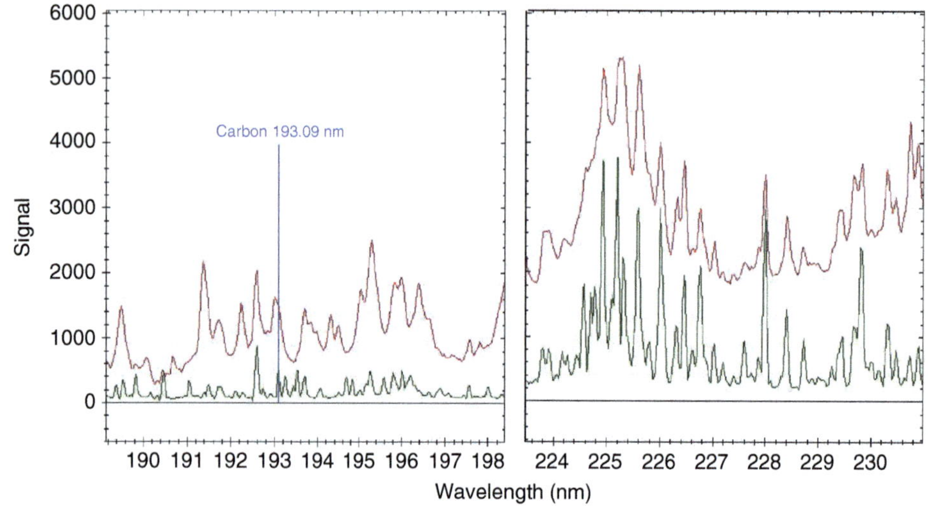

Figure 13.16 LIBS spectra of 86L20 steel, argon environment, using an ungated laser (red, 270 μJ, 140 Hz,300 mS integration), and gated laser (green, 5 mJ, 650 nS gate delay).

laser spectrum exhibits a carbon emission line that is reasonably isolated from neighboring peaks while the microjoule laser spectrum exhibits a shoulder to the left of the carbon emission line. Both techniques are capable of measuring differences in carbon concentrations in steel, but the shoulder in the nongated microjoule laser spectrum makes it difficult to analyze lower concentration levels. The lowest carbon concentration shown in Figure 13.17 for the ungated microjoule laser was for 304H stainless near 0.05% carbon concentration. The smallest carbon peak from the gated millijoule laser was observed from the certified reference material "CP-Iron/IARM 27E" with a carbon concentration of 0.002% (https://www.armi.com/hubfs/Certificates%20of%20Analysis/IARM-27E.pdf).

13.2.7 Calibration

HHLIBS "calibration" can refer to spectrometer wavelength or intensity, and it can also refer to a particular application algorithm that yields elemental concentrations based on the observed emission lines in the LIBS spectrum.

13.2.7.1 Wavelength and Intensity Calibration

For HHLIBS to function well, the wavelength calibration (wavelength vs. pixel number) and intensity response (digital-to-analog converter [DAC] counts/photon vs. pixel number) must either remain stable over time and temperature, or the instrument must have some method to compensate for temperature-related drifts. All spectrometers experience a temperature-dependent shift in wavelength simply due to thermal expansion. To first order, this can be calibrated and built into the software while at the factory. However, in practice, manufacturers typically use known material standards that can be tested in the field. Through measurement of standard's emission line locations and intensity values, corrections can be applied to the factory stored wavelength/intensity calibration. The correction values are, thus, updated every time the reference standards are checked. At least, one commercial HHLIBS can make measurements from an internal calibration surface at the desired intervals based on time or temperature, which frees the user from needing an external calibration standard. These techniques can result in wavelength reproducibility to approximately 0.01 nm despite large temperature changes or severe instrument shock.

13.2.7.2 Application Calibration

The term "calibration" also refers to the algorithm by which LIBS spectral data is converted into elemental concentration results. Calibration techniques include traditional univariate methods, such as those used by OES

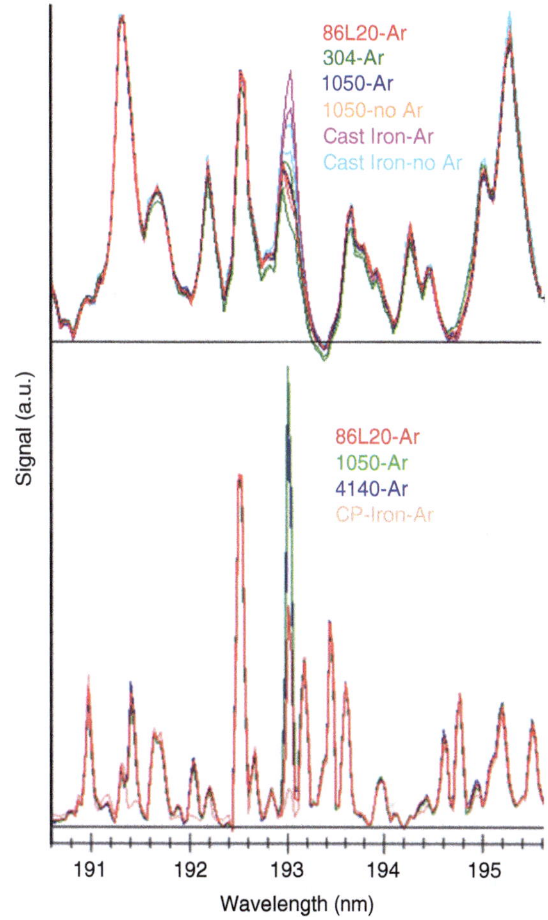

Figure 13.17 Comparison of carbon emission peak at 193.09 nm of various steels. Data processed for baseline removal and normalization to 192.5 nm iron peak. Top: 270 μJ, ungated, 140 Hz, 300 ms integration, argon and no argon (labeled); bottom: 5 mJ, gated, 650 ns gate delay, argon.

(ASTM 1996) and more modern multivariate methods utilizing chemometric modeling (Guang et al. 2015; Zhang et al. 2018) and artificial neural networks (D'Andrea et al. 2014).

The most common method of application calibration in current HHLIBS is similar to conventional OES which creates a linear or polynomial function that relates an element's emission line intensity to the element's concentration. One difficulty with LIBS is that overall plasma intensity can vary significantly from laser shot to shot depending on the material under test. In these cases, to compensate for intensity variation, a matrix element which makes up the bulk of the material is often used as an internal standard. For example, when analyzing copper in steel, the copper emission line intensity will be divided by a nearby iron (matrix element) emission line intensity. The calibration that results is a polynomial that predicts elemental concentration based on that element's "intensity ratio." Another method for intensity compensation is to use the entire integrated spectrum over a selected region as an internal standard rather than a single matrix element peak.

Figure 13.18 shows a typical calibration in this case for molybdenum in iron. In this example, two Mo peaks (553.3 and 550.7 nm) were summed and referenced to the iron line at 537.1 nm. In addition to the straight summing of multiple peaks, signed multiplication factors may be applied to compensate for certain matrix effects or interference effects resulting from an overlapping peak, all techniques common to traditional OES.

Chemometric and machine learning algorithms (Kramer 1998; Mark and Workman 2018) are a rich area for LIBS data analysis. In recent years, some papers using HHLIBS-collected data have been published on PCA (principal component analysis), PLS (partial least squares) and PLS-DA (partial least squares discriminant analysis). PCA

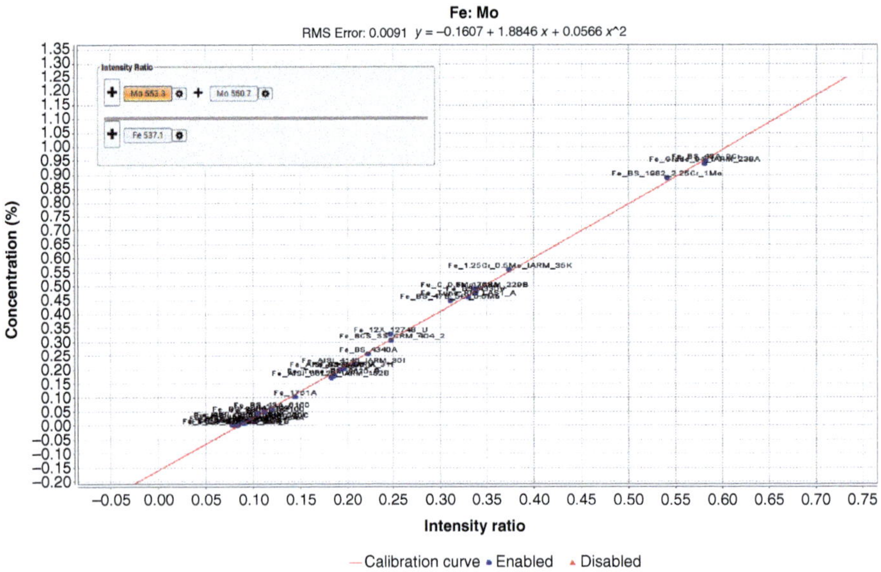

Figure 13.18 Second-order polynomial calibration curve for prediction of molybdenum in iron based on measurement of assayed standards. The intensity ratio on the horizontal axis is calculated from the peak ratios defined in the upper left equation.

and PLS-DA are utilized for material classification and several examples of their use with HHLIBS data applied to geochemical materials have been demonstrated (Harmon et al. 2017, 2018). PLS is a quantitative prediction technique, and its use with HHLIBS data for determining elemental concentrations in steel has been successfully demonstrated (Afgan et al. 2017). Artificial neural network algorithms have also successfully extracted accurate elemental concentration in steels from HHLIBS data (X. Tan, personal communication).

There can be several drawbacks to chemometric and machine learning calibration methods. The data sets required for accurate classification or calibration can be quite large, and, thus, time-consuming to create. A second classical issue is that the models are only as good as the data contained within the model data set and, thus, predictions made on samples falling outside the model data set can yield erroneous results. Finally, chemometric model transfer between instruments is extremely difficult for any measurement technology, but especially so for LIBS due to the large range of variables associated with the LIBS technique (absolute laser power, optics and spectrometer transfer function, detectors) that are difficult to reproduce exactly between instruments.

Improved calibration transfer algorithms for both univariate and multivariate calibration schemes are needed for HHLIBS. Future developments along these lines should contribute significantly to HHLIBS performance enhancement and cost reduction for market-driven commercial applications.

13.2.8 Compact Electronics and Power Supply

Cell phone technology evolution is largely responsible for the availability of small powerful processors, high-capacity memory, high-resolution displays, Wi-Fi–Bluetooth–universal serial bus (USB) communications, global positioning sytem (GPS), and miniature cameras. All of these subunits are now incorporated in most HHLIBS as well as other handheld analyzers, such as X-ray fluorescence (XRF), Raman, NIR, and mid-infrared (MIR) instrumentation. These off-the-shelf components are combined with custom electronics for spectrometer communication, laser drive, and motor drive submodules.

Lithium-ion batteries are used to power HHLIBS instruments. With current battery technology, a 230 g battery yields 50 Wh of energy. Depending on operating characteristics, a 5 or 10 mJ DPSS laser will consume on the order to 10–15 W average while firing. The rest of the HHLIBS system might continuously consume several watts, depending on implemented options. An HHLIBS under heavy usage might run the laser with a 40% duty cycle. Assuming around 8 W of average power consumption, a single 50 Wh 230 g battery can run for about six hours which is more than adequate for normal applications. Various operation times are listed by the different vendors, which obviously depends on actual battery size and internal electronics; however, they are all along these lines. All current commercial HHLIBS enable quick battery change-out in the event that even longer run times are required.

13.3 Commercial HHLIBS Specifications

The currently available commercial HHLIBS have a variety of capabilities and features, many of which have been described in this chapter. The differences arise from the technology utilized, features included, and desired applications for which the products are intended. Table 13.1 shows a brief summary of the basic HHLIBS features.

13.4 HHLIBS Applications

Completely untethered HHLIBS have now been in use for over five years and they have found their way into a variety of diverse application areas (David and Omenetto 2012). The main market-driving application for HHLIBS is metal analysis in the areas of positive material identification (PMI), nondestructive testing (NDT), scrap sorting, and quality control. A secondary market has also been recently developing for a variety of geochemical applications.

Applications just starting or still in development, but not yet driving the HHLIBS market cover a wide range of areas including lithium brine analysis (https://www.sciaps.com/wp-content/uploads/2016/07/SciAps-ApNote_Lithium-Concentrations-in-Brines_Feb2017.pdf?x24702), monoatomic salt PMI (https://www.clinicaltrialsarena.com/products/nanolibs-handheld-libs-analyzer), organic carbon in soils (https://www.sciaps.com/wp-content/uploads/2018/12/SciAps_ApNote_Carbon-in-SoilsFinalRGB.pdf?x24702), soil contamination (Kumar et al. 2019), archeological analysis (Botto et al. 2019), and forensic analysis (Hark and East 2014), as a few examples.

Figure 13.19 shows a rough breakdown of the current HHLIBS market, several examples of which are described herein.

Table 13.1 Commercial HHLIBS instrument characteristics as compiled from Noll et al. (2018), information generally available on the Internet, and personal communications.

	BRUKER	BWTEK	Hitachi	RIGAKU	SCIAPS	THERMO	TSI	VELA
Laser pulse energy	<1 mJ	<1 mJ	<1 mJ	<1 mJ	5 mJ	10 mJ	>5 mJ	up to 200 μJ
Laser pulse rep rate (Hz)	5000	1000–5000	7500	–	50	20	–	up to 4000
Laser ave range power (mW)	100	<200	<450	–	250	200	–	Up to 400
Laser wavelength (nm)	1064	1064	1064	1064	1064	1064	1574	1064
Raster	1D	1D	1D	1D	2D	No	No	1D
Argon purge	No	No	No	No	Yes	Yes	No	No
Detector gating	No	No	No	No	Yes	Yes	–	No
Spectral range (nm)	170–720	200–800	240–510	220–480	188–950	186–420	–	–
Weight (kg, with battery)	2.4	1.8	1.5	1.5	2.2	2.9	–	1.2
Measurement speed (s)	–	0.3	1	~2	2–3, ~10 carbon	~10 carbon	1–3	1

Source: Based on Noll et al. (2018).

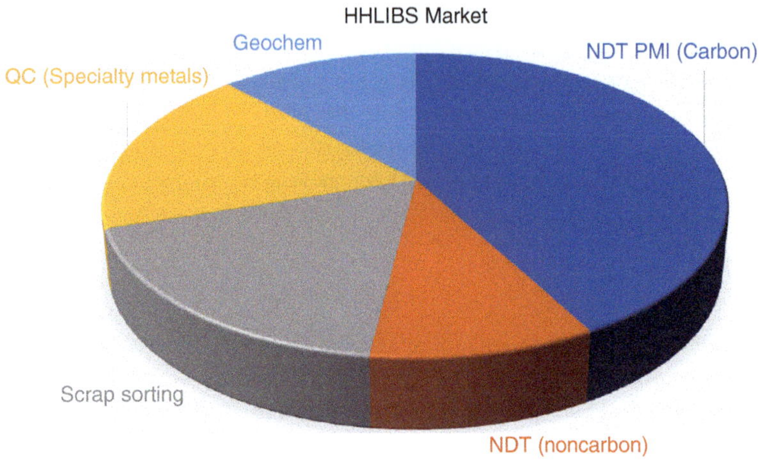

Figure 13.19 Estimated market breakdown of current handheld LIBS sales (D. Sackett and K. Smith, personal communications).

13.4.1 Scrap Sorting

For the last two decades, handheld XRF (HHXRF) has been the dominant instrument of choice for scrap sorting and it continues to be. HHLIBS, however, has found its way into certain scrap applications where HHXRF has a difficulty – for example, where measurement of lighter elements, such as silicon, magnesium, lithium, beryllium and boron are required. Aluminum is the largest scrap-sorting application for LIBS, and can differentiate the aluminum series from 1000 to 8000. One example of this is for scrap produced from the use of 5000 and 6000 series aluminum alloys in the automotive industry where magnesium and silicon concentrations are measured to determine the alloy type (https://www.recycling-magazine.com/2016/10/17/new-sorting-system-for-separating-aluminium-alloys, TSI: https://www.tsi.com/getmedia/859f93cc-4296-46be-a81d-f048d7788a64/LIBS_Automated_Aluminum_Scrap_Sorting_LIBS-028-US?ext=.pdf).

HHLIBS can also differentiate between grades within a series including 6061/6063, 3003/3004, and 7050/7075 alloys (https://hha.hitachi-hightech.com/en/blogs-events/blogs/2019/05/30/identify-aluminium-lithium-alloys-in-a-second-with-the-new-vulcan-optimum-handheld-libs-analyser). The differentiation of these very similar alloy pairs relies on HHLIBS ability to make accurate measurements of Cu, Mg, Si, Zn, Zr, and Cr which may only be present in a few percent down to several tenths of a percent (https://en.wikipedia.org/wiki/Aluminium_alloy). However, the recent improvement in X-ray detection technology, specifically for Mg and Si, are moving several of these tough differentiations back into the HHXRF application arena (https://www.sciaps.com/aluminum-sorting-with-x-ray-or-libs-sciaps-has-the-answers).

The aircraft/aerospace industry makes use of lithium-containing aluminum for the combined purpose of lower density by up to 10% and higher stiffness by up to 15% (Prasad et al. 2014). Lithium is an extremely bright LIBS emitter, and can easily be detected to less than 10 ppm levels making the technique very useful for these types of alloy separations.

Other HHLIBS-sorting applications claimed by the various manufacturers include Co, Cu, Fe, Mg, Ni, Pb, Sn, Ti, and Zn alloys. Most HHLIBS manufacturers list scrap-sorting as a major application for their commercial instruments:

http://mhmp.bruker.com/acton/attachment/18602/f-0036/1/-/-/-/-/EOS%20500%20HH-LIBS%20brochure.pdf.

https://hha.hitachi-hightech.com/en/product-range/products/handheld-xrf-libs-analysers/handheld-libs-analysers.

https://www.rigaku.com/en/products/libs/kt100s/app-scrapmetal.
https://tsi.com/solutions/solutions-by-application/scrap-metal-sorting-and-recycling.
https://www.sciaps.com/industries/scrap-metal-and-alloys.
https://www.thermofisher.com/us/en/home/industrial/spectroscopy-elemental-isotope-analysis/portable-analysis-material-id/industrial-elemental-radiation-solutions/niton-apollo-handheld-libs-analyzer.html.
http://velalibs.com/index.php?lang=en.

13.4.2 Carbon Analysis

One of the most important examples for HHLIBS NDT and PMI is for carbon analysis in low alloy and stainless steels, which previously could only be done by larger arc/spark OES instrumentation. L-grade stainless steels, used in critical applications for petrochemical, food processing, pharmaceutical, and medical equipment, can have carbon concentrations as low as 100 ppm. Improvements in HHLIBS capabilities just during the last three years have yielded limits of detection (LODs) down to 80 ppm (SciAps 2018) enabling the differentiation of L-grade (<300 ppm carbon) and H-grade (>400 ppm carbon) stainless steels (https://www.astm.org/Standards/A240.htm), which is critical for correct material confirmation in these various applications that require either high temperature or high strength operation.

Making measurements of such low carbon concentrations can be difficult due to contamination on the sample surface. Just as in classical OES methods of carbon measurement, HHLIBS steel samples are typically ground to remove any dirt and corrosion buildup. Figure 13.20 shows an example of an HHLIBS stainless-steel calibration which utilizes the 193 nm carbon emission line ratioed to a nearby iron emission line.

13.4.3 Equivalent Carbon for Weldability

Another sizable NDT application is the measurement of Carbon Equivalents (CE), which is related to "weldability" of the material and is obviously important to know or confirm before pipes are welded or repaired. There are several

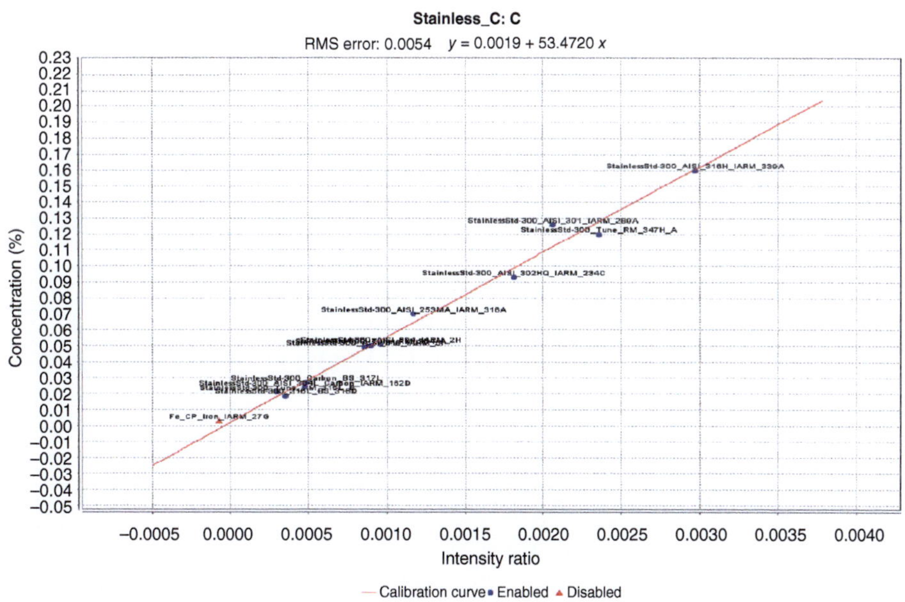

Figure 13.20 Carbon in stainless calibration data and fitted calibration equation. Assayed concentration vs. HHLIBS intensity ratio of carbon 193.1 nm peak intensity to nearby iron emission line. Error bars represent ± one standard deviation of the mean of five measurements.

formulas used for CE, but one example specified by the International Institute of Welding, or IIW, is:

$$CE = \%C + \frac{\%Mn + \%Si}{6} + \frac{\%Cr + \%Mo + \%V}{5} + \frac{\%Cu + \%Ni}{15}$$

All of these elements can be analyzed with HHLIBS, enabling CE determination and verification that the resulting value is less than the IIW threshold of 0.40%, thus reducing the risk of future cracking near the weld joint.

13.4.4 Corrosion

Flow-accelerated corrosion (FAC) in carbon steels from water flow is of large concern in nuclear power plants and highly dependent on chromium concentrations in steel (Poulson 2014). If the chromium concentration decreases to below 0.1%, the rate of corrosion in carbon steel flow systems increases rapidly. HHLIBS can make the required measurement within three seconds to a confidence level of 0.015% Cr, making it a desirable alternative to the HHXRF and the radiation hazards associated therewith (SciAps 2016). Figure 13.21a shows a plot of HHLIBS measured versus assayed chrome content.

Sulfidic corrosion in refineries results from naturally occurring sulfur found in crude oil. Greatly accelerated at temperatures above 260 °C, pipes and pressure vessel walls are slowly thinned over time. The sulfidic corrosion rate is related to several factors including sulfur content, temperature, and silicon content of the carbon steel in use (https://inspectioneering.com/airs/2017-02-13/6123/a-primer-on-sulfidation-corrosion [accessed 1 November 2019]). The American Petroleum Institute's (API's) guidance on avoiding sulfidation corrosion in oil-refining operations is to confirm the silicon content of piping and pressure vessels is above 0.1% (API RP 939-C 2009). HHLIBS testing of silicon on a variety of refinery components is shown in Figure 13.21b with a working LOD of 0.02% and an accuracy of 0.02–0.03% using spark-OES as a reference.

13.4.5 Additional Positive Material Identification (PMI) Applications

PMI is critical for the aerospace, petrochemical and power-generation industries (Rigaku Inc. 2018; SciAps 2019). While HHXRF has traditionally been used for this application, HHLIBS excels in cases where light elements are involved, such as B, Be, Li, Si, and Mg. HHLIBS is particularly well-suited for measurement of Be in Bronze, Si and Mg in Al, and Al and Mg in Ti alloys (Patterson 2015). Measurement of boron-containing nickel and stainless aerospace alloys is also well-suited for HHLIBS. Li, B, and Be measurement down to 10 ppm in Aluminum alloys is also possible (SciAps 2019).

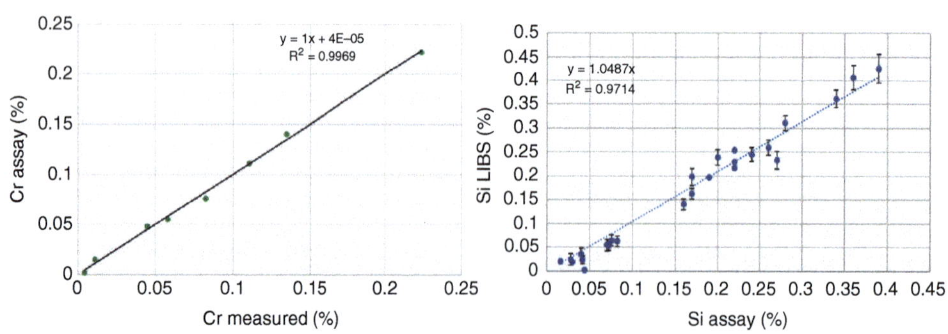

Figure 13.21 (a) Chrome assay vs. HHLIBS-measured values in various carbon steels for flow-accelerated corrosion threshold (0.1% Cr) evaluation; (b) HHLIBS silicon concentration vs. assay made on various refinery components for sulfidic corrosion threshold (0.1% Si) evaluation using the Si 212.4 nm emission line.

13.4.6 Other Applications

There are other applications for HHLIBS, some of which were mentioned in Section 13.4 and many that were not mentioned, too numerous to list here. Some of these are covered elsewhere in this book. While they do not yet account for significant commercial sales, there is a great potential for many of them to become commercially significant in the coming years as technology and algorithms develop further.

13.5 Summary and Future Expectations

Technology advances and size reductions in pulsed lasers, high-power drive electronics, spectrometers, and data processing/communication components over the last decade have led to the current revolution in handheld LIBS instrumentation. It is expected that there is still progress to be made with laser size reduction and pulse power enhancement through improved optical components and optical coating technology. Along with continued spectrometer size reduction, there are new technologies in development, such as spatial heterodyne spectroscopy (Allen and Angel 2018) enabling very small and very high-resolution instruments and which should work well for laser ablation molecular isotopic spectrometry or LAMIS (Bol'shakov et al. 2016). Echelle spectrometers, with recent modeling showing miniature HHLIBS size-compatible versions are possible, are also an excellent candidate for future HHLIBS enabling very large spectral range and extremely high-resolution measurements (B. Beardsley, personal communication). As data processing power continues to increase, advanced chemometric algorithms, neural network and machine learning algorithms should also find their way into standard commercial products. While the current market-driving applications are mainly in metals analysis, the future looks bright for a large host of other applications that can benefit from this exciting field.

References

Afgan, M.S., Hou, Z., and Wang, Z. (2017). Quantitative analysis of common elements in steel using a handheld μ-LIBS instrument. *Journal of Analytical Atomic Spectrometry*. https://doi.org/10.1039/C7JA00219J.

Allen, A. and Angel, S. (2018). Miniature spatial heterodyne spectrometer for remote laser induced breakdown and Raman spectroscopy using Fresnel collection optics. *Spectrochimica Acta Part B: Atomic Spectroscopy.*. https://doi.org/10.1016/j.sab.2018.07.013.

API RP 939-C (2009). *Guidelines for Avoiding Sulfidation (Sulfidic) Corrosion Failures in Oil Refineries*, 1e. American Petroleum Institute.

Asimellis, G., Hamilton, S., Giannoudakos, A., and Kompitsas, M. (2005). Controlled inert gas environment for enhance chlorine and fluorine detection in the visible and near-infrared by laser-induced breakdown spectroscopy. *Spectrochimica Acta Part B*. http://doi.org/10.1016/j.sab.2005.05.035.

ASTM E158-86 (1996). *e1 Standard Practice for Fundamental Calculations to Convert Intensities into Concentrations in Optical Emission Spectrochemical Analysis*. West Conshohocken, PA: ASTM International www.astm.org.

Aubert, J.J. (2001). Microchip lasers. In: *Encyclopedia of Materials: Science and Technology*, 5594–5598. https://doi.org/10.1016/B0-08-043152-6/00978-5.

Bol'shakov, A., Mao, X., Gonzalez, J., and Russo, R. (2016). Laser ablation molecular isotopic spectrometry (LAMIS): current state of the art. *Journal of Analytical Atomic Spetroscopy* https://doi.org/10.1039/C5JA00310E.

Botto, A., Campanella, B., Legnaioli, S. et al. (2019). Applications of laser-induced breakdown spectroscopy in cultural heritage and archaeology: a critical review. *Journal of Analytical Atomic Spectrometry*. https://doi.org/10.1039/C8JA00319J.

Connors, B. and Day, D. (2018). Screening food packaging for fluorinated compounds using laser-induced breakdown spectroscopy. *International LIBS Conference* 2018, Atlanta, GA (21–26 October 2018).

Connors, B., Somers, A., and Day, D. (2016). Application of handheld laser induced breakdown spectroscopy (LIBS) to geochemical analysis. *Applied Spectroscopy*. https://doi.org/10.1177/0003702816638247.

Cremers, D.A. and Radziemski, L.J. (1983). Detection of chlorine and fluorine in air by laser-induced breakdown spectroscopy. *Analytical Chemistry* https://doi.org/10.1021/ac00259a017.

Cremers, D.A. and Radziemski, L.J. (2006). *Laser-Induced Breakdown Spectroscopy*. Chichester: Wiley.

Crocombe, R.A. (2018). Portable spectroscopy. *Applied Spectroscopy* https://doi.org/10.1177/0003702818809719.

Czerny, M. and Turner, A.F. (1930). Über den astigmatismus bei spiegelspektrometern. *Zeitschrift für Physik*. http://dx.doi.org/10.1007/BF01340206.

D'Andrea, E., Pagnotta, S., Grifoni, E. et al. (2014). An artificial neural network approach to laser-induced breakdown spectroscopy quantitative analysis. *Spectrochimica Acta Part B: Atomic Spectroscopy* https://doi.org/10.1016/j.sab.2014.06.012.

Darmont, A. (2009). Spectral response of silicon image sensors. White Paper. https://britastro.org/sites/default/files/attachments/SpectralResponse_WhitePaper_April09.pdf (accessed 1 November 2019).

David, H.W. and Omenetto, N. (2012). Laser-induced breakdown spectroscopy (LIBS), part II: review of instrumental and methodological approaches to material analysis and applications in different fields. *Applied Spectroscopy* 66 (4): 347–419.

Day, D.R. (2018). Analyzer alignment, sample detection, localization, and focusing method and system. US Patent 9,939,383 B2, filed 5 February 2016 and issued 10 April 2018.

Day, D.R., Sackett, D.W. (2017). Libs analyzer sample presence detection system and method. US Patent 9,664,565, filed 26 February 2015 and issued 30 May 2017.

Day, D.R., Derman, K., Egan, J.E., and Soucy, P.E. (2016). Handheld libs analyzer end plate purging structure. US Patent 9,395,243 B2, filed 29 January 2015 and issued 19 July 2016.

Gornushkin, I.B., Amponsah-Manager, K., Smith, B.W. et al. (2004). *Microchip Laser-Induced Breakdown Spectroscopy: A Preliminary Feasibility Investigation*. https://doi.org/10.1366/0003702041389427.

Guang, Y., Shujun, Q., Pengfei, C. et al. (2015). Rock and soil classification using PLS-DA and SVM combined with a laser-induced breakdown spectroscopy library. *Plasma Science and Technology*. https://doi.org/10.1088/1009-0630/17/8/08.

Häkkänen, H.J. and Korppi-Tommola, J.E. (1995). UV-laser plasma study of elemental distributions of paper coatings. *Applied Spectroscopy* 49 (12): 1721–1728.

Hark, R. and East, L.J. (2014). Forensic application of LIBS. In: *Springer Series in Optical Sciences*. https://doi.org/10.1007/978-3-642-45085-3-14.

Harmon, R.S., Hark, R.R., Throckmorton, C.S. et al. (2017). Geochemical fingerprinting by handheld laser-induced breakdown spectroscopy. *Geostandards and Geoanalytical Research*. https://doi.org/10.1111/ggr.12175.

Harmon, R.S., Throckmorton, C.S., Hark, R.R. et al. (2018). Discriminating volcanic centers with handheld laser-induced breakdown spectroscopy (LIBS). *Journal of Archaeological Science* https://doi.org/10.1016/j.jas.2018.07.009.

Kramer, R. (1998). *Chemometric Techniques for Quantitative Analysis*. Marcel Dekkter: Basel.

Kumar, R., Devanathan, A., Mishra, N.L., and Rai, A.K. (2019). Quantification of heavy metal contamination in soil and plants near a leather tanning industrial area using libs and TXRF. *Journal of Applied Spectroscopy* https://doi.org/10.1007/s10812-019-00919-w.

Lyngnes, O., Traggis, N., Dessau, K., and Myatt, C. (2006). Coating technologies for high-damage-threshold optics. *Optics & Photonics News* https://doi.org/10.1364/OPN.17.6.000028.

Mark, H. and Workman, J. (2018). *Chemometrics in Spectroscopy*. https://doi.org/10.1016/B978-0-12-805309-6.09994-3

Miziolek, A.W., Palleschi, V., and Schechter, I. (2006). *Laser Induced Breakdown Spectroscopy*. http://dx.doi.org/10.1017/CBO9780511541261.

Mukhopadhyay, P. (2011). *Development and Characterization of Diode Pumped Solid State Lasers*. Saarbrucken, Germany: LAP LAMBERT Academic Publishing GmbH & Co. KG.

Musazzi, S. and Perini, U. (2014). *Laser-Induced Breakdown Spectroscopy: Theory and Applications*. Berlin Heidelberg: Springer-Verlag.

Noll, R. (2012). *Laser-Induced Breakdown Spectroscopy: Fundamentals and Applications*. Berlin Heidelberg: Springer-Verlag.

Noll, R., Frick-Begemann, C., Connermann, S., and Sturm, V. (2018). LIBS analyses for industrial applications – an overview of developments from 2014 to 2018. *Journal of Analytical Atomic Spectroscopy* https://doi.org/10.1039/c8ja00076j.

Patterson, J.I. (2015). *HH-XRF and HH-LIBS for Alloy Analysis: Choosing the Right Tool for the Right Job*. Portable Analytical Technologies https://pdfs.semanticscholar.org/3346/41a9f180db4a11cb1ffbe6d59805e0be1923.pdf (accessed 1 November 2019).

Poulson, B. (2014). Predicting and preventing flow accelerated corrosion in nuclear power plant. *International Journal of Nuclear Energy*. http://dx.doi.org/10.1155/2014/423295.

Prasad, N.E., Gokhale, A.A. and Wanhill, R.J. (2014). *Aluminum-Lithium Alloys: Processing, Properties, and Applications*. https://doi.org/10.1016/C2012-0-00394-8.

Quant, F., Farmer, K.R., Tan, P.V. et al. (2016). Handheld laser induced breakdown spectroscopy device. US Patent 9,506,869 B2, filed 14 October 2014 and issued 29 November 2016.

Rakovsky, J., Cermak, P., Musset, O., and Veis, P. (2014). *A Review of the Development of Portable Laser Induced Breakdown Spectroscopy and its Applications*. http://dx.doi.org/10.1016/j.sab.2014.09.015.

Rigaku Inc. (2018). Positive Material Identification (PMI) of Metal Alloys. https://www.rigaku.com/en/products/libs/kt100s/app003 (accessed 1 November 2019).

Scheeline, A. (2017). How to design a spectrometer. *Applied Spectroscopy* 71 (10): 2237–2252.

SciAps (2016). *Fast, Laser-Based Cr Analysis for Flow Accelerated Corrosion (FAC) Applications*. Application Note. https://www.sciaps.com/fast-laser-based-cr-analysis-for-flow-accelerated-corrosion-fac-applications (accessed 1 November 2019).

SciAps (2018). *Carbon Analysis in Stainless and Carbon Steels with Handheld LIBS*. Application Note. https://www.sciaps.com/wp-content/uploads/2018/04/SciAps_ApNote_LIBSCarbonMARCH2018.pdf?x24702 (accessed 1 November 2019).

SciAps (2019). NDT and PMI. https://www.sciaps.com/industries/ndt-pmi (accessed 1 November 2019).

Tashuk, M.T., Cravetchi, I.V., Tsui, Y.Y., and Fedosejevs, R. (2007). Micro-LIBS. In: *Laser-Induced Breakdown Spectroscopy* (ed. T. Page), 173–196. Amsterdam: Elsevier.

Wang, S.X.. (2017). Laser induced breakdown spectroscopy (libs) apparatus based on high repetition rate pulsed laser. US Patent 9,797,776 B2, filed 27 May 2016 and issued 24 October 2017.

Zayhowski, J.J. and Mooradian, A. (1989). *Single-Frequency Microchip Nd Lasers*. https://doi.org/10.1364/OL.14.000024.

Zhang, T., Tang, H., and Hua, L. (2018). Chemometrics in laser-induced breakdown spectroscopy. *Journal of Chemometrics*. https://doi.org/10.1002/cem.298.

14

Miniaturized Mass Spectrometry – Instrumentation, Technology, and Applications

Dalton T. Snyder

Resource for Native Mass Spectrometry Guided Structural Biology, The Ohio State University, Columbus, OH, USA

14.1 Introduction

Mass spectrometry (MS) is a highly sensitive, selective, and versatile analytical tool for complex mixture analysis. Despite its overwhelming utility, mass spectrometry has usually been constrained to conventional analytical laboratories. Benchtop mass spectrometers are bulky, heavy, power-hungry, and expensive, akin to the first electronic computers of the mid-1900s or cell phones of the late twentieth century, which were not widely used, yet viewed promising. Even so, just as researchers and manufacturers reduced the size and power consumption of electronic devices while increasing their ease of use, so too is mass spectrometry being miniaturized to the point of portability, and so the scope of applications that can be investigated by mass spectrometry is being expanded.

At its core, mass spectrometry is the science and technology of ions, i.e. charged particles. The goal of most mass spectrometry experiments is to measure, ideally in an unbiased manner, the mass-to-charge ratios (m/z) of ions generated from a particular sample, and, in doing so, deduce the chemical makeup of the sample's constituents. For example, mass spectrometry can determine whether an apple is organic by determining the presence or absence of pesticides, herbicides, and fungicides on the apple. Foliar phytochemicals can be monitored in response to some external stimulus, e.g. herbivory or volatile signaling cues from nearby plants. Trace amounts of explosives and narcotics can be detected from suspicious packages or residues during traffic stops or at an airport. Mass spectrometry can also detect chemical warfare agents CWAs on the battlefield in order to protect soldiers or civilians during times of war.

With conventional benchtop mass spectrometers, samples from each of these situations would have to be transported back to the lab for analysis. While there are many analytical questions that favor benchtop systems, which typically have higher sensitivity, resolution, and mass accuracy than their portable counterparts, in the cases listed above, portable mass spectrometers with modest performance would be more fitting. Removing suspicious packages from X-ray machines at the airport and moving or shipping them to a proper analytical laboratory is at best inconvenient and at worst logistically impractical. If endogenous metabolites or externally applied chemicals (pesticides) are to be measured from plant foliage, the chemical makeup of the sample may change by the time it reaches the laboratory. In situations where lives are at risk – detecting CWAs or explosives, for example – the laboratory often *must* be brought to the sample rather than the sample being brought to the laboratory. There is, therefore, an increasing need and desire for portable analytical instrumentation, with mass spectrometry at the forefront as one of the most powerful tools in the analytical chemist's arsenal.

The first step in the mass spectrometry experiment is to produce ions from a sample of typically neutral molecules. Once ions are produced by the ionization source and transported to the mass analyzer in a vacuum, their m/z values can be measured with high accuracy and precision (indicating the molecular formulas of the

Portable Spectroscopy and Spectrometry 1: Technologies and Instrumentation, First Edition.
Edited by Richard A. Crocombe, Pauline E. Leary and Brooke W. Kammrath.

analytes) and, if coupled with tandem mass spectrometry, the arrangement of atoms of each of the sample's constituents can be determined. In order for the ion's m/z to be measured successfully by the analyzer, the ion must be transported from ambient air outside the vacuum chamber to the mass analyzer far below ambient pressure, either directly so through a capillary inlet or via an intermediate interface (e.g. a membrane inlet for the transport of dissolved gases and volatile organics). This represents a change in pressure of over seven orders of magnitude! While the ion is in vacuum, its trajectory can be controlled very precisely by electric (e.g. ion trap or quadrupole) and/or magnetic (e.g. magnetic sector) fields without substantial deviations in the ion motion that would otherwise be caused by collisions with background gas molecules. Vacuum systems are by far the most difficult subsystems to miniaturize and are usually the power and size bottleneck for portable MS systems.

14.2 Instrumentation

14.2.1 Ion Introduction and Vacuum Systems

The chemical makeup of a sample is determined by measuring the m/z values of the sample's components using a mass analyzer. In order for the ions' trajectories to be manipulated accurately, precisely, and *predictably*, the ions must be analyzed in a vacuum, normally at pressures where collisions with background gas have a negligible effect on ion motion. Generally, it is best if the mean free path of the ions is ~5–10 times the length of ion travel in the mass analyzer. For quadrupole ion traps, which can function at the highest pressure of all analyzers, the typical operating pressure is 10^{-5} to 10^{-3} Torr. In exceptional cases, ion traps can be operated at pressures around 1–10 Torr, but this is not typical and mass spectral resolution and sensitivity are severely compromised. Quadrupole mass analyzers operate at 10^{-6}–10^{-5} Torr, time-of-flight (TOF) mass analyzers require at most 10^{-6} Torr, and sectors require similarly stringent pressure regimes. For TOF and sector mass analyzers, collisions are more detrimental toward resolution than for ion traps and quadrupoles. Orbitrap and Fourier transform ion cyclotron resonance mass analyzers are not discussed here as they are not appropriate analyzers for miniature systems due to their requirement for ultrahigh vacuum (problematic for terrestrial applications but feasible in space) and their complexity.

In order to achieve the necessary vacuum conditions for mass analysis, the mass analyzer needs to be seated in a vacuum chamber, typically a hollow block of stainless steel or aluminum. Various feedthroughs for vacuum pumps and control voltages are machined into a chamber that is otherwise sealed gas-tight with o-rings. A critical consideration in the design of a vacuum chamber is how ions will be introduced into the system. This decision determines which pumps are chosen to maintain the requisite mass analyzer pressure, and the size and power consumption of the pumps regulate the portability of the entire mass spectrometer.

In general, there are three strategies for ion introduction, shown in Figure 14.1. The first option (Figure 14.1a), which matches the strategy employed for benchtop mass spectrometers, is to couple differential pumping with a continuous atmospheric pressure interface (API). Ions from the ion source are introduced through an "inlet," usually a small hole in the vacuum chamber or a capillary that protrudes from the chamber, which allows the ions to pass from ambient pressure (760 Torr) to a first vacuum region at ~1–10 Torr. The modest pressure in the first region of differential pumping is often achieved in portable systems with a single diaphragm or scroll pump. The first region of differential pumping must usually have an ion optical element, such as an radio-frequency (RF) multipole or an ion funnel as well as a skimmer in order to keep the desolvating ions from being defocused as they undergo free jet expansion. In rare cases, e.g. Figure 14.1a, the inlet capillary stretches the length of the first pressure region. This strategy is simple but sacrifices sensitivity as ions are lost between the stages of differential pumping. The ions are then transported to the second stage of differential pumping which is held at high vacuum pressure and which houses the mass analyzer. Although differential pumping is ideal for benchtop systems, which do not have size and power constraints, it is challenging to implement effectively on portable spectrometers coupled with a continuous ion source due to the limitations in pumping capacity. Even so, a high acquisition rate

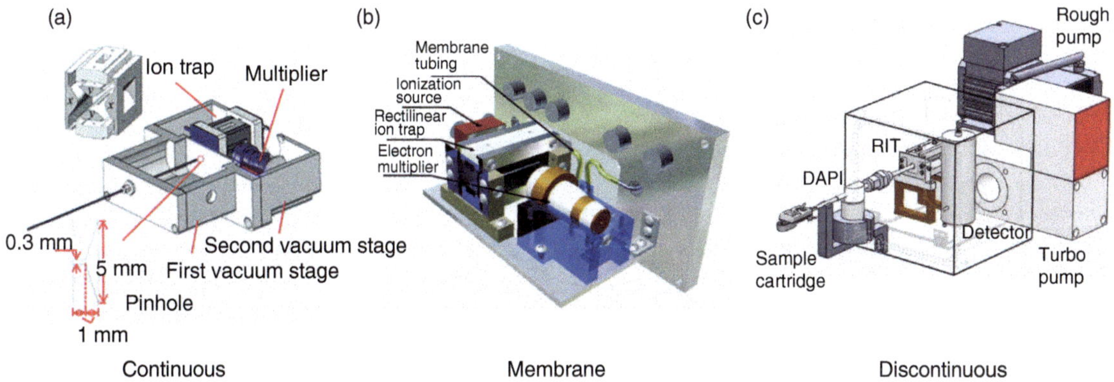

Figure 14.1 Atmospheric pressure interfaces for portable mass spectrometers: (a) continuous atmospheric pressure interface (*Source*: Zhai, Y., Feng, Y., Wei, Y., Wang, Y., and Xu, W. (2015). Development of a miniature mass spectrometer with continuous atmospheric pressure interface. *Analyst* 140: 3406–3414. © 2015 Royal Society of Chemistry); (b) membrane inlet (*Source*: Gao, L., Song, Q., Patterson, G.E., Cooks, R.G., and Ouyang, Z. (2006).Handheld rectilinear ion trap mass spectrometer. *Anal. Chem.* 78: 5994–6002. © 2006 American Chemical Society); and (c) discontinuous atmospheric pressure interface (DAPI) (*Source*: Li, L., Chen, T.C., Hendricks, P.I., Cooks, R.G., and Ouyang, Z. (2014). *Anal. Chem.* 86: 2909–2916. © 2014 American Chemical Society).

(# of scans/second) is maintained because the ion source and mass analyzer can be operated continuously, and any external ionization source may be coupled with the system.

A second, particularly popular, option is to place the ion source inside the vacuum chamber ("internal" ionization) and introduce the sample through a membrane inlet (Figure 14.1b). The membrane, typically polydimethylsiloxane, limits gas conductance into the instrument and is simple to implement. Liquid and gas samples can both be introduced and heated so that volatiles and semivolatiles are desorbed and pass through the membrane and into the ionization region. Usually, the ion source is a variant of electron impact (EI) ionization, which generates molecular cations for mass spectrometry analysis. Like the continuous atmospheric interface, the system pressure is maintained throughout the analysis, and so the mass analyzer and ion source can be operated continuously, resulting in a high acquisition rate. The chemical composition of the membrane can be tailored for molecules of a particular size, polarity, or affinity; albeit, characteristics that increase the membrane's selectivity also constrain its applicability (the main disadvantage). Membrane introduction mass spectrometry (MIMS) nonetheless remains a viable option for many targeted portable MS applications (e.g. monitoring volatiles in the environment), and the lax pumping requirements of MIMS systems are attractive for reducing system weight and power (SWaP).

The third API strategy is to pulse ions into the vacuum chamber, that is, implement a discontinuous atmospheric pressure interface (DAPI, Figure 14.1c). A compromise of sorts between the continuous and membrane interfaces, DAPI allows the use of smaller and lower power pumps while also maintaining compatibility with external ionization sources. Similar to the continuous API, ions travel through an aperture in the vacuum chamber or through a capillary inlet. However, in DAPI implementations, the ion introduction is pulsed rather than continuous. This is typically accomplished by replacing a portion of the metal inlet capillary with a soft silicone section that is constricted by a pinch valve. To introduce ions (and neutrals), the pinch valve is opened for ~10–20 ms, during which the pressure in the vacuum chamber increases by two to three orders of magnitude. Ions are trapped in the mass analyzer or some other trapping device while the neutrals are pumped away until the requisite pressure for mass analysis is attained, a period of time on the order of 500 ms. Mass analysis can then proceed as usual. The main compromise when using a discontinuous interface is the loss of acquisition rate. Because the mass analyzer must be pumped down to operating pressure after ions and neutrals are introduced, the typical acquisition rate of DAPI systems is ~1 Hz. The variable pressure in the vacuum chamber also limits DAPI systems to ion traps as mass analyzer candidates. Even so, DAPI is a promising ion introduction technology because the use of small vacuum pumps enables greater portability and simplifies the MS design since a single vacuum stage is all that is needed, and the ability to couple with external ambient ionization sources is an attractive proposition.

The API design drastically affects both the size of the inlet aperture and the choice of backing and turbopumps. While a larger diameter inlet allows more ions into the mass analyzer per unit time, a greater number of neutral gas molecules are also introduced. Larger inlets, therefore, must also be accompanied by higher pumping capacity (read: bulkier and higher power pumps), which, in turn, limits instrument portability. So while a continuous API may seem the ideal option between the three choices discussed above, in practice, the apertures between pressure stages must often be made quite small to limit gas introduction, which limits ion fluence. For systems which use non-trapping mass analyzers (e.g. quadrupoles and magnetic sectors), it is usually not efficient to couple a DAPI interface with the analyzer without differential pumping, and so either the continuous or membrane API (or another API with limited gas conductance) is the only option.

The miniaturization of mass spectrometry would certainly not be possible without the ability to maintain a high vacuum with sufficiently small pumps. Figure 14.2 shows several of the most common backing and high vacuum pumps used in miniature and portable mass spectrometers, and Table 14.1 lists some of their specifications. KNF Neuberger, Pfeiffer, and Creare are the main manufacturers of small vacuum pumps. Usually, the pumping system of a miniature mass spectrometer consists of a ~10–80 L/s turbopump backed by a ~5 L/min diaphragm or scroll pump. The Pfeiffer line of high vacuum pumps includes the HiPace 10 (10 L/s), HiPace 30, HiPace 60 P, and HiPace 80, with higher capacity being accompanied by higher power consumption and instrumental bulk. For example, while the HiPace 80 specifications indicate a pumping capacity of up to 67 L/s (for nitrogen), its maximum 110 W power consumption and 2.4 kg mass are noticeably higher than the HiPace 10's 28.8 W maximum power consumption and 1.8 kg mass. Creare also manufactures specialized miniature vacuum pumps for portable mass spectrometers, including a 550 g turbopump and a 130 g drag pump. Both offer >4 L/s pumping capacity and a low power consumption of 12 W while maintaining a remarkably small footprint. Regardless of which high vacuum pump is chosen, it must be coupled with a backing pump, most often a ~5 L/min diaphragm pump akin

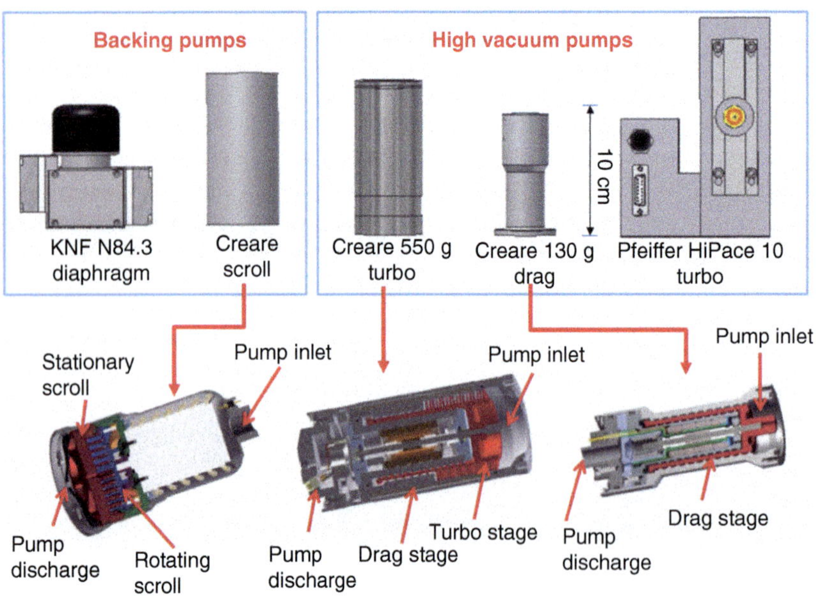

Figure 14.2 Representative backing and high-vacuum pumps used in portable mass spectrometers. *Source*: Chen, C., Chen, T., Zhou, X., Kline-Schoder, R., Sorenson, P., Cooks, R.G., and Ouyang, Z. (2014). *J. Am. Soc. Mass Spectrom.* 26, 240–247. © 2014 American Chemical Society.

Table 14.1 Characteristics of the vacuum pumps in Figure 14.2.

Pump	Weight (g)	Pumping speed	Compression ratio	Max. foreline pressure (Torr)	Max. power (W)	Rotor speed (rpm)	Inlet ID (mm)
KnF N84.3 diaphragm pump	900	5 L/min (at 1 atm)		760	18		
Creare 130 g scroll pump	350	1 L/min (at 1 torr)		760	6		
HiPace 10 turbopump	1800	10 L/s (at 0.1 mtorr)	3×10^6	18	28	90k	24
Creare 550 g turbopump	550	>4 L/s (at 0.1 mtorr)	1×10^9	10	12	100k	53
Creare 130 g drag pump	130	>4 L/s (at 0.1 mtorr)	1×10^5	10	12	200k	25

Source: Chen, C., Chen, T., Zhou, X., Kline-Schoder, R., Sorenson, P., Cooks, R. G., and Ouyang, Z. Design of portable mass spectrometers with handheld probes: aspects of the sampling and miniature pumping systems. *J. Am. Soc. Mass Spectrom.* 26: 240–247. © 2014 American Chemical Society.

to the KNF N84.3, which weighs 900 g and consumes up to 18 W of power. KNF offers similar pumps with small variations in pumping capacity. The Pfeiffer MVP line of diaphragm pumps (not shown) consists of the MVP 010 which has a pumping capacity of up to 10 L/min and weighs only 2 kg. Meanwhile, Creare manufactures a 350 g scroll pump with a pumping capacity of 1 L/min and maximum power consumption of 6 W. The ultimate pressure of these backing pumps is usually on the order of ~1 Torr, with final pressures below 10^{-5} Torr when coupled with a turbopump. This pressure is sufficiently low for quadrupole mass filters and ion traps which make up the majority of miniature mass analyzers.

14.2.2 Sampling and Ionization

Because mass spectrometry uses electric and magnetic fields to control the trajectories of charged particles in the gas phase, the first step in the analysis procedure is usually to generate ions from the corresponding neutral molecules that make up a sample and to transfer those ions into the vacuum chamber. If internal ionization is utilized instead of external ionization, then the neutral molecules are moved into the vacuum chamber and then ionized. Ionization and sampling are, therefore, critical steps to consider in the mass spectrometry workflow since they affect overall instrument design and can either limit or augment the capabilities of a portable mass spectrometer.

Ionization can be broadly classified as either external or internal, depending on where ions are generated from the corresponding neutral molecules. In the former case, sampling and ionization take place outside the vacuum chamber, and often sampling and ionization are conducted simultaneously (e.g. with ambient ionization sources). Sampling and external ionization can also be decoupled. One example is electrospray ionization (ESI), wherein the sample is dissolved in an ESI-compatible solvent to accomplish the sampling step. Ionization is effected by transporting the solution to the sharp end of a metal needle or glass capillary and applying a high potential difference between the solution and the inlet capillary of the instrument, generating a mist of charged droplets and eventually bare ions. In contrast, internal ionization is accomplished *inside the vacuum chamber*, and so the neutral sample (solid, liquid, or gas) must first be volatilized and transported into vacuum from atmospheric pressure. As discussed in the next paragraph, internal ionization is limited in scope, yet it remains a powerful tool in combination with well-established spectral EI ionization libraries.

Internal ionization occurs inside the vacuum chamber; the most common method is EI ionization in which an energetic (70 eV) electron beam from a resistively heated filament is directed at a volatilized sample. Figure 14.3a shows an arrangement of an EI ionization source coupled to a 3D ion trap. EI produces molecular cations (i.e. the neutral molecule loses an electron) but is also "hard" enough to cause fragmentation. As such, a single analyte generally produces an array of mass spectral peaks including the molecular cations and fragments from unimolecular dissociation. Fortunately, EI is highly reproducible across instrument platforms and is ubiquitous. EI mass spectra acquired on a portable instrument can be directly compared to publicly available or internal databases – the National Institutes for Standards and Technology (NIST) EI database, for example – and sample components can be identified with high accuracy so long as they have undergone prior separation by gas chromatography (GC) to limit spectral contamination. Drawbacks of EI include limited filament lifetime at high pressure, especially in the presence of oxygen, and high power consumption. Glow discharge EI ionization can produce EI-like spectra without the use of a filament and at a fraction of the power consumption. A potential difference is applied between opposing plates in a pressurized (~0.5 Torr) region inside the vacuum chamber, generating cations, anions, and electrons. The electrons are then extracted through an aperture in one of the plates and are used for ionization.

As just described, for systems which use internal ionization, the sample must be converted to the gas phase (volatilized), moved inside the vacuum chamber, and finally ionized. Hence, the sample must be volatile, which severely limits the scope of molecular classes that can be observed to small, mostly nonpolar analytes such as volatile organic compounds (VOCs) and explosives. Even so, one of the primary benefits of internal ionization – often coupled with membrane inlets or other gas-constricting manifolds – is that less pumping capacity can be used to maintain vacuum compared to spectrometers that use external ionization. Several sampling methodologies can be happily married with internal ionization sources. Common suitors are headspace sampling, swabs, and solid-phase microextraction probes. Headspace sampling involves injecting a gaseous sample at equilibrium with its headspace and is particularly common for analysis of VOCs. Usually, internal ionization is coupled with GC to analyze headspace vapors. A drawback of this method is its limited sensitivity since the sample is not preconcentrated, yet its simplicity is attractive. Similarly, swabs made of paper, nylon, or other porous materials can also be used for sample collection. The swab is then inserted into a slot on the instrument and heated, thereby desorbing volatile and semivolatile analytes which are ionized and detected. Solid-phase microextraction (SPME) probes are common sampling devices (Figure 14.3b). Using SPME, analytes can instead be preconcentrated on a hydrophobic sorbent bed prior to injection onto a GC. SMPE devices are typically fibers coated with liquid or solid polymeric films. The fiber is dipped into a sample of interest, wherein the sample adsorbs to the fiber material. The SPME probe is then inserted into the mass spectrometer and the sample-containing fiber is exposed and heated to desorb analytes from the fiber. SPME devices are small and convenient for trace sampling, are reusable (like some swabs), and have increased sensitivity due to sample concentration prior to injection into the instrument.

External ionization sources generate ions outside the vacuum chamber. Whereas for internal ionization sampling and ion generation are distinct, external ionization sources may entail only ionization or both ionization and sampling. The most common sources in which sampling and ionization are still distinct are ESI, nanoelectrospray ionization (nESI), and atmospheric pressure chemical ionization (APCI). To generate ions using ESI or nESI, a sample is dissolved in a solvent and the resulting solution is placed into a metal or glass capillary with a sharp tip. With the application of a high voltage between the solution and the instrument inlet (Figure 14.3c, voltage not shown), charged microdroplets are formed which eventually fission and desolvate to leave bare protonated or salt-adducted ions. Spray-based ionization is generally only suitable for polar analytes, yet it remains the most popular ionization source in mass spectrometry due to high sensitivity, ease of use at atmospheric pressure, and "softness" (i.e. very little fragmentation is induced in the source). APCI, another "soft" ionization source, uses a corona discharge to generate ions from a sample that has been heated and volatilized. APCI is suitable for polar and some nonpolar compounds so long as they are not thermally labile, and the ions observed are similar to those from ESI. In general, instruments which utilize external ion sources tend to have broader applicability in exchange for requiring higher pumping capacity to remove excess solvent and gas that is introduced through the instrument inlet. Matrix-assisted laser desorption/ionization (MALDI) and laser desorption ionization (LDI) are worth mentioning as alternative ion sources that use lasers to effect ionization. Because lasers increase overall

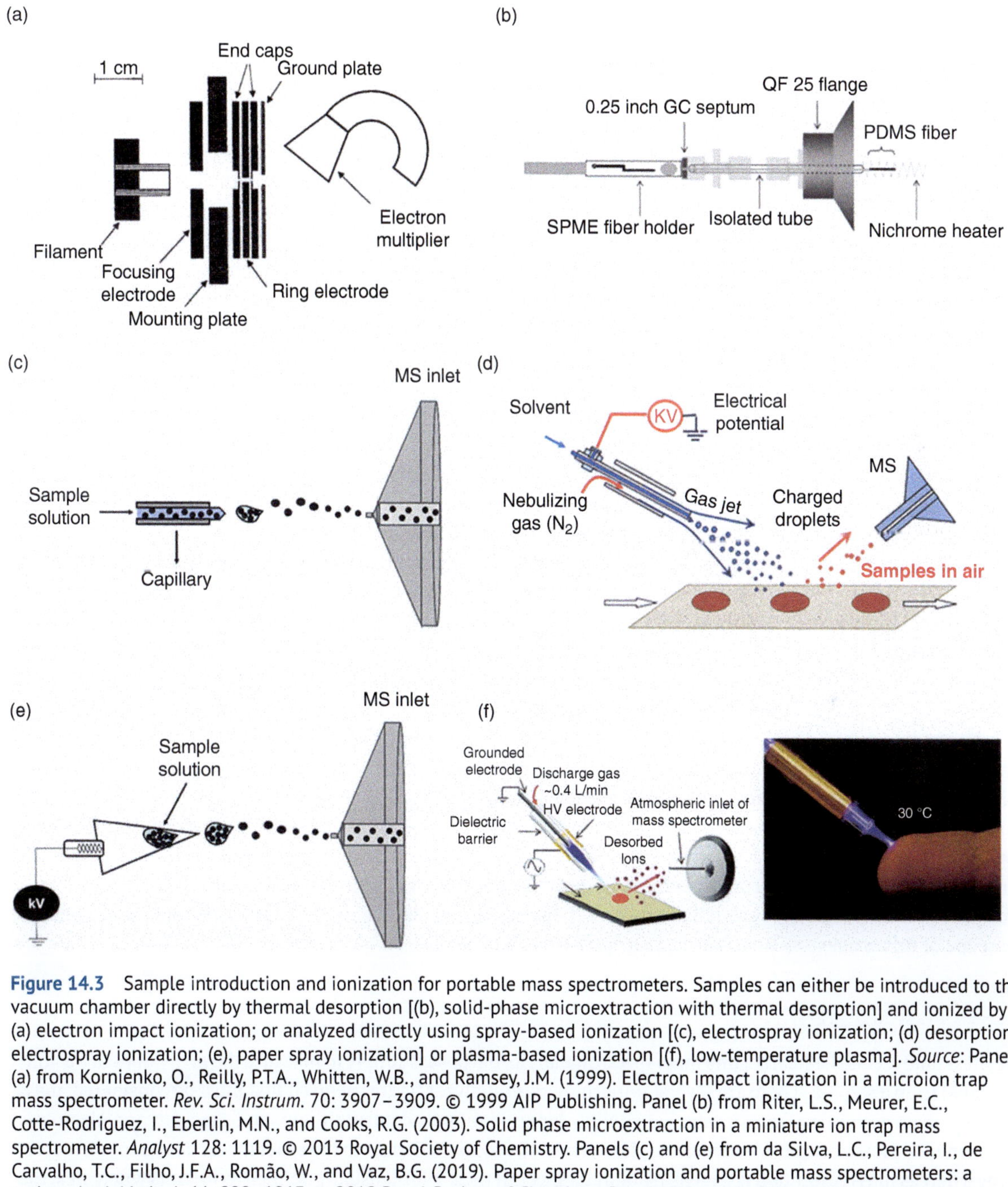

Figure 14.3 Sample introduction and ionization for portable mass spectrometers. Samples can either be introduced to the vacuum chamber directly by thermal desorption [(b), solid-phase microextraction with thermal desorption] and ionized by (a) electron impact ionization; or analyzed directly using spray-based ionization [(c), electrospray ionization; (d) desorption electrospray ionization; (e), paper spray ionization] or plasma-based ionization [(f), low-temperature plasma]. *Source*: Panel (a) from Kornienko, O., Reilly, P.T.A., Whitten, W.B., and Ramsey, J.M. (1999). Electron impact ionization in a microion trap mass spectrometer. *Rev. Sci. Instrum.* 70: 3907–3909. © 1999 AIP Publishing. Panel (b) from Riter, L.S., Meurer, E.C., Cotte-Rodriguez, I., Eberlin, M.N., and Cooks, R.G. (2003). Solid phase microextraction in a miniature ion trap mass spectrometer. *Analyst* 128: 1119. © 2013 Royal Society of Chemistry. Panels (c) and (e) from da Silva, L.C., Pereira, I., de Carvalho, T.C., Filho, J.F.A., Romão, W., and Vaz, B.G. (2019). Paper spray ionization and portable mass spectrometers: a review. *Anal. Methods* 11: 999–1013. © 2019 Royal Society of Chemistry. Panel (d) from Ifa, D.R., Wu, C., Ouyang, Z., and Cooks, R.G. (2010). Miniaturized mass spectrometry – instrumentation, technology & applications. *Analyst* 135: 669–681. © 2010 Royal Society of Chemistry. Panel (f) reproduced from Harper, J., Charipar, N.A., Mulligan, C.C., Zhang, X., Cooks, R.G., and Ouyang, Z. (2008). Low-Temperature plasma probe or ambient desorption ionization. *Anal. Chem.* 23: 9097–9104. © 2008 American Chemical Society.

instrument volume and power consumption, they are not ideal for portable mass spectrometers, but they may be useful for targeted applications and for mass spectrometry imaging.

Ambient ionization is used to generate ions from samples in their native environment without any kind of sample pretreatment or preparation. Because sampling and ionization steps take place simultaneously, ambient ionization is the preferred method of ion generation for portable instruments. Ambient ionization was coined at the advent of desorption electrospray ionization (DESI), a technique in which a spray of charged microdroplets is aimed at a sample (Figure 14.3d). The sample is dissolved in a thin film of solvent and subsequently desorbed by further impacts of microdroplets. This generates a beam of fissioning/desolvating microdroplets that can be collected in the inlet of a mass spectrometer for analysis. The second ambient ionization technique reported in the literature was direct analysis in real time (DART), in which excited state gaseous species (helium, nitrogen, or argon) formed from exposure to a high potential difference cause a cascade of ion/molecule reactions which eventually produces pseudomolecular (protonated) ions. While DESI is most appropriate for polar analytes, DART is applicable to polar and some nonpolar analytes. Both DESI and DART use compressed gases, in DESI to generate the microdroplet spray and in DART to create excited-state atoms/molecules, which adds to overall SWaP. In addition, the extra gas load may increase system pressure, yet these drawbacks may be overshadowed by the versatility of combining sampling and ionization. Paper spray ionization is another ambient ionization source that replaces the glass or steel capillary of ESI with a simple paper triangle to which sample and high voltage are applied (Figure 14.3e). In this case, the paper substrate can also conveniently act as a swabbing material for sample collection. The paper can be modified in various ways to increase its sensitivity and/or selectivity (e.g. by increasing or decreasing its hydrophobicity) or to decrease the voltage needed to generate a spray (e.g. by doping in nanowires or carbon nanotubes). Other raw materials may be used as substrate, not just paper. For example, leaves and other plant material can be cut to a point and solvent and high voltage can be applied to generate ions ("leaf spray ionization"). Hence, paper spray is remarkably simplistic and easy to use, yet also quite versatile.

Plasma-based ambient ionization sources also abound. For example, a low-temperature plasma (LTP, Figure 14.3f) can be generated from an inert carrier gas (He, N_2, Ar) at atmospheric pressure by applying a high alternating current (AC) voltage across a dielectric barrier. The plasma interacts directly with the sample for ionization and desorption. It is limited to analytes of low molecular weight (particularly those that are also volatile) but is applicable to both polar and nonpolar analytes and has low power consumption (~5 W). Many other plasma ionization sources exist, namely desorption atmospheric pressure chemical ionization (DAPCI), flowing afterglow atmospheric pressure glow discharge, plasma-assisted desorption ionization, and dielectric barrier discharge ionization. The differences in device configuration, AC vs. direct current (DC) discharge, gas substrate, plasma properties (e.g. temperature and "hardness"), flow rate, and power consumption are beyond the scope of this article.

14.2.3 Mass Analyzers

The mass analyzer is the heart of any mass spectrometer; its purpose is simply to measure the m/z values of the ions generated from a sample. There are several important mass analyzer figures of merit that must be considered: mass accuracy, mass resolution, mass range, sensitivity, dynamic range, acquisition rate, and pressure and electronic requirements. Mass accuracy describes the closeness of the measured mass to the exact theoretical (monoisotopic) mass, whereas mass resolution is a measure of mass spectral peak width, usually defined as $m/\Delta m$ (where m is the measured m/z and Δm is the peak width measured as the full-width at half-maximum [FWHM] intensity). Sensitivity is the slope of the calibration curve (measured ion intensity plotted against concentration), dynamic range is a measure of the ratio of the highest concentration over the lowest concentration that can be measured simultaneously, and acquisition rate is the number of scans per second. Portable mass spectrometers should strive to achieve 10^2 ppm (parts per million) mass accuracy, unit mass resolution (i.e. resolving successive integer m/z values), $\sim 10^3$ dynamic range, and >1 Hz acquisition rate (maintaining compatibility with chromatographic methods).

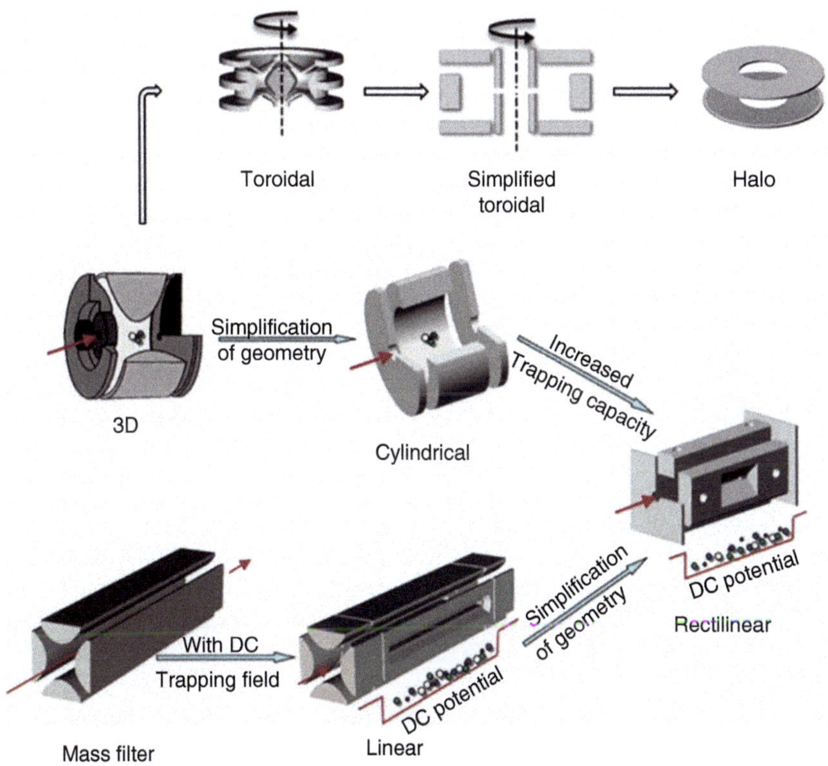

Figure 14.4 Common mass analyzers in portable mass spectrometers. All are ion traps except for the quadrupole mass filter. *Source*: Snyder, D.T., Pulliam, C.J., Ouyang, Z., and Cooks, R.G. Miniature and fieldable mass spectrometers: recent advances. *Anal. Chem.* 88(1): 2–29. © 2015 American Chemical Society.

Due to the limitations imposed by the vacuum system, orbitrap, Fourier transform ion cyclotron resonance, and TOF mass analyzers – all of which function best at ultrahigh vacuum pressures ($<10^{-7}$ Torr) – are generally not found in portable mass spectrometers. Instead, quadrupole ion traps and quadrupole mass filters operating at more modest pressure (ion trap: 10^{-3}–10^{-5} Torr; quadrupole mass filter: $<10^{-5}$ Torr) are the practical choices. Notably, ion trap resolution and sensitivity conveniently benefit from the additional collision cooling at higher pressures! Figure 14.4 shows a sampling of the most prominent analyzer geometries. Quadrupole mass filters and ion traps satisfy the figures of merit qualifications listed above. They typically obtain mass errors <0.1 Da as well as unit mass resolution and good dynamic range and sensitivity, all while having an acquisition rate on the order of 1–10 Hz.

A quadrupole mass filter consists of a set of four steel rods with hyperbolic cross-sections arranged in a square configuration. With the application of an RF voltage to one pair of rods opposite each other and the opposite phase of RF to the other pair of rods, a quadrupolar potential (or quadrupolar "field," a common misnomer) is created inside the device. It is beyond the scope of this chapter to discuss the details of quadrupolar potentials and the Mathieu parameters that govern both quadrupolar mass filters and quadrupole ion traps, so only a brief description of the operation of these devices will be given here. If only an RF potential is applied to the quadrupole rods, then the quadrupole is said to be in "transmission" or "RF-only" mode, in which case ions of a large range of m/z values are transmitted. However, if a *resolving DC* is also applied (i.e. a positive DC potential on one pair of rods and the opposite polarity on the other pair), then the quadrupole mass filter is said to operate in "RF/DC" or "mass-selective stability" mode. That is, it is acting as an m/z resolving quadrupole by allowing only a small range of m/z values through the device. If the RF and DC potentials applied to the rods are ramped simultaneously while keeping their ratio near-constant, then a mass spectrum may be obtained in order of increasing m/z.

A linear quadrupole ion trap can be thought of as a quadrupole mass filter with endcap electrodes whose applied DC potentials prevent ions from escaping from the device along its axis, thus trapping the ions. In contrast to the quadrupole which operates in mass-selective stability mode, the ion trap operates in "mass-selective instability" mode, wherein the ions are made unstable in order of m/z and are ejected toward an ion detector, generating a mass spectrum. To accomplish this task, only the RF voltage is applied to the trap rods, and the amplitude of the RF voltage is ramped linearly so as to make the trajectories of ions of successively higher m/z unstable so that they are ejected into a charged particle detector. The ions can be ejected at the Mathieu stability boundary at Mathieu $q = 0.908$, where

$$q = 4V_{RF}m/zer_0{}^2\Omega^2 \tag{14.1}$$

V_{RF} is the zero-to-peak amplitude of the RF voltage, e is the elementary charge, r_0 is the inscribed radius of the device (half distance between opposite rods), and $\Omega = 2\pi f$ is the angular RF frequency (f is the RF frequency in Hz). The ions can instead be ejected at any arbitrary Mathieu q parameter (frequency) by applying a supplemental dipolar AC frequency of low voltage to the ion trap rods, a method known as "mass-selective instability with resonance ejection." This change in "operating point" can benefit resolution, sensitivity, and mass range, though the optimum ejection point must be carefully determined experimentally. Unlike the quadrupole mass filter, the ion trap's performance metrics (save for dynamic range) fortunately do not inherently depend on device size, a characteristic also not shared by TOF and sector mass analyzers. Ion traps are thus the ideal candidates for portable mass spectrometers, with quadrupole mass filters coming in second place. Resolution and mass accuracy ought to be maintained in miniature versions of ion traps, and any loss in dynamic range can be compensated by making arrays of ion traps, as discussed below. As the size of an ion trap decreases, so does the path length of the ions, and so the analyzer can be operated at higher pressure while maintaining resolution. The drive RF frequency can be increased to compensate for any resolution losses since

$$m/\Delta m\ \alpha\ \Omega_{RF}/P = \Omega_{RF}\tau/4\sqrt{(3)} \tag{14.2}$$

where $m/\Delta m$ is the mass resolution, Ω_{RF} is the angular RF frequency in radians/s, P is the buffer gas pressure, and τ is the relaxation time of an ion caused by ion-neutral collisions.

The linear ion trap has evolved from the 3D quadrupole ion trap, which consists of a ring electrode and two endcap electrodes, all with hyperbolic cross-sections. The RF voltage is usually only applied to the ring electrode, and auxiliary resonance frequencies may be applied in a dipolar manner to the endcaps for resonance excitation or ejection. The 3D ion trap shares many of the same qualities of the linear ion trap, but the linear trap sports greater ion injection efficiency and trap capacity. As a result, 3D ion traps are becoming less prominent. Through the use of resonance excitation – applying a low-voltage auxiliary AC waveform with frequency matching an ion's *secular frequency* (the ion's m/z-dependent and, hence, q-dependent frequency of motion in the trap) – product ion MS/MS spectra may be generated. All types of quadrupole ion traps have the ability to perform single analyzer MS/MS experiments – tandem mass spectrometry – in which an ion is isolated, activated through collisions with intentionally introduced background gas molecules, and fragmented, and the fragments are subsequently mass analyzed. Product ion spectra can be used to confirm molecular identity through spectral library matching or to probe the connectivity of atoms in an ion whose structure is unknown and whose MS/MS spectrum is not already found in spectral libraries. Precursor ion scans, neutral loss scans, two-dimensional tandem mass spectrometry scans, and other unusual yet promising scanning methodologies are more recent innovations but have yet to see wide acceptance. Even so, the wide array of capabilities of quadrupole ion traps is yet another attractive feature for portable spectrometers.

Quadrupole ion traps come in many additional sizes and geometries in addition to the linear ion trap and 3D trap just described. For example, the cylindrical ion trap is a simplified 3D quadrupole ion trap in which the cross-sections of the electrodes are rectangular and the edges are flat instead of hyperbolic. Similarly, the rectilinear ion trap is a simplified linear ion trap with rectangular cross-sections. While this electrode geometry compromises the quadrupolar character of the electric field, unit resolution and good mass accuracy can still easily be obtained in a package that is far easier to manufacture than traps designed with hyperbolic cross-sections. Moreover, quadrupole ion traps of all geometries are highly tolerant to electric field imperfections and to pressure variations, more so than other instruments. This is especially important as manufacturing tolerances become more stringent with smaller analyzer dimensions. The cylindrical ion trap and rectilinear ion trap retain all of the characteristics and capabilities of the 3D trap and linear trap, respectively, and they can be manufactured as serial or parallel arrays of mass analyzers that can enhance the capabilities of the analyzer, decrease power consumption by reducing the operating voltage, or increase the dynamic range. Maintaining consistency in size and electric field across the analyzers in the array remains a challenge and has prevented widespread adoption, yet arrays remain promising configurations.

The toroidal ion trap can be generated by rotating a 3D ion trap about an axis parallel to the line through the endcaps but tangent to the outside of the ring electrode. The result is a "donut" or toroidal shape, and the electric field inside the device is largely quadrupolar if an RF potential is applied. The main advantage of the toroidal trap in comparison to 3D and linear ion traps is the increase in dynamic range and sensitivity with a small increase in device size. The trapping capacity of the toroidal trap is perhaps orders of magnitude higher than a 3D trap of comparable inscribed radius. Like the cylindrical and rectilinear ion traps, toroidal traps with rectangular cross-sections have been investigated, and an elegant "halo ion trap" consisting of two ceramic plates with ring electrodes patterned on the plates has also been demonstrated. The toroidal traps have all the capabilities of other ion traps and portable versions have even been commercialized by TORION (now part of PerkinElmer). Improvements to the mass analyzer geometry are still ongoing, as the electrode curvature warps the electric field inside the device, which can have deleterious side effects.

While the main focus of this section has been quadrupole mass filters and quadrupole ion traps, by far the most common analyzers in portable mass spectrometers, there are other analyzers worth mentioning. Both sectors and TOF instruments suffer from performance degradation as the mass analyzer is shrunk, and so miniaturizing these devices comes at a steep cost in terms of mass spectral resolution. Even so, TOF instruments, which simply consist of plates with high voltages applied, offer theoretically unlimited mass range as well as excellent mass accuracy and resolution. Yet their requirement for ultrahigh vacuum limits their adoption in portable systems. Miniature magnetic sectors have been an interesting topic of research in recent years, with reports of coded aperture systems maintaining mass spectral resolution without sacrificing ion throughput at smaller dimensions. That is, the coded aperture design uses the Felgett advantage in a cycloidal mass analyzer geometry to increase signal-to-noise ratio without sacrificing resolution.

14.2.4 Separations Prior to Mass Analysis

Analysis of samples with portable mass spectrometers is preferably conducted without any extraneous subsystems to minimize SWaP. The sampling interface, ion source, vacuum system, mass analyzer, detector, and data systems can all be considered essential. Separation capabilities prior to mass analysis, such as ion mobility (IM) spectrometry, gas or liquid chromatography (LC), and capillary electrophoresis (CE), which are auxiliary subsystems, can, however, be useful and sometimes essential for targeted applications or when the complexity of the sample is too high for unit resolution mass analyzers.

GC and LC, both conducted prior to analyte ionization, involve the injection of a sample into a "carrier" stream which pushes the sample components through a stationary phase (the "column") where the constituents are separated in space and eluted usually in order of polarity. For GC, the carrier is an inert gas such as helium or nitrogen and the stationary phase is a polymer (e.g. dimethylsiloxane), whereas in LC, the carrier is a combination of polar and nonpolar solvents (water, methanol, acetonitrile, etc.) and the stationary phase consists of glass beads with a polar or nonpolar (e.g. C_{18}) coating. Because the stationary phase is generally nonpolar, sample constituents elute from the column starting with the most polar compounds. Both gas and liquid chromatography can be coupled to portable mass spectrometers and are useful for analyzing particularly complex mixtures where species overlapping in m/z space are problematic. GC is the more common method for portable systems and is usually coupled with internal EI ionization. Its most significant drawback is that only volatile and semivolatile analytes will be observed because the sample must be heated and vaporized prior to GC injection. In addition, only a few ionization sources can be coupled with GC, usually EI or glow discharge ionization. For LC, usually coupled with ESI, the analyte must be soluble in the mobile phase, and solvent reservoirs and liquid-handling systems add to SWaP.

Although CE has not been coupled with commercialized portable mass spectrometers, it is worth noting as an alternative to GC and LC. CE separates analytes by their electrophoretic mobility in an aqueous buffer. A potential difference is applied across some length of silica capillary tubing, and as the injected sample plug moves along the length of the capillary, the analytes separate by size and charge. In general, ions that are small and more highly charged will tend to migrate the quickest, and so they will be ionized (via ESI) and detected by the mass analyzer sooner than larger ions in low charge states, generating "chromatograms" similar to GC-mass spectrometry (GC–MS) and LC–MS. In contrast to GC, CE is an attractive separation technique for polar, nonvolatile analytes.

GC, LC, and CE are separations that are conducted prior to analyte ionization, whereas IM is a *post-ionization* separation scheme that takes place in the gas phase. IM involves the injection of ionized analytes into a drift gas with an electric field established across the pressurized IM chamber. The electric field pushes ions through the drift gas, but collisions between the ions and the drift gas cause the analytes to exit from the mobility cell in order of their gas-phase mobility. The ion's mass, charge, size, and shape all affect the observed "drift time" (i.e. when the ions exit the cell). IM is most promising for larger molecules such as polypeptides and proteins, yet can also be useful for small molecules. For example, it can be used to distinguish isomeric species that are otherwise identical to the mass analyzer, though most IM systems do not have the resolution to do so. Like the other separation techniques, SWaP is increased by adding IM as a subsystem, and having a pressurized IM cell necessitates increased pumping capacity and likely more stages of differential pumping. Like TOFs and magnetic sector analyzers, the resolution of the IM separation depends on device size. As such, IM is rarely coupled to portable mass spectrometers (aside from the commercial Excellims MC3100 discussed later), but its adoption is likely to increase in the near future, particularly for forensic applications which often require two orthogonal dimensions of compound identification (e.g. IM and MS, GC and MS, or LC and MS).

14.2.5 Detectors

After the ions are sorted into m/z bins by the analyzer, they must be converted into a measurable signal (a current or voltage) that is then digitized and processed by a computer. As the number of ions sorted by the mass analyzer per second is generally less than 100 000 ($\sim 10^{-14}$ A), the signal must be amplified prior to digitization. The most common charged particle detectors in mass spectrometry are the electron multiplier, the Faraday cup, and the microchannel plate. An electron multiplier, the most common mass spectrometry detector, uses the kinetic energy of incident ions – accelerated to high energies via a large attractive voltage applied to the detector – to cause collision with a material having good secondary emission properties (e.g. Cu–BeO) and cause the release of secondary electrons. The electrons are then further accelerated and multiplied in stages, with each stage providing additional amplification. Typical gain is six to seven orders of magnitude. Discrete dynode detectors have 10–20 stages of amplification. A second type, the continuous dynode electron multiplier (or channeltron), uses a continuous lead-doped glass surface to generate secondary electrons from ion collisions. The channeltron device first converts incident positive or negative ions into electrons using a conversion dynode, and these secondary electrons then proceed through the cascading effect of the multiplier. Overall, electron multipliers provide high gain (sensitivity) and dynamic range (10^4–10^6) with fast response time, though their lifetime is limited due to oxidation, surface contamination, and electron depletion. Mass discrimination is observed with these detectors since the gain is velocity dependent. Hence, lower m/z ions tend to generate larger signals due to their higher velocities, though this effect can be somewhat mitigated by increasing the voltage on the conversion dynode. A microchannel plate is a type of electron multiplier with many parallel channels. By biasing one side of the plate relative to the other, a cascading effect can be induced after a first collision of an analyte ion onto the device, generating amplification per channel of 10^2–10^4 while ensuring fast rise times due to the minimal travel distance. Electron multipliers and microchannel plates can be made sufficiently small for most portable mass spectrometers and require high-voltage power supplies for operation.

In contrast to the cascading effects of electron multipliers, Faraday cups inherently provide no amplification, yet they also have less mass discrimination. A Faraday cup is simply a metal cylinder, cup, or plate onto which ions are neutralized. A current is generated which proceeds through a resistor and is amplified with external circuitry. Because there is no cascade of electrons within the detector, Faraday cups have much lower sensitivity than electron multipliers and their signals are more prone to be overwhelmed by background noise. They also provide slower rise times which can be improved by sacrificing sensitivity. As a result, these devices are not commonly used in portable mass spectrometers.

14.2.6 Data Acquisition, Control, and Interpretation

After ions are sorted by the mass analyzer and converted to an amplified analog signal by the detector and its associated electronics, they must then be digitized, processed, and interpreted by a computer. For a portable system, the computer is ideally a rugged laptop or touchscreen tablet. Data processing can include signal averaging, background subtraction, peak annotation, molecular formula determination, MS/MS interpretation, isotope analysis, charge state assignment, quantitation (via internal or external standards), and more. The computer then displays the data to the user, who must either further interpret the data, act on it, or initiate further processing or mass spectrometry analyses.

The user interacts with the mass spectrometer through a GUI (graphical user interface), of which there are often "expert" and "novice" modes. Expert modes of operation may allow the user to redefine the instrument's scan function, manually calibrate the m/z or intensity axis, and various other tasks. For most users, who are not likely to have backgrounds in mass spectrometry, a novice interface presents options to select particular analyses depending on the situation of interest and to optimize signal intensity through tuning of instrument parameters (e.g. timing, voltages, pressures, etc.). For example, the mass spectrometer can focus on detecting drugs, explosives, or VOCs, for which internal and external libraries of spectra are readily available. The instrument may guide these users through particular tasks such as calibration, sampling, ionization, and MS/MS analysis. Ideally, the instrument is capable of performing all of these tasks with minimal user intervention, and the instrument should identify the components in the mixture and also report a measure of confidence in the results. For most applications, the capabilities of the computer should be entirely internal and should not rely on external connections (e.g. the Internet) which are less reliable in unpredictable field environments. However, cloud integration presents a promising alternative for future data analytics in which data is acquired locally, processed in the cloud, and returned to be displayed to the user rather than being interpreted locally on the computer itself. Potential advantages of this methodology are access to faster and more sophisticated processing techniques which would otherwise be resource intensive on a local machine, comparison to a greater number of inhouse or Internet-accessible spectral libraries, remote instrument control, and the ability to monitor long-term instrument performance.

14.2.7 Commercially Available Systems

While miniature mass analyzers, new sampling and ionization strategies, and smaller and lower power vacuum pumps and electronics have been evolving over the course of several decades, only more recently have these technologies converged to a point where portable mass spectrometers could feasibly be commercialized. Table 14.2 lists currently available *portable or fieldable* mass spectrometers for which specifications are readily available from the manufacturers. Note that some systems which are *miniature* but not *fieldable* or *portable* have been excluded from this list, but are discussed briefly at the end of this section.

Three of the systems described in Table 14.1 contain quadrupole mass filters as analyzers, namely the FLIR G510, the Inficon Hapsite, and the Q Technologies AQUA MMS (miniature mass spectrometer). The G510 and the Hapsite both couple GC to EI ionization, with several auxiliary sampling strategies such as SPME, headspace sampling, a preconcentrator, and so on. They have relatively modest mass range that sufficiently covers volatile and semivolatile analytes compatible with these sampling methodologies, and they can operate on battery power for two to four hours. Intended applications of these devices include forensics, CWA and explosive detection, VOC monitoring, and more. The AQUA MMS similarly uses a quadrupole mass filter with EI ionization but has a more limited mass range that is suitable for detection of small organics and gases dissolved in water.

Most of the commercialized portable mass spectrometers understandably utilize ion traps as the mass analyzer. These include the 1st Detect TRACER 1000, the 908 Devices MX908, the Bayspec Continuity and Portability, Excellims' MC3100, the MassTech MT Explorer 50, the Torion (now PerkinElmer) T-9, and the Mini β from PURSPEC. Most of these systems use linear ion traps, with the exception of the MX908 which uses an unknown geometry, the MT Explorer 50 (which uses a 3D quadrupole ion trap), and the Torion T-9 (the only system to include a unique toroidal ion trap geometry). Many of these systems have atmospheric pressure interfaces that allow them to couple with ambient ionization sources, notably ESI, APCI, DART, MALDI, and/or paper spray. Like the FLIR and Inficon instruments, the T-9 has an internal ionization source and utilizes technologies such as SPME and GC. Most operate in both positive and negative ion modes, have upper m/z limits around m/z 1000, and can operate on battery power for at least two hours. Common applications include explosive and CWA detection, narcotics analysis, and environmental monitoring.

There are other commercial systems that are *miniature* but not strictly portable. These include several single quadrupole mass spectrometers, including the 4500 MiD from Microsaic Systems, the DART QDa, ACQUITY, and SQ Detector 2 from Waters, and the Advion expression CMS (Compact Mass Spectrometer). Usually, these are intended as compact and simple benchtop detectors for pharmaceutical research, often coupled with LC. They are compatible with a variety of external ionization sources. Compact triple quadrupole systems do exist, e.g. the Agilent Ultivo and the Waters Xevo TQ-S micro, with sizes comparable to an LC stack. Lastly, TOFWERK develops custom-fieldable TOF mass spectrometers for a variety of purposes, including VOC detection and elemental analysis, and ZeteoTech has developed a "digital" MALDI-TOF platform for detection of high molecular weight biological species such as bacteria and proteins.

14.3 Applications

Portable instruments have expanded the utility of mass spectrometry beyond the analytical laboratory so that *in situ* measurements can be made in the field in real time. Though the virtually limitless potential of portable MS has yet to be realized in its entirety, several important applications have been described in detail in the literature and are emerging as practical cases.

Detection of illicit or dangerous substances is one case where it is logistically impractical or disadvantageous to collect the sample and bring it to a laboratory setting. Police officers, for example, have several tools at their

Table 14.2 Commercially available miniature and portable mass spectrometers.

Instrument	Affiliation	Sample introduction	Ionization source(s)	Polarity	Analyzer	Mass range (*m/z*)	Resolution	MS/MS	Power	Mass (kg)	Intended application
TRACER 1000	1st Detect	Teflon-coated fiberglass swabs; sampling wand	CI	+/−	Linear ion trap	Unknown	Unknown	Unknown	110 VAC/230 VAC, 50 Hz/60 Hz	25	Explosives; Narcotics
MX908	908 Devices	Thermal desorption swab for solids/liquids; direct gas/vapor analysis	Internal	+/−	Ion trap	50–460	Unknown	Yes	>3 h of continuous operation	3.9	Chemical warfare agents; Narcotics; Emerging threats; Explosives; Precursors
Continuity	Bayspec	AI	TD-ESI; APCI; EI	+/−	Linear ion trap	50–1200	0.49 amu FWHM	Yes	72 W	20	Explosives; Narcotics; Pesticides; Mycotoxins
Portability	Bayspec	AI	TD-ESI; APCI; EI	+/−	Linear ion trap	50–650	0.49 amu FWHM	Yes	72 W	10	Explosives; Narcotics; Pesticides; Mycotoxins
MC3100	ExcellIMS	AI; Thermal desorption; Continuous liquid introduction	ESI; Corona discharge	+	Linear ion trap + Ion mobility	20–2800	Unit	Yes	250 W	35	Narcotics; Food & Drug safety; Pharmaceuticals
G510	FLIR	Gas chromatography; Heated sample probe; Membrane introduction for vapors; Internal dual-bed preconcentrator with split/splitless injector port; direct liquid sampling via syringe; SPME; PSI	EI	+	Quadrupole mass filter	15–515	0.7 amu FWHM	No	2–4 h on battery	16.3	Narcotics; Explosives; Environmental contaminants
HAPSITE	Inficon	Gas chromatography; headspace sampling; SituProbe; SPME; Thermal desorption	EI	+	Quadrupole mass filter	41–300	Unknown	No	2–3 h on battery	19	VOCs; SVOCs; Chemical warfare agents; Toxic industrial materials

(Continued)

Table 14.2 (Continued)

Instrument	Affiliation	Sample introduction	Ionization source(s)	Polarity	Analyzer	Mass range (*m/z*)	Resolution	MS/MS	Power	Mass (kg)	Intended application
MT Explorer 50	MassTech	Direct sampling	ESI, AP MALDI; APCI; sESI; DART	+/−	3D ion trap	30–2500	Up to 6000 @ 2000 Da	Yes	100–300 W	34	Narcotics; Explosives; Pesticides;
TORION T-9	PerkinElmer (Torion)	SPME; Needle trap; Gas chromatography	EI	+	Toroidal ion trap	41–500	Unit	Unknown	2.5 h on battery	14.5	Environmental volatiles and semivolatiles; Explosives; Chemical warfare agents; Hazardous substances; Food safety
Mini β	PURSPEC	AI (noncontinuous API)	AI (ESI, paper capillary spray)	+/−	Linear ion trap	50–2000	<1 amu	Yes	100 W	20	Narcotics; Food and drug regulation; Biomedical & clinical
AQUA MMS	Q Technologies	Membrane inlet	EI	+	Quadrupole mass filter	Up to 200	Unit	No	<50 W	Unknown	Oil/hydrocarbons in water; dissolved gases

APCI = atmospheric pressure chemical ionization; API = atmospheric pressure interface; AI = ambient ionization; CI = chemical ionization; DART = direct analysis in real time; EI = electron impact; ESI = electrospray ionization; FWHM = full-width at half-maximum; LTP = low-temperature plasma; PSI = prepless sample introduction; sESI = secondary electrospray ionization; SPME = solid-phase microextraction; TD = thermal desorption.

disposal for detection of illicit substances (drugs or alcohol), including chemical and electrochemical tests, but with the emergence of drug analogs and more potent alternatives (e.g. fentanyl and its derivatives), it is increasingly of value for officers to have immediate access to highly sensitive and selective, yet broadly applicable, analytical detectors. Figure 14.5a, for example, shows a mass spectrum of 10 ng each of three cathinones, 3,4-methylenedioxypyrovalerone (*m*/*z* 276, MDPV), methylone (*m*/*z* 208), and mephedrone (*m*/*z* 178), deposited on a Teflon slide and detected using a Griffin AI-MS 1.2 cylindrical ion trap mass spectrometer. The analytes were sampled from the slide using DESI. This experiment demonstrates the high sensitivity and broad detection range that mass spectrometry brings to drug analysis in the field, as all three cathinones, though structurally diverse, can be detected in a single scan and in trace amounts. Portable mass spectrometers can be used for the detection of a range of nefarious substances, not just drugs. Figure 14.5b shows detection of gunshot residues methyl centralite (*m*/*z* 241) and ethyl centralite (*m*/*z* 269) via MS/MS, using rayon swabs to collect the material from a suspected shooter's hand and also to serve as the ionization substrate ("swab touch spray" ionization). A Mini 12 rectilinear ion trap mass spectrometer with DAPI was used for this experiment. This forensic application could provide an on-site test to determine whether a suspect has recently used a firearm.

Police officers are not the only public servants that may benefit from the adoption of MS as a primary field analytical methodology. Portable MS could be invaluable for warfighters who are constantly exposed to unknown, dangerous, and foreign substances. Figure 14.5c shows mass spectra of CWA VX (*m*/*z* 267, ethyl ({2-[bis(propan-2-yl)amino]ethyl}sulfanyl) (methyl)phosphinate) and its degradation product (*m*/*z* 175) recorded using a Griffin 450 GC–MS via EI ionization coupled to quadrupole mass analyzer detection, and Figure 14.5d shows detection of explosives Tetryl, TNT, and HMX collected using DESI on a Mini 10 mass spectrometer. One could imagine affixing a miniature mass spectrometer to a rover or drone that could be wirelessly controlled and driven to a sample for detection. This strategy would mitigate risk to personnel, particularly if the sample is believed to be acutely toxic, explosive, or otherwise dangerous. Even if personnel are directly involved in the mass spectrometry analysis rather than robots, in contrast to laboratory-based mass spectrometry, portable mass spectrometry offers immediate answers to pressing questions (is this substance dangerous?) in potentially life-threatening situations.

At the moment, the most common applications of portable mass spectrometry are related to first responders and warfighters, but there are other uses that have garnered interest, namely water quality monitoring and detection of VOCs. In both cases, the technologies of choice are membrane inlets (which gases and volatile analytes may freely diffuse through) combined with EI ionization and quadrupole mass detection. Single photon ionization (irradiating the sample with an ultraviolet laser) is another useful source for VOCs and can prevent spectral overlap (compared to EI) since fragments are not produced. Portable mass spectrometry can measure trace concentrations of dissolved gases, such as oxygen, carbon dioxide, and nitrogen, volatile organics, such as benzene and toluene, polycyclic aromatic hydrocarbons, polychlorinated biphenyls, and dioxins, to name a few. An example where these detection capabilities would be useful is the detection of oil or pharmaceutical contamination in water after an industrial disaster. The concentrations of VOCs in drinking water may also be monitored to ensure that chemical levels are below regulated levels. More generally, any form of environmental pollution monitoring can benefit from online or offline mass spectrometry detection, and onsite measurements necessitate portable instrumentation.

Though the current list of applications for portable mass spectrometry is somewhat limited, it is expanding quite rapidly as portability opens new avenues of analysis. Figure 14.6 illustrates several more recent demonstrations of portable mass spectrometry for food quality screening, point-of-care clinical analysis, and disease diagnostics. One could imagine a world in which supermarket shoppers use pocket-sized portable mass spectrometers to determine what chemicals (fungicides, pesticides, and herbicides) are on the produce they intend to purchase, answering questions such as "is this apple really organic?". The mass spectrum in Figure 14.6a results from detection of fungicide diphenylamine (*m*/*z* 170) directly from the surface of an apple using LTP ionization coupled to a Mini 12 mass

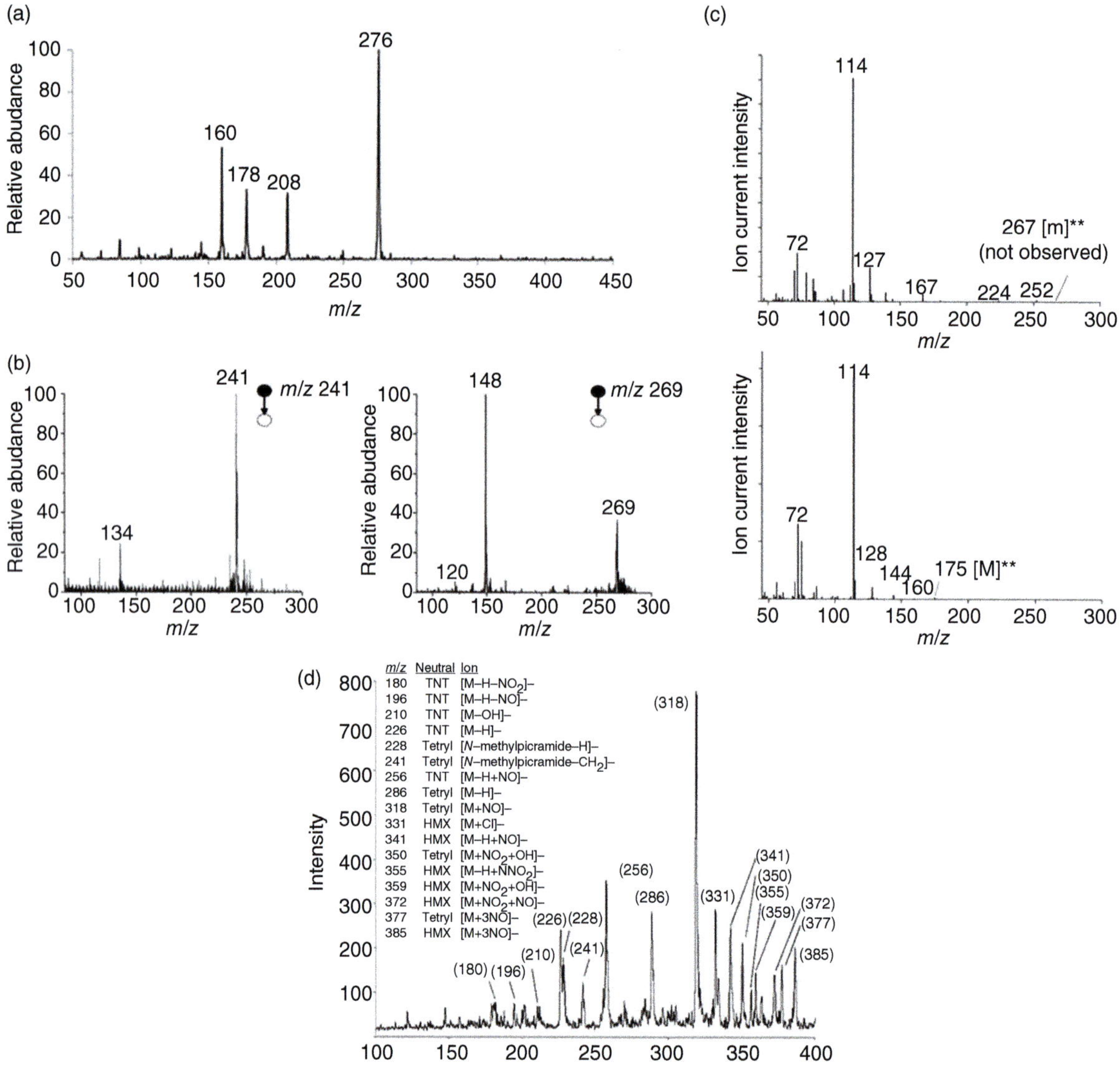

Figure 14.5 Detection of narcotics, gunshot residues, chemical warfare agents, and explosives using miniature mass spectrometers: (a) desorption electrospray ionization mass spectrum of 10 ng each of 3,4-methylenedioxypyrovalerone (*m/z* 276, MDPV), methylone (*m/z* 208), and mephedrone (*m/z* 178) on a Teflon slide using a Griffin AI-MS 1.2 cylindrical ion trap mass spectrometer; (b) swab touch spray product ion spectra of methyl centralite (*m/z* 241) and ethyl centralite (*m/z* 269) detected by a Mini 12; (c) quadrupole mass spectrum after electron impact ionization of chemical warfare agent VX (*m/z* 267) and its degradation product (*m/z* 175) recorded using a Griffin 450 GC–MS; and (d) negative ion mass spectrum of explosives Tetryl, TNT, and HMX collected using desorption electrospray ionization on a Mini 10 mass spectrometer. *Source*: Panel (a) from Vircks, K.E. and Mulligan, C.C. (2012). Rapid screening of synthetic cathinones as trace residues and in authentic seizures using a portable mass spectrometer equipped with desorption electrospray ionization. *Rapid Commun. Mass Spectrom.* 26: 2665–2672. © 2012 John Wiley & Sons. Panel (b) from Fedick, P.W., Bain, R.M. Swab touch spray mass spectrometry for rapid analysis of organic gunshot residue from human hand and various surfaces using commercial and fieldable mass spectrometry systems. *Forensic Chem.* 5: 53–57. © 2017 Elsevier. Panel (c) from Smith, P.A., Lepage, C.J., Lukacs, M., Martin, N., Shufutinsky, A., and Savage, P.B. Field-portable gas chromatography with transmission quadrupole and cylindrical ion trap mass spectrometric detection: chromatographic retention index data and ion/molecule interactions for chemical warfare agent identification. *Int. J. Mass Spectrom.* 295: 113–118. © 2010 Elsevier. Panel (d) reproduced from Sanders, N.L., Kothari, S., Huang, G., Salazar, G., Cooks, R.G. Detection of explosives as negative ions directly from surfaces using a miniature mass spectrometer. *Anal. Chem.* 82: 5313–5316. © 2010 American Chemical Society.

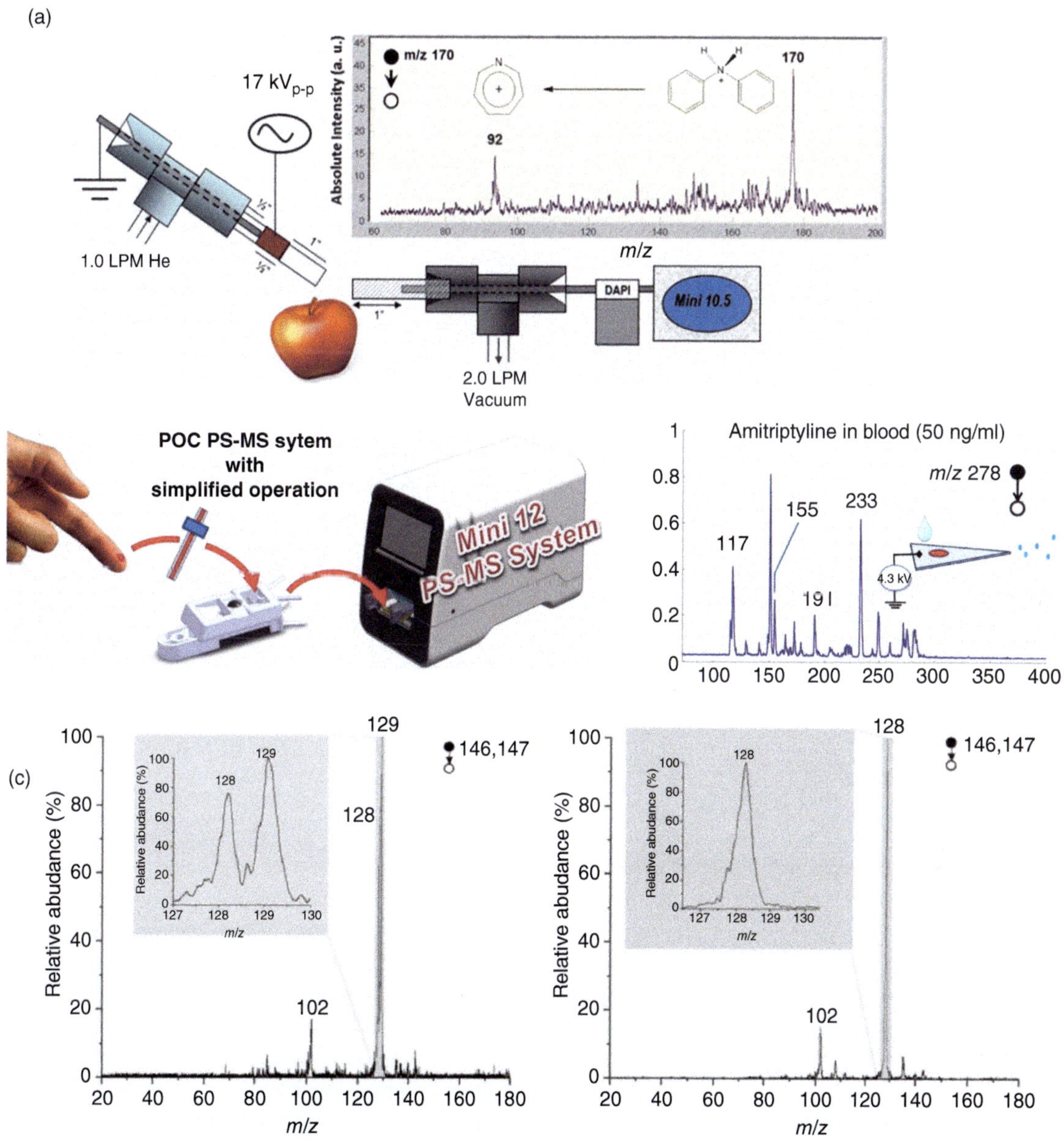

Figure 14.6 Miniature mass spectrometers for detection of agrochemicals, point-of-care blood monitoring, and tissue analysis: (a) detection of fungicide diphenylamine directly from an apple using low-temperature plasma ionization on a Mini 10.5; (b) MS/MS spectrum of amitriptyline in blood using paper spray on a Mini 12; and (c) MS/MS spectra of deprotonated forms of 2-hydroxyglutarate (*m/z* 146) and glutamate (*m/z yy* 147) from an IDH (isocitrate dehydrogenase) mutant glioma (left) and IDH wild-type (right). The peak at *m/z* 129 occurs only in the IDH mutant. *Source*: Panel (a) from Soparawalla, S., Tadjimukhamedov, F.K., Wiley, J.S., Ouyang, Z., and Cooks, R.G. In situ analysis of agrochemical residues on fruit using ambient ionization on a handheld mass spectrometer. *The Analyst* 136(21): 4392. © 2011 Royal Society of Chemistry. Panel (b) reproduced from Li, L., Chen, T.C., Hendricks, P.I., Cooks, R.G., and Ouyang, Z. (2014). Miniature mass spectrometer for clinical and other applications—introduction and characterization. *Anal. Chem.* 86: 2909–2916. © 2014 American Chemical Society. Panel (c) from Pu, F., Alfaro, C.M., Pirro, V., Xie, Z., Ouyang, Z., and Cooks, R.G. (2019). Rapid determination of isocitrate dehydrogenase mutation status of human gliomas by extraction nanoelectrospray using a miniature mass spectrometer. *Anal. Bioanal. Chem.* 411: 1503–1508. © 2019 Springer Nature.

spectrometer operated in the MS/MS mode. The product ion MS/MS scan of protonated diphenylamine is shown. Point-of-care analysis is a particularly promising direction, as it may offer money and time savings while providing immediate answers for many medical questions. It is reasonable to suppose that in the future, portable mass spectrometers would be used by doctors and nurses for drug and metabolite monitoring, early disease diagnosis, or bacteria detection onsite at the doctor's office. Figure 14.6b illustrates point-of-care drug-in-blood monitoring by paper spray mass spectrometry on a Mini 12 mass spectrometer. A drop of blood was placed on a paper triangle and solvent and high voltage were applied to generate ions that were detected by a rectilinear ion trap. Amitryptyline (*m*/*z* 278, a prescription antidepressant) was detected via MS/MS at a concentration of 50 ng/mL. This experiment could be used to determine whether patients have been taking their prescribed mood-altering medications, and, because quantitative results can be obtained, to identify any changes in dosage that may be beneficial to them. This experiment may also be useful for therapeutic drug monitoring. Figure 14.6c shows negative ion mode MS/MS spectra of *m*/*z* 146 and 147 (using a 2 *m*/*z* wide isolation window), which correspond to deprotonated forms of 2-hydroxyglutarate and glutamate, respectively, from brain smears of (left) an isocitrate dehydrogenase (IDH) mutant glioma and (right) an IDH wild type. It is evident from the MS/MS spectra that only the IDH glioma gives the abundant product ion at *m*/*z* 129 (a fragment of 2-hydroxyglutarate which has been shown to be indicative of glioma), whereas the wild type gives only fragments of glutamate, an indication of healthy tissue. In other words, MS/MS on a portable mass spectrometer can be used to detect cancerous tissues.

It is beyond the scope of this text to discuss every imaginable application of portable mass spectrometry. The field is advancing rapidly and new ideas are emerging every day; so, the topics included here should be understood as a subset of the applications that are either the most common or those which are exciting new developments. Space science, in particular, has been omitted from this text yet has been one of the primary drivers of miniature mass spectrometers and continues to push boundaries while also benefiting from new inventions. For those interested in the topic, reviews of extraterrestrial mass spectrometry are readily available in the literature.

14.4 Summary and Outlook

While mass spectrometry as a field has matured over the last several decades of research, only recently (~last 20 years) have portable mass spectrometers been the topic of intense research interest, culminating in the commercialization of several notable miniature and portable MS systems within the last decade. The advances in vacuum system, mass analyzer, sampling, ionization, and data acquisition/manipulation technologies will continue to drive mass spectrometry forward, resulting in substantially smaller, more power-efficient, and higher performing systems. As the overall system size decreases, the range of applications for portable MS will continue to increase rapidly, particularly if versatile methods of sampling and ionization (ambient ionization) are incorporated into newer designs. At the moment, internal ionization combined with membrane inlets (and often GC) and quadrupole mass filters remain popular choices for miniature systems, yet miniature ion traps coupled with ambient ionization and unique vacuum system configurations (e.g. DAPI) are emerging as the most sensible choice.

Acronyms

AI	ambient ionization
APCI	atmospheric pressure chemical ionization
API	atmospheric pressure interface
CI	chemical ionization
CWA	chemical warfare agent

DAPI	discontinuous atmospheric pressure interface
DART	direct analysis in real time
DESI	desorption electrospray ionization
EI	electron impact (ionization)
ESI	electrospray ionization
FWHM	full-width at half maximum (a measure of mass spectral resolution)
GC	gas chromatography
IDH	isocitrate dehydrogenase
LDI	laser desorption ionization
LTP	low-temperature plasma (ionization)
MALDI	matrix-assisted laser desorption ionization
MS	mass spectrometry (or mass spectrometer, depending on context)
MS/MS	tandem mass spectrometry (two stages of mass selection and/or analysis, typically with an intermediate isolation and fragmentation step)
nESI	nanoelectrospray ionization
POC	point-of-care
PSI	prepless sample introduction or paper spray ionization, depending on the context
sESI	secondary electrospray ionization
SPME	solid-phase microextraction
TD	thermal desorption
Tetryl	2,4,6-trinitrophenylmethylnitramine (an explosive)
TNT	trinitrotoluene (an explosive)
HMX	1,3,5,7-tetranitro-1,3,5,7-tetrazoctane (an explosive)
VOC	volatile organic compound
VX	ethyl ({2-[bis(propan-2-yl)amino]ethyl}sulfanyl)(methyl)phosphinate (a chemical warfare agent, i.e. a nerve agent)

Further Reading

Blakeman, K.H., Wolfe, D.W., Cavanaugh, C.A., and Ramsey, J.M. (2016). *Anal. Chem.* 88: 5378–5384.

Chen, C., Chen, T., Zhou, X. et al. (2015). *Journal of the American Society for Mass Spectrometry.* 26: 240–247.

de Hoffman, E. and Stroobant, V. (2007). *Mass Spectrometry: Principles and Applications*, 3e. Wiley.

Jiang, T., Zhang, H., Tang, Y. et al. (2017). *Anal. Chem.* 89: 5578–5584.

Liu, X., Wang, X., Bu, J. et al. (2019). *Anal. Chem.* 91: 1391–1398.

Snyder, D.T., Pulliam, C.J., Ouyang, Z.O., and Cooks, R.G. (2015). *Anal. Chem.* 88: 2–29.

Tian, Y., Higgs, J., Barney, B., and Austin, D.E.J. (2014). *Mass Spectrom.* 49: 233–240.

15

Portable Gas Chromatography–Mass Spectrometry: Instrumentation and Applications

Pauline E. Leary[1], Brooke W. Kammrath[2,3] and John A. Reffner[4]

[1] Federal Resources, Stevensville, MD, USA
[2] Department of Forensic Science, Henry C. Lee College of Criminal Justice and Forensic Sciences, University of New Haven, West Haven, CT, USA
[3] Henry C. Lee Institute of Forensic Science, West Haven, CT, USA
[4] Department of Sciences, John Jay College of Criminal Justice, City University of New York, New York, NY, USA

15.1 Introduction

Gas chromatography–mass spectrometry (GC–MS) is a valuable laboratory method for the analysis of a wide range of analytes. It is extensively used and has even been called the gold standard for the identification of numerous chemical substances (Fogerson et al. 1997; Schecter et al. 1999; Tilstone et al. 2006). GC–MS systems integrate two independent technologies. Separation of mixtures into their components is performed using the GC; identification of these components is established based upon the GC and MS data. Independently, each method has value. GC is a premier analytical technique for the separation of volatile compounds. It combines rapid analysis, high resolution, excellent quantitative results, and moderate costs. MS systems are among the most information-rich detector types available. They require only micrograms or less of a sample, but provide data for both qualitative and quantitative analyses (McNair and Miller 2009). When combined, the value of the integrated system is dramatically increased over what either method can offer independently. For instance, the addition of MS detection to GC analysis enables confirmation of each chemical detected and presumptively identified in the GC. Likewise, the ability to separate complex mixtures on the front end using GC makes it possible to concentrate and purify components and then recover spectra of these purified components at low concentrations, even when masked within complex matrices.

There are many applications that benefit from the use of GC–MS in the field, but the development of portable versions of these systems has been relatively slow compared with the development of other types of portable spectrometers, such as infrared and Raman systems. Challenges faced when miniaturizing and extending GC–MS to field operators have been significant. In some instances, these challenges are based upon hardware limitations; in others, they are based upon challenges to the ability to effectively use these systems to generate results at the sample site. An example of the former is the need to use vacuum pumps. The availability of pumps to achieve the vacuum levels required to maintain good MS performance has been a limiting factor that continues to impact instrument design and portability. An example of the latter is the challenge of data interpretation in the field. GC–MS data are complex, and for many of the applications where portable GC–MS applications may have the largest potential benefit, operators frequently have a limited scientific and/or GC–MS background. As a result, extracting meaningful results from GC–MS data can be difficult and has limited the use of these systems in some applications. It seems fair to say that as the technology has progressed, vendors of these systems have worked to bridge the gap between generating powerful, but complex, GC–MS data and the translation of these data into meaningful results at the scene.

Portable Spectroscopy and Spectrometry 1: Technologies and Instrumentation, First Edition.
Edited by Richard A. Crocombe, Pauline E. Leary and Brooke W. Kammrath.

This chapter explores the current state of commercially available portable GC–MS technology. After a brief review of the history of these systems, the components critical to the performance of portable systems will be discussed followed by a review of the applications of this technology. The chapter will end with a discussion of what the future might hold for these systems as they continue to evolve to meet the needs of end users. This chapter is specific to integrated GC–MS systems. For a thorough discussion of standalone portable MS, see the chapter in this volume by Snyder; for a thorough discussion of standalone high-pressure MS, see the chapter in this volume by Blakeman and Miller.

15.2 History of Portable GC–MS

One of the earliest reports that demonstrate the value of being able to perform GC–MS analysis at the scene is the inclusion of GC–MS capability on the 1976 Viking voyage to Mars. Plans for the voyage were reported in 1969 (National Aeronautics and Space Administration 1970). By 1970, scientists were performing experiments to provide reference data intended to aid in the interpretation of soil-organics experiments planned for the voyage (Simmonds 1970). GC–MS was successfully deployed on the voyage in 1976 and the instrumentation details are documented (Biemann et al. 1977). The Viking voyage involved the deployment of GC–MS capability for field analysis, but the GC–MS technology used was not as much portable as it was transportable. The system was designed specifically for the mission and was, essentially, part of a fixed mobile laboratory.

Even by the year 2000, portable GC–MS instrumentation was still considered cumbersome and broadly categorized as either vehicle-portable or man-portable. Vehicle-portable GC–MS systems were described as transportable, fixed-laboratory systems, while man-portable systems were described as luggable (Meuzelaar et al. 2000). It seems the characteristics that made GC–MS instruments portable were not necessarily their size and weight, but rather their independent power and carrier gas supplies (Henry 1997). Today, it is reasonable to expect that portable systems offer weights that enable them to be hand-carried to the sample site.

The first truly portable GC–MS, the Inficon (East Syracuse, NY) HAPSITE was introduced in 1996. This system operated on a battery with an independent carrier gas supply. It offered a quadrupole MS and used a non-evaporative getter (NEG) pump to achieve vacuum. This product went through a series of iterations through the years. The most recent, the HAPSITE ER, was introduced to the market in 2008 and is designed to support the needs of the emergency-response community (Crume 2009). This system weighs 42 lbs (Inficon 2015). Most portable GC–MS competitors to the original HAPSITE, and for much of the life cycle of the HAPSITE product line, were either vehicle-portable or man-portable systems. In recent years, though, a number of vendors have presented alternatives to the HAPSITE family of products that are more comparable in size and weight. Today, the PerkinElmer (Waltham, MA) TORION T-9 and the FLIR (Wilsonville, OR) Griffin G510 are two such examples. The TORION T-9 uses a toroidal ion trap MS and weighs 32 lbs (PerkinElmer Inc 2020); the Griffin G510 uses a quadrupole MS and weighs 36 lbs (FLIR Systems, Inc. 2019). Details about the specifications of these systems and their intended applications are presented throughout this chapter. There are also vendors who provide GC–MS systems that are deployed to the sample site but are a bit heavier. One example is the Bruker (Billerica, MA) E^2M which weighs 72 lbs (Bruker 2020). Another option for performing GC–MS at the sample site can occur when GC–MS systems that are not necessarily portable are taken to the sample site as part of a fixed mobile laboratory (Mississippi National Guard 2015; United States Environmental Protection Agency 2020). Figure 15.1 shows images of the HAPSTIE ER, TORION T-9, and Griffin G510.

Aside from the reduction in size and weight, there are other aspects of portable GC–MS systems that have evolved and improved through the years that are making these systems more field-friendly. Some systems now offer software interfaces that are intuitive with easy-to-select options, even when the operator is wearing personal protective equipment. Modes of operation and required accessories are minimized. This is important not only for logistical

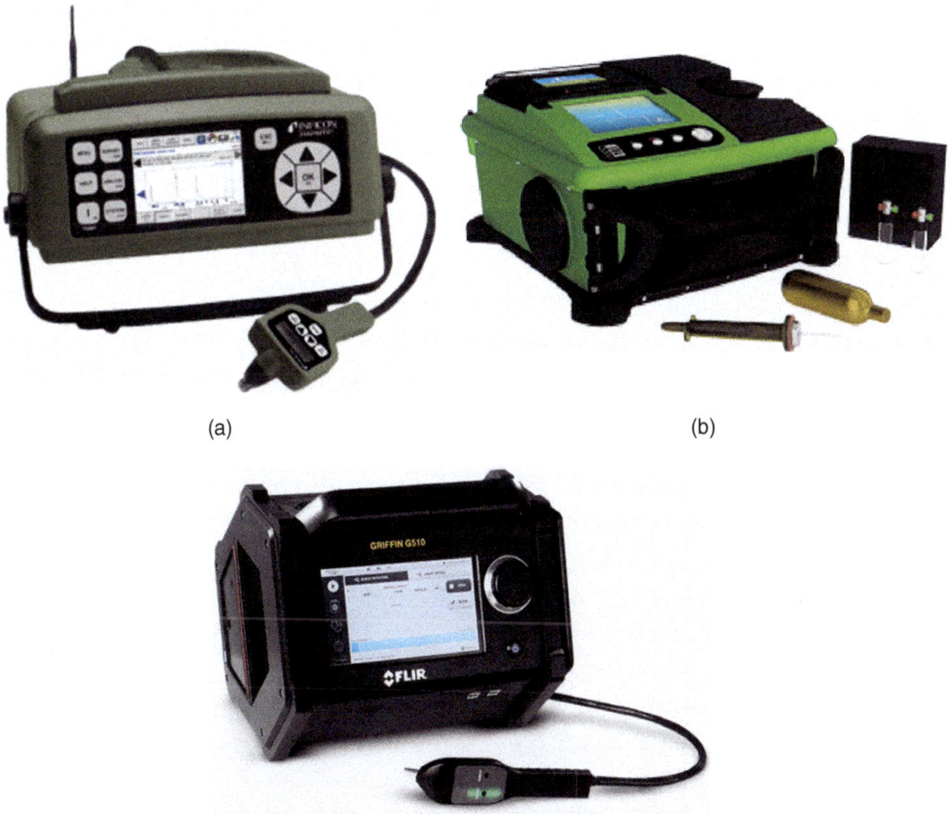

Figure 15.1 Images of the Inficon HAPSITE ER (a) (*Source*: INFICON), PerkinElmer TORION T-9 (b) (*Source*: PerkinElmer Inc.), and FLIR Griffin G510 (c) (*Source*: FLIR Systems, Inc.).

reasons, but also because when GC–MS systems are used in the laboratory, it is typical for the operator to have some experience and educational background in the sciences. They also will likely have a general understanding of the fundamentals of the technology, as well as how it works. This is not always the case with field users. While some field users have formal education in the sciences, this type of background is not necessarily typical, nor is it always expected. Therefore, ease of use, library-search capabilities, data interpretation, and training requirements must be considered when the technology will be deployed for field use (Leary et al. 2016). In addition (and especially for many military and emergency-response applications), the end user will frequently be under considerable mission stresses and may have very little time to collect samples in an extremely tense, hostile, and dangerous environment. The instrumentation and software interface must be as simple and intuitive as possible so that the operator will have the best chance for success under these conditions.

The approach to deployment has also evolved and can impact whether the technology is successfully deployed by an organization. To relieve some of the burden of deploying portable GC–MS systems, an organization may focus the use of the technology on either a specific type of response or for detection, identification, and/or quantitation of a specific set of target chemicals. While this approach can limit the potential value of the technology that could be realized, it can significantly ease the burden of education and training users and significantly improve the chances for success in the field for the detection and identification of specific types of target chemicals.

15.3 Critical Components for Portability

During GC–MS analysis, a sample is introduced to the system using one of a number of different sample-introduction methods. The sample is vaporized and carried through the GC column in the presence of a carrier gas. During its transfer through the column, the individual components of the sample separate based, essentially, upon boiling point. As each purified chemical elutes from the GC column, it enters the MS where it is ionized, fragmented, and sorted based upon mass-to-charge (m/z) value. By using a combination of both the GC and MS data, an identification of each chemical in the sample is made.

When designing portable GC–MS systems, the components that are important for the performance of lab-based GC–MS systems, such as sample collection and introduction, column and carrier gas, and mass analyzer and vacuum requirements, are also important for the performance of portable systems. When designing portable systems, though, optimization of performance is balanced with the goal of minimizing size, weight, and power consumption. Sample collection and introduction, GC (column and carrier gas), MS (mass analyzer, ionization source), and vacuum components will be considered separately.

15.3.1 Sample Collection and Introduction

Sampling for portable GC–MS analysis is important. If an appropriate sampling method is not available for a specific analyte, the ability to analyze the sample at the scene may not be possible. Sampling devices are also important because they can increase instrument size, weight, and deployment logistics. Solid-phase microextraction (SPME) (Pawliszyn 1997) is commonly considered to be the easiest sampling method for field analysis. SPME provides a solvent-free method for sample collection. The SPME sampling device is small and weighs only a few ounces so it can be carried to the sample (Leary et al. 2016). It is, essentially, a cartridge containing a fiber coated with a polymer or sorbent onto which analytes selectively adsorb/absorb and concentrate. There are a variety of commercially available fibers that may be used. Fiber selection is based upon molecular weights and polarity of the intended analytes. Although useful in many applications, SPME is limited for a number of reasons including its selective sorption of chemicals which can significantly bias results, as well as its inability to sample solid samples. Other methods are routinely employed including direct and dynamic air sampling, headspace analysis, purge-and-trap options, and solid-sampling accessories. Different portable GC–MS systems offer different sampling options, and depending upon the application, one sampling method over another may be preferred. Factors that influence the desired sampling method include sample type, sample phase of matter, desired limits of detection, quantitative capabilities, and environmental conditions. Table 15.1 summarizes sampling options for some of the commercially available portable GC–MS systems.

Recently, PerkinElmer introduced coiled microextraction (CME) as a sampling method. The appearance of the device is similar to the SPME sampling device in size, weight, and appearance, but the internal SPME fibers are replaced by a CME insert. The insert is a treated wire that is finely coiled to trap liquid samples, including those with suspended solids, by capillary action. Once the solvent has dried, the dissolved compounds remain on the coiled wire ready for introduction into the heated injection port of the TORION T-9 for thermal desorption. The CME is intended to be a fast, easy, and reliable method for sample collection and introduction (PerkinElmer, Inc. 2018). Figure 15.2 shows a stereophotomicrograph of the sampling area of the CME insert and the SPME fiber. The length of the sampling area for both the CME and SPME devices is approximately 1 cm.

A device that has been used in the laboratory to improve sample collection and introduction and can be extended to use in the field is the MicroShot (see Figure 15.3) by Merlin Instrument Company (Centennial, CO). The MicroShot enables a simplified collection of samples for GC analysis. A GC syringe is mounted in the MicroShot injector and a spring-driven plunger slide holds and supports the syringe plunger, making it easy to rinse and fill the syringe without damaging the plunger. The sample volume is controlled by a calibrated

Table 15.1 Sampling options for portable GC–MS systems.

Instrument	Sampling options
Hapsite ER	Air probe Headspace (option) SituProbe (option) SPME (option) Thermal desorption (option)
Torion T-9	SPME Coiled microextraction (CME) Air sampling with trap (option)
Griffin G510	Heated air probe Direct liquid sampling via syringe Liquid extraction via SPME fiber or PSI-Probe w/Gerstel Twister™ (option) Solid PSI-probe™ thermal separation via TAG™ (option)

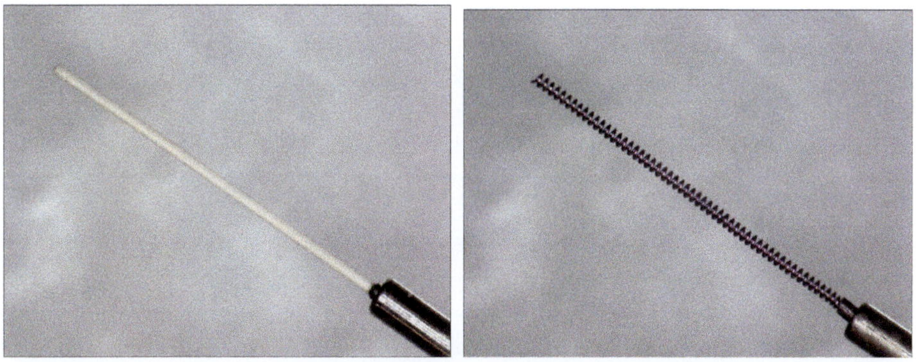

Figure 15.2 Sampling area of the CME (left) and the SPME fiber sampling area.

volume rod. When the injector is filled and armed, pushing the syringe needle into the injection port triggers the sample injection, reducing the injection process to a simple motion. Advantages to its use include improvement in injection precision, easy manipulation of the GC syringe plunger, fast injection (minimizing needle residence time in the injection port), and convenient sampling from a wide variety of sample containers (Merlin Instrument Company 2019).

15.3.2 Gas Chromatography

GC describes all chromatographic methods in which the mobile phase is a gas and may involve either a solid or liquid stationary phase retained on a solid sorbent or column wall. Gas–solid chromatography (GSC) comprises all techniques with an active solid as a stationary phase, while gas–liquid chromatography (GLC) is comprised of those techniques involving a liquid stationary phase. With the exception of a few specialized areas, such as the analysis of inorganic gases, GLC systems are more commonly used (Robards et al. 1994).

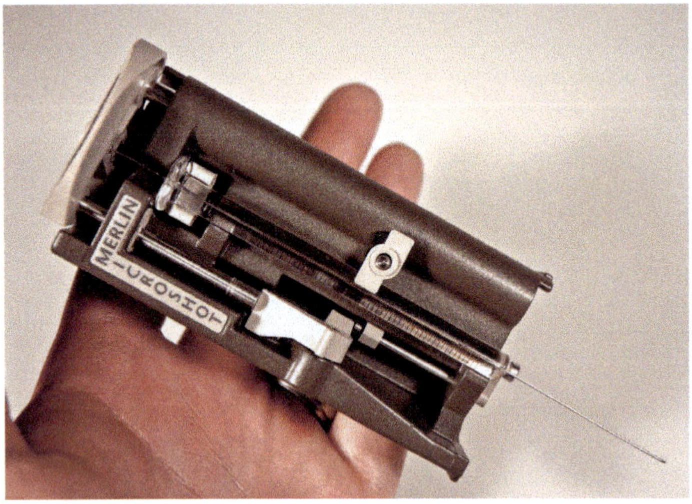

Figure 15.3 MicroShot GC Sampling Device. *Source*: Merlin Instrument Company.

Chemicals travel through a GC at a rate that is characteristic of the chemical and based upon properties such as boiling point and vapor pressure. Separation of different chemicals occurs because each chemical partitions differently between the stationary and mobile phases. The time it takes for a chemical to travel through the GC at a specified flow rate is related to this partitioning behavior and is known as its retention time. Retention time can be used to identify a specific chemical within a mixture sample. Further analytical evaluation of each purified analyte that elutes from the GC may be performed when an MS is linked to the GC.

In addition to detecting and identifying a single analyte within matrices using GC, the chromatographic "pattern" generated when all of the chemicals in a sample elute from a column may also be used for identification, not of an individual analyte, but rather of the sample as a whole. This type of chromatographic-pattern analysis is frequently performed to identify ignitable liquids such as gasoline, diesel, and kerosene (Stauffer et al. 2008).

The two primary components of a GC that impact performance are the GC column and the carrier gas. Open-tubular (capillary) columns are the column of choice for the most portable GC–MS systems. These columns are well-suited for field-portable systems because the developments that led to the widespread adoption of open-tubular columns in the 1980s, primarily for chromatographic performance, coincidentally made this important component strong and lightweight. The low thermal mass of an open-tubular GC column allows for temperature-program analysis using relatively little power when heating is limited to the column itself. Using such a column assembly, aggressive temperature ramp rates are possible even when an instrument is operated using battery power. The resulting rapid analysis speeds improve the ability to quickly complete GC analyses in the field (Smith 2015). Open-tubular GLC columns with nonpolar stationary phases are the standard column option for most commercially available portable GC–MS systems. There have been portable systems that offer columns with alternative stationary phases to support specified applications such as those used in the cannabis and petrochemical industries (Restek n.d.), but these options are not widely available on most portable systems (Table 15.2).

Other properties of open-tubular columns that impact analytical performance include column length, stationary-phase film thickness, and diameter (Agilent Technologies Inc 2012). Increasing column length will increase resolution, column backpressure, and analysis time. Therefore, the choice of column length is a compromise between efficiency, operating pressure, and analysis time. The shortest column capable of generating the separating efficiency needed should be used (Robards et al. 1994). Stationary-phase film thickness has an

Table 15.2 Summary of GC component types for portable GC–MS systems.

Instrument	Column	Length (m)	Film thickness (μm)	Stationary phase	Inner diameter (mm)	Temperature limit (°C)	Carrier gas
Hapsite ER	DB-1MS	15	1	100% dimethylpolysiloxane (GLC)	0.25	200	Nitrogen
Torion T-9	MXT-5	5	0.4	Crossbond diphenyl dimethyl polysiloxane (GLC)	0.1	300	Helium
Griffin G510	DB-5MS	15	0.25	Phenyl arylene polymer (GLC)	0.18	300	Helium (hydrogen capable)

influence over retention, resolution, bleed, inertness, and capacity. Column diameter has an influence over efficiency, retention, pressure, carrier gas flow rate, and column capacity (Agilent Technologies, Inc. 2012).

The temperature limit of the column is also an important specification for portable GC–MS systems. Like with lab-based systems, a sample is amenable to GC analysis if it is thermally stable with an appreciable vapor pressure at the column temperature. This allows the sample components to vaporize in, and move with, the gaseous mobile phase (Robards et al. 1994). Some chemicals of interest to field-portable GC–MS users, such as low-volatility compounds, may need an elevated column temperature to achieve the vapor pressure required for GC–MS analysis. The temperature range of a system's GC column in these instances is an important consideration.

There are a number of different types of low-volatility chemicals of interest to users of portable GC–MS. Two examples include chemicals used in warfare, such as the V-series agents and incapacitating agents (Pitschmann 2014). The V-series agents are part of the group of persistent agents that can remain on skin, clothes, and other surfaces for long periods of time due to low-volatility characteristics (Keyes et al. 2015). A majority of V-series agents are listed in Schedule 1 of the Chemical Weapons Convention (CWC). Exposure to solids, liquids, or vapors from these agents can result in death within minutes (Ellison 2008). Military incapacitating agents are the third- and fourth-generation chemical warfare agents (CWAs) that became popular during the Cold War. In addition to military-specific agents, materials in this class encompass a wide variety of commercially available medicinal drugs that interfere with the higher functions of the brain, such as attention, orientation, perception, memory, motivation, conceptual thinking, planning, and judgment (Ellison 2008). BZ (3-quinuclidinyl benzilate) is an example of an incapacitating agent and is listed in Schedule 2 (Toxic Chemicals) of the CWC. It has a boiling point of 320 °C (Incapacitating Agents 2007) and, like similar low-volatility compounds, can be a challenge for some GC–MS systems. The ability to run the column at higher temperatures makes analysis of this type of substance possible.

Not only is the ability to run at a higher column temperature important for some applications, but the system must also maintain good temperature control throughout the analysis path of the sample through the GC–MS. If good temperature control is not achieved across joints, such as between the GC and the MS, and transfer lines, such as in the sampling interface, analytes in the vapor phase may be lost during transfer through the system and will not be detected. For portable systems, the ability to heat the column to high temperatures and maintain good thermal control while minimizing power consumption so the system can run on battery is important.

Different portable GC–MS systems use different carrier gases. The carrier gas employed for use should be inert to analytes and compatible with other instrument components. Carrier gases selected for most GC applications include hydrogen, helium, and nitrogen. The choice of carrier gas among these three is important, primarily because it balances the column resolution required with analysis time. It impacts the resolution of a system through its effect on column efficiency (Robards et al. 1994). Efficiency is observed chromatographically as narrow and well-resolved peaks. The efficiency of an open-tubular column, measured in plates (N) or plates per meter (N/m), increases as the inner diameter of the column decreases (Sigma Aldrich 2017). The efficiency a gas imparts upon column performance is exemplified in its van Deemter curve, which is a plot of the column efficiency as a function of the average linear velocity of the carrier gas. The most efficient separations, i.e., those with the lowest N values, are usually achieved with nitrogen as a carrier gas. However, nitrogen is only the most efficient gas at low flow rates. At higher flow rates, which are required for faster analysis and typical for open-tubular columns used in portable GC–MS systems, the efficiency of nitrogen drops off quickly. Both helium and hydrogen are significantly more efficient than nitrogen at higher flow rates.

When comparing hydrogen with helium, hydrogen is likely the best choice from a performance perspective for GC applications, especially with regard to speed of analysis under otherwise identical system configurations (de Zeeuw 2011). For mobile laboratories, hydrogen generators can be used to produce high-purity hydrogen which can be used as the carrier gas. However, helium is probably used in 90% of GC applications (Grob 1997). Leland Gas Technologies (South Plainfield, NJ) offers disposable helium canisters that are used for both the Griffin G510 and TORION T-9 systems. For both the systems, a single canister can perform over 100 analyses. There are logistics considerations since the canisters are compressed gases, but these canisters have extremely long shelf lives, which is a significant advantage when maintaining inventory.

15.3.3 Mass Spectrometry

Mass spectrometry is one of the most generally applicable of all the analytical methods, providing qualitative and quantitative information about the atomic and molecular composition of inorganic and organic materials. The MS produces charged particles that consist of the parent ion and ion fragments of the original molecule. It sorts these ions according to their m/z value. The mass spectrum is the record of the relative numbers of different kinds of ions and is characteristic of every compound, including isomers (Willard et al. 1988).

Primary components of an MS include the ionization source, mass analyzer, and detector. With regard to portable GC–MS systems, the use of pumps to achieve the required operating pressures is also an important consideration because this component limits the size and weight of handheld forms of the technology. It also drives the mass analyzer type and design used in a system. Table 15.3 summarizes mass analyzer details and specifications of commercially available portable GC–MS systems described within this chapter.

Once an analyte has been vaporized and successfully transported through the GC column, it enters the ion source of the MS. The ion source charges the molecule to form an ion. This process is required to control the path of the substance in a magnetic or oscillating electric field. Fragmentation of the species that enters the ion source is also expected to occur when electron ionization (EI) is used. This fragmentation is characteristic of the starting compound and the ionization conditions.

Table 15.3 Mass analyzer details and specifications of portable GC–MS.

Instrument	Ion source	GC/MS interface	Mass analyzer	Mass range	Membrane-interfaced MS-only mode	Operating pressure (Torr)	Vacuum pump
Hapsite ER	EI	Membrane	Quadrupole	41–300	Yes	~10^{-6} or lower	NEG
Torion T-9	EI with dynamic ionization	Direct	Toroidal ion trap	41–500	Not available	~10^{-4}	Turbomolecular (10 l/s)
Griffin G510	EI	Direct	Quadrupole	15–515	Yes	~10^{-6}	Turbomolecular (80 l/s)

All portable GC–MS systems described within this chapter use EI. Unlike with other ionization methods, such as electrospray ionization (ESI), EI does not require the analyte to be polar. The success of EI for the ionization of organics led to its adoption as the standard ion source for use in organic mass spectrometry. The vapor pressure of many organic substances is sufficient, under the reduced pressure conditions of the MS source chamber, to require that the ion source act only as an ionizer, i.e., it does not have to volatilize the sample. This method was useful to replace gas-discharge sources for a number of reasons. First, gas-discharge sources (such as glow-discharge sources) operate at a higher pressure relative to the MS requirements, requiring heavy differential pumping. Second, the discharge, particularly the high-voltage type, can exhibit instability with a resultant fluctuating ion beam. Third, a critical factor of gas dischargers was the often-large ion-energy spread produced. Fourth, the energy of the ionizing electrons generated in the gas discharge is not always adequately controlled (Coburn and Harrison 1981).

EI is referred to as a hard-ionization method because it produces a substantial portion of the ionized molecules with such high energies that they fragment before leaving the ion source. The masses of these fragment ions provide the basic structural information used in interpretation (McLaffery and Turecek 1993). During EI, gas analyte molecules are bombarded by energetic electrons (typically 70 eV), which leads to the generation of a molecular radical ion (M^+) that can subsequently generate the ionized fragments. EI spectra are highly reproducible and, therefore, can be compared with mass spectral libraries so identification of unknowns can be achieved (Santos and Galceran 2003). GC–MS systems that use EI are most popular because they frequently provide both of these molecular and fragment ions. All the commercially available portable GC–MS systems described within this chapter use EI as an ionization source.

Once ions are created, the ion beam can be separated according to the respective m/z values using a variety of mass analyzer types. Magnetic deflection, quadrupole filter, ion trap, orbitrap, time of flight, and cyclotron resonance are the separating techniques most commonly used in commercial MS systems (McLaffery and Turecek 1993). The list of MS types used in commercially available portable GC–MS systems is much smaller and primarily includes quadrupole filter and ion trap systems. It seems the greatest practical advantage of quadrupole mass analyzers when used in field-portable GC–MS systems is their ability to generate mass spectra that can be more easily interpreted to determine the identity of a true unknown. They also are usually directly comparable with spectra in the commercially available NIST (National Institute of Standards and Technology) MS database, which is an extensive collection of searchable MS data. With regard to field-portable ion trap systems, a significant benefit is that resolution does not inherently depend upon trap size. Also, they run at higher operating pressures than quadrupole systems so they require less aggressive pumping. Pumping requires a significant amount of power, which is not ideal for portable systems that need to operate on batteries and are intended to be small and lightweight. The need to pump aggressively challenges deployment of portable GC–MS in the field.

Quadrupole mass analyzers require high-vacuum (low-pressure) environments to achieve good performance and typically operate at vacuum levels on the order of 10^{-6} Torr. They work by creating a quadrupole field using four electrically conducting parallel rods. Opposite pairs of electrodes are electrically connected. One diagonally opposite pair of rods is held at a positive direct-current (DC) potential, and the other is held at the negative of this DC potential. A radio-frequency (RF) oscillator supplies a specific signal to the first pair of rods, and an RF signal retarded by 180° to the second pair. The equipotential surfaces in the region between the four rods appear as oscillating hyperbolic potentials. Ions from the ion source are injected into the quadrupole array through an aperture. As the ions proceed down the longitudinal z-axis, they undergo transverse motion in the x- and y-planes perpendicular to the longitudinal axis. The DC electric fields tend to focus positive ions in the positive plane and defocus them in the negative plane. As the superimposed RF field becomes negative during part of the negative half-cycle of the alternating field, positive ions are accelerated toward the electrodes and achieve a substantial velocity. The following positive half-cycle has an even greater influence on the motion of the ion, causing it to reverse its direction (away from the electrode) and accelerate even more. The ions exhibit oscillations with increasing amplitudes until they finally collide with the electrodes and become neutral particles. The lighter the ion in mass, the smaller the number of cycles before it is collected by the electrode. By controlling the ratio of DC/RF, the field can be established to pass ions of a narrow m/z range down the entire length of the quadrupole array (Willard et al. 1988). The smaller the range of stable m/z values, the better the resolution, but sensitivity is decreased. By simultaneously ramping the DC and RF amplitudes, ions of various m/z values are allowed to pass through the mass filter to the detector, i.e., mass-selective stability, and an entire mass spectrum can be recorded (Willard et al. 1988).

A benefit of using quadrupole mass analyzers in portable GC–MS systems is the ability of some of these systems to perform MS-only analysis. This is also sometimes referred to as survey mode. During GC–MS analysis, a sample is analyzed by separation through the GC, followed by the generation of mass spectra of each eluted sample component. During analysis in MS-only mode, the air sample collected bypasses the GC and enters the MS through a membrane that allows the MS system to, among other things, maintain the required vacuum for MS analysis. The instrument response is independent of the amount of time required for GC analysis and, therefore, is almost real time. The mass spectra of individual components are not separated and isolated for conclusive identification when using MS-only mode, so the data may be very complex. But there are some purported benefits of this capability. First, the intensity of the MS-only signal may be used to deduce an approximate concentration of volatile organic compounds (VOCs) at the point of sampling so that optimum positioning for GC–MS sampling can be determined. Once this optimum position is located, the system can then be switched to the standard mode and air sampling for GC–MS analysis can be performed (Inficon 2018). In systems that do not offer MS-only mode for analysis such as the TORION T-9, optimum sampling position for air samples can be determined by measuring the concentration at the anticipated sampling location using a photoionization detector (PID). PIDs are handheld devices commonly used by first responders and the military to measure the concentration of VOCs and other gases on scene. PIDs can reliably measure in the ppm (parts per million) and ppb (parts per billion) range (Rae Systems n.d.). A second purported benefit of MS-only analysis is that it may be used to detect ions associated with one or two specific target analytes (Beckley et al. 2013). A challenge to MS-only analysis is that some chemicals of interest are not of appropriate volatility for analysis in this mode. In addition, some chemicals of interest may be of appropriate volatility but are not detected because they cannot penetrate the membrane and enter the MS.

Ion trap mass analyzers are also used in portable GC–MS systems. The operating pressure for ion traps is higher than for other forms of MS systems such as quadrupoles, allowing for less stringent pumping requirements (Lammert et al. 2006). In addition, ion traps perform better at higher pressures due to collisional cooling (Tolmachev et al. 2000). This can be beneficial for portable systems. Commercially available ion trap mass analyzers employ three electrodes – two endcap electrodes that are normally at ground potential, and between them a ring electrode to which an RF voltage, often in the megahertz range, is applied – to generate a quadrupole electric field. This type of quadrupole ion trap was used to store ions and characterize their properties in the isolated state (Cooks et al. 1991). In addition to the confinement of ions, developments allowed for the ion trap to function as an MS in that the m/z values of the trapped ion species could be measured. The principal method for measuring the m/z values of confined ions is to tip the potential well of the ion trap so that the ions leave the ion trap in order of ascending m/z value, i.e., mass-selective instability (March 1997). Using mass-selective ejection, an ion trap MS takes a sample, ionizes it, and then traps ions over a large mass range of interest simultaneously. Once trapped, ions can be sequentially ejected in order of mass by adjusting the strength of the quadrupole field holding the ions. This is usually accomplished by ramping the RF voltage applied to the ring electrode. The ejected ions are detected by an external electron multiplier to produce a mass spectrum (Stafford Jr. 2002).

A primary challenge for these ion trap MS systems is the management of ion–ion repulsions in the trap. Ion trap repulsions can impact the mass spectrum due to space charge as well as ion–ion reactions and ion–molecule reactions. Space charge is a consequence of Coulomb's law, which quantifies the force (F) between two point charges (q and q') separated by a distance r as:

$$F = k\left(\frac{qq'}{r^2}\right)$$

When space charge occurs, shifts in m/z values may be observed in the spectrum, or sensitivity may be compromised (Busch 2004). Ion chemistry occurs when ions in the trap react with each other to form ion complexes such as dimers. Although this behavior is not always observed, when it is, it can challenge the library-search results of the system. For these chemicals, the resulting mass spectrum may not only contain fragments of the ions initially formed during ionization but may also contain fragment ions that are characteristic of the ion complexes. In extreme cases, all ions initially formed during the ionization process may be consumed during reactions in the trap due to ion chemistry. The mass spectrum, under these extreme conditions, may only contain fragments that are characteristic of the ion complexes.

It is usually possible to address space charge and minimize the occurrence of ion chemistry by controlling the number of ions in the trap. This may be achieved a number of different ways including (i) increasing trap volume, (ii) adjusting instrument settings such as filament current to reduce the number of ions created, and (iii) optimizing the number of ions generated in real time using dynamic ionization. While these approaches can help in controlled settings, in situations where the operator is under significant mission and time constraints analyzing samples of unknown identity and concentration, they are not always effective.

The TORION T-9 uses a toroidal ion trap MS. A primary benefit of the toroidal design is the increase in trapping volume achieved when compared with a conventional ion trap design of similar inscribed radius. This system also uses dynamic ionization to manage the number of ions entering the trap. Overloading the trap is the principal obstacle to obtaining reproducible spectra for an ion trap system (Adams 1989) and, therefore, it is critical that overload of the trap be prevented. Alternatively, too few ions in the trap will result in poor spectral quality. Optimizing the number of ions in real time is critical for good performance but can be challenging because the amount of analyte eluting from the column is continually changing. For these reasons, this toroidal ion trap system performs dynamic-ionization control which varies ionization time to control the trap's ion population. As the concentration entering the MS increases, the ionization time decreases; as the concentration entering the MS decreases, ionization time increases.

The TORION T-9 attempts to address the issue of ion chemistry using onboard library capabilities. During the development of its proprietary libraries, compounds of interest may be analyzed using the system under a variety of conditions, including at various concentrations. Then, library entries can be created in accordance with how the chemical behaves when analyzed by the system under these different conditions. For samples that behave differently at different concentrations, for example, more than one library entry may be present so that, regardless of the concentration of the sample encountered in the field, an identification will be made. This enables effective library-matching results when deployed in the field.

The TORION T-9 also addresses the issue of space charge during library searching. As previously mentioned, when space charge occurs, shifts in m/z values may be observed in the spectrum. For example, when too many ions of bromoform are generated in the trap, the 171 fragment that is expected may not be observed; it will shift to 172, or even 173 in extreme cases. If the potential of space charge is not accounted for during the library-search process, this behavior may result in a missed identification. Figure 15.4 shows the impact of space charge on the mass spectrum of bromoform. The top MS shows the fragments 171, 173, and 175 and their relative abundances when space charge does not occur during the analysis. The bottom MS shows what can happen to the mass assignment and relative abundance of these fragments when space charge occurs. In spite of the fact that space charge is evident in the bottom spectrum, this toroidal ion trap system will still accurately identify the substance, although the match factor is lower (88.2) than for the spectrum collected under normal conditions (99.6). The ability to

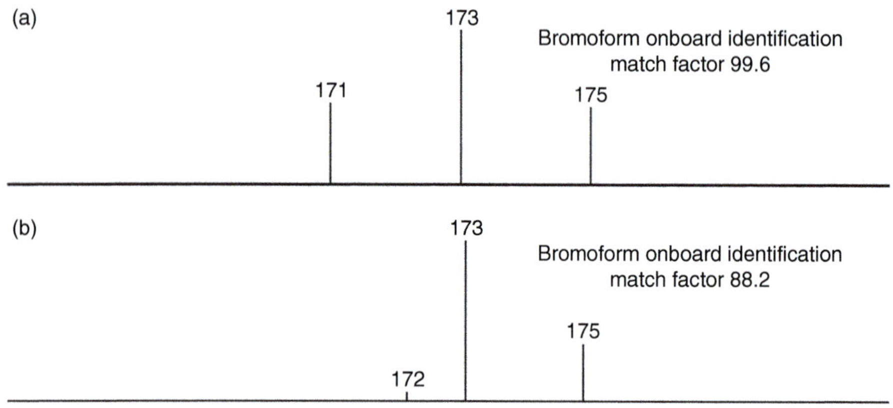

Figure 15.4 The mass spectra of bromoform when it is free from space charge (a) and when space charge occurs (b).

identify the target chemical in the field, even in cases when space charge occurs, is important. This is because it is frequently difficult or even impossible for users to control the amount of sample introduced to the system when analysis is performed in the field. Therefore, the availability of libraries of ion trap spectra can be critical for these portable systems.

15.3.4 Pumping and Vacuum Requirements

As previously mentioned, a primary challenge to the miniaturization of GC–MS systems is the need to achieve vacuum to perform the analysis. Pumps used to achieve vacuum may not only be bulky but may also be heavy. Depending upon the type of mass analyzer, the vacuum levels required for analysis may be high and, therefore, the performance requirements of pumps required to achieve such vacuum may be aggressive.

High vacuum is required for many portable GC–MS systems because the distance ions need to travel to reach the detector can be relatively large. These ions must reach the detector without undergoing collisions with other gaseous species along this path. The mean free path of the ion needs to be at least as large as the distance the ion must travel to reach the detector. Indeed, collisions would produce a deviation of the trajectory and the ion would lose its charge against the walls of the instrument before detection (de Hoffmann and Stroobant 2007).

The mean free path of the gas molecules can be calculated, and this parameter serves to define the conditions of viscous and molecular flow. When the mean free path of the gas molecules exceeds the dimensions of the vacuum container, the system is under molecular-flow conditions. Under such conditions, the residual gas molecules move without colliding with other gas molecules, instead ultimately colliding within the vacuum chamber (Busch 2001). The mean free path, ℓ, of a particle is dependent upon pressure, p, in accordance with the equation:

$$\ell = \hbar T/(\sqrt{2}\,\pi\, d^2 p\, \pi)$$

where $\hbar$ is the Boltzmann constant and d is the diameter of the particle. Depending upon the size and geometry of the mass analyzer used, i.e., the distance ions must travel to the detector, high vacuum may be a requirement for effective analysis. Due to the ability to miniaturize ion trap systems, the mean free path required for ions generated in ion trap systems can be lower than for quadrupole systems, i.e., the smaller the ion trap, the higher vacuum it can withstand to maintain unit-mass resolution. Therefore, pumping requirements for these systems are lower.

Pumping is accomplished when a net (nonrandom) direction to the movement of residual gas molecules in the vacuum chamber is attained (Busch 2001). Different pumps achieve this goal differently. There are a number of different types of pumps used to achieve vacuum in portable GC–MS systems. Historically, quadrupole field-portable systems required the use of a NEG pump to achieve the necessary vacuum required (Crume 2009).

Although quadrupole portable GC–MS systems like the HAPSITE ER that employ NEG pumps are still commercially available for sale and offer the advantage of lighter weight than other types of pumps, turbomolecular pumps capable of pumping 80 L/s are now available as an option for use with quadrupole systems. Ion trap systems use either diaphragm-backed turbomolecular pumps or even less aggressive pumps to achieve the required vacuum. Both the NEG and turbomolecular pump will be described in more detail as their use in field-portable systems is important.

NEG pumps remove or pump active gases by gettering (chemically combining) gases with metal alloys (Physical Electronics Inc. 2001). A primary advantage of the NEG pump, especially considering its ability to achieve the high vacuum required for quadrupole systems, is its lightweight (approximately 5 lbs) (Crume 2009). This feature is important for field-portable GC–MS systems. While the use of NEG pumps is important, especially considering they enabled the introduction of the field-portable (rather than transportable) quadrupole GC–MS systems to the field as far back as the year 1996 (Crume 2009), there are some challenges NEG pumps present when applied to field-portable GC–MS analysis. These challenges are both analytical and practical.

Analytically, the use of a NEG pump requires the placement of a membrane at the interface between the GC and MS. This membrane helps maintain vacuum while selectively allowing for the flow of organic compounds to the MS while blocking the flow of inorganic gases to the MS (Inficon 2008). In some instances, this membrane may also unwittingly prevent the introduction of some chemicals to the MS, challenging the ability of the system to analyze these substances (Bier and Cooks 1987). Practically, the NEG pump requires the use of a service module to activate the NEG pump (Inficon 2008). This activation process is required regularly in order for the NEG pump to remain functional, and can present logistical challenges for some field users. In addition, the service module itself is quite heavy (Crume 2009), including both a roughing pump and a turbomolecular pump to help achieve activation (Inficon 2016). While the service module is not required for analysis at the sampling point, it is required for sustainment of the portable GC–MS in the field. Activation is a process required for NEGs because when the getter material is exposed to air, the surface will "skin over" with reacted gases. This means the NEG will be totally enclosed in an envelope of oxides, nitrides, and other reacted gases. Additionally, the bulk of the material will be saturated with dissolved hydrogen gas. Since under these conditions, the getter material will be essentially inert, it will not provide an active getter-pumping surface. Activation, then, is the process to prepare the getter surface for pumping. This is done in situ by heating under vacuum (Danielson n.d.).

Turbomolecular pumps achieve vacuum by repeated collisions of residual gas molecules with rotating blades of a turbine motor. The edge speed of the rotors approaches the velocities of the residual molecules themselves. When a collision occurs, a direction is imparted to the motion of the residual gas molecules toward a region of higher pressure and toward the pump exhaust (a roughing pump) (Busch 2001). In portable GC–MS systems, the turbomolecular pump is frequently the limiting component that defines the ruggedness of the system.

Although the use of turbomolecular pumps may be considered generally advantageous, there are some disadvantages of turbomolecular pumps compared with NEG pumps. First, although these pumps are not necessarily large, they are heavy, which is undesirable in field-portable systems. In addition, they are sensitive to particulates and sediments and need to be protected from these contaminants. In modern field-portable GC–MS systems that are commercially available, pumps are contained within the system so that risk of exposure to particulates and sediments under normal wear-and-tear conditions is minimized.

15.4 Applications

Portable GC–MS is used to detect, identify, and quantify many different types of samples in many different industries. The value of being able to perform analysis at the scene is different for different applications. For instance, the ability to provide a confirmatory analysis of CWAs and other toxic industrial chemicals (TICs) during military operations is valuable because it allows the unit to independently maintain its initiative and momentum while at the same time enabling adjustments to both operational and strategic decisions in real time (Leary et al. 2019). On the other hand, performing analysis of ignitable liquid residues (ILRs) at the scene of a fire investigation is valuable because it preserves sample integrity. ILRs are a mixture of volatile chemicals. By the time samples are received in the laboratory, a significant amount of the chemicals present in the ILR have likely evaporated (Lentini 2013). Analysis at the scene, therefore, provides a more accurate analysis of the sample at the time it was recovered.

There are a number of established applications where portable GC–MS is frequently used including environmental assessments for the analysis of both hazardous air pollutants and other VOCs, as well as for the analysis of CWAs and TICs. As the technology continues to evolve, the value of these systems within these disciplines continues to grow. In addition, there are a number of emerging applications where the technology offers value including for fire investigations, the analysis of illicit drugs, forensic investigations of explosives, cannabis analysis as related to medical marijuana, and the petrochemical industry. There are also a number of niche research projects that demonstrate the versatility of the application of this technology in the field.

15.4.1 Environmental Applications

Portable GC–MS systems, at least historically, were most widely used in environmental applications (Henry 1997). There are a number of different types of environmental investigations where portable GC–MS analysis has value including site investigations for the analysis of ground contaminants, worker-safety assessments, and vapor-intrusion studies. Site investigations typically are performed by analyzing ground contaminants in soil, soil gases, or water. Worker-safety assessments might be undertaken if, for example, workers are digging in a known contaminated area where the air around the workers needs to be monitored to detect the off-gassing of dangerous chemicals. Vapor-intrusion studies are usually performed to verify contaminants in the ground are not penetrating a building structure (D. Schenk, 20 November 2019, personal communication). Regardless of the type of investigation, environmental assessments are usually focused on the analysis of hazardous air pollutants and other VOCs.

The first focused application for the use of portable GC–MS was for the analysis of hazardous air pollutants. In fact, the name HAPSITE is an acronym for Hazardous Air Pollutants on SITE (Crume 2009). The ability to detect and identify hazardous air pollutants is important because these chemicals can be dangerous, and amendments to the United States Clean Air Act (42 US Code §§7401-7671q 1970) make it a requirement that Environmental Protection Agency (EPA) regulate emissions of these pollutants. The original list included 189 pollutants. Since 1990, EPA has modified the list through rulemaking to include 187 pollutants. The hazardous air pollutants' list not only contains many VOCs, but also includes other hazardous compounds such as chlorine, titanium tetrachloride, and phosphorus (United States Environmental Protection Agency 2016).

VOCs are also detected and identified using portable GC–MS (Beckley et al. 2013; Eckenrode 2001; Fair et al. 2009; Gorder and Dettenmaier 2011; Henry 1997; Ho et al. 2001). These substances are members of a large group of organic chemicals that include any compound of carbon (excluding carbon monoxide, carbon dioxide, carbonic acid, metallic carbides or carbonates, and ammonium carbonate). Their detection and identification are important for a number of reasons. They participate in atmospheric photochemical reactions that contribute to ozone formation. They also play a role in the formation of secondary organic aerosols, which are found in airborne particulate matter. Finally, many individual VOCs are known to be harmful to human health (United States Environmental Protection Agency 2017).

15.4.2 CWAs and TICs

As previously mentioned, the military's ability to detect and identify CWAs using a confirmatory method like GC–MS (Eckenrode 2001; Parrish 2005; Seto et al. 2005; Smith et al. 2004) is important because it enables the unit to independently maintain its initiative and momentum while at the same time enabling adjustments to both operational and strategic decisions in real time (Leary et al. 2019). Portable GC–MS systems are used for the detection and identification of CWAs. A chemical agent is defined as a chemical substance that is intended for use in military operations to kill, seriously injure, or incapacitate mainly through its physiological effects (United States Department of Defense Joint Chiefs of Staff 2016). When these toxic chemicals and their precursors are prohibited under the CWC (Organisation for the Prohibition of Chemical Weapons n.d.), they are classified as CWAs. CWAs include choking, nerve, blood, blister, and incapacitating agents (United States Army, Marine Corps, Navy, Air Force 2005). Compared to conventional weapons, relatively small amounts of modern chemical agents may cause high number of casualties. Therefore, CWAs have been classified as weapons of mass destruction (WMD) (Szinica 2005).

Portable GC–MS systems are also used for the detection and identification of TICs (Bowerbank et al. 2009; Fair et al. 2009). TICs are industrial chemicals that when released into the atmosphere can have severe toxic effects on people and the environment. Examples include ammonia, chlorine, hydrogen chloride, hydrogen cyanide, and phosgene. The Occupational Safety and Health Administration (OSHA) of the United States Department of Labor classifies TICs into two groups, chemical and physical hazards, based on the type of risk they pose (Occupational Safety and Health Administration n.d.).

The military has spent significant resources to develop capabilities to detect and identify these dangerous chemicals. Ion mobility spectrometers continue to be a critical screening technology used by the military for exactly this purpose. (For a thorough discussion of portable ion mobility spectrometry, see the chapter in this volume by DeBono and Leary, and the chapter in volume 2 by Leary and Joshi.) However, ion mobility spectrometers are used for presumptive rather than confirmatory identifications. Before military units had the ability to perform confirmatory identifications of gases and vapors at the scene, it was necessary for them to operate under conditions that were not ideal. In most instances, tactics, techniques, and procedures dictated that decisions were made based upon information gathered using methods and instruments designed to classify or presumptively identify these substances. These methods included wet-chemical methods, colorimetric tests, electrochemical sensors, photoionization detectors, and ion mobility spectrometers (Detection and Measurement of Chemical Agents 1999; Murray 2013; Sun and Ong 2005). In other situations, samples were sent to remote laboratories, if access to such laboratories was available, for confirmatory testing. It could take days, weeks, or even months for results of testing to be returned. These limitations required that, in many instances, potentially critical intelligence be abandoned in theater with operations continuing as if the intelligence did not exist at all. In cases of personnel exposure to hazardous chemical agents, the inability to conclusively identify the agent causing the illness frequently required that the symptoms of the exposure be treated rather than the root cause (Leary et al. 2019).

Another benefit of portable GC–MS for this application is the large breadth of chemicals that can be characterized using this method. Even if a chemical has never been encountered or previously analyzed using GC–MS, the ability to interpret the GC–MS data collected in the field can help discern chemical identity and provide critical information about the threat. Essentially, the ability to analyze a substance using GC–MS is dependent upon (i) whether the chemical can be sampled and introduced to the system, (ii) has an appreciable vapor pressure at the column temperature so that it can pass through the GC, and (iii) whether the ionization method used in the mass analyzer is capable of generating ion fragments in the atomic-mass-unit (amu) range of the instrument. For modern portable GC–MS systems, this means that there are almost limitless numbers of chemicals that portable GC–MS systems can be used to help identify. In these situations, even though the sample will not alarm or be identified at the sample site, the data generated can be provided to scientific support teams for interpretation, providing potentially valuable intel. This is especially important as CWAs continue to evolve. New chemicals synthesized and developed for use in chemical warfare can be designed to circumvent detection and identification capabilities in the field, and the requirements of treaties like the CWC (Mirzayanov 2009). These substances can be tailored to meet toxicological or pharmacological goals. In addition, manufacturing, storage, transport, and delivery methods can be considered and optimized (Leary et al. 2019). The use of portable GC–MS is important in these situations because the data can be interpreted to identify new threats that might go undetected or uninterpreted using other methods.

The ability to analyze a sample using portable GC–MS is not possible if the sample cannot be introduced to the system. As such, the phase of matter of the sample can be critical. A portable GC–MS system capable of identifying chemicals in multiple phases of matter is currently under development through mid-2022. In October 2018, FLIR Systems, Inc. announced it had received an award from the United States Department of Defense (DOD) Joint Program Executive Office for Chemical, Biological, Radiological and Nuclear Defense (JPEO-CBRND) in support of the multiphase chemical agent detector (MPCAD) to develop a capability that will provide the ability to identify chemicals at low levels in aerosol, gas, liquid, and solid phases of matter. This instrument was proposed as a portable GC–MS solution that will provide the warfighter with field confirmation capabilities and better enable the Joint Forces to combat and interdict chemical weapons of mass destruction (WMD) (FLIR Systems, Inc. 2018).

15.4.3 Fire Investigation

There are some emerging applications where industries will likely benefit from the use of portable GC–MS system including the field of fire investigation. The use and value of GC–MS for the analysis of fire debris to establish the presence and identity of ignitable liquids is well-established (ASTM International 2014; Newman 2004; Stauffer et al. 2008). More recently, this technology has been taken to the field so that analysis may be performed at the scene of a fire investigation (Visotin and Lennard 2016). This capability is important. Ignitable liquids are comprised of different mixtures of volatile chemicals. As a result, fire debris samples containing residues of these liquids may change over time, e.g., between sample collection in the field and analysis in the laboratory. Analysis in the field would, therefore, more accurately represent the chemical composition of the sample recovered at the scene rather than after some components may have evaporated during transport to and storage in the laboratory. In instances where analysis at a laboratory is required, portable GC–MS may be used to determine the best location from which to recover samples for laboratory submission. This helps minimize the potential of sending samples containing no residue of ignitable liquids to the laboratory. Negative samples not only cost money to perform the analytical testing, but also increase the potential for laboratory backlog.

The environmental and human health aspects of fire scenes were recently evaluated using portable GC–MS. Portable GC–MS was used for the on-site analysis of water samples for the identification of hazardous organic compounds at fire scenes. This research showed that portable GC–MS was capable of detecting and identifying a range of volatile and semivolatile organic compounds in fire water runoff, and could be used in conjunction with conventional laboratory analysis methods for a comprehensive understanding of hazardous organics released at fire scenes. Deployment of this portable instrumentation provides first responders with a rapid, on-site screening tool to appropriately manage the runoff water from firefighting activities. This ensures that environmental and human health is proactively protected (Lam et al. 2019).

15.4.4 Illicit Drugs

The analysis of illicit drugs in the field by law enforcement and other first responders is an important capability that has become routine. An analytical scheme for such testing may include the use of colorimetric tests for screening, followed by either infrared or Raman spectroscopy for identification. Although portable GC–MS is more difficult to deploy than either of these methods, a primary advantage of using portable GC–MS for identification at the scene is that GC–MS can conclusively identify trace levels of drugs in street samples, even when these drugs are adulterated with large amounts of cutting agents. Figure 15.5 shows a chromatogram of an illicit drug sample recovered by law enforcement. Both portable infrared and Raman spectrometers identified this sample as quinine. Analysis of the sample using portable GC–MS (FLIR Systems, Inc. Griffin G510, Wilsonville, OR) showed the sample also contained the illicit drugs ketamine and fentanyl, as well as the drug levimasole.

These types of samples challenge other spectroscopic methods that struggle to detect drug samples at low concentrations, which is where GC–MS excels. This capability is especially advantageous in situations where high-potency opioids like fentanyl and its analogues are present at very low concentrations (Leary et al. 2017).

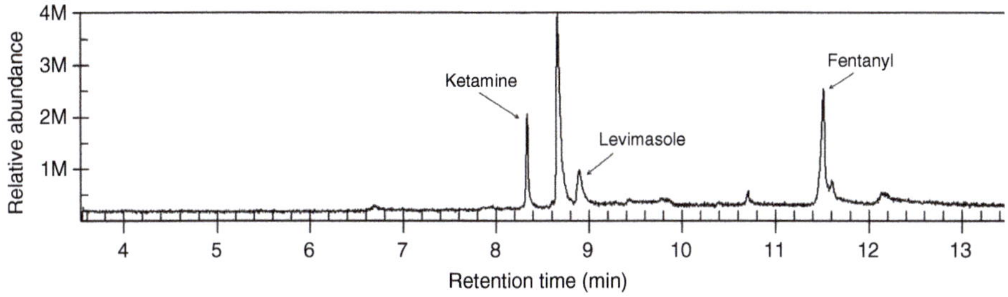

Figure 15.5 Chromatogram of illicit drug substance (identified using portable infrared and Raman as quinine).

15.4.5 Forensic Investigations of Explosives

The field of explosives analysis may also benefit from the use of field-portable GC–MS. Explosives are detected in the field using methods, such as ion mobility spectrometry, which can detect particulate residues of these substances at very low levels. While this type of analysis is important, the use of field-portable GC–MS can extend the analytical capabilities in the field to not only provide confirmation of initial ion mobility spectrometry and other testing results, but also to establish provenance of explosives when chemical markers used during explosive manufacture are detected and identified (Leary et al. 2016). Chemical markers may remain in a sample at trace levels and can be used as a point of differentiation among samples (Leary 2014). This is especially important in situations where homemade explosives such as hexamethylene triperoxide diamine (HMTD) and triacetone triperoxide (TATP) are analyzed. The ability to detect trace residues in these samples can provide insight into the history of the explosive sample including solvents and processes used for manufacture, providing useful investigative information in real time. Figure 15.6 shows a chromatogram of a sample of the explosive Composition C-4 tagged with DMNB (2,3-dimethyl-2,3-dinitrobutane) analyzed using portable GC–MS with SPME headspace analysis (Smiths Detection GUARDION, Edgewood, MD). The sample was purported to contain 0.1% DMNB which was automatically detected and identified using the system. Other advantages of performing GC–MS analysis of explosives at the scene of explosive events is because (i) it enables the development of render-safe procedures based upon real-time threat identification and assessment; (ii) intelligence turnaround time to support offensive operations is improved so Commanders are

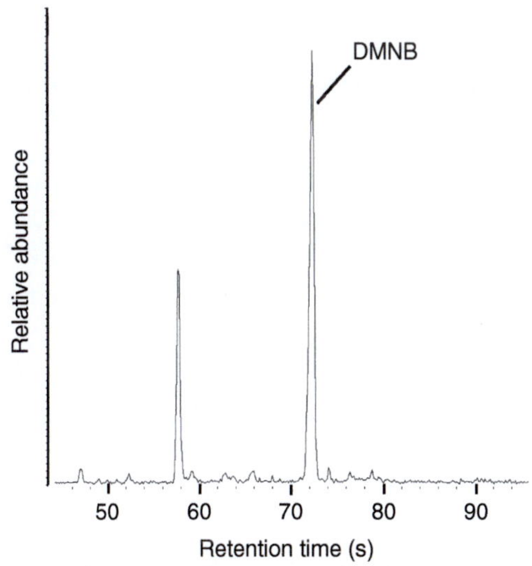

Figure 15.6 Chromatogram of composition C-4 tagged with DMNB.

empowered to make actional decisions using reliable, confirmed information; (iii) real-time results help guide and optimize scene processing; and (iv) analysis at the scene enables the most accurate evaluation of the scene; results are representative of the scene at the time of analysis, not at some later time when samples are received and analyzed at the laboratory (Moquin et al. 2020).

15.4.6 Cannabis Analysis

In 1970, the United States Congress placed marijuana (cannabis) into Schedule I of the Controlled Substances Act because they considered it to have "no accepted medical use." As of December 2020, 33 of 50 states in the United States and Washington, D.C. have legalized the medical use of marijuana. Proponents of medical marijuana argue that it can be a safe and effective treatment for the symptoms of cancer, AIDS, multiple sclerosis, pain, glaucoma, epilepsy, and other conditions (ProCon.org n.d.). More than 60 cannabinoid species have been isolated from cannabis and identified (Brenneisen 2007). The ability to detect and identify some of these chemicals is important. For instance, Δ-9 tetrahydrocannabinol (THC) and cannabidiol (CBD) are used for determining a product's total potency. In addition, the ratio of THC to CBD can determine the end usage of products as well as pricing and associated state tax revenue (908 Devices, Inc. 2017). For these and other reasons, the ability to detect, identify, and quantify these species in the field using portable GC–MS has become important.

15.4.7 Petrochemicals

There are a number of different ways portable GC–MS systems may provide value to the petrochemical industry including the ability to establish the quality of crude oil coming from a well, as well as its use to troubleshoot processing. The composition of crude oil is never consistent, creating the opportunity for operational problems to arise, such as raw material feedstock changes between incoming crude oil batches. Production pressures require the source of the problem to be identified as quickly as possible. The best way to achieve this is often to use GC–MS to identify the presence and type of chemical responsible and determine at what point it entered the process (Harrison 2011).

15.4.8 Others

There are other industries and applications that may benefit from the use of GC–MS in the field. A review of the literature shows a variety of different types of analysis that have already been attempted including the use of portable GC–MS to detect boar taint from samples on the slaughter processing line (Verplanken et al. 2015), to differentiate damaged and undamaged yellow Starthistle flower heads (Beck et al. 2015), and to detect and quantify munitions constituents in groundwater (Bednar et al. 2012). Portable GC–MS was also used to evaluate the volatility profiles from stockpiled almonds and compare results with data collected from benchtop systems. Despite the portable GC–MS system detecting fewer volatiles than the benchtop system, both systems resolved humidity treatments and identified potential fungal biomarkers at extremely low water activity levels. This ability to resolve humidity levels suggests that volatile profiles from germinating fungal spores could be used to create an early warning, nondestructive, portable detection system of fungal growth (Beck et al. 2016). While these applications and others that are currently being explored for viability may not be as widely deployed as others, the ability to perform these types of analyses in the field is testament to the ruggedness and versatility of modern portable GC–MS platforms.

15.5 The Future of Portable GC–MS

A primary value of portable GC–MS when compared with other field technologies is its ability to conclusively identify chemicals, even when present at low concentration within complex sample matrices. Depending upon the application, the features of a system that are important can be different. For instance, for military users "smaller, lighter, faster" is a standard mantra and especially appropriate for field-portable GC–MS systems. These systems can be deployed to remote locations within hostile environments where they must be discretely hand-carried to the scene. In these situations, the addition of even a few ounces of weight can be critical. Further development to extend each system's current capabilities to be able to analyze a wider breadth of chemicals would be valuable. When deployed in other applications, though, development of flexibility in sampling methods and adding quantitative capabilities to a system may be more important than size and weight. These features may be important for applications such as environmental analysis, fire investigation, and illicit-drug analysis.

To achieve systems that are smaller, until the Griffin G510 was introduced to the market in 2017, innovation in mass analyzer design was the road taken. The use of ion trap designs has been important to achieving smaller and lighter systems due to the higher operating pressures of these systems. Vendors such as Smiths Detection and PerkinElmer developed portable ion trap GC–MS systems, while 908 Devices developed a variety of high-pressure MS products including standalone ion trap MS systems. For all of these ion trap systems, overcoming the consequences of trapping ions either through improvements in mass spectral design or in data processing capabilities would be beneficial. While not discussed in this chapter (see chapter in this volume by Blakeman and Miller), the high-pressure MS system by 908 Devices known as the MX908 has become an important capability that continues to grow into different markets. The ability to improve the spectral resolution of this high-pressure MS system and couple it with front end separation technologies like GC would provide an impressive portable capability that would be quite field-friendly.

Acknowledgments

The authors would like to acknowledge Dalton T. Snyder for his review and suggestions for improvement to this chapter, and support provided by the Connecticut State Police K9 Unit, and specifically TFC Jon Naples and TFC Brian Faughnan for samples provided for analysis.

References

42 US Code §§7401-7671q (1970). *United States Clean Air Act*. Office of the Law Revision Counsel of the U.S. House of Representatives.

908 Devices Inc. (2017). Application Note 1.0: Ultrafast Total Potency Analysis of Cannabindoids in Decarboxylated Oil Extracts Using G908. 908 Devices.com. http://908devices.com/wp-content/uploads/2017/09/TotalPotency_AppNote_2017_v8.pdf (accessed 28 October 2017).

Adams, R.P. (1989). *Identification of Essential Oils by Ion Trap Mass Spectroscopy*. San Diego, CA: Academic Press, Inc.

Agilent Technologies Inc. (2012). *Agilent J&W GC Column Selection Guide*. Agilent. https://www.agilent.com/cs/library/catalogs/public/5990-9867EN_GC_CSG.pdf (accessed 4 October 2017).

ASTM International (2014). *ASTM E1618–14, Standard Test Method for Ignitable Liquid Residues in Extracts from Fire Debris Samples by Gas-Chromatography-Mass Spectrometry*. West Conshohocken: ASTM International https://doi.org/10.1520/E1618-14.

Beck, J.J., Porter, N., Cook, D. et al. (2015). In field volatile analysis employing a hand-held portable GC-MS: emission profiles differentiate damaged and undamaged yellow starthistle flowers. *Phytochem. Anal.* 26: 395–403. https://doi.org/10.1002/pca.2573.

Beck, J.J., Willett, D.S., Gee, W.S. et al. (2016). Differentiation of volatile profiles from stockpiled almonds at varying relative humidity levels using benchtop and portable GC-MS. *J. Agric. Food Chem.* https://doi.org/10.1021/acs.jafc.6b04220.

Beckley, L., McHugh, T., Gorder, K. et al. (2013). *Use of on-Site GC/MS Analysis to Distinguish between Vapor Intrusion and Indoor Sources of VOCs*. Environmental Security Technology Certification Program (ESTCP) https://clu-in.org/download/issues/vi/VI-ER-201119-FR.pdf.

Bednar, A., Russell, A., Hayes, C. et al. (2012). Analysis of munitions constituents in groundwater using field-portable GC-MS. *Chemosphere* 87 (8): 894–901. https://doi.org/10.1016/j.chemosphere.2012.01.042.

Biemann, K., Oro, J., Toulminn, P. III, et al. (1977). The search for organic substances and inorganic volatile compounds in the surface of Mars. *J. Geophys. Res.* 82 (28): 4641–4658. https://doi.org/10.1029/JS082i028p04641.

Bier, M. and Cooks, R. (1987). Membrane interface for selective introduction of volatile compounds directly into the ionization chamber of a mass spectrometer. *Anal. Chem.* 59: 597–601. https://doi.org/10.1021/ac00131a013.

Bowerbank, C.R., Lee, E.D., Sadowski, C.S. et al. (2009). Rapid field detection of chemical warfare agents and toxic industrial chemicals using a hand-portable GC-TMS system. *LCGC* 27 (4). http://www.chromatographyonline.com/rapid-field-detection-chemical-warfare-agents-and-toxic-industrial-chemicals-using-hand-portable-gc?id=&sk=&date=&pageID=5 (accessed 18 April 2020).

Brenneisen, R. (2007). Chemistry and analysis of phytocannabinoids and other cannabis constituents. In: *Marijuana and the Cannabinoids* (ed. M.A. ElSohly), 17. Totowa, NJ: Humana Press.

Bruker (2020). *Technical Details of the Bruker E2M Mobile Gas Chromatograph Mass Spectrometer*. Bruker. https://www.bruker.com/products/cbrne-detection/gc-ms/e2m/technical-details.html (accessed 14 April 2020).

Busch, K.L. (2001). High-vacuum pumps in mass spectrometers. *Spectroscopy* 16 (5): 14–18.

Busch, K.L. (2004, June). Space charge in mass spectrometry. *Spectroscopy* 19 (6): 35–38.

Coburn, J. and Harrison, W. (1981). Plasma sources in analytical mass spectrometry. *Appl. Spectros. Rev.* 17 (1): 95–164. https://doi.org/10.1080/05704928108060402.

Cooks, R.G., Gilsh, G.L., McLuckey, S.A., and Kaiser, R.E. (1991). Ion trap mass spectrometry. *Chem. Eng. News (C&EN)* 69 (12): 25–41.

Crume, C. (2009). *History of Inficon HAPSITE: Maintenance Management, Support, and Repair*. KD Analytical. http://www.kdanalytical.com/instruments/inficon-hapsite-history.aspx (accessed 5 June 2016).

Danielson, P. (n.d.). How to use getters and getter pumps. R&D Magazine (February 2001). http://www.normandale.edu/departments/stem-and-education/vacuum-and-thin-film-technology/vacuum-lab/articles/how-to-use-getters-and-getter-pumps (accessed 9 November 2017).

Detection and Measurement of Chemical Agents (1999). *In Chemical and Biological Terrorism: Research and Development to Improve Civilian Medical Response*. Washington, DC: National Academies Press.

Eckenrode, B. (2001). Environmental and forensic applications of field-potable GC-MS: an overview. *J. Am. Soc. Mass Spectrosc.* 12 (6): 683–693. https://doi.org/10.1016/S1044-0305(01)00251-3.

Ellison, D.H. (2008). *Handbook of Chemical and Biolgical Warfare Agents*, 2e. Boca Raton, FL: Taylor & Francis Group, LLC.

Fair, J.D., Bailey, W.F., Felty, R.A. et al. (2009). Method for rapid on-site identification of VOCs. *J. Environ. Sci.* 21 (7): 1005–1008. https://doi.org/10.1016/S1001-0742(08)62375-X.

FLIR Systems, Inc. (2019). *Person-Portable GC-MS Chemical Identifier FLIR Griffin™ G510*. FLIR Systems, Inc. https://www.flir.com/globalassets/imported-assets/document/griffin-g510-datasheet.pdf (accessed 14 April 2020).

FLIR Systems, Inc. (2018). *FLIR Systems Receives Award Totaling $28.7M for the Next Generation Chemical Detector (NGCD) from the United States Army*. FLIR Systems, Inc. https://investors.flir.com/news-releases/news-release-details/flir-systems-receives-award-totaling-287m-next-generation (accessed 16 April 2020).

Fogerson, R., Schoendorfer, D., Fay, J., and Spiehler, V. (1997). Qualitative detection of opiates in sweat by EIA and GC-MS. *J. Anal. Toxicol.* 21: 451–458.

Gorder, K.A. and Dettenmaier, E.M. (2011). Portable GC/MS methods to evaluate sources of cVOC contamination in indoor air. *Ground Water Monit. Remed.* 31 (4): 113–119. https://doi.org/10.1111/j.1745-6592.2011.01357.x.

Grob, K. (1997). *Carrier Gases for GC*. Restek http://www.restek.com/Technical-Resources/Technical-Library/Editorial/editorial_A017.

Harrison, S. (2011). How gas chromatography-mass spectrometry can be utilised as a tool for forensic analysis. *Hydrocarbon Eng.* 16 (10): 33–36.

Henry, C. (1997). Product review: take the show on the road: portable GC and GC/MS. *Anal. Chem.* 69 (5): 195A–200A. https://doi.org/10.1021/ac9715682.

Ho, C.K., Itamura, M.T., Kelley, M., and Hughes, R.C. (2001). *Review of Chemical Sensors for in-Situ Monitoring of Volatile Contaminants*. Albuquerque and Livermore: Sandia National Laboratories.

de Hoffmann, E. and Stroobant, V. (2007). *Mass Spectrometry: Principles and Applications*, 3e. West Sussex: John Wiley & Sons Ltd.

Incapacitating Agents (2007). *Compendium of Chemical Warfare Agents* (ed. S.L. Hoenig). New York: Springer https://doi.org/10.1007/978-0-387-69260-9_4.

Inficon (2015). *Specifications of the HAPSITE® ER Chemical Identification System*. Inficon.

Inficon (2016). *Operating Manual, HAPSITE Service Module*. Inficon. http://products.inficon.com/getattachment.axd/?attaname=service-module-operating-manual (accessed 9 November 2017)

Inficon (2008). *Hapsite Smart Plus Operators Manual*. Inficon. http://products.inficon.com/GetAttachment.axd?attaName=9243f3a0-981e-435b-aa6b-412124af0900 (accessed 11 March 2017).

Inficon (2018). *HAPSITE ER Chemical Identification System*. Inficon. http://products.inficon.com/GetAttachment.axd?attaName=b0ddf534-db3e-4920-b9c1-ec872bc28a4d (accessed 13 December 2020).

Keyes, D.C., Benitez, F.L., Velez-Daubon, L.I., and Talavera, F. (2015). *E-Medicine Drugs and Diseases, Emergency Medicine*. CBRNE – Nerve Agents, V-series - Ve, Vg, Vm, Vx. http://emedicine.medscape.com/article/831760-overview#showall (accessed 9 March 2017).

Lam, R.L., Kingsland, G., Johnstone, P. et al. (2019). Rapid on-site identification of hazardous organic compounds at fire scenes using person-portable gas chromatography-mass spectrometry (GC-MS) - part 2: water sampling and analysis. *Foren. Sci. Res.*: 1–15. https://doi.org/10.1080/20961790.2019.1662648.

Lammert, S.A., Rockwood, A.A., Wang, M. et al. (2006). Miniature Toroidal radio frequency ion trap mass analyzer. *J. Am. Soc. Mass Spectrom.* 17 (7): 916–922. https://doi.org/10.1016/j.jasms.2006.02.009.

Leary, P.E. (2014). Counterfeiting: a challenge to forensic science, the criminal justice system, and its impact on pharmaceutical innovation. Ph.D. thesis. Graduate Center of the City University of New York, New York.

Leary, P.E., Kammrath, B.W., Lattman, K.J., and Beals, G.L. (2019). Deploying portable gas chromatography-mass spectrometry (GC-MS) to military users for the Identificaiton of toxic chemical agents in theater. *Appl. Spectrosc.* 73 (8): 841–858. https://doi.org/10.1177/0003702819849499.

Leary, P.E., Dobson, G.S., and Reffner, J.A. (2016). Development and applications of portable gas chromatography-mass spectrometry for emergency responders, the military, and law-enforcement organizations. *Appl. Spectrosc.* 70: 888–896. https://doi.org/10.1177/0003702816638294.

Leary, P.E., Gayle, B.A., Gross, D.L., and Frunzi, M. (2017). The fentanyl epidemic. *Fire Eng.* 170 (10): 26–32.

Lentini, J.J. (2013). Analysis of ignitable liquid residues. In: *Scientific Protocols for Fire Investigation*, 2e (ed. J.J. Lentini). Boca Raton, FL: Taylor & Francis Group, LLC.

March, R.E. (1997). An introduction to quadrupole ion trap mass spectrometry. *J. Mass Spectrom.* 32: 351–369.

McLaffery, F.W. and Turecek, F. (1993). *Interpretation of Mass Spectra*, 4e. Sausalito, CA: University Science Books.

McNair, H.M. and Miller, J.M. (2009). *Basic Gas Chromatography*, 2e. John Wiley & Sons, Inc.

Merlin Instrument Company (2019). *Merlin Microshot*. Merlin Instrument Company. https://merlinic.com/merlin-microshot (accessed 17 April 2020).

Meuzelaar, H.L., Dworanski, J.P., Arnold, N.S. et al. (2000). Advances in field-portable mobile GC/MS instrumentation. *Field Anal. Chem. Technol.* 4 (1): 3–13.

Mirzayanov, V.S. (2009). *State Secrets: An Insider's Chronicle of the Russian Chemical Weapons Program*. Denver: Outskirts Press, Inc.

Mississippi National Guard (2015). *Frequently Asked Question's*. Mississippi National Guard. https://ms.ng.mil/aboutus/units/66tc/47cst/Pages/FAQs.aspx (accessed 14 April 2020).

Moquin, K., Higgins, A.G., Leary, P.E., and Kammrath, B.W. (2020, May). Optimized explosives analysis using portable gas chromatography-mass spectrometry for battlefield forensics. *Curr. Trends Mass Spectr.* 18 (2): 1–8.

Murray, G.M. (2013). Detection and screening of chemicals related to the chemical weapons convention. *Encycl. Anal. Chem.*: 1–37. https://doi.org/10.1002/9780470027318.a0403.pub2.

National Aeronautics and Space Administration (1970). *Astronautics and Aeronautics, 1969*. Washington, D.C.: Scientific and Technical Information Division, Office of Technology Utilization.

Newman, R. (2004). Interpretation of laboratory data. In: *Fire Investigation* (ed. N. Nic Daéid). Boca Raton: CRC Press, LLC.

Occupational Safety and Health Administration (n.d.). *Toxic Industrial Chemicals (TICs) Guide*. United States Department of Labor. https://www.osha.gov/SLTC/emergencypreparedness/guides/chemical.html (accessed 21 February 2020).

Organisation for the Prohibition of Chemical Weapons (n.d.). *Genesis and History of the Chemical Weapons Convention*. Organisation for the Prohibility of Chemical Weapons. https://www.opcw.org/chemical-weapons-convention/genesis-and-historical-development (accessed 1 November 2016).

Parrish, L.D. (2005). *Application of Solid Phase Microextraction with Gas Chromatography – Mass Spectrometry as a Rapid, Reliable, and Safe Method for Field Sampling and Analysis of Chemical Warfare Agent Precursors*. Bethesda: Disseration – Uniformed Services University of the Health Sciences.

Pawliszyn, J. (1997). *Solid Phase Microextraction*. Waterloo: Wiley-VCH, Inc.

PerkinElmer Inc. (2018). *Product Note: Custodion CME Syringes for Easy Liquid Sample Collection and Injection*. PerkinElmer. https://www.perkinelmer.com/lab-solutions/resources/docs/PRD_Custodion_CME_Syringe_014110_01.pdf (accessed 25 April 2020).

PerkinElmer Inc. (2020). *Torion T-9 Portable GC/MS*. PerkinElmer Inc. https://www.perkinelmer.com/product/torion-t-9-portable-gc-ms-instrument-ntsst090500 (accessed 14 April 2020).

Physical Electronics Inc (2001). *Technicians Non-Evaporative Getter Pump Component Manual*. Physical Electronics Inc. http://www.pascaltechnologies.com/files/Pumps/Ion%20Pumps/Other%20Ion%20Pumps/Physical%20Electronics%20NEG%20Technicians%20Manual.pdf (accessed 9 November 2017).

Pitschmann, V. (2014). Overall view of chemical and biochemical weapons. *Toxins* 6: 1761–1784. https://doi.org/10.3390/toxins6061761.

ProCon.org (n.d.). Should Marijuana Be a Medical Option? https://medicalmarijuana.procon.org (accessed 25 March 2018).

Rae Systems (n.d.). *Photoionization Detectors (PIDs)*. Rae Systems: http://www.raesystems.com/solutions/photoionization-detectors-pids (accessed 25 March 2018).

Restek (n.d.). *MXT-Msieve 5A Columns, Technical Description*. Restek. http://www.restek.com/catalog/view/8578 (accessed 26 October 2017).

Robards, K., Haddad, P.R., and Jackson, P.E. (1994). *Principles and Practice of Modern Chromatographic Methods*. London: Academic Press Limited.

Santos, F. and Galceran, M. (2003). Modern developments in gas chromatography-mass spectrometry-based environmental analysis. *J. Chromatogr. A* 1000: 125–151. https://doi.org/10.1016/S0021-9673(03)00305-4.

Schecter, A.J., Sheu, S.U., Birnbaum, L.S. et al. (1999). A comparison and discussion of two differing methods of measuring dioxin-like compounds: gas chromatography-mass spectrometry and the calux bioassay - implications for health studies. *Organohalogen Compd.* 40: 247–250.

Seto, Y., Kanamori-Kataoka, M., Tsuge, K. et al. (2005). Sensing technology for chemical-warfare agents and its evaluation using authentic agents. *Sens. Actuators B* 108: 1930197. https://doi.org/10.1016/j.snb.2004.12.084.

Sigma Aldrich (2017). *How to Choose a Capillary GC Column*. Sigma Aldrich. https://www.sigmaaldrich.com/analytical-chromatography/gas-chromatography/column-selection.html (accessed 8 November 2017).

Simmonds, P.G. (1970). Whole microorganisms studied by pyrolysis-gas chromatography-mass spectrometry: significance for extraterrestrial life detection experiments. *Appl. Microbiol.* 20 (4): 567–572.

Smith, P.A. (2015). Portable gas chromatography. In: *Analytical Separation Science*, vol. 3 (eds. J. Anderson, A. Berthod, V. Pino and A.M. Stalcup). Wiley https://doi.org/10.1002/9783527678129.assep031.

Smith, P.A., Jackson Lepage, C.R., Koch, D. et al. (2004). Detection of gas-phase chemical warfare agents using field-portable gas chromatography-mass spectrometer systems: instrument and sampling strategy considerations. *Trends Anal. Chem.* 23: 296–306. doi: 10.106/S0165-9936(04)00405-4.

Stafford, G. Jr., (2002). Ion trap mass spectrometry: a personal perspective. *J. Am. Soc. Mass Spectrosc.* 13: 589–596.

Stauffer, E., Dolan, J.A., and Newman, R. (2008). *Fire Debris Analysis*. Elsevier, Inc.

Sun, Y. and Ong, K.Y. (2005). *Detection Technologies for Chemical Warfare Agents and Toxic Vapors*. Boca Raton, FL: CRC Press LLC.

Szinica, L. (2005, October 30). History of chemical and biological warfare agents. *Toxicology* 214 (3): 167–181. https://doi.org/10.1016/j.tox.2005.06.011.

Tilstone, W.J., Savage, K.A., and Clark, L.A. (2006). *Encyclopedia of Forensic Science: An Encyclopedia of History, Methods and Techniques*. Santa Barbara: ABC-CLIO, Inc.

Tolmachev, A.V., Udseth, H.R., and Smith, R.D. (2000). Radial stratification of ions as a function of mass to charge ratio in collisional cooling radio frequency multipoles used as ion guides or ion traps. *Rapid Commun. Mass Spectrom.* 14 (20): 1907–1913. https://doi.org/10.1002/1097-0231(20001030)14:20<1907::AID-RCM111>3.0.CO;2-M.

United States Army, Marine Corps, Navy, Air Force (2005). *Potential Military Chemical/Biological Agents and Compounds*. http://www.marines.mil/Portals/59/MCRP%203-37.1B%20z.pdf (accessed 23 April 2016).

United States Department of Defense Joint Chiefs of Staff (2016). Department of Defense Dictionary of Military and Associated Terms. http://www.dtic.mil/doctrine/new_pubs/jp1_02.pdf (accessed 23 April 2016).

United States Environmental Protection Agency (2016). *Initial List of Hazardous Air Pollutants with Modifications*. United States Environmental Protection Agency. https://www.epa.gov/haps/initial-list-hazardous-air-pollutants-modifications (accessed 25 March 2018).

United States Environmental Protection Agency (2017). Volatile Organic Compounds' Impact on Indoor Air Quality. https://www.epa.gov/indoor-air-quality-iaq/volatile-organic-compounds-impact-indoor-air-quality (accessed 11 December 2020).

United States Environmental Protection Agency (2020). *EPA Region 9 Laboratory Field Services Summary*. United States Environmental Protection Agency. https://www.epa.gov/regionallabs/epa-region-9-laboratory-field-services-summary#mobilelab (accessed 14 April 2020)

Verplanken, K., Wauters, J., Van Dirme, J. et al. (2015). Development and validation of a rapid detection method for boar taint by means of solid phase microextraction and a person-portable GC-MS. In: *Belgian Association for Meat Science and Technology (BAMST) Symposium*, 29–30. Belgium: Melle. http://hdl.handle.net/1854/LU-7067733 (accessed 23 April 2016).

Visotin, A. and Lennard, C. (2016). Preliminary evaluation of a next-generation portable gas chromatograph mass spectrometer (GC-MS) for the on-site analysis of ignitable liquid residues. *Aust. J. Forensic Sci.* 48 (2) https://doi.org/10.1080/00450618.2015.1045554.

Willard, H.H., Merritt, L.L. Jr., Dean, J.A., and Settle, F.A. Jr., (1988). *Instrumental Methods of Analysis*, 7e. Belmont: Wadsworth Publishing Company.

de Zeeuw, J. (2011). *Fast(Er) GC: How to Decrease Analysis Time Using Existing Instrumentation? Part IV: Using Hydrogen as the Carrier Gas*. doi: https://blog.restek.com/?p=3520

16

Development of High-Pressure Mass Spectrometry for Handheld and Benchtop Analyzers

Kenion H. Blakeman and Scott E. Miller

908 Devices Inc., Boston, MA, USA

16.1 Introduction

Mass spectrometry (MS) is commonly referred to as the gold standard for chemical analysis, given its sensitivity, selectivity, and wide application ranges [1]. Single-molecule detection is possible for some applications and resolving powers in excess of 1 000 000 have been reported [2, 3]. While mass spectrometers can be extraordinarily powerful, they are typically large, complex, and expensive; require regular maintenance; have limited robustness; are confined to a laboratory environment; and require trained users to operate [4–6]. For decades, there has been strong interest in miniaturizing mass spectrometers, and in moving them outside the laboratory environment. A key set of parameters to consider for miniaturization and portability are size, weight, and power (SWaP) [7, 8]. A miniature mass spectrometer design must be robust enough to survive transport to a new location and operate on limited power supplies such as batteries or generators. Since mass spectrometers are classically optimized for performance in a particular location and only moved on the rarest of occasions by trained personnel, the design challenges are unique for portable instruments [9–11].

The transition from laboratory tool to field-portable mass spectrometer was first demonstrated by the Inficon HAPSITE system in 1996 [12]. While not intended to be a comprehensive list, several examples of field-portable mass spectrometers that have been released since then are summarized in Table 16.1. Some key differences include the type of mass analyzer that is used, including quadrupole mass filters (QMFs) and quadrupole ion traps (QITs). While a gas chromatograph can add additional analytical capabilities, it adds additional size, weight, and complexity to a system and may not be included depending on the intended use case. Despite multiple decades of field-portable MS development, considerable diversity in how systems are designed is seen.

As the mass spectrometer miniaturization has progressed, there has been a trend toward commercializing systems with weights in the 15–35 kg range. A key challenge for miniaturization has been the vacuum requirements associated with performing mass spectrometry. Vacuum requirements below $<10^{-5}$ Torr are common for mass spectrometers and are typically achieved using large turbomolecular pumps that introduce considerable weight to the system. Ion trap mass analyzers can operate at higher pressures (e.g. 10^{-3} Torr) but this is still well into the molecular flow regime [13, 14]. For technologies without such pumping requirements, such as ion mobility spectrometry (IMS) and optical methods, such as Fourier transform infrared spectroscopy (FTIR) and Raman, smaller analyzers have been commercialized [11, 15, 16]. For example, Smiths Detection released the LCD 3.3 handheld IMS in 2009, which has dimensions of 10.54 × 17.3 × 4.65 cm and weighs 0.65 kg [17]. The FirstDefender was a Raman spectrometer by Ahura Scientific, which was released in 2005 with dimensions of 19.3 × 10.7 × 4.4 cm and a weight of 1.8 kg [18].

Portable Spectroscopy and Spectrometry 1: Technologies and Instrumentation, First Edition.
Edited by Richard A. Crocombe, Pauline E. Leary and Brooke W. Kammrath.

Table 16.1 Examples of commercially available field-portable mass spectrometers highlighting key specifications.

Name and developer	Year introduced	Mass analyzer	GC	Weight (kg)	Dimensions (cm)	Reference
Inficon HAPSITE	1996	QMF	Yes	19	46 × 43 × 18	[106]
Bruker E2M/MM2	2004	QMF	Yes	37.7	39 × 39 × 28	[107]
908 Devices M908	2014	QIT	No	2.0	22 × 18.5 × 7.6	[108]
Torion T-9	2015	QIT	Yes	14.5	47 × 36 × 18	[109]
Griffin G510	2017	QMF	Yes	16.3	33.7 × 33.7 × 40	[110]
908 Devices MX908	2017	QIT	No	3.9	29.8 × 21.6 × 12.2	[111]
Bayspec Continuity	2017	Dual QIT	No	20	33 × 33 × 43	[112]
1st Detect Tracer 1000 MS-ETDM	2019	QIT	No	24	36 × 45 × 53	[113]

The disparity between mass spectrometers in the 15 kg weight range at best and optical and IMS detectors in the 0.5–2 kg range leads to questions on how to bridge the weight gap. Since the late 1990s, one research approach has been to increase the operating pressure of an ion trap mass spectrometer. If the operating pressure was high enough (fractions of a Torr and above), the turbomolecular pump could be eliminated and only a highly miniaturized roughing pump would be required, a significant departure from conventional methods which utilize two large vacuum pumps.

In this chapter, the background of ion trap operation and its evolution over time will be summarized. Key developments on scaling an ion trap for higher pressure operation will be covered. Experimental results from high-pressure mass spectrometry (HPMS) experiments will be covered in detail. Finally, devices based on HPMS and their applications across different disciplines will be described.

16.2 Ion Trap Development for HPMS

16.2.1 Trapped Ion Motion Characterization

The development of HPMS analyzers centers on the development and analytical performance of QITs. The QIT mass analyzer was first reported by Paul and Steinwedel in 1953 [19]. It consisted of a central ring electrode separated from two endcap electrodes by insulating material. Trap dimensions are defined as the internal radius of the ring electrode (r_0) and the distance from the center of the trap to the endcap electrodes (z_0). Ions are introduced and ejected from the ion trap through apertures in the endcap electrodes.

Since the initial work by Paul and Steinwedel, there has been considerable research into understanding the motion of trapped ions. While a complete mathematical analysis is beyond the scope of this summary, several detailed analyses are cited here [20–22]. Central to the understanding of trapped ion motion are stability diagrams (Figure 16.1). Stability diagrams are determined by the Mathieu equation for ion motion and describe conditions where ions are stable in the ion trap based on dimensionless parameters a and q. The parameters a and q are further defined as a_z and q_z to define trajectories in the z-dimension, and a_r and q_r to define motion in the r-dimension. The equations for each of these dimensionless parameters are defined in Eqs. (16.1) through (16.4):

$$a_z = -\frac{16\,\text{eU}}{m(r_0^2 + 2z_0^2)\Omega^2} \tag{16.1}$$

$$a_r = \frac{8\,\text{eU}}{m(r_0^2 + 2z_0^2)\Omega^2} \tag{16.2}$$

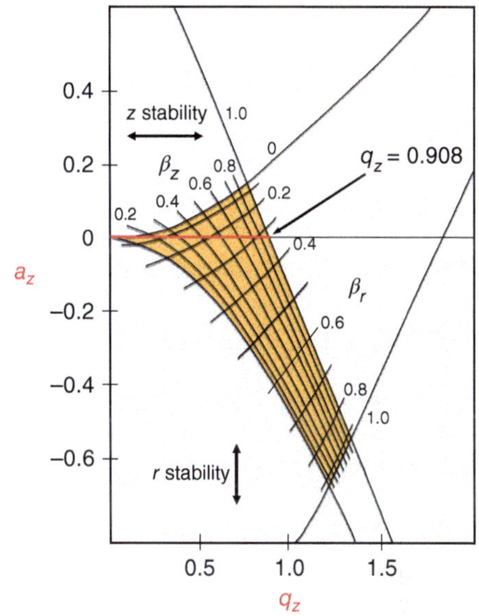

Figure 16.1 Stability diagram for a QIT. Ions are trapped with stable ion trajectories in the shaded orange regions. In typical mass-selective instability runs, no DC voltage is applied (red line) and ions are ejected when the RF voltage is tramped and q_z reaches 0.908.

$$q_z = \frac{8\,\mathrm{eV}}{m(r_0^2 + 2z_0^2)\Omega^2} \tag{16.3}$$

$$q_r = \frac{-4\,\mathrm{eV}}{m(r_0^2 + 2z_0^2)\Omega^2} \tag{16.4}$$

where U is the direct current (DC) voltage applied to the ring electrode, V is the zero-to-peak (0–p) radio frequency (RF) voltage applied to the ring electrode, Ω is the angular frequency of the RF signal applied to the ring electrode, e is the elementary charge, m is the mass of the ion, r_0 is the trap radius, and $2z_0$ is the distance between the two endcap electrodes. The coefficients assume a purely QIT geometry and are approximate values that will vary for different geometries [23, 24]. Theory predicts that areas of ion stability exist in both the radial and axial dimensions of the parameter space. Where these two regions overlap, the conditions are consistent with ions being stable in the ion trap [19, 22].

Additional insight into ion trap operation was gained through the work of Dehmelt, who modeled the frequency of motion of trapped ions as a simple harmonic oscillator in the z-dimension. The Dehmelt pseudopotential well depth (D_z) approximates the maximum energy for an ion to remain stably trapped. Equation (16.5) defines D_z under the Dehmelt approximation conditions where $q_z < 0.4$.

$$D_z = \frac{\mathrm{eV}^2}{4mz_0\Omega^2} = \frac{q_z V}{8} \tag{16.5}$$

Building off the D_z calculation, Dehmelt also described a relationship between the charge capacity of an ion trap (N_{max}) relative to D_z and the z_0 critical dimension of the QIT (Eq. 16.6).

$$N_{\mathrm{max}} = 2.8 \times 10^7 D_z z_0 \tag{16.6}$$

Both the D_z and N_{max} calculations proved critical in future decades when ion trap critical dimension scaling was investigated [25, 26].

16.2.2 Mass-Selective Detection

While the theoretical characterization of ion motion developed by Paul, Steinwedel, and March has proven robust, methods for mass analysis have significantly changed over time. Methods developed by Paul and Steinwedel and then Dawson and Whetten were inefficient and limited to spectroscopy-type experiments [19, 27, 28]. Contemporary ion trap mass analysis is derived from mass-selective instability operation, developed by Stafford and coworkers at Finnigan MAT Corporation and the University of Kent in the early 1980s [29, 30]. In this mode of operation, ions are injected and trapped with a quadrupolar field for an indefinite time frame. Ions are ejected in order of increasing mass-to-charge ratio (m/z) by ramping the RF voltage applied to the ring electrode with fixed Ω until they are ejected when $q_z > 0.908$. Using this mode of operation, the mass range for an ion trap (m/e) can be described on the basis of Dehmelt theory as a function of m/z by varying V, Ω, and the QIT dimensions until an unstable q_z value is reached (Eq. 16.7) [25].

$$\frac{m}{z} = \frac{8\,\mathrm{V}}{q_z(r_0^2 + 2z_0^2)\Omega^2} \tag{16.7}$$

While ion ejection could also be achieved by ramping the RF frequency in a similar fashion to the RF voltage, it has proven uncommon since resonant circuits are required to produce the 1–10 kV (p–p) RF voltages commonly used for this application [22].

The performance of mass-selective instability scans (mass range, sensitivity, and mass resolution) has been improved by introducing resonance ejection to the mass instability scans. Each m/z ion has a unique three-dimensional trajectory within the ion trap that occurs at a characteristic frequency, known as the secular frequency and defined in the z-dimension as (ω_z) [22]. The frequency of motion is commonly expressed by the trapping parameter β_z, which is a complex dimensionless function of the previously described a_z and q_z dimensional parameters when q_z is <0.4 (Eq. 16.8).

$$\beta_z^2 \cong a_z + \frac{{q_z}^2}{2} \tag{16.8}$$

The β_z parameter can then be correlated back to the ion's real secular motion by Eq. (16.9).

$$\omega_z = \frac{\beta_z \Omega}{2} \tag{16.9}$$

Since the β_z value varies between 0 and 1 for a trapped ion, the ion's secular frequency is less than half of the drive RF frequency. During mass-selective instability scans with resonance ejection, an axial RF signal with a voltage on the order of hundreds of millivolts is applied across the endcap electrodes at the axial secular frequency of ion motion. This results in ions being ejected from the ion trap at a lower q_z and extends the mass range of the ion trap. The most common mode of operation uses a resonant RF frequency just below ½ of the drive RF frequency [22].

16.2.3 Mass Resolution for QITs

In all previously discussed methods of ion trap operation, the gas background pressures were maintained below 10^{-5} Torr to minimize collisions between ions and neutral molecules, which are potentially detrimental to mass resolution and sensitivity [31, 32]. Commercial interest in the ion trap as a detector for gas chromatography led to experiments with the addition of helium buffer gas to the vacuum chamber. Stafford and coworkers increased the helium buffer gas pressure in their $r_o = 1$ cm (cm) ion trap from approximately 10^{-6} to 10^{-2} Torr and determined that optimal performance was achieved at a pressure near 10^{-3} Torr [29]. Mass resolution was studied as a function of increasing helium pressure focusing on the PFTBA m/z 69 and m/z 502 peaks. The mass resolution ($m/\Delta m$) at full-width at half-maximum (FWHM) for the m/z 69 peak improved from 70 to 185 with the helium gas at 10^{-3} Torr, while m/z 502 was not detected without buffer gas but was detected at a mass resolution of 1700 with the buffer gas. They suggested that increased helium gas pressure caused collisional dampening of ion motion toward the center of the QIT, where the ions experience a more uniform trapping field, resulting in better defined secular motion of the trapped ions and ejection from the ion trap in a tighter time span, improving mass resolution. The performance in mass resolution and sensitivity of the relatively high operating pressure (10^{-3} Torr) opened the door for the ion trap to be used as a gas chromatography detector [33].

While the work by Stafford and coworkers significantly increased the operational pressure regime for ion traps, the vacuum requirements still confined these instruments to laboratory use. A theoretical breakthrough in ion trap scaling was realized by Goeringer and coworkers in 1992. While increasing the pressure from 10^{-5} to 10^{-3} Torr was successful, the ion neutral collisions with the buffer gas continued to be a concern due to their expected deleterious effect on resolution. Goeringer et al. created a model that assumed the motion of an ion could be approximated by a damped harmonic oscillator where the motion of the ion is determined from the RF frequency while being damped by ion neutral collisions. A dampening factor was added to the Mathieu equations for ion motion to account for these ion neutral collisions. They developed a new relationship (Eq. 16.10) for mass resolution ($m/\Delta m$) as a function of buffer gas pressure (P), collisional relaxation time (τ), and the RF frequency (Ω) [34].

$$\frac{m}{\Delta m} \propto \frac{\Omega}{P} = \frac{\Omega\tau}{4\sqrt{3}} \tag{16.10}$$

This relationship shows peak widths (Δm) scale linearly with increased pressure and inversely with RF drive frequency. For a trap operated at a constant RF frequency, when pressure increases, the mean free path decreases proportionally, the ions experience more frequent ion neutral collisions, and mass resolution is reduced. However, if the RF frequency is scaled with the operating pressure, ions now experience more RF cycles per ion neutral collision and mass resolution can be maintained.

Building off the work by Goeringer et al., another model for ion trap motion was developed by Arnold and coworkers at the University of Utah [35]. Arnold et al. studied the dispersion of ions at the point of trap ejection rather than ions confined within a trap. They developed a relationship (Eq. 16.11) for a mass resolution that also includes additional variables including scan rate (a_m), secular frequency ($\omega_0 = 0.5\,\Omega$), thermal energy (kT), collisional relaxation time (τ), and neutral mass (M).

$$\frac{m}{\Delta m} \leq \frac{\Omega/2}{\sqrt{\frac{\pi a_m \Omega z_0}{m} + 8/\tau^2}\sqrt{\frac{kT}{m r_0^2 \omega_0^2} + \frac{1}{3}\frac{M}{m}}} \tag{16.11}$$

There are several significant differences between the models worth mentioning. The Arnold et al. mode assumes mass resolution scaled as scan rate to the ½ power under high scan speed conditions regardless of operating pressure. Since sensitivity and mass resolution are often a direct trade-off, this is a key consideration for detecting low-concentration analytes. The Goeringer et al. model did not consider the effect of scan rate on mass resolution. The Arnold et al. model predicts that mass resolution will be dependent on the relative masses of the analyte and neutral molecules, while the Goeringer et al. model does not explicitly consider these masses. However, consideration of the masses of the neutral molecules is important since there are multiple buffer gas options.

Whitten et al. addressed the possibility of scaling the RF frequency and trap size by a factor of 1000 to accommodate ion trap operation in the low-Torr pressure range [36]. They added a drag coefficient to the equations describing ion motion and investigated the effects on the ion trap stability diagram, concluding that mass resolution *could* be maintained at higher pressures, and discussed multiplexing ion traps for improved sensitivity and charge capacity at higher pressures.

16.2.4 Ion Trap Dimensional Scaling and Simplified Geometries

Once the potential advantages of miniaturizing ion traps were recognized, the fabrication of small traps needed to be addressed. The hyperbolic QIT was developed with an emphasis on producing trapping fields that are as close to a pure quadrupole as possible [19]. However, higher order fields are created in ion traps by incorporating entrance and exit apertures required to introduce and detect the analyte. They are also introduced by any trap asymmetry, which becomes difficult to avoid as trap dimensions are reduced. The quadrupolar is the strongest field and dominates trap behavior by providing a linear restoring force for trapped ions with distance from the center of the trap. Higher order components such as hexapole, octapole, decapole, and dodecapole fields are strongest near the boundaries of the trap and decrease nonlinearly with distance toward the center where ions predominantly reside [37, 38]. Since ions mainly reside close to the center of the ion trap, they mainly experience the quadrupolar fields, although the other field strengths are significant and must be addressed during trap development [24, 39].

While the hyperbolic electrodes of the QIT were effective at the 1 cm dimension scale, researchers quickly realized fabricating these precise features would become infeasible when trap dimensions were decreased in size. The cylindrical ion trap (CIT) design was introduced in 1977 by Bonner and coworkers [40]. The hyperbolic shapes of the three trap electrodes are replaced with cylindrical planar electrodes, which are amenable to both conventional machining and microfabrication techniques with miniaturized dimensions. Starting in the late 1990s, multiple groups studied the performance of miniaturized CITs operating in the 10^{-3} Torr pressure regimes [5, 41–43]. Immediate concerns surrounded the performance of the CIT geometry, which simplifies fabrication but adds non-quadrupolar field characteristics to the ion trap geometry and influences performance [2, 44–46].

The challenges associated with nonideal ion trap geometries and higher order fields had been encountered in the earlier work by Finnigan Corporation. They noticed a mass shift in their mass spectra where ions were detected at the wrong m/z value depending on the chemical structure of the ion. They solved the problem extending the trap by 10.8% in the z_0-dimension to correct for field perturbations caused by apertures in the endcap electrodes used to inject and eject ions from the trap. They believed the endcap electrodes added an octapolar field component to the trapping fields and stretching the z_0-dimension offset the effect and corrected the mass shift [47]. Multiple groups have theoretically and experimentally studied the effects of changing CIT critical dimensions on ion trap performance [48–51]. While there are different trade-offs to consider, no set of critical dimensions for the CIT became standard as different groups prioritized different performance trade spaces depending on specific applications and system design requirements.

Another challenge of small ion traps is charge capacity. As noted earlier in Eq. (16.6), the charge capacity of an ion trap scales linearly with the z_0-dimension as charge repulsions limit the number of ions that can be stored in a trap. One potential solution is planar arrays of ion traps to maintain charge capacity [25, 26]. While charge capacity decreases linearly with the z_0-dimension, volume scales as z_0^3. Miniaturized ion trap arrays have the potential to maintain the charge capacity of a traditional 1 cm ion trap while storing that amount of charge in a lower volume of space. Early experiments into ion trap arrays focused on arrays of less than 10 individual traps with critical dimensions in the low mm dimensions [52–54]. The smallest trap arrays that have successfully performed mass analysis were reported by Pau and coworkers [51]. A trap consisting of $r_o = 20\,\mu m$ CIT arrays (256 and 2304 elements) was fabricated from p-doped polysilicon with silicon dioxide insulating layers. Xenon gas was detected at 10^{-4} Torr of helium pressure with xenon concentration of 1%. Blaine and coworkers fabricated arrays with up to 10^6 elements of $r_o = 1\,\mu m$ CITs [48, 49]. However, when operated as a mass analyzer, the positively charged electrical current measured from the trap array could not be completely attributed to trapped ions and mass spectra were not presented. A major concern for ion trap arrays is uniformity across array elements that could result in a loss of mass resolution when the signals from multiple individual traps are detected simultaneously in time. Given this concern, there has been interest in developing ion trap designs that do not require multiple trap elements but still give the sensitivity benefits and high charge capacity that are predicted from trap arrays.

An alternative approach to limited charge capacity in miniature ion traps adapts linear ion trap (LIT) designs developed by Hager and Schwartz for miniature MS operation [55, 56]. The LIT consists of four hyperbolic rods similar to a QMF, except the rods are cut into three sections each set at unique DC potentials. Ions are contained along the length of the rods by applying carefully selected RF voltages to the rods. Trapping ions along the length of the rods improves charge capacity and minimizes space charge effects without loss in mass resolution. Similar to the QITs, the hyperbolic features of the LIT are not amenable to miniaturization. Ouyang and coworkers developed a simplified geometry based on the LIT where the four central hyperbolic rods are replaced with four planar electrodes and the two end segments of the LIT are replaced with two planar electrodes known as the rectilinear ion trap (RIT) in 2004 [57]. The RITs have been fabricated with critical dimensions on the single mm scale in a variety of materials and their performance has been thoroughly characterized [58–60].

Another approach that uses elements of the LIT and CIT was demonstrated by Schultze and Ramsey in 2014 in the development of the stretched length ion trap (SLIT) [61]. The SLIT design replaces the cylindrical features found in CITs with a slot that stretches along the y-plane of the trap electrodes and increases sensitivity up to an order of magnitude compared to similarly sized CITs. Related designs such as a serpentine pattern along the y-plane of the trap electrodes have been investigated, demonstrating the versatility of this design. SLIT traps have been demonstrated with critical dimensions in the hundreds of μm, a size scale that is amenable to operating at pressures of up to 1 Torr [61].

Several additional simplified geometries that have been reported in the literature include multi-ring stacks [62], the coaxial ion trap [63], the toroidal ion trap [64–66], and the halo ion trap [67, 68]. Considerations for different trap designs include fabrication techniques, optimal critical dimensions, operating pressure, mass resolution optimization, and material selection.

16.2.5 Vacuum Pump Requirements for Miniature MS

There are multiple parameters related to the vacuum system that need to be jointly considered for a functional miniature mass spectrometer:

1. Minimize the power consumption of the pumping system for its intended operating pressure – this reduces the required battery payload (volume and weight) for useful uptime.
2. Maximize the pumping speed at operating pressure – the gas throughput of the mass spectrometer critically defines its response time, and the efficiency of ion transport from atmospheric ionization sources.
3. Minimize the time required to pump the MS down from atmospheric pressure – handheld tools are generally expected to be "shelf-ready" so time-to-readiness of a few seconds is ideal.
4. Minimize the cost of the pumping system – it is particularly critical to avoid exotic materials and eliminate extreme precision parts including precision bearings.
5. Minimize the size/weight of the pumping system.
6. Maximize the lifetime of the pumping system – eliminate high-loading bearings or wearing surfaces.
7. Ensure that the pumping system is capable of servicing the required vacuum and flow across the operating temperatures of interest in fieldwork (usually −20 to 50 °C) and at different elevations.
8. The pumping system must survive and function through physical shock (e.g. drops) and vibration.

Among the groups who have investigated ion trap miniaturization, different approaches have been used to handle the vacuum requirements of the mass spectrometer. A typical mass spectrometer consists of at least two stages of pumping including a roughing pump and turbopump. Depending on inlet gas flow, the roughing pump can achieve pressures ranging from 10^{-2} Torr to single-digit Torr values, while the turbopumps are backed by roughing pumps and reach operating pressures below 10^{-2} Torr [69].

Sample introduction can take several forms. The simplest approach is to consistently sample an ambient environment through a restricted atmospheric inlet or capillary. This approach can provide fast response times but has high pumping requirements. A disadvantage is the potential for debris to obstruct the capillary, leading to significant downtime. Semipermeable membranes can be implemented to provide the needed pressure drop. These membranes may also be made selective to the analytes of interest while keeping other components outside of the mass spectrometer [70–72]. However, membrane inlets are typically considered a consumable and logistical burden, and in use can negatively impact response and cleardown time. An approach to minimize pumping requirements uses a valve to sample for a limited time, after which the valve closes, the vacuum pumps achieve the desired operating pressure, and the sample is mass analyzed. This method, known as discontinuous atmospheric pressure interface (DAPI), minimizes pumping requirements but increases analysis time since the sample is only being analyzed for a fraction of the total operational time, while also increasing mechanical complexity [73, 74].

16.2.6 High-Pressure Scaling Approaches

As noted above, almost all transportable mass spectrometers operate at pressures below 10^{-3} Torr with turbomolecular pumping. Running at such low pressures means that no matter how much miniaturization of the mass analyzer is achieved, the SWaP and ruggedness of the overall instrument will be determined entirely by the pumping mechanism [36] – turbomolecular pumping is a catastrophe by every figure of merit. Accordingly, several efforts have been made to increase buffer gas pressure beyond the traditional molecular flow/10^{-3} Torr regime.

In one approach to high-pressure operation, Xu et al. used theoretical modes to study the effects of ion neutral collisions for ions trapped in air buffer gas up to 1 Torr operating pressure with a 5 mm RIT instrument operating at 1.1 MHz, matching the RF frequency regime used with traditional QIT instruments. Their theoretical model predicted that FWHM peak widths would increase from roughly 1 Da at 4.1×10^{-3} Torr to 9.5 Da at 0.25 Torr [75]. Experimental results by Song et al. using this RIT configuration did not investigate pressures above 5×10^{-2} Torr, but demonstrated FWHM peak widths of 0.98 Da at 4.1×10^{-3} Torr air buffer gas with an increase to 2.00 Da at 5×10^{-2} Torr [76]. Based on this FWHM vs. pressure trends, an FWHM of nearly 40 Da would have been projected at 1 Torr, making MS identifications quite difficult.

Jiang et al. developed a miniature mass spectrometer based on a 4 mm LIT and investigated several ion ejection methods as a function of air buffer gas pressure. Ions were trapped at an RF frequency of 1 MHz and then mass analyzed by two different ion trap modes known as dipolar resonance ejection and QE-dipolar resonance ejection modes where QE stands for quadrupole enhanced. Peak widths for a methionine–arginine–phenylalanine–alanine peptide (MRFA) peak at m/z 524 were found to increase from 3.5 Da at 2.95×10^{-3} Torr to 4.23 Da at 6.44×10^{-3} Torr when the dipolar resonance ejection method was used. In comparison, the same peak widths increased from 1.58 Da at 2.95×10^{-3} Torr to 2.19 Da at 6.44×10^{-3} Torr with the QE-dipolar resonance ejection method. The pressure regime studied was limited by the operating pressure range of the electron multiplier detector [77].

Another approach to higher pressure operation was taken by Jiang et al. This group used a 1 cm CIT and two different RF frequency and pressure regimes. First, the CIT was operated at an RF frequency of 0.84 MHz and an operating pressure of 5×10^{-6} Torr to benchmark performance in the high-vacuum regime. Next, the operating pressure was increased to approximately 1.5×10^{-2} Torr and the RF frequency was increased to 2.4 MHz. Using this configuration, mass spectra of methyl salicylate were acquired. The pressure regime studied was limited by the operating range of the electron multiplier detector used [78].

Recent work by Decker and coworkers in 2019 has focused on miniaturizing a microscale planar ion trap (PLIT) for higher pressure operation. A microscale 800 μm PLIT was fabricated with lithographic methods operated with an RF frequency of 2.71 MHz. Mass spectra of toluene were acquired from 2.5×10^{-3} to 4.2×10^{-2} Torr of helium buffer gas and peak widths at the highest pressure were estimated to be between 2.3 and 2.7 Da. Performance trade-offs such as power requirements for the RF driver, ion trap pseudopotential well depth, and mass resolution compared to a larger 2.5 mm critical dimension trap were also analyzed [79].

The elevated pressure operating strategies described so far focused on scaling RF frequency by less than a factor three compared to conventional ion traps. One challenge in the development of higher pressure MS is the generation of an appropriate RF waveform for driving the ion trap. The amplitude of the RF signal must be sufficiently stable to reliably eject ions without peak broadening. While the theory is clear that increasing the RF frequency should counteract increased RF collisions at higher pressures, increasing the RF frequency is not a trivial task. To produce an RF waveform used to trap and analyze ions, the required power scales as the RF frequency squared. Furthermore, power also scales with the capacitance squared. Smaller ion traps have smaller gaps between the electrodes, increasing the trap capacitance. Finally, electrical discharges must be avoided despite the presence of ions generated from the ionization source and the high voltages produced at high RF frequencies [80–82].

The approach of operating ion traps at higher RF frequencies was taken by Ramsey and coworkers starting in the late 1990s. In the seminal paper, a 500 μm CIT was used to record the mass spectra of xenon atoms and PFTBA.

The RF frequency was 6.5 MHz for xenon and 5.0 MHz for PFTBA [42]. With the ion trap dimensions reduced by 20 times compared to conventional traps and the RF frequency increased by a factor of five and higher, this scaling method was unique and predicted to yield a high-quality mass spectra at higher pressures than had previously been achieved. Following the initial demonstration, several ion trap geometries were studied and used to measure the mass spectra of toluene, chlorobenzene, and 1-chloronaphthalene at nominally 10^{-3} Torr helium [41].

Once the microscale ion trap operation was demonstrated, several methods for improving performance were characterized. To improve mass resolution and sensitivity of the microscale CITs, ions were ejected by mass-selective instability scans with resonance ejection utilizing nonlinear resonances. This method was termed double resonance ejection since trapped ions are resonant with both the quadrupolar (ion trapping) field, as well as the supplemental (axial) RF voltage applied to the endcap electrode at an RF frequency selected to match a nonlinear field resonance [83]. The axial RF voltage is applied during the RF ramp to eject the ions at a lower q_z value. Multiple resonance ejection techniques such as this are common in ion trap instruments; however, higher order field components are typically minimized or well-controlled [84]. Microscale CITs have large components of high-order field strength that are difficult to control because of their geometrical simplification and small critical dimensions [85, 86]. This presents a unique opportunity and need for methods such as using higher order field components to improve performance.

The large components of higher order fields in the microscale CIT designs revealed significant opportunities for performance improvements from nonlinear fields. The stability diagram was experimentally mapped out for six different resonances that demonstrated ions were ejected at a lower q_z value. Without the use of the double resonance ejection, the mass resolution ($m/\Delta m$) for xenon atoms was found to be 44. With optimized double resonance ejection, mass resolutions up to 660 were measured for xenon atoms [83].

Since an obvious application for microscale traps was the analysis of volatile organic components in air (VOCs), a semipermeable membrane sample inlet was added to the MS instrument. Limit of detection (LOD) values were evaluated for six analytes and varied from 0.26 parts per million (ppm) for xylene to 13.8 ppm for acetone at operating pressures of 1.5×10^{-3} to 3.8×10^{-3} Torr in a helium carrier gas. Tandem mass spectrometry (MS/MS) was performed on chlorobenzene and improved the detection limit by a factor of 2.1 [87]. The effect of sample pressure on signal intensity was evaluated in the microscale CIT. Xenon peak intensities were characterized as a function of xenon pressure at different helium pressures. Increasing the helium pressure shifted the xenon concentration at which the signal strength was largest to a higher value, after which signal strength decays with pressure (all pressures below 10^{-3} Torr). It was postulated that this effect was caused by scattering of xenon ions by xenon neutrals within the trap [88].

16.2.7 High-Pressure (~1 Torr) Trap Operation

Following optimization of the 500 μm CIT in the low pressure (<10^{-3} Torr) regime, the focus turned to strategies for operating the microscale ion trap as a HPMS instrument, defined as pressures above 0.1 Torr. While Whitten et al. addressed the need to scale the ion trap to accommodate higher pressure operation, the next obstacle to higher pressure operation became the ion detector [36]. Electron multipliers are widely used in mass spectrometers due to their high sensitivity and sampling bandwidth [89–91]. However, they are pressure-intolerant due to their surface chemistry and even the most robust electron multipliers are only rated to approximately 10^{-2} Torr.

Initial work separated the development of the ion source and microscale CIT from the detector by designing a differentially pumped mass spectrometer. The ionization source and microscale ion trap were housed in one chamber while the electron multiplier detector was housed in a separate vacuum chamber. The CIT and mounting scheme limited the gas flow between the two chambers – the CIT served as a small diameter orifice. The pressure drop inherent in this design made it possible to analyze ion source and trap performance at ≈1 Torr while still using an electron multiplier at $\approx 10^{-2}$ Torr. The two vacuum chambers were connected through a vacuum hose with an on/off valve to equalize the pressure between the two chambers when a high-pressure detector was being analyzed [92].

MS at approximately 1 Torr was demonstrated by Blakeman et al. in the early 2010s in helium buffer gas using a 500 μm CIT operated in a differentially pumped mass spectrometer at RF frequencies ranging from 6.73 to 9.43 MHz. The ionization source was a thermionic filament (operated outside of the rated pressure range) while the detector was an electron multiplier operated around 10^{-2} Torr. Filament parameters were optimized to detect xenon isotope peaks at a pressure of 1.2 Torr.

Three analytes were characterized over pressure ranges from approximately 5×10^{-2} Torr to 1 Torr, including xenon, 2-chloroethyl ethyl sulfide (CEES), and octane. Average peak FWHM for the five most abundant xenon isotopes were found to increase from 0.49 to 1.19 Da (an increase of 0.7 Da) when the pressure was increased from 4.3×10^{-2} to 1.0 Torr. Peak widths for CEES and octane were compared to a theoretical model [35]. Peak widths for CEES *m/z* 75 peak were found to increase by 0.52 ± 0.02 Da-Torr^{-1} while the theoretical model predicted peaks to broaden by 0.60 Da-Torr^{-1}. For octane *m/z* 41, the peak widths increased by 0.52 ± 0.05 Da-Torr^{-1} compared to the theoretical mode of 0.55 Da-Torr^{-1} [92].

Following the demonstration of HPMS with helium buffer gas operation, the focus moved toward HPMS operation with ambient air as the buffer gas. While helium is traditionally used in laboratories for its favorable collisional properties as a 4 Da species, a truly portable instrument would benefit from eliminating the need for a carrier gas, with the attendant gas cylinders, regulators, and logistical burden for users. The previous strategy of using a thermionic filament as an ionization source was not appropriate for use in ambient air, since filaments have poor tolerance to oxidation. Glow discharge ionization (GDI) was chosen as the replacement option since it operates well in the pressure regime around 1 Torr, draws low power (<0.5 W), and can be used as an electron or ion source. Emission currents for glow discharge as an electron source were characterized across a pressure regime from approximately 0.5 to 2 Torr of both nitrogen and air to ensure ionization stability when coupled to a microscale CIT at these operating pressures. While the target buffer gas was air, nitrogen was also characterized to confirm that additional air constituents did not detrimentally affect system performance [93].

In addition to ionization, air buffer gas operation presented new ion trap performance challenges since O_2 and N_2 have roughly seven times larger mass than helium, increasing the ion neutral collisional energy considerably. Mass spectra for *p*-xylene were compared between helium, nitrogen, and air buffer gases at 1 Torr buffer gas pressure for a 500 μm CIT operated at 10 MHz RF frequency. Both the nitrogen and air mass spectra were acquired with the glow discharge electron source, while the helium mass spectra were acquired with a tungsten filament. Signal intensity was better for nitrogen and air since their ionization source was better suited for the high-pressure operation. Peak widths for *p*-xylene *m/z* 91 and *m/z* 106 peaks were compared between helium and air buffer gases as a function of operating pressure and are summarized in Table 16.2. Despite wider peak widths than traditionally observed with mass spectrometers at this mass range, each constituent of a four-component mixture of VOCs could be distinguished by the mass spectra acquired in 1 Torr of air [93].

Table 16.2 Peak broadening vs. pressure for *p*-xylene in helium vs. air buffer gases.

p-Xylene peak	Peak broadening rate (Da-Torr^{-1})	Peak width at 1 Torr (Da)
m/z 91 in Helium	0.51 ± 0.07	0.62
m/z 106 in Helium	0.41 ± 0.06	0.62
m/z 91 in Air	2.0 ± 0.1	4.78
m/z 106 in Air	1.6 ± 0.2	5.04

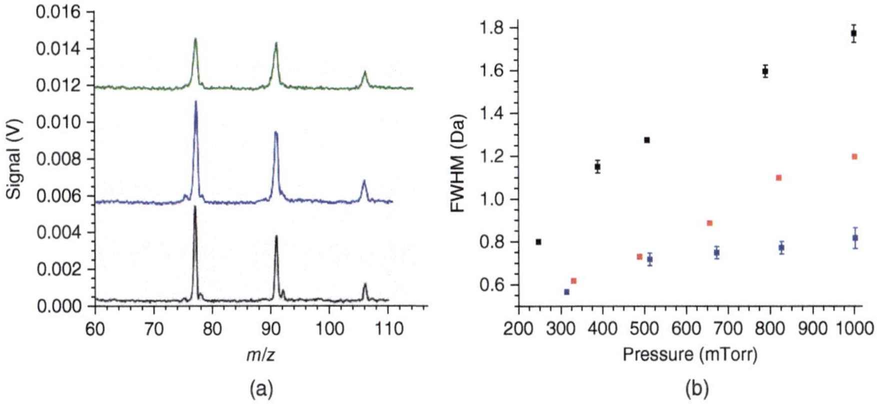

Figure 16.2 (a) Mass spectra of *p*-xylene with $r_0 = 100\ \mu m$ at an RF drive frequency of 59.44 MHz with ambient air pressures of 0.313 Torr (– –), 0.672 Torr (– –, blue), and 1.02 Torr (– –, green); and (b) peak widths for *p*-xylene 106 Da with increasing ambient air pressure for 30.34 MHz, $r_0 = 165\ \mu m$ (■), 47.24 MHz, $r_0 = 100\ \mu m$ (■, red), and 59.44 MHz, $r_0 = 100\ \mu m$ (■, blue).

The Ramsey group followed the theoretical trends predicted by Goeringer and Arnold and worked to increase the frequency of the drive RF, and, thereby, the mass spectral resolution. CITs with critical dimensions between 500 and 100 μm were analyzed (Figure 16.2). As the trap size decreased, higher RF frequencies could be evaluated while maintaining an operational mass range of approximately 140 *m/z* using maximum RF voltage of approximate 1 kV (p–p). The same analyte *p*-xylene was used as the analyte for each CIT and the peak width for the *m/z* 106 peak was quantitated. With an RF frequency of 59.44 MHz, the peak width was 0.8 Da and subunit resolution was achieved despite the elevated operating pressure. At 59.44 MHz, dropping the pressure from 1.02 Torr of ambient air to 0.313 Torr further improved peaks from 0.8 to 0.56 Da. Peak widths for a total of two traps and three RF frequencies are characterized to show the trade space between peak widths, RF frequency, and air buffer gas pressure [94].

With ionization and mass analysis demonstrated at 1 Torr of air, the last step to consider was ion detection. Previous work had been performed using electron multiplier detectors which are not compatible with operation above 10^{-2} Torr. In contrast, Faraday cup detectors are capable of operation regardless of gas pressure, but do not offer the same sensitivity levels as electron multipliers. Two Faraday cup detectors were acquired and evaluated on HPMS platforms at pressures up to 1 Torr [95]. Mass spectra of four different aromatic benzene derivatives were acquired with these Faraday cup detectors in a series of experiments which demonstrated the full process of the mass spectrometer (ionization, mass analysis, detection) operated at 1 Torr of nitrogen buffer gas [96].

16.3 Commercialization and Applications

The first commercially available mass spectrometer operating at approximately 1 Torr came in 2014 with the launch of the M908 by 908 Devices, Inc. (Boston, MA). The M908 is a handheld mass spectrometer with dimensions of 22 cm × 18.5 cm × 7.6 cm and weighing 2.0 kg (Figure 16.3). It uses a GDI source, a microscale ion trap as the mass analyzer, and the detector is a Faraday cup [97]. Pumping is performed by a small, custom, mechanical scroll pump to achieve an operating pressure in the 1 Torr range in under 10 seconds. A notable feature of the M908 is the absence of an expensive, fragile, turbomolecular pump. The use of a mechanical pump as the sole source of vacuum increases the overall ruggedness of the device. The M908 is compliant with military (MIL-STD-810G) standards for shock/drop/vibration.

The miniature ion trap, ionizer, and Faraday cup are contained in an "MS core" (Figure 16.4). The core is user-replaceable in the event of severe contamination on-scene in order to reduce downtime to a few minutes (Figure 16.5). Post-event, the core can be easily cleaned and returned to service.

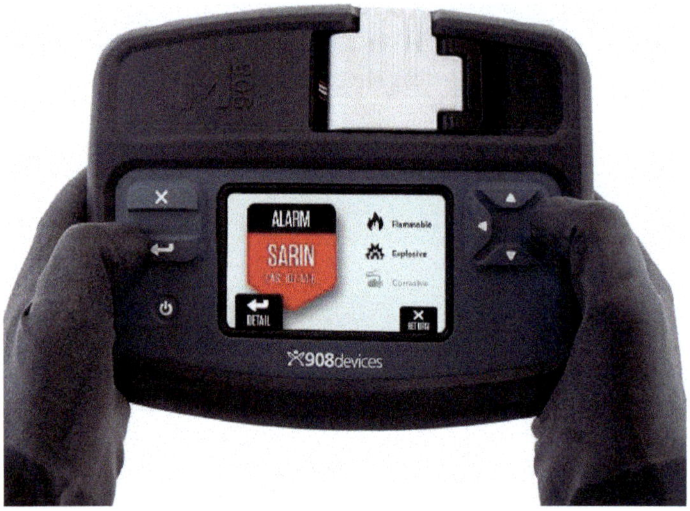

Figure 16.3 M908 handheld mass spectrometer. *Source*: Figure courtesy of 908 Devices Inc.

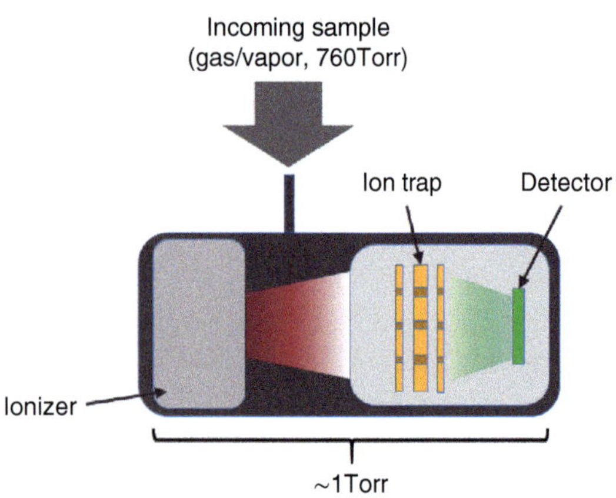

Figure 16.4 Diagram showing the components of the M908 MS Core. The entire core is at a pressure of ~1 Torr.

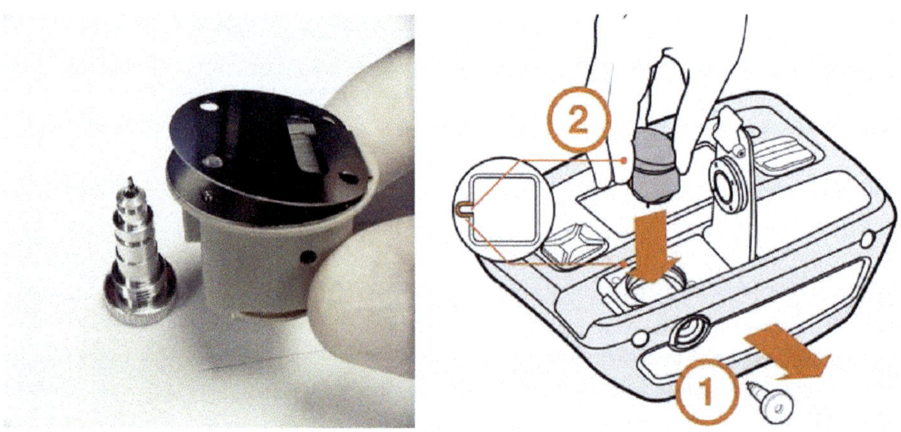

Figure 16.5 The M908 MS core and inlet capillary (a). *Source*: Figure courtesy of 908 Devices Inc.. Replacement of the MS core (b).

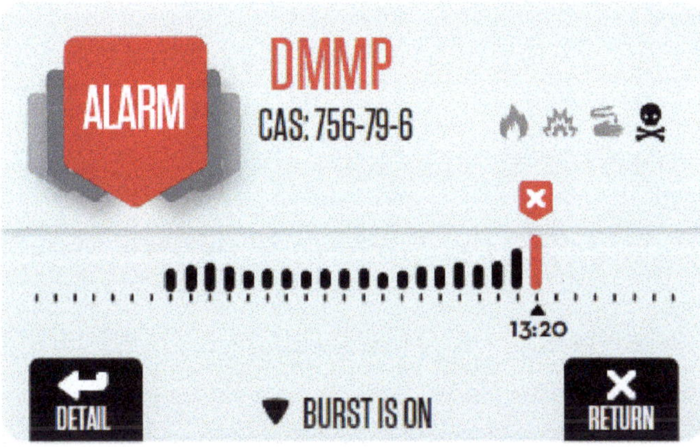

Figure 16.6 M908 alarm screen in vapor mode. Each bar represents a mass spectrum – black bars indicate ion current detected, red bars indicate ion current detected and a match to the target library. *Source*: 908 Devices.

The M908 was designed specifically for the detection and identification of chemical threats at the point of need. It is battery operated, with a battery life of 2–4 hours depending on the mode of use. Vapor phase threats are detected by direct sampling through a small inlet capillary. The device measures ambient air directly and requires no additional carrier gas supply or dopant gas. The target library includes a variety of toxic industrial chemicals (TICs), solvents, and chemical warfare agents (CWAs).

Operationally, the M908 measures a mass spectrum of the surrounding air about once every second, detecting positively charged ions generated by the GDI. If the measured chemical signal (the total ion current) rises above a set threshold, the spectrum is automatically tested against the target library. If the spectrum matches one or more targets in the library, the M908 displays an alarm on the screen, along with the identity of the chemical or chemicals (Figure 16.6). All spectral analysis is done by the software. By design, the user is not required to assess the mass spectrum, making the M908 suitable for use by civilian users, first responders, and military teams.

Solid- and liquid-phase threats, as well as trace samples, are analyzed by swab sampling using a thermal desorption (TD) mode. Sampling swabs, made of Teflon-coated fiberglass, are used to deliver minute amounts of condensed phase material into the M908. The device rapidly heats the swab and any present threat material to over 200 °C, producing vapor which is measured by HPMS. TD mode targets include explosives and low-volatility CWAs, such as VX. Detection/identification limits vary by chemical: in the vapor phase ranging from sub-ppm for CWAs to 10s of ppm for some TICs, and in TD mode from sub-microgram to 10s of micrograms.

In 2017, a second-generation handheld mass spectrometer, the MX908 was launched by 908 Devices (Figure 16.7). The MX908 utilizes a dual-polarity atmospheric pressure chemical ionization (APCI) source driven by fast switching electronics that are capable of scanning both positive and negative mode polarities in approximately 0.1 seconds. The microscale ion trap design is optimized for the APCI source and the Faraday cup detector electronics are optimized for the fast switching dual-polarity MS scans (Figure 16.8). While slightly larger than the M908 with a weight of 3.9 kg and dimensions of 29.8 cm × 21.6 cm × 12.2 cm, the design updates improved the sensitivity of the platform by a factor of up to a thousand. The MX908 operates on battery power with a battery life of four hours and is MIL-STD 810G compliant.

The MX908 has both vapor and thermal desorption capability. The target library, which covers both positive- and negative-mode analytes, comprises over 100 threats, including TICs, CWAs, explosives, and narcotics (including fentanyls). Limits of detection/identification are summarized in Table 16.3.

Ionization via APCI is known to produce less fragmentation than the more common electron impact (EI) ionization [98–100]. Mass spectra produced by the MX908 exhibit large fractions of protonated molecules and higher *m/z* fragments. Additional fragments can be generated using the process of collision-induced dissociation (CID)

Figure 16.7 MX908 handheld mass spectrometer. *Source*: Figure courtesy of 908 Devices Inc.

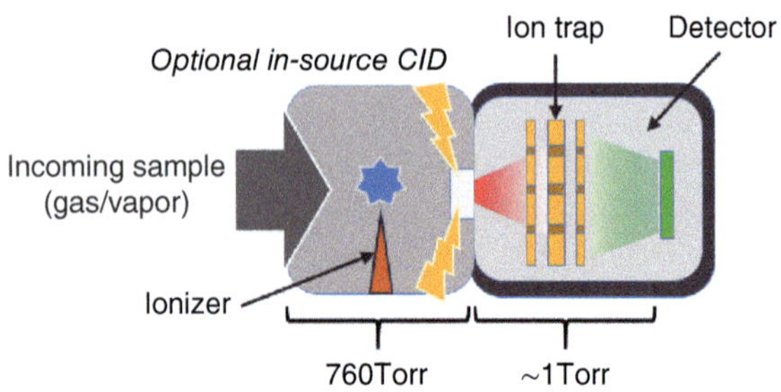

Figure 16.8 Diagram showing the components of the MX908 MS Core. The ionization chamber on the left operates at atmospheric pressure. The chamber containing the ion trap and detector operates at ~1 Torr. A small aperture separates the two chambers.

Table 16.3 Estimated LOD for key target classes in MX908.

Target class	Mode	Estimated LOD
Explosives – organonitrates	TD	<100 ng
Explosives – peroxides	Vapor/TD	<10 ppm/<3000 ng
Narcotics	TD	<100–1000 ng
Fentanyls	TD	<100 ng
CWA: nerve agents (G)	Vapor	<100 ppb
CWA: nerve agents (V)	Vapor/TD	<100 ppb/<100 ng
CWA: blister agents (HD)	Vapor	<2000 ng

[101, 102]. Ions generated in the ionization chamber (in-source CID) can be accelerated through the aperture with controlled levels of force, providing a well-defined amount of kinetic energy to the ions and fragmenting them to specific degrees. The MX908 uses multiple levels of CID to improve chemical specificity.

An example is shown in Figure 16.9. Two fentanyl analogs with identical molecular weights, 3-methylfentanyl and butyryl fentanyl (350.5 amu) are analyzed at low, medium, and high levels of CID voltage. At low CID voltage (left), no fragmentation is observed and the two fentanyls are indistinguishable. At a moderate CID voltage, two fragments are observed for both 3-methyl fentanyl (orange) and butyryl fentanyl (blue). The lower mass fragment is 105 m/z for both targets, but the higher mass fragment observed is 189 m/z for butyryl fentanyl and 202 for 3-methylfentanyl. The matching algorithms on the MX908 utilize both the mass/charge values of the ions observed

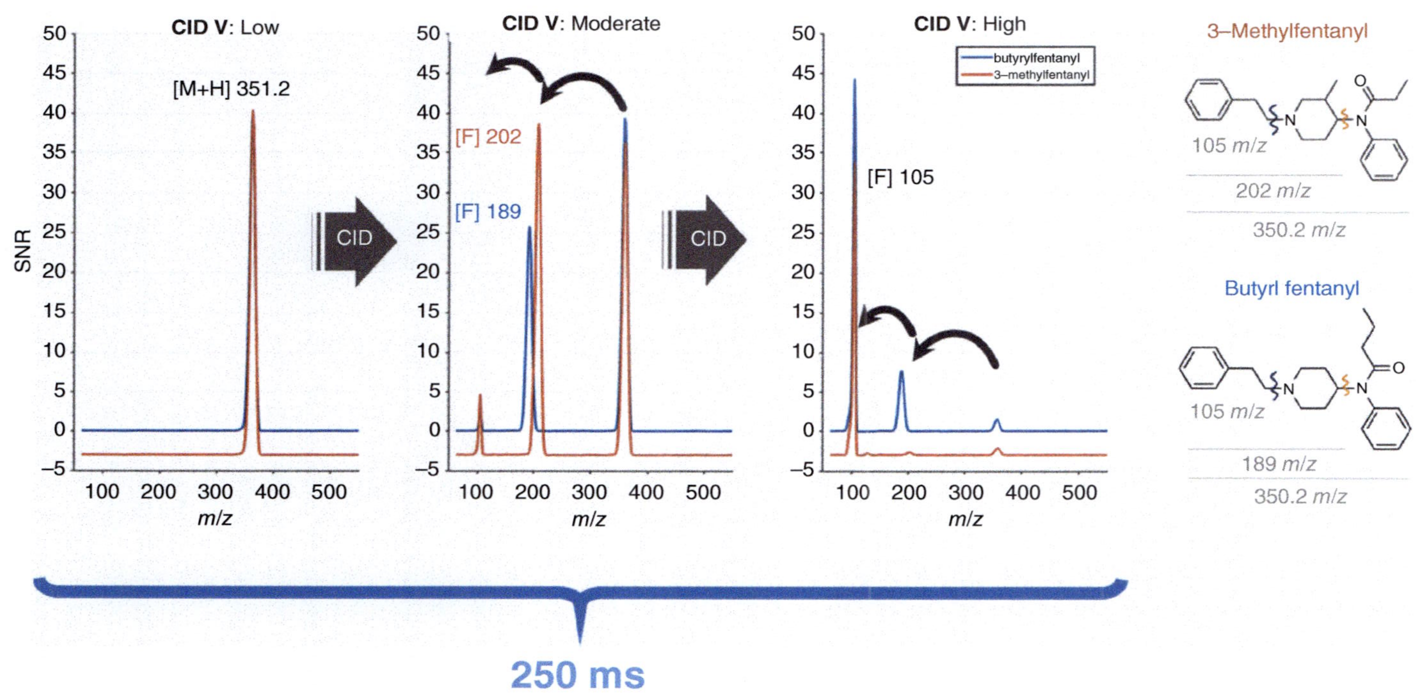

Figure 16.9 Use of CID to distinguish isobaric fentanyl analogs.

Figure 16.10 The Rebel amino acid analyzer uses CE-ESI-HPMS to detect and quantify 33 metabolic targets from samples of cell culture media. *Source*: Figure courtesy of 908 Devices Inc.

and the CID energy level at which they are observed to identify targets, reducing false alarm rates even in the presence of large amounts of cutting agents or other interferents [103].

HPMS has been used as the detector in conjunction with chemical separation techniques to improve the analysis of complex mixtures. An example is the recently released benchtop amino acid analyzer, the Rebel (Figure 16.10). Rebel was launched by 908 Devices in 2019, and utilizes HPMS detection in conjunction with a microchip-based capillary electrophoresis–electrospray ionization (CE–ESI) method to separate and quantify amino acids in complex biological samples, such as cell culture media [104, 105].

The Rebel's Spent Media assay is designed to automatically detect and quantify 33 metabolites, including the 20 common amino acids and 6 vitamins, in mammalian and fermentation media. The built-in autosampler gives the capability for running over 200 samples in a batch with no user intervention after initiating the batch. The entire process of analyzing a sample from the autosampler loading a sample through the separation, MS analysis, and automated data analysis takes under 10 minutes per sample. The separation and MS analysis itself take just under three minutes.

The Rebel is a benchtop instrument, not intended to be mobile. However, the use of HPMS as the detector provides considerable advantages over more typical liquid chromatography-MS (LC-MS) systems. Operating the system at ~1 Torr allows all vacuum pumps to be contained within the device so that the system sits entirely on the benchtop. No carrier or curtain gases are necessary. All of the required reagents are provided in the sample kit – no additional solvents are needed. The HPMS detector is essentially maintenance-free allowing the Rebel to be operated by analysts and biomanufacturing personnel, rather than analytical chemists.

The microfluidic chip used in Rebel is shown in Figure 16.11. The built-in autosampler delivers the sample and the background electrolyte (BGE) to the respective wells. A sequence of pneumatic pressures is applied to the wells to deliver a precise amount of sample into the separation channel. Typical volumes injected are around 4 nl. A sequence of applied high voltages induces an electrophoretic separation as the sample passes through the channel, and the sample is emitted from the corner of the chip as an electrospray. The chip is precisely positioned to deliver the sample to the inlet of the HPMS.

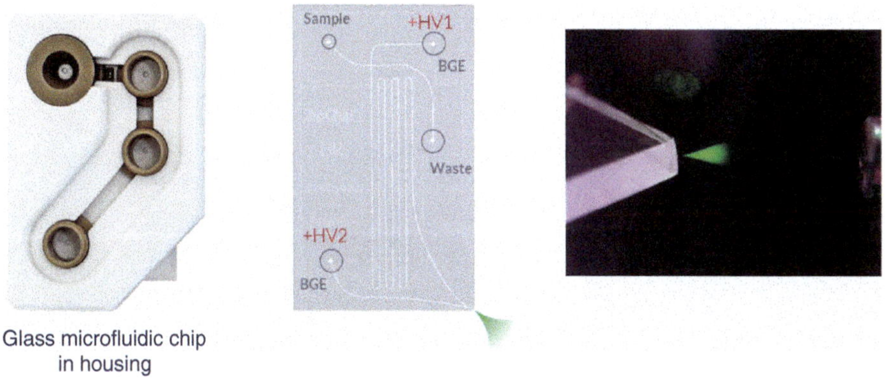

Figure 16.11 (a) The Rebel microfluidic chip. *Source*: Figure courtesy of 908 Devices Inc. (b) Microfabricated glass layer of the chip showing sample and electrolyte wells, separation channel, and position of electrospray in green. (c) The positioning of the chip relative to the inlet of the HPMS. *Source*: Figure courtesy of 908 Devices Inc.

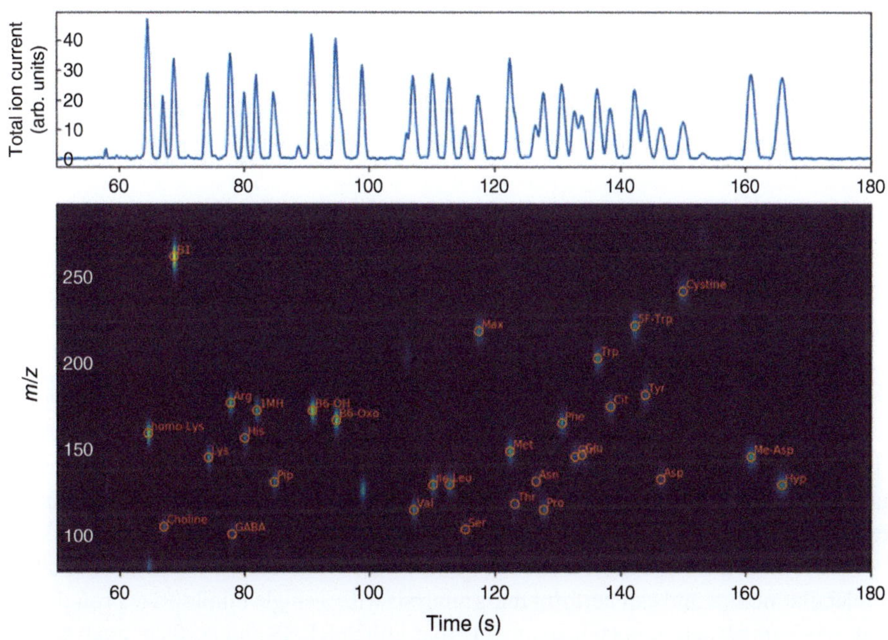

Figure 16.12 Electropherogram and *m/z* map of the standard solution of analytes, all at a concentration of 25 μM. Thirty-one components are identified and quantified in three minutes. The total ion current is shown in the top trace.

An example of Rebel data is shown in Figure 16.12. A mixture of amino acids, biogenic amines, and vitamins, all at concentrations of 25 micromolar (μM), was analyzed. The sample volume injected was 3.6 nanoliters (nl). Thirty-one components were identified over a CE run of just under three minutes (min). The observed levels of total ion current signal for each component correspond to minute quantities of material: the electrospray contains 9 femtomoles of the component. The estimated limits of quantitation for Rebel are <5 μM, corresponding to below 1–4 pg (depending on the specific target) leaving the chip in the electrospray and demonstrating the sensitivity of the HPMS detector.

Applications for CE–ESI–HPMS include screening of raw materials and monitoring metabolite concentrations at different time points during cell culture growth. The separation and detection methods are tolerant to common

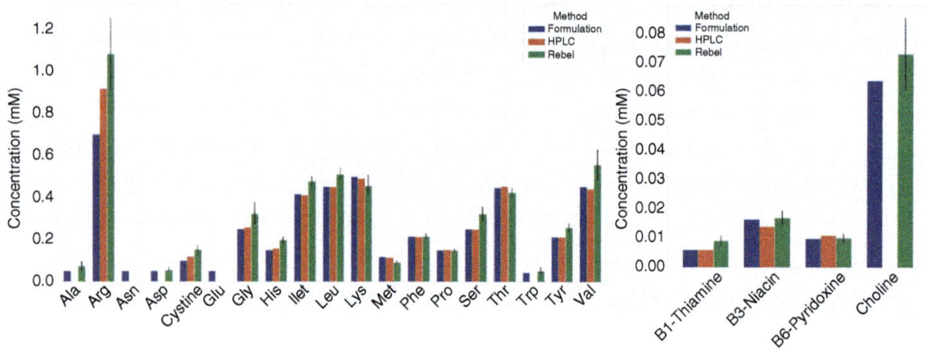

Figure 16.13 Concentration values for amino acids and vitamins in DMEM-F12 cell culture media as determined by HPLC analysis ($n = 1$, orange) and Rebel analysis ($n = 5$, green). The manufacturer-reported values are shown in blue.

LC-MS interferents such as salts, sugars, antibodies, and proteins commonly found in cell culture media samples. Figure 16.13 shows the results of the analysis of fresh cell culture media (DMEM/F12, P/N D8437, Sigma-Aldrich). The media was run on both the Rebel and via HPLC, and the results were compared to the manufacturer's reported values. For the HPLC analysis, the sample was diluted 10×, then derivatized to add UV absorbent tags for detection. For the Rebel analysis, the sample was diluted 10× and analyzed directly.

The Rebel ran the sample and quantified all components in a single seven-minute run. The HPLC method required two separate runs – one for amino acids and one for vitamin analysis. Data were aligned very well between the Rebel ($n = 5$) and HPLC ($n = 1$). Note: the standard B vitamin method for HPLC did not cover choline; so, it was not detected.

16.4 Conclusions

Ion trap mass spectrometers have been workhorse instruments in laboratories since the 1980s. Miniaturization efforts have sought to move these mass spectrometers from laboratories where they are maintained and operated by trained personnel to portable units that no longer require advanced MS knowledge to operate. Ion traps have been miniaturized and analyzed using a variety of geometries by both academic and commercial groups. A key step toward truly handheld, portable units has been the introduction of HPMS. HPMS had led to MS instruments that no longer require turbomolecular pumps and can perform mass analysis with a single miniaturized roughing pump. The M908, MX908, and Rebel are all commercial instruments that utilize HPMS and perform analyses at the point of use for applications, such as TICs, CWAs, drugs, explosives, and metabolites.

References

1 Ouyang, Z., Noll, R.J., and Cooks, R.G. (2009). Handheld miniature ion trap mass spectrometers. *Anal. Chem.* https://doi.org/10.1021/ac900292w.

2 Denisov, E., Damoc, E., Lange, O., and Makarov, A. (2012). Orbitrap mass spectrometry with resolving powers above 1,000,000. *Int. J. Mass Spectrom.* https://doi.org/10.1016/j.ijms.2012.06.009.

3 Pelander, A., Decker, P., Baessmann, C., and Ojanperä, I. (2011). Evaluation of a high resolving power time-of-flight mass spectrometer for drug analysis in terms of resolving power and acquisition rate. *J. Am. Soc. Mass Spectrom.* https://doi.org/10.1007/s13361-010-0046-z.

4 Snyder, D.T., Pulliam, C.J., Ouyang, Z., and Cooks, R.G. (2016). Miniature and fieldable mass spectrometers: recent advances. *Anal. Chem.* https://doi.org/10.1021/acs.analchem.5b03070.

5 Badman, E.R. and Cooks, R.G. (2000). Miniature mass analyzers. *J. Mass Spectrom.* https://doi.org/10.1002/1096-9888(200006)35:6<659::AID-JMS5>3.0.CO;2-V.

6 Verbeck, G.F. and Bierbaum, V.M. (2015). Focus on harsh environment and field-portable mass spectrometry: editorial. *J. Am. Soc. Mass Spectrom.* https://doi.org/10.1007/s13361-014-1057-y.

7 Palmer, P.T. and Limero, T.F. (2001). Mass spectrometry in the U.S. space program: past, present, and future. *J. Am. Soc. Mass Spectrom.* https://doi.org/10.1016/S1044-0305(01)00249-5.

8 Ecelberger, S.A., Cornish, T.J., Collins, B.F. et al. (2004). Suitcase TOF: a man-portable time-of-flight mass spectrometer. *Johns Hopkins APL Tech. Dig. (Appl. Phys. Lab.)* 25: 14–19.

9 Gnyba, M., Smulko, J., Kwiatkowski, A., and Wierzba, P. (2011). Portable Raman spectrometer - design rules and applications. *Bull. Polish Acad. Sci. Tech. Sci.* https://doi.org/10.2478/v10175-011-0040-z.

10 Ho, C.K., Itamura, M.T., Kelley, M., and Hughes, R.C. (2001). Review of chemical sensors for in-situ monitoring of volatile contaminants. *Contract* https://doi.org/10.2172/780299.

11 Cottingham, K. (2003). Product review: ion mobility spectrometry rediscovered. *Anal. Chem.* https://doi.org/10.1021/ac031304h.

12 Harris, C.M. (2002). GC to go. *Anal. Chem.* 21: 585A–589A.

13 Glish, G.L. and Vachet, R.W. (2003). The basics of mass spectrometry in the twenty-first century. *Nat. Rev. Drug Discov.* https://doi.org/10.1038/nrd1011.

14 Haag, A.M. (2016). Mass analyzers and mass spectrometers. *Adv. Exp. Med. Biol.* https://doi.org/10.1007/978-3-319-41448-5_7.

15 Crocombe, R.A. (2018). Portable spectroscopy. *Appl. Spectrosc.* https://doi.org/10.1177/0003702818809719.

16 Sorak, D., Herberholz, L., Iwascek, S. et al. (2012). New developments and applications of handheld Raman, mid-infrared, and near-infrared spectrometers. *Appl. Spectrosc. Rev.* https://doi.org/10.1080/05704928.2011.625748.

17 Smiths Detection. LCD 3.3 Person worn CWA and TIC detector. https://www.smithsdetection.com/products/lcd-3-3 (accessed 23 October 2019).

18 ThermoFisher Scientific. FirstDefender™ RMX Handheld Chemical Identification. https://www.thermofisher.com/order/catalog/product/FIRSTDEFENDERRMX (accessed 23 October 2019).

19 Paul, W. and Steinwedel, H. (1953). Ein Neues Massenspektrometer Ohne Magnetfeld. *Zeitschrift fur Naturforschung – Sec. A J. Phys. Sci.* https://doi.org/10.1515/zna-1953-0710.

20 March, R. E.; Todd, J. F. J. *Practical Aspects of Trapped Ion Mass Spectrometry*; 2016. https://doi.org/10.1201/9781420083743.

21 Bharti, A., Ma, P.C., and Salgia, R. (2007). *Mass Spectrom. Rev.* https://doi.org/10.1002/mas.20125.

22 March, R.E. (1997). An introduction to quadrupole ion trap mass spectrometry. *J. Mass Spectrom.* https://doi.org/10.1002/(SICI)1096-9888(199704)32:4<351::AID-JMS512>3.0.CO;2-Y.

23 Drewsen, M. and Brøner, A. (2000). Harmonic linear Paul trap: stability diagram and effective potentials. *Phys. Rev. A – At. Mol. Opt. Phys.* https://doi.org/10.1103/PhysRevA.62.045401.

24 Alheit, R., Kleineidam, S., Vedel, F. et al. (1996). Higher order non-linear resonances in a Paul trap. *Int. J. Mass Spectrom. Ion Process.* https://doi.org/10.1016/0168-1176(96)04380-7.

25 Dehmelt, H.G. (1968). Radiofrequency spectroscopy of stored ions I: storage. *Adv. At. Mol. Opt. Phys.* https://doi.org/10.1016/S0065-2199(08)60170-0.

26 Major, F.G. and Dehmelt, H.G. (1968). Exchange-collision technique for the RF spectroscopy of stored ions. *Phys. Rev.* https://doi.org/10.1103/PhysRev.170.91.

27 Dawson, P.H. and Whetten, N.R. (1968). Ion storage in three-dimensional, rotationally symmetric, quadrupole fields. II. A sensitive mass spectrometer. *J. Vac. Sci. Technol. A* https://doi.org/10.1116/1.1492570.

28 Dawson, P.H. and Whetten, N.R. (1968). Ion storage in three-dimensional, rotationally symmetric, quadrupole fields. I. Theoretical treatment. *J. Vac. Sci. Technol. A* https://doi.org/10.1116/1.1492569.

29 Stafford, G.C., Kelley, P.E., Syka, J.E.P. et al. (1984). Recent improvements in and analytical applications of advanced ion trap technology. *Int. J. Mass Spectrom. Ion Process.* https://doi.org/10.1016/0168-1176(84)80077-4.

30 Stafford, G. (2002). Ion trap mass spectrometry: a personal perspective. *J. Am. Soc. Mass Spectrom.* https://doi.org/10.1016/S1044-0305(02)00385-9.

31 Dawson, P.H. (1977). The effect of collisions on ion motion in quadrupole fields. *Int. J. Mass Spectrom. Ion Phys.* https://doi.org/10.1016/0020-7381(77)80082-X.

32 Blatt, R., Schmeling, U., and Werth, G. (1979). On the sensitivity of ion traps for spectroscopic applications. *Appl. Phys.* https://doi.org/10.1007/BF00894999.

33 Borman, S.A. (1983). New gas chromatographic detectors. *Anal. Chem.* https://doi.org/10.1021/ac00258a735.

34 Goeringer, D.E., Whitten, W.B., Ramsey, J.M. et al. (1992). Theory of high-resolution mass spectrometry achieved via resonance ejection in the quadrupole ion trap. *Anal. Chem.* https://doi.org/10.1021/ac00037a023.

35 Arnold, N.S., Hars, C., and Meuzelaar, H.L.C. (1994). Extended theoretical considerations for mass resolution in the resonance ejection mode of quadrupole ion trap mass spectrometry. *J. Am. Soc. Mass Spectrom.* https://doi.org/10.1016/1044-0305(94)85008-9.

36 Whitten, W.B., Reilly, P.T.A., and Ramsey, J.M. (2004). High-pressure ion trap mass spectrometry. *Rapid Commun. Mass Spectrom.* https://doi.org/10.1002/rcm.1549.

37 Wang, Y., Franzen, J., and Wanczek, K.P. (1993). The non-linear resonance ion trap. Part 2. A general theoretical analysis. *Int. J. Mass Spectrom. Ion Process.*

38 Franzen, J., Gabling, R.H., Schubert, M., and Wang, Y. (1995). Nonlinear ion traps. *Pract. Asp. Ion Trap Mass Spectrom.*

39 Eades, D.M., Johnson, J.V., and Yost, R.A. (1993). Nonlinear resonance effects during ion storage in a quadrupole ion trap. *J. Am. Soc. Mass Spectrom.* https://doi.org/10.1016/1044-0305(93)80017-S.

40 Bonner, R.F., Fulford, J.E., March, R.E., and Hamilton, G.F. (1977). The cylindrical ion trap. Part I. general introduction. *Int. J. Mass Spectrom. Ion Phys.* https://doi.org/10.1016/0020-7381(77)80034-X.

41 Kornienko, O., Reilly, P.T.A., Whitten, W.B., and Ramsey, J.M. (1999). Micro ion trap mass spectrometry. *Rapid Commun. Mass Spectrom.* https://doi.org/10.1002/(SICI)1097-0231(19990115)13:1<50::AID-RCM449>3.0.CO;2-K.

42 Kornienko, O., Reilly, P.T.A., Whitten, W.B., and Ramsey, J.M. (1999). Electron impact ionization in a microion trap mass spectrometer. *Rev. Sci. Instrum.* https://doi.org/10.1063/1.1150010.

43 Wells, J.M., Badman, E.R., and Cooks, R.G. (1998). A quadrupole ion trap with cylindrical geometry operated in the mass-selective instability mode. *Anal. Chem.* https://doi.org/10.1021/ac971198h.

44 Xu, W., Chappell, W.J., Cooks, R.G., and Ouyang, Z. (2009). Characterization of electrode surface roughness and its impact on ion trap mass analysis. *J. Mass Spectrom.* https://doi.org/10.1002/jms.1512.

45 Chaudhary, A., Van Amerom, F.H.W., and Short, R.T. (2014). Experimental evaluation of micro-ion trap mass spectrometer geometries. *Int. J. Mass Spectrom.* https://doi.org/10.1016/j.ijms.2014.06.032.

46 Lee, W.W., Oh, C.H., Kim, P.S. et al. (2003). Characteristics of cylindrical ion trap. *Int. J. Mass Spectrom.* https://doi.org/10.1016/j.ijms.2003.08.001.

47 Louris, J.N., Cooks, R.G., Syka, J.E. et al. (1987). Instrumentation, applications, and energy deposition in quadrupole ion-trap tandem mass spectrometry. *Anal. Chem.* https://doi.org/10.1021/ac00140a021.

48 Austin, D.E., Cruz, D., and Blain, M.G. (2006). Simulations of ion trapping in a micrometer-sized cylindrical ion trap. *J. Am. Soc. Mass Spectrom.* https://doi.org/10.1016/j.jasms.2005.11.020.

49 Cruz, D., Chang, J.P., Fico, M. et al. (2007). Design, microfabrication, and analysis of micrometer-sized cylindrical ion trap arrays. *Rev. Sci. Instrum.* https://doi.org/10.1063/1.2403840.

50 Tian, Y., Higgs, J., Li, A. et al. (2014). How far can ion trap miniaturization go? Parameter scaling and space-charge limits for very small cylindrical ion traps. *J. Mass Spectrom.* https://doi.org/10.1002/jms.3343.

51 Pau, S., Pai, C.S., Low, Y.L. et al. (2006). Microfabricated quadrupole ion trap for mass spectrometer applications. *Phys. Rev. Lett.* https://doi.org/10.1159/000089267.

52 Badman, E.R. and Cooks, R.G. (2000). A parallel miniature cylindrical ion trap array. *Anal. Chem.* https://doi.org/10.1021/ac000109p.

53 Tabert, A.M., Griep-Raming, J., Guymon, A.J., and Cooks, R.G. (2003). High-throughput miniature cylindrical ion trap array mass spectrometer. *Anal. Chem.* https://doi.org/10.1021/ac0346858.

54 Ouyang, Z., Badman, E.R., and Cooks, R.G. (1999). Characterization of a serial array of miniature cylindrical ion trap mass analyzers. *Rapid Commun. Mass Spectrom.* https://doi.org/10.1002/(SICI)1097-0231(19991230)13:24<2444::AID-RCM810>3.0.CO;2-F.

55 Hager, J.W. (2002). A new linear ion trap mass spectrometer. *Rapid Commun. Mass Spectrom.* https://doi.org/10.1002/rcm.607.

56 Schwartz, J.C., Senko, M.W., and Syka, J.E.P. (2002). A two-dimensional quadrupole ion trap mass spectrometer. *J. Am. Soc. Mass Spectrom.* https://doi.org/10.1016/S1044-0305(02)00384-7.

57 Ouyang, Z., Wu, G., Song, Y. et al. (2004). Rectilinear ion trap: concepts, calculations, and analytical performance of a new mass analyzer. *Anal. Chem.* https://doi.org/10.1021/ac049420n.

58 Erickson, B.E. (2004). Miniaturized rectilinear ion trap. *Anal. Chem.* 17: 305A.

59 Hendricks, P., Duncan, J., Noll, R.J. et al. (2011). Performance of a low voltage ion trap. *Int. J. Mass Spectrom.* https://doi.org/10.1016/j.ijms.2011.05.009.

60 Wu, G., Cooks, R.G., Ouyang, Z. et al. (2006). Ion trajectory simulation for electrode configurations with arbitrary geometries. *J. Am. Soc. Mass Spectrom.* https://doi.org/10.1016/j.jasms.2006.05.004.

61 Ramsey, J.M. and Schultze, K.P. (2014). Miniature charged particle trap with elongated trapping region for mass spectrometry. US8878127B2.

62 Pau, S., Whitten, W.B., and Ramsey, J.M. (2007). Planar geometry for trapping and separating ions and charged particles. *Anal. Chem.* https://doi.org/10.1021/ac0706269.

63 Peng, Y., Hansen, B.J., Quist, H. et al. (2011). Coaxial ion trap mass spectrometer: concentric toroidal and quadrupolar trapping regions. *Anal. Chem.* https://doi.org/10.1021/ac200600u.

64 Lammert, S.A., Plass, W.R., Thompson, C.V., and Wise, M.B. (2001). Design, optimization and initial performance of a toroidal RF ion trap mass spectrometer. *Int. J. Mass Spectrom.* https://doi.org/10.1016/S1387-3806(01)00507-3.

65 Lammert, S.A., Rockwood, A.A., Wang, M. et al. (2006). Miniature toroidal radio frequency ion trap mass analyzer. *J. Am. Soc. Mass Spectrom.* https://doi.org/10.1016/j.jasms.2006.02.009.

66 Higgs, J.M., Petersen, B.V., Lammert, S.A. et al. (2016). Radiofrequency trapping of ions in a pure toroidal potential distribution. *Int. J. Mass Spectrom.* https://doi.org/10.1016/j.ijms.2015.11.009.

67 Austin, D.E., Wang, M., Tolley, S.E. et al. (2007). Halo ion trap mass spectrometer. *Anal. Chem.*

68 Wang, M., Quist, H.E., Hansen, B.J. et al. (2011). Performance of a halo ion trap mass analyzer with exit slits for axial ejection. *J. Am. Soc. Mass Spectrom.* https://doi.org/10.1007/s13361-010-0027-2.

69 Chen, C.H., Chen, T.C., Zhou, X. et al. (2015). Design of portable mass spectrometers with handheld probes: aspects of the sampling and miniature pumping systems. *J. Am. Soc. Mass Spectrom.* https://doi.org/10.1007/s13361-014-1026-5.

70 Hoch, G. and Kok, B. (1963). A mass spectrometer inlet system for sampling gases dissolved in liquid phases. *Arch. Biochem. Biophys.* https://doi.org/10.1016/0003-9861(63)90546-0.

71 Davey, N.G., Krogh, E.T., and Gill, C.G. (2011). Membrane-introduction mass spectrometry (MIMS). *TrAC* https://doi.org/10.1016/j.trac.2011.05.003.

72 Giannoukos, S., Brkić, B., Taylor, S., and France, N. (2015). Membrane inlet mass spectrometry for homeland security and forensic applications. *J. Am. Soc. Mass Spectrom.* https://doi.org/10.1007/s13361-014-1032-7.

73 Gao, L., Cooks, R.G., and Ouyang, Z. (2008). Breaking the pumping speed barrier in mass spectrometry: discontinuous atmospheric pressure Interface. *Anal. Chem.* https://doi.org/10.1021/ac800014v.

74 Gao, L., Sugiarto, A., Harper, J.D. et al. (2008). Design and characterization of a multisource hand-held tandem mass spectrometer. *Anal. Chem.* https://doi.org/10.1021/ac801275x.

75 Xu, W., Song, Q., Smith, S.A. et al. (2009). Ion trap mass analysis at high pressure: a theoretical view. *J. Am. Soc. Mass Spectrom.* https://doi.org/10.1016/j.jasms.2009.06.019.

76 Song, Q., Xu, W., Smith, S.A. et al. (2010). Ion trap mass analysis at high pressure: an experimental characterization. *J. Mass Spectrom.* https://doi.org/10.1002/jms.1684.

77 Jiang, T., Xu, Q., Zhang, H. et al. (2018). Improving the performances of a "brick mass spectrometer" by quadrupole enhanced dipolar resonance ejection from the linear ion trap. *Anal. Chem.* https://doi.org/10.1021/acs.analchem.8b03332.

78 Jiang, P., Zhou, Z., Wu, Z., and Zhao, Z. (2018). Low-vacuum cylindrical ion trap mass spectrometry. *Instrum. Sci. Technol.* 46 (6): 614–627.

79 Decker, T.K., Zheng, Y., Ruben, A.J. et al. (2019). A microscale planar linear ion trap mass spectrometer. *J. Am. Soc. Mass Spectrom.* https://doi.org/10.1007/s13361-018-2104-x.

80 Jones, R.M., Gerlich, D., and Anderson, S.L. (1997). Simple radio-frequency power source for ion guides and ion traps. *Rev. Sci. Instrum.* https://doi.org/10.1063/1.1148297.

81 Schaefer, R.T., MacAskill, J.A., Mojarradi, M. et al. (2008). Digitally synthesized high purity, high-voltage radio frequency drive electronics for mass spectrometry. *Rev. Sci. Instrum.* https://doi.org/10.1063/1.2981691.

82 O'Connor, P.B., Costello, C.E., and Earle, W.E. (2002). A high voltage RF oscillator for driving multipole ion guides. *J. Am. Soc. Mass Spectrom.* https://doi.org/10.1016/S1044-0305(02)00700-6.

83 Moxom, J., Reilly, P.T.A., Whitten, W.B., and Ramsey, J.M. (2002). Double resonance ejection in a micro ion trap mass spectrometer. *Rapid Commun. Mass Spectrom.* https://doi.org/10.1002/rcm.635.

84 Snyder, D.T. and Cooks, R.G. (2016). Successive resonances for ion ejection at arbitrary frequencies in an ion trap. *J. Am. Soc. Mass Spectrom.* https://doi.org/10.1007/s13361-016-1473-2.

85 Decker, T.K., Zheng, Y., McClellan, J.S. et al. (2018). Double resonance ejection using novel radiofrequency phase tracking circuitry in a miniaturized planar linear ion trap mass spectrometer. *Rapid Commun. Mass Spectrom.* https://doi.org/10.1002/rcm.8267.

86 Zhou, X., Liu, X., Chiang, S. et al. (2018). Stimulated motion suppression (STMS): a new approach to break the resolution barrier for ion trap mass spectrometry. *J. Am. Soc. Mass Spectrom.* https://doi.org/10.1007/s13361-018-1995-x.

87 Moxom, J., Reilly, P.T.A., Whitten, W.B., and Ramsey, J.M. (2003). Analysis of volatile organic compounds in air with a micro ion trap mass analyzer. *Anal. Chem.* https://doi.org/10.1021/ac034043k.

88 Moxom, J., Reilly, P.T.A., Whitten, W.B., and Ramsey, J.M. (2004). Sample pressure effects in a micro ion trap mass spectrometer. *Rapid Commun. Mass Spectrom.* https://doi.org/10.1002/rcm.1389.

89 Collins, R.D. (1969). The use of electron multipliers in mass spectrometry. *Vacuum* https://doi.org/10.1016/s0042-207x(69)90023-2.

90 Dietz, L.A. (1965). Basic properties of electron multiplier ion detection and pulse counting methods in mass spectrometry. *Rev. Sci. Instrum.* https://doi.org/10.1063/1.1719460.

91 Tuithof, H.H., Boerboom, A.J.H., and Meuzelaar, H.L.C. (1975). Simultaneous detection of a mass Spectrum using a Channeltron electron multiplier array. *Int. J. Mass Spectrom. Ion Phys.* https://doi.org/10.1016/0020-7381(75)87040-9.

92 Blakeman, K.H., Wolfe, D.W., Cavanaugh, C.A., and Ramsey, J.M. (2016). High pressure mass spectrometry: the generation of mass spectra at operating pressures exceeding 1 Torr in a microscale cylindrical ion trap. *Anal. Chem.* https://doi.org/10.1021/acs.analchem.6b00706.

93 Blakeman, K.H., Cavanaugh, C.A., Gilliland, W.M., and Ramsey, J.M. (2017). High pressure mass spectrometry of volatile organic compounds with ambient air buffer gas. *Rapid Commun. Mass Spectrom.* https://doi.org/10.1002/rcm.7766.

94 Blakeman, K.H. (2015). *Development of high pressure mass spectrometry for handheld instruments*. The University of North Carolina at Chapel Hill.

95 Amptek. A250CF CoolFET® Charge Sensitive Preamplifier. https://www.amptek.com/products/charge-sensitive-preamplifiers/a250cf-coolfet-charge-sensitive-preamplifier#Documentation (accessed 31 October 2019).

96 Schultze, K.P. (2014). *Advanced System Components for the Development of a Handheld Ion Trap Mass Spectrometer*. The University of North Carolina at Chapel Hill.

97 Brown, C.D., Knopp, K.J., Krylov, E., and Miller, S. (2014). Compact Mass Spectrometer. US9099286B2.

98 Chen, H., Gamez, G., and Zenobi, R. (2009). What can we learn from ambient ionization techniques? *J. Am. Soc. Mass Spectrom.* https://doi.org/10.1016/j.jasms.2009.07.025.

99 Holčapek, M., Jirásko, R., and Lísa, M. (2010). Basic rules for the interpretation of atmospheric pressure ionization mass spectra of small molecules. *J. Chromatogr. A* https://doi.org/10.1016/j.chroma.2010.02.049.

100 Herrera, L.C., Grossert, J.S., and Ramaley, L. (2008). Quantitative aspects of and ionization mechanisms in positive-ion atmospheric pressure chemical ionization mass spectrometry. *J. Am. Soc. Mass Spectrom.* https://doi.org/10.1016/j.jasms.2008.07.016.

101 Marquet, P., Venisse, N., Lacassie, E., and Lachâtre, G. (2000). In-source CID mass spectral libraries for the "general unknown" screening of drugs and toxicants. *Analusis* https://doi.org/10.1051/analusis:2000280925.

102 Rodgers, M.T. and Armentrout, P.B. (1998). Statistical modeling of competitive threshold collision-induced dissociation. *J. Chem. Phys.* https://doi.org/10.1063/1.476754.

103 Falconer, T., Kern, S., and Voelker, S. (2018). Rapid, in situ detection of synthetic opioids. *66th ASMS Conference on Mass Spectrometry and Allied Topics*, San Diego, CA.

104 Gilliland, W.M., Mellors, J.S., and Ramsey, J.M. (2017). Coupling microchip electrospray ionization devices with high pressure mass spectrometry. *Anal. Chem.* https://doi.org/10.1021/acs.analchem.7b03484.

105 Gilliland, W.M. and Ramsey, J.M. (2018). Development of a microchip CE-HPMS platform for cell growth monitoring. *Anal. Chem.* https://doi.org/10.1021/acs.analchem.8b03708.

106 Inficon. HAPSITE® ER Chemical Identification System. https://products.inficon.com/en-us/nav-products/product/detail/hapsite-er-identification-system (accessed 23 October 2019).

107 Bruker. E^2M the Mobile Enhanced Environmental Mass Spectrometer. https://www.bruker.com/products/cbrne-detection/gc-ms/e2m/overview.html (accessed 23 October 2019).

108 908devices. M908 Spec Sheet. http://908devices.com/wp-content/uploads/2017/04/SpecSheet-908_2016_v3.pdf (accessed 31 October 2019).

109 PerkinElmer. Torion T-9 Portable GC/MS. https://www.perkinelmer.com/product/torion-t-9-portable-gc-ms-instrument-ntsst090500 (accessed 23 October 2019).

110 FLIR. Person-Portable GC-MS Griffin G510. https://www.flir.com/products/griffin-g510 (accessed 23 October 2019).

111 908devices. MX908 Spec Sheet. https://908devices.com/wp-content/uploads/2017/09/MX908_SpecSheet_EOD_.pdf (accessed 31 October 2019).

112 Continuity™ Transportable High-Sensitivity Mass Spectrometer. https://www.bayspec.com/spectroscopy/continuity-transportable-high-sensitivity-mass-spectrometer (accessed 23 October 2019).

113 TRACER 1000. https://www.1stdetect.com/products/tracer-1000-ms-etd (accessed 23 October 2019).

17

Key Instrumentation Developments That Have Led to Portable Ion Mobility Spectrometer Systems

Reno F. DeBono[1,3] and Pauline E. Leary[2]

[1] Trace Detection Scientist, Formerly of Barringer Instruments, Warren, NJ, USA
[2] Federal Resources, Stevensville, MD, USA
[3] Smiths Detection, Edgewood, MD, USA

17.1 Background and History

Ion mobility spectrometry (IMS) is a gas-phase separation technique which works by predictably and reproducibly creating ions from neutral analyte molecules and separating these ions based on their size and shape under the influence of an electric field in the presence of a buffer gas. IMS is likely the most widely used trace detection method deployed in the field for testing of dangerous substances. It is selective, offers low detection limits, provides reliable identifications, and fast analysis times. IMS systems are size, weight, and power (SWaP) friendly, and perform at ambient pressure. These features are important for portability and amenable to miniaturization, with a robust deployment of portable and handheld systems in a variety of different markets. Systems have been extensively used by the military, as well as in aviation security, border inspection, and by first responders.

In chapter one of this volume, Leary, Crocombe, and Kammrath provide a general classification scheme of portable spectrometers in accordance with SWaP expectations. This classification scheme generally holds true for portable IMS systems and categorizes person-portable spectrometers as those that weigh between 3 and 20 kg, handheld systems as those that weigh between 0.5 and 3 kg, and wearable systems as those that weigh less than 0.5 kg. This chapter includes discussions of person-portable, handheld, and wearable IMS systems. Most of the largest portable IMS systems deployed would be classified as person portable; the smallest would be classified as wearable. For portable IMS systems, especially within markets, such as aviation security, where high sample throughput at a centralized location, such as a security checkpoint, is expected, end users are willing to accept person-portable systems that may not operate on battery power. These IMS systems, sometimes referred to as benchtop systems, are classified as portable since they are deployed to field users for analysis on site. On the other hand, users of the smallest IMS detectors have very strict requirements for battery operation. Therefore, significant resources are expended during design and development of handheld and wearable IMS systems like those used by the military to provide personal monitoring for chemical warfare agents (CWAs) and toxic industrial chemicals (TICs) to verify that these systems are as lightweight as possible and can run on commercial AA batteries for extended periods of time.

IMS is particularly effective at screening for the presence of hazardous chemicals, such as explosives, narcotics, TICs, CWAs, and opioids, including fentanyl analogs. IMS systems can be deployed either independently as a primary screening tool, or as a secondary complement to another screening method, such as X-ray, for alarm resolution or orthogonal detection. Targets that alarm using the initial screening method can be further interrogated using IMS. This two-tiered approach can be especially useful in situations where high throughput is required.

Portable Spectroscopy and Spectrometry 1: Technologies and Instrumentation, First Edition.
Edited by Richard A. Crocombe, Pauline E. Leary and Brooke W. Kammrath.

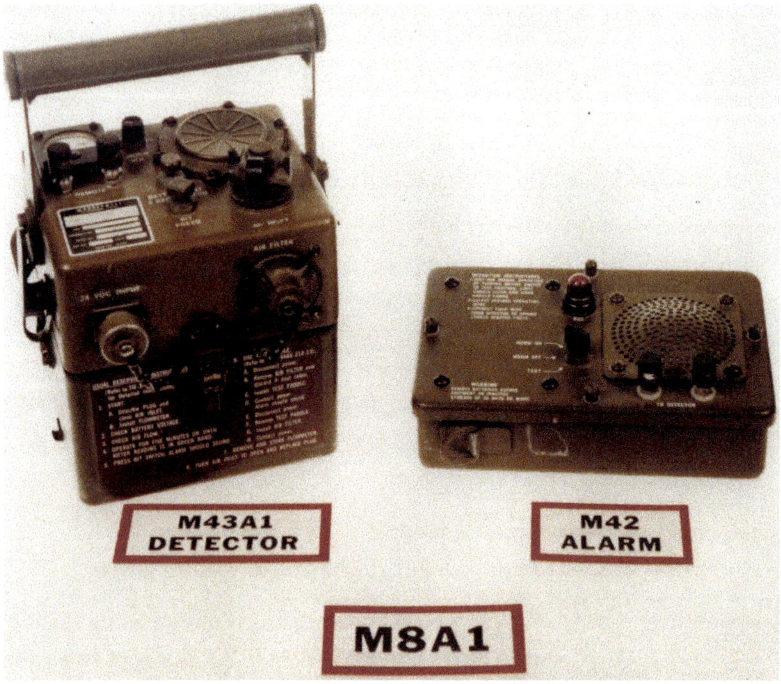

Figure 17.1 M8A1 detector standardized by the United States Army in 1981. *Source*: Image courtesy of the United States Army.

Acceptance of IMS in the field is largely driven by its versatility regarding deployment, and the ease with which it fits within an organization's optimized concept of operations. Although IMS systems are deployed to a variety of different end users, the military and aviation industry have been critical to the success of portable IMS systems.

Initial development of portable IMS systems was due to military interest in the detection of CWAs. One of the earliest IMS detectors was a product manufactured by Intellitec [1] and known as the M8A1, which was standardized by the United States Army in 1981. The M8A1 was configured with an M43A1 ion detector linked with the standard M42 alarm. Figure 17.1 shows the configured M8A1 system. The M43A1 used an ^{241}Am alpha radioactive ionization source. A pump drew air samples into the detector unit and then through a heater and filter. Contaminated air passed over the ionization source causing the ions of the chemical agent to cluster. A geometrically configured cell collected the clusters as electric current while a monitor sensed any voltage change which activated the alarm. The M43A1 detector could be placed up to 400 m from the M42 alarm to provide an early warning of a chemical cloud. The M8A1 alarm became the United States Army's single most important chemical detection capability, and by 1987, over 32 000 units were in the field. During Operation Desert Shield/Storm in 1990–1991, the United States Army used over 12 000 M8A1 systems as the main detection capability for chemical defense [2–5]. The military continues to actively support the development and deployment of IMS systems for CWA, TIC, and explosives detection. For instance, in 2019, Smiths Detection reported it provided more than 91 000 joint chemical agent detector (JCAD) units over the past 14 years [6].

The aviation industry has also been critical to the development of IMS detectors for the detection of explosives. In 2004, more than 10 000 explosives detectors based on IMS had been used at security checkpoints in airports worldwide [7]. Today that number is significantly higher due to the fact that over the last 25 years, IMS technology has sufficiently matured and continues to meet the well-defined and evolving aviation detection standards for explosives detection within regulatory bodies, such as the Transportation Security Administration (TSA) in the United States, the Department for Transport (DfT) in the United Kingdom, the Transport Canada in Canada, the European

Civil Aviation Conference (ECAC), and the Civil Aviation Administration of China (CAAC). Requirements of Customs, border control, correctional facilities, and law enforcement have been critical to the development of IMS detectors for trace narcotics detection, and more recently for detection of fentanyl and its analogs. For a thorough discussion on the applications of IMS, see chapter 8 in volume two of this text by Leary and Joshi.

17.2 Principles of Ion Mobility Spectrometry

Over the last 50 years, two variants of IMS systems have developed. The conventional form is based on a fixed-mobility property associated with the ion, which is not a function of the electric field. The second is based on changing ion mobility as a function of a changing electric field [8] (p. 1). The former is the basis of both drift-tube IMS (DTIMS) and aspiration IMS (AIMS), while the latter is the basis of differential mobility spectrometry (DMS). DMS is often referred to as field asymmetric waveform IMS (FAIMS), and over the years, has also been called other names including nonlinear drift spectrometry, differential IMS (DIMS), field ion spectrometry, and radio frequency IMS (RF IMS). Following brief discussions of AIMS and FAIMS systems, this chapter will focus on DTIMS.

17.2.1 Aspiration Ion Mobility Spectrometry

AIMS is based on the principle that ions move in air, and if an electric field is applied perpendicular to the flow of air through a drift tube, the ions will be pushed toward the walls of the drift tube [9]. In these systems, the ionization occurs at ambient conditions in the presence of a relatively high amount of water from the ambient environment, which affects the kinetics of ionization reactions [10]. At high-enough water-vapor concentrations, sensitivity can be compromised. Ions continuously enter the drift tube carried by a laminar flow of ambient air. No gating grid is used to pulse the ions, so instrument resolution is dependent on the inlet aperture dimensions. Theoretically, 100% of the ions are injected into the analyzer, so the signal-to-noise ratio (SNR) can be significantly improved compared with DTIMS systems [11] (p. 128). If the electric field polarity is switched, both positive and negative ions can be detected. A series of ion detectors, such as Faraday plates, are placed along the path of the deflected ions to generate the signal. For IMS detectors, ion detection is achieved by converting ions striking the detector into a current by neutralization at a collector electrode referred to as a Faraday plate [11] (pp. 155–156). As the ion swarm is neutralized, current flows in the metal Faraday plate: For positive ions, current flows to the plate and for negative ions, current flows away from the plate. If the electric field is fixed, then the distance the ions travel before they strike a detector is proportional to their mobility. In these fixed-field situations, ions of faster velocity collide with the detector earlier than ions with slower velocity. Poor spatial ion separation, due to space-charge effects and diffusion, results in poor resolving power. Both effects can be reduced by increasing the flow rate so that ion concentration and drift time decrease; however, the flow rate is limited since laminar flow is required.

Peaks associated with variable ion mobilities can be broad and overlapping. For this reason, the use of pattern recognition for detection algorithms in these systems is typical. Sensitivity is adjustable since ionization efficiency is flow-rate dependent [10]. AIMS is intrinsically more sensitive but less selective than DTIMS. AIMS systems are simple, offer high sensitivity, low cost, and are small. Figure 17.2 shows a schematic demonstrating the operating principle of an AIMS instrument that uses a radioactive ^{241}Am ionization source. The pattern generated as the output signal is used for identification. Environics Oy (Finland) has been a primary manufacturer of AIMS systems.

17.2.2 Field-Asymmetric Ion Mobility Spectrometry

To separate ions, FAIMS exploits the differences in their mobilities in low and high fields. Separations performed based upon ion mobility are typically classified as low or high field depending upon whether the ion gains translational energy between collisions (high field) or achieves a steady-state drift velocity (low field) [12]. As previously

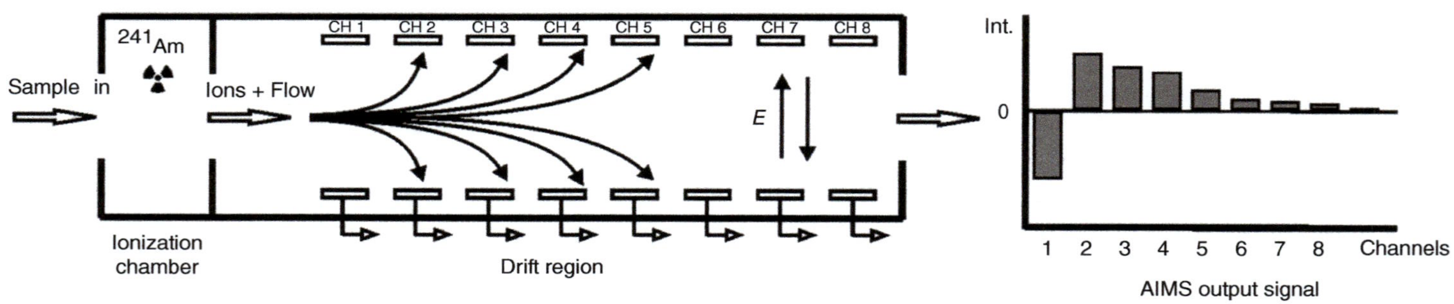

Figure 17.2 Schematic demonstrating the operating principle of an AIMS instrument. *Source:* Mäkinen et al. [10]. © 2010, American Chemical Society.

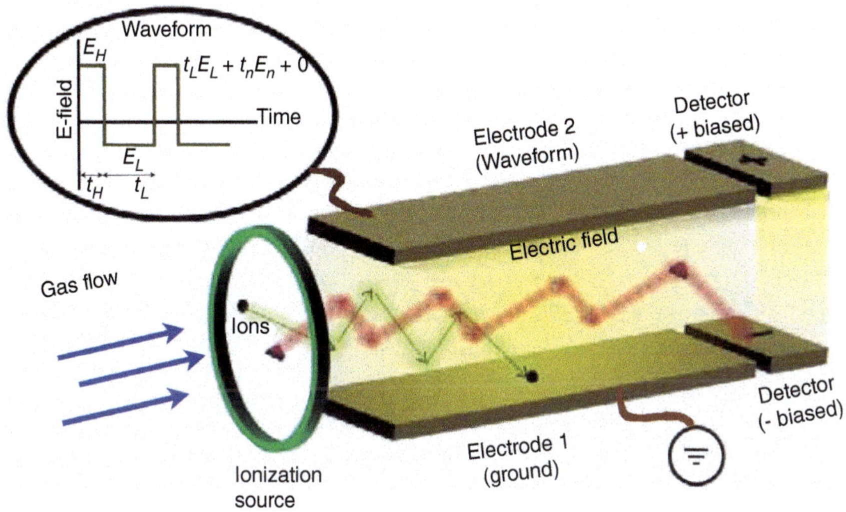

Figure 17.3 Schematic demonstrating the operating principle of a FAIMS instrument. *Source*: Zrodnikov and Davis [13]. Licensed under CC BY 4.0.

described by Zrodnikov and Davis, during a FAIMS measurement, an electric field is created by applying a waveform between two electrodes; one that is grounded and one with an applied high-frequency alternating current (AC) square waveform. One cycle of the waveform induces an electric field which is an initially high strength, and it is then switched to a low-strength field in the opposite direction. Since an ion will have different mobility in the high- and low-strength regimes, the ion will have a net displacement toward one of the electrodes over each cycle of the waveform. This displacement is counterbalanced by the application of a direct current (DC) compensation voltage (CV). Each CV value allows ions with that corresponding specific differential mobility to pass through the drift region and reach the ion detector. The CV can also be specified to a value characteristic of a certain ion of interest to achieve an ion-filtering regime, or it can be swept across a range to record the abundance of all ion species that are present in a sample. Dispersion plots, which show ion abundance for scans of CV at different values of the asymmetric waveform amplitude, are commonly used to display sampled data [13]. The peak voltage of the waveform is referred to as the dispersion voltage (DV). Figure 17.3 shows a schematic demonstrating the operating principle of FAIMS.

FAIMS systems offer good separation power, but they have historically had power consumption issues associated with electric field requirements. High power consumption requirements are not a significant problem for lab-based systems connected to shore power, but are not well-suited for field instruments intended to operate on battery power. Recently, power issues have been resolved resulting in the development of a miniaturized, lightweight, handheld DMS system [14]. In the past 15 or so years, FAIMS has had some success penetrating security markets, but has not yet been widely adopted.

17.2.3 Drift-Tube Ion Mobility Spectrometry

17.2.3.1 Theory

IMS works by predictably and reproducibly creating ions from neutral analyte molecules or atoms and separating these ions according to their size and shape. The first DTIMS, the configuration which formed the fundamental basis of current designs utilized in most field-deployed systems, was constructed about 50 years ago in 1970 [15] (p. 96). In this configuration, a neutral analyte is introduced to the system via a carrier gas and vaporized (if not already in the gas phase). The gas-phase analyte is then introduced to the ionization chamber where it is ionized, usually in the presence of dopants, using atmospheric pressure chemical ionization (APCI). Ionization is selective producing either positive or negative ions which are injected into the drift tube via an ion shutter, which in most commercial IMS instruments is a gating grid. The ion shutter opens briefly for 50–200 μs and then closes, allowing a pulse of either positive or negative ions to enter the drift tube, depending on the drift-tube polarity. Ions travel the length of the drift tube, which is maintained at constant voltage and temperature, through a gas known as the buffer gas. In the drift tube, the ions separate based on their size, shape, and collisional cross-section, and are then detected using a Faraday plate. The amount of time it takes the ion to travel from the ion shutter to the detector is measured and used for identification (under specified operating conditions). Figure 17.4 shows the schematic of the operation of a typical DTIMS system. Figure 17.5 shows an IMS spectrum. Table 17.1 is a summary of many of the commercially available IMS systems currently on the market and widely deployed for the analysis of threats. Most of the deployed platforms are based on DTIMS due to their selectivity, which allows good material identification coupled with a low false-alarm rate.

Many factors impact the formation and movement of ions through an IMS. During the ionization process, dopants may be added to purposely form stable ion adducts or create weak interactions with the ion. The size and shape of the ions formed, to first order, are proportional to the molecular weight (MW) of the ion. However, the IMS signal also captures three-dimensional structural information from the average cross-sectional area. The average cross-sectional area is a function of the sweep volume of the ion as it rotates and vibrates. It enables the separation of structural isomers and ions of the same mass (isobaric ions). The ability to separate isobaric ions is particularly important since many lower MW compounds, those at less than about 600 atomic mass units (amu), have very similar masses, but are differentiable in IMS due to this capability. The basis of separation is a result of the interaction of the ion with the buffer gas. As a result, it is not surprising that properties of the buffer gas impact the effective cross-sectional area of the ion enabling separation of two ions of the same mass but different cross-sectional areas. Typically, for portable instruments, clean dry air, which is created in situ by drying ambient air and filtering organic material, is used as the buffer gas. The moisture content of the air is critical and must be taken into account by either controlling it or compensating for changes. Typically, water-vapor levels in the buffer gas are in the range of 1–10 parts per million (ppm).

The ions associated with field-deployed IMS platforms are generally molecular ions with either positive or negative charge. An electric field is applied to the drift tube to move an ion swarm from the ionization chamber through a buffer gas down the length of the drift tube. The buffer gas is typically air at atmospheric pressure. The buffer gas is at sufficiently high pressure (>~1 Torr, typically at atmospheric pressure for field-deployable platforms) that the ion mean free path is small and, therefore, the ions undergo multiple collisions and rapidly achieve a constant velocity. The velocity (v) of the ion through the buffer gas is equal to the length of the drift tube (L) divided by the time it takes an ion to move from the gating grid to the detector, which is known as its drift time (t_d) (Eq. (17.1)).

$$v = \frac{L}{t_d} \tag{17.1}$$

An ion's drift velocity depends on the applied electric field strength, E, and its mobility constant, K. In the low-field regime of <~500 Volts/centimeter (V/cm), which is the regime within which most commercial DTIMS systems operate, K is essentially independent of E and can be treated as a constant [34]. The mobility K is a joint property of the ions and the buffer gas through which the motion occurs. Dividing v by E defines the mobility of the ion in terms of ion velocity per unit field (cm^2/Vs), which normalizes the velocity of the ion to E.

$$K = \frac{v}{E} = \frac{L}{t_d E} \tag{17.2}$$

As air pressure increases, the number of collisions of the ion with neutral air molecules (78% N_2, 21% O_2) increases, and the mobility of the ion decreases. If the air is heated, air density decreases and mobility increases. As a result, reduced mobility, K_0, is typically used to account for changes in temperature (T) and pressure (P) by normalizing to 273 Kelvin and 760 Torr as shown in Eq. (17.3). To first order, K_0 is independent of temperature

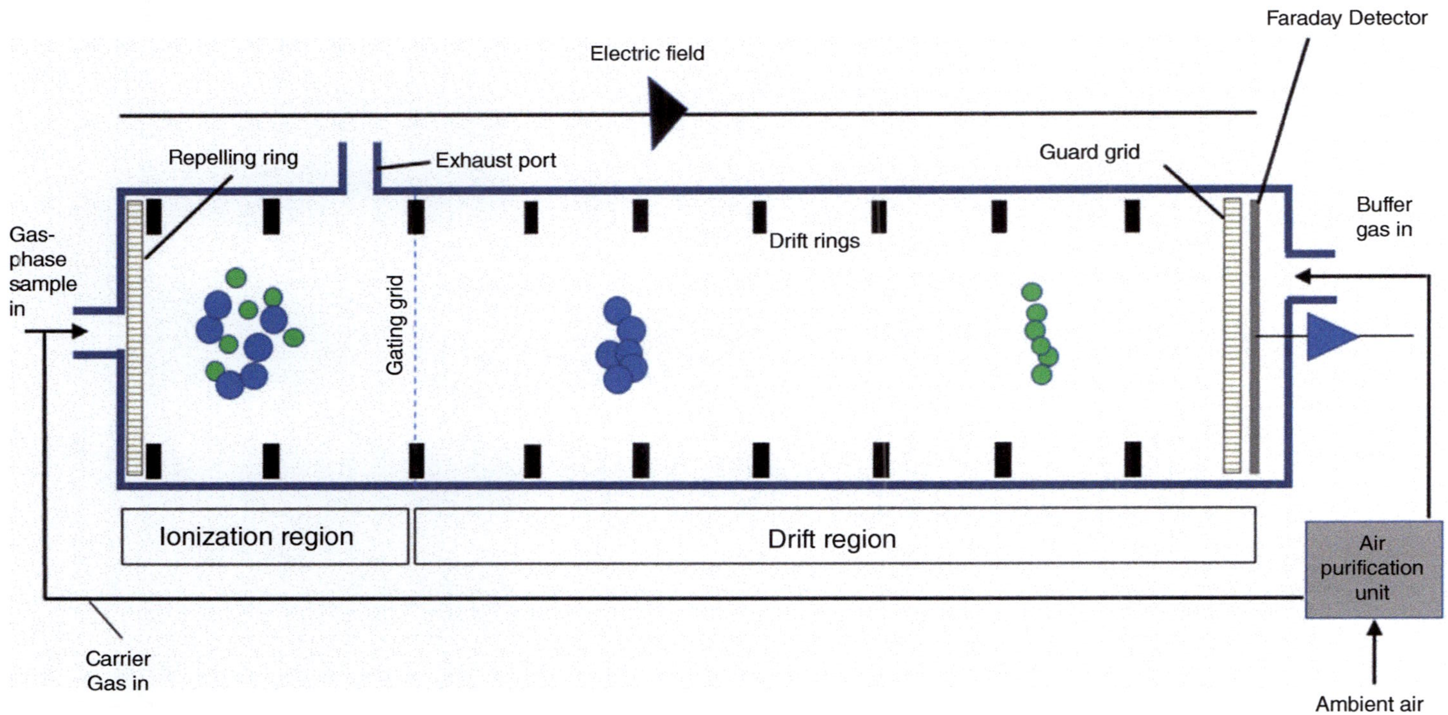

Figure 17.4 Schematic of drift-tube IMS where the sample is brought in with a clean carrier flow.

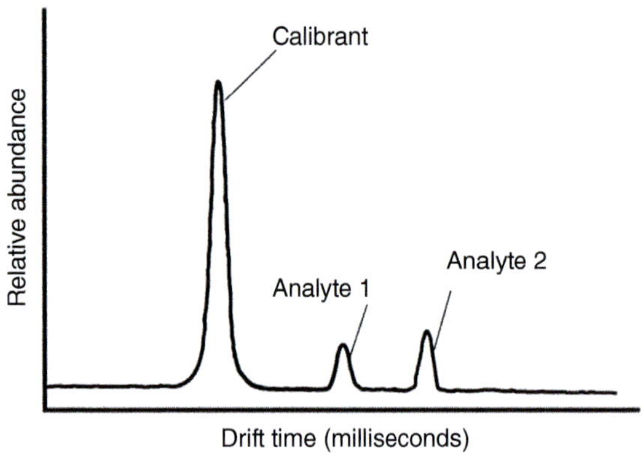

Figure 17.5 IMS spectrum of a two-component sample.

and pressure.

$$K_0 = \frac{L}{t_d E}\left(\frac{P}{760}\right)\left(\frac{273}{T}\right) \quad (17.3)$$

As the ions travel through the drift tube, they are separated based upon their different mobilities which are a function of their size and shape. The greater the number of collisions and interactions the ion has with the buffer gas at a given electric field strength, the slower it will move. Larger ions move more slowly than smaller ions. IMS does not measure the properties of a single ion, but a group of gaseous ions referred to as a swarm [11] (p. 1). It is a swarm of ions with similar mobility that move through the drift tube with average mobility. The effective cross-sectional area is approximately proportional to the mass of the ion, and generally, a good correlation to first order is observed for $1/K_0$ as a function of MW. Figure 17.6, which was compiled from data previously published by Karpas [35] and Metternich et al. [36], shows the relationship between $1/K_0$ and MW. Linear regression was performed for the combined data sets as if they were one series, and the R^2 value calculated was 0.9656. Since molecules with the same MW can have different average cross-sectional areas based on structure, it is not surprising that this relationship is not exactly one to one. Table 17.2 provides a general sense of how mobility scales with MW, detailing drift times measured for different substances along with their K_0, $1/K_0$ and MW. With regard to size and shape, reduced mobility coefficients are influenced by ion mass in cases in which a linear relationship exists within a homologous series. However, ions of the same mass but different functional groups, or ions of the same functional groups but different geometrical arrangements (isomers), often exhibit different K_0 values, reflecting the influence of size and shape on mobility [15].

Ion swarm peaks will have a full width at half maximum (FWHM) of between 150 and 300 μs for a gating pulse of 150 μs. Figure 17.7 shows an IMS spectrum with three peaks with K_0 values of 1.8000, 1.2000, and 1.000 associated with molecular ions with amu values of approximately ~121, 262, and 346, respectively. FWHM for the peaks are 200, 250 and 280 μs, respectively.

The exact relationship between mobility, K, and chemical identity in IMS is not completely understood. So, the expected mobility of an ion cannot be accurately determined from first principles. The mobility of an ion is dependent on conditions used (electric field strength, pressure, temperature, buffer composition) while for mass spectrometry the mass-to-charge ratio (m/z) is independent of measuring conditions. Historical attempts to apply theories regarding mobility indicated modifications to these theories were necessary. The evolution of many of these modifications, starting with Langevin's initial theoretical proposal in 1903 and leading to Eq. (17.4), which evolved from studies of mono- or diatomic ions in pure gases at subambient pressure, are summarized in Karasek's 1974 paper "Plasma Chromatography" [15, 37].

$$K = \frac{3e(2\Pi)^{\frac{1}{2}}(1+\alpha)}{16N(\mu k T_{\text{eff}})^{\frac{1}{2}}\Omega_D(T_{\text{eff}})} \quad (17.4)$$

Table 17.1 Field-deployed IMS systems categorized by IMS type and system weight.

IMS type	Vendor	≤1.2 kg	≤3.5 kg	≤6 kg	≤13 kg	≤23 kg
DTIMS	Smiths Detection	LCD3.3 [16]	SABRE 5000 [17]	GID-3 [18]	IONSCAN 600 [19]	IONSCAN 500DT [20]
	Rapiscan			Hardened Mobile Trace [21]	Itemiser DX and Itemiser 4DX [22]	
	Leidos				H150 [23]	B220 [24]
	Bruker [25]	uRAIDplus	RAID M100 RoadRunner		DE-tector flex	DE-tector
	Nuctech			TR1000DB-A [26]	TR2000DB-A [27]	
AIMS	RAE Systems	ChemRAE[a] [28]				
	Environics Oy	ChemPro100i [29], ChemProX [30]				
FAIMS	Owlstone Inc.			OwlSens-C [31]	OwlSens-T [32]	
	Chemring Sensors and Electronic Systems	Juno [14]				

a) This product is an original equipment manufacturer version of the Environics ChemPro 100.
Source: Based on MRIGlobal [33].

where,

e = electron charge
α = correction factor
N = number density of neutral gas molecules
μ = reduced mass ($1/u = 1/m + 1/M$), where m is the mass of the ion, M is the mass of the neutral gas molecules
k = Boltzmann constant
T_{eff} = effective temperature of an ion based on thermal energy and energy acquired by an electric field
$\Omega_D(T_{eff})$ = effective cross-section of the ion which is a function of the T_{eff}

Since drift velocities are sensitive to the exact identity of ions in the swarm, which is affected by the composition of the gas atmosphere and by changes in temperature, Eq. (17.4) is limited. Thus, the relationship between K and Ω_D is inaccurately described or incomplete for large organic ions where Ω_D is influenced by polar molecules or

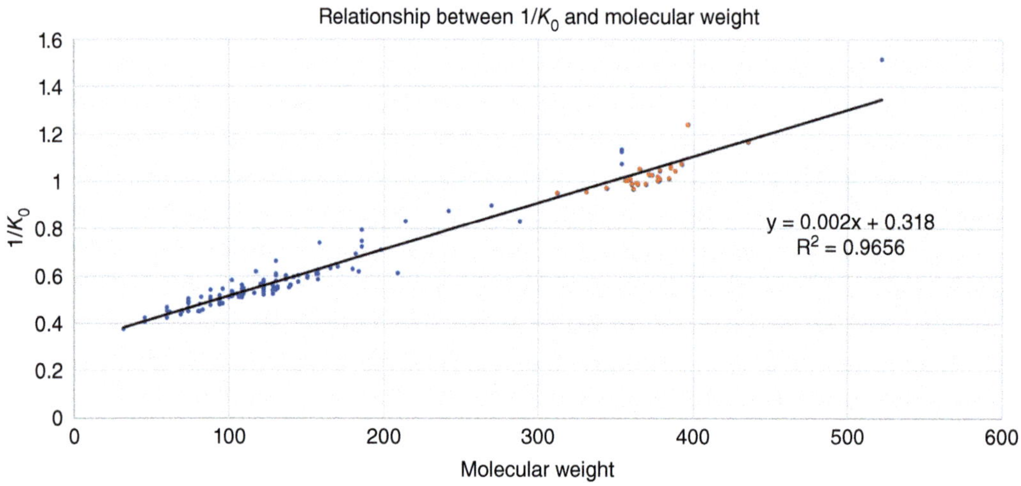

Figure 17.6 Relationship showing the dependence of $1/K_0$ on MW from previously published data for aliphatic/aromatic amines (blue) and synthetic cannabinoids (orange) in air or nitrogen between 150 and 250 °C. *Source*: Karpas [35]; Metternich et al. [36].

Table 17.2 Drift times associated with K_0 values for an 8-cm drift tube operated at 200 °C at atmospheric pressure in the air.

K_0	$1/K_0$	t_d (ms)	~MW
2.400	0.4167	7.696	50 ± 8
2.200	0.4545	8.395	69 ± 8
2.000	0.5000	9.235	92 ± 9
1.800	0.5556	10.261	121 ± 10
1.600	0.6250	11.543	156 ± 11
1.400	0.7143	13.192	201 ± 13
1.200	0.8333	15.391	262 ± 15
1.000	1.000	18.469	346 ± 18
0.800	1.250	23.087	473 ± 22

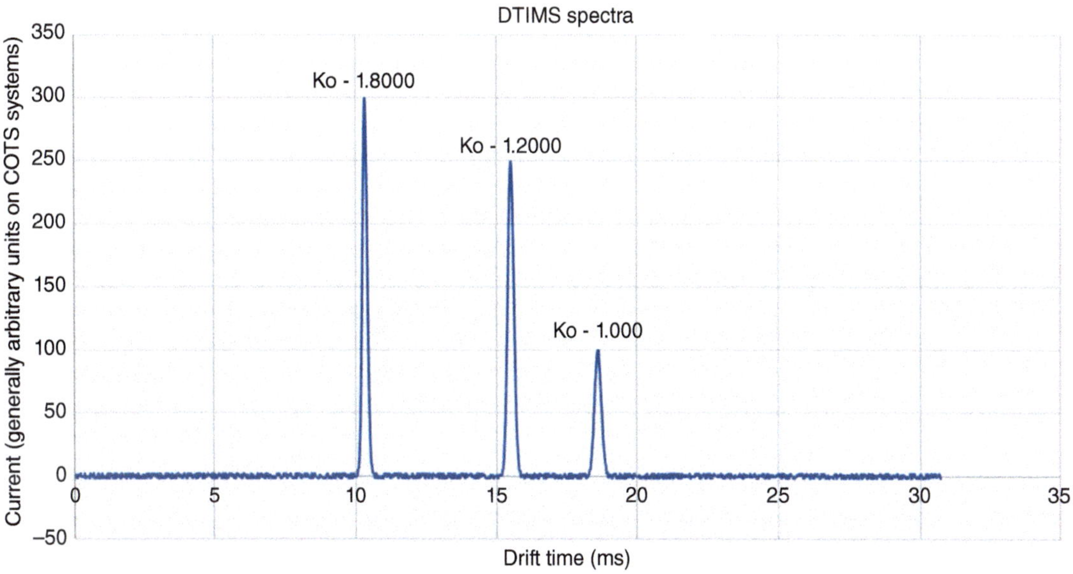

Figure 17.7 Example IMS spectrum for ions with K_0 values of 1.800, 1.200, and 1.000.

moisture in the drift region. There is no comprehensive model that includes ion-neutral associations in polarizable gas atmospheres and the role of temperature in controlling Ω_D and, thus, K. The distribution of charge within a large organic ion also affects the observed mobility values, once again through the formation of clusters. This is not included in current formulas for mobility. Consequently, Eq. (17.4) and other formulas for mobility are incomplete descriptions of mobility measurements in the air at atmospheric pressure, and the correlation of ion structure, collision cross-section, and mobility over a range of experimental conditions should be considered an underdeveloped topic in IMS.

Electric field strength also has an impact on mobility, which is a function of the normalized field strength, E/N, where N is the gas number density. Field-deployable IMS systems operate at atmospheric pressure. So, N at 1 atm and 0 °C is approximately 2.69×10^{19} cm^{-3} for an ideal gas at standard temperature and pressure. The units for E/N are Townsends (Td), where 1 Td = 10^{-17} Vcm^2. This means for $E = 250$ V/cm at STP, $E/N = 0.93$ Td. The relationship between K_0 and E/N is shown in Eq. (17.5) [8] (p. 57).

$$K_0\left(\frac{E}{N}\right) = (K_0)\left(1 + a\left(\frac{E}{N}\right)^2 + b\left(\frac{E}{N}\right)^4 + c\left(\frac{E}{N}\right)^6 \ldots\right) \tag{17.5}$$

Why does E/N impact K_0? Above a critical normalized field strength, $(E/N)_c$, the amount of energy imparted to the ion by the electric field due to collisions of the ions with the neutral buffer gas is non-negligible compared with the thermal energy of the buffer gas, and is not quickly dissipated. As a result, the effective temperature of the ion increases. $(E/N)_c$ is dependent on the ion, the buffer gas, and temperature of the buffer gas. Under low-field conditions (at which most traditional DTIMS and AIMS systems operate (E < 500 V/cm, E/N < 3 Td)), K_0 is independent of E and E/N. Changes in temperature and pressure result in changes to N, which will result in variations of E/N. Table 17.3 shows a variation in E/N for $E = 250$ V/cm for different temperatures and pressures. The Torr value of 760 corresponds to sea level and 525 Torr corresponds to 10 000 ft. For an IMS operating at 50 °C, the E/N field can vary from 1.10 to 1.59 Td when going from sea level to 10 000 ft. Since this is the low-field limit, K_0 is not impacted.

Between $(E/N)_c$ and about 100 Td (~20 000 V/cm at standard temperature and pressure), only terms up to $(E/N)^4$ are significant [38]. Three possible behaviors are observed for K_0 as a function of increasing E/N. These are

Table 17.3 Variation in *E/N* (Td) as a function of temperature and pressure for $E = -250$ V/cm.

Temperature (°C)	760 Torr	525 Torr
0	0.93	1.35
50	1.10	1.59
100	1.27	1.84

referred to in the literature as types A, B, and C behaviors. Type-A behavior is when K_0 increases, i.e., the effective cross-sectional area of the ion decreases. This behavior is the most common and is where clustering of a component in the buffer gas, i.e., water or added modifier with the ion, is reduced with increasing electric field strength. Type-B behavior is when K_0 first decreases and then increases. Type-C behavior is when K_0 decreases, i.e., the effective cross-sectional area of the ion increases. The differences in mobilities at high and low electric field strength are used in FAIMS systems to differentiate between different types of ions. The design of field-deployable instruments requires an understanding of how K_0 varies as a function of drift-tube conditions. Table 17.4 summarizes the impact of different operating conditions on mobility.

17.2.3.2 IMS Components

17.2.3.2.1 Sample Introduction

Samples for IMS analysis are either collected as vapors or microscopic particles. In either case, the sample matrix can be complex, and most commercially available portable IMS systems do not offer front-end separation mechanisms to isolate the target analyte prior to introduction into the IMS. Therefore, both the analyte and the sample matrix are introduced to the system during analysis. The sample matrix has the potential to impact ion formation and detection results and may enhance or suppress the signal of the analyte. To prevent this from happening, dopants are used to control the ionization reactions so that reproducible, stable analyte ions are formed regardless of the sample matrix. Without the use of dopants, the matrix of the sample or trace contaminants in the buffer gas will control the ionization chemistry. Fortunately, most of the common chemicals detected by IMS are selectively and preferentially ionized compared to the common real-world background material encountered during analysis.

The sample matrix can also cause unwanted contamination of the system. IMS particle-detection systems which thermally desorb samples will typically have hot inlets and reaction regions which minimize dirt ingression and maintain unit sensitivity. From a system-maintenance perspective, the IMS front end is ideally designed to minimize the accumulation of dirt and allow easy purging of the inlet and reaction region. This is particularly important for high throughput applications like those within the aviation-security industry, such as checkpoint and checked-baggage screening. Many systems used for high throughput swab analysis have built-in, automatic bakeout procedures which are executed during low-use periods. Bakeout for IMS systems is a term that is applicable to both vapor- and particle-detection systems and refers to a process where the temperature of the drift tube and other system components are increased for a specified time period to allow contamination and other debris to volatilize and pass through the system. Automatic bakeouts keep these systems working effectively at optimum sensitivity.

Sample introduction using vapor sampling is performed when the target analyte has a reasonable vapor pressure such that ppb (parts per billion) levels of material are introduced to the system during the sampling cycle. To prevent overload, some IMS systems offer a feature that causes automatic sampling to shut down if an overload is suspected. IMS systems are trace detectors with sensitivities in the low ppb range (low ng–pg range). In the real world where sample matrix and concentration are frequently unknown, the potential of the system analyzing high-concentration samples is real and can overload the system. Well-designed IMS systems have factored this in

Table 17.4 Impact of conditions on IMS response for DTIMS and FAIMS systems.

Condition	DTIMS: Low field (<500 V/cm, @ 1 atm, <3 Td)	FAIMS: Electric field is switched between high- and low-field values during the analysis
Ion mobility (K_0)	Independent of E	The function of E, can increase, decrease or increase/decrease
Energy from an electric field	Negligible	Non-negligible, resulting in high effective temperature for ion during high-field condition which can decrease K_0 under high-field operation
Solvation by neutral gas molecules (water)	Increases effective cross-sectional area of the ion which will decrease K_0	Can help increase differences in K_0 between high and low electric fields
Temperature	Drift times of ions shift to lower values due to reduced air density as drift tube temperature increases. Reduced mobility may stay constant if there is no impact on clustering. Clustering is more significant at lower temperatures <150 °C	Can impact the differences in K_0 observed between high and low fields
Pressure	As pressure decreases, drift times of ions decrease	
Ion size	As ion size increases, K_0 decreases	With larger ions, a charge may be delocalized more and differences in K_0 may not be as large between high and low fields
Impact of clustering dopant	Ion size ↑ due to clustering and adduct formation, ↓ mobility	The ion charge is highly delocalized to minimize solvation effects; Declustering occurs which ↓ ion size and ↑ mobility
Adduct formation with ion (e.g., chloride or nitrate adduct formation)	Leads to new ion being formed, e.g., RDX*Cl^-, RDX*NO_3: ↑ ion size ↓ mobility	If electric field strength is high enough to dissociate adduct from the neutral analyte, then large mobility ↑. Otherwise, only minor changes to mobility are observed

during development. In these instances, the sampling cycle, which usually is preset for a specified amount of time for a given analysis, will automatically stop the sampling, but allow the IMS analysis to continue. Vapor sampling works well and is routinely deployed for the IMS detection and identification of both CWAs and TICs.

For semivolatile compounds, such as drugs and explosives, trace amounts of microscopic particles are collected by swabbing a surface. The swab substrate may be made of a number of different materials including paper, cotton, Nomex ®, Teflon®-coated glass fiber, or other materials. The swab is heated at the IMS inlet to drive the sample into the vapor phase for analysis. Some modern IMS systems designed for trace-particle sampling use low-thermal-mass inlets with sophisticated inlet designs. These inlets and designs help minimize contact between the sample swab and the sample-inlet mechanism to reduce contamination, prevent system overload, and even lower power requirements. Some systems also offer variable-temperature desorption which can help optimize selectivity.

In most cases, with commercial portable IMS systems, sample flow into the system is performed either in the presence of ambient air or as a diluent in a clean carrier gas. Figure 17.4 shows a schematic of a single-tube DTIMS where a clean carrier gas is used. In this configuration, ambient air is dried and purified and it serves as both carrier and buffer gas. If the water content in the sample stream is not controlled, variable sensitivity may be observed for some compounds.

17.2.3.2.2 Ionization and Ionization Sources

17.2.3.2.2.1 Ionization All commercially available IMS systems use a soft ionization process based on APCI. When performing APCI, the sample enters the ionization region where primary reactant ions created by the ionization source in the presence of the carrier gas and chemical dopants selectively ionize the sample components. The carrier gas is typically clean dry air (<10 ppm H_2O) which consists of nitrogen, oxygen, CO_2, water vapor, and other trace gases. The selective and reproducible production of analyte ions in the air is critical to the success of IMS [39]. Also critical to its success is the ability to perform the analysis at ambient pressure with preferred ionization for analytes of interest over general interferents found in the sampling environment. Chemical dopants that enter the ionization chamber can be used to control the reaction chemistry by creating reproducible and stable analyte ion species and suppressing ionization of interferent species. They may also be used to create calibrant peaks.

To produce reproducible and stable ions, instruments have been designed to provide a controllable ionization environment. During the ionization process, the primary reactant ions formed depend on the ionization source and the composition of the carrier gas: Subsequently, the types of analyte ions produced depend on the nature of the primary reactant ions formed and dopants present. Of course, sample-matrix components are also part of the process. Dopant chemistry is a complicated but very significant aspect of instrument design and method development [39, 40]. Higher-proton-affinity dopants, such as ammonia [41], nicotinamide [42], or isobutyramide [43] can be added to suppress ionization of low-proton-affinity species so that selectivity is improved, and false alarms are minimized. The addition of high-electron-affinity chlorinated dopants, such as hexachloroethane and dichloromethane [41], are added to produce chloride ions to create an environment for the most effective ion chemistry. If dopants are not added to the system, reactant ions produced will be uncontrolled and what is present in the ambient background of the sample or the air stream. Typical ions formed in negative and positive modes along with relevant examples of ion complexes are shown in Tables 17.5 and 17.6, respectively. RDX, PETN, TNT, EGDN, and DNT are all explosive substances.

Table 17.5 Typical ions formed in negative-ion mode.

Basis of formation	Short-form	Example	Comment
Dissociative charge transfer	X^-	Cl^-	
Adducts with react-ions (typically Cl^-)	$(M + Cl)^-$	$(RDX + Cl)^-$	Source dopant added
Adducts with nitrates NO_3^-	$(M + NO_3)^-$	$(PETN + NO_3)^-(HNO_3 + NO_3)^-$	Source typically ambient nitrates
Adducts with NO_2^-	$(M + NO_2)^-$	$(RDX + NO_2)^-$	
Adducts with O_2^-	$(M + O_2)^-$	$(RDX + O_2)^-$	Typically suppressed in the presence of nitrates and chlorides
Proton abstraction	$(M - H)^-$	$(TNT\text{-}H)^-$, $(DNT\text{-}H)^-$	
Dimers	$(M2 + X)^-$	$(2RDX + Cl)^-$	Unstable at higher drift tube temperatures
Decomposition	$(NO_3)^-$	EGDN, NG	

Source: Based on Kozole et al. [43].

Table 17.6 Typical ions formed in positive-ion mode.

Basis of formation	Short-form	Example	Comment
Protonated molecules	$(M+H)^+$	$(\text{cocaine}+H)^+$	Proton affinity
Adduct formation	$(M+A)^+$	$(\text{tabun}+NH_4)^+$	
Proton-bound dimers	$(2M+H)^+$	$(2\ \text{tabun}+H)^+$	High temperatures can destabilize dimers

Source: Based on Puton and Namieśnik [44].

In addition to controlling reaction chemistry, when present in the ionization chamber, dopants may also be added to the buffer gas to modify its composition. As an example, although not commonly considered a dopant, water vapor is a chemical that is used to modify buffer-gas composition to optimize IMS analysis.

17.2.3.2.2.2 Ionization Sources Virtually, all IMS systems produced prior to the year 2000 used a radioactive source. These sources played an important role in the fundamental studies of IMS [15] (pp. 22–23, 38–39). Radioactive sources, such as the commonly used ^{63}Ni source, are regulated in the United States by the Nuclear Regulatory Commission (NRC), and in other countries by similar national organizations. Nonradioactive sources based on plasma discharge, photoionization, and high-energy photoionization (HEPI) have become more favorable options from regulatory and maintenance perspectives. Due to the ability to perform IMS without the use of a radioactive source, the burdens historically associated with the deployment of IMS systems containing a radioactive component have been removed.

17.2.3.2.2.2.1 Radioactive Sources Radioactive ionization sources, such as ^{63}Ni and ^{241}Am, used in IMS systems have many operational benefits. Typically, they are coated metal foils inserted into the reaction chamber: They are simple, stable, have long lifetimes, and require no power supply. They emit high energy particles, such as beta particles from ^{63}Ni, and alpha particles from ^{241}Am. The kinetic energy of these emitted particles exceeds the ionization thresholds of most analytes of interest and background air components [45–47]. These sources usually have high ionization efficiency, but low ionization selectivity [39, 40, 48, 49].

The most favored radioactive ionization source is 10 mCi of ^{63}Ni coated as a thin layer on a metal strip, generally nickel or gold. [50] The half-life of ^{63}Ni is about 100 years. The maximum energy of the electrons emitted from the ^{63}Ni is 67 keV, with average energy near 17 keV. Almost all the energy of this source is dissipated in the air at an ambient pressure within 10 to 15 mm from the surface of the metal. A common geometry of this source in IMS systems is cylindrical [15, 51] (p. 137). ^{241}Am has also been used as a source in IMS, usually in AIMS platforms. ^{241}Am was used as the ionization source in the M43A1 detector, one of the earliest portable IMS systems, which was deployed to military users for CWA detection [4]. Its half-life is about 432 years. Alpha particles emitted from ^{241}Am are highly energetic, with energies above 5.4 MeV. They have a short effective range in the air so that ionization is efficient in small-volume sources [52].

During the ionization process, beta particles emitted from the ^{63}Ni source collide with neutral molecules of O_2 and N_2 in the air causing electron ejections as they travel. This results in the formation of a plasma consisting of O_2^+, N_2^+, and thermalized electrons [15] (pp. 22–23, 38–39). Similar species are produced for ^{241}Am, except the high-energy alpha particles strip electrons from a large number of O_2 and N_2 molecules, producing a plasma consisting of thermalized electrons along its travel path, as well as N_2^+ and O_2^+.

These primary reactant ion species then interact with air to create reactant ions as follows:

(1) The thermalized electrons react with O_2 to create O_2^-, which is hydrated as $O_2^-(H_2O)_n$: The degree of hydration depends upon temperature and water-vapor levels.

Table 17.7 Common ionization sources used in commercially available portable IMS systems.

Manufacturer	Instrument	Ionization source
All	Most pre-2000 instruments	Radioactive ^{63}Ni
Bruker [53]	RAID-M	Radioactive ^{63}Ni
Bruker [54, 55]	DE-tector, DE-tector-flex Raid-M_{NR}, RoadRunner	High Energy Photoionization (HEPI)
Environics Oy	AIMS platforms	Radioactive ^{241}Am
Leidos	B220	Photoionization (pulsed UV ion source, photonic ionization) [56]
Rapiscan	Itemiser DX	Radioactive ^{63}Ni [57]
Rapiscan	Itemiser 4DX	Photoionization (krypton lamp) [56, 58]
Smiths Detection	Ionscan 600	Dielectric barrier discharge [56]
Smiths Detection	LCD	Pulsed corona
Smiths Detection	Ionscan 500DT	Radioactive ^{63}Ni

(2) The positive ions N_2^+, N_4^+, O_2^+, O_4^+ are created and react with water vapor to generate H^+ with water clusters of $H^+(H_2O)_n$, where $n = 1$–10: The degree of clustering depends upon temperature and water-vapor levels.

Even though radioactive sources offer operational advantages, and the behavior of the radioactive sources used in IMS systems is well understood, the regulatory burdens associated with radioactive sources have resulted in the movement away from these sources and the development of nonradioactive sources for use in IMS systems.

17.2.3.2.2.2.2 Nonradioactive Sources Nonradioactive sources used in commercially available IMS systems include corona discharge and barrier discharge ionization sources, and photoionization sources. Table 17.7 details some of the available portable systems and the different radioactive and nonradioactive sources used in each. The details of the exact operation of these nonradioactive sources are usually not disclosed by the vendors beyond what is available in the patent literature. A few have been described below. It is important to recognize among instrument platforms that there may be minor-to-significant differences in performance based on the primary ionization source used.

17.2.3.2.2.2.2.1 Corona Discharge A corona discharge is, by definition, a gas discharge where the geometry confines the gas-ionizing processes to high-field ionization regions around the active electrode. To form a corona-discharge IMS source, a sharp needle or thin wire is placed 2–8 mm from a metal plate or discharge electrode with a voltage difference of 1–3 kV between the needle and the plate. An electric discharge develops in the gap between the needle or the wire and the opposing conductor [15] (p. 138).

Corona sources use a strong electric field to accelerate electrons toward the opposite electrode. The electrons collide with the neutral gas molecules, ejecting more electrons along the way, creating a plasma of O_2^+, N_2^+, and thermalized electrons. In positive-ion mode, these primary reactant species subsequently form similar reactant ions to those formed when using a radioactive source (N_2^+, N_4^+, O_2^+, O_4^+), along with similar water clusters of $H^+(H_2O)_n$ [59]. In negative-ion mode, ionization behavior is different from that of radioactive sources due to the formation of NO_3^- [15, 60] (p. 137), [61]. To achieve a more similar ionization profile to that observed in radioactive sources, a pulsed voltage across the two electrodes is imposed, which causes an ion avalanche with enough

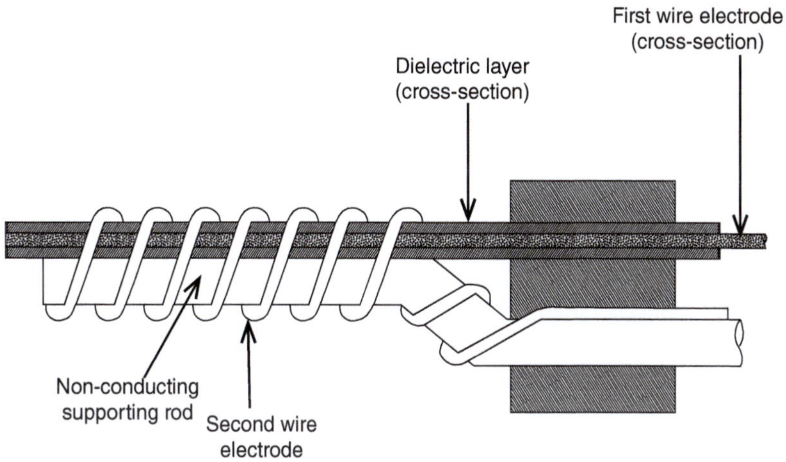

Figure 17.8 Schematic of a DBD source. *Source*: Kubelik et al. [65].

energy to produce $[(H_2O)_nO_2]^-$ in analytically useful quantities, without producing significant levels of NO_2. Further, by pulsing the corona, the lifetime of the electrodes, and, hence, long-term stability of the system, is enhanced [61]. Commercially, corona discharge as an ionization source in field-deployed IMS was first proposed by Graseby Dynamics (now Smiths Detection) [62]. Subsequently, the first commercial IMS system using a corona discharge ionization source was likely the Smiths Detection LCD in 2002 for the government of the United Kingdom, and, subsequently, for the government of the United States as the JCAD in 2002 [11] (p. 28). There are some disadvantages to corona discharge sources including the need for an external power supply, corrosion of components, maintenance of the discharge, and the formation of corrosive chemical vapors, such as NO_x and ozone. Stability may be degraded by corrosion of the needle [62].

17.2.3.2.2.2.2.2 Dielectric Barrier Discharge Dielectric barrier discharge (DBD) sources are discharge ionization sources which have an insulating (dielectric) material in between metal electrodes, typically made from glass, quartz, ceramics, enamel, mica, plastics, silicon rubber, or Teflon with a gap distance between 0.1 and 10 mm [63, 64]. DBDs have low power consumption, long lifetimes, and work at ambient pressures. For these reasons, they are favorable for incorporation into field-deployable IMS instruments. The DBD source is inserted into the ionization region of the IMS.

DBDs are self sustaining in electrode configurations containing an insulating material in the discharge path. This so-called dielectric barrier is responsible for a self-pulsing plasma operation and, thus, the formation of nonthermal plasma at normal pressure [64]. The discharge creates electrons with high average kinetic energy (1–10 eV), metastable species, and high energy photons. The ionization profile for DBDs similar to that of corona discharge sources, and like with the corona, an advantage of pulsing is a reduction in production NO_3^-. A primary analytical benefit of a DBD ionization source is a larger and more controlled ionization region.

DBD ionization devices may come in planar and cylindrical configurations. With cylindrical configurations, the plasma can be generated in the volume between the electrodes or at the surface of the dielectric. With surface-based DBD sources, microdischarges are generated at the surface of the dielectric, resulting in a more homogenous plasma source than the volume-based DBD sources. A surface-based DBD source was recently proposed [65]. Figure 17.8 shows a schematic of this DBD which consists of a metal wire electrode coated with a dielectric layer and a second wire electrode wrapped around the coated metal wire electrode. Ionization starts with a short burst of a high-voltage pulse which initiates a plasma discharge.

17.2.3.2.2.2.2.3 Photoionization Most organic molecules will ionize when exposed to ultraviolet (UV) light in the photon range of 8–12 eV (155–103 nm) [49]. Application of an electrical discharge in the lamp in the presence of a noble gas at low pressure will result in the emission of photons. The formation of positive ions with photons has been described as direct ionization through the reaction described in Eq. 17.6 where $h\nu$ is the photon energy and M is the neutral molecule.

$$M + h\nu = M^{+} + e^{-} \tag{17.6}$$

The wavelength emitted by a UV lamp depends on the filler noble gas (Ar, Kr, H, Kr, or Xe) used and the lamp window material (LiF, MgF_2, CaF_2, CaF_2, or Al_2O_3). The Kr-filled UV lamp (10.6 eV, 116 nm) with an MgF_2 window is the most common lamp with the strongest output and longest lifetime. A Kr lamp can ionize most hydrocarbons, alcohols, aldehydes, ketones, and esters if they do not contain chlorine, fluorine, or bromine. It can also ionize all amines and sulfides [66]. The probability of direct ionization is very low, especially for analytes present in a small proportion compared to other components of the carrier gas. Therefore, to promote the ionization efficiency of analyte molecules, a photoionization dopant, such as toluene or acetone, can be used. The Rapiscan Itemiser 4DX uses anisole as the photoionization dopant to create M^+ or MH^+ reactant ions in positive ion mode, and low energy electrons in negative mode. A metal surface can also be used with a UV lamp using the photoelectric effect to release low energy electrons from the metal surface. The major advantages of UV lamps are their small size, simplicity of use, and commercial availability. The disadvantage of UV lamps is a short operating lifetime. For instance, the Kr UV lamp used in the Itemizer 4DX offers a three-month lifetime [67]. These lamps also suffer a loss of 50% intensity through every 1.5 mm of purified air [68].

17.2.3.2.3 Ion Injection

In commercial DTIMS systems, ion injection is performed using gating-grid ion shutters. Ion gating currently is based on the presence of two closely spaced wire grids which span the cross-section between the ionization region and the drift tube. If the wire grids are in the same plane, this type of ion shutter is known as a Bradbury–Nielson (BN) gate. If the wire grids are in different planes, it is known as a Tyndall–Powel (TP) gate. The wire used in making BN and TP gates are typically less than 100 μm in diameter with wire spacing of ~1 mm. For the TP gate, this distance refers to the distance of the nearest wire in the adjacent plane. TP gates are typically used in commercial equipment due to their robustness as well as their ease and cost of manufacture.

Gate pulsing is important to system performance. Application of an electric field gradient across the wire grids results in an orthogonal force on the ions traversing the gating grids. This electric field is modulated to open or close the gate. In the closed state, the field gradient is strong enough to cause the ions to strike the gating grids; in the open state, the ions traverse the gate into the drift region. The gating pulse is a primary factor that impacts the FWHM of the ion peak which will be discussed later in the "Resolving Power, Resolution, Duty Cycle, and Coadding Scans" section of this chapter. In addition to the impact on FWHM, the timings of the opening and closing of the gate become important as ion mobility decreases because the slowest moving ions need to pass through the gate before it closes. As gate-pulse widths go below 100 μs, this can be a problem. For conventional gates and peak mobilities between 0.5 and 2.5 cm^2/Vs, gating pulses should be 100 μs or greater. Otherwise, slower mobility ions will be filtered out resulting in very small ion signals [69]. Typically, ions are gated between 100 and 300 ms every 15–30 ms; hence, only ~1% of ions are injected into the drift tube. For AIMS and FAIMS systems, where ions are continuously injected, theoretically, 100% of the ions could be injected, depending on inlet aperture design.

17.2.3.2.4 Drift Tube

The drift tube is the most important part of the IMS; it is where separation occurs. The objective of the IMS design is to create a homogenous electric field inside the drift tube [70]. Ions enter the drift tube in the presence of a carrier gas, which in commercial instruments is typically clean dry air. In addition, in properly designed DTIMS systems, a counterflow of clean dry air enters at the end of drift region near the detector to provide counterflow drift gas. The counterflow is used to ensure no neutral molecules from the ionization enter the drift tube. Prior to the introduction of air to the IMS, it is cleaned to remove volatile organics and reach water-vapor levels down to approximately 1–10 ppm. Control of air quality is important to ensure reliable ion formation and mobility. Air purification is one of the key development areas critical to portable systems and will be discussed later. It is also possible to use other carrier gases, such as nitrogen, helium, or argon, but their use requires the presence of a pressurized tank which is not desirable for field-portable systems. The drift tube is constructed of a series of ring electrodes creating a uniform field usually held between 200 and 300 V/cm. The positioning of the ring electrodes controls the uniformity of the field. The closer and greater the number of ring electrodes, the more uniform the field. Uniformity of the electric field is important because nonuniform fields act like a defocusing lens that degrades resolution. The mobility of the ion is dependent upon the electric field voltage in accordance with Eq. (17.7) [71] which shows the relationship among the ion mobility, K, the length of the drift tube, L, the electric field gradient E, and the measured drift time t_d. This relationship can also be stated in terms of the applied voltage (V) across the drift tube.

$$K = \frac{L}{Et_\mathrm{d}} = \frac{L^2}{Vt_\mathrm{d}} \tag{17.7}$$

Typical drift tube lengths are 10 cm or less. Very short drift tube lengths require very fast ion shutters and amplifiers for acceptable separation capability. Typically, drift times for most commercial instruments with drift tubes between 4 and 10 cm is between 2 and 20 ms. The mobility of ions measured by field-portable equipment has molecular masses between 30 and 500 amu and mobilities between 2.5 and 0.8 cm^2/Vs. Table 17.2 shows the expected drift times for peaks for K_0 values between 2.4 and 0.8 cm^2/Vs for an 8-cm drift tube at 200 °C with a gating pulse of about 150 μs.

17.2.3.2.5 Guard Grid

The guard grid is located very close to the Faraday plate detector and has an applied potential such that the field gradient between the guard and detector is much higher than the drift tube field gradient. The guard grid is required to capacitively uncouple the approaching ion swarm from the Faraday plate. If the guard grid was not present, a current would be induced on the Faraday plate before the ion swarm came into contact with it, resulting in a non-Gaussian peak with tailing on the short-drift-time side. With the guard grid, the ion swam is rapidly accelerated to minimize this effect resulting in Gaussian peak shape.

17.2.3.2.6 Ion Detector – Faraday Plate

The ion detector at the end of the drift tube is a critical component of overall IMS sensitivity, and in portable IMS systems, it is a Faraday plate. A Faraday plate is used, as opposed to a Faraday cup, since the Faraday plate operates at atmospheric pressure. The detector converts the swarm of gas-phase ions into an electrical signal that provides both time and amplitude information. To ensure sharp and properly formed peaks, a guard grid is placed before the Faraday plate and held at the potential to quickly accelerate the ions to the plate, negating peak broadening due to capacitive charging.

The ion mobility spectrum is a plot of current as a function of time. Typical IMS peaks have FWHM values between 100 and 500 μs, and the Faraday plate offers a fast rise time (<10 μs), and fast recovery times, which are essential for minimizing peak broadening. Ion currents detected at the Faraday plate are on the order of a few nanoamps. For a typical IMS peak which is 200-us wide, Johnson noise ($\sim 1 \times 10^{-12}$ A at 25 °C) limits the detection. For a single scan, 100 ions need to reach the detector to generate a signal [15] (pp. 155–158). The diameter of the Faraday plate, compared to the diameter of the drift tube, can impact peak width if the electric field near the outer diameter of the tube is not homogenous. This can be mitigated by using a Faraday detector with a smaller diameter than the drift tube. In these situations, a compromise between signal and resolution is required [70].

The Faraday plate is ideally designed and manufactured to have a long life and no maintenance issues. Some IMS systems will heat the Faraday plate to ensure neutralized ions are removed. Otherwise, samples which are thermally desorbed and release semivolatile species can condense on the detector degrading sensitivity over time. Instrument designers will put a lot of careful attention into the preamplifier electronics associated with the Faraday plate in order to reduce electronic noise and remove high-frequency noise. Care must also be taken to minimize any time delay introduced by the electronics or data processing which can artificially increase the measured drift time.

17.2.3.2.7 Air-Flow and Air-Purification Systems

The key air flows in the IMS are the drift flow, sample flow, and exhaust flow. The drift flow should be countered to the ion flow in the drift tube to prevent neutrals in the ionization region from entering the drift tube. If neutrals do enter the drift tube, then uncontrolled ion clustering will occur which can result in unacceptable drift-time variability. The sample-flow medium must be clean and dry to prevent ion clustering and produce reliable and stable drift times. The exhaust flow is used to remove the unionized sample from the reaction region.

Clean, dry air is the primary support medium for these flows. Ambient air is typically pulled into the system and cleaned and dried internally. The ability to use ambient air is a very valuable field-deployment feature since gas accessories are, therefore, not required; however, keeping the airflow through the system clean and dry can be challenging. Medium- and high-performance IMS systems typically require water content to be <10 ppm for good performance. Ambient air water-vapor content can vary from 1000 to 40 000 ppm (0.02–36 g/m^3 of water vapor over the typical environmental conditions encountered from −20 to 50 °C).

To clean and dry the air, a variety of approaches are used. For small handheld systems, disposable molecular sieve packs are used and need to be replaced regularly. For these systems, regenerative air purification units currently are not an alternative due to power restrictions. Since most of the water vapor that enters the system will occur during sampling, disposable-sieve lifetime may be extended by minimizing the amount of external air which enters the system during sampling, or by recirculating the existing air supply. Historically, thin permeable membranes were used across the inlet to allow the sample to get through while minimizing water vapor ingression. Challenges to this approach, though, are that the membrane will create a delay in sample entry, and can also prevent some substances from entering the system. Smiths Detection developed an alternative method to limit the amount of water that enters the sample. In their LCD handheld IMS system, a small-volume pulse of sample vapor is pulled into the IMS through a pinhole [11] (pp. 48–49).

For larger systems, initially, charcoal and Drierite desiccant were used to clean and dry the ambient air. In these situations, the lifetime of the drying material could be extended by predrying the air using thermoelectric coolers or Nafion™ drying systems. Eventually, small regenerative air purification systems were developed. Typical regenerable systems have two sieve towers which can be operated in alternating sequence to provide clean air and to regenerate the sieve tower. This is particularly useful for systems which are used 24/7 at checkpoints, such as encountered in aviation security. Regenerable air purification systems also have the advantage of providing more consistent water-vapor levels, which reduces the width of detection windows, thus improving system performance.

17.2.3.2.8 Outgassing

IMS systems are trace detection systems and very careful attention must be paid to ensure all the components used to build these systems are clean at the trace level. If not, volatile interferents can degas from these components, which can impact the ionization process and ion interactions in the drift tube. Introduction of trace levels of contamination can have significant impacts on performance. The ability to build such systems generally requires the development of cleaning processes for all components, which consist of washing, baking, and testing to ensure no outgassing of contaminants that degrade detection performance will occur. The cleaning processes that instrument manufacturers have developed are a critical core competency which is required for the mass production of portable systems.

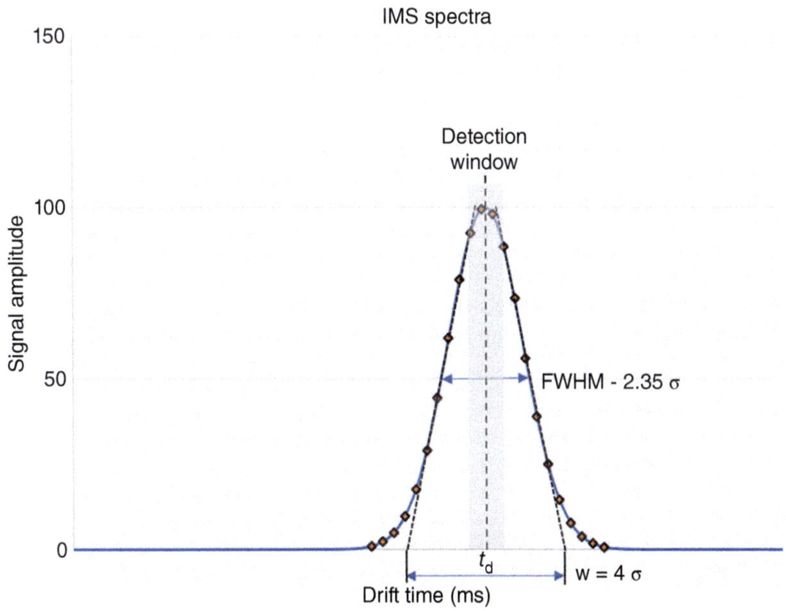

Figure 17.9 IMS peak characteristics including drift time (t_d), FWHM, sigma (σ), peak width (W) detection window, and digitization points.

17.2.3.3 Figures of Merit

17.2.3.3.1 IMS Peaks, Peak Windows, Calibration, and Informing Power

An IMS peak is typically Gaussian in shape with a drift time associated with its maximum amplitude. A peak's FWHM, base width (Δ W), and standard deviation (σ) are shown in Figure 17.9. During data acquisition, the analog signal is digitized based on the digitization rate of the instrument. Digitization is shown in Figure 17.9 as discrete points across the peak. The instrument can estimate the peak position by (i) using the digitization point at the maximum amplitude or interpolating the maximum based on the points around this maximum, or (ii) fitting the peak. Fitting the peak usually generates a more accurate peak-position value. Each instrument manufacturer uses its own method for determining peak position, which may be proprietary.

The expected peak position of the analyte peak is determined from the library K_0 value. For DTIMS systems, this can be converted into drift-time space using two methods. In method 1, the system is designed to measure E, P, and T in real time and calculate the expected drift time using Eq. (17.8). L is the length of the drift in cm, P is the pressure in Torr, T is temperate in K, t_d(analyte) is the drift time of the analyte, K_0(analyte) is the mobility of analyte, and E is the strength of the electric field.

$$\text{Method 1}: t_d(\text{analyte}) = \frac{L}{K_0(\text{analyte})E}\left(\frac{P}{760}\right)\left(\frac{273}{T}\right) \tag{17.8}$$

Uncertainty in the drift time of the peak arises from the precision and accuracy of the sensors used to measure the values and the slight variations associated with the manufacture of the drift tube. Note this equation is valid for large variations of P, variations of E in the low-field limit, and a narrow range of temperatures for a fixed IMS drift-tube temperature. If the drift temperature is not fixed and is dependent on the ambient environment, which could range from −20 to 50 °C, then the detection library will factor this into the K_0 value used by the above equation to predict the expected ion peak drift time. This method has been used very effectively by some of the small handheld platforms with an acceptable false-alarm rate.

A more accurate method using an internal calibrant of defined K_0 and measured drift time dispenses the need to measure T, P, and E. This method is based on the use of Eq. (17.9) to predict the expected position of the analyte peak. The t_d(calibrant) is the measured drift time of the calibrant, K_0(calibrant) is the known reduced mobility of the calibrant, K_0(analyte) is the known reduced mobility of the analyte peak and t_d(analyte) is the calculated expected position of the analyte.

$$\text{Method 2}: t_d(\text{analyte}) = \frac{K_0(\text{calibrant})}{K_0(\text{analyte})} t_d(\text{calibrant}) \tag{17.9}$$

This is achieved typically by using either an external or internal calibrant substance to generate a calibrant peak to compensate for slight differences between manufactured instruments (drift-tube length, electric field, temperature) and ambient-pressure impacts. For a list of some internal calibrants used, see Kaur-Atwal; [72] these are typically specific to a manufacturer. A key criterion that must be considered when using this method is how often the calibration is performed. For some platforms, it is done constantly; for others, it may be done every 5–15 minutes or at even longer intervals. The interval which is chosen is determined by how much drift may occur in P, T, and E and maintain an acceptable detection window.

There is sometimes a discrepancy between K_0 values established using applied detection platforms and those established in academic settings. This is particularly true if the K_0 value of the calibrant in the applied detection platform is different from the academic literature. The K_0 values used by the field-deployable instrument will still be internally consistent, and it is possible to convert from an instrument's K_0 scale to a literature-corrected K_0 scale by using a well defined literature value, such as shown in Eq. (17.10).

$$K_0(\text{corrected}) = K_0(\text{vendor}) \times K_0(\text{TNT} - \text{literature})/K_0\ (\text{TNT} - \text{vendor}) \tag{17.10}$$

Once peak position is determined, the measured drift time is compared to the expected drift time of the peak from the detection library. All instruments have variability associated with the drift-time measurement, which creates variability in the peak-position measurement. This uncertainty is typically determined experimentally by characterizing the peak under a range of conditions, e.g., at different loading levels and environmental conditions at which the instrument will be operated in the field. In addition to individual instrument variability, unit-to-unit variability must be accounted for in the calibration and detection algorithm. This can be achieved in two ways. The first way is to optimize the calibration either during the calibration stage of manufacturing or during subsequent service. The second is to account for instrument-to-instrument variability, generally, by expanding detection windows slightly. This second method is preferred since this means a common detection algorithm can be loaded into every instrument and no additional unit calibration is required. It also makes library expansion much easier in the field.

The determined uncertainty is then used to define the detection-window width (shown in Figure 17.9). It is important to note that uncertainty is also dependent on instrument design. Although attempts are made to reduce the uncertainty during development to ensure a minimal variation of critical parameters that impact drift time, uncertainty is inevitable. Drift time of an IMS peak is a function of the drift-tube temperature, pressure, electric field, and water-vapor content. If the measured peak position is within the detection window, the first peak-alarm criterion is met. Detection-window width impacts the false-rate performance of the instrument which to a first order scales with detection-window size. It also impacts the ability to uniquely identify compounds with similar K_0 values. For typical fixed-temperature commercial systems, detection width windows in the range of $\pm$25–75 μs, or approximately $\pm$0.001 K_0 are expected. If the drift tube temperature is not fixed, then detection windows can be significantly larger. Alarm criteria will usually also consider other peak characteristics including amplitude and width.

Another measurement that helps establish a system's performance is informing power [73]. The informing power of an IMS correlates with false-alarm rate and the ability of the system under defined conditions to uniquely identify an analyte. Informing power of the scan period can be measured as the scan period divided by the average detection-window width associated with analyte peak position. For example, if the IMS system had a scan period of 20 ms and a typical peak window width of 0.100 ms, the informing power is 20/0.1 = 200. If the system is monitoring in both positive- and negative-ion modes, the overall informing power is 400. The informing power correlates inversely with false-alarm rate, i.e., the higher the information content, the lower the false-alarm rate. It is also a measure of the ability of the system to uniquely identify what was detected.

17.2.3.3.2 Resolving Power, Resolution, Duty Cycle, and Coadding Scans

The resolving power, R_p, for a DTIMS is defined for a single peak as the ratio of the drift time (t_d) to the FWHM [74]. Eq. (17.11) shows this relationship.

$$R_p = \frac{t_d}{\text{FWHM}} \tag{17.11}$$

We have discussed the measurement of drift time, but have not yet considered the FWHM measurement. The FWHM of the ion swarm is given by Eq. (17.12) where t_g is the gating pulse width and t_{diff} is diffusional broadening [74]. Diffusional broadening is a function of the temperature and is determined using Eq. (17.13), [74] where q is the charge on the ion. These equations are assuming a Gaussian shape of the ion swarm peak, which is a good assumption.

$$\text{FWHM} = \sqrt{t_g^2 + t_{diff}^2} \tag{17.12}$$

$$t_{diff} = \left[\frac{16\ kT\ \ln 2}{Vq}\right]^{\frac{1}{2}} t_d = \left[\frac{16\ kT\ \ln 2}{Vq}\right]^{\frac{1}{2}} \frac{L^2}{KV} \tag{17.13}$$

The key takeaway points from these equations are that FWHM is controlled by the gating pulse width, while diffusional broadening is controlled by temperature, electric field strength, and drift-tube length. Other factors that can impact the FWHM of the ion through their effect on either drift time or diffusional broadening are columbic repulsion (space charge) for intense peaks and electric field inhomogeneities. For most field-portable IMS instruments, which have drift tube lengths in the range of 3–10 cm and electric field strengths in the range of 200–300 V/cm, pulse widths of 100–300 μs are used with resolving power between 20 and 60 [75]. Resolving power can be increased by reducing the gating pulse width, but this reduces the signal. Therefore, a compromise between resolving power and sensitivity is necessary. Generally, a gating pulse of 200 μs is used. When comparing resolving power of instruments, peaks with similar K_0 values should be used. The higher the resolving power, the higher the ability of the instrument to resolve two closely spaced peaks. This ability can impact the false-negative rate of the instrument when complex spectra are present. Note that in the literature, there is a tendency to confuse terminology of resolving power and resolution, and, sometimes, resolving power is referred to as resolution.

Resolution (R) for an IMS is the ability to separate two peaks with drift times t_1 and t_2 and peak widths w_1 and w_2 as defined using Eq. (17.14). Resolution is particularly important when attempting to detect a peak in the presence of other peaks with similar drift times to minimize false negatives. There are mathematical tools which can be used to deconvolve closely overlapping peaks, but these can induce artifacts which must be considered when using this approach.

$$R = \frac{2(t_2 - t_1)}{(w_1 + w_2)} \tag{17.14}$$

The scan period for DTIMS systems is the time the signal is monitored after the gating grid is opened before opening the gate again. The scan period encompasses the range of mobilities in the detection library, which are characteristic of the mobilities of the detection substances over the anticipated operating environmental conditions. If the MW of the ions observed is less than 500 amu ($K_0 \sim 0.80$), then for the DTIMS systems outlined in Table 17.1, scan periods are on the order of 23 ms. In order to improve the SNR, scans can be coadded. When scans are coadded, the signal increases with the number, n, of coadded scans. Noise, on the other hand, increases as a function of the square root of n. The anticipated SNR improvement as a function of the number of coadded

Table 17.8 Anticipated signal-to-noise ratio improvement by coadding scans.

Number of scans coadded	1	2	4	16	32	100	256
Signal-to-noise (S/N) improvement	1.0	1.4	2.0	4.0	5.7	10.0	16.0

scans is shown in Table 17.8. Typically, the number of coadded scans required to achieve optimal results is determined during method development. The amount of time required to collect a spectrum is, therefore, equal to the scan period times the number of coadded scans. The total analysis time is equal to this value, times the number of spectra collected during the measurement. For the example of a 23-ms scan period with 20 added scans, an IMS spectrum is measured in 460 ms. Further, if the total analysis time is 10 seconds, ~22 average spectra each of 0.46 seconds duration would be measured during this analysis. To speed up processing, detection algorithms are typically applied to the IMS spectra rather than the individual scans.

The duty cycle of DTIMS system is the ratio of the gating pulse to the scan period. Typical gating-pulse widths of 0.200 ms are used. If the scan period is 23 ms, then the duty cycle is approximately 1%. When the gate is open, ions are injected into the drift tube; when closed, the ions generated in the reaction are neutralized when they strike the closed gate. From a practical perspective for DTIMS systems, 98–99% of ions are thrown away. For AIMs and DIMS, 100% of the ions are injected into the analyzer so the SNR can be improved by as much as 50 [11] (p. 128).

17.2.3.3.3 Verification and Sensitivity Testing

It is typical for field-deployed IMS systems to provide a test sample referred to as either a verification sample or a confidence-test sample. Differently configured systems will use this sample to test different things, but generally, it will be used to establish that (i) the components required to get the sample into the ionization chamber of the system are functioning properly, (ii) the system is calibrated, and (iii) the detection algorithm is being applied properly and the system is alarming as intended. The verification sample should be different from the calibration sample in order to assure the system is properly calibrated.

17.2.3.3.4 Sensitivity, Limit of Detection, and Limit of Alarm

IMS sensitivity is a function of many factors including sample introduction efficiency, sample ionization efficiency, gating efficiency, the efficiency of ion transfer through the drift tube, and conversion of the ions into an electronic signal as a function of its arrival time. The sample matrix can also impact sensitivity. For IMS platforms, there are generally two figures of merit used to evaluate sensitivity; (i) the limit of detection (LOD) which is usually calculated as baseline plus three times the instrument noise, and (ii) the limit of alarm (LOA) which is usually calculated as baseline plus ten times the instrument noise. LOA may be set higher if high background chemical noise is expected. To minimize false alarms, IMS platforms typically operate with detection libraries set at the LOA rather than the LOD.

17.2.3.3.5 Battery Requirements

The battery lifetime of portable systems varies as a function of size and weight, i.e., systems that are smaller and lighter have longer lifetimes than those that are larger and heavier. Usually, attempts are made to minimize the use of components that consume power through engineering design. For small handheld detectors that may also be worn, expected battery life can be on the order of 75 hours. For handheld systems, battery lifetimes of four hours are reasonably expected and although hot-swappable batteries are frequently desirable for portable spectrometers, this is not usually a critical issue for IMS systems. Some modern benchtop systems can operate for up to one hour on battery. Since IMS systems perform at ambient pressure, they have an advantage with regard to battery life over other trace detectors, such as mass spectrometers, that require power-hungry and somtimes-difficult-to-maintain pumps to achieve vacuum.

17.3 Current Innovations and Future Directions

The current state and future direction of IMS seem to be focused in a few areas. Increasing the breadth of chemicals that can be detected in a single analysis is important, and is possible using modern systems capable of performing fast-switching polarities. The ability to detect both positive and negative ions simultaneously is important because there are many situations where screening for substances that generate both positive and negative ions is necessary. For example, standard military and commercial explosives are generally detected in the negative mode. However, homemade explosives, such as HMTD and TATP, that have become more commonly encountered in recent years, are detected in the positive mode. CWAs and TICs also fall into this category. For instance, many nerve agents generate positive ions, while blister agents, such as mustard gas, generate negative ions. The ability to screen for these different types of ions simultaneously in these situations is significant.

There are generally two approaches to performing simultaneous detection of positive and negative ions. The first is by polarity switching. Historically, switching polarities was slow – on the order of about ten seconds [76]. Fast stabilization of the electric field in the reaction and drift regions was difficult, but many of these stabilization challenges have been addressed. In 2001, Rapiscan (then GE Ion Track) developed a fast-switching-polarity IMS that detected both positive and negative ions by switching the drift tube at 20 Hz [77]. More recently, IMS systems capable of a 1–2 ms stabilization are achievable [78]. The second approach to simultaneous detection is the use of two drift tubes. An advantage to this approach is that the operating conditions, and specifically the drift-tube temperature, can be optimized for each mode of operation. A single sample enters the system and is split between two IMS systems – one run in the positive mode and one run in the negative mode. An example of this type of system is the Smiths Detection 500DT (DT being an acronym for dual tube). This product has been on the market since about the year 2005 and, although it is still a viable trace detection system that offers good performance, it is competing in today's market with systems that are cheaper, smaller, lighter, and require less power.

Improvement of detection algorithms is also an important area of interest and development for IMS systems. The size of detection libraries needs to increase due to an increasing breadth of threats, but these systems need to maintain or reduce false-alarm rates and minimize coalarms. Coalarms occur when two different threats are programmed into a detection library, but the presence of one may cause both threats to alarm. To date, very robust detection algorithms have been developed based on peak finding and/or the use of logic, which can be combined with desorption time for samples which use variable-temperature desorption profiles. Typically, a system's detection algorithm will look at a peak's position, width, amplitude, and desorption profile to generate an alarm. Some analytes may produce multiple ion peaks which can be helpful in reducing false positives. But there is information in peak shape, desorption profiles, and patterns associated with common interferences that cannot easily be used. This is where the use of artificial intelligence (AI) may help. Currently, none of the commercial platforms appears to be actively using this approach.

Other innovations look at improving resolving power and reducing drift-tube size. In 2013, Kirk et al. [79] published a paper which examined the use of increased electric field strength in the drift tube with short gating pulses to reduce diffusional broadening. This paper took an entirely different look at the fundamental equations of IMS and suggested a nonconventional way of operating the IMS, showing it was possible to develop short drift-tube systems that were capable of achieving better resolution than larger laboratory systems.

The authors of this chapter used the classical equations associated with IMS to examine three different scenarios to establish the potential for IMS performance improvement: Table 17.9 shows the results predicted for a conventional IMS configuration, a higher-resolution IMS configuration, and a shorter-drift-tube configuration with normal resolution. The impact on TNT peak of K_0 1.54 is examined.

In scenario one, a drift tube of length 5 cm and gating pulse of 200 μs was used. A TNT peak with K_0 1.54 would have a drift time of 11.328 ms and FWHM of 266 μs. The FWHM according to Eq. (17.12) is a function of the gating pulse and the diffusional broadening and both are comparable. The resolving power of the system is 43 and peaks

Table 17.9 Impact of changing electric field and gating pulse on DTIMS performance.

	Variable	Scenario 1[a] (Conventional (normal) resolution)	Scenario 2[a] (Higher resolution)	Scenario 3[a] (Shorter drift tube with normal resolution)
IMS configuration ($T = 40\,°C$, $P = 760$ Torr)	E (V/cm)	250	2000	825
	E/N (Td)	1.06	8.5	3.5
	Length of drift tube (cm)	5	5	2
	Gating pulse (μs)	200	5	20
IMS peak at K_0 1.54	Drift time (ms)	11.328	1.416	1.133
	Diffusional broadening (μs)	175	7.75	13.8
	FWHM (μs)	266	9.22	24.0
	Resolving power	43	153	46
	Resolution (difference in K_0 required to fully resolve closely space peaks for $R = 1.0$)	0.064	0.017	0.059
Impact on # of coadded scans	Max scan period at K_0 0.6 (ms)	21.81	2.726	2.181
	Signal: Max coadded scans in 0.50 s	18	147	183
	Noise: Square root of # of scans	4.3	12.1	13.5
	s/n improvement[b]	1	2.83	3.16

a) Calculations based on $T = 40\,°C$, $P = 760$ Torr.
b) Normalized to 1.

with differences greater than 0.064 K_0 can be fully resolved. The typical scan period is 22 ms. To improve the SNR, 18 scans could be coadded over 0.5 second.

In scenario two, the resolution of the system was improved by increasing the electric field from 250 to 2000 V/cm and reducing the gating pulse from 200 to 5 μs. The TNT peak position shifted from 11.328 to 1.416 ms due to the much higher electric field, and E/N increased from 1 to 8.5 Td, outside the low-field limit. The overall FWHM is 9.22 ms which has increased the resolving power almost four times from 43 to 153. The resolution has also increased by ~4. In this scenario, peaks with ΔK_0 values greater than 0.017 can be fully resolved. This will help improve identification and minimize false negatives. With the much shorter scan period of 2.7 ms, 147 scans can be coadded over 0.5 second which improves the SNR over the conventional system by a factor of 2.83.

In scenario three, it was possible to reduce the length of the drift tube by 60% while maintaining the same performance as the conventional system. The drift-tube length is reduced from 5 to 2 cm, the electric field strength is increased from 250 to 825 V/cm, and the gating pulse is reduced from 200 to 20 μs. The TNT peak position shifted from 11.328 to 1.133 ms due to the higher electric field and short drift-tube length. E/N increased from 1.05 to 3.5 Td. The overall FWHM is now 24 μs for the TNT peak and the resolving power is 46. Resolution is similar to the conventional scenario, but the system is more compact. With the much shorter scan period of 2.1 ms, 183 scans can be coadded which improves the SNR over the conventional system by a factor 3.16.

Attempts at ion modification are another area of innovation. Mass spectrometry has always offered the ability to fragment ions to get an extra dimension of information about the analyte which can improve identification and reduce false alarms. Historically, it has been difficult to fragment ions at atmospheric pressure for IMS. With the advent of DMS and the ability to apply high AC fields between metal grids, high fields can be inserted into the reaction region or part way down the drift tube of an IMS. Terms such as ion modification and field-induced fragmentation (FIF) have been used to describe the results of this approach. Ions at ambient pressure can undergo fragmentation in high-frequency electric fields >100 Td (~14 000 V/cm) [80]. Smiths Detection has described methods to combine ion modification with IMS drift tubes [81]. Gary Eiceman's group [80] has demonstrated the ability to cleave bonds for alcohols and six-member ring rearrangements for acetates. The ability to use new ionization process, such as electrospray ionization or derivatizations to ionize high-melting-point compounds that could not otherwise be desorbed and introduced to the system, is also an area of interest.

17.4 Conclusions

It is clear that IMS is a powerful analytical technique which has been used effectively in the field for many decades to assess dangerous threats including CWAs, TICs, drugs, and explosives. These systems are valuable to field users for a number of reasons, including their fast analysis times, simplified user interfaces that provide yes/no results, and sensitivity to threats of interest. Their relative ease of deployment is due, in part, to the fact that they operate at atmospheric pressure and use ambient air as both the carrier gas and buffer gas. The use of these systems by the military and within the aviation-security markets is extensive, and there are a number of other field applications where these systems have proven value. Recent attempts to extend the successful vapor-phase JCAD IMS system to the analysis of solids and liquids [82] is testament to the desire to even further expand the use of these systems to the adapting list of threats faced by the military and other security experts. Systems have evolved over the years to their current state where dual-ion-mode analysis is possible using lightweight battery-operated systems, and a review of the current innovations and future directions for this technology suggests they will continue to evolve to meet the ever-changing threat-detection needs.

Acronyms

AC	Alternating current
AI	Artificial intelligence
AIMS	Aspiration ion mobility spectrometry
amu	Atomic mass units
APCI	Atmospheric pressure chemical ionization
BN	Bradbury–Nielson
CAAC	Civil Aviation Administration of China
CV	Compensation voltage
CWA	Chemical warfare agent
DBD	Dielectric barrier discharge
DC	Direct current
DfT	Department for Transport in the United Kingdom
DIMS	Differential ion mobility spectrometry
DMS	Differential mobility spectrometry
DNT	Dinitrotoluene
DTIMS	Drift-tube ion mobility spectrometry
DV	Dispersion voltage
ECAC	European Civil Aviation Conference
EGDN	Ethylene glycol dinitrate
FAIMS	Field asymmetric waveform ion mobility spectrometry
FIF	Field-induced fragmentation
FWHM	Full width at half maximum
HEPI	High-energy photoionization
HMTD	Hexamethylene triperoxide diamine
IMS	Ion mobility spectrometry
JCAD	Joint chemical agent detector
LOA	Limit of alarm
LOD	Limit of detection
MW	Molecular weight
NRC	National Regulatory Commission
ppb	Parts per billion
PETN	Pentaerythritol tetranitrate
ppm	Parts per million
SWaP	Size, weight, and power
RDX	Research department explosive
RF IMS	Radio frequency ion mobility spectrometry
TATP	Triacetone triperoxide
TIC	Toxic industrial chemical
TNT	Trinitrotoluene
TP	Tyndall Powell
TSA	Transportation Security Administration
UV	Ultraviolet

Abbreviations and Symbols

α	Correction factor
^{241}Am	Americium-241 radioactive ionization source
C	Celsius
cm	Centimeter
e	Electron charge
E	Electric field
$(E/N)_c$	Critical normalized field strength
eV	Electronvolt
g	Grams
Hz	Hertz
hν	Photon energy
k	Boltzmann constant
K	Mobility constant
K_0	Reduced mobility
keV	Kiloelectronvolt
kg	Kilogram
kV	Kilovolts
L	Length of the drift tube
m	Meters
mCi	Millicurie
mm	Millimeter
ms	Millisecond
N	Number density of neutral gas molecules
ng	Nanogram
^{63}Ni	Nickel-63 radioactive ionization source
$\Omega_D(T_{eff})$	Effective cross-section of the ion which is a function of the T_{eff}
P	Pressure
pg	Picogram
q	Ion charge
R	Resolution
R_p	Resolving power
R^2	Coefficient of determination
SNR	Signal-to-noise ratio
STP	Standard temperature and pressure
s	Seconds
σ	Standard deviation
T	Temperature
T_d	Townsend
t_1	Drift time analyte one (Eq. (17.14))
t_2	Drift time analyte two (Eq. (17.14))
t_d	Drift time
t_{diff}	Diffusional broadening

t_g	Gating pulse width
T_{eff}	Effective temperature of an ion based on thermal energy and energy acquired by an electric field
μ	Reduced mass (1/u = 1/m + 1/M), where m is the mass of the ion, M is the mass of the neutral gas molecules
μs	Microseconds
μm	Micrometers
v	Velocity
V	Volts
w_1	Peak width analyte one (Eq. (17.14))
w_2	Peak width analyte two (Eq. (17.14))
ΔW	Base width

References

1 Fatah, A.A., Barrett, J.A., Arcilesi, R.D. Jr., et al. (2000). *Guide for the Selection of Chemical Agent and Toxic Industry Material Detection Equipment for Emergency First Responders*, B1. Rockville: National Law Enforcement and Corrections Technology Center.

2 Smart, J. (2009). *History of U.S. Army Research & Development, Chemical and Biological Detectors, Alarms, and Warning Systems*. Aberdeen Proving Ground: U.S. Army Research, Development and Engineering Command.

3 Preparedness Directorate Office of Grants and Training (2007). *Guide for the Selection of Chemical Detection Equipment for Emergency First Responders*, 3rde. United States Department of Homeland Security.

4 Headquarters Department of the Army (1982. Change 2). *Technical Manual 43-0001-26-1, Change 2, 30 December 1985. Army Equipment Data Sheets: Chemical Defense Equipment*, 1–3. Headquarters, Department of the Army.

5 Office of the Special Assistant for Chemical Biological Defense & Chemical Demilitarization Programs (2006). *Department of Defense Chemical and Biological Defense Program Annual Report to Congress*, A3–A4. United States Department of Defense.

6 Galford, C. (2019). Smiths Detection Named Developer of DOD's Next Chemical Agent Detector. *Homeland Preparedness News* (16 January 2019).

7 Eiceman, G.A. and Stone, J.A. (2004). Ion mobility spectrometers in national defense. *Analytical Chemistry* 76: 390A–397A.

8 Shvartsburg, A.A. (2008). *Differential Ion Mobility Spectrometry: Nonlinear Ion Transport and Fundamentals of FAIMS*. CRC Press, Taylor & Francis Group 9781420051063, 1420051067.

9 Utriainen, M., Kärpänolga, E., and Paakkanen, H. (2003). Combining miniaturized ion mobility spectrometer and metal oxide gas sensor for the fast detection of toxic chemical vapors. *Sensors and Actuators B: Chemical* 93 (1–3): 17–24.

10 Mäkinen, M.A., Anttalainen, O.A., and Sillanpää, M.E.T. (2010). Ion mobility spectrometry and its applications in detection of chemical warfare agents. *Analytical Chemistry* 82 (23): 9594–9600.

11 Eiceman, G.A., Zarpas, Z., and Hill, H.H. Jr., (2014). *Ion Mobility Spectrometry*, 3ee. Boca Raton: CRC Press, Taylor & Francis Group. ISBN: 978-1-4398-5997-1.

12 McDaniel, E.W. and Mason, E.A. (1973). *Mobility and Diffusion of Ions in Gases*, 372. John Wiley & Sons, Inc.

13 Zrodnikov, Y. and Davis, C.E. (2012). The highs and lows of FAIMS: predictions and future trends for high field asymmetric waveform ion mobility spectrometry. *Journal of Nancomedicine & Nanotechnology* 3: e109.

14 Chemring Sensors & Electronic Systems (2018). Juno point trace vapor detection of CWA and TICs. *Chemrimg Sensors & Electronic Systems* [Online]. www.chemring.co.uk/~/media/Files/C/Chemring-V3/documents/sensors/datasheet-junomar-2018.pdf.

15 Eiceman, G.A. and Karpas, Z. (2005). *Ion Mobility Spectrometry*, 2e. Boca Raton: CRC Press, Taylor & Francis Group. ISBN: 978-0-8493-2247-1.

16 Smiths Detection (2017). Technical Information, LCD 3.3, Compact Wearable CWA Identifier and TIC Detector. *LCD 3.3 Person Worn CWA and TIC Detector* [Online]. https://www.smithsdetection.com/products/lcd-3-3 (accessed 26 December 2018).

17 Smiths Detection (2012). Technical Information SABRE 5000 Handheld Trace Detector for Explosives, Chemical Agents and Toxic Industrial Chemicals or Narcotics. *Smith Detection* (28 November) [Online]. https://www.smithsdetection.com/products/sabre-5000 (accessed 26 September 2019).

18 Smiths Detection (2011). GID-3 Chemical Warfare Agent Detector. *Smith Detection* (15 August) [Online]. https://www.smithsdetection.com/products/gid-3 (accessed 30 April 2020).

19 Smiths Detection (2017). *Technical Information – IONSCAN 600 Explosives and Narcotics Trace Detector*. Hemel Hempstead: Smiths Detection.

20 Smiths Detection (2011). *IONSCAN 500DT Explosives and Narcotics Trace Detection. Smith Detection* (15 February) [Online]. https://www.smithsdetection.com/products/ionscan-500dt-2 (accessed 8 August 2013).

21 MRIGlobal (2018). CBRNE Tech Index Hardened Mobile Trace. *CBRNE Tech Index* [Online]. https://www.cbrnetechindex.com/p/3330/Morpho-Detection-Inc/Hardened-MobileTrace-HMT (accessed 30 April 2020).

22 MRIGlobal. CBRNE Tech Index Itemiser DX Series. *CBRNE Tech Index* [Online]. https://www.cbrnetechindex.com/Print/3332/rapiscan-systems/itemiser-dx-series (accessed 30 April 2020).

23 MRIGlobal (2018). CBRNE Tech Index H150 and H150E Handheld Trace Detectors. *CBRNE Tech Index* [Online]. https://www.cbrnetechindex.com/p/6177/L3Harris/H150-and-H150E-Handheld-Trace-Detectors (accessed 30 April 2020).

24 MRIGlobal (2018). CBRNE Tech Index B220 Series. *CBRNE Tech Index* [Online]. https://www.cbrnetechindex.com/P/6179/L3Harris/B220-Series (accessed 30 April 2020).

25 MRIGlobal (2018). CBRNE Tech Index Bruker IMS Compare. *CBRNE Tech Index* [Online]. https://www.cbrnetechindex.com/Compare (accessed 30 April 2020).

26 Nuctech Company Limited (2011). TR1000DB-A Handheld Explosives and Narcotics Trace Detector. *Nuctech* [Online]. http://www.nuctech.com/en/SitePages/ThDetailPage.aspx?nk=PAS&k=BHFCEF&pk=FGAFHB (accessed 30 April 2020).

27 Nuctech Company Limited (2017). Nuctech TR2000DB-A Desktop Explosives and Narcotics Trace Detector. *PT. Ekacitra Bumikarya* [Online]. http://ekacitra.com/brosur/NUCTECH%20TR2000DB-A.pdf (accessed 3 May 2020).

28 RAE Systems. ChemRAE Chemical Warfare Agent Detection. *RAE Systems* [Online]. https://www.raesystems.com/sites/default/files/content/resource/FeedsEnclosure-ChemRAE_US_DS.pdf (accessed 29 April 2020).

29 Environics Oy (2019). ChemPro100i Handheld Chemical Detector. *Environics Oy* [Online]. https://www.environics.fi/wp-content/uploads/2020/01/US_ChemPro100i-Handheld-Chemical-Detector.pdf (accessed 29 April 2020).

30 Environics Oy (2019). ChemProX New Generation Handheld Chemical Detector. *Environics Oy* [Online]. https://www.environics.fi/wp-content/uploads/2020/01/ChemProX_new_generation_handheld_chemical_detector-2.pdf (accessed 29 April 2020).

31 Owlstone Inc (2017). OwlSens-C - Advanced Chemical Detector. *Owlstone Inc* [Online]. https://www.owlstoneinc.com/media/uploads/files/OwlsensC_Slick_2017_FINAL.pdf (accessed 29 April 2020).

32 Owlstone Inc (2017). OwlSens-T - Advanced Chemical Sensing. *Owlstone Inc* [Online]. https://www.owlstoneinc.com/media/uploads/files/OwlSensT_TIC_Slick_2017_FINAL.pdf (accessed 29 April 2020).

33 MRIGlobal. CBRNE Tech Index: ChemRAE. CBRNE Tech Index [Online]. https://www.cbrnetechindex.com/Print/3422/rae-systems-inc/chemrae (accessed 3 May 2020).

34 Prox, T., Prieto, M., Bryant, J. and Yost, R.A. (2012). Partial Ovoidal FAIMS Electrode. 8,237,118 B2. USA.

35 Karpas, Z. (1989). Ion mobility of aliphatic and aromatic amines. *Analytical Chemistry* 61: 684–689.

36 Metternich, S., Zörntlein, S., Schönberger, T., and Huhn, C. (2019). Ion mobility spectrometry as a fast screening tool for synthetic cannabinoids to uncover drug trafficking in jail via herbal mixtures, paper, food, and cosmetics. *Drug Test Analyst* 11 (6): 833–846.

37 Karasek, F.W. (1974). Plasma chromatography. *Analytical Chemistry* 46: 710A–720A.

38 Shvartsburg, A.A., Li, F., Tang, K., and Smith, R.D. (2006). High-resolution field asymmetric waveform ion mobility spectrometry using new planar geometry analyzers. *Analytical Chemistry* 78 (11): 3706–3714.

39 Puton, J., Nousiainen, M., and Sillanpää, M. (2008). Ion mobility spectrometers with doped gases. *Talanta* 76 (5): 978–987.

40 Waraksa, E., Perycz, U., Namieśnik, J. et al. (2016). Dopants and gas modifiers in ion mobility spectrometry. *TrAC Trends in Analytical Chemistry* 82: 237–249.

41 Kozole, J. et al. (2012). Characterizing the gas phase ion chemistry of an ion trap mobility spectrometry based explosive trace detector using a tandem mass spectrometer. *Talanta* 99: 799–810.

42 Keller, T. et al. (2006). Application of ion mobility spectrometry in cases of forensic interest. *Forensic Science International* 161 (2–3): 130–140.

43 Kozole, J. et al. (2015). Gas phase ion chemistry of an ion mobility spectrometry based explosive trace detector elucidated by tandem mass spectrometry. *Talanta* 140: 10–19.

44 Puton, J. and Namieśnik, J. (2016). Ion mobility spectrometry: current status and application for chemical warfare agents detection. *TrAC Trends in Analytical Chemistry* 85 (Part B): 10–20.

45 Karasek, F.W. and Denney, D.W. (1974). Role of nitric oxide in positive reactant ions in plasma chromatography. *Analytical Chemistry* 46 (6): 633–637.

46 Spangler, G.E. and Lawless, P.A. (1978). Ionization of nitrotoluene compounds in negative ion plasma chromatography. *Analytical Chemistry* 50 (7): 884–892.

47 Spangler, G.E. and Collins, C.I. (1975). Reactant ions in negative ion plasma chromatography. *Analytical Chemistry* 47 (3): 393–402.

48 Kim, S.H., Karasek, F.W., and Rokushika, S. (1978). Plasma chromatography with ammonium reactant ions. *Analytical Chemistry* 50 (1): 152–155.

49 Chuang, C., Dandan, J., and Haiyang, L. (2019). UV photoionization ion mobility spectrometry: fundamentals and applications. *Analytica Chimica Acta* 1077: 1–13.

50 Simmonds, P.G., Fenimore, D.C., Pettitt, B.C. et al. (1967). Design of a nickel-63 electron absorption detector and analytical significance of high-temperature operation. *Analytical Chemistry* 39 (12): 1428–1433.

51 Siegel, M.W. (1983). Rate equations for prediction and optimization of chemical ionizer sensitivity. *International Journal of Mass Spectrometry and Ion Physics* 46: 325–328.

52 Paakanen, H. (2001). About the applications of IMCELLTM MGD-1 detectors. *International Journal of Ion Mobility Spectrometry* 4: 136–139.

53 Bruker Daltronics (2016). Chemical Threat Mitigation, Compact of CWA/TIC Detector: Bruker μRAIDplus. *Bruker* [Online]. https://www.bruker.com/fileadmin/user_upload/8-PDF-Docs/CBRNE_Detection/Literature/1815859_uRAIDplus.pdf (accessed 3 May 2020).

54 Bruker Daltronics (2015). Explosives Trace Detection, Explosives Threat Mitigation: Bruker RoadRunner. *Bruker* [Online]. https://www.opecsystems.com/persistent/catalogue_files/products/brukerroadrunnerbrochure.pdf (accessed 3 May 2020).

55 Bruker Daltronics (2017). Explosives Trace Detection, Explosives Threat Mitigation: Bruker DE-tector flex. *Bruker* [Online]. https://www.bruker.com/fileadmin/user_upload/8-PDF-Docs/CBRNE_Detection/Literature/1854797_DE-tector_flex_brochure.pdf (accessed 3 May 2020).

56 Verkouteren, M. and Lawrence, J. (2017). IMS Measurements and Preliminary Peak Assignments for Trace Explosives in non-RAD ETDs. *Trace Explosives Detection (TED) Workshop*. Santa Fe, NM.

57 United States Regulatory Commission (2014). License Number 20-023904-01E. *Nuclear Materials License, NRC Form 274.*

58 Smiths, N., McLain, D., and Steeb, J. (2017). *Ion Mobility Spectrometer Field Test, Results from the Former New Brunswick Facility*. Argonne National Laboratory.

59 Dzidic, I., Carroll, D.I., Stillwell, R.N., and Horning, E.C. (1976). Comparison of positive ions formed in nickel-63 and corona discharge ion sources using nitrogen, argon, isobutane, ammonia and nitric oxide as reagents in atmospheric pressure ionization mass spectrometry. *Analytical Chemistry* 48 (12): 1763–1768.

60 Tabrizhi, M., Khayamian, T., and Taj, N. (2000). Design and optimization of a corona discharge ionization source for ion mobility spectrometry. *Review of Scientific Instruments* 71 (6): 2321–2328.

61 Hill, C.A. and Thomas, C.L.P. (2003). A pulsed corona discharge switchable high resolution ion mobility spectrometer-mass spectrometer. *Analyst* 128: 55–60.

62 Turner, R.B., Taylor, S.J., Clark, A. and Arnold, P.D. (2001). Corona Discharge Ion Source for Analytical Instruments. 6,225,623, USA.

63 Hu, J., Li, W., Zheng, C., and Hou, X. (2011). Dielectric barrier discharge in analytical spectrometry. *Applied Spectroscopy Reviews* 46 (5): 368–387.

64 Brandenburg, R. (2018). Dielectric barrier discharges: progress on plasma sources and on the understanding of regimes and single filaments. *Plasma Sources Science and Technology* 26 (5): 1–29.

65 Kubelik, I., Feldberg, S., Atamanchuk, B., et al. (2017). Dielectric Barrier Discharge Ionization Source for Spectrometry. 9,778,224 B2 USA.

66 TSI Incorporated (2013). Photoionization Detection (PID) Technology, Application Note TSI-147 Rev B. *TSI Incorporated* [Online]. https://www.tsi.com/getmedia/e6812861-60a7-4dee-82a7-09b1ea9f734c/TSI-147_Photo_Ionization_Detection_Technology?ext=.pdf (accessed 1 May 2020).

67 TSA Trace (2020). Product Accessory Kit, with Lamps and Dopants, Itemiser 4DX. *TSA Trace* [Online]. https://www.tsatrace.com/Product-Accessory-Kit-with-Lamps-and-Dopants-Itemiser-4DX_p_87.html (accessed 3 May 2020).

68 Nazarov, E.G., Miller, R.A., Eiceman, G.A., and Stone, J.A. (2006). Miniature differential mobility spectrometry using atmospheric pressure photoionization. *Analytical Chemistry* 78 (13): 4553–4563.

69 Kirk, A.T. and Zimmermann, S. (2014). Bradbury-Nielsen vs. field switching shutters for high resolution drift tube ion mobility spectrometers. *International Journal of Ion Mobility Spectrometry* 17: 131–137.

70 Soppart, O. and Baumbach, J.I. (2000). Comparison of electric fields within drift tubes for ion mobility spectrometry. *Measurement Science and Technology* 11 (10): 1473–1479.

71 Davis, E.J., Grows, K.F., Siems, W.F., and Hill, H.H. Jr., (2012). Improved ion mobility resolving power with increased buffer gas pressure. *Analytical Chemistry* 84 (11): 4858–4865.

72 Kaur-Atwal, G., O'Connor, G., Aksenov, A.A. et al. (2009). Chemical standards for ion mobility spectrometry: a review. *International Journal of Ion Mobility Spectrometry* 12: 1–14.

73 Committee on Assessment of Security Technologies for Transportation (2004). *Opportunities to Improve Airport Passenger Screening with Mass Spectrometry*. Washington, DC: National Research Council of The National Academies.

74 Siems, W.F., Wu, C., Tarver, E.E. et al. (1994). Measuring the resolving power of ion mobility spectrometers. *Analytical Chemistry* 66 (23): 4195–4201.

75 Cottingham, K. (2003). Product review: ion mobility spectrometry rediscovered. *Analytical Chemistry* 75 (19): 435A–439A.

76 Jenkins, A. and McGann, W.J. (2004). Enhancements to Ion Mobility Spectrometers. 6,765,198 B2 USA.

77 McGann, W.J., Geodecke, K., Neves, J., and Jenkins, A. (2001). Simultaneous, dual mode IMS system for contraband detection and identification. *International Journal of Ion Mobility Spectrometry*: 144–147.

78 Zaleski, H., et al. (2017). Fast-Switching Dual-Polarity Ion Mobility Spectrometry. 9,709,530 B2 USA.

79 Kirk, A.T., Allers, M. et al. (2013). A compact high resolution ion mobility spectrometer for fast trace gas analysis. *Analyst* 138 (18): 5159–5504.

80 Shokri, H. et al. (2020). Field induced fragmentation (Fif) spectra of oxygen containing volatile organic compounds with reactive stage tandem ion mobility spectrometry and functional group classification by neural network analysis. *Analytical Chemistry* 92 (8): 5862–5870.

81 Atkinson, J.R., Clark, A., Taylor, S.J., and Munro, W.A. (2011). Detection Apparatus. 7,932,489 B2 USA.

82 Smiths Detection Inc. (2018). *Smiths Detection Inc. to Develop Enhanced Chemical Explosives Detection Capability for United States Department of Defense*. Edgewood, MD: Smiths Detection.

18

X-Ray Sources for Handheld X-Ray Fluorescence Instruments

Sterling Cornaby

Moxtek Inc., Orem, UT, USA

18.1 Background

X-ray fluorescence (XRF) is used to identify elements within a sample, as well as quantify the amounts of each element from bulk percentages to parts per million. The major advantage of portable or handheld X-ray fluorescence (HHXRF) is that the instrument can be taken to the sample, rather than the sample to the instrument [1]. Portability has proven to be a very big advantage, demonstrated by the rapid growth of the HHXRF market in 2004–2020. HHXRF is taken directly to mines, on-site in mining investigations, used directly in scrap metal sorting yards, taken into homes to test lead in paint, to name just as a few examples. The development of miniature, lightweight X-ray sources was the critical component needed to shift XRF spectroscopy from the lab to the field. The X-ray source, within HHXRF instruments is small, on the order of 15 cm in length, lightweight on the order of 300 g, and low powered, consuming 5–10 W [2]. Additionally, all HHXRF instruments have single-element energy-dispersive (ED) detectors such as a silicon drift detector (SDD) or a PIN (p-type–intrinsic–n-type) diode. All of the qualities of these small components are needed to make the HHXRF instrument portable, battery powered, and about the size of a handheld cordless drill. The tube has to be very small, consume a small amount of power, and robust enough to withstand shocks from dropping and temperature extremes found in the field (Figure 18.1). These constraints made the miniature X-ray source what it is today, with an "X-ray source" defined to be composed of a miniature X-ray tube combined with a miniature high-voltage power supply (HVPS). The technical advancements in small X-ray sources are a major advancement that makes HHXRF possible.

Previously to about the year 2004, radioisotopes were used in the HHXRF instruments due to their lightweight, small size, and the monochromatic radiation with energies capable of exciting the K-lines in most materials of interest [3, 4]. The handling of radioactive materials severely limited HHXRF based on radioisotopes due to strict regulations on their handling and transportation.

In 1999, Moxtek developed its first few miniature X-ray tube-based X-ray sources, funded by a NASA SBIR for the CHEMIN XRD/XRF instrument for Mars missions [5]. This work developed into X-ray sources for HHXRF in the next few years. Having a miniature X-ray tube-based source was a huge advantage over radioisotopes, because it freed up legal restrictions, since in an X-ray source based on an X-ray tube, the radiation can be turned on and off. Other advantages included having the flux output level adjustable to maximize the signal from the detector, and the high voltage (HV) being adjustable, which is useful in changing the range of detected elements.

The rapid growth of the HHXRF market in 2004–2020 required that the X-ray sources be mass produced. In the second decade of the twenty-first century, the global demand for the miniature X-ray sources reached over 10 000 units per year level, far surpassing the market for the benchtop and laboratory analytical XRF instruments. Market-driven applications put HHXRF instruments in scrap yards, mines, deserts, high mountains, and arctic conditions in the hands of non-XRF experts.

Portable Spectroscopy and Spectrometry 1: Technologies and Instrumentation, First Edition.
Edited by Richard A. Crocombe, Pauline E. Leary and Brooke W. Kammrath.

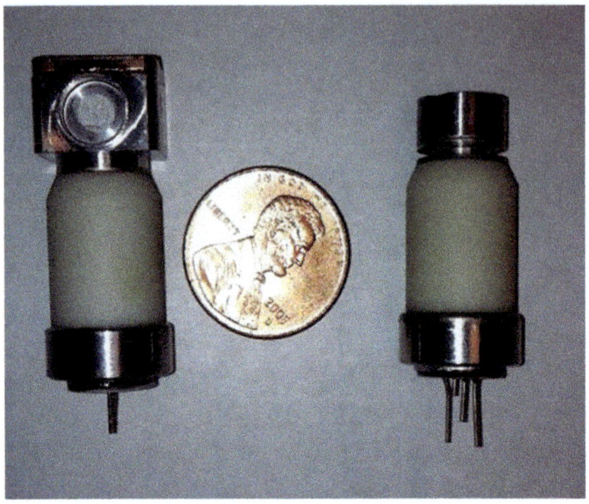

Figure 18.1 Transmission window (right) and side window (left) tubes. The tube body is ceramic; the cathode and anode assemblies are metal alloys. *Source*: Courtesy of Moxtek, Inc.

Year	Weight	Voltage & Current	Tube Power
2001	(763 gm)	35 kV 100 μA	3.5 W
2004	(320 gm)	40 kV 100 μA	4 W
2006	(533 gm)	50 kV 200 μA	10 W
2009	(335 gm)	50 kV 200 μA	4 W
2012	(250 gm)	50 kV 200 μA	4 W
2013	(750 gm)	70 kV 1000 μA	12 W

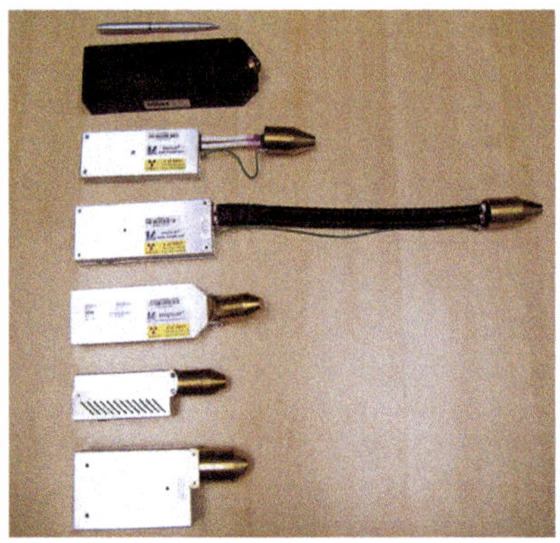

Figure 18.2 X-ray source weight and size reduction in 2001–2013 and a transition from cabled to monolithic design. The pen at the top of the image is shown as a size reference. *Source*: Courtesy of Moxtek, Inc.

From the start, there has been a demand for reducing the size and weight of both the tubes and HVPSs. Figure 18.2 shows the reduction is size and weight over the period from 2001 until 2015. It also shows the increasing high voltage HV and power of the miniature X-ray sources. From about 2015 to today, miniature X-ray sources have stayed the same size. The emphasis on technological advances has currently shifted to performance enhancements.

Over the last five years, one major optimization desired by manufacturers of HHXRF instruments is possessing X-ray sources which are spectroscopically identical. A spectroscopically identical X-ray source allows one to switch out an X-ray source in an instrument with no other changes. For example, the MOXI™ source gives the same XRF spectra source-to-source with minimal or no recalibration of the instrument. This allows for swapping of X-ray sources in an instrument, without requiring instrument recalibration. This benefit is very much desired by HHXRF manufacturers because it reduces manufacturing complexity (the time required for calibration), and reduces service time for these instruments. This will be discussed in detail in Section 18.4.5.

18.2 The Miniature X-Ray Source

An X-ray source is composed of a miniature X-ray tube and a miniature HVPS; sometimes the HVPS is also called a generator. Figure 18.3 gives a schematic of the main parts of both an X-ray tube and the HVPS. The next two

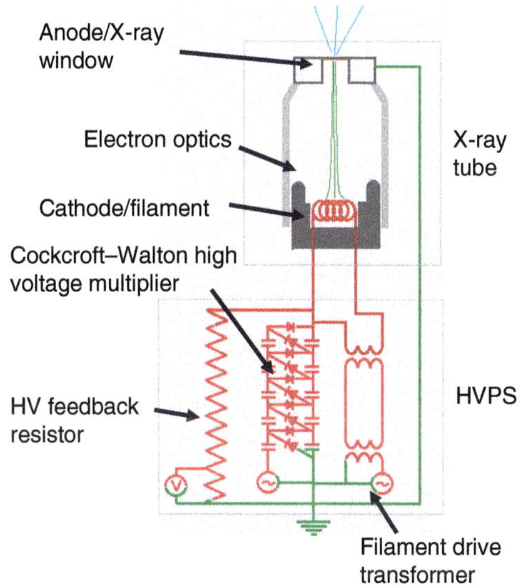

Figure 18.3 A schematic of an X-ray source with all its major parts labeled.

sections will describe the X-ray tube and the HVPS. In the final section, we will go over some specific examples of miniature X-ray sources.

18.2.1 The X-Ray Tube

The miniature X-ray tube used in HHXRF is more or less the standard hot-cathode tube design used since William Coolidge developed the original X-ray tube [6]. The main difference is the extremely small size achieved by the use of advanced materials and processing methods (Figure 18.1). The tube has a tungsten filament cathode which provides a source of electrons, an anode target for the electrons to decelerate in, which generates X-rays, and passive electron optics to guide the electron beam to the anode.

Some important aspects of the miniature tube include:

- A very small filament, which takes a low amount of input power to heat the filament to about 1800 °C. A low-power filament is needed to extend the time of battery operation in a portable instrument.
- A ceramic/metal vacuum envelope, providing robustness over a glass vacuum envelope. This helps to prevent the tube from breaking in the event of both mechanical and thermal shocks experienced by the instrument when used in the field.
- A "simple diode" or unipolar X-ray tube design, allowing for a single variable HV to run the X-ray tube. There are no other components, such as a "gate" for turning the tube on and off, or steering the electron beam, which are fairly common on many X-ray tubes [6]. The passive electron optics design is challenging, and is simply achieved by shaping of metal components of the tube. The passive electron optics need to work over a large range of 4–70 kV, and keep the electron beam at nearly the same location and size on the anode, with no active feedback. All of this effort simplifies the X-ray tube and the HVPS, so that the entire X-ray source can be smaller.
- The most popular "transmission-window" design combines the anode of the tube and the X-ray window of the tube into the same part. The beryllium window (usually 100–250 μm thick) is coated with a thin, ~1 μm, layer of the desired anode material. Details on the anode layer thickness are very application dependent. This allows for closer source-to-sample distance, which is a critical parameter for miniature X-ray sources in HHXRF.

All of these efforts result in a very small and robust miniature X-ray tube, well-suited for HHXRF.

Functionally, an X-ray tube produces X-rays with an electron beam impinging on the anode generating two general types of interactions: bremsstrahlung radiation and electron beam-induced characteristic X-ray radiation [7–11]. Bremsstrahlung radiation, also known as "braking" radiation, is caused by the impinging electrons interacting with the nucleus of an atom. As the electron approaches the atom, it is deflected and possibly stopped by the strong electric field surrounding the atom. This deceleration of the electron produces the bremsstrahlung radiation. This broadband radiation produces fewer photons at higher X-ray energies, with the intensity of bremsstrahlung X-rays approaching zero (in keV) at the tube's HV setting (in kV).

Characteristic radiation is based on impinging electrons interacting with the electrons surrounding an atom. Kossel [8–11], Moseley [12, 13], and Barkla [14] presented the theoretical and experimental foundation for the quantum theory of producing characteristic X-rays. For induced characteristic radiation, an electron of high energy may ionize an atom by displacing an electron from one of the inner atomic shells. An electron from an outer shell may fall into the shell vacancy, emitting either another electron from the atom, or a photon, the photon being the desired reaction for the X-ray tube. The energy of the X-ray is dependent on the energy levels of the shells participating in the process, and therefore is "characteristic" to the material of the target.

The spectrum of X-rays from two different materials, which shows both bremsstrahlung radiation and electron-induced characteristic radiation, is shown in Figure 18.4, with no X-ray filters and a vacuum flight path. To tailor a spectrum for a specific XRF application, the HV setting on the tube is changed, or a variety of filters are put in front of the anode, to filter out lower energy X-rays from the source, and this is described in more detail below.

18.2.2 The High-Voltage Power Supply

The HVPS used is based upon the Cockcroft–Walton HV generator [15]. It generates the HV, and also provides the emission current needed to drive the electron current through the X-ray tube. It generates this DC HV based upon an oscillator, coupled with a ladder of capacitors and diodes (HV multiplier), using a half-wave rectifier design, as shown in Figure 18.3. To achieve the needed HV of 4 up to 70 kV, the input oscillating transformer has an HV oscillation amplitude of a few hundred volts to a few thousand volts, and a frequency of several kHz. The capacitor–diode ladder consists of between 8 and 15 stages.

The next critical component is the HV feedback resistor, which allows for monitoring the HV, and for controlling the HV on the Cockcroft–Walton HV generator. XRF requires a very stable high voltage, which in turn requires

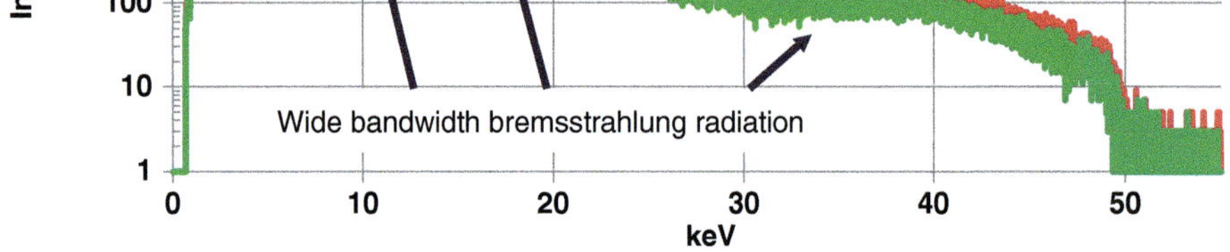

Figure 18.4 An X-ray spectrum from both a tungsten and rhodium anode X-ray tube which shows both the bremsstrahlung and the characteristic X-ray radiation.

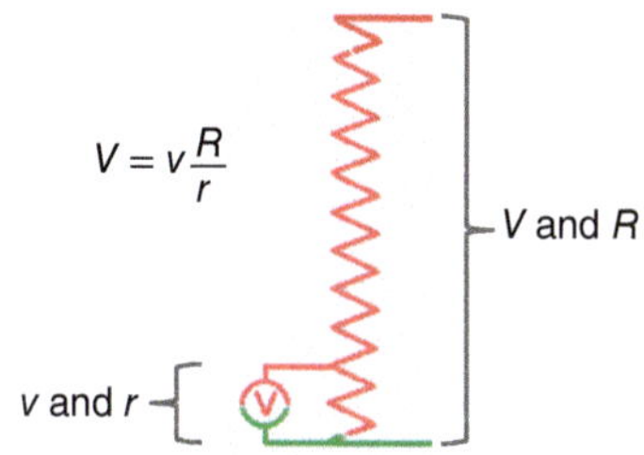

Figure 18.5 A schematic of the HV feedback resistor. "*V*" is the HV of the HVPS calculated by knowing "*R*" the resistance of the entire resistor, "*r*" is the resistance of the resistor subsection, and "*v*" is the measured voltage of this resistor subsection.

having a very stable and a predicable HV feedback resistor. Figure 18.5 outlines the simple circuit that shows how the HV is measured, by a simple ratio of the resistance of the entire resistor to just a smaller subsection of the resistor.

The final component of the HVPS is the filament drive transformer. Miniature X-ray sources are run with the anode at ground, and the cathode is at HV. The filament is driven with an alternating current (AC) current provided by a series of transformers. Since the filament is at HV, the AC drive is achieved by a series of two or more transformers to isolate the HV.

One other critical design issue is achieving HV in small sizes without arcing. Both the tubes and the HVPSs are potted in a highly insulating material to prevent this. The layout of the HV components is also optimized to reduce the field gradients and provide enough HV standoff in the very small packages need for HHXRF.

18.2.3 Physical Examples of Miniature X-Ray Sources

Table 18.1 and Figure 18.6 show examples of some commercial X-ray sources currently being produced [10, 11]. The MOXI source, which is the latest design, is meant specifically for handheld applications, being very small, very light, and of lower power. The MAGPRO® X-ray source is meant for applications which need some more power, and are often put in small benchtop XRF instruments.

There are several characteristics of the X-ray source that are important. As mentioned earlier, for HHXRF the X-ray source has to be light and small, and consume low amounts of power. Other key properties described below include flux stability, target material selection, and source-to source spectra differences.

18.2.4 X-Ray Flux and Spectra

The flux coming from a tube-based X-ray source is often described by stating the highest operating voltage (spectral information), the emission current (flux information), and the anode material (spectral information),

Table 18.1 This table shows the basic parameters of both the MOXI® and the MAGPRO X-ray sources.

Parameters	MOXI™	MAGPRO
HVPS output kV range (kV)	5–50	5–70
HVPS emission current range (μA)	5–200	10–1000
Anode material	W, Rh and others	W, Rh and others
HVPS power limit on X-ray tube (W)	4	12
HVPS power input voltage (V DC)	9	24
Total weight (g)	310	700

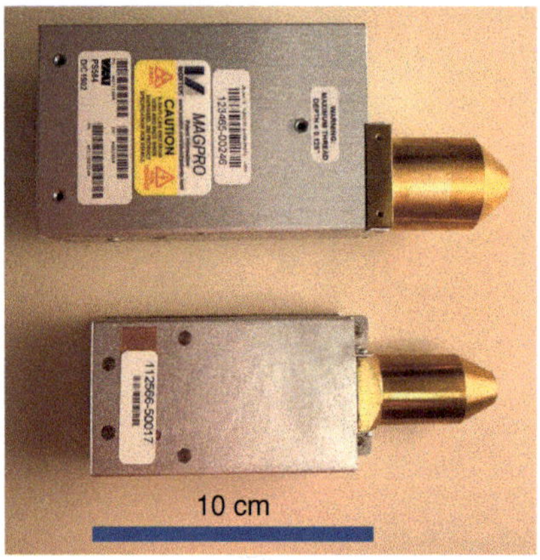

Figure 18.6 Image of the MAGPRO (top) and MOXI (bottom) X-ray source. *Source*: Courtesy of Moxtek, Inc.

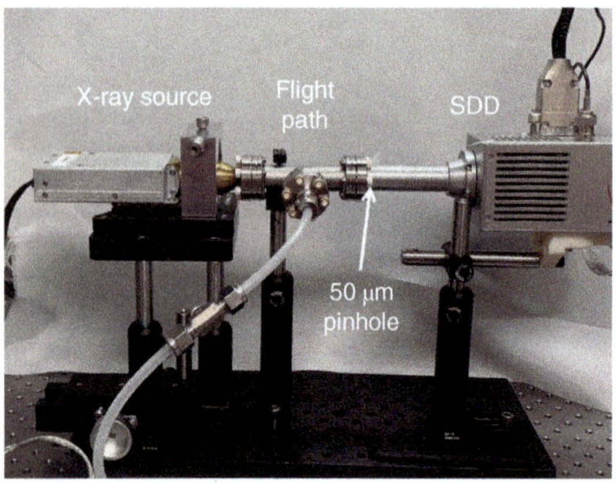

Figure 18.7 This is an image of a setup to measure the flux. The vacuum flight path is specifically needed to eliminate air absorption for better detection of X-rays under about 5 keV. The windows on the flight path are 8 μm thick, the flight path is 9.5 cm long. The pinhole was placed directly on the SDD window and has a 50 μm diameter in a 0.5 mm thick tungsten foil. *Source*: Courtesy of Moxtek, Inc.

e.g. a 50 kV, 0.2 mA, tungsten anode X-ray source. This is sensible for many X-ray applications, but it is not as useful for comparing miniature X-ray sources to larger X-ray tubes [16].

One reasonably direct way to determine the X-ray radiation emitted from an X-ray source is with an ED detector, such as a PIN diode or a SDD. The radiation directly from an X-ray source is much too intense to measure the spectrum, but that can be overcome by putting a small pinhole of 10–200 μm in front of the detector to decrease the flux hitting the detector. The flux measurement setup shown in Figure 18.7 gives results similar to the spectra shown in Figure 18.4.

It is exceedingly difficult to focus or collimate X-rays, especially in a handheld format. Therefore the designs of HHXRF instruments all feature the X-ray tube, the sample, and the detector in closest possible proximity to optimize the signal-to-noise ratio (SNR) of the spectra. The miniature sources shown in Figure 18.6 can get to within 2–3 cm from the sample. In this tight geometry, with just a few watts of tube power, the intensity of the emitted fluorescent X-ray radiation in most cases will reach the maximum count rate of a majority of existing SDDs [16]. This tight anode-to-sample distance is due to two major design features, the small X-ray tube size and the X-ray transmission anode on the X-ray tube.

18.2.5 Radiation Dose from Miniature X-Ray Sources

In an enclosed XRF system, the sample chamber is designed to be radiation-tight for radiation safety. Additionally, these instruments have interlocks which shut off the X-ray source if the sample chamber is opened to keep X-ray exposures from happening. For a HHXRF, there is more of a chance for X-ray exposure, since there is no radiation

Table 18.2 Table of radiation dose from two tungsten anode miniature X-ray sources, at two different distances of 2 cm, and at 30 cm away.

Tube type	Radiation dose in mRem/s							
	No X-ray filter				1 mm Al filter			
	At 30 cm		At 2 cm		At 30 cm		At 2 cm	
	40 kV	60 kV	40 kV	60 kV	40 kV	60 kV	40 kV	60 kV
Magnum W anode at 4 W	18 at 100 μA	– –	3000 at 100 μA	– –	4 at 100 μA	– –	600 at 100 μA	– –
MAGPRO W anode at 12 W	55 at 300 μA	78 at 200 μA	9100 at 300 μA	12 500 at 200 μA	15 at 300 μA	20 at 200 μA	1800 at 300 μA	3100 at 200 μA

This is the dose with no additional filtration, just the native beryllium window.

enclosure, and they are open beam devices. Commercial instruments include a number of safety protocols, but they are not discussed here.

Focusing on just the X-ray source, it is considered radiation-safe if the radiation is strictly coming out of the window opening; radiation should not come out in other directions through other structures of the X-ray source. Repeating this in another way, if the window opening of the X-ray source is blocked from emitting radiation, the X-ray source will emit less than 0.2 mRem/h. (2 μGray/h) in any direction when turned on at the maximum kV and power. For reference, natural background radiation is about 0.01 mRem/h.

Another frequent question is "What is the radiation dose if the X-ray source window opening is not covered?" Table 18.2 outlines the radiation does for both the Magnum® tube, used in the MOXI source, and the MAGPRO source. These values were measured with a Black Piranha radiation meter, which is a common radiation dose meter used in the medical field. At the distance of 30 cm, the dose rate is covering a circular area with a diameter of about 25 cm. Often, a filter is used in conjunction with the X-ray source. The table includes the radiation dose both with a 1 mm aluminum filter and without the filter.

The full body limit of radiation per year is less than 5000 mRem (50 mGray) for a radiation worker. An unfiltered miniature X-ray source can reach this limit in a localized area in just an hour or two at a 30 cm distance. At the distance of 2 cm, covering a very limited area of a few cm^2, you could cause a very localized radiation burn on your skin if you held the tube right on your skin and left it on for a few seconds at full power. These small low-power X-ray sources need to be respected when it comes to radiation safety, especially at very close distances.

18.3 The Selection of a Target Anode Material for XRF

The basic premise of XRF is that high-energy X-rays are aimed at a sample and then the atoms in the sample may absorb an X-ray. This ionizes the atom by displacing an electron from one of the shells close to the atom. The next step is for an electron from an outer shell to fall into the shell vacancy, emitting the excess energy in the form of characteristic X-ray radiation. The energy of the X-ray is dependent on the energy levels of the shells participating in the process and therefore is "characteristic" to the material of the sample. More details in using XRF for material identification may be found in Van Grieken [3], Jenkins [17, 18].

A critical detail for the X-ray source used for XRF is that X-ray source needs to produce X-rays at energies higher than absorption edge of the element in order to ionize the atom. Every element has different absorption edges, just as they have different characteristic X-ray lines. The X-ray source needs to be tuned in various ways for effective detection of different elements.

There are three basic methods for tuning the spectrum shape from the X-ray source:

- Setting the HV on the X-ray tube.
- Putting an X-ray filter between the X-ray source and the sample.
- Selection of the anode material on the X-ray tube.

The next few sections will go through these items in detail, especially on the selection of the anode material since this is a central aspect of the miniature X-ray source.

18.3.1 Setting the High Voltage on the X-Ray Tube

From an XRF instrument perspective, setting the HV on the X-ray source is an easy change to make by changing the HV setting on the source's HVPS. Therefore, for any particular application, the HV can be adjusted to optimize for an XRF application quite easily. Increasing the HV does reach a maximum; the X-ray sources in HHXRF are limited to 40 or 50 kV. This limit is partially due to the efficiency of the PIN detectors and SDDs, which get quite low above 20 keV. At higher energies, the X-rays are progressively not absorbed by the detector and are therefore not detected. At 20 keV, 0.50 mm thick silicon absorbs roughly 30% of X-rays at that energy, and the absorption gets lower at higher X-ray energies. Also, the HV limit on the X-ray source is influenced by radiation safety; it is increasingly difficult to shield X-rays at higher energies. Therefore X-ray sources for HHXRF are not capable of exciting the K emission lines of elements with larger atomic numbers than lanthanum ($Z = 57$), and so in HHXRF the L emission lines (lower energy) are used to detected elements heavier than lanthanum.

In practice, HVs like 50 kV are used in the more expensive and versatile HHXRF instruments. The HVs in these higher end HHXRF instruments are used for applications such as detecting cadmium for hazardous element detection, or in mining applications, whereas 35 or 40 kV is a typical maximum voltage setting for alloy identification in scrap metal sorting. In these less expensive, single-application HHXRF instruments, the X-ray sources are typically fixed to a single voltage.

Lower voltage settings are a benefit for light element XRF detection. Often, for detecting light elements emitting fluorescent X-rays lower than 3 keV, the HV is set to a voltage from about 8 to 15 kV. These HV settings maximize the L-line X-ray intensity of the tube's Ag or Rh anode, both of which are close to 3 keV, which excites elements with an absorption edge lower than 3 keV. Additionally, keeping the voltage lower on the tube effectively turns off the detection of elements with characteristic X-ray lines above 10–15 keV, as well as reducing background at lower keV.

18.3.2 X-Ray Filters

The use of X-ray filters in front of the X-ray source is very application specific. The details for using a particular filter can be wide ranging, depending on exactly what element or elements are of interest. Most often the filter's main function is to eliminate lower energy X-rays from the spectrum, thereby improving the SNR. For example, in metal detection in scrap metal sorting, a filter such as a 0.5 mm filter of aluminum is often used. This filter eliminates most of the X-rays below about 10 keV from the X-ray source. This reduces the background between 4 and 15 kV, which is the range where fluorescence from common metals are like iron, nickel, copper, and zinc are located.

Another more complicated filter example is one used for the Restriction of Hazardous Substances (RoHS) application. RoHS specifically is looking for the elements Cr, Br, Hg, Pb, and Cd, whose fluorescence spans a wide range of X-ray energies from 5.4 keV for Cr to 23 keV for Cd. For this application, a tungsten anode on the X-ray source is desirable. A copper filter 270 μm thick blocks all the energy X-rays just above 9.0 keV, reducing the background

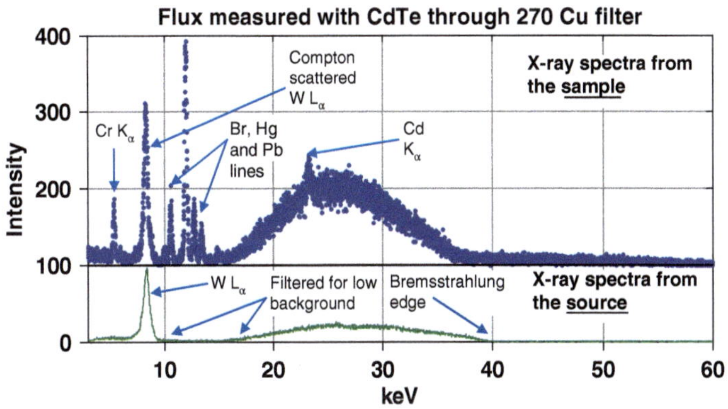

Figure 18.8 This figure shows two resulting spectra using a tungsten anode source at 40 kV with a 270 μm thick copper filter with a Cd–Te detector. The lower graph shows the spectrum from the X-ray source through 270 μm filter combination. The upper graph shows the spectrum from a polyethylene sample with 750 ppm Cr, 400 ppm Br, 75 ppm Hg, 250 ppm Pb, and 100 ppm Cd. These figures show that the copper filter both lets the W L_α-line from the source through to excite Cr, and suppresses the background under the Br, Hg, and Pd lines.

from 9 to about 15 keV, which is the region where the X-ray emission lines from Br, Hg, and Pb occur. Additionally, the tungsten L_α-line from the X-ray tube is at 8.3 keV, which is just below the absorption edge at 9.0 keV of the copper foil; therefore, the sample is exposed to X-rays at 8.3 keV. The tungsten L_α-line at 8.3 keV is effective at exciting the chromium K_α-line at 5.4 keV, thereby enhancing the signal from the chromium in the sample. Figure 18.8 shows an example of using a copper filter for a (RoHS) XRF application.

From an XRF instrument perspective, changing the filter is independent from the X-ray source; therefore changing a filter is not too hard to do. Very often several filters are mounted on a wheel, after the X-ray tube, in the front of the instrument. Depending on the exact XRF application, the wheel is turned to the filter appropriate for that particular application. Often a filter selection is done simultaneously with a change in the X-ray source's HV for a particular XRF application.

18.3.3 X-Ray Anode Selection

Unlike the HV or the filter, the anode of the X-ray tube is fixed; you are unable to change an anode material due to this material being on the inside the vacuum of the X-ray tube. Getting the anode material correct is critical for both a single-XRF application instrument, as well as instruments for multiple XRF applications, because, in many ways, it defines the application capabilities of the instrument. Therefore, the choice of anode material strongly depends on the particular XRF application.

For HHXRF, there are two major classes of X-ray tube anode types: "light element" anodes, specifically for exciting light elements with absorption edges below about 3 keV, and "general use" anodes for exciting elements higher than about 3 keV. Each class of anode will be discussed in the two sections below.

18.3.3.1 Anodes for Light Element XRF Detection

In HHXRF, light elements are defined as being from Mg ($Z = 12$) to Cl ($Z = 17$) in the periodic table. Light element XRF detection is a very technically demanding application, especially for HHXRF, and it determines most of the design limitations on X-ray sources for HHXRF, as well as on the X-ray detectors.

X-ray tube anodes for light element detection in HHXRF are most commonly made of either rhodium or silver. These elements are ideal for light element excitation in a sample because they both have L-lines which are near 3 keV: silver has L-lines between 2.9 and 3.4 keV and rhodium has L-lines between 2.7 and 3.0 keV. The L-line strongly excites fluorescent lines just under 3 keV, such as from S, P, Si, Al, and Mg, which are the specific elements of interest for light element XRF. HHXRF X-ray sources designed for light element analysis use a very thin (~125 μm, or thinner) beryllium window on the X-ray tube. This thinner window is for the purpose of minimizing X-ray absorption of the Rh or Ag L-lines in the X-ray window itself. In these light element applications, there is no X-ray filter. Air is often an unwanted filter, absorbing roughly 25% of the X-rays between 2.5 and 3 kV in about 1 cm of air (for reference, a 3 kV X-ray has a wavelength of about 0.4 nm). For lower energy X-rays subsequently emitted from the sample, the air absorption gets much stronger, and the air path is especially difficult to eliminate in HHXRF instruments. Some benchtop instruments used for light element XRF incorporate a helium purge, or a vacuum chamber, to eliminate absorption due to the air. These fundamental limitations of the HHXRF technique for light elements have led to the development of handheld laser-induced breakdown spectroscopy instrumentation – see the chapter in this volume by Day for details.

Often, X-ray sources used for light element detection are used for other XRF applications as well, which require running the X-ray source at a HV of 40–50 kV. This presents a design constraint in the X-ray source: a source optimized for 10–20 kV operation is not optimized for 40–50 kV. For HHXRF almost all the tubes are a transmission anode design, which consists of the target material deposited directly on the X-ray window with a thickness of about 1 μm. The anode thickness is optimally set to be thick enough to stop the X-ray tube's electron beam, and then thin enough to not cause any additional filtering of the generated X-rays. In the design of the anode thickness, this optimum is specific for a single HV (Figure 18.9).

For light element detection, getting a large L-line from the X-ray source onto the sample to excite the light elements is paramount. Yet the same X-ray source is needed for XRF applications with higher voltages as well. This leads to a compromise in the design between the two applications.

The rhodium or silver anode layer has a few target thickness options for operation between 20 and 40 kV. This range of target thicknesses is split between the two application needs, for light element XRF and other XRF applications. An X-ray tube manufacturer will have a selection of proprietary thicknesses optimized over this range for HHXRF, with the exact target thickness picked by the instrument manufacturer (Tables 18.3 and 18.4). Figure 18.10 shows the X-ray flux advantage and disadvantage of three different rhodium anode thicknesses at different voltage settings. Since the light element XRF detection is so much more challenging, any HHXRF instrument made specifically for light element detection will almost always use the thinner rhodium target, RH3. The RH3 target thickness increases the L-line X-ray signal below 20 kV, which is exactly the reason that this thickness

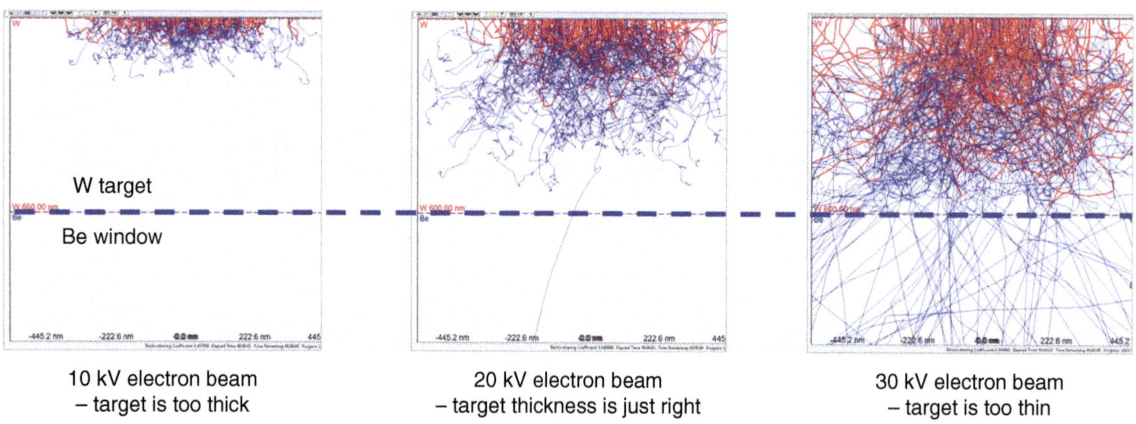

Figure 18.9 This figure shows the electron interaction of a 1 μm W foil with a 10, 20, and 30 kV electron beam. This thickness will have an optimal performance at 20 kV. Below 20 kV, the extra thickness of the foil acts as an X-ray filter, and above 20 kV, electrons get through the tungsten target into the beryllium window and do not generate the wanted X-rays. The electron interaction simulations are done in CASINO [19]. *Source*: Based on CASINO, Introduction to CASINO https://www.gel.usherbrooke.ca/casino.

Table 18.3 This table gives the calculated and measured X-ray flux output of three Rh anode thicknesses produced and Moxtek, optimized for 20 kV (RH3), 25 kV (RH2), and 30 kV (RH7).

Target ID	Target material	Optimized kV		At 10 kV	At 20 kV	At 30 kV	At 40 kV	At 50 kV
RH3	Rh	20.0	e-Anode Sim.	0.95	1.00	0.87	0.62	0.38
			X-ray L-lines	0.58	1.00	0.81	0.53	0.36
RH2	Rh	25.0	e-Anode Sim.	0.90	0.99	0.99	0.90	0.70
			X-ray L-lines	0.45	0.84	1.00	0.82	0.60
RH7	Rh	30.0	e-Anode Sim.	0.87	0.99	1.00	0.96	0.82
			X-ray L-lines	0.43	0.77	1.00	0.85	0.63

For each anode thickness, the line labeled *"e⁻ anode Sim"* shows the calculated emission of the X-ray L-lines from the anode. For each anode thickness, the line labeled *"X-ray L-lines"* shows the experimentally measured X-ray intensity in the L-lines, normalized to the X-ray tube power. The electron modeling trends line up well with the experimentally measured X-ray intensity. The differences between the two are due to the modeling not taking everything into effect such as self-anode absorption and X-ray window absorption.

Table 18.4 This table gives the calculated and measured X-ray flux output of two Ag anode thicknesses produced and Moxtek, optimized for 20 kV (AG2) and 40 kV (AG1).

Target ID	Target material	Optimized kV		At 10 kV	At 20 kV	At 30 kV	At 40 kV	At 50 kV
AG2	Ag	20.0	e-Anode Sim.	0.94	1.00	0.91	0.70	0.45
			X-ray L-lines	0.56	1.00	0.82	0.53	0.36
AG1	Ag	40.0	e-Anode Sim.	0.83	0.98	0.97	1.00	0.93
			X-ray L-lines	0.36	0.70	0.95	1.00	0.83

For each anode thickness, the line labeled *"e⁻ anode Sim"* shows the calculated emission of the X-ray L-lines from the anode. For each anode thickness, the line labeled *"X-ray L-lines"* shows the experimentally measured X-ray intensity in the L-lines, normalized to the X-ray tube power. The electron modeling trends line up well with the experimentally measured X-ray intensity. The differences between the two are due to the modeling not taking everything into effect such as self-anode absorption and X-ray window absorption.

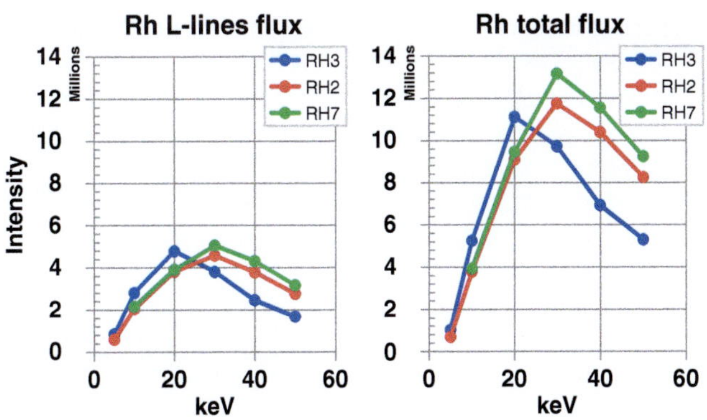

Figure 18.10 The figure on the left shows the flux intensity for the Rh L-lines as a function of the voltage on an X-ray source, normalized to the X-ray tube power of 1 W. For light element detection, the target of choice is RH3, a thinner anode, which gives roughly a 20% increase in the L-line intensity at settings between 5 and 20 keV. The disadvantage of the RH3 anode thickness is that there is a loss in overall X-ray intensity of roughly 40% at settings between 30 and 50 kV at the high end, which most HHXRF manufacturers find acceptable.

is most often chosen. The X-ray flux drop-off for settings 30–50 kV is not wanted, but applications using the higher voltages still work reasonably well, even with the large reduction in X-ray intensity.

All the arguments for the rhodium anode apply equally well to the silver anode. The choice between the two target types is often to the decision of a particular instrument manufacturer. A downside to the rhodium target is the rhodium line at 2.70 keV is very close to the K-line for Cl at 2.62 keV. Several of the Rh L energies do not excite Cl, and the peaks are close together causing peak overlap with Cl. In this particular case of Cl detection, a silver anode is a better choice.

18.3.3.2 Anodes for Other XRF Applications

General use anodes for HHXRF are most commonly made of tungsten, and work with many XRF applications, with light element XRF being a notable exception. Tungsten anodes produce more flux per unit current, the reason being that the bremsstrahlung radiation production scales with the atomic number "*Z*" of the target [7]. Additionally tungsten is a very robust material, being a refractory metal with a high melting point, which often gives it the edge over other materials. Sources that have a tungsten anode have a beryllium window thickness of 250 μm or more. Tungsten anodes are almost never used for light element XRF, so a thinner window is of no value. The tungsten anode thickness is optimized for the highest kV the source typically produces. For example an HHXRF instrument for scrap metal sorting has an X-ray source fixed at 35 kV. The tungsten anode thickness on the transmission anode source would likewise be optimized for 35 kV (Figure 18.11). Table 18.5 gives a list of typical tungsten anode thicknesses as well as the kV the anode is optimized for.

XRF applications that use a tungsten anode are very often used in conjunction with a filter exterior to the tube. Two notable filters are an aluminum filter of about 0.5 mm thickness for scrap metal sorting, and a copper filter for hazardous element detection (see Section 18.3.2).

For applications outside XRF, tungsten is the anode of choice for X-ray imaging due to tungsten's mechanical robustness and higher X-ray output. For X-ray diffraction, a tight single energy of X-rays is desired. Common anode materials are chromium (5.4 K_α), copper (8.0 K_α), and molybdenum (17.5 K_α). For X-ray diffraction, only a single characteristic line is desired, and the wide band of bremsstrahlung is not wanted because it contributes to the background noise. Often, sources for XRD are run at a kV that is three to four times the keV of the desired-ray emission line, so a copper source with a character line at 8 kV will be run at 30–40 kV on the X-ray tube.

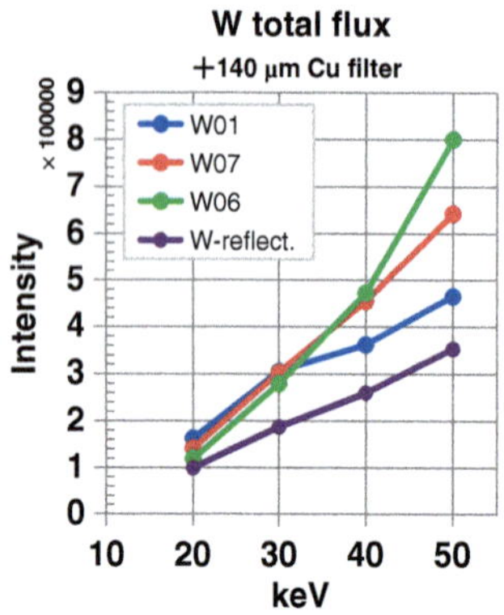

Figure 18.11 This figure shows the X-ray intensity of several different tungsten anode types optimized for different high voltages, normalized to the X-ray tube power of 1 W. This is the flux through a copper filter, which is often how such a source would be used.

Table 18.5 This table gives the calculated and measured X-ray flux output of 3 W anode thicknesses produced and Moxtek, optimized for 25 kV (W01), 40 kV (W07), and 60 kV (W06).

Target ID	Target material	Optimized kV		At 10 kV	At 20 kV	At 30 kV	At 40 kV	At 50 kV	At 60 kV	At 70 kV
W01	W	25.0	e-Anode Sim.	0.88	0.97	0.96	0.82	0.58	0.39	0.26
			X-ray 3.5–10 keV	0.25	0.75	1.00	0.92	0.72	—	—
W07	W	40.0	e-Anode Sim.	0.77	0.87	0.98	0.99	0.95	0.78	0.58
			X-ray 3.5–10 keV	0.15	0.54	0.84	1.00	0.96	—	—
W06	W	60.0	e-Anode Sim.	0.37	0.69	0.90	0.96	0.98	1.00	0.94
			X-ray 3.5–10 keV	0.01	0.32	0.52	0.68	0.81	0.94	1.00
W–R	W	—	e-Anode Sim.	—	—	—	—	—	—	—
			X-ray 3.5–10 keV	0.37	0.79	0.95	1.00	0.83	—	—

W–R is a thick refection anode, like more traditional X-ray tube designs. For each anode thickness, the line labeled "*e^- anode Sim*" shows the calculated emission of the X-ray L-lines from the anode. For each anode thickness, the line labeled "*X-ray L-lines*" shows the experimentally measured X-ray intensity between 3.5 and 10 kV, normalized to the X-ray tube power. The electron modeling trends line up well with the experimentally measured X-ray intensity. The differences between the two are due to the modeling not taking everything into effect such as self-anode absorption and X-ray window absorption.

18.4 Functionality of X-Ray Sources for HHXRF

X-ray sources for XRF have to be extremely stable and reproducible. More specifically, the **spectrum shape** (example in Figure 18.4) should not change in an XRF instrument for a particular HV setting over time, or any other variables like temperature. XRF manufacturers and XRF instrument users want their instrument to be repeatable. They all want the XRF instrument to give the same stoichiometric result each time it is used. If there is even a minor drift in the spectrum shape from the X-ray source, this has a good chance of likewise affecting the XRF stoichiometry results.

For many lab-based instruments, recalibration procedures are used to keep the results consistent over time, dealing with any drifts within the instrument, including the X-ray source over time. Recalibration can also be done within a HHXRF instrument, but in practice many instruments are calibrated a single time in production, and are never fully recalibrated again. Most HHXRF units do have a single reference sample to check the calibration, but the HHXRF device is more or less expected to give the same result in the stoichiometry for the lifetime of the device, without a full recalibration.

From an X-ray source perspective, this means that HHXRF manufactures want X-ray sources in which the spectrum shape, at a particular HV setting, is very repeatable and stable over the lifetime of the X-ray source. Keeping this spectrum shape the same over many different conditions for miniature X-ray sources for HHXRF is the major theme of the rest of this section. It is central to the design of sources for HHXRF.

From an X-ray source perspective, here is a list of major ways one can change the spectrum shape on an X-ray source:

1. **Change the HV**: This is very changeable, by design. The HV stability and repeatability are at the core of this section.

2. **Change the emission current**: This does change the X-ray intensity, but should not change the spectrum shape.
3. **Change the anode target material or the anode thickness**: The anode material and thickness are stable for a single X-ray source in normal operation.
4. **Change the X-ray filter**: In a HHXRF device the filters on the filter wheel are unlikely to change.

The HV stability of the X-ray source is extremely critical for XRF. Changing the HV on the X-ray source changes the spectrum shape, which changes the ratio of X-ray counts between each element's X-ray lines from a sample. The ratio between the different characteristic lines gives the stoichiometry of these elements, so if this ratio changes the XRF calibration is compromised (Figure 18.12). Because HV stability is so critical to the source's spectrum shape stability, and the spectrum shape stability is so critical to the XRF results, we look directly at the HV stability (see Section 18.4.1) from the spectrum from the source, by measuring the bremsstrahlung edge to insure the X-ray source has a stable HV (see Section 18.4.1).

The emission current changes the intensity of the X-rays from the source; it does not change the spectrum shape. Therefore, changing the emission current does not change the resulting ratios between X-ray lines in XRF from a sample. The emission current is important; setting it allows maximizing the counts on the XRF, and HHXRF manufactures need the flux to be linear with the emission current (Section 18.4.4). But the emission current is not as nearly critical an issue, compared to a stable HV which affects the spectrum shape.

The anode thickness in the X-ray tube also affects the spectrum shape, in that the anode stops the electrons generating X-rays, and also acts as an X-ray filter for the generated X-rays. Tuning the thickness of the anode has several benefits needed for technically demanding applications such as light element XRF, as discussed in Section 18.3.3.1. Having tightly controlled thickness on the anode material is important to maintain an analogous spectrum shape between two different sources. Getting two different sources to have the same spectrum shape will be discussed below in Section 18.4.4.

Changing an X-ray filter also drastically changes the source's spectrum shape. In practice the filters are changed via a filter wheel on the front of the detector, based on the application (Section 18.3.2). If a HHXRF instrument is calibrated with a specific source and set of X-ray filters, the anode and filters for this instrument should not change, so the calibration should be stable. The designed filters will not change, but unwanted filters, such as

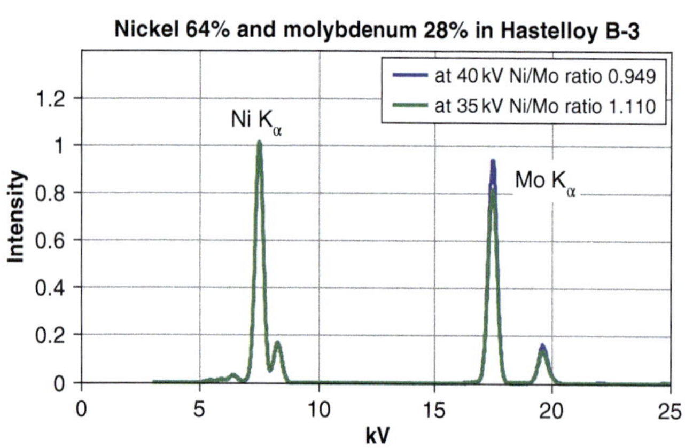

Figure 18.12 This figure shows two resulting spectra from a Hastelloy B-3 sample excited with a silver anode tube with a 70 µm Cu filter. The two spectra where taken with the tube set to 40 and 35 kV. This high-voltage shift changed the ratio between the peaks by 16%. In this setup, the nickel-to-molybdenum ratio can be changed by 0.5% by changing the voltage by just 200 V. This demonstrates that a stable high voltage is needed to maintain a consistent ratio between elemental X-ray lines, and to maintain a HHXRF calibration.

debris getting on the X-ray window, of the X-ray tube, or anywhere else along the beam path, can compromise the spectrum shape.

In the following sections, we will be exploring each of these factors, with a major emphasis on keeping the HV repeatable and stable over time and other variables. Substantial effort has been taken to fully characterize the miniature X-ray sources under these categories: environmental stress tests, X-ray flux and spot stability and repeatability tests, and multi-settings tests. The intent of this test is to provide confidence for those using these X-ray sources in HHXRF.

18.4.1 Environmental Stress Testing

One consideration HHXRF has over lab-based XRF instruments is the environment [20]. HHXRF is meant to be used in a large range of field environments, and give the same stoichiometry in each of these environments. This translates into having the X-ray source functioning over an extended temperature range. Not only do the components need to stay operational, large efforts are made to keep the X-ray output stable over extended temperature ranges as well.

The typical temperature specification is a storage temperature range of −40 to 80 °C and an operational range of −10 to 60 °C. While developing a miniature X-ray source, the design is considered to be robust when the X-ray source can survive a 120 hour environmental test during which the temperature is ramped between the temperature extremes repeated 20 times. Each testing cycle is meant to find any weaknesses in the design, and when failure points are found they are fixed.

On recent X-ray source designs, the tests over these temperature ranges have also been done in conjunction with measuring the spectral output, in effort to minimize the spectrum shape changes over extended temperature ranges. Recent designs of the X-ray sources are made to vary less than 100 V in the HV over the operational temperature range of −10 –60 °C, to make the HHXRF system stable as possible, and give nearly identical results independent of the temperature of the environment.

18.4.2 X-Ray High-Voltage Stability

Of all items on the list of ways to change the spectrum shape, the X-ray source's HV stability and repeatability are the most critical. This section's focus is on verifying and monitoring an X-ray source to make sure the HV is stable over time.

A stable spectrum shape from an X-ray source requires that the HV on that X-ray source must remain constant over time, hours to days to months, as well as over on and off cycles. Measuring the HV stability is accomplished by evaluating the X-ray spectrum directly, specifically the bremsstrahlung edge, over and over, in a setup like the one shown in Figure 18.7. A spectrum's bremsstrahlung edge is fit with a proprietary algorithm, giving the value of the bremsstrahlung edge in keV. This measurement provides the "spectroscopic" voltage in keV. The bremsstrahlung edge in keV coincides with the X-ray source's actual HV in kV (Figure 18.13).

If the bremsstrahlung edge moves over time, it means that the HV has shifted on the HVPS. Any changes in the HV over time in the X-ray source are very undesirable. The bremsstrahlung edge technique is very sensitive; changes in the HV on the order of 50–100 V can be seen. From an X-ray source's perspective, this measurement technique is extremely valuable because it detects voltage shifts semidirectly in units of keV, decoupling this key issue from other causes of flux drift, such as emission current changes or X-ray spot movement on the anode.

Figure 18.14 shows the results of the bremsstrahlung edge detection from a series of spectra measured every minute for eight hours. The HV stability results shown in Figure 18.12 are exactly what an XRF instrument needs from its X-ray source: the HV is stable to within ±100 V for HHXRF over and extended time frame of hours to the life time of the source.

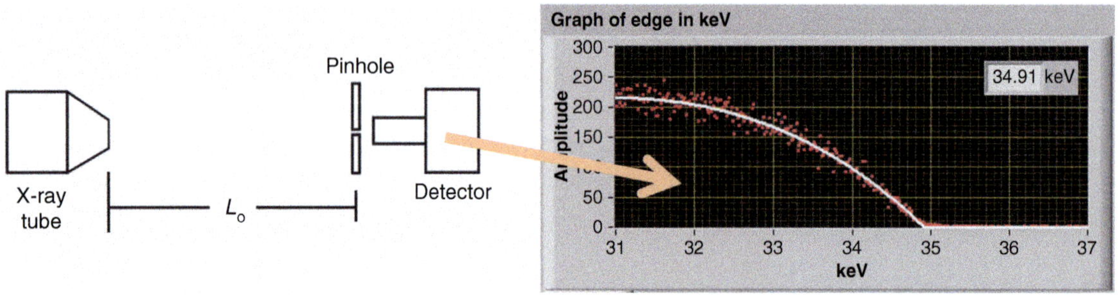

Figure 18.13 A simple schematic on the right shows the setup used to measure the bremsstrahlung edge from an X-ray source. The graph on the right shows a spectrum measurement, zoomed in on the bremsstrahlung edge. In this case, the high-voltage calibrated source was set to 35 kV and the bremsstrahlung edge was measured to be at 34.9 keV, giving a small 0.10 kV offset between the two values.

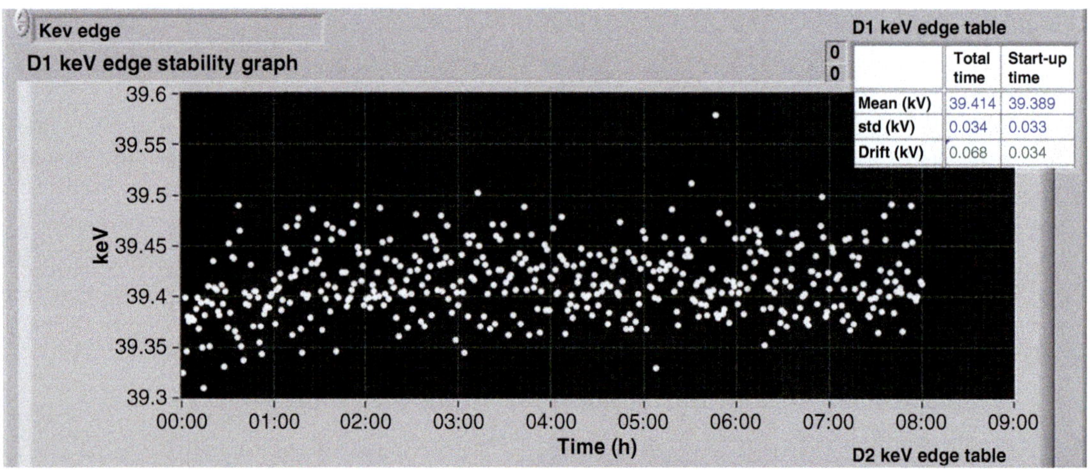

Figure 18.14 A high-voltage repeatability test from a miniature X-ray source for HHXRF. An algorithm finds the bremsstrahlung edge in a spectrum, and records it for each of the spectra taken over eight hours. This source achieved a voltage of 39.4 kV (set at 40 kV), with a standard deviation of 34 V. This level of high-voltage stability is good for getting a stable HHXRF instrument.

18.4.3 X-Ray Flux Stability and Repeatability

The HV stability is just one of the most critical elements to having a stable flux, and having a stable and repeatable X-ray flux from the X-ray source is also needed for XRF. Types of issues that affect the flux stability are changes in the HV, spot movement on the X-ray tube, and changes in the emission current. As an X-ray source warms up when used, there are inevitable small changes in both the X-ray tube and the HVPS which will shift the X-ray flux around. The aim in designing these X-ray sources is to make the source as stable as possible throughout any changes.

The stability and repeatability test evaluates how the X-ray source operates over time. For the stability test, the X-ray source is left on for the entire test. For the repeatability test, the X-ray source is periodically cycled on an off: typical off cycles are 2–10 seconds, and typical on cycles are 30–120 seconds. The test durations can be from a time frame of several minutes to days. Testing the stability and repeatability source involves several detectors and instruments simultaneously measuring and evaluating the X-ray source, and Figure 18.15 shows an X-ray source testing setup, which incorporates several detectors. Having all these instruments evaluating the X-ray source simultaneously effectively pinpoints any issues on an X-ray source if the X-ray source is unstable.

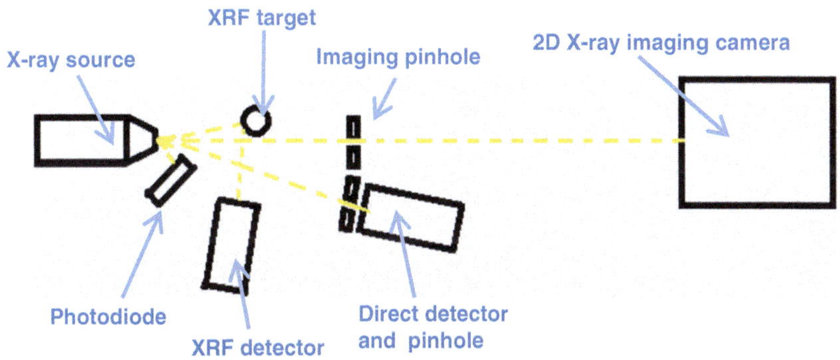

Figure 18.15 A simple schematic of the stability and repeatability test showing all the major instruments, three flux detectors and an imaging detector, simultaneously measuring and evaluating the X-ray source. *Source*: Based on [10].

For both the stability and repeatability tests, the flux is measured directly from the source, which gives the flux stability (Figure 18.16), spectrum shape stability as well as the HV stability (Section 18.4.2). XRF flux is measured a second way using a secondary XRF target, simulating XRF instruments' stability. The flux output is measured a third way with a photodiode, which is very good for monitoring fast flux changes over tenths of a second. A time series of pinhole X-ray images of the X-ray tube's anode is also measured, which gives X-ray spot position stability (Figure 18.17). Each of these methods gives flux stability information, and differences between each measurement give vital clues as to what the causes any flux instability, if there is any. In addition to the four ways used to evaluate the flux from the source, several thermocouples are placed on different locations of the source's tube and HVPS,

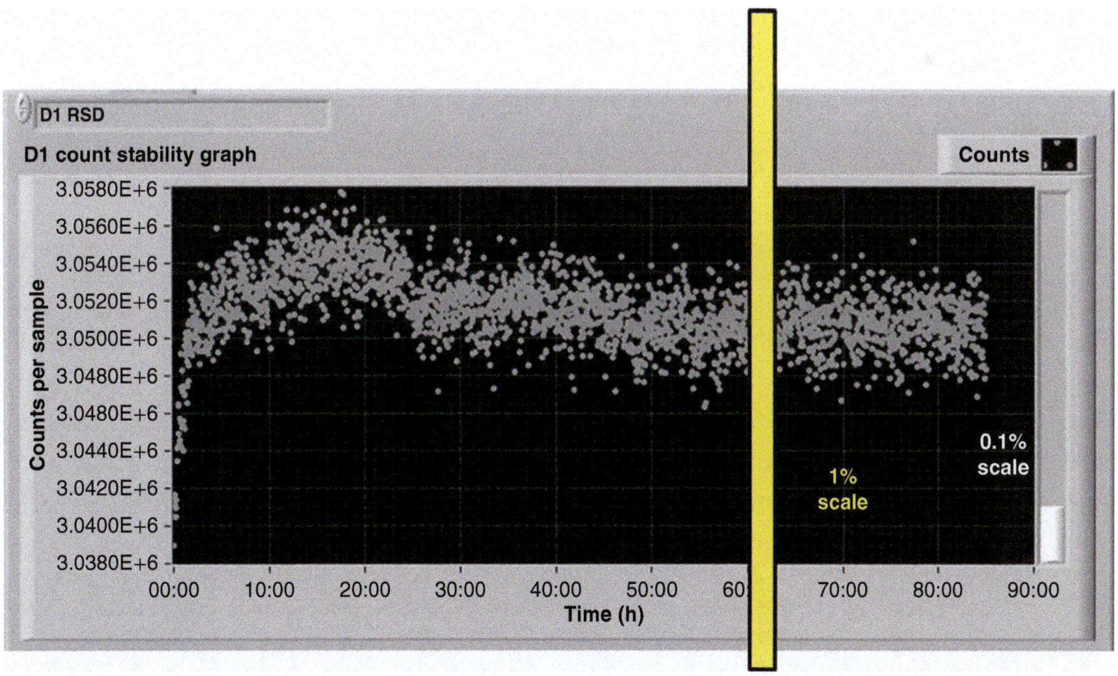

Figure 18.16 An X-ray flux measured directly from ULTRA-LITE source, set at 50 kV and 80 μA (the full 4 W) during a repeatability test, cycled on for 120 seconds and off for 5 seconds for 2400 cycles. This test measured an X-ray flux repeatability of 0.10% RSD over 85 hours. During the source's warm-up time of 30–45 minutes, the X-ray flux drifted by 0.45% [10].

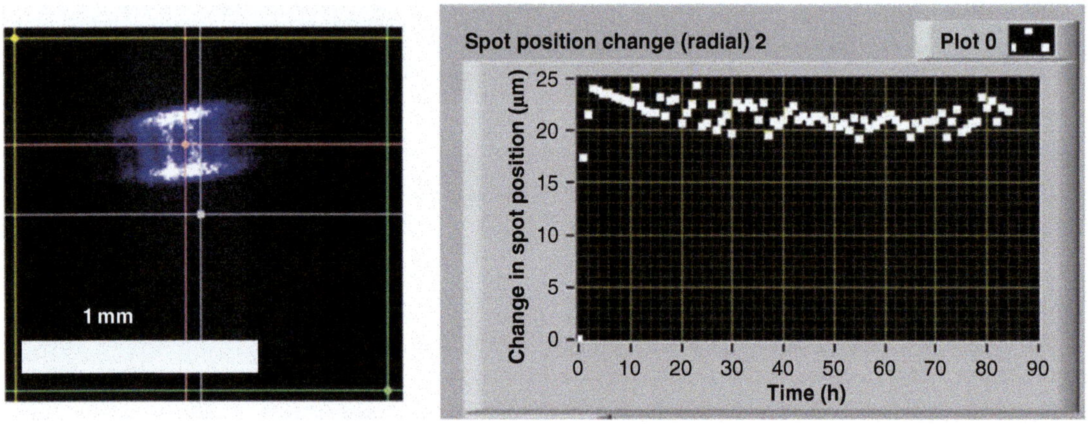

Figure 18.17 An X-ray spot repeatability test from ULTRA-LITE source, taken simultaneously with information in Figures 18.16. To the left is a single pinhole camera image of the X-ray spot, with a spot size of 390 × 340 μm FWHM. The small white dot in the center of the image represents the physical center of the anode. The graph to the right shows the drift of the spot over time; it drifts ~20 μm as the source warms up over the first 30–60 minutes and has a constant position thereafter over 85 hours [21, 22].

and every signal and power line going into and out of the HVPS is monitored. Collecting all this information simultaneously permits finding correlations in the data, which is critical for troubleshooting issues, as problems are found. This stability and repeatability testing verifies that the X-ray source has a stable HV, a stable X-ray flux, and a stable X-ray spot needed for a HHXRF. At our company, this level of testing for sources has been used in development for several years [10]. The basics of this testing platform have also been moved to the production level, because it has been very effective at both detecting and diagnosing any issues in the X-ray sources produced.

18.4.4 Multi-Settings Testing and Flux Linearity

The stability and repeatability testing verifies that an X-ray source is stable at one setting over time, and the multi-settings test determines the X-ray source functions over its entire range of voltage and emission current settings. In the multi-settings test, the X-ray source is turned on and off for a few seconds over a few hundred to thousands of different HV and emission current settings. The flux value is recorded with a photodiode, and every signal and power line going into and out of the HVPS is monitored. All of this information provides several ways to evaluate the X-ray source. Here we present data collected with the just photodiode which give information on:

- The flux turn-on time.
- The flux stability for a few seconds at each setting.
- The flux linearity over many settings.

For HHXRF, a quick turn-on time of under a second is desired. Figure 18.18 shows the signal flux, measured by the photodiode, as the X-ray source is turned on. The turn-on time can be measured over all the settings evaluated.

The flux stability for a few seconds at each setting is very valuable as well, showing if there are any settings that have an unstable flux. Figure 18.19 shows photodiode flux stability over a few seconds at each setting on an early prototype X-ray source. This was a very extensive test, with the source tested in 1 μA steps (ranging from 0 to 200 μA) and 0.5 keV steps (ranging from 4 to 50 data keV) resulting in data taken over 13 000 individual settings, with the entire test taking about 24 hours. A low standard deviation in the flux signal represents the desired stable flux output, and is represented in blue dots in the plot of Figure 18.19. A high standard deviation in the flux signal represents the undesired unstable flux output, and is represented in either red or yellow dots in the plot

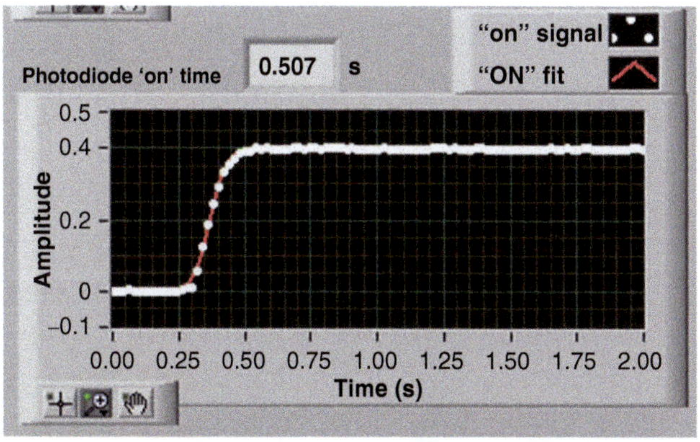

Figure 18.18 This shows the X-ray flux signal measured over time by the photodiode as the source turns on. For this setting of 50 kV and 80 μA, to the flux is fully on after 0.51 seconds [10].

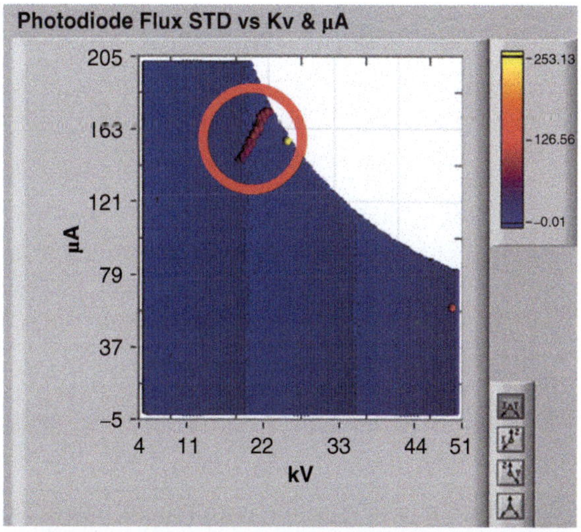

Figure 18.19 An example of data collected from the multi-settings test where the set voltage and the set current are plotted against the standard deviation of the X-ray flux measured, with a photodiode. An unstable flux output region is clearly seen inside the red circle [10].

of Figure 18.19. The settings with unstable flux stability are clearly identified in the region circled in red. This mapping procedure insures that the X-ray source is functional with a stable flux output over all the available HV and emission current settings that may be needed for a given XRF application.

Lastly, the flux linearity of the X-ray source is measured over several HVs. The flux output should be linear with the emission current on the source when the source is held at a constant HV (Figure 18.20). Measuring the X-ray flux at several emission current settings verifies this relationship. For HHXRF, flux linearity as a function of current is desired to insure the instrument's calibration.

18.4.5 X-Ray Spectrum Stability Between Different X-Ray Sources

The previous sections focused on making sure the spectrum shape from a single X-ray source was stable and consistent over time and other variables. Spectrum shape stability is critical for an X-ray source within a HHXRF instrument, so the HHXRF instrument can maintain its calibration for months to years. In this section, the focus shifts from the same spectrum shape from over time on a single source, to obtaining the same spectrum shape across several different X-ray sources.

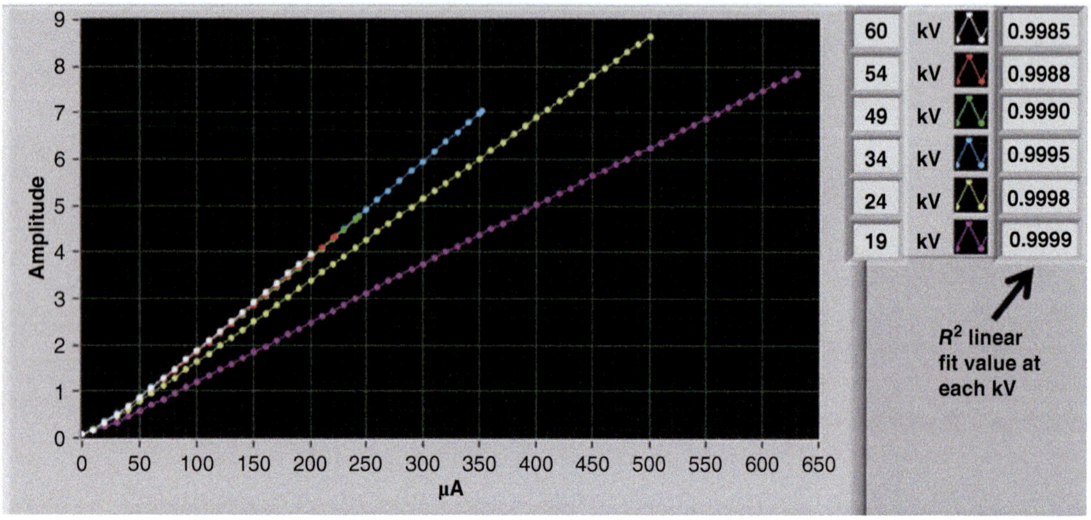

Figure 18.20 The flux linearity from a tungsten X-ray source at a few selected high voltages. For settings from 60 to 19 kV the source flux is very linear, with R^2 values all under 0.998. Several of the higher voltage linear curves are on top of each other due to the photodiode not detecting higher energy X-rays.

One feature desired by manufacturers of HHXRF instruments is having X-ray sources which are spectroscopically identical. A spectroscopically identical X-ray source allows one to switch out the source in an instrument with no other required changes: the new source will generate the same XRF spectra as the old source, with minimal or no recalibration of the instrument. This allows swapping of X-ray sources in an instrument, without requiring instrument recalibration, and is a major plus for instrument manufacturers, in both their manufacturing and service areas.

One of the major obstacles in making sources that are close to spectroscopically identical is having a tight and repeatable control over the voltage of the X-ray source's HVPS, which is discussed extensively in this section. A tight control is also needed within the anode thickness material, as well as any filters that are used. Controlling these features will allow for an XRF equivalent X-ray source with the same spectrum shape, so that one X-ray source can be exchanged for another X-ray source and the calibration on the XRF instrument is unaffected.

A proven solution for spectroscopically identical sources is the digitally controlled Moxi X-ray source. These handheld X-ray sources are spectroscopically identical for several XRF applications. Each digital X-ray source is calibrated to within ±250 V over all settings between 4 and 50 kV, and this voltage accuracy is tight enough to allow for XRF equivalence for several XRF applications. We will outline how the new digital sources are HV calibrated, and then show how this calibration leads to a spectroscopically identical X-ray source.

18.4.5.1 High-Voltage Calibration of X-Ray Sources

In order to calibrate the HV on an X-ray source, two basic requirements are needed:

- A meaningful and accurate way to measure the HV.
- A way for each X-ray source to store the HV calibration information, offsetting source-to-source differences due to variance in electrical components which make up the source.

The X-ray source's HV is measured by collecting a spectrum from the tube's anode, as discussed in Section 18.4.2. The bremsstrahlung edge measurement provides the "spectroscopic" voltage in keV, while the electronic monitors on the source's HVPS give an electronically measured voltage in kV (Figure 18.13). The difference between the two gives a correction factor, which can be used to tighten the accuracy between these two values.

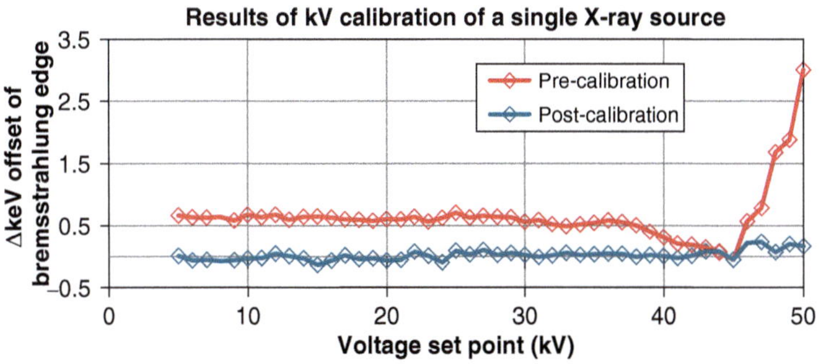

Figure 18.21 This shows the offset between the tube's set voltage and bremsstrahlung edge measurements in 1 kV increments. The red line is the offset before the X-ray source is high-voltage calibrated, and the blue curve shows the offset after the digital X-ray source is high-voltage calibrated. Before calibration, the offset was close to 600 V, and gets as high as 3 kV at 50 kV. After calibration, the source is within ±250 V, compared to the bremsstrahlung edge measurements.

The X-ray sources are designed to be operated over a range from 5 to 50 kV, and therefore spectra are taken at several HV. The bremsstrahlung edge is recorded over several tube voltage settings, generating a keV to kV offset for each voltage setting. In Figure 18.21, the red line shows the offset between the spectroscopically measured bremsstrahlung edge in keV and the pre-calibrated tube's set voltage in kV. One key idea is the pre-calibrated offset is nonlinear after about 40 kV. The electronic components do not have a strictly linear relationship with HV. This nonlinearity is due to variance in the HVPS board components, specifically the HV feedback resistor (Figure 18.5).

For an analog X-ray source, without any onboard memory, the variance in the electronic components in aggregate is the major driving factor for the HV offset of any particular X-ray source. This represents the limit in HV accuracy for an analog source. In a digital source, with onboard memory, a digital correction factor is stored for each kV setting, compared to bremsstrahlung edge. Most of the HV variance due to analog board component variations can be removed by the combination between measuring several bremsstrahlung edge spectra and storing a correction factor on the source's memory.

Manufacturers need every X-ray source produced to output the same tube spectrum shape to gain the benefits they want. Figures 18.22 and 18.23 show the HV offset between the bremsstrahlung edge measurement and the electronic HV setting over a 100 different X-ray sources. Figures 18.22 and 18.23, which intentionally have the same scale, show both the accuracy and precision of the HV between X-ray sources is drastically increased by the HV calibrating procedure. It is easily seen that the calibration done on the digital tube greatly reduces the voltage offset variation between multiple X-ray sources.

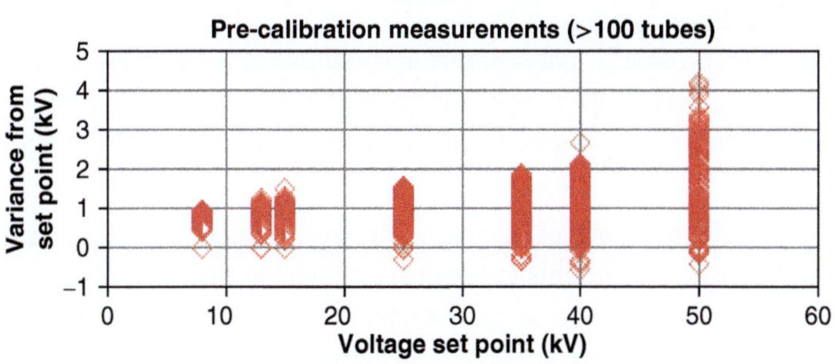

Figure 18.22 This graph shows the offset between the tube's set voltage and bremsstrahlung edge over seven settings pre-calibration, on over 100 separate X-ray sources. This is the offset to be expected on a strictly analog X-ray source. Pre-calibration, the high voltage is typically offset by 800 V, and can vary by up to ±1000 V (at 35 kV).

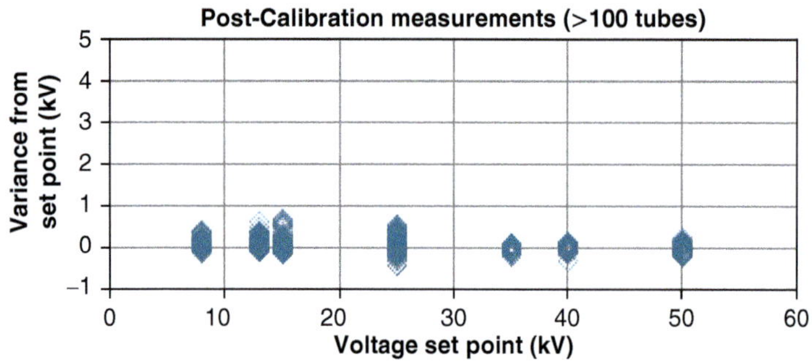

Figure 18.23 This graph shows the offset between the tube's set voltage and bremsstrahlung edge over seven settings post-calibration, on over 100 separate X-ray sources. Post-calibration, the high voltage is typically offset by 100 V, and can vary by up to ±250 V (at 35 kV).

The voltage accuracy and precision achieved in the digital X-ray source are much more tightly controlled. The benefit of this increased voltage accuracy has resulted in reducing manufacturing complexity in the calibration, and service time for XRF handheld instruments manufactures. The next section will quantify what this voltage accuracy permits in terms of the resultant stoichiometry from an XRF instrument.

18.4.5.2 X-Ray Source High-Voltage Dependence on XRF Stoichiometry

Several experiments were performed exploring how voltage variation on the X-ray source impacts the calculated stoichiometry from XRF for two different applications: nonaluminum metal sorting and aluminum metal sorting. The basic outline of the experiment is as follows:

- Five spectra were collected from a sample with over 800 k counts at a single HV setting.
- The spectra where then converted into elemental stoichiometry percentages with SinerX fundamental parameters software. The resulting stoichiometry percentages were averaged from the five spectra, giving a basic measurement variance.
- The HV of the tube was adjusted by a step size ranging from 200 to 500 V, depending on the sample, and another five spectra were collected.
- This was done over a range of voltage from 500 to 2000 V, depending on the sample.

Two samples were investigated:

- Hastelloy, with the tube nominally set to 35 kV to represent a "harder" case for nonaluminum metal alloy sorting (Figure 18.12). Hastelloy was chosen due to the Ni and Mo fluorescent peaks being far apart in energy. This large difference in the characteristic X-ray energies (7.4 keV Kα for Ni and 17.4 keV Kα for Mo) maximizes the change in the calculated stoichiometry when the tube's voltage is changed.
- Al 7075, with the tube nominally set to 12 kV, to represent a "harder" case for aluminum metal alloy sorting. Al 7075 was chosen due to the Al and Zn peaks being far apart in energy. This large difference in the characteristic X-ray energies (1.5 keV Kα for Al and 8.6 keV Kα for Zn) maximizes the change in the calculated stoichiometry, when the tube's voltage is changed.

From the data collected, the effects of changing the HV on the X-ray source on the stoichiometry could be quantified. As expected, as the voltage was changed the number of counts in the peaks nearest to the bremsstrahlung edge changed the most, Mo for the Hastelloy and Zn for the Al7075. The calculated Mo stoichiometry percentage was much more sensitive to tube voltage changes than the Ni percentage in the Hastelloy, and likewise with the Zn compared to the Al in the Al7075 sample.

If the Mo and Ni (or Al and Zn) calculated stoichiometry from an XRF spectra stays constant from a given XRF measurement, then we expect spectral XRF equivalence. The question becomes what X-ray source HV variance

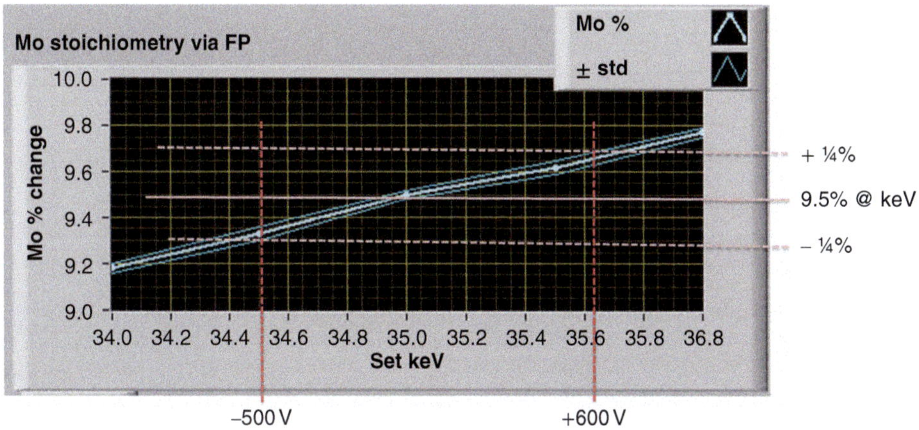

Figure 18.24 This graph shows the stoichiometry change of the Mo as the high voltage of the tube is changed. The standard deviation of the measurements is represented by the thin lines just above and below the main center line. Holding the Mo peak stoichiometry to under ±¼% error requires a high-voltage accuracy on the order of ±500 V. Our current high-voltage calibration is under ±250 V at 35 kV.

keeps the Mo and Ni (or Al and Zn) calculated stoichiometry small enough for XRF equivalence? To "quantify" this, if the stoichiometry of the elements stayed under 0.25%(0.5%), then the result is considered to be "XRF equivalent." This quantification has been verified in practice, in that it works as intended in next-generation handheld X-ray sources sold today. Figure 18.24 shows the results for the stoichiometry of the Mo changing as a function of the tube's HV for the Hastelloy sample at 35 kV, and likewise Figure 18.25 for the Zn lines for the Al7075 sample at 12 kV. From these graphs, it is fairly easy to see what the voltage range on the X-ray source needs to be to keep the calculated stoichiometry at or under ¼% (Figures 18.24 and 18.25).

Nonaluminum metal sorting sources may have a ±500 V variance in HV; this limit will produce "equivalent" sources, meaning that the stoichiometry should not vary more than ¼%, tube-to-tube. The test case with the Hastelloy sample is a more demanding example for this application, due to the zinc and molybdenum lines being very far apart in X-ray energy. Therefore, with most other typical samples in which the element X-ray lines are closer in energy to each other, the difference a voltage change will have on the stoichiometry will tend to be lower.

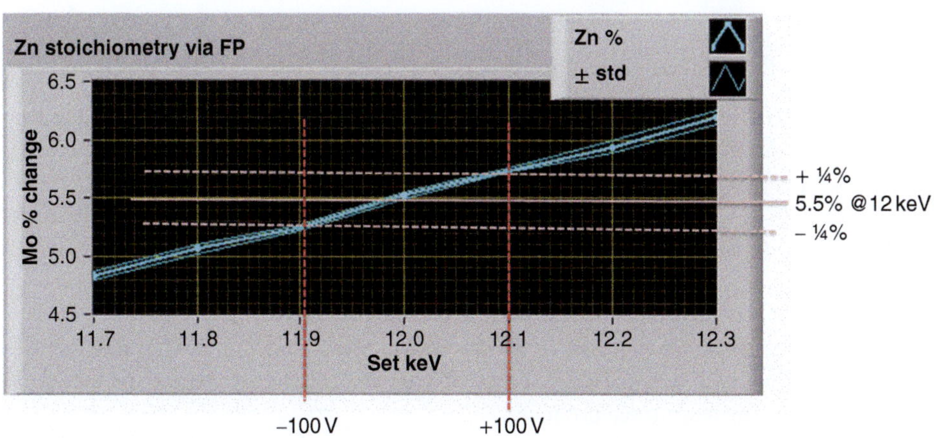

Figure 18.25 This graph shows the stoichiometry change of the Zn as the high voltage if the tube is changed. Holding the Mo peak stoichiometry to under ±¼% error requires a high-voltage accuracy on the order of ±100 V. Our current high-voltage calibration is under ±250 V at 12 kV.

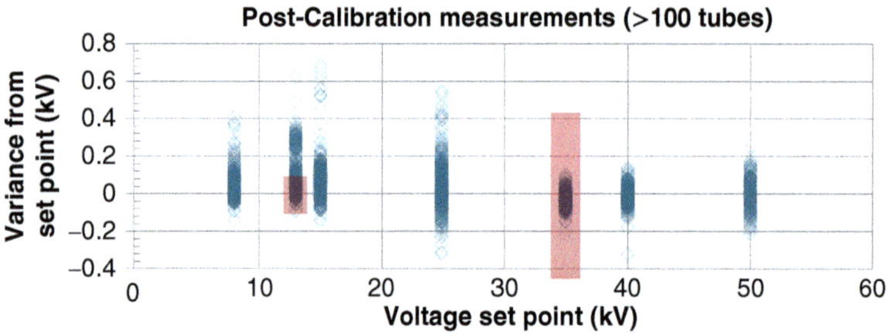

Figure 18.26 This graph shows the offset between the tube's set voltage and bremsstrahlung edge over seven settings post-calibration, on over 100 separate X-ray sources. This is the exact same graph as Figure 18.23, with the *y*-axis zoomed in too show more detail. The red bars show the source's high-voltage accuracy needed to keep calculated stoichiometry from XRF spectra under ¼%, for two different experimental setups, aluminum metal testing at 12 keV and nonaluminum metal sorting at 35 keV.

This test case can be extended to other XRF methods, such as ppm detection for RoHS/WEEE (waste electrical and electronic equipment), in which the X-ray source is at or above 35 kV. For these XRF techniques, keeping the voltage of the source to ±250 V should keep the measurements very close to equivalent.

Aluminum metal sorting sources may have a ±100 V variance in HV; this limit will produce "equivalent" sources, meaning that the stoichiometry should not vary more than ¼%, tube-to-tube. Similarly, the test case with the Al7075 sample is a more demanding example for this application, due to the aluminum and zinc lines being very far apart in X-ray energy. Therefore, with most other typical samples in which the element X-ray lines are closer in energy to each other, the difference a voltage change will have on the stoichiometry will tend to be lower. For low-voltage XRF techniques, keeping the voltage of the source to ±250 V is not enough to keep the stoichiometry to within ¼%, but it is almost enough to keep it within ½%. Even so, this level of improvement in voltage variation is drastically smaller than for an analog X-ray source counterpart.

Figure 18.26 replots the variance of over 100 digitally calibrated sources, along with red bars giving the range of voltages required for "XRF equivalent," keeping the stoichiometry variation under ¼%. From this, we can see that the higher voltage XRF applications are well within the voltage tolerance needed to keep the voltage variance down to be "XRF equivalent." For the more demanding XRF application at low kV, the control is not to the "XRF equivalent" level, but is closer to this ideal with the calibrated digital sources than without it.

18.5 Conclusion

The major advantage of portable or HHXRF is that the instrument can be taken to the sample, rather than the sample to the instrument. As a few examples, these instruments are taken directly to mines, on-site in mining investigations, used directly in scrap metal sorting yards, and taken into homes for lead in paint investigations. The development of miniature, lightweight, X-ray sources was the critical component needed to shift XRF spectroscopy from the lab to the field. This portability has proven to be a very big advantage, demonstrated by the rapid growth of the HHXRF market in 2004–2020.

For HHXRF, keeping the X-ray source small is very important, with a major emphasis on keeping the HV repeatable and stable over time and other variables (e.g. ambient temperature). Full characterization of these miniature X-ray sources provides confidence in the stability of the X-ray source within the HHXRF instrument, and therefore the qualitative and quantitative results the instrument produces.

"XRF equivalent" X-ray sources for HHXRF can be produced by careful HV calibration of the digital X-ray source. This HV calibration is achieved by bremsstrahlung edge measurements, and storing HV offset information on each source. The stored offset corrects for HV discrepancies due to the variance in the electronic components of the HVPS.

References

1 Potts, P.J. and West, M. (eds.) (2008). *Portable X-Ray Fluorescence Spectrometry, Capabilities for in Situ Analysis*. Cambridge, UK: RSC Publishing.

2 Cornaby, S. and Kozaczek, K. (2016). Sources for handheld X-ray fluorescence instruments. In: *Encyclopedia of Analytical Chemistry* (ed. R. Meyers). John Wiley & Sons.

3 Van Griken, R.E. and Markiwicz, A.A. (2002). *Handbook of X-Ray Spectroscopy*, 433. New York: Marcel Dekker.

4 Bosco, G. (2013). Development and application of portable, hand-held X-ray fluorescence spectrometers. *TrAC Trends Anal. Chem.* 45: 121–134.

5 Reyes-Mena, A., Knight, L.V., and Cornaby, S. (2000). *An XRD/XRF Instrument for the Microanalysis of Rocks and Minerals*. SBIR contract number NAS2-00010.

6 Coolidge, W.D. (1913). A powerful Röntgen ray tube with a pure electron discharge. *Phys. Rev.* 2: 409.

7 Lee, S.M. and McDonald, C. (2009). *Handbook of Optics*, 3e, vol. V (ed. M. Bass). New York: McGraw-Hill Companies, Inc., Chapter 54.

8 Kossel, W. (1916). Bemerkungen zum Seriencharakter derRongtgenspektren. *Verh. Deut. Phys. Ges.* 18 (339): 396.

9 Kossel, W. (1917). Zum Ursprung der Υ-Strahlenspektren und Rontgespektren. *Phys. Z.* 18: 240.

10 Kossel, W. (1920). Zum Bau der Rontgespektren. *Z. Phys.* 1: 119.

11 Kossel, W. (1920). Über die Ausbildung der Röntgenserien mit wachsender Ordnungszahl. *Z. Phys.* 2: 470.

12 Moseley, H.J.G. (1913). The high frequency spectra of the elements part I. *Philos. Mag.* 26 (6): 1024.

13 Moseley, H.J.G. (1914). The high frequency spectra of the elements part II. *Philos. Mag.* 27 (6): 703.

14 Barkla, C.J. (1915). X-ray fluorescence and the quantum theory. *Nature* 95: 7.

15 Cockcroft, J.D. and Walton, E.T.S. (1932). Experiments with high velocity positive ions. (I) Further developments in the method of obtaining high velocity positive ions. *Proc. Roy. Soc. London Ser.* A136: 619.

16 Cornaby, S. and Kozaczek, K. (2012). Flux comparisons of miniature X-ray source: XRF wattage-maps by application. *58th Denver X-Ray Conf. Adv. X-Ray Anal.* 56: 202–208.

17 Jenkins, R., Gould, R.W., and Gedcke, D. (1995). *Quantitative X-Ray Spectrometry*. New York: Marcel Dekker.

18 Jenkins, R. (1999). *X-Ray Fluorescence Spectrometry*, 2e. Chichester, UK: Wiley-Interscience.

19 CASINO. https://www.gel.usherbrooke.ca/casino (accessed 10 October 2020).

20 Cornaby, S., Morris, S., Smith, J. et al. (2012). Moxtek's new ultra-lite X-ray sources: performance characterizations. *58th Denver X-ray Conf. Adv. X-ray Anal.* 56: 202–208.

21 Caruso, D., Dinsmore, M., and Cornaby, S. (2009). Miniature X-ray sources and the effects of spot size on system performance. *58th Denver X-Ray Conf. Adv. X-Ray Anal.* 53: 228–233.

22 Uchida, H., Hasuike, K., Torii, K., and Tsunemi, H. (2006). Quantitative method of measuring spot size of microfocus X-ray generator. *Jpn. J. Appl. Phys.* 45 (6A): 5277–5279.

19

Semiconductor Detectors for Portable Energy-Dispersive XRF Spectrometry

Andrei Stratilatov

Moxtek, Inc., Orem, UT, USA

19.1 Introduction

Historically, the primary sensors used in energy-dispersive X-ray Fluorescence (EDXRF) spectrometers [1] were cryogenic cooled Si(Li) detectors. These detectors had a good energy resolution of 135 eV peak full-width at half-maximum (FWHM) using the 5.895 keV Mn-K_α line emitted by an ^{55}Fe isotope source [2]. However, because of their size, cost, and complexity they are used only in analytical laboratories. Significant progress in semiconductor technology allowed silicon wafers of very low concentration of structural defects (dislocations) and impurities to be manufactured. That, together with planar technology introduced by J. Kremer [3], resulted in fabrication of Si detectors that do not require deep cryogenic cooling to achieve acceptable energy resolution. However, the fabrication of semiconductor detectors is a complex technology that uses multiple techniques, including doping, deposition, patterning, and surface passivation. The first thermoelectrically cooled Si-PIN detector developed at Amptek, Inc. in 1991 had a resolution of 850 eV [1]. Continuous refinement of planar technology, and progress in development of low noise front end electronics, led to significant improvements in the resolution of Si-PIN diodes over the following three decades. Today, the resolution of Si-PIN detectors with 10 mm^2 size, is about 150 eV at a temperature of −55 °C. This temperature is easily achieved with a cooling system based on a two-stage thermoelectric Peltier cooler (TEC), at a reasonable power consumption of $\leq$0.5 W. However, the resolution of Si-PIN detectors degrades rapidly with increasing sensitive area, because of the increased detector capacitance.

The revolutionary invention of silicon drift detector (SDD), made by Emilio Gatti and Pavel Rehak in 1982 [4], gave a new push to the development of semiconductor detectors for x-ray spectrometry, and provided remarkably better detector performance parameters than Si-PIN. The capacitance of SDDs is very small, and almost independent from detector sensitive area, which yields excellent energy resolution around 125 eV, a high counting rate, a peak-to-background ratio $\geq$ 10 000, and a higher working temperature up to −10 °C, or even higher, when high resolution is not a primary requirement.

Suitable low-noise-readout electronics have to be designed to exploit the SDD's intrinsic parameters, and to meet compactness requirements of portable XRF instruments. The integration of these electronics in custom application-specific integrated circuits (ASICs) was successful, and today readout ASICs for semiconductor detectors are commercially available [5].

The development of fast digital pulse processors (DPPs) played a significant role in the improvement of the performance of semiconductor detector systems. Today, DPPs have practically supplanted analog systems for output signal filtering and shaping. DPPs apply appropriate algorithms to filter the pulse and extract the pulse amplitude. DPPs offer several advantages over traditional analog systems: very high resolution (low signal-to-noise ratio), reduced ballistic deficit, higher throughput, and better stability.

Portable Spectroscopy and Spectrometry 1: Technologies and Instrumentation, First Edition.
Edited by Richard A. Crocombe, Pauline E. Leary and Brooke W. Kammrath.

All of these components, a PIN or SDD detector, cooled with a Peltier cooler, as well as the readout electronics, are small enough to easily fit on a small handheld XRF instrument. Let us briefly review the basic physical principles of semiconductor detectors.

19.2 Semiconductor Detector Fundamentals: Signal Formation

There are several fundamental textbooks which detail the relevant aspects of semiconductor detector physics and signal processing [2, 6, 7]. In addition, many research articles have been published over the decades of development of semiconductor detectors, fabrication technology, and associated electronics. Here we give just a brief review of the principles behind semiconductor detectors and readout electronics.

Semiconductor detectors are basically solid-state ionization chambers that have several advantages over gas ionization chambers.

- The average energy for creating electron–hole pair in Si is one order of magnitude (3.65 eV) smaller than those of gases (≈30 eV).
- The high density (2.33 g/cm^3) of Si permits the construction of detectors of relatively low thickness with sufficient quantum efficiency (high stopping power).
- In Si, electrons and holes have high mobility (μ_n = 1450 cm^2/V·S, μ_h = 450 cm^{2}S·V), which leads to short collection time, i.e. high counting rate.
- Excellent mechanical rigidity simplifies a device design.
- Elements of front-end electronics and a detector can be integrated into a single device.

In the simplest configuration, an absorbing medium is capped by a pair of electrodes with an applied voltage, as illustrated in Figure 19.1. Absorbed radiation liberates charge in the form of electron–hole pairs.

$$Q_s = \frac{E}{E_i} e \tag{19.1}$$

where E is absorbed energy, E_i is the energy required to form a charge pair, and e is the electronic charge [6]. In the presence of electric field, carriers move (drift) parallel to the field. Due to interaction with crystal lattice, the velocity does not depend on the time during which the carrier is accelerated. This is because the carrier is always in equilibrium with the crystal lattice since the characteristic times of phonon excitation are much smaller than the transport time. So, the carrier velocity is only a function of electric field

$$\vec{v}(x) = \mu \vec{E}(x) \tag{19.2}$$

The moving charge induces an electrical current in the external circuit. It is important to understand that signal current begins instantaneously and reaches its maximum when electrons reach the anode. The instant current induced by moving charge is stated by Ramo's (or Ramo–Schokley) theorem [6].

$$i_k = -q\vec{v}\ \vec{E}_Q \tag{19.3}$$

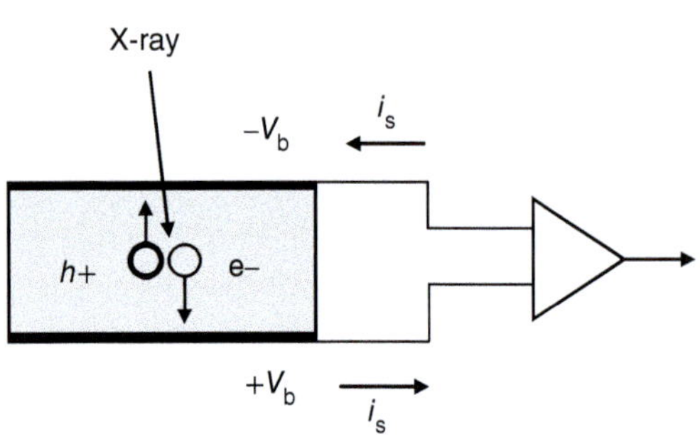

Figure 19.1 The detector and amplifier.

Or, in scalar form, that current i induced on a given electrode due to the motion of a charge is given by:

$$i = E_v q v \tag{19.4}$$

where q is the moving electric charge; v is its instantaneous velocity, and E_v is the component of the electric field (weighting field) in the direction of v at the charge's instantaneous position, under the following conditions: charge removed, given electrode raised to unit potential, and all other conductors grounded.

19.2.1 Parallel Plate Geometry with Uniform Field

A semiconductor detector with very large overbias can be approximated by uniform field. The bias voltage V_b is applied across the electrode spacing d. The electric field

$$E = \frac{V_b}{d} \tag{19.5}$$

determines the motion of the charge carrier in the detector. The carrier's velocity

$$v = \mu E = \mu \frac{V_b}{d} \tag{19.6}$$

The weighting field is obtained by applying unit potential to the collection electrode and grounding the other:

$$E_v = \frac{1}{d} \tag{19.7}$$

So induced current

$$i = qvE_v = q\mu \frac{V_b}{d}\frac{1}{d} = q\mu \frac{V_b}{d^2} \tag{19.8}$$

Since the electric field and the weighting field are uniform throughout the detector, the current is constant until the charge reaches its terminal electrode.

Assume that the charge is created at the opposite electrode and transverses the detector thickness d. The required collection time, i.e. the time required to traverse the distance d

$$t_c = \frac{d}{v} = \frac{d}{\mu \frac{V_b}{d}} = \frac{d^2}{\mu V_b} \tag{19.9}$$

The induced charge

$$Q = it_c = q\mu \frac{V_b}{d^2}\frac{d^2}{\mu V_b} = q \tag{19.10}$$

Next assume an electron–hole pair formed at coordinate x from the positive electrode. The collection time for the electron

$$t_{ce} = \frac{x}{V_c} = \frac{xd}{\mu_e V_b} \tag{19.11}$$

and the collection time for the hole

$$t_{ch} = \frac{d-x}{V_c} = \frac{(d-x)d}{\mu_h V_b} \tag{19.12}$$

Since electrons and holes move in opposite directions, they induce current of the same sign at a given electrode, despite their opposite charge. The induced charge due to the motion of the electron

$$Q_e = e\mu_e \frac{V_b}{d^2}\frac{xd}{\mu_e V_b} = e\frac{x}{d} \tag{19.13}$$

Correspondingly, the hole contributes

$$Q_h = e\mu_h \frac{V_b}{d^2}\frac{(d-x)d}{\mu_h V_b} = e\left(1 - \frac{x}{d}\right) \tag{19.14}$$

Assume that $x = d/2$. After the collection time for the electron

$$t_{ce} = \frac{d^2}{2\mu_e V_b} \tag{19.15}$$

The induced charge is $e/2$. At this time the hole, due to its lower mobility $\mu_h \approx \mu_e/3$, has induced $e/6$, yielding a cumulative induced charge of $2e/3$. After the additional time for the holes collection, the remaining charge $e/3$ is induced, yielding the total charge e. The measured charge depends on the integration time. An integration time larger than the collection time of all charge carriers yields the full charge. A shorter integration time yields a fractional charge.

19.2.2 Detectors' Resolution: Physical Limit and Electronic Noise Influence

19.2.2.1 Fluctuation in Signal Charge: Fano Factor

A key characteristic of signal sensors is not just the magnitude of the signal, but also the fluctuation of the signal for a given absorbed energy. Both determine the minimum signal threshold and the relative resolution $\Delta E/E$. One of the remarkable features of the signal fluctuations in semiconductor detectors is that they are smaller than the simple statistical variance

$$\sigma_Q = \sqrt{N_Q} \tag{19.16}$$

A detailed calculation of this phenomenon is quite complicated, so the following derivation introduces some simplifications to illustrate the basic mechanism.

Two mechanisms contribute to the mean ionization energy. First, conservation of momentum requires excitation of lattice vibrations, and second, many modes are available for momentum and energy transfer with excitation energy less than the bandgap. Energy can be absorbed by either lattice vibration, i.e. phonon production without ionization (electron–hole pair production), or ionization, i.e. formation of a mobile charge pair. Let us assume that absorbed energy produced N_x excitations (phonons) and N_{ion} ionizations charge pairs. The sum of the energies going into excitation and ionization is equal to the energy deposited by incident radiation

$$E_0 = E_{ion}N_{ion} + E_x N_x \tag{19.17}$$

where E_{ion} and E_x are the energy required for a single ionization or excitation. In semiconductors, E_{ion} is the bandgap and E_x is the average phonon energy. Assuming Gaussian statistics, the variance in the number of excitations $\sigma_x = \sqrt{N_x}$

and the variance in the number of ionizations $\sigma_x = \sqrt{N_x}$. For a single event, the energy deposited in the detector is fixed (although can vary from one event to the next). If the energy required for excitation is much smaller than required for ionization, sufficient degree of freedom will exist for some combination of ionization and excitation process to dissipate precisely the deposited energy. Hence, for a given energy deposited in the sample, a fluctuation in excitation must be balanced by an equivalent fluctuation in ionization:

$$E_x \Delta N_x + E_{ion}\Delta N_{ion} = 0 \tag{19.18}$$

If for a given event more energy goes into ionization, less energy will be available for excitation. Averaging over many events, this means that the variance in the energy allocated to the two types of processes must be equal: $E_{ion}\sigma_{ion} = E_x\sigma_x$, so,

$$\sigma_x = \frac{E_x}{E_{ion}}\sqrt{N_x} \tag{19.19}$$

From the total energy (19.18) we can extract

$$N_x = \frac{E_0 - E_{ion}N_{ion}}{E_x} \tag{19.20}$$

And insert this into (19.19) to obtain

$$\sigma_i = \frac{E_x}{E_{ion}}\sqrt{\frac{E_0}{E_x} - \frac{E_{ion}}{E_x}N_{ion}} \tag{19.21}$$

Overall, the number N_Q of charge pairs formed is the total deposited energy E_0 divided by the energy deposition required to produce a charge pair. Since each ionization forms a charge pair that contributes to the signal,

$$N_{ion} = N_Q = \frac{E_0}{E_i} \tag{19.22}$$

Thus, the variance in ionization process

$$\sigma_{ion} = \frac{E_x}{E_{ion}}\sqrt{\frac{E_0}{E_x} - \frac{E_{ion}}{E_x}\frac{E_0}{E_i}} \tag{19.23}$$

which can be rewritten as

$$\sigma_{ion} = \sqrt{\frac{E_0}{E_i}}\sqrt{\frac{E_x}{E_{ion}}\left(\frac{E_i}{E_{ion}} - 1\right)} \tag{19.24}$$

The second factor on the right-hand side is called the Fano factor F. Since σ_{ion} is proportional to the variance in signal charge Q and the number of charge pairs is $N_Q = E_0/E_i$,

$$\sigma_Q = \sqrt{FN_Q} \tag{19.25}$$

In silicon $E_x = 0.037$ eV, $E_{ion} = E_g = 1.1$ eV, and $E_i = 3.6$ eV for which the above expression yields $F = 0.08$, in reasonable agreement with the measured value $F = 0.1$. Thus, the variance of the signal charge is smaller than naively expected, $\sigma_Q \approx 0.3\sqrt{N_Q}$.

A similar treatment can be applied if the degrees of freedom are much more limited, and Poisson statistics are necessary. However, when applying Poisson statistics to the situation of a fixed energy deposition, which imposes an upper bound on the variance, one cannot use the usual expression for variance: $var\, N = N$. Instead, the variance is $(N - \overline{N})^2 = F\overline{N}$, as shown by Fano in the original paper. An accurate calculation of the Fano factor requires a detailed accounting of the energy-dependent cross-sections and the density of states of the phonon modes. One can use this result to calculate the intrinsic resolution of semiconductor detectors

$$\Delta E_{FWHM} = 2.35 \cdot E_i\sqrt{FN_Q} = 2.35 \cdot E_i\sqrt{F\frac{E}{E_i}} = 2.35 \cdot \sqrt{FEE_i} \tag{19.26}$$

where in Si $E_i = 3.62$ eV, and in Ge $E_i = 2.9$ eV. For both the Fano factor $F \approx 0.11$. So, in well-designed detector systems, detectors have sufficiently small capacitance that, together with low noise readout electronics, allows electronic noise of ≤80 eV FWHM; so, the variance of the detector signal is a significant contribution. For example, Si detector intrinsic noise at 5895 eV (Mn Kα line emitted by the ^{55}Fe isotope source) is $\Delta E = 114$ eV, which is the theoretical limit for detector system energy resolution at this X-ray photon energy.

19.2.2.2 Recombination and Trapping

One might expect that electrons and holes generated by absorbed X-ray photon will quickly recombine. However, it is not the case for Si (and Ge) because of several reasons. First, electrons and holes are dispersing by diffusion and second, they are separating by applied electric field. Furthermore, silicon is an indirect semiconductor

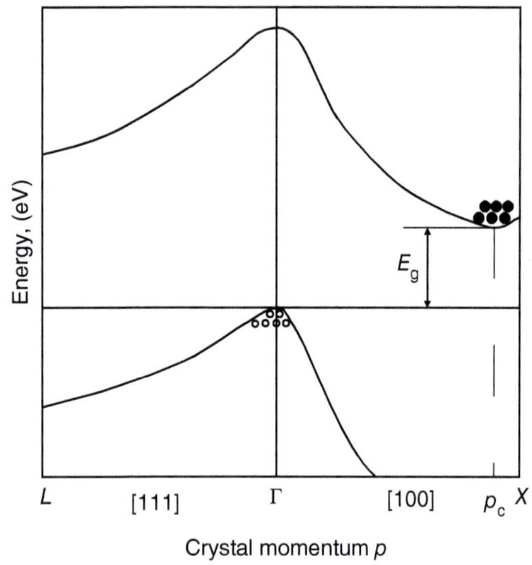

Figure 19.2 Schematic presentation of the bond structure versus crystal momentum in indirect semiconductors (Si, Ge). In Si, the minimum bandgap associated with nonzero crystal momentum (wavevector).

and annihilation of electrons at the bottom of the conduction band and holes at the top of the valence band by direct transition is suppressed. This is because the bottom of the conduction band and the top of the valence band are offset in momentum space, and large momentum transfer to the crystal lattice is required for electron–hole annihilation to take place (see Figure 19.2).

This momentum offset needs to be found from another source, for example from lattice vibrations – phonons. Thus, in indirect semiconductors, annihilation is a two-step process. This leads to carrier's lifetime being significantly larger (ms) than collection time (ns). Impurities and crystal defects (vacancies, dislocations) can introduce additional energy levels (states) in the bandgap, which can trap charge. That drastically decreases the carrier's lifetime, and increases thermally generated current due to "stepping stone effect." This leads to increased collected charge fluctuation and also increased background at low energy.

Impurities and crystal structure imperfections can act as traps for charge carriers. Traps can capture and remove mobile carriers available for signal formation. On the other hand, thermal excitation and applied electric field can release the carrier from the trap, resulting in delayed charge collection. In X-ray or gamma spectroscopy, these effects can lead to smear the signal distribution to lower energies. Even small amounts of trapping can lead to significantly low energy tails that adversely affect the ability to separate adjacent lines.

19.2.2.3 Thermal Noise

At any nonzero temperature, some thermal energy is shared by the electrons in the crystal. It is possible for a valence electron to gain sufficient thermal energy to be elevated across the bandgap into the conduction band, and leave a vacancy in the valence band, resulting in creation of the electron–hole pair. The motion of both of these charges contributes to the observed conductivity of the material. The probability per unit time that an electron–hole pair is thermally generated is given by

$$p(T) = CT^{3/2} \exp\left(-\frac{E_g}{2kT}\right) \tag{19.27}$$

where T is the absolute temperature,
E_g is the bandgap energy,
k is the Boltzmann constant,
C is the proportionality constant characteristic of the material.

As reflected in the exponential term, the probability of the thermal excitation is critically dependent on the ratio of the bandgap energy and absolute temperature, and can be decreased drastically if the material is cooled.

Thermal excitations cause not only fluctuations in the number of electron–hole pair, but also charge carrier velocity fluctuations. The spectral density of the noise power caused by thermal fluctuations can be derived directly as the long wavelength limit of in Planck's theory of blackbody radiation.

19.2.2.4 Leakage Current

PIN diodes and SDDs are reverse-biased devices. Ideally, reverse bias removes all mobile carriers from the junction volume, so no current can flow. However, thermal excitation can promote electrons across the bandgap, and a current flows even in the absence of radiation, hence the terms "dark current" or "leakage current." The probability of electrons surmounting the bandgap is increased strongly by the presence of impurities in the lattice, as they introduce intermediate energy states in the gap that serve as "stepping stones," as well as for recombination transitions. The reverse bias current depends exponentially on temperature

$$I_l \propto T^2 \exp\left(\frac{E_g}{2kT}\right) \tag{19.28}$$

where E_g is the bandgap energy and k the Boltzmann constant, so cooling the detector can reduce leakage substantially. The ratio of leakage currents at temperatures is:

$$\frac{I_l(T_2)}{I_l(T_1)} = \left(\frac{T_2}{T_1}\right)^2 \exp\left[-\frac{E_g}{2k}\left(\frac{T_1 - T_2}{T_1 T_2}\right)\right] \tag{19.29}$$

In Si ($E_g = 1.12\,\text{eV}$), this yields a 10-fold reduction in leakage current when the temperature is lowered by 14 °C from room temperature. In some situations surface leakage current can be as high as bulk leakage current or even greater. This can be prevented by use of guard rings. Leakage current is one of the most important factors that affect detectors resolution. So, only high-purity Si (low impurity and defect concentration) should be used for detector manufacturing. In general, in high-quality detectors, leakage current should not be higher than 1 nA/cm^2.

19.2.2.5 Signal Formation and Energy Resolution

Once an X-ray photon is detected and current pulse is formed by sensor, it needs to be amplified. Several types of amplifiers can be used for this purpose. However, the most popular preamplifier which is used with semiconductor detectors in laboratory and portable spectrometers is the charge-sensitive preamplifier invented by Emilio Gatti in 1956. It is represented in Figure 19.3.

The basic building block is an inverting voltage amplifier with high input resistance. Typically in the first amplification stage, a junction gate field-effect transistor (JFET) or metal-oxide-semiconductor field-effect transistor (MOSFET) is used. For the inverting amplifier, the voltage gain is:

$$\frac{dV_{out}}{dV_i} = -A \text{ and } V_{out} = -AV_i. \tag{19.30}$$

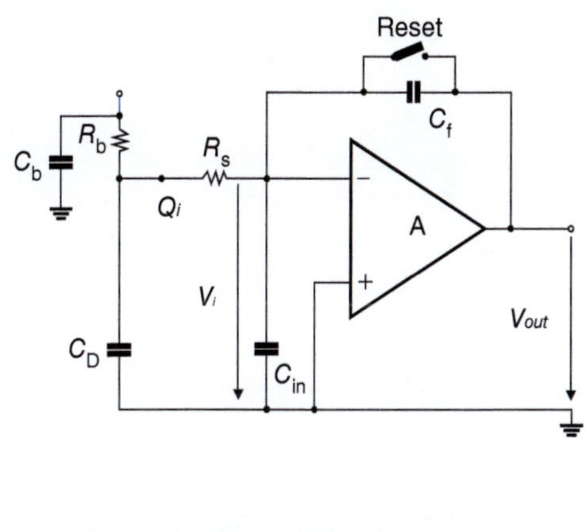

Figure 19.3 Principle of charge-sensitive amplifier. C_d represents a detector capacitance, R_s series resistance, C_{in} total input capacitance that includes amplifier input capacitance and stray capacitance, and C_f feedback capacitance. The reset switch is used for bringing the system into its operation condition.

Thus, the charge deposited on C_f is:

$$Q_f = C_f V_i = C_f(A + 1)V_i \quad (19.31)$$

We can assume an infinite input resistance, i.e. input current cannot flow into the amplifier, and all input current must charge up C_f and $Q_f = Q_i$. In this case the amplifier "dynamic" input capacitance is

$$C_i = \frac{Q_i}{V_i} = C_f(A + 1) \quad (19.32)$$

The voltage output per unit input charge

$$A_q = \frac{V_{out}}{Q_i} = \frac{AV_i}{C_iV_i} = \frac{A}{C_i} = \frac{A}{A+1} \cdot \frac{1}{C_f} \approx \frac{1}{C_f} \ (A >> 1) \quad (19.33)$$

Or output voltage is

$$V_{out} \simeq \frac{Q_i}{C_f} \quad (19.34)$$

Consequently, arriving X-ray photons will form a "step like" output signal until the reset switch brings the system back to its initial operational condition (when maximum output voltage will be reached), as in shown in Figure 19.4.

As already discussed above, the ionization process results in statistical signal peak broadening given by

$$\sigma_Q = \sqrt{FN_Q} \quad (19.35)$$

Repetitive measurements of the signal charge yields a Gaussian distribution, where standard deviation equals the root mean square (RMS) noise level. This allows calculation of the equivalent energy broadening (when the

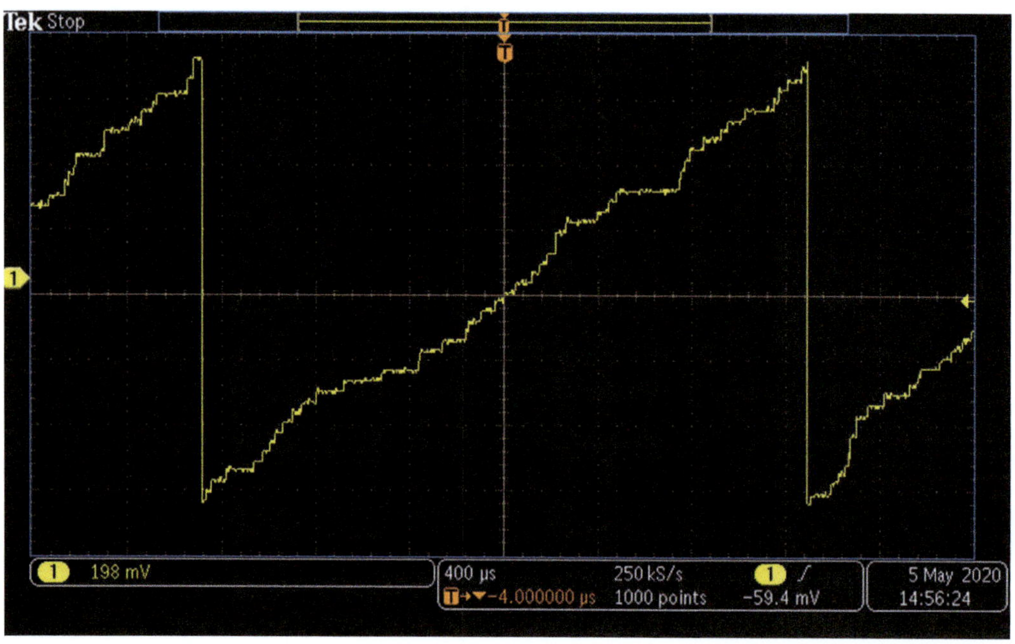

Figure 19.4 Output signal of the charge-sensitive amplifier between two resets. Small steps correspond to X-ray photon detection events.

electronic noise is zero), which is known as detector dispersion or "Fano noise," that is the detector resolution limit

$$\frac{\sigma(E)}{E} = \frac{\sigma(N_Q)}{N_Q} = \sqrt{\frac{F}{N_Q}} = \sqrt{\frac{FE_i}{E}} \tag{19.36}$$

$$\sigma(E) = \sqrt{FE_iE} \tag{19.37}$$

It is common practice to express energy peak broadening ("detector resolution") in terms of the full-width at half-maximum (FWHM) as shown in Figure 19.5, although the correct definition for energy resolution is $\Delta E/E$

$$R = \Delta E = 2.355\sigma(E) = 2.35\sqrt{FE_iE} \tag{19.38}$$

In addition to signal fluctuations originating in the sensor, the detector resolution (and detection limit) is affected by fluctuations introduced by electronics. Electronics noise introduces baseline fluctuations, which are superimposed on the signal and alter the peak amplitude. Electronics noise and "Fano noise" are not correlated and they sum in quadrature

$$R^2 = \Delta E^2 + N_{el}^2 \tag{19.39}$$

The typical detector front-end equivalent scheme used for noise analysis is represented in Figure 19.6.

The detector is represented by capacitance C_D, the current noise generator i_{nd} models the "shot noise" caused by the detector leakage current. Resistors exhibit noise due to thermal velocity fluctuations of the charge carriers. This noise can be modeled as a voltage or a current generator. Resistors shunting the detector input act as a noise

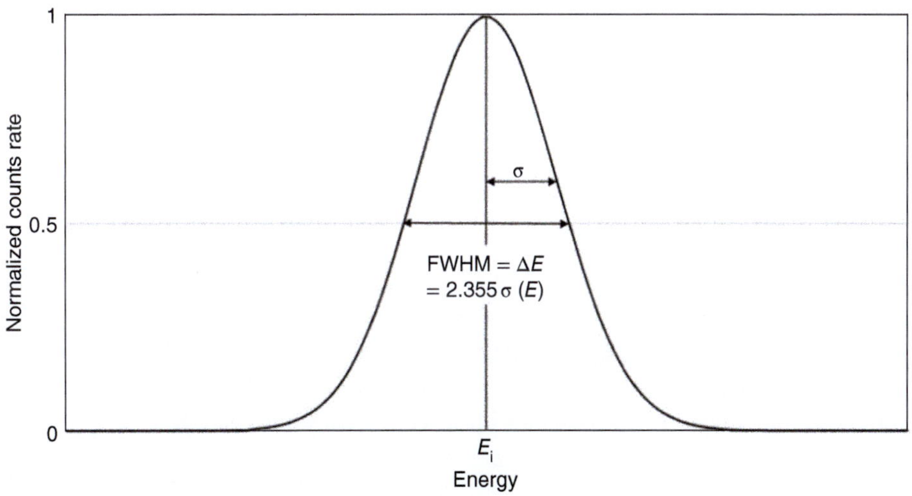

Figure 19.5 Definition of resolution.

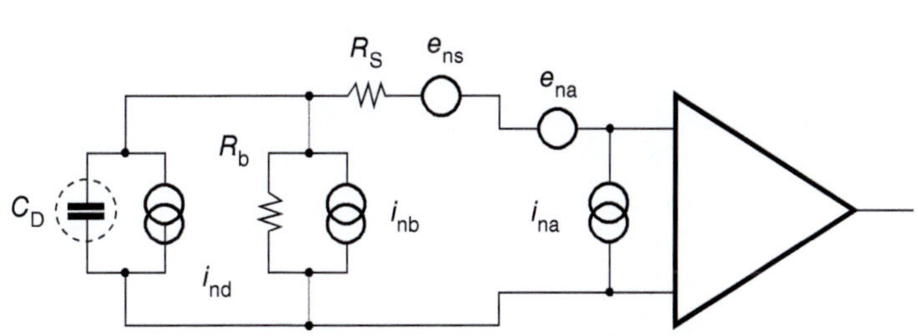

Figure 19.6 An equivalent detector front end electronics circuit for noise calculations.

current generator, while resistors in series with the input act as voltage noise sources. Since the bias resistor effectively shunts the input because C_b (see Figure 19.3) passes current fluctuations to the ground, it acts as a current generator i_{nb}, and its noise has the same effect as the shot noise current from the detector. The series resistor R_s acts as a voltage generator. In-depth analysis of the detector's and preamplifier's noise is covered in several fundamental textbooks, for example [1, 4, 5]. Here we give a brief summary of different noise contributions, which are important for understanding of the detectors performance characteristics and their limitations.

19.2.2.6 Johnson (Thermal) or Voltage Noise

When a current flows in resistors and semiconductors, electrons experience scattering on phonons (atomic vibrations), that leads to fluctuations in the electrons' velocities [8]. The degree of scattering depends on the energy kT of the vibrating atoms and one can write

$$e_n^2 = \frac{dV^2}{df} = 4kTR \tag{19.40}$$

Similarly, for the current

$$i_n^2 = \frac{di^2}{df} = \frac{4kT}{R} \tag{19.41}$$

The largest component of voltage noise is usually the JFET channel resistance, because of the coupling to the gate by gate-channel capacitance. This resistance is proportional to the inverse of the JFET transconductance, g_m, and RMS voltage noise density can be written as

$$e_n^2 = \frac{K4kT}{g_m} \tag{19.42}$$

Another contributor to thermal noise is a resistance of the back layer of the sensor and physical contact between the sensor and JFET gate that are in series with the JFET channel resistance. So, R_s will contribute to the voltage noise as

$$e_n^2 = 4kTR_s \tag{19.43}$$

19.2.2.7 Current or Shot Noise

The current that flows in sensors is the result of many individual moving charges and experiences fluctuations due to different physical factors that result in noise. The spectral density of current or shot noise can be given as

$$i_n^2 = 2eI \tag{19.44}$$

where e is electronic charge and I is the average current.

Trapping and detrapping processes in resistors, dielectrics, and semiconductors can introduce additional fluctuations whose noise power frequently exhibits a $1/f$ spectrum with spectral density of its voltage as:

$$e_{nf}^2 = \frac{A_f}{f} \tag{19.45}$$

The amplifier's noise is described fully by a combination of voltage and current sources at its input, shown in Figure 19.6 as e_{na} and i_{na}. So, for the circuit in Figure 19.6, one can write four noise components as

$$i_{nd}^2 = 2eI_d \tag{19.46}$$

$$i_{nb}^2 = \frac{di^2}{df} = \frac{4kT}{R_b} \tag{19.47}$$

$$e_{ns}^2 = 4kTR_s \tag{19.48}$$

$$e_{nf}^2 = \frac{A_f}{f} \tag{19.49}$$

Since individual noise contributions are random and uncorrelated, so that they add in quadrature.

19.2.2.8 Equivalent Noise Charge

Above, the noise contributions have been expressed in terms of voltage and current. It is often more useful to express them in the quantity to be measured. The noise level is then specified as the signal level for which the signal-to-noise ratio (S/N) equals one. So, detector systems that measure signal charge can be characterized in terms of equivalent noise charge (ENC). This is the signal charge that yields a signal-to-noise ratio (S/N) of one.

$$\text{ENC} = Q_\text{n} = \frac{Q_\text{s}}{\text{S/N}} \tag{19.50}$$

where Q_s is the signal charge. For a given detector material, the signal charge is determined by absorbed energy, then the noise also can be expressed in terms of energy – eV, keV, or in electrons. For example in Si $Q_\text{n} = 1e$ corresponds to 3.64 eV. Then, energy resolution for ionization energy E_i is

$$\Delta E_\text{n} = E_\text{i} \cdot \text{ENC} \tag{19.51}$$

Equivalent noise charge is the most convenient measure of system noise when designing or characterizing detector systems. Nevertheless, in analyzing individual noise contributions, the basic noise parameters – voltage and current – are more useful.

The signal produced by the detector usually is not used directly; it is further amplified and shaped. The purpose of these procedures is to optimize the signal-to-noise ratio, to reduce interference between subsequent signals, to minimize the ballistic deficit effect, and to form signal shape for more accurate amplitude measurements. Mathematically, the pulse shaper can be presented in the form of a weighting function (see for details [6, 9]). In this case the ENC is given as

$$Q_\text{n}^2 = \frac{1}{2} i_\text{n}^2 \int_{-\infty}^{\infty} [W(t)]^2 \text{d}t + \frac{1}{2} C^2 e_\text{n}^2 \int_{-\infty}^{\infty} [W'(t)]^2 \text{d}t \tag{19.52}$$

The weighting function can be expressed in terms of characteristic time usually as "shaping time" or "peaking time." This leads to a general formulation of the equivalent noise charge

$$Q_\text{n}^2 = i_\text{n}^2 F_\text{i} T_\text{S} + e_\text{n}^2 F_\text{v} \frac{C^2}{T_\text{S}} + F_\text{vf} A_\text{f} C^2 \tag{19.53}$$

where F_i, F_v, and F_vf depend on the shape of the pulse determined by the shaper for a step unit and T_S is the characteristic time, whose choice, in general, depends on the type of the shaper.

For the circuit shown in Figure 19.6

$$Q_\text{n}^2 = \left(2eI_\text{d} + \frac{4kT}{R_\text{b}} + i_\text{na}^2\right) F_\text{i} \tau_\text{S} + (4kTR_\text{s} + e_\text{na}^2) F_\text{v} \frac{C_\text{d}^2}{\tau_\text{S}} + F_\text{vf} A_\text{f} C_\text{d}^2 \tag{19.54}$$

where e is the electronic charge; I_d is the detector leakage current; k the is the Boltzmann constant; i_na is the JFET leakage current (typically negligible); R_b is the bias resistor; R_s is the total series resistance; C_d is the detector capacitance; e_na is the amplifier noise parameter; A_f is the device-specific noise parameter that determines $1/f$ noise contribution; F_i, F_v, and F_vf are the shaper parameters; and τ_s is the shaping time.

The typical dependence of the ENC as a function of shaping time and basic noise parameters is shown in Figure 19.7.

In real detector system analysis, the detector capacitance needs to be substituted with the sum of all capacitance shunting the input, i.e. the sensor capacitance, the preamplifier input capacitance, and stray capacitance (wire, etc.).

The summary:

- ENC strongly depends on leakage current that depends on temperature.
- ENC strongly depends on the capacitance at the preamplifier input.
- ENC strongly depends on voltage noise parameter of the preamplifier.

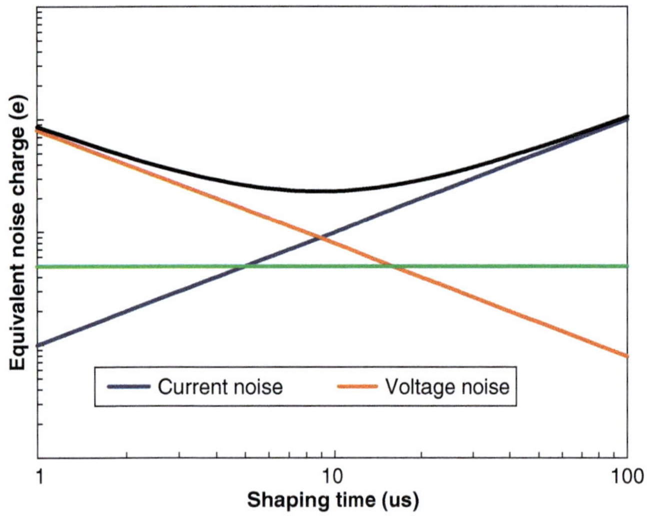

Figure 19.7 Equivalent noise charge as a function of shaping time. The green line is $1/f$ noise contribution and the black line is the resulting ENC.

- Signal-to-noise ratio can be improved by the optimal shaper design.
- The current and voltage contribution to ENC depend on shaping time: the current noise increases with shaping time, but the voltage noise decreases with τ_s, and the optimal shaping time corresponds to the equal values of these contributions.
- The $1/f$ noise contribution can be minimized by optimization of the shaper frequency band.

In general, to improve the detector system energy resolution, the developer should to strive to:

- Improve quality of start material and fabrication processes to decrease leakage current.
- Design the sensor and preamplifier with as low as possible capacitance.
- Design preamplifier with low voltage noise, mainly first amplification stage.
- Design an optimal shaper.

19.3 Detectors for Portable Spectrometers: Design and Performance

19.3.1 Si-PIN Diode Detectors

A silicon PIN diode, employing a simple planar geometry, is the basic style of semiconductor X-ray detectors. From the early 1980s until today, most have been produced using slightly n-doped, 400–5000 μm thick, <111> silicon wafer, of 5 kΩ resistance. Figure 19.8 shows the cross-section of a typical PIN diode detector fabricated using the planar technology introduced by J. Kemmer in 1980. Typical wafer thickness is 400–625 μm and resistance 5–10 kΩ.

The p+ and n+ layers are formed by introducing dopants. The similarly doped area shown on the left and right sides of the entrance window contact ("cathode") forms a guard ring that isolates the detector diode from the edge of the chip. The guard ring is necessary because mechanical damage of the edge leads to unacceptably high leakage current. The guard ring can be grounded or floating. Electrical contacts are formed by deposited metal layers, typically aluminum. The intermediate surface between electrodes is protected by a dielectric layer of SiO_2.

When sufficient reverse voltage V_b is applied, the detector gets fully depleted. The depletion width is

$$w_d = \sqrt{\frac{2\varepsilon(V_b + V_{bi})}{Ne}} \tag{19.55}$$

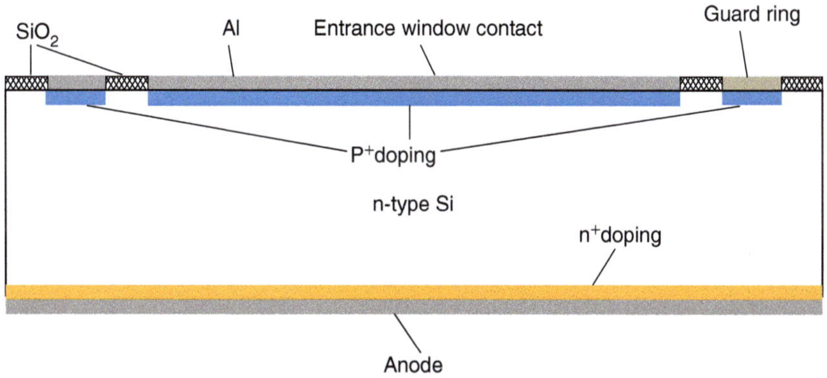

Figure 19.8 Cross-section of a PIN-diode detector. Typically the entrance window connected to the gate electrode of the JFET.

where ε is the dielectric constant (for Si $11.9\varepsilon_0$), V_{bi} is the inbuilt junction potential ($\approx$0.5 V for Si), and N is the dopant concentration. When the detector is fully depleted, the leakage current at given temperature and the detector capacitance

$$C = \varepsilon \frac{A}{w_d} \tag{19.56}$$

is minimal. For example, the capacitance of the XPIN-6 detector from Moxtek, Inc., with 6 mm^2 sensitive area, is about 1.8 pF.

Figure 19.9 presents the typical design of the PIN diode detector, packaged in a TO-8 can, which is the most common design for portable X-ray spectrometers. The sensor chip, together with JFET and thermistor (or a diode), is mounted on the ceramic substrate that is then placed on the two-stage Peltier TEC. The detector sensitive surface is apertured by an internal collimator to eliminate edge effects coursed by incomplete charge collection. These affects can lead to poor peak-to-background ratio and formation of "ghost" peaks in the XRF spectra. All the necessary connections between metalized contacts and case pins are done by wire bonding. The detector assembly is protected with a nickel cap with a thin X-ray transparent window. The detector volume is evacuated and sealed. In most cases, manufacturers install getters inside of the sealed enclosure which allow the pressure inside to be kept low enough for several years.

The first amplification stage is typically a discrete JFET, with integrated feedback capacitor of about 50 fF, although several groups have reported fabrication of PIN diode detectors with integrated "on-chip" field-effect transistors (FETs) [10]. An example of the typical PIN diode detector design in TO-8 package is shown in Figures 19.9 and 19.10.

The entrance vacuum-tight and light-tight window is a very important part of the detector package. The window must provide good sealing and low absorption of the X-ray in a wide energy range. The most frequently used windows are made from vacuum-tight beryllium (Be) foils of 8 and 25 μm thickness.

For better resistance to harsh environment, some manufacturers protect the Be window with a thin coating. For example, the DuraBerillium window from Moxtek is protected with DuraCoat® coat: a proprietary refractory material consisting of elements lighter than sodium that ensure beryllium windows stay vacuum tight in harsh environments. This product is highly resistant to moisture, chemicals, and abrasion, making it ideal for XRF applications. The Be window limits detector sensitivity at low energies, below 1.5 keV. When enhanced X-ray transmission at low energies is important, materials other then Be windows can be used, which will be discussed below.

The detectors for portable and handheld spectrometers are made as single modules together with a small low-noise preamplifier. The preamplifier is placed into aluminum housing that provides electromagnetic (EMI) shielding, and serves as a heatsink for the TEC's hot side. The typical design of the detector for handeld EDXRF

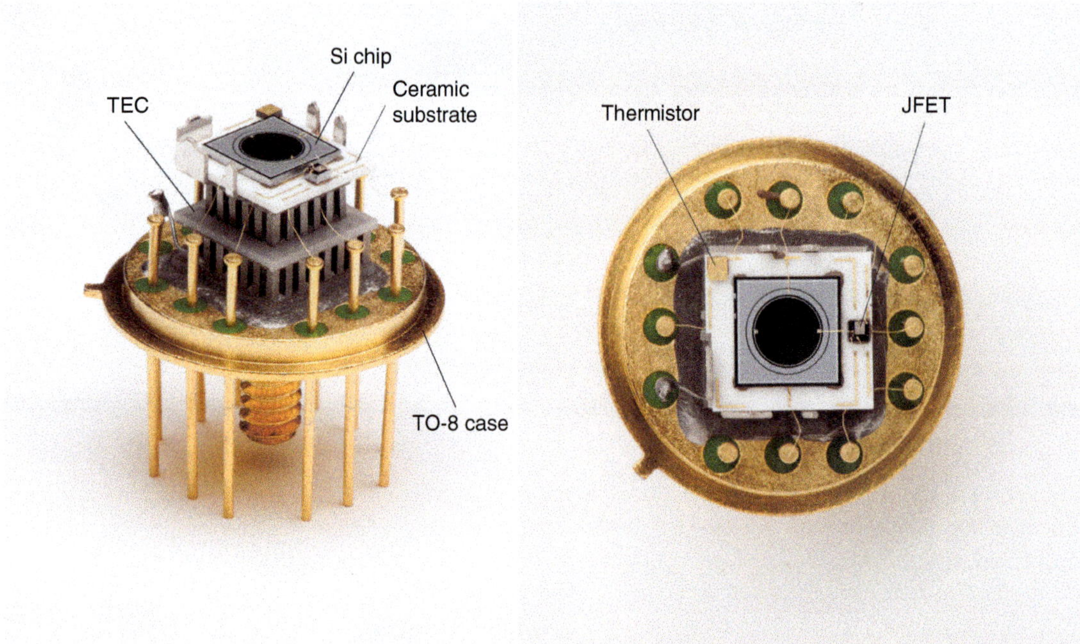

Figure 19.9 Typical design of the SI-PIN detector in TO-8 case. Moxtek, Inc., XPIN-6 detector. *Source*: Courtesy of Moxtek, Inc.

Figure 19.10 Typical design of the PIN diode detector module for handheld EDXRF spectrometer. XPIN-6 and XPIN-13 from Moxtek, Inc. *Source*: Courtesy of Moxtek, Inc.

spectrometers is presented in Figure 19.10. The two-stage TEC is effective enough to provide ΔT between the detector chip and the detector housing up to 80 °C at reasonable power consumption of about 0.5 W. So, the detector can operate at −35 °C, or even lower when higher resolution is required.

The typical resolution of PIN diode detectors lies within an interval of 150–200 eV at 5895 eV (Mn K_α line obtained from ^{55}Fe isotope source) and strongly depends on the detector sensitive area, i.e. the detector capacitance, as is shown in Figure 19.11. However, PIN diode detectors produce fast signals and demonstrate good stability with counting rate. The counting rate mainly is limited by shaping time of the pulse processor.

Despite the fact that SDDs today have better performance parameters, Si-PIN detectors are used in portable XRF spectrometers when price is the primary consideration, for example, in some handheld spectrometers from Bruker Nano Analitics and Olympus NDT. The major Si-PIN diode detector manufacturers are currently Moxtek, Inc., and Amptek, Inc.

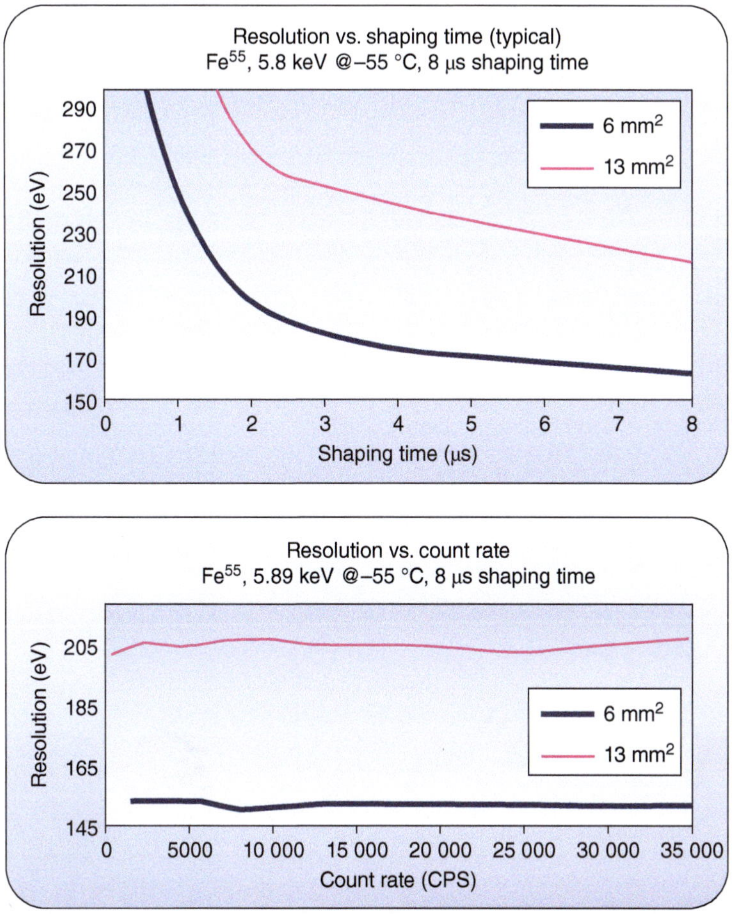

Figure 19.11 Resolution of Si-PIN detectors as a function of peaking time, taken with detectors with different sensitive areas (upper part), and resolution stability against counting rate taken at a constant peaking time. XPIN-6 and XPIN-13 Si-PIN detectors from Moxtek, Inc.

19.4 Silicon Drift Detectors

In 1984 E. Gatti and P. Rehak presented a brilliant invention – a semiconductor drift chamber. Initially this device was designed and used as a particle-tracking position-sensitive detector. Later, the silicon drift chamber or SDD was optimized for X-ray spectrometry.

The main idea underlying the principle of the SDD is to create an electric field in a fully depleted bulk silicon wafer that drives electrons to the same small n+ doped electrode (anode), regardless of the position in which the electron was generated by thermal excitation or by X-ray ionization. In this case, the anode can be very small. In other words, in this structure the charge generated in a large area is collected in small collection electrode: the capacitance of this electrode can be very small (1–50 fF) and almost independent from the detector sensitive area. There are various drift detector topologies optimized for different detector applications [7]. Radial devices (Figure 19.12) are most suitable and interesting for energy measurements [11].

The cross-section of a drift detector is shown in Figure 19.13. The X-ray entrance detector window (top) is formed by a large, uniformly doped p+ area (diode). The opposite side (bottom) is segmented into concentric p+ doped rings (diodes), and at the center of the device there is small n+ doped anode. The drift field is formed by applying

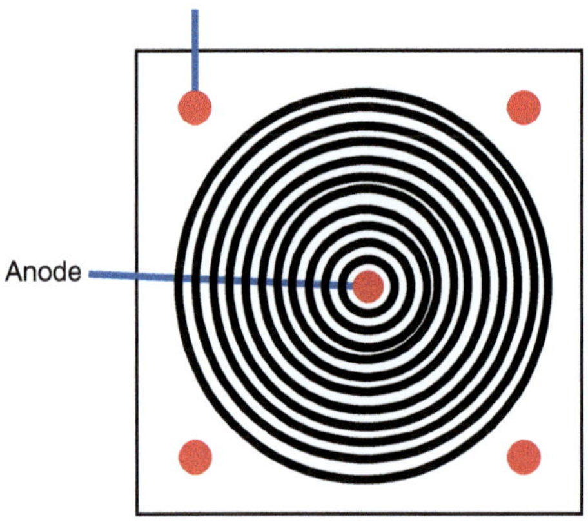

Figure 19.12 Radial SDD with point-type anode in the center.

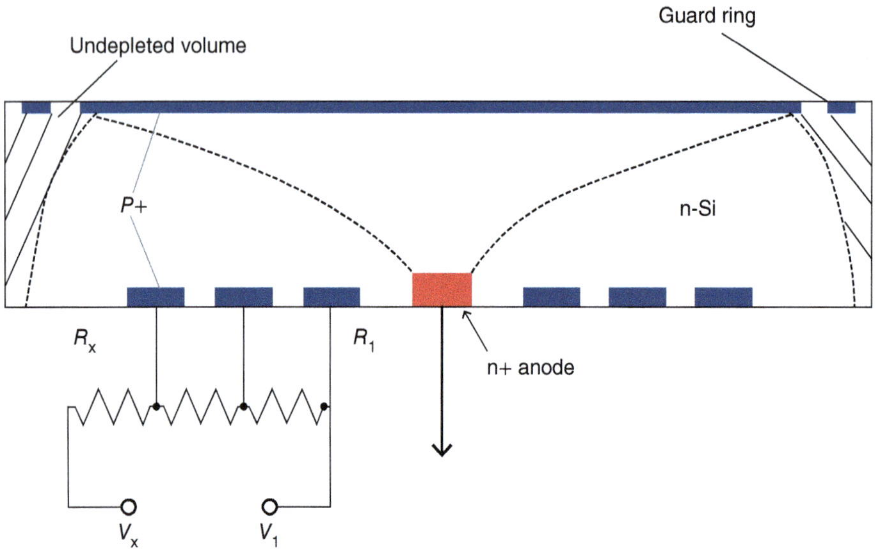

Figure 19.13 Cross-section of a drift detector. R_1 is the innermost ring, R_x is the outermost ring. The voltage of the drift rings V_1, V_x, and X-ray entrance window is provided by an external source.

different voltages to n+ rings. By applying suitable voltages to the ring electrodes, one can create the drift field as a potential valley (or funnel) for majority carriers (electrons). So, the field continuously slopes down from the outer edge of the detector sensitive area on the top to the collecting electrode on the bottom.

Such a drift field can be formed by applying different voltages to each electrode from external power supply, but in most cases an integrated voltage divider is used to supply these voltages, which is more practical. However, other methods of rings biasing, for example "self-biasing," can be used too [2].

The depletion in this type of SDD is achieved by applying voltage to the entrance window electrode that starts to deplete the bulk from the top to the bottom and to the concentric rings, which deplete the bulk from the ring toward the anode. There is a limitation on the attainable drift field, which comes from the condition that the voltage between diodes formed opposite each other on the two wafer surfaces has to be lower than the wafer's full-depletion voltage. So, the maximum drift field is obtained when the voltage between R_1 and R_x is slightly lower than the doubled depletion voltage, and the potential of the X-ray entrance window is about halfway in between. Another design replaces concentric rings with a continuous resistive spiral [12].

The anode can be connected to a discrete JFET transistor by a bond pad in the same way as a PIN diode. The negative side of this method is the added stray capacitance of bonding pad and flying wire. The detector with on-chip electronics was developed to overcome this problem [13]. The cylindrical geometry JFET was implanted into the center of the anode. However, the presence of the transistor in the center of the sensitive area leads to some problems with incomplete charge collection that decreases the peak-to-background ratio. To overcome this, perfect symmetry was sacrificed by moving the anode and the JFET to the edge of the device. In this design, the drift field forming electrodes have a droplet shape, and the anode and JFET are protected from X-ray exposure by an internal collimator [14].

Today, the most advanced detector readout electronics are amplifiers integrated in ASICs. ASICs have remarkably low ENC (1–3 electrons RMS) and very low input capacitance. That allows for very good resolution at low temperatures, down to 122 eV at 5895 eV X-ray photon energy, or good resolution <140 eV at moderate cooling or even at room temperature [5]. Improved semiconductor technology results in fabrication of wafers of very high quality. The current density at the anode of 25 pA/cm^2 has been reported at room temperature. That results in resolution from 148 eV at +30 °C, down to 124 eV at −30 °C [15]. Using ASICs also enables low dead time and high count rate, since excellent resolution can be achieved at sub-microsecond peaking times [5, 16].

There are different types of commercially available SDDs for portable spectroscopy on today's market from different manufacturers, such as KETEK GmbH (Germany), PNDetector (Germany), XGLab-Bruker (Italy), and Amptek, Inc. (USA). They offer SDDs with different types of front-end electronics, different sensitive areas ranging from 7 mm^2 up to 150 mm^2, and different shapes: round, droplet, hexagonal, and oval. Typical performance parameters are resolution better than 130 eV within temperature range from −60 to −25 °C. Some detectors can be used at higher temperatures with a trade-off of poorer resolution. The maximum achievable counting rate is up to 2 Mcps, depending on peaking time.

Traditionally, Be windows of 8–25 μm thickness were used in X-ray detector packages limiting detector sensitivity to low-energy X-rays (<1.5 keV). Now, practically all detector manufacturers have introduced SDDs with windows having good X-ray transparency down to sub-keV energies. Low-noise detector systems equipped with such windows allow detection of light elements.

19.5 Si Detectors' Quantum Efficiency: X-Ray Entrance Windows

The Si detectors' quantum efficiency I_{reg}/I_0 is determined by absorption of X-rays in the detector volume that is given by

$$I = I_0 \exp(-\mu_a \rho x) \tag{19.57}$$

where I_0 is the X-ray intensity at the surface, ρ is the Si density, μ_a is the mass absorption coefficient that decreases generally with energy, and x is the penetration depth. So, Si detector's quantum efficiency is limited from the high energy side by a decreased value of μ_a that can be compensated only by increased thickness of the detector, as is shown in Figure 19.14.

A second important part of semiconductor detector design that affects detector efficiency (or one can call it sensitivity since this parameter includes quantum efficiency of the detector and window transparency) is the entrance X-ray window. The detector X-ray entrance window for portable or handheld EDXRF spectrometers must satisfy following requirements:

- High X-ray transparency within the X-ray energy range of interest.
- Light-tightness.
- Vacuum-tightness.

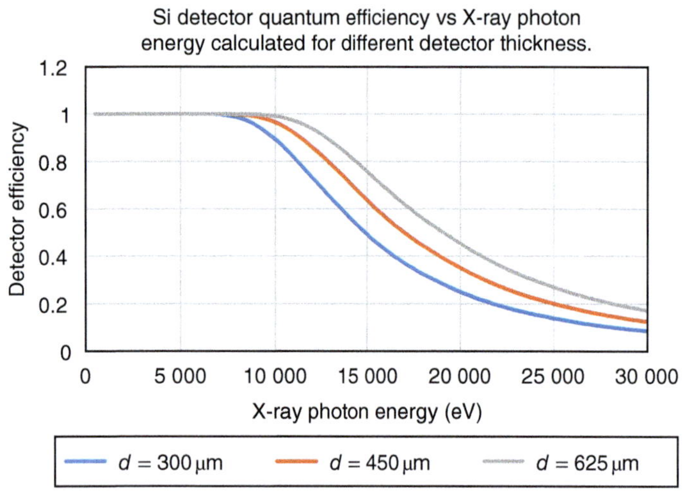

Figure 19.14 Si detector quantum efficiency as a function of X-ray energy, calculated for detectors of different thickness.

- Must withstand a pressure difference of at least slightly higher than 1 bar and multiple pressure cycles.
- Must withstand multiple temperature cycles typically from approximately −20 to +80 °C.
- Should have a conducting layer for detector EMI shielding.

In many instances these requirements are in opposition to each other. For example, for high X-ray transparency, a window must be made from materials that consist of only light elements, have as low as possible density, and be very thin. Obviously these requirements can be satisfied only for the price of mechanical strength, i.e. can be more easily damaged by pressure differences.

The light-tightness often requires deposition of an additional light-absorbing layer, which also will attenuate low-energy X-rays. Thus, an optimal entrance window design can be achieved by series of reasonable compromises between different parameters. Today all major semiconductor detector manufacturers offer detectors with entrance window that provides enhanced sensitivity to low atomic number (Z) elements down to lithium. Typically the low-energy window designs include support structures. Mechanically, the structure supports the film over pressure changes of up to 1.5 atm, and maintains the integrity of the film over thousands of pressure cycles. Optically, the support structure needs a large open area percentage, thereby maximizing the window's X-ray transmission. For example, Moxtek has traditionally made silicon rib support structures, made from silicon wafers. The silicon rib structures are available in an assortment of sizes ranging from 10 to 500 mm^2.

PNdetector GmbH advertises "pnWindow" used in their SDD^{GL} – Genius Line detectors, which allow detection of light elements down to Li with excellent peak-to-background information. The "pnWindow" design is kept as company proprietary information. Amptek, Inc., patented "C-Series" X-ray windows that utilize silicon nitride (Si_3N_4) with an Al coating. This window extends SDD sensitivity range down to boron. The transparency of the C1 window is around 65% at 1000 eV, and 25% at 500 eV X-ray energy. In their VITUS product line, KETEK GmbH uses an ultrathin AP3.3 polymer window from Moxtek, Inc. Also, in 2015, KETEK reported successful development of a new window based on graphenic carbon [17].

Moxtek's AP3 X-ray windows are specifically made to transmit low-energy X-rays, from about 50 eV and above. This ultrathin polymer membrane allows for detection of X-ray fluorescence of light elements such as boron, carbon, nitrogen, and oxygen, as well as heavier elements. The windows are vacuum tight (less than 5×10^{-10} mbarL/s of He), and can withstand up to 1.5 atm of pressure without breaking. These windows are found in a variety of X-ray detectors throughout the world, including XRF detectors mounted in scanning electron microscopes (SEMs). The AP3 product line windows consist of an ultrathin polymer membrane of 3000 Å, layered with a proprietary boron hydrate of about 200 Å, and coated with an Al layer of 300 Å thickness. The window's X-ray transmission was calculated and tested at the Advanced Light Source within an energy range of 108–1300 eV. Calculated and experimentally obtained X-ray transmission of the AP3 windows is presented in Figure 19.15. Moxtek has also developed hexagonal support structures for AP3 ultrathin films. Examples of both support structures are shown in Figure 19.16.

The calculated optical transmission of the silicon ribs support structure is 76.4%, and of the hexagonal diamond support structure is 72.2%. Unfortunately, the presence of support structures makes the optical transmission of the entrance window dependent on the incidence angle of incoming X-ray photon with regard to the window plane.

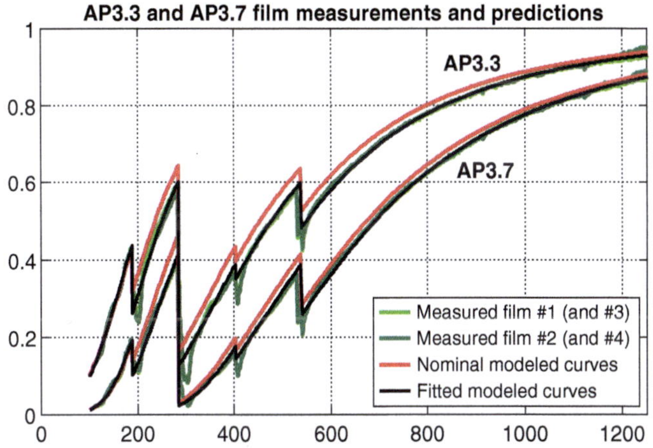

Figure 19.15 This figure shows the measured and modeled X-ray transmission for both the AP3.3 and AP3.7 films, without any support structures. The light and dark green curves show the measured transmission for two AP3.3 films and two AP3.7 films. The red curves are modeled curves using the nominal film thicknesses for the AP3.3 and AP3.7 films. The black curves are also modeled, but using thicknesses which result from fits to the measured transmission curves. Discontinuities in these curves correspond to K-edges of windows material constituents.

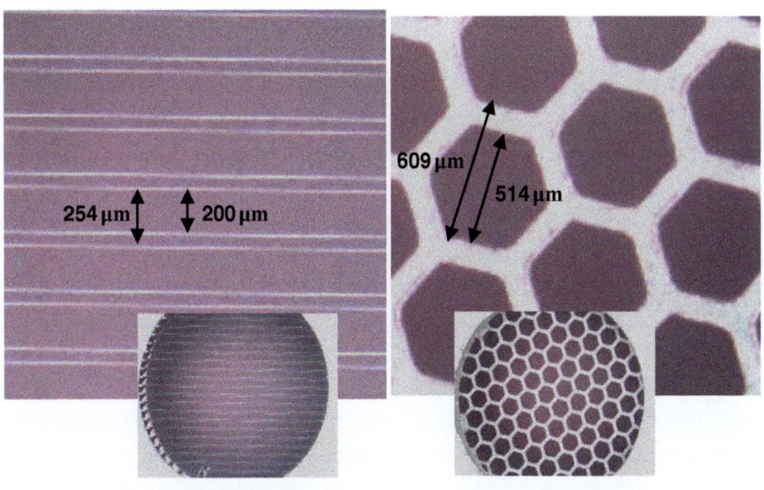

Figure 19.16 Images of two different support structures. The images on the left show the traditional silicon rib structure. The silicon structure has a thickness of 381 μm, a rib width of 54 μm, a rib pitch of 254 μm, and a gap width of 200 μm. On the right is the hexagonal structure with a thickness of 90 μm, a rib width of 95 μm, a hexagonal pitch of 609 μm, and a gap width of 514 μm. Overall, both of these support structures have a 6.4 mm ID (32 mm^2 active area). *Source*: Courtesy of Moxtek Inc.

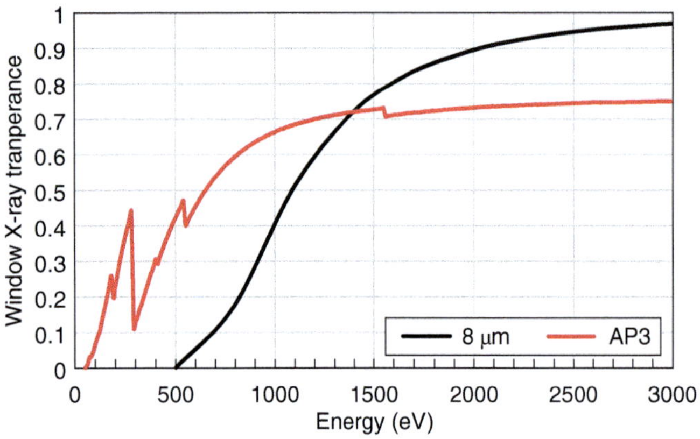

Figure 19.17 X-ray transmission of AP3 window and Be window of 8 μm thickness within 50–3000 eV energy range.

That decreases X-ray transmission of the window at low energies for the fluorescence radiation emitted from a small irradiated point on a sample surface into a wide solid angle. The X-ray transmission of the AP3 window at normal beam incidence against those for the thinnest beryllium window (8 μm) is shown in Figure 19.17.

19.5.1 Influence of Environmental Noises on the Semiconductor Detector's Performance

A negative attribute of very sensitive detector electronics is that they can be easily affected by external disturbances, such as electromagnetic interference (EMI), microphonic effects caused by environmental electromagnetic fields, ground loops, and mechanical vibrations. These sources of disturbances are often encountered in close proximity to these semiconductor detectors, especially in portable energy-dispersive spectrometers, when the X-ray tube and detector are mounted very close to each other. In addition, handheld spectrometers are always at the risk of picking up environmental interference from vibrating equipment, cell phones, etc.

Different methods of the microphonic noise filtering have been proposed and analyzed in several publications [18, 19]. A study of the influence of these external interferences on the detector energy resolution was carried out by the Moxtek research group. This research work has been done focusing on periodic external interference disturbances associated with mechanical vibrations (microphonic noise), and electrical noise that affects the detector–preamplifier assembly. Because of this, we prefer to use term periodic "interference," rather than "noise," which implies a random disturbance.

All measurements were done using standard Moxtek PIN diode detectors, along with standard data acquisition electronics: charge-sensitive preamplifier and DPP. The external electrical disturbance was entered into the detector assembly by a simple connection of a function generator output and the detector housing, connected to the preamplifier's ground. Mechanical vibrations were applied to the detector by a piezoelectric transducer connected to a function generator, with the transducer pressed to the detector housing for a good acoustic contact. All experiments were done by measuring the detector energy resolution as a function of the detector peaking time, within a peaking time range between 4 and 100 μs, at different constant generator frequencies within 1–100 kHz range. The generator output signal amplitude was found experimentally, and the amplitude of the interference was set at a level high enough to observe the external disturbance in the detector, but without causing unreasonably high resolution degradation. The detector resolution measurements were done with 5895 eV X-ray photons, using a ^{55}Fe isotope source. When either a mechanical or electrical periodic external interference is introduced to the detector, it drastically changes the resolution-peaking time dependence shape. The example of the resolution curve when the detector is under influence of the mechanical vibrations of 100 kHz is presented in Figure 19.18.

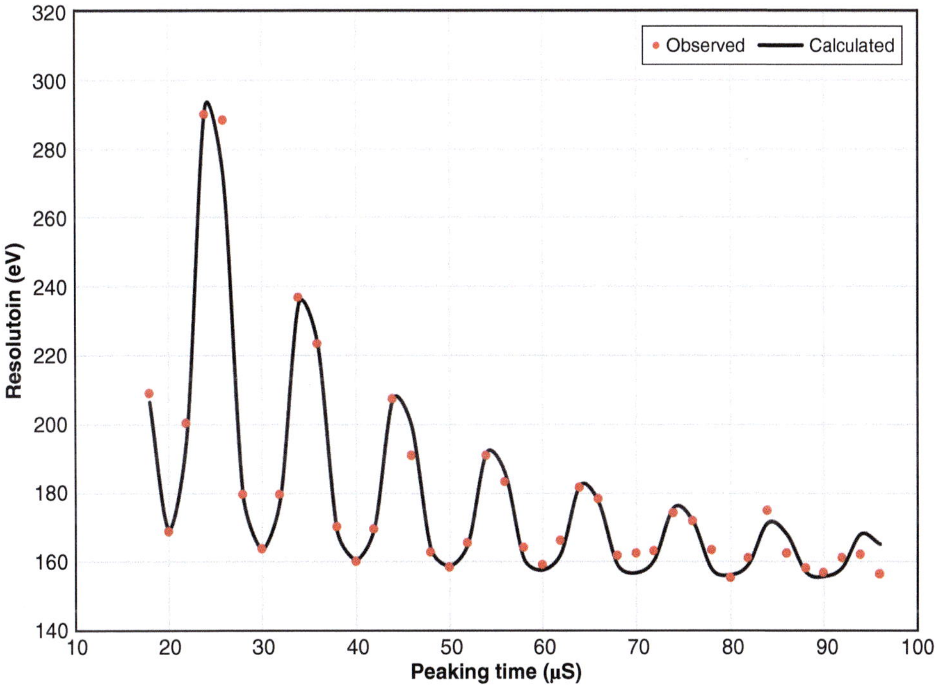

Figure 19.18 The detector energy resolution as a function of the peaking time obtained with the PIN diode detector affected by the acoustic disturbance of 100 kHz. The experimental observations are shown as points. Solid line represents the best LMS fit obtained using (66).

As one can see, when the detector is affected by periodic interference, the resolution-peaking time dependence has several minimums and maximums, i.e. represented by the oscillating curve in Figure 19.18. As was expected, electrical and microphonic interference have the same effect on the resolution-peaking time dependence. That means the external interferences, regardless of if they affect the detector itself, or the detector's preamplifier, will result in adding a periodical wave into the preamplifier output signal (see Figure 19.19). Thus, when developing a mathematical model for the effect of the periodic interference influence on the detector resolution-peaking time dependence, we do not need to consider the interference nature, but the interference periodicity will play the key role in this model.

The full experimental peak width at half-maximum is given by the very well-known formula:

$$\mathrm{FWHM} = \sqrt{\Gamma_{\mathrm{Fano}}^2 + \Gamma_{\mathrm{ENC}}^2 + \Gamma_{\mathrm{others}}^2} \tag{19.58}$$

where:

$\Gamma_{\mathrm{Fano}} = (2.35)\sqrt{\varepsilon F E}$ is the Fano noise,
Γ_{ENC} is the equivalent noise charge (ENC)
Γ_{others} is any other factor that may affect resolution such as incomplete charge collection, traps concentration, etc.

The second term gives the dependence of ENC (random noise) from the peaking time (19.53). We can rewrite (19.53)in a form more convenient for the LMS fit:

$$Q_{\mathrm{n}}^2 = A_1 \cdot \tau + A_3 \cdot \frac{1}{\tau} + A_3 \tag{19.59}$$

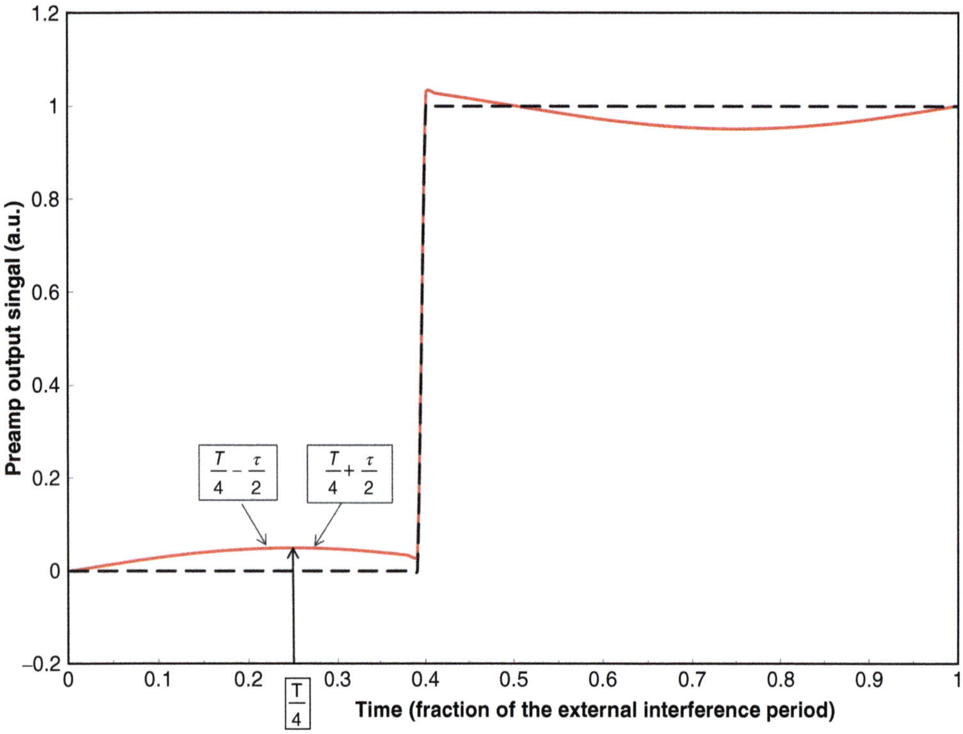

Figure 19.19 Schematic presentation of the detector signal superimposed with the external periodical interference. Dashed line represents the detector signal as a step function, red solid line represents the sinusoidal wave that models the external interference.

where $A_1 - A_3$ are parameters that will be refined during least median of squares (LMS) procedure. Let us assume that the detector works under the influence of some external periodic disturbance; that means, we have to add one more term into (19.58):

$$\mathrm{FWHM} = \sqrt{\Gamma^2_{\mathrm{Fano}} + \Gamma^2_{\mathrm{ENC}} + \Gamma^2_{\mathrm{Ext}} + \Gamma^2_{\mathrm{others}}} \tag{19.60}$$

Assuming the other noise sources and the external interference are significantly greater than $\Gamma^2_{\mathrm{others}}$, we can omit this term. To find out the explicit form of Γ^2_{Ext}, let us consider the preamplifier output signal as a step function, corresponding to a detected X-ray photon, superimposed with a sinusoidal signal with frequency ω, corresponding to a signal of external interference, as shown in Figure 19.4. Thus, for the external interference voltage, we can write down:

$$\langle U_{\mathrm{t}} \rangle^2 = \frac{1}{T} \int_0^T {D_{\mathrm{m}}}^2 \cdot \sin{(\omega t)}^2 \mathrm{d}t \tag{19.61}$$

where D_{m} is the external disturbance signal peak amplitude, T is a period of external interference signal.

The DPP forms output trapezoidal pulse from moving average calculations as [6]:

$$S_0(n) = \sum_{k=0}^{N-1} W(k) \cdot S_{\mathrm{i}}(n-k) \tag{19.61}$$

where S_0 and S_{i} are the output and input signals, and W is the weighting function that yields the desired pulse shape. Therefore, the deviation from the average baseline value between two X-ray events, or deviation of the output signal from the true value caused by external interference, is equal to the external interference average

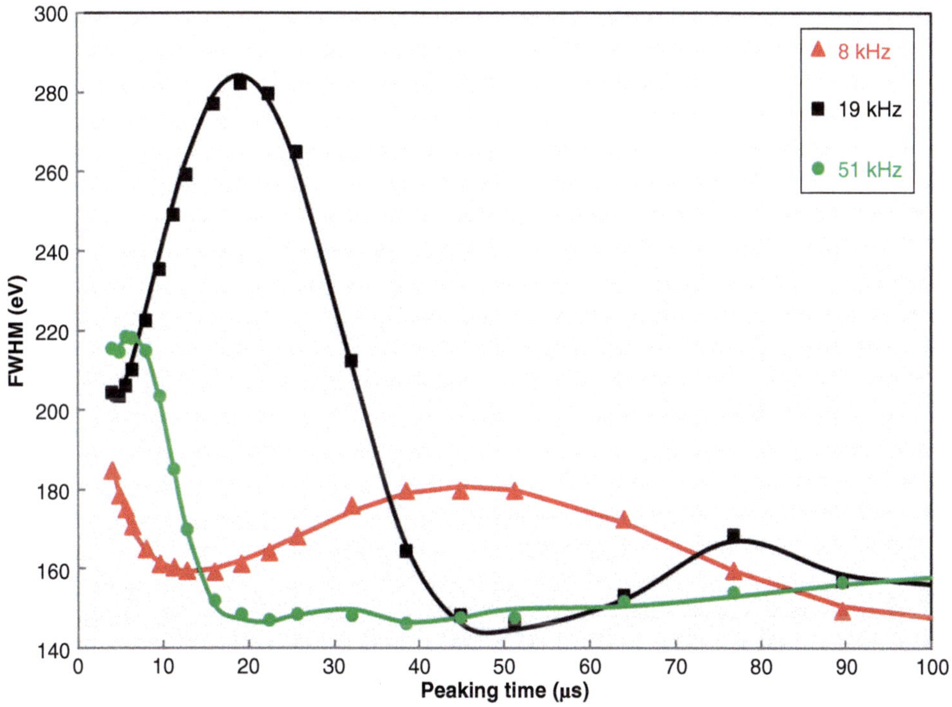

Figure 19.20 The detector energy resolution-peaking time dependencies obtained when the detector was under influence of external interferences of different frequencies: triangles – 8 kHz, squares – 19 kHz, dots – 51 kHz. Solid lines correspond to the best LMS fits.

voltage calculated according to (19.61) within peaking time interval. The maximum amplitude of this deviation will be reached when the interval of averaging is centered around the maximum (or minimum) of the interference signal, as is shown in Figure 19.20. Thus, we can write down the deviation peak amplitude as:

$$D_{\mathrm{m}} = \frac{1}{\tau}\int_{\frac{T}{4}-\frac{\tau}{2}}^{\frac{T}{4}+\frac{\tau}{2}} A_{\mathrm{m}} \sin(\omega t)\mathrm{d}t = A_{\mathrm{m}} \frac{2\cdot\sin\left(\frac{2\pi T}{4T}\right)\cdot\sin\left(\pi\cdot\frac{\tau}{T}\right)}{2\pi\cdot\frac{\tau}{T}} = A_{\mathrm{m}}\frac{\sin(\pi\cdot p)}{\pi\cdot p} \tag{19.62}$$

where

A_{m} is the external interference signal amplitude;
T is the external interference period;
τ is the peaking time;

$$p = \frac{\tau}{T}.$$

One can note that when $p = n$, where n is an integer, the external interference will have no effect on the detector resolution, and its effect on the resolution will decrease with peaking time increase; this is exactly what we observe in experiment (Figure 19.20). Now, for the external interference, combining (19.61) and (19.62) and neglecting the integration on phase since the external interference is not correlated with photon detection events, we can write down:

$$\Gamma_{\mathrm{Ext}}^2 = \langle U_{\mathrm{t}}\rangle^2 = \sum_{k=0}^{N-1}[W(k)]^2\left[\frac{A_{\mathrm{m}}}{\sqrt{2}}\cdot\frac{\sin(\pi p)}{\pi p}\right]^2 \tag{19.63}$$

and rewrite (19.58) in a form convenient for the LMS fit if the weighting function is known explicitly. Since in our work we used DPPs from different manufacturers, we did not know $W(k)$ in explicit form. However, it was found that the best agreement between the calculated dependence of FWHM as a function of the peaking time and the experimental observations can be reached when Γ^2_{Ext} is given as:

$$\Gamma^2_{\text{Ext}} = K \cdot \frac{(\sin(\pi p))^4}{(\pi p)^2} \tag{19.64}$$

That can be explained by form of the weighting function that can vary from one DPP manufacturer to another, but we found that (19.64) works well for at least two DPPs from different manufacturers. Finally, the expression:

$$\text{FWHM} = \sqrt{((2.35)\sqrt{\varepsilon FE})^2 + \left(A_1 \cdot \tau + A_2 \cdot \frac{1}{\tau} + A_3\right) + \left(A_4 \frac{(\sin(\pi p))^4}{(\pi p)^2}\right)} \tag{19.65}$$

was used for LMS data fit, where coefficients A1–A5 were refining parameters. The results of the best fits of the resolution versus peaking time dependences are shown in Figure 19.20. As one can see, (19.65) gives an adequate description of the external interference effect on the detectors resolution regardless of the nature of this interference, and, we believe these considerations are applicable not only to PIN diode detectors, but to other types of detectors as well.

It was demonstrated that periodic interference and white noise that might influence the detectors in practice affect the detector energy resolution-peaking time dependence in different ways: white noise causes the energy resolution degradation within a wide range of the peaking times, but the periodic interference significantly changes resolution-peaking time dependence shape. Because of this, it is always useful to take measurements of the resolution-peaking time dependence in the wide range for proper detection system diagnostic, because just a single point measurement taken at one peaking time can lead to erroneous conclusion when the detector is under influence of the periodic interference. The detector resolution degradation depends not only on the external interference amplitude, but also on the peaking time-to-interference period ratio, i.e. on the interference frequency. Thus, the frequency of the external interference can be found from the resolution-peaking time dependence measurements using proposed model. That might help, for example, to develop proper digital filter when dealing with fixed frequency interference and not avoidable external disturbance.

19.6 Conclusion

Within the last decade, semiconductor detector technology has made significant progress. Currently, SDDs cooled with Peltier elements have a typical energy resolution between 125 and 140 eV, within detector temperature ranges between −35 and 15 °C. The maximum count rate is about one million counts per second, and the detector sensitive area is up to 150 mm^2. Modern detectors are compact, have low power consumption, and are mechanically robust and reliable devices, which can work within a wide range of ambient temperatures and under harsh conditions. They were successfully used, for example, in recent NASA missions on Mars. ASIC preamplifiers played a significant role in the improvement of detector performance parameters such as increased detector counting rate and improved sensitivity at low energies (below 1 keV). Thus, SDDs and (PIN diode detectors, when cost is the major concern) are the most suitable detectors for handheld XRF spectrometers, and their performance is equivalent to that of detectors made for in-lab use.

However, Si detectors have some fundamental limitations. They exhibit low quantum efficiency at high X-ray photon energy (above 15 keV), need to be cooled down, and their operational temperatures have to be stabilized. The demand for detectors with higher quantum efficiency at high energy, and with higher working temperatures (lower power consumption), will continue to increase for applications like basic research, for the space explorations, security, and medical industries, to name a few. Several research groups at universities and companies are working to satisfy this demand, not only by improving existing silicon detector technology, but also by engaging in a search for alternatives to silicon materials. Significant progress has been achieved in the development of detectors based on cadmium telluride – CdTe (Amptek, Inc), and the cadmium zinc telluride CZT (Kromek Group), which have high quantum efficiency and good energy resolution at the incident photon energy up to 130 keV (see for example [20]). CdTe and CZT detectors are now commercially available. Another potential candidate for use in X-ray detectors is silicon carbide – SiC. Prof. G. Bertuccio and his group demonstrated that SiC detectors can operate at high and unstable temperatures up to 100 °C [21]. However, these types of detectors are still not being fully developed and more efforts are required to make them suitable for use in portable EDXRF spectrometers.

Acronyms and Abbreviations

ASIC	application specific integrated circuit
CZT	cadmium zinc telluride
DPP	digital pulse processor
ENC	equivalent noise charge
EMI	electromagnetic interference
FWHM	full width half maximum
JFET	junction gate field-effect transistor
MOSFET	metal-oxide-semiconductor field effect transistor
LMS	least median of squares
RMS	root mean square
SDD	silicon drift detector
Si(Li)	lithium-doped silicon detector
Si-PIN	silicon p-type–intrinsic–n-type detector
S/N	signal-to-noise ratio
TEC	thermoelectric cooler
TO-8	a specific metal can design, containing electronics

References

1 Pantazis, T., Pantazis, J., Huber, A., and Redus, R. (2010). The historical development of the thermoelectrically cooled X-ray detector and its impact on the portable and hand-held XRF industries. *X-Ray Spectr.* 39: 90–97.

2 Lowe, B.G. and Saren, R.A. (2014). *Semiconductor X-Ray Detectors*. Taylor & Francis Group, Ltd.

3 Kemmer, J. (1984). Improvement of detector fabrication by the planar process. *Nucl. Instr. Methods Phys. Res. A* 226 (1): 89–93.

4 Gatti, E. and Rehak, P. (1985). Semiconductor drift chambers for position and energy measurements. *Nucl. Instr. Methods* A 235: 224–234.

5 Bombelli, L., Fiorini, C., Frizzi, T., Alberti, R., and Longoni, A. (2011). "CUBE", a low noise CMOS preamplifier as alternative to JFET-nd for high-count rate spectroscopy. *Proceedings, 2011 IEEE Nuclear Science Symposium and Medical Imaging Conference (NSS/MIC 2011),* Valencia, Spain (23–29 October 2011).

6 Spieler, H. (2005). *Semiconductor Detector Systems*, Series on Semiconductor Science and Technology (Book 12). Oxford University Press.

7 Lutz, G. (1999). *Semiconductor Radiation Detectors*. Springer.

8 Sze, S. and Kwok, K. (2007). *Physics of Semiconductor Devices*. John Willey & Sons, Inc.

9 Radeka, V. (1972). Trapezoidal filtering of signals from large germanium detectors at high rates. *Nucl. Instr. Methods* 99 (3): 525–539.

10 Betta, G. and Dalla, F. (2000). Silicon PIN radiation detectors with on-chip front-end junction field effect transistors. *IEEE Trans. Nucl. Sci.* 47 (3): 829–833.

11 Rawlings, K. (1986). Radial semiconductor drift chambers. *Nucl. Instr. Methods* A256: 297–304.

12 Rehak, P., Gatti, E., Longoni, A. et al. (1989). Spiral silicon drift detector. *IEEE Trans. Nucl. Sci.* 36 (1): 203–209.

13 Hartmann, R., Hauff, D., Krisch, S. et al. (1994). Design and test at room temperature of the first silicon drift detector with in-chip electronics. *Proceedings of 1994 IEEE International Electron Device Meeting*, San Francisco, CA, USA, pp. 535–538. doi: https://doi.org/10.1109/IEDM.1994.383476.

14 Fiorini, C. and Longoni, A. (2008). Semiconductor drift detectors for X-ray and gamma-ray spectroscopy and imaging. *Nucl. Instr. Methods B* B266: 2173–2181.

15 Bertuccio, G., Ahangarianabhari, M., Graziani, C. et al. (2015). A silicon drift detector-CMOS front-end system for high resolution X-ray spectroscopy up to room temperature. *J. Instr.* 10: 001002.

16 Bombelli, L., Fiorni, C., Frizzi, T., et al. (2010). Low-noise CMOS charge preamplifier for X-ray spectroscopy detector. *IEEE Nuclear Science Symposuim & Medical Imaging Conference*, Knoxville, TN, pp. 135–138. doi: https://doi.org/10.1109/NSSMIC.2010.5873732.

17 Huebner, S., Miyakawa, N., Kapser, S. et al. (2015). High performance X-ray transmission window based on graphenic carbon. *IEEE Trans. Nucl. Sci.* 62 (2): 588–593.

18 Fontanelli, F., Gatti, F., and Nostro, A. (1996). An adaptive system for the reduction of the microphonic noise in cryogenic detectors signal. *Nucl. Instr. Methods A* 370: 218–219.

19 Morales, J. and Garda, E. (1992). Filtering microphonics in dark matter germanium experiments. *Nucl. Instr. Methods Phys. Res. A* 321: 410–414.

20 Wilson, M.D., Cernik, R., Chen, H. et al. (2011). Small pixel CZT detector for hard X-ray spectroscopy. *Nucl. Instr. Methods Phys. Res. A* 652 (1): 158–161.

21 Bertuccio, G., Caccia, S., Puglisi, D., and Macera, D. (2011). Advances in silicon carbide X-ray detectors. *Nucl. Instr. Methods Phys. Res. A* 652 (1): 193–196.

20

Field-Deployable Utility of Benchtop Nuclear Magnetic Resonance Spectrometers

Koby L. Kizzire and Griffin Cassata

Department of Forensic Science, Henry C. Lee College of Criminal Justice and Forensic Sciences, University of New Haven, West Haven, CT, USA

20.1 Introduction

Soon after the close of World War II, Felix Bloch and Edward Purcell reported independent discoveries that demonstrated the absorption of radio frequency light by nuclei in a magnetic field [1, 2]. By the early 1950s, Varian produced the first commercially available instrument based on this principle [3], and by 1952, both investigators won the Nobel prize in physics [4]. The unconventional speed of these events demonstrated what most modern chemists will still affirm about nuclear magnetic resonance (NMR) – NMR spectroscopy is one of the most powerful analytical techniques for discovering and understanding inorganic and organic structures of compounds.

The physics of the NMR measurement favors higher magnetic fields and, thus, the subsequent pattern of instrument development led to a trend of bigger and better NMR spectrometers. Alongside advancements in user-friendly interfaces, instruments began gaining bulk quickly [5]. Larger magnets with greater magnetic fields were introduced to improve resolution and sensitivity and, in turn, the elucidative capabilities of the instrument. The fixed magnets of many of these instruments posed a restriction as the highest achievable homogeneous magnetic fields were quickly reached due to limitations in existing materials. However, after discoveries in superconduction and the recent rise of hybrid instruments, this trend has continued into the present day [6]. Where instruments once routinely operated on as low as a 0.5 T magnet [6], company announcements are now reporting the release of instruments that operate at 28.2 T [7], and NMR spectroscopy experiments have been conducted in fields as high as 35.2 T [8].

While the capabilities of modern instruments are unquestionably incredible, they have come at substantial monetary cost. Prices vary depending on an overall package, but high-field NMR spectrometers (i.e. possessing field strengths of 7.05 T or greater) cost anywhere from $100 000 USD to well over $10 000 000 USD [9]. Rooms and even small buildings have been constructed or outfitted to receive new high-field NMR spectrometers, and the superconducting materials require continual replenishment of liquid He and N_2 over the lifetime of the instrument. Inevitable repairs or maintenance require the aid of a technician at the installation site and routinely cost thousands of dollars. The value of information gained from these instruments has clearly been worth the cost to those who can afford it.

In the face of these financial hurdles, it is not difficult to conceive of why the instruments of yesteryear are gaining attention. Scientists in the early decades of NMR were successfully utilizing data generated on instruments operating cryogen-free with field strengths of 1–2 T [10], and not every experiment requires high-resolution measurements. One notable company that moved to fill this niche, Anasazi Instruments, began producing 1.4 and 2.1 T fixed-magnet systems with a primary interest from academia. These instruments provide a superior affordability and ruggedness to modern high-field instrumentation that alleviate some concerns of allowing

Portable Spectroscopy and Spectrometry 1: Technologies and Instrumentation, First Edition.
Edited by Richard A. Crocombe, Pauline E. Leary and Brooke W. Kammrath.

inexperienced students to use them. However, they also contain traditional magnets of greater mass with the 2.1 T magnet weighing 918 kg [11].

Advances in several key components of NMR spectrometers have allowed a level of miniaturization of the modern versions of these units, and current commercial versions of these systems are often referred to as benchtop instruments. Although structural elucidation of simple molecules can be achieved by these lower field NMR spectrometers, the key uses of these instruments are often related to confirmation or quantitation of a known structure. Additional reasons for interest vary depending on the demands of various consumers. In addition to academia, marketing attention has especially focused on quality assurance and control, research, and the possibility of use in fieldwork [12]. The reduction in mass and external needs for some of these units has been so dramatic that they can be easily moved by one person to a site of greater benefit to the user.

The change in NMR spectrometer development trends toward instrumentation with lower field strengths has caused some understandable nomenclatural confusion. It is clear from a quick browse of review literature and manufacturer websites that there is no unified convention concerning what to call NMR spectrometers that have potential as field-deployable instruments for structure identification. Some common examples of descriptors include the more widely accepted benchtop [12–15] along with desktop [10, 16], portable [17–20], compact [21–23], and combinations of these and additional terms. The use of the term portable, which exists directly or in indirect public claims by manufacturers, must be approached carefully. While techniques like Raman and infrared spectroscopy (IR) have been reduced to a handheld scale, portable NMR spectroscopy is more akin to the development of the desktop computer versus the immovable, room-sized counterparts of the mid-twentieth century. A comparison of various NMR spectrometers is shown in Figure 20.1.

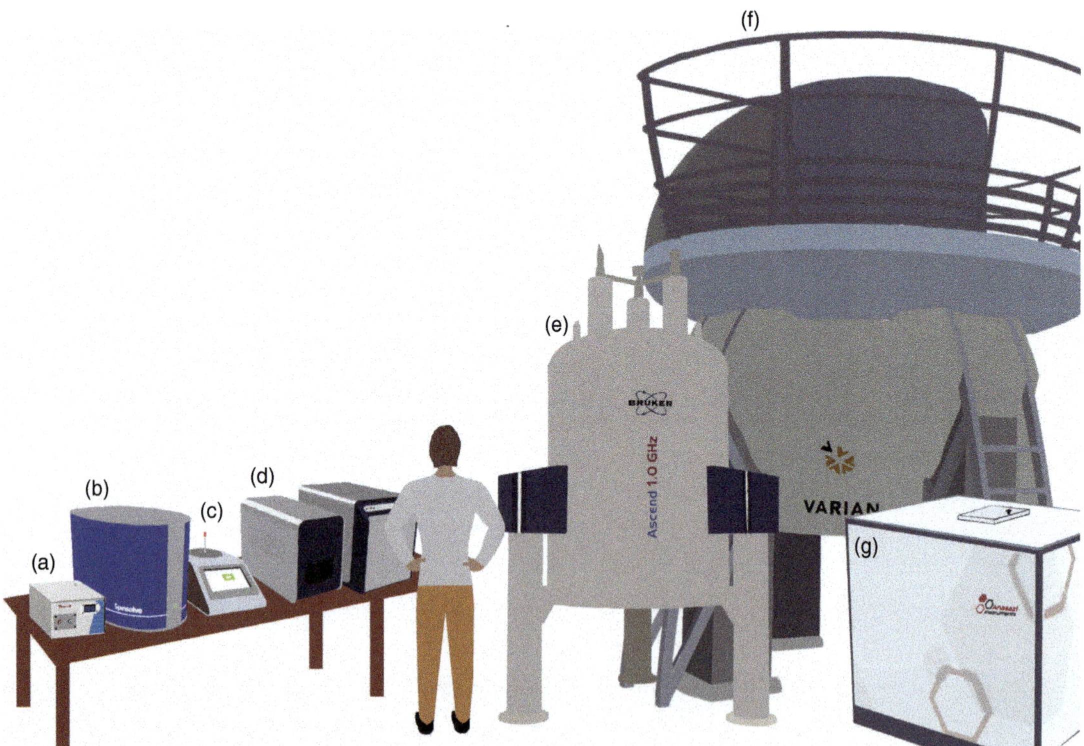

Figure 20.1 Size comparison of various NMR spectrometers. These instruments are (a) Thermo Scientific's picoSpin 40, (b) Magritek's Spinsolve 60 Carbon, (c) Nanalysis' NMReady-60PRO, (d) Oxford Instrument's X-Pulse, (e) Bruker's Ascend 1.0 GHz spectrometer, (f) Vairan's Unity Inova a 900 MHz Spectrometer, and (g) Anasazi's Eft-90 NMR Spectrometer. Instruments (a–d) are bNMR spectrometers. Instrument (g) is a low-field NMR with similar capabilities to instruments (a–d), and instruments (e) and (f) are ultrahigh-field spectrometers.

For the sake of clarity, the instruments that will be addressed herein operate using permanent magnets of approximately 1.0–2.3 T, provide structural data useful for the identification of a substance, and are of a sufficiently small footprint and external power and/or resource requirement to allow potential field use. These instruments will be referred to as benchtop NMR (bNMR) spectrometers. In addition to the level of field-deployment capability achieved by these instruments, brief background of NMR theory and key advancements that have allowed the commercially available state of the art in bNMR spectrometers will be addressed.

20.2 NMR Theory

NMR is measured spectroscopically, and like other spectroscopic instruments, NMR instruments use electromagnetic waves to elicit a response from the analyte. More specifically, radio waves force some of the nuclei being measured to transition to higher energy states, and as nuclei return to states of lower energy, they release energy that is detected by the instrument. The detected energy is displayed on NMR spectra and analyzed for key structural elements in the analyte. NMR spectroscopy, however, requires the atom being analyzed to be NMR active. This intrinsic activity of certain atoms is directly associated with the quantum spin states of their respective nuclei.

Nuclear spin is the sole quality of a nucleus that leads to the possibility for NMR analysis. Spin is based on the electromagnetic principle that postulates that moving charges create magnetic fields. Taking the entire atom into consideration, the electrons circulate around the nucleus causing a magnetic field on the atom. Inside the nucleus, though, the protons and neutrons also have their own spin producing their own separate magnetic field. The magnetic field produced by these nuclear particles acts as any magnetic field would and can be manipulated with external fields. The term spin, describing movement in subatomic particles, is typically used in the classical descriptions of quantum physics. Subatomic particles do not spin as one would imagine a top or a coin to spin. Nuclear spin is more of an energy shift and a concentration of that energy throughout the particle.

The nuclear spin quantum number, I, is a measure of the magnetic properties of the particles in the nucleus. The I values possessed by an atom are interpretations of how the energy is concentrated in that atom, and the various energetic states the nucleus can occupy [24]. When $I = ½$, the energy concentration is symmetrical and spherical about the particle. Because of this, the energy can only occupy one of two confirmations, either a (i) spin left or a (ii) spin right as described in the classical model of nuclear spin. NMR analysis of nuclei with $I = ½$ is simplified because of this limitation, and this accounts for most nuclei that are commonly investigated with this technique (e.g. ^{1}H, ^{13}C, and ^{19}F).

When the atom is exposed to an external magnetic field, the nucleus' spin, which is magnetic, will either align with (i.e. parallel to) or against (i.e. antiparallel to) the magnetic field if it has $I = ½$. Whether the spin is with or against the magnetic field, the resulting spin axis on the nucleus is never exactly parallel. The axis is always askew of true parallel to the magnetic field and instead precesses about the axis of the applied magnetic field. This precession is measured in terms of the Larmor frequency, which is different for each atom in each respective magnetic field [25]. Equation (20.1) shows the relationship of the Larmor frequency to the magnetic field.

$$\nu = \frac{\gamma}{2\pi} B_0 \tag{20.1}$$

In this equation, ν is the Larmor frequency, γ is the gyromagnetic ratio of a given nucleus, and B_0 is the strength of the magnetic field supplied by the instrument. The gyromagnetic ratio (γ) can be thought of as the affinity of an atom to be manipulated by an applied magnetic field. The Larmor frequency needs to be matched by a radio frequency (RF) pulse in order to influence the nuclear precessional orientation. Because of its natural abundance relative to other isotopes, desirable spin state, ease of interpretation due to limited bonding, and prevalence in many organic and inorganic compounds, the ^{1}H nucleus was most studied by NMR spectroscopy historically and is arguably still of greatest analytical interest. One common practice for classifying the performance of a given NMR spectrometer is the Larmor frequency of ^{1}H. Most commonly encountered frequencies in bNMR spectrometers are

43, 60, 80, and 100 MHz corresponding to approximate magnetic field strengths of 1, 1.4, 2, and 2.3 T respectively.

The energy gap between spin states in a nucleus of $I = ½$ is quantized as indicated by the discussion above. Equation 20.2 shows the relationship of the energy separation between the spin states for these nuclei [24].

$$\Delta E = \gamma \hbar B_0 \tag{20.2}$$

Planck's constant, h, divided by 2π is $\hbar$. Given that the gyromagnetic ratio is an apparent constant inherent to each nucleus using standard operating conditions, it is clear that the magnetic field of the instrument is the manipulatable feature that can influence the magnitude of this gap.

To obtain a signal in NMR spectroscopy, the nucleus' spin must absorb and release energy as it transitions between spin states. Most common NMR active nuclei have positive spin quantum numbers. In these nulcei, the lower energy level is the one that aligns the nucleus' spin with the magnetic field, and the higher energy level is opposed to that field. There must be a greater number of nuclei attributed to the lower level than in the higher energy level in order to allow for a net absorption of energy from the RF pulse. This ratio of high-energy spin to low-energy spin at thermal equilibrium is called the Boltzmann distribution [26], which is defined in Eq. 20.3.

$$\frac{N_{\text{high}}}{N_{\text{low}}} = e^{-\frac{\Delta E}{kT}} = e^{-\frac{\gamma \hbar B_0}{kT}} \tag{20.3}$$

Boltzmann's constant is k and T is the temperature. Increasing the magnetic field strength of the instrument leads to an increase in the gap between spin states, and as expected, the Boltzmann distribution expressed here ($N_{\text{high}}/N_{\text{low}}$) will decrease indicating a higher population of nuclei attributed to the lower energy state.

The movement of charged, subatomic particles produces a local spin-induced magnetic field. These subatomic magnetic fields are far weaker than the field produced by the instrument's magnet, but they are close enough to other subatomic magnetic fields in the molecule to show measurable effects. This effect is called shielding. When a nucleus of interest is surrounded by other nuclei and electrons producing magnetic fields, which is sometimes referred to as the magnetic environment, the overall magnetic field experienced by the nucleus is different from the one produced by the instrument's magnet, which is expressed in Eq. 20.4 [27].

$$\nu = \frac{\gamma}{2\pi} B_0 (1 - \sigma) \tag{20.4}$$

The shielding constant, σ, can be either positive or negative. The difference in magnetic field experienced by each nucleus occurs in a micro-environmentally dependent manner, which greatly aids the identification of structural features of compounds. As expressed in Eq. (20.1), the Larmor frequency will change based on the magnetic field experienced by the nucleus. The minor variations of the Larmor frequency lead to minor changes in signal output, which are expressed as parts-per-million shifts, or ppm shifts, relative to a standard signal (e.g. tetramethylsilane). Typically, on an NMR spectrum these shifts are represented as δ. A more shielded nucleus will be shifted upfield and display a lower δ than a nucleus that is deshielded and shifted downfield. Shift values reflect Larmor frequency differences as expressed in Eq. 20.5.

$$\Delta \nu = \frac{(\Delta \delta) \gamma B_0}{2\pi} \tag{20.5}$$

While the shifts of any two given signals are reported as the same ppm value between instruments of differing magnetic field strengths, the Larmor frequency difference of the signals is much greater on the instrument with the stronger field. Thus, stronger magnets not only mean stronger signal due to the population differences noted in Eq. 20.3 but also greater resolution.

The effect of magnetic field sensitivity of the instrument can also be approached by considering an approximation of the signal-to-noise ratio (S/N), which is expressed in Eq. 20.6 [6].

$$\frac{\text{signal}}{\sqrt{\text{noise}}} \propto \frac{B_0^2}{\sqrt{\alpha B_0^{\frac{1}{2}} + \beta B_0^2}} \tag{20.6}$$

In this equation, α and β relate to the cofactors of sample efficiency. If the RF coil is lossless, i.e. all input energy is converted to output energy, and the sample is conductive, then α approaches zero. If the RF coil is lossy, i.e. some input energy is lost to other processes, then the sample is nonconductive, and β approaches zero. If α is 0, the S/N is directly related to B_0, and if β is 0, the S/N is related to $B_0^{\frac{7}{4}}$. In either case, greater B_0 results in higher S/N [6].

Even the most fundamental aspects of NMR theory presented here provide insight into the current state of NMR technology, especially with regard to the climb in magnetic field strength. Many of the aforementioned equations also indirectly hint at another vital need in any NMR spectrometer. Changes in B_0 as a result of lack of homogeneity within the sample volume of the instrument's magnetic field would at the very least contribute to band broadening of signals. S/N and resolution are greatly improved by the homogeneity of the field, and the struggle to achieve both strength and homogeneity in a permanent magnet has characterized the development of today's bNMR spectrometers.

20.3 Magnet Miniaturization

When searching for literature on portable or mobile NMR spectrometers, one might be surprised to find decades-old examples. Many of these are based on earth's field NMR and include low-resolution techniques with very specialized uses (e.g. well logging) [28]. The magnets were often large, and creating a compact system was severely limited by the state of electronic components and computers. It was not until the mid-1990s that a truly compact and portable NMR spectrometer was developed – the NMR Mobile Universal Surface Explorer [29]. This instrument is used to make diffusion and relaxometry measurements as well as to develop images that provide information on surface characteristics of materials [10]. The low resolution of this instrument and others like it makes them unsuitable for spectroscopic measurements needed to identify substances and will not be further addressed in this chapter.

In the 2000s, intensive research efforts eventually culminated in the first commercially available bNMR spectrometer in 2009. Many technological hurdles were overcome by advancements in non-NMR technologies. The ever-improving abilities of the computer manufacturing industry to reduce the footprint of processors and storage hardware while improving data capacity provided an unarguable advantage in the reduction of the NMR spectrometer's size. Another marked improvement was the reduction in the size of RF coils thanks in part to the cell phone industry [26]. However, there is no question that the size of any NMR spectrometer is governed chiefly by the magnet. To achieve an acceptable volume of homogeneous magnetic field at a given field strength, large volumes (~1 m^3) of heavy metallic material had to be used in permanent magnet systems of the 1960s and 1970s [5], and the field strengths and general properties of the materials used imposed severe limitations on the state and utility of NMR data [19].

In the search for magnetic materials with higher field strengths, two key properties are typically investigated – coercivity and remanence. Coercivity is the ability of a material to resist changes in its permanent magnetic field when an external magnetic force is applied. Remanence is the magnitude of a material's magnetic field in the presence of no applied magnetic force. In 1966, a rare earth cobalt alloy was found to have higher coercivity and remanence when compared with previous generations of alloys [30]. The continued research of rare earth alloys since that time has produced NdFeB-based materials, which are now commonly employed in many technological applications [31]. While NdFeB-based magnets have greatly increased the ability to achieve strong and highly stable magnetic fields, the amount of material required to make an effective NMR spectrometer using traditional magnet designs would be too heavy to achieve portability of any kind [5].

One key discovery reported in 1980 eventually led to the smaller magnetic systems of modern bNMR spectrometers. Following the development of rare earth cobalt magnets, Halbach reported that permanent magnets had the capability to generate strong, homogeneous fields when multiple magnets of a given shape and magnetization orientation were properly arranged [32]. This discovery generated little interest in the NMR community for over a decade in part due to the interest in field strengths that far exceeded the use of this technique. However, the sky-rocketing costs of high-resolution instrumentation and advancements made by the Blümich and Blümler groups have made the Halbach-type array a staple of modern NMR spectrometer miniaturization efforts. The low cost of materials needed to produce these instruments has even led to a rise in researchers building their own units, which are sometimes referred to as "home-built" instrumentation [17, 19, 33].

Current use of the Halbach-type array coupled with improvements in modern magnetic alloys has resulted in experimental field strengths as high as 5.16 T [34], and the highest commercially available field strength currently used in bNMR spectrometers is ~2.3 T (~100 MHz). The exact arrangement, shape, and alloy composition of the components in the array vary between vendors, but in modern bNMR spectrometers, they generally consist of concentrically oriented magnetic units such that a high magnetic field is directed across a gap in the center of a cylindrical shape and very little stray field occurs externally [35]. Not only does this dramatically reduce the footprint of the magnetic system, but also it allows placement of the completed device near other magnetically sensitive laboratory equipment.

20.4 Improvements in Sensitivity and Resolution

The size of NMR spectrometers has been greatly reduced, and efforts to further minimize the instruments are ongoing, but as has been made clear in NMR theory, any NMR instrument based on permanent magnets cannot be expected to outperform high-field instrumentation in terms of sensitivity and resolution. Many researchers and manufacturers are currently more focused on improving the utility of data in order to broaden the capabilities and applications of low-field instruments. Additionally, the ^{1}H shift range is one of the narrowest among commonly analyzed nuclei. For the relatively low fields of bNMR spectrometers, other nuclei have more desirable resolution in terms of frequency [26], which explains manufacturer interest in providing heteronuclear capabilities. These currently include ^{7}Li, ^{11}B, ^{13}C, ^{15}N, ^{19}F, ^{23}Na, ^{29}Si, and ^{31}P [36].

A notable technical hurdle for fixed magnetic materials operated at ambient temperatures is thermal drift, which can affect both resolution and sensitivity. While this has been mitigated in part by the development of new alloys, temperature fluctuations in a magnetic material can cause changes related to solid-state expansion, and these materials have coefficients that directly associate temperature changes with magnetic field strength [35]. As is demonstrated by Eq. 20.1, changes in field strength will cause proportional changes in the Larmor frequency, which will lead to broadening of the signal when the data of multiple acquisitions are averaged to improve S/N. This is especially true of spectra obtained over long periods of time. Thermal drift effects are less of a problem in superconducting instruments because the magnet temperature is maintained by boiling helium, but they have been observed immediately following helium fills or when data are collected over several weeks [37].

For low-resolution instruments, thermal drift presents a strong concern for lower concentration samples and many two-dimensional (2D) coupling experiments as these require longer acquisition periods. Many techniques used in high-resolution experiments can be utilized on these instruments. Field locking is performed using a signal related to a standard internal or external to the sample, and deuterated solvents are most typically used to this end [38]. Measurement of the fluctuation of this signal allows for drift adjustments to be made in the values observed in sample output prior to signal averaging.

Many modern bNMR spectrometer manufacturers have installed temperature control units to more directly address the environmental concern. Instead of modulating a near ambient temperature expected in most laboratories, bNMR spectrometers are designed to operate with a temperature in the magnetic environment around 30–40 °C. The internal temperature must remain relatively stable, and consequently, operating specifications recommend external environments of 18–26 °C with no more than 1.5 °C fluctuations per hour [39, 40]. As should be expected, operation of any of these instruments requires a minimum of 30 minutes to achieve a thermally equilibrated readiness state. The level of thermal stability that would be required to obtain acceptable signal with sole use of environmental measures far exceeds the capacity of any temperature control units [41], so locking is used in conjunction with this method.

NMR spectrometry is a measurement of low sensitivity when compared with other forms of spectroscopy. Most measurements are performed at thermal equilibrium, where the gap induced in spin energy levels is known as polarization [42]. This leads to a defined nuclear population difference noted in Eq. 20.3. Sensitivity for these acquisitions is demonstrated in bNMR spectrometers by the need for either highly concentrated samples (e.g. 100 mM or greater concentrations for ^{1}H experiments) or high acquisition times over which field drift can be of concern.

Hyperpolarization techniques have been investigated for decades and are now finding popularity in low-field measurements. As the name suggests, these methods are designed to induce greater differences in nuclear spin alignment than observed under thermal equilibrium alone [42], and have been shown to produce increases in signal that are orders of magnitude greater than thermally equilibrated acquisitions. Some of the most popular are varieties of dynamic nuclear polarization (DNP) and signal amplification by reversible exchange (SABER), but broad applicability of these techniques will require overcoming cost and the incorporation of complex equipment among other hindrances [15].

20.5 Current bNMR Spectrometers

The utility of today's bNMR spectrometers can be limited in a field setting. Original designs centered on the use of these instruments in academic, pharmaceutical, and quality control laboratory settings [12] where stable ambient conditions and limited movement of the device were expected. All bNMR spectrometer manufacturers have made claims of portability [43–46]. Before an examination of these claims, one should note that the term portability does not have a strict application or meaning even when applied to instruments, and there are no clear regulations or restrictions on the use of this term in the context of field instrumentation. This is due in part to the fact that each end user has unique needs that can be satisfied by various levels of instrument mobility.

All modern bNMR instruments can be operated solely on standard portable power supplies or generators without additional accessory needs. In the past, permanent magnet NMR spectrometers required sample spinning using air streams to provide ample homogeneity of the magnetic environment. These air streams would also be used to hold or transition a sample to the appropriate position for analysis. The air had to be supplied by portable tanks or a forced air line fitted at the site of installation. Due to the current state of autoshimming functions and other field homogeneity compensations and improvements, bNMR spectrometers do not have this requirement.

According to Gałuszka et al., instruments have been categorized for field use as transportable, portable, or handheld, where weight is one primary consideration in the division of these categories. Transportable instruments are recommended to be between 10 and 20 kg and portable instruments between 0.5 and 10 kg [47]. The United States' National Institute for Occupational Safety and Health (NIOSH) has similarly categorized field instruments as personal (i.e. wearable), portable, and transportable [48] but has not published weight ranges to accompany these terms. NIOSH does provide an ergonomic recommendation for the maximum weight of a single two-handed lift or carry at 23 kg, and additional movements and duration of the lift are expected to reduce that maximum [49]. While the size of bNMRs have been greatly reduced, most are still prohibitively heavy for a single individual to carry in view of these guidelines.

The current bNMR spectrometer market is dominated primarily by four manufacturers – Magritek, Nanalysis, Oxford Instruments, and Thermo Scientific. The differences in spectrometers supplied by each reflect slightly different capabilities and end user focus. Some specifications are listed in Table 20.1. Separate educational models of these instruments supplied by Magritek and Nanalysis have limited function and available nuclei

Table 20.1 Instrument specifications of bNMR spectrometers.

Manufacturer and model	Field strength	Weight	Dimensions	Linewidth at % maximum peak height	Nuclei available[a]
Magritek [36, 50]					^{1}H, ^{7}Li, ^{11}B, ^{13}C, ^{15}N, ^{19}F, ^{23}Na, ^{29}Si, and ^{31}P
Spinsolve 43	1 T	55 kg	58 × 43 × 40 cm	0.5 Hz at 50% and 20 Hz at 0.55%	
Spinsolve 60	1.4 T	60 kg	58 × 43 × 40 cm	0.5 Hz at 50% and 20 Hz at 0.55%	
Spinsolve 80	2 T	72.5 kg	58 × 43 × 40 cm	0.5 Hz at 50% and 20 Hz at 0.55%	
Spinsolve ULTRA	1 or 1.4 T	55 or 60 kg	58 × 43 × 40 cm	0.2 Hz at 50%, 6 Hz at 0.55%, and 12 Hz at 0.11%	
Nanalysis [40, 51]					^{1}H, ^{7}Li, ^{11}B, ^{13}C, ^{19}F, and ^{31}P
NMReady-60PRO	1 T	25 kg	30 × 28 × 49 cm	1.0 Hz at 50%	
100PRO	2.3 T	97 kg	37 × 41 × 65 cm	1.0 Hz at 50%	
Oxford Instruments [39] X-Pulse[b]					^{1}H, ^{7}Li, ^{11}B, ^{13}C, ^{19}F, ^{23}Na, ^{29}Si, and ^{31}P
Magnetic Unit	1.4 T	150 kg	39 × 54 × 43 cm	0.35 Hz at 50% and 10 Hz at 0.55%	
Electronics Unit		22 kg	30 × 61 × 42 cm		
Thermo Scientific [46, 52]					^{1}H and ^{19}F
picoSpin 45	1.1 T	4.8 kg	18 × 15 × 29 cm	1.8 Hz at 50%	
picoSpin 80	2 T	19 kg	43 × 26 × 25 cm	1.8 Hz at 50%	

a) Listed nuclei are available for all models from a given manufacturer. Magritek, Nanalysis, and Oxford Instruments advise consultations for nuclei not listed.
b) The X-Pulse consists of two separate units. The units are separated to show weights and dimensions of each.

(i.e. ^{1}H and/or ^{19}F only). These instruments are not specifically addressed herein. Most of the units discussed here have masses greater than most would consider acceptable weights for portable instrumentation, but might be suitable for mobile laboratories. This is true of many bNMR spectrometers, but the focus herein will be on the previously mentioned manufacturers. The reader should be aware that other manufacturers of instrumentation with similar size and field strengths (e.g. Bruker's Fourier 60 and 80 models) to the largest of these are also available.

Of all bNMR manufacturers, Magritek presents the largest variety of optional accessories and instrument capabilities including a base educational model. Recent advances in shimming technology have led to the production of the Spinsolve ULTRA series of bNMRs, which possess line widths of less than 0.2, 6, and 12 Hz at 50, 0.55, and 0.11% peak heights, respectively [50]. These are the narrowest line widths of commercially available bNMRs and are achieved without the need for the resolution enhancement capability that Magritek provides on Spinsolve series instruments.

Nanalysis is recognized in the United States for its NMReady-60 professional and educational spectrometers, which offer similar nuclear and accessory capability to Magritek but with wider signal linewidths. The company is poised to have the highest field strength bNMR upon the release of the 100PRO [51], which also has a corresponding educational model, the 100e. Consequently, the 100 series instruments will have the highest signal resolution by virtue of the 2.3 T (100 MHz) field strength.

The X-Pulse by Oxford Instruments was also very recently announced and replaces the former Pulsar spectrometer [39]. Where Nanalysis' and Magritek's spectrometers allow the option of a probe for a single nucleus in addition to standard available nuclei, the X-Pulse's advantage is an optional broadband tuner that allows analysis of ^{7}Li, ^{11}B, ^{13}C, ^{23}Na, ^{29}Si, and ^{31}P without the need for multiple probes. The bulk of separate magnet and electronic units, which make this instrument the largest of bNMRs in both weight and volume, are a drawback in field use as they remove already limited space in mobile laboratories and may pose weight-related restrictions. However, the bulk of the magnet becomes an advantage for flow reaction monitoring at temperatures between 20 and 70 °C and provides a moderate level of thermal drift resistance relative to other spectrometers.

Thermo Scientific produces the picoSpin Series II spectrometers with 1.1 and 2 T field strengths, which are based on the first commercially available bNMR spectrometer. Developed originally for academic use, these instruments are limited in terms of multinuclear capabilities and, in contrast to their competitors, do not have 2D experiment options. The picoSpin instruments also use a less conventional capillary cartridge with sample injection via syringe in which sample viscosity and air introduction can be of some concern. This design allows small volumes of approximately 40 μL to be sufficient for analysis [52]. There is also a soft-lock capability that allows acceptable performance of these instruments in relatively harsh environments. Manufacturer claims include the operation of the picoSpin 45 "on mountains and aboard NASA's Weightless Wonder zero-gravity aircraft" [53]. Of particular note, the limited consumable requirements and footprint, durability, and 5 kg mass of the picoSpin 45 allow classification of this instrument as portable, not simply field-deployable.

Overall, bNMR spectrometers allow a level of portability that is inconceivable in standard low-field instruments and any high-field instrument. If employed in a field setting, all of these instruments would be best suited to an environmentally controlled mobile laboratory in a context where time constraints are limited if turned on at the operational site. They perform best when infrequently moved or shut down. In terms of general portability, the picoSpin 45 most clearly fits constraints.

20.6 Applications

The data that a bNMR spectrometer can provide are similar in nature to what are gained from IR spectra. The signals can be compared to a library standard spectrum to identify a compound of interest and are related to the structural features of the compound analyzed. However, without a standard for comparison, the information that can be gleaned from spectra is somewhat limited. Some applications of Fourier transform-based experiments on higher field instruments apply to bNMR spectrometers. To a degree, a moderate level of structural elucidation can be achieved on these instruments. Structural elucidation is a process whereby NMR data, either alone or in conjunction with other forms of spectroscopy and/or mass spectrometry, allow an analyst to ascertain the chemical structure of a compound.

Elucidative capability may be generally defined by an estimated molecular weight maximum. However, even moderate signal overlap in compounds of lower molecular weight can greatly reduce the ability of bNMR units to identify an unknown compound for which no standard or library comparison can be made. Some manufacturers offer 2D capabilities, which correlate two 1D hetero- or homonuclear spectra shift patterns, in order to expand the capacity to solve structures. For example, an elucidation of strychnine on a 1.0 T (43 MHz) instrument has been reported and utilizes multiple 1D and 2D experiments [54]. This study was an analysis and assignment of peaks using purchased standards and comparisons to data generated on a 9.4 T (400 MHz) instrument. Although adulterant identification was claimed utilizing the 1.0 T instrument, some peak assignments made utilizing signals well below acceptable S/N were clearly informed by the foreknowledge of structures of the compounds involved.

The most promising qualitative and quantitative applications of bNMR instruments relate to analyses where the structure is known in whole or in part. Compound identifications are reasonably made in situations where reaction progress is being monitored or targeted analysis is performed. Synthetic processes are analyzed by examining the loss and/or gain of expected signals in reagents and products. Recent developments in these methods center on the development of continuous flow techniques especially for monitoring changes in hazardous reagents within the confines of a fume hood [38]. These techniques have seen some success in specialized applications where few components are in the reaction mixture, but they are hampered by sensitivity issues previously discussed [15]. Targeted analysis, which is of greater importance in conventional use of field deployable instrumentation, relies on expectations of a limited number of structures and spectral library comparisons to achieve identification of an analyte.

Similar to other non-separatory techniques like IR, the success of identification or accuracy of quantitation is greatly aided by a sample where the target analyte is a major constituent or nearly pure. While targeted analysis with a bNMR spectrometer can have broad applications in materials [55], pharmaceutics [56, 57], forensics [58–60], food [61], and fossil fuels [62] among others, the analysis of mixtures can pose limitations on its use as it does for high-field instruments. This is especially true for 1D-^{1}H NMR, which has the greatest potential for rapid analysis in field situations.

A direct approach to the problem of mixtures is the use of hyphenated techniques. These have been conceived of for decades, especially with regard to liquid chromatography (LC)-NMR, but initially suffered from inherent difficulties related to sensitivity and interfering solvent signals. Advances in solvent suppression capabilities of NMR instruments in general have attracted new attention to liquid-based separations [63], but most reported applications still use high-field instrumentation. Successful separation and analysis of polymers has been reported using size exclusion chromatography and a 62 MHz bNMR spectrometer [64], but the general cost and expansion of the instrumental footprint of hyphenated instruments currently limits their utility in field settings.

The more conventional approach to the analysis of mixtures is the use of spectral deconvolution in 1D and 2D spectra. In addition to visual and manual approaches with reference spectra [60], several types of algorithms and methodologies are already available in high-field applications and are suitable for bNMR data. Deconvolution algorithms currently exist for many third-party NMR data processing programs such as Mestrelab's Mnova, ACD-Labs' ACD/NMR Workbook Suite, and Nucleomatica's iNMR. In bNMRs, deconvolution techniques have shown recent successes in the compositional analysis of fruit juices [65] and the differentiation of metabolites in urine and serum samples [66], and newer techniques like pure shift analysis have also been investigated [21]. As an alternative to deconvolution of acquired data, the measurement process can also be altered to improve deconvolution capabilities. An example of this approach is resolution enhancement, wherein the free induction decay signal is artificially enhanced to make spectral lines appear narrower [26].

20.7 Conclusion

While bNMR spectrometers cannot provide the sensitivity and resolution of high-field instruments, they excel in affordability, convenience, and relative mobility. These instruments already fill an important niche in academic, industrial, and government laboratory settings by making NMR spectroscopy a more accessible technique. Most facilities that use high-field NMR spectroscopy can only afford one instrument, which often means significant demand and delays in getting results. bNMR instruments are frequently advertised as a way to alleviate that burden and free up their high-field counterparts for experiments that actually need the resolution.

In recent years, efforts of further miniaturization have seen greatest success in instruments designed for relaxometric measurements, but current research in bNMR spectroscopy is largely focused on exploring current capability and improvements in data acquisition and processing. This trend is expected to continue as reduction in magnet volume necessitates a reduction in sample volume that reinforces sensitivity issues [67]. If broadly applicable versions of hyperpolarization techniques become available, it is possible that the next generation of bNMR spectrometers could all become feasible candidates for portable use. To date, only one can make that claim. Regardless, the small footprint and limited infrastructural needs of current bNMR spectometers currently make them suitable for field use in mobile laboratories.

References

1 Purcell, E.M., Torrey, H.C., and Pound, R.V. (1946). Resonance absorption by nuclear magnetic moments in a solid. *Physical Review* 69: 37–38.

2 Bloch, F. (1946). Nuclear induction. *Physical Review* 70: 460–474.

3 Rabenstein, D.L. (2001). NMR spectroscopy: past and present. *Analytical Chemistry* 73: 214A–223A.

4 NobelPrize.org (2019). The Nobel Prize in Physics 1952. *Nobel Media AB*. https://www.nobelprize.org/prizes/physics/1952/summary (accessed 14 September 2019).

5 Danieli, E., Perlo, J., Blümich, B., and Casanova, F. (2010). Small magnets for portable NMR spectrometers. *Angewandte Chemie International Edition* 49: 4133–4135.

6 Moser, E., Laistler, E., Schmitt, F., and Kontaxis, G. (2017). Ultra-high field NMR and MRI – the role of magnet technology to increase sensitivity and specificity. *Frontiers in Physics* 5: 33–47.

7 Wishart, D.S. (2019). NMR metabolomics: a look ahead. *Journal of Magnetic Resonance* 306: 155–161.

8 Gan, Z., Hung, I., Wang, X. et al. (2017). NMR spectroscopy up to 35.2 T using a series-connected hybrid magnet. *Journal of Magnetic Resonance* 284: 125–136.

9 Schwalbe, H. (2017). New 1.2 GHz NMR spectrometers – new horizons? *Angewandte Chemie International Edition* 56: 10252–10253.

10 Blümich, B. and Singh, K. (2018). Desktop NMR and its applications from materials science to organic chemistry. *Angewandte Chemie International Edition* 57: 6996–7010.

11 Anasazi Instruments. *The Eft-60 & Eft-90 Spectrometers*. Anasazi Instruments. https://aiinmr.com/high-quality-durable-nmr-spectrometers-pmnmr-instruments (accessed 14 January 2020).

12 Frost & Sullivan (2018). *Benchtop NMR Spectroscopy Market Analysis: Ease of Use and Low Cost Boosts Growth*. https://static1.squarespace.com/static/5707ede0d210b8708e037a1e/t/5b733ac14ae237dd6d8ede6e/1534278340205/FS_Benchtop-NMR-Nanalysis.pdf (accessed 11 July 2019).

13 Blümich, B. (2019). Low-field and Benchtop NMR. *Journal of Magnetic Resonance* 306: 27–35.

14 Gouilleux, B., Charrier, B., Akoka, S. et al. (2016). Ultrafast 2D NMR on a Benchtop spectrometer: applications and perspectives. *Trends in Analytical Chemistry* 83: 65–75.

15 Grootveld, M., Percival, B., Gibson, M. et al. (2019). Progress in low-field Benchtop NMR spectroscopy in chemical and biochemical analysis. *Analytica Chimica Acta* 1067: 11–30.

16 Zhong, Y., Huang, K., Luo, Q. et al. (2018). The application of a desktop NMR spectrometer in drug analysis. *International Journal of Analytical Chemistry* 2018: 3104569.

17 Louis-Joseph, A. and Lesot, P. (2019). Designing and building a low-cost portable FT-NMR spectrometer in 2019: a modern challenge. *Comptes Rendus Chimie* 22: 695–711.

18 Lei, K.M., Ha, D., Song, Y.Q. et al. (2020). Portable NMR with parallelism. *Analytical Chemistry* https://doi.org/10.1021/acs.analchem.9b04633.

19 Hugon, C., Aguiar, P.M., Aubert, G., and Sakellariou, D. (2010). Design, fabrication, and evaluation of a low-cost homogeneous portable permanent magnet for NMR and MRI. *Comptes Rendus Chimie* 13: 388–393.

20 Doğan, N., Topkaya, R., Subaşı, H. et al. (2009). Development of Halbach magnet for portable NMR device. *Journal of Physics: Conference Series* 153: 012047.

21 Cataing-Cordier, T., Bouillaud, D., Bowyer, P. et al. (2019). Highly resolved pure-shift spectra on a compact NMR spectrometer. *ChemPhysChem* 20: 736–744.

22 Halse, M.E. (2016). Perspectives for hyperpolarisation in compact NMR. *Trends in Analytical Chemistry* 83: 76–83.

23 Blümich, B. (2018). Beyond compact NMR. *Microporous and Mesoporous Materials* 269: 3–6.

24 Macomber, R.S. (1998). Chapter 2: Magnetic properties of nuclei. In: *A Complete Introduction to Modern NMR Spectroscopy*. New York: Wiley.

25 Warwick, A. (1993). Frequency, theorem and formula: remebering Joseph Larmor in electromagnetic theory. *Notes and Records of the Royal Society of London* 47: 49–60.

26 Blümich, B. (2016). Introduction to compact NMR: a review of methods. *Trends in Analytical Chemistry* 83: 2–11.

27 Harris, R.K., Becker, E.D., Cabral de Menezes, S. et al. (2001). NMR nomenclature: nuclear spin properties and conventions for chemical shifts (IUPAC recommendations 2001). *Pure and Applied Chemistry* 73: 1795–1818.

28 Blümich, B., Perlo, J., and Casanova, F. (2008). Mobile single-sided NMR. *Progress in Nuclear Magnetic Resonance Spectroscopy* 52: 197–269.

29 Eidmann, G., Savelsberg, R., Blümler, P., and Blümich, B. (1996). The NMR MOUSE, a Mobile universal surface explorer. *Journal of Magnetic Resonance, Series A* 122: 104–109.

30 Halbach, K. (1981). Physical and optical properties of rare earth cobalt magnets. *Nuclear Instruments and Methods in Physics Research* 187: 109–117.

31 Fischbacher, J., Kovacs, A., Gusenbauer, M. et al. (2018). Micromagnetics of rare-earth efficient permanent magnets. *Journal of Physics D: Applied Physics* 51: 193002.

32 Halbach, K. (1980). Design of permanent multipole magnets with oriented rare earth cobalt material. *Nuclear Instruments and Methods* 169: 1–10.

33 Yu, P., Xu, Y., Wu, Z. et al. (2018). A low-cost home-built NMR using Halbach magnet. *Journal of Magnetic Resonance* 294: 162–168.

34 Kumada, M., Iwashita, Y., Aoki, M., and Sugiyama, E. (2003). The strongest permanent dipole magnet. *Proceedings of the 2003 Particle Accelerator Conference,* vol. 3, 1993–1995. IEEE.

35 Blümler, P. and Casanova, F. (2016). Chapter 5: Hardware developments: Halbach magnet arrays. In: *Mobile NMR and MRI: Developments and Applications* (eds. M.L. Johns, E.O. Fidjonsson, S.J. Vogt and A. Haber). Cambridge: The Royal Society of Chemistry.

36 Magritek Ltd. *Spinsolve Benchtop NMR*. Magritek Ltd. http://www.magritek.com/products/spinsolve (accessed 15 January 2020).

37 Najbauer, E.E. and Andreas, L.B. (2019). Correcting for magnetic field drift in magic-angle spinning NMR datasets. *Journal of Magnetic Resonance* 305: 1–4.

38 Singh, K. and Blümich, B. (2016). NMR spectroscopy with compact instruments. *Trends in Analytical Chemistry* 83: 12–26.

39 Oxford Instruments PLC (2019). *X-Pulse: Benchtop NMR Spectrometer Technical Data Sheet*. Oxford Instruments PLC, xp-spec-09-19.

40 Nanalysis Corp (2019). *NMReady-60 User Manual*. Nanalysis Corp.

41 Manz, B. (2014). *The Lock and the Importance of Field Stability*. Magritek Ltd. http://www.magritek.com/2014/08/12/the-lock-and-the-importance-of-field-stability (accessed 27 December 2019).

42 Nikolaou, P., Goodson, B.M., and Chekmenev, E.Y. (2015). NMR hyperpolarization techniques for biomedicine. *Chemistry* 21: 3156–3166.

43 Magritek Ltd. *Magritek Home Page*. Magritek Ltd. http://www.magritek.com (accessed 15 September 2019).

44 Oxford Instruments PLC. *Pulsar*. Oxford Instruments PLC. https://nmr.oxinst.com/products/nmr-spectrometers/pulsar (accessed 15 September 2019).

45 Nanalysis Corp. *Nanalysis Home Page*. Nanalysis Corp. https://www.nanalysis.com (accessed 15 September 2019).

46 Thermo Fisher Scientific. *picoSpin 80 Series II NMR Spectrometer*. Thermo Fisher Scientific. https://www.thermofisher.com/order/catalog/product/912A0832 (accessed 15 September 2019).

47 Gałuszka, A., Migaszewski, Z.M., and Namieśnik, J. (2015). Moving your laboratories to the field – advantages and limitations of the use of field portable instruments in environmental sample analysis. *Environmental Research* 140: 593–603.

48 DHHS (NIOSH) Publication (2012). *Components for Evaluation of Direct-Reading Monitors for Gases and Vapors*. DHHS (NIOSH) Publication, No. 2012–162. pp. 1–99.

49 Waters, T.R., Putz-Anderson, V., Garg, A., and Fine, L.J. (1993). Revised NIOSH equation for the design and evaluation of manual lifting tasks. *Ergonomics* 36: 749–776.

50 Magritek Ltd (2019). *Spinsolve Ultra* [Brochure]. Magritek Ltd.

51 Nanalysis Corp (2019). *100PRO* [Brochure]. Nanalysis Corp.

52 Thermo Scientific (2014). *picoSpin User Guide*. Thermo Scientific, No. 269–298002, Revision A.

53 Thermo Scientific (2016). *Thermo Scientific picoSpin Benchtop NMR Spectrometers* [Brochure]. Thermo Scientific, No. BR52832_E 04/16M.

54 Singh, K. and Blümich, B. (2017). Desktop NMR for structure elucidation and identification of strychnine adulteration. *Analyst* 142: 1459–1470.

55 Pulst, M., Golitsyn, Y., Reichert, D., and Kressler, J. (2018). Ion transport properties and ionicity of 1,3-dimethyl-1,2,3-triazolium salts with fluorinated anions. *Materials* 11: 1723–1731.

56 Kern, S., Wander, L., Meyer, K. et al. (2019). Flexible automation with compact NMR spectroscopy for continuous production of pharmaceuticals. *Analytical and Bioanalytical Chemistry* 411: 3037–3046.

57 Assemat, G., Gouilleaux, B., Bouillaud, D. et al. (2018). Diffusion-ordered spectroscopy on a Benchtop spectrometer for drug analysis. *Journal of Pharmaceutical and Biomedical Analysis* 160: 268–275.

58 Duffy, J., Urbas, A., Niemitz, M. et al. (2019). Differentiation of fentanyl analogues by low-field NMR spectroscopy. *Analytica Chimica Acta* 1049: 161–169.

59 Antonides, L.H., Brignall, R.M., Costello, A. et al. (2019). Rapid identification of novel psychoactive and other controlled substances using low-field ^{1}H NMR spectroscopy. *ACS Omega* 4: 7103–7112.

60 Pagès, G., Gerdova, A., Williamson, D. et al. (2014). Evaluation of a Benchtop cryogen-free low-field ^{1}H NMR spectrometer for the analysis of sexual enhancement and weight loss dietary supplements adulterated with pharmaceutical substances. *Analytical Chemistry* 86: 11897–11904.

61 Chen, Y.F., Singh, J., Midgley, J., and Archer, R. (2020). Influence of time-temperature cycles on potato starch retrogradation *In Tuber* and starch digestion *In Vitro*. *Food Hydrocolloids* 98: 105240.

62 Voigt, M., Legner, R., Haefner, S. et al. (2019). Using fieldable spectrometers and chemometric methods to determine RON of gasoline from petrol stations: a comparison of low-field ^{1}H NMR @ 80 MHz, handheld Raman and Benchtop NIR. *Fuel* 236: 829–835.

63 Bhavyasri, K., Sindhu, K., and Rambabu, D. (2019). A review: liquid chromatography-nuclear magnetic resonance spectroscopy (LC-NMR) and its applications. *IOSR Journal of Pharmacy* 9: 54–62.

64 Botha, C., Höpfner, J., Mayerhöfer, B., and Wilhelm, M. (2019). On-line SEC-MR-NMR hyphenation: optimization of sensitivity and selectivity on a 62 MHz Benchtop NMR spectrometer. *Polymer Chemistry* 10: 2205–2346.

65 Matviychuk, Y., Yeo, J., and Holland, D.J. (2019). A field-invariant method for quantitative analysis with Benchtop NMR. *Journal of Magnetic Resonance* 298: 35–47.

66 Percival, B.C., Grootveld, M., Gibson, M. et al. (2019). Low-field, Benchtop NMR spectroscopy as a potential tool for point-of-care diagnostics of metabolic conditions: validation, protocols and computational models. *High-Throughput* 8: 2.

67 Schwarz, I., Rosskopf, J., Schmitt, S. et al. (2019). Blueprint for nanoscale NMR. *Scientific Reports* 9: 6938–6948.

21

Rapid DNA Analysis – Need, Technology, and Applications

Claire L. Glynn and Angie Ambers

Department of Forensic Science, Henry C. Lee College of Criminal Justice and Forensic Sciences, University of New Haven, West Haven, CT, USA

21.1 Need for Speed

21.1.1 Addressing Industry Needs

The traditional forensic DNA (deoxyribonucleic acid) typing/profiling process involves a multistep procedure that includes DNA extraction, DNA quantitation, short tandem repeat (STR) amplification via polymerase chain reaction (PCR), separation and detection via capillary electrophoresis (CE), and profile interpretation (Figure 21.1). These processes, which technically could all be completed in 10–12 hours, realistically take weeks or months to complete in operational crime laboratories, depending upon the current backlog of the laboratory.

In order to expedite the process, integration of all traditional STR DNA profiling steps into a single instrument was conceived, and the development of Rapid DNA technologies emerged. While there are ongoing efforts to speed up the individual traditional in-laboratory processes, e.g. Rapid PCR protocols, there exists a need for portable, field-based Rapid DNA technologies whereby all the processes occur within one instrument, and which require reduced need for expert technical and scientific personnel. The hypothesis was that Rapid DNA technology could not only be an effective tool to eliminate DNA backlogs within accredited DNA laboratories, but also offer potential for application in field-based environments outside of the laboratory, including booking stations, border controls, disaster sites, and more.

Rapid DNA is defined by the Federal Bureau of Investigation (FBI) as follows: "Rapid DNA, or Rapid DNA analysis, is a term used to describe the fully automated (hands free) process of developing a DNA profile from a reference sample buccal (cheek) swab without human intervention" [1].

In 2010, under the direction of the FBI, the Rapid DNA Program Office was established in response to the growing need for the development and oversight of Rapid DNA technologies. The directive of this office was to strategize and facilitate research and development of Rapid DNA technology, with eventual integration, for use by law enforcement agencies. The Rapid DNA Program office brings together input and experts from a variety of agencies, some of which include the Department of Defense (DOD), the National Institute of Standards and Technology (NIST), the National Institute of Justice (NIJ), and other federal agencies.

The development of Rapid DNA technologies began with several aims:

- To establish a tool which could produce a DNA profile in a much-reduced time-period, as compared to current traditional DNA profiling methods employed in operational crime laboratories.
- To take this instrumentation out of the laboratory and into the field where it could be employed in field-based applications such as police booking stations, the battlefield, border controls, and disaster sites.
- To develop a technology that would be housed within one unit and could be operated by nontechnical users.

Portable Spectroscopy and Spectrometry 1: Technologies and Instrumentation, First Edition.
Edited by Richard A. Crocombe, Pauline E. Leary and Brooke W. Kammrath.

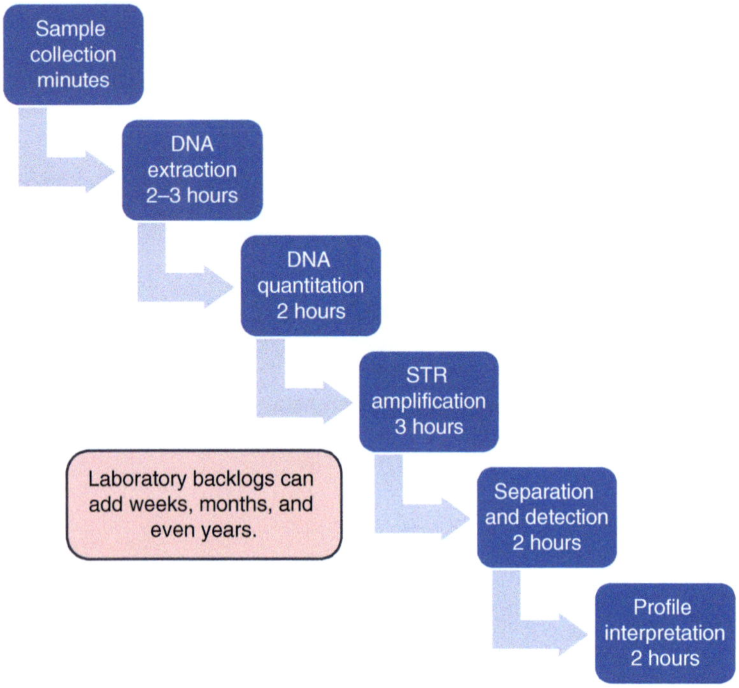

Figure 21.1 Traditional forensic DNA profiling workflow.

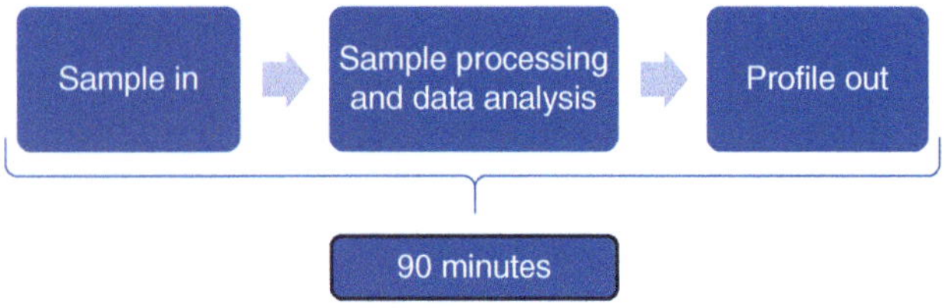

Figure 21.2 Rapid DNA technology workflow.

Current Rapid DNA platforms combine the multistep process involved in traditional STR DNA profiling into one single instrument, with a "sample in, profile out" approach in design (Figure 21.2).

The platforms typically incorporate microfluidic and molecular biology technologies together to generate a DNA profile from a biological sample beginning with DNA purification, rapid thermal cycling, highly multiplexed amplification, focused DNA sequencing, optical detection of DNA sequences, and automated data analysis. The end result (the interpreted DNA profile/data) is simplified. The system first produces an indication of whether a full, partial, or no profile was obtained, and then subsequent built-in software analyses can compare Questioned versus Known samples and conduct kinship matching (if applicable). Rapid DNA policies have been developed by the National DNA Index System (NDIS) and the Scientific Working Group on DNA Analysis Methods (SWGDAM), and developmental validation studies were designed to ensure that Rapid DNA technologies generate results equivalent to those from an accredited DNA laboratory.

Within the last decade, there have been two companies that have emerged as the main players in the Rapid DNA technology market. These include ANDE® Corporation (Waltham, MA), and the Applied Biosystems division of Thermo Fisher Scientific (Waltham, MA) who acquired IntegenX (Pleasanton, CA).

21.1.1.1 ANDE (Accelerated Nuclear DNA Equipment) Corporation

The founding director of ANDE corporation, Dr. Richard Selden, established ANDE Corporation in 2004 in the Greater Boston area. Together with Dr. Selden, pioneering research was performed at MIT's Whitehead Institute to develop the first Rapid DNA platform. Following several years of successful pilot studies, in 2009 ANDE was awarded a competitive research and development grant which was sponsored by a variety of Federal agencies, including the DOD, FBI and the Department of Homeland Security (DHS). The ANDE Rapid DNA™ Identification System was the first to receive NDIS approval from the FBI. This approval represented a major milestone in the integration of Rapid DNA technology as it allows accredited NDIS laboratories to profile DNA samples with the ANDE Rapid DNA Identification System, and use the resulting profiles for upload to the FBI's Combined DNA Index System (CODIS), without the need for manual interpretation or technical review. ANDE's current platform, the ANDE 6C system utilizes a custom FlexPlex™ PCR chemistry, which produces a DNA profile generated from 27 STR loci within 94 minutes.

21.1.1.2 Applied Biosystems

In 2011, Northrop Grumman partnered with IntegenX, and was awarded a contract to produce a rapid human DNA identification system to the Army's Biometrics Identity Management Agency. In 2018, IntegenX was purchased by Thermo Fisher and now operates as part of the Applied Biosystems brand. Applied Biosystems produces the RapidHIT™ systems, including the RapidHIT 200 and RapidHIT ID platforms which utilize the GlobalFiler® Express PCR chemistry, and produces a DNA profile generated from 20 STR loci within two hours for the RapidHIT 200 platform, and 90 minutes for the RapidHIT ID platform. To date, the RapidHIT 200 System has been used to upload more than 1000 STR profiles into national DNA databases, both in the United States and in the United Kingdom.

It is of course essential that the evolution and ultimate integration of Rapid DNA technologies is closely monitored and overseen by organizations such as the NIST. NIST has been involved with the testing of various platforms since 2012.

In order for Rapid DNA technologies to be fully integrated into current forensic processes, it is necessary that the profiles produced be compatible with the original CODIS 13 core STR loci and the expanded CODIS 20 core STR loci (in effect as of 1 January 2017). This ensures that profiles generated with Rapid DNA instruments can be compared to those generated through traditional in-house DNA profiling methodologies. NIST has been performing maturity assessments since 2013 to assess the performance of these platforms (see Section 21.2.3).

21.1.2 The Rapid DNA Act of 2017

On 18 August 2017, the Rapid DNA Act (Public Law 115-50) was signed into US Federal law. The act was passed unanimously by the US House and Senate. The Rapid DNA Act amended the original DNA Identification Act of 1994, thereby allowing the DNA profiling of arrestees in booking station environments onsite using Rapid DNA technology platforms. The aim of this act was to: "establish a system for integration of Rapid DNA instruments for use by law enforcement to reduce violent crime and reduce the current DNA analysis backlog" [2]. With issuance of this law, US law enforcement agencies are given the "green light" by the Director of the FBI to process single-source reference buccal samples. There are two important features of the Rapid DNA Act. The first is that the generation and analysis must be fully automated, without human intervention or human data review. The second is that the FBI approves the Rapid DNA systems for use, and also that the FBI establishes the standards and procedures for use. Since 2017, the FBI has been preparing policies, the necessary Information Technology (IT) infrastructure, and modifications to CODIS in order to implement this new legislation. It is important to recognize that the Rapid DNA Act of 2017, the DNA Identification Act of 1994, and the FBI's Quality Assurance Standards only apply to those agencies and laboratories operating under the NDIS and CODIS agreements.

Table 21.1 Comparison of Rapid DNA Analysis and modified Rapid DNA Analysis.

	Rapid DNA analysis	Modified Rapid DNA analysis
Defined as:	Automated extraction, amplification, separation, detection, and interpretation with *no human interpretation*	Automated extraction, amplification, separation, detection, *with human interpretation* and technical review
Instrumentation:	Rapid DNA system (includes interpretation software)	Rapid DNA instrument
Approval by:	FBI (referred to as NDIS-approved)	Formal approval not required if validated in accordance with the FBI Director's Quality Assurance Standards (QAS)
DNA profiles eligible for NDIS:	Yes, for accredited forensic DNA laboratories using an NDIS-approved Rapid DNA system	Yes, for accredited forensic DNA laboratories using Rapid DNA instruments

21.1.3 Rapid DNA Analysis

In December 2014, an addendum was issued to the FBI's Quality Assurance Standards (QAS) allowing the implementation of Rapid DNA technologies in accredited crime laboratories with single-source reference buccal samples. Two modalities of analysis were defined within the addendum Rapid DNA analysis and modified Rapid DNA analysis. The term "Rapid DNA analysis (RDA)" is defined as the fully automated process of developing a DNA profile *without* human intervention, while the term "modified Rapid DNA analysis (mRDA)" is defined as the fully automated process of developing a DNA profile, but *with* human interpretation and technical review of the resulting profile (Table 21.1).

21.2 Technology

21.2.1 ANDE Rapid DNA System

The first prototype Rapid DNA instrument was released in 2012 and is known as the DNAscan™ Rapid DNA Analysis System, from GE Healthcare Life Sciences and NetBio. NetBio later (in 2017) changed their name to ANDE Corporation. The DNAscan Rapid DNA Analysis System evolved into the ANDE 4C System, which was NDIS-approved by the FBI in 2016. The system has since undergone additional evolution, with the ANDE 6C System now being the latest generation platform.

21.2.1.1 DNAscan™/ANDE® 4C Rapid DNA Analysis System

In order to satisfy the FBI's definition of Rapid DNA analysis, the DNAscan™/ANDE 4C system utilizes a fully automated profile interpretation and expert system software package (Figure 21.3).

The DNAscan™/ANDE 4C System consists of three primary components: the BioChipSet Swab, the BioChipSet Cassette, and the DNAscan™/ANDE 4C instrument. Within the BioChipSet cassette (Figure 21.4), DNA is purified, PowerPlex® 16 STR loci are amplified, and PCR products are separated by electrophoresis through a capillary system. All necessary reagents are preloaded onto the cassette, e.g. extraction reagents, STR amplification reagents, buffers, separation polymer. Within the cassette, there are electrophoretic channels molded into the cassette. Pneumatic pressure drives the reagents through the cassette, with a sophisticated thermal subsystem for multiplexed STR amplification [3].

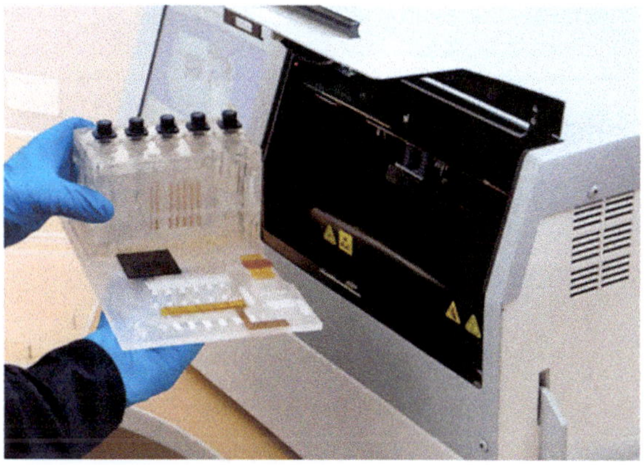

Figure 21.3 DNAscan™/ANDE 4C Rapid DNA System. *Source*: French et al. [3].

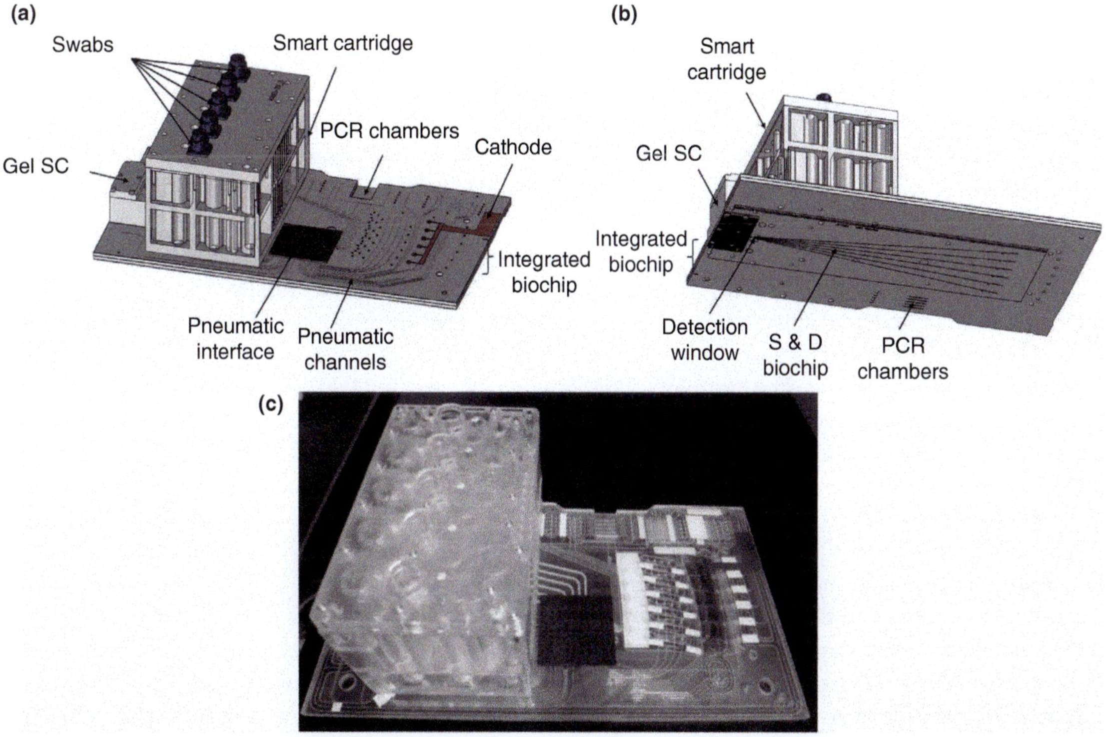

Figure 21.4 Schematic of DNAScan™/ANDE® 4C BioChipSet Cassette (a: top view, b: bottom view, and c: photograph). *Source*: Tan et al. [4], 2020 Springer Nature. Licensed under CC BY 4.0.

While the PowerPlex 16 system had previously been approved by NDIS in a traditional laboratory setting, the use of the system in a Rapid DNA instrument required full developmental validation [3].

Integrated expert system software analyzes the data and provides the user with results via a checkmark system (depending on the user account type and security settings for each user). A green checkmark represents that a full profile was obtained and suitable for database searching. A red "X" represents that no profile was obtained.

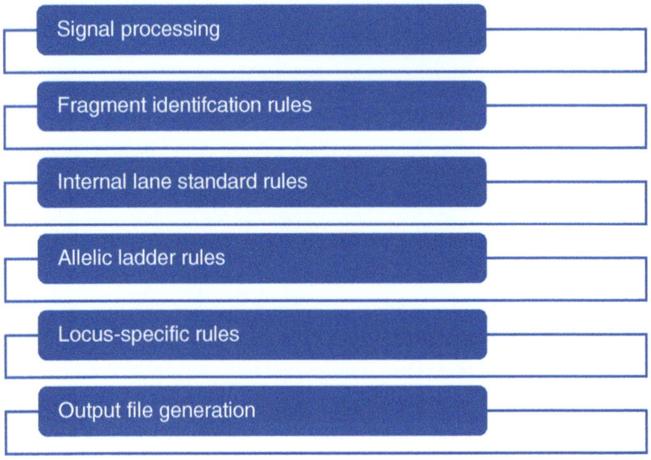

Figure 21.5 Steps in DNAscan™/ANDE 4C expert system.

The expert system software processes the raw data, assigns allele designations, and interprets the DNA profiles. A series of steps are automatically followed within the expert software system (Figure 21.5).

If desired, the data can be reviewed by a forensic scientist and all raw data files can be exported. The genotypes are exported as CODIS Common Message Format (.cmf) files with allele tables listing all passing allele calls; the electropherogram is viewed as a bitmap; the .xml or .cmf file can be uploaded to CODIS; and the .fsa files are exported and can be analyzed in Genemapper® /Genemarker® STR analysis software.

The DNAscan™/ANDE 4C Rapid DNA Analysis System underwent extensive developmental validation, in collaboration with many external laboratories. The system provided high-quality, concordant results for reference buccal samples. The automated data analysis and the 2016 NDIS approval paved the way for its implementation in accredited DNA crime laboratories. The FBI Rapid DNA Program Office stated that "the system could confidently be used in an accredited laboratory for the rapid analysis of known exemplars and could be used by non-scientists in non-laboratory settings with appropriate protocols, quality assurance, and quality control measures in place" [5]. However, the increase of CODIS loci from 13 to 20 in 2017 caused the NDIS approval of the 4C system to be retracted by the FBI.

21.2.1.2 ANDE 6C Rapid DNA Analysis System

On 1 January 2017, the original CODIS 13 core loci were updated and expanded to the new CODIS 20 core loci, in order to facilitate global data sharing efforts. While the ANDE 4C Rapid DNA System had proven successful for processing known buccal reference samples, it was hoped that other forensic specimens could also be processed using Rapid DNA technology. As a result, the ANDE 4C System underwent significant upgrades with expanded capabilities and resulted in the production of the ANDE 6C Rapid DNA System (Figure 21.6).

On 1 June 2018, the ANDE 6C System received NDIS approval for use in accredited DNA laboratories. The approval was granted following extensive multi-laboratory developmental validation studies [6]. The process is still based on the "Swab In, Profile Out" principle.

The ANDE 6C Rapid DNA System is based upon the same principles of previous generations with single-use microfluidic chips, rapid thermal cycling, and electrophoretic detection of fluorescently labeled PCR products, and automated data analysis, however with several enhancements made throughout the system. The instrument was specifically engineered for field deployment and on-site testing operations. In addition to its relatively small size (dimensions 75 cm × 45 cm × 60 cm) and weight (117 lbs), it is operational at altitudes as high as 3048 m (10,000 ft) and was ruggedized to account for vibration and shock during transportation, meeting the strict requirements of US Military Standard 810G (MIL-STD-810G) [7].

Figure 21.6 ANDE 6C Rapid DNA System. *Source*: www.ande.com.

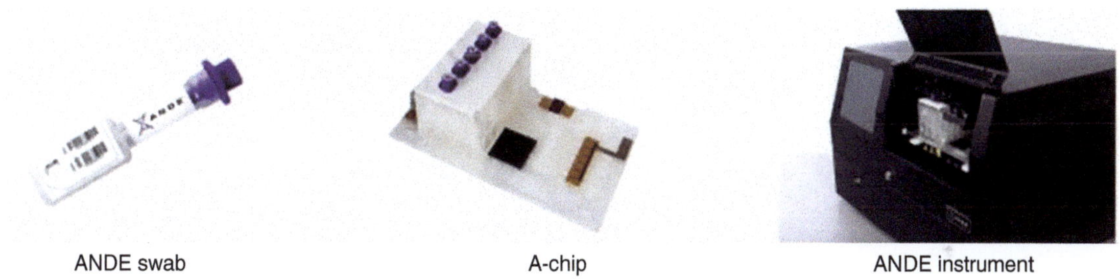

Figure 21.7 ANDE 6C System components; swab, A-Chip, instrument. *Source*: Carney et al. [6].

The ANDE 6C Rapid DNA System consists of three components: the ANDE swab, the A-Chip or I-Chip, and the ANDE 6C Instrument (Figure 21.7). The automated Expert System software could be considered a fourth component.

The ANDE swab is a self-contained sterile swab loaded with a desiccant which facilitates sample drying and prevents sample degradation during storage. There is an embedded RFID tag on the top surface of the swab handle for sample tracking within the instrument.

The ANDE 6C System can be used with two different single-use microfluidic FlexPlex Chips. The A-Chip (A = Arrestee) and the I-Chip (I = Investigative). The A-Chip is used for known reference buccal samples, while the I-Chip is used for more challenging, lower DNA concentration samples such as blood, other forensically relevant body fluids, touch DNA, tissues, bone samples, etc. (Table 21.2).

The major difference between the two chips is that the I-Chip runs four samples, while the A-Chip runs five samples. Within the I-Chip (Figure 21.8), a concentration step is performed after DNA purification in order to maximize the quantity of DNA for STR amplification.

The chips do not need to be refrigerated, and all reagents onboard the chip are stable for up to six months. The possibility of sample cross-contamination is eliminated as each chip has dedicated channels.

Within each chip, the DNA is extracted and purified using silica-based guanidinium salt extraction, 27 STR loci (including the CODIS 20 core loci) are PCR-amplified using a custom FlexPlex assay. The amplified PCR products, which are fluorescently tagged, are electrophoretically separated, and then detected using laser-induced fluorescence. The samples travel through the chip under pneumatic pressure.

Table 21.2 Characteristics of A-Chips and I-Chips.

	A-Chip	I-Chip
Sample types	Buccal reference samples	Forensic samples (blood, semen, tissue, etc.)
Number of samples per run	five	four
Analysis time	94 minutes	106 minutes
Chip storage	Room temp (5–25 °C) for six months	
Operating conditions	10–40 °C 20–80% humidity	

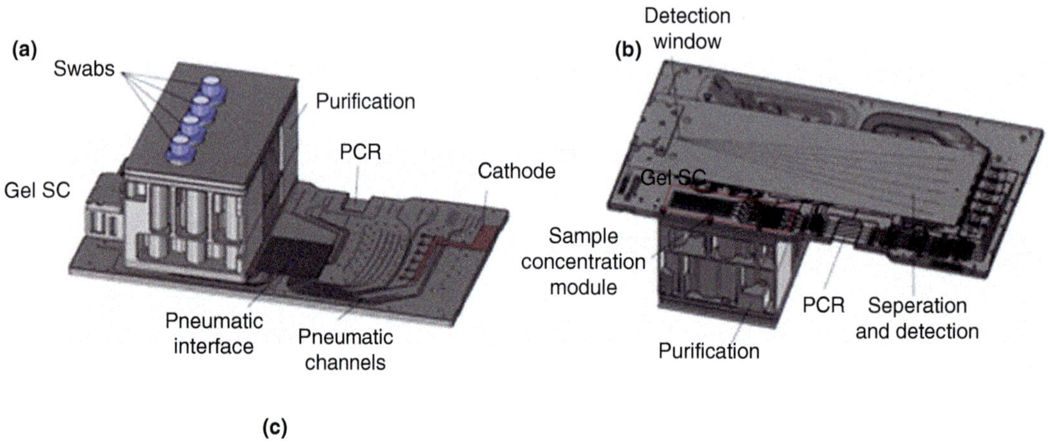

Figure 21.8 Schematic of the ANDE I-Chip. *Source:* Turingan et al. [8]. © 2020, John Wiley & Sons.

The custom FlexPlex assay (based on the Promega Fusion 6C System) utilizes 27 STR loci and a six-dye fluorescent system (Figure 21.9). The assay contains:

- CODIS core 20 loci, plus Amelogenin (sex-determining marker)
- Y chromosome STRs (Y-STRs)
- Additional autosomal STRs (PENTA E, D6S1043, SE33).

The custom assay was created in order to allow the use of this assay on a global scale. The loci selected were chosen from the "cores" or "sets" used across the globe (Figure 21.10).

Therefore, the assay is compatible for comparisons with the CODIS 20 core loci (United States), the UK core set, the European Standard Set (ESS), Germany's core loci, and the Australian core, in addition to the D6 locus, which is commonly used in China. The accuracy and precision of the allele calls made using the allelic ladder and the integrated expert system software in the ANDE 6C Rapid DNA System were investigated in several studies. Precision was determined by calculating the standard deviation of base pairs from each allele in the ladder and confirming that it was below the acceptable target value of 0.5 base pairs. Accuracy and precision of allelic calls was confirmed [6, 9] .

At the end of the chip run, as the user need not be a technical expert, the integrated touchscreen will report the success of each sample via a simplified symbol system. A green checkmark signifies that the DNA profile obtained satisfies NDIS requirements for database searching. A yellow checkmark signifies that the DNA profile obtained

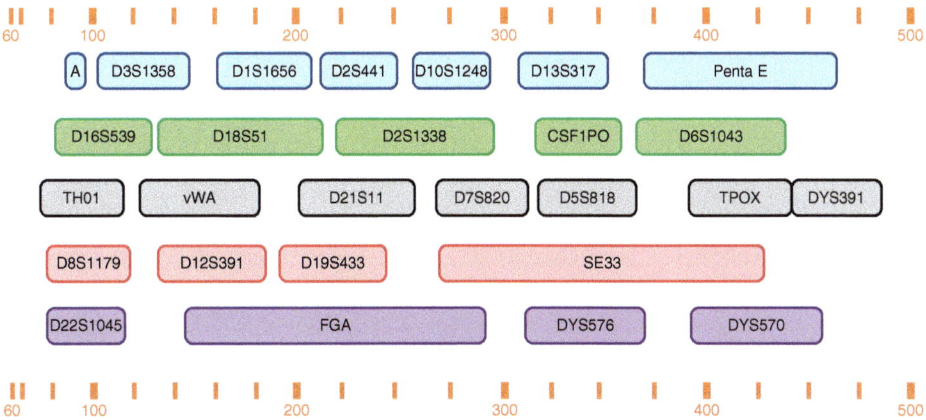

Figure 21.9 Loci included in ANDE's FlexPlex assay. *Source*: Turingan et al. [8]. © 2020, John Wiley & Sons.

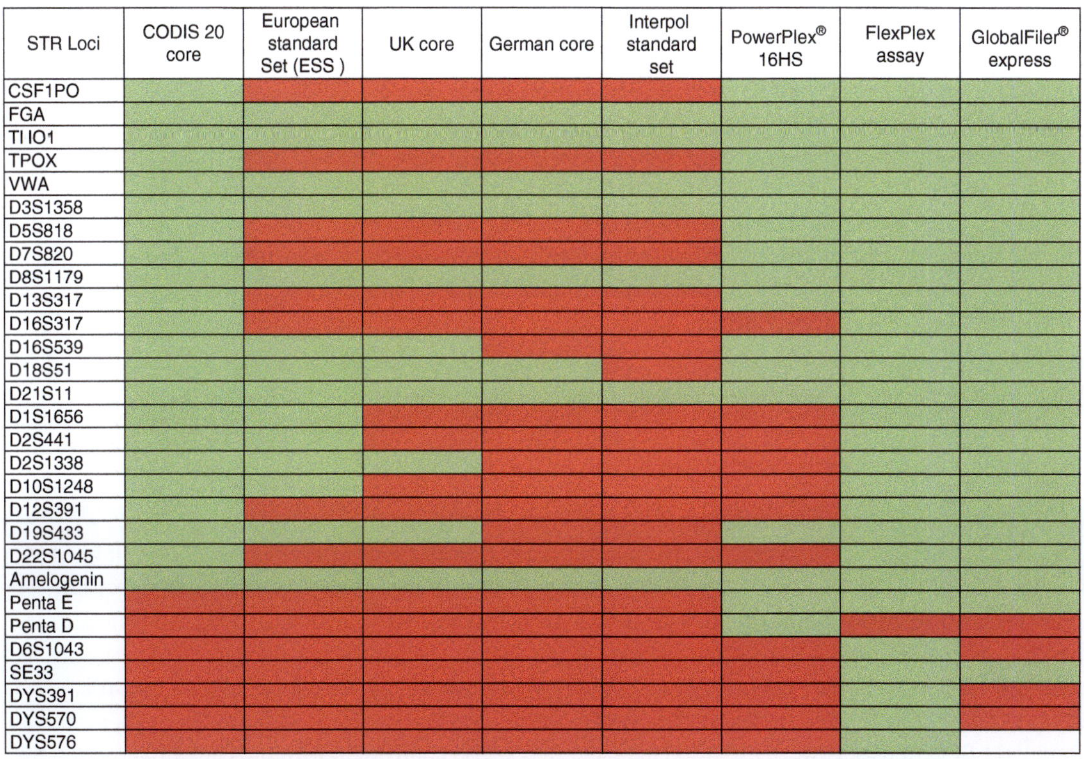

Figure 21.10 Global comparison of STR loci (green cells = included, red cells = not included).

satisfies NDIS requirements for database searching, but should be reviewed by a DNA analyst, or repeat the sample run. A red "X" signifies that the DNA profile obtained does not meet NDIS requirements for database searching and the sample run should be repeated.

The resulting electropherogram (EPG) for each sample can be exported and reviewed by a DNA analyst. All encrypted output files can be exported and decrypted using the FAIRS™ expert software package (ANDE Corporation). FAIRS allows the user to create and manage databases from both samples processed using the ANDE Rapid DNA System and conventionally produced DNA profiles that can be imported manually. The user can search and match samples within the databases, and also perform kinship analysis. The FAIRS Export Service allows all run data to be exported to external sources. The following data can be exported:

- .xml and .cmf files for database searching.
- .png files of electropherograms.
- .fsa files of raw data for import into genotype software, e.g. GeneMapper®/GeneMarker®.

In 2019, a comprehensive developmental validation study was published investigating the ANDE 6C Rapid DNA System for use with arrestee and reference buccal swab processing. The study reported successful interpretation of over 2000 reference buccal samples tested, with 99.99% concordant alleles. Furthermore, the study evaluated several other important operational aspects of the system. Species specificity was investigated by testing 10 different species, some of which included micro-organisms found to be prevalent in the oral cavity, other commonly encountered micro-organisms, primates, and domestic animals. No DNA profiles considered "passing" or suitable for database searching were generated for any of the 10 species tested. Some peaks were detected in the mouse and primate samples, however, were failed by the expert system. The limit of detection was investigated by testing 30 reference swabs collected with one, three, or six swipes. All samples generated full profiles, with average peak height ratios of 82, 81, and 78%, for six swipes, three swipes, and one swipe, respectively. In addition, the researchers further investigated the use of already purified DNA as a starting sample and subjected it to a full Rapid DNA run. A range of quantities were evaluated which include 0.1, 0.25, 0.5, 1.0, and 2.0 μg. Only the 2.0 μg samples yielded full profiles. All others produced only partial profiles and were flagged/failed by the system.

Stability was investigated and demonstrated that freshly collected moist swabs could be stored at 4 °C or room temperature for seven days and would still produce full DNA profiles. During the collection of buccal reference samples, it is possible that other substances present in the mouth may also be collected on the swab which could act as PCR inhibitors. Ten potential PCR inhibitors were investigated, some of which included mint, beer, cigarette, coffee, gum, and mouthwash. All of the samples tested produced full profiles, and these inhibitors did not appear to have a negative impact on the PCR amplification.

The results of this comprehensive validation study reveal the ANDE 6C Rapid DNA System to provide results that are reliable, reproducible, and robust, therefore revealing its suitability for industry implementation.

21.2.2 Applied Biosystems Rapid DNA Systems

21.2.2.1 RapidHIT System/RapidHIT 200 System

In 2013, the first generation of the RapidHIT Human Identification System was developmentally validated [10]. Developmental validation studies demonstrated the system to have a high level of precision and accuracy. The system was reported to be "fit for purpose" with regard to the success rates, sensitivity, and robustness to PCR inhibitors. Further validation studies were soon after reported [11, 12]. The RapidHIT System is described as a fully integrated sample-to-profile system, which is capable of processing five samples, plus three controls, simultaneously (Figure 21.11).

The RapidHIT System uses PowerPlex 16HS chemistry and is also compatible with the GlobalFiler Express kit chemistry. Within the system, the traditional DNA laboratory workflow is housed within one single instrument, with DNA purification, PCR amplification, and fragment separation/detection occurring automatically, in approximately 90 minutes using the PowerPlex 16 chemistry, and 120 minutes using the GlobalFiler Express chemistry. The resulting DNA profiles are reviewed by DNA analysts and exported as CODIS searchable files. As the resulting profiles require manual review by a DNA analyst, this method falls under the modified Rapid DNA analysis method.

The RapidHIT System employs a single-use kit, which contains a series of individual cartridges: a sample cartridge (Figure 21.12), a control cartridge, an anode cartridge containing a linear polyacrylamide separation gel, and a buffer cartridge for CE separation. The system uses pneumatic pressure to drive the samples and reagents

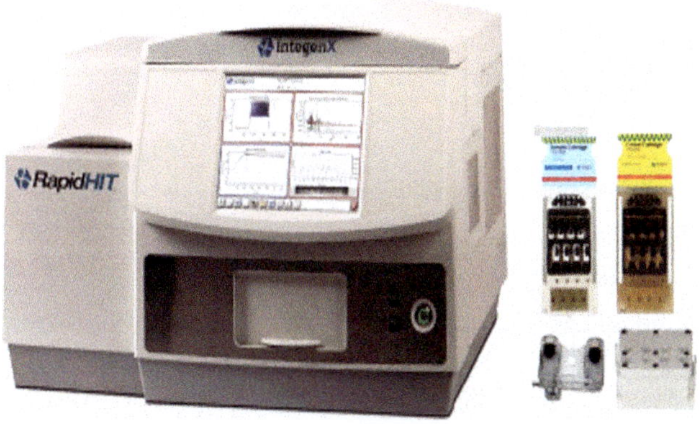

Figure 21.11 RapidHIT Human DNA Identification System. *Source:* Jovanovich et al. [11].

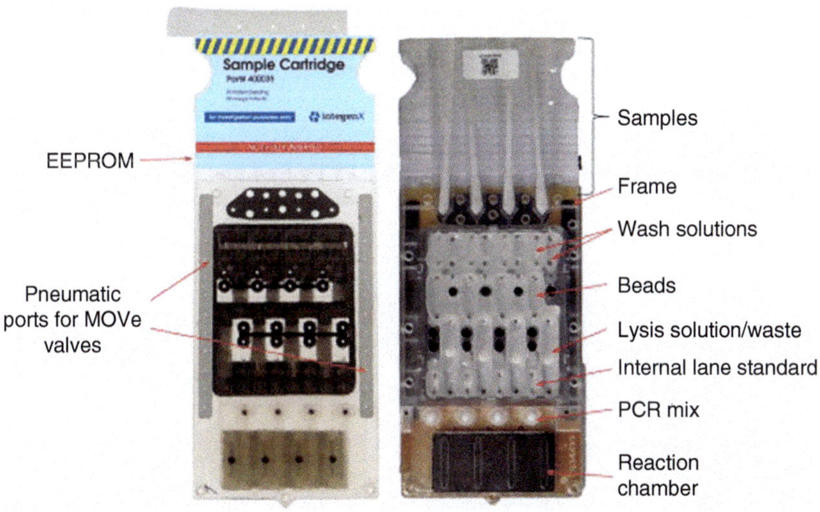

Figure 21.12 RapidHIT system sample cartridge. *Source:* Jovanovich et al. [11].

through the cartridges. Up to four sample swabs are processed in the sample cartridge. One sample swab and three controls are processed in the control cartridge. The three controls are: a positive amplification control, a negative amplification control, and an allelic ladder.

The RapidHIT instrument is composed of four main subsystems which perform the required processes:

1. Sample preparation subsystem
   - The sample and control cartridges are inserted.
   - The samples are lysed, extracted, and purified using the DNA IQ System.
   - PCR amplification occurs using the PowerPlex 16 HS chemistry.
   - The Internal Lane Standard (ILS) is added (ILS 60, Promega).
2. Separation subsystem
   - Receives the STR amplified PCR products, plus the allelic ladder.
   - Eight separation capillaries are injected with the separation polymer.
   - The PCR products are injected into the capillaries using the separation buffer.
   - DNA fragments are separated by size via capillary array gel electrophoresis.
   - The capillaries are cleaned for the next sample run.

3. Detection subsystem
   - The fluorescently labeled STR products are captured as they pass through the detection window.
   - 20 mW, 488 nm solid-state laser is employed to excite the fragments.
   - The excited fragments are detected using a CCD detector.
4. Control and analysis subsystem
   - Utilizes an embedded computer system with the required processor and software.
   - Image data from the CCD detector is captured.
   - The data are processed and analyzed.

The RapidHIT System was shown to produce highly reproducible results with an 88% first pass success rate for buccal samples processed producing all 16 out of 16 loci called [11]. Other interlaboratory system studies were performed at the FBI laboratory, NIST, and the United States Army Criminal Investigation Laboratory (USACIL), which also produced full concordant profiles [13], while other internal validation studies produced similar results [14, 15]. The positive results generated from these studies encouraged further development of the system and revealed great potential for its use in both accredited DNA laboratories and booking stations for reference buccal reference sample processing. In 2014, IntegenX received approval from the FBI to upload directly to NDIS samples that have been processed using the RapidHIT System with the GlobalFiler Express Kits.

21.2.2.2 RapidHIT ID System

In 2016, the next generation of RapidHIT System was developed, the RapidHIT ID System (Figure 21.13). This new system was generated to specifically meet the needs of a field-based, decentralized environment. Significant changes were made to the system to increase its efficiency, increase its capacity, reduce its footprint, and require less user intervention.

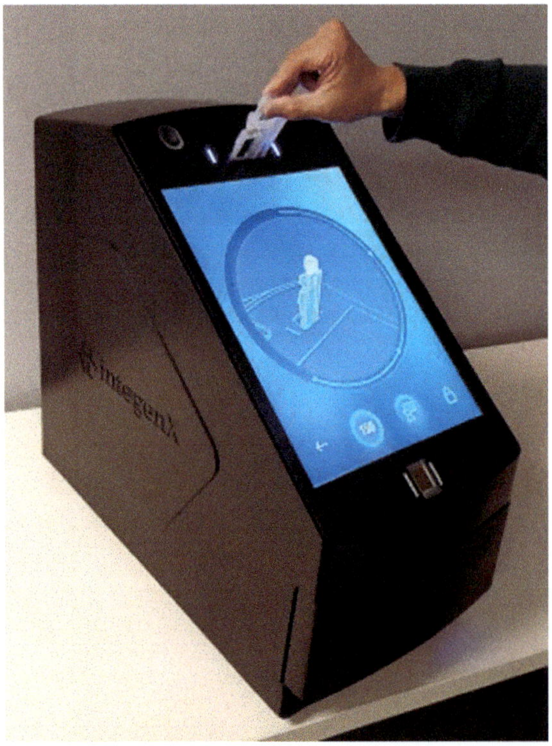

Figure 21.13 RapidHIT ID System. *Source:* Salceda et al. [16].

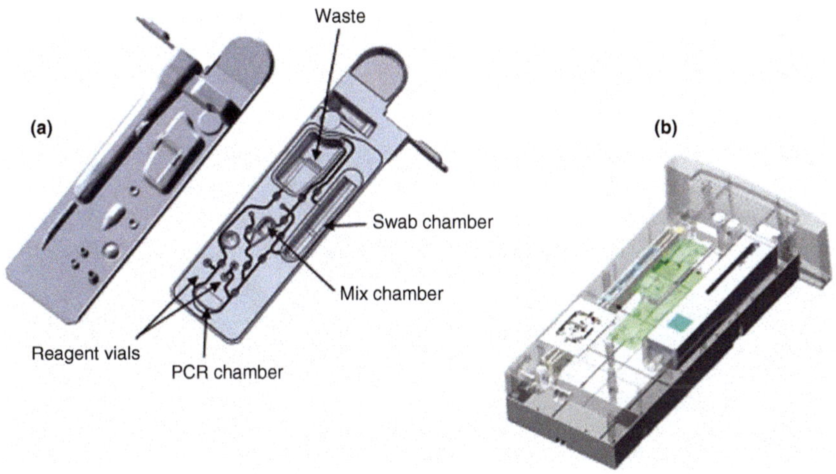

Figure 21.14 RapidHIT ID (a) sample cartridge and (b) primary cartridge. *Source*: Salceda et al. [16], © 2017, Elsevier.

The previous cartridge, with a four subsystem configuration, was completely redesigned and replaced with a low-cost single-use cartridge for each sample, a control cartridge, and a primary cartridge housing the bulk reagents, and the capillary electrophoresis module suitable for running 150 samples (Figure 21.14).

The sample and primary cartridges are reported to have a reasonable shelf life and can be stored for six months at temperatures up to 25 °C. The sample cartridge contains the GlobalFiler Express primer set, master mix, and the size standard in separate vials. The primary cartridge contains the lysis buffer and electrophoretic reagents for 150 samples. After the DNA is extracted from the sample, the lysate moves from the swab chamber to the PCR chamber where the lysate is captured on a paper disc. The primer and master mixes then enter the PCR chamber, and thermal cycling occurs with 28 cycles, following the manufacturer's protocol. The size standard then enters the PCR chamber containing the PCR products causing them to flow into the mix chamber where the size standard and PCR products combine. This mixture is denatured and enters the primary cartridge for capillary electrophoresis to separate the DNA fragments, which are then optically detected. The raw DNA data are processed by the system with the resulting STR profile/electropherogram displayed on the screen. The .fsa files can be exported to GeneMarker HID for further analysis. An allelic ladder is not included in the sample processing, instead a "digital" allelic ladder is applied during the data analysis stages imported from a library of allelic ladders.

Another substantial upgrade and addition to the RapidHIT ID System was the development of the new RapidLINK software (IntegenX) for data analysis and DNA profile interpretation. The RapidHIT ID instruments transfer the output data to a central computer that operates the RapidLINK software. The RapidHIT ID System has a sophisticated security process with a fingerprint reader and camera for authenticating access. Furthermore, predictive analytics are utilized to monitor and manage consumable inventory across geographically distributed RapidHIT ID Systems.

21.2.3 Maturity Assessment

Since the release of the first prototype Rapid DNA instruments in 2012 (DNAscan Rapid DNA Identification System; ANDE Corporation/NetBio, and RapidHIT™® 200 Human Identification System; Applied Biosystems/IntegenX), the Applied Genetics Group of NIST has been investigating and evaluating their performance.

In August 2013, the first maturity assessment was performed across three federal laboratories [13]. This assessment investigated the DNAscan (ANDE/NetBio), and RapidHIT 200 (Applied Biosystems/IntegenX) instruments, using buccal reference samples. In this interlaboratory study, the success rate was measured by calculating the percentage of complete and correct/concordant profiles utilizing the CODIS 13 core loci. With the ANDE Rapid DNA instrument, the study reported 84% success, with 168 of 200 samples tested providing full concordant profiles. With the RapidHIT Rapid DNA instrument, the study reported 94% success, with 141 of 150 samples tested providing full concordant profiles. The combined results for both instruments report 88% success, with a total of 309 of 350 samples tested providing full concordant profiles. The results of this interlaboratory study provided the necessary success rates to further perform developmental validation studies.

In October 2014, the second maturity assessment was performed across seven independent laboratories, spanning US federal, state, and private laboratories [17]. The aim of this study was to assess the current state of the technology for the CODIS 13 core loci as software and hardware updates had been implemented. Both automated (Rapid DNA analysis) and manual (modified Rapid DNA analysis) review of the data sets were conducted. In addition, two STR PCR chemistries were employed: PowerPlex 16 (Promega Corporation, Madison, WI) and GlobalFiler Express (Applied Biosystems). The results of the study reported success for the CODIS 13 core loci ranged from 76% with automated data review to 80% with manual data review. The peak height ratios and stutter percentages were also found to be within observed developmental validation ranges using both PCR chemistries; PowerPlex 16 and GlobalFiler Express.

In January 2017, the original CODIS 13 core loci were expanded to 20. Therefore, a third maturity assessment was performed to assess the performance of Rapid DNA instruments with the expanded CODIS 20 core loci. In addition, there were significant updates to the existing platforms, and new platforms were developed. In 2018, the third, and most recent, maturity assessment was performed to evaluate the implementation of Rapid DNA outside of a laboratory environment for use only on single-source reference buccal samples [18]. This maturity assessment was performed using the ANDE 6C System, the RapidHIT 200 System, and the RapidHIT ID System. The participants in the study utilized the instruments in US federal, state, and local laboratories, police agencies, and commercial vendors. Each participant processed unknown buccal epithelial reference samples and sent the resulting data to NIST for evaluation. The success rate was measured by calculating the percentage of complete and correct profiles generated by the various Rapid DNA technologies, when compared to those generated using traditional STR DNA profiling methods in a laboratory environment. As the ANDE 6C system was the only platform at the time that was fully approved for NDIS upload, it was the only instrument eligible for Rapid DNA analysis. Therefore, the others underwent modified Rapid DNA analysis whereby the data were manually interpreted. The results of the study with the Rapid DNA analysis reported average success of 85% when analyzing the CODIS 20 core STR loci, ranging from 60 to 100% for the ANDE 6C instruments. The results of the study with the modified Rapid DNA analysis reported average success of 90% when analyzing the CODIS 20 core STR loci, ranging from 60 to 100% for all instruments. Remarkably, concordance between the different chemistries across the instruments (custom FlexPlex and GlobalFiler Express) was observed for 99.98% of the STR allele calls compared. A summary of the results from the three maturity assessments is provided in Table 21.3.

The continued efforts by NIST over the last eight years to support the development, validation, and implementation of Rapid DNA technologies has provided encouragement to law enforcement agencies that these instruments meet the standards and capabilities needed.

Table 21.3 Summary of results from Rapid DNA maturity assessments.

Year	Instrument	STR typing chemistry	Samples tested per Lab	Total samples	Average % success
2013 [13]	ANDE 4C	PowerPlex® 16	200	350	88.3%
	RapidHIT 200		150		
2014 [17]	ANDE 4C	PowerPlex® 16	100	280	75.25%
	RapidHIT 200	PowerPlex® 16	60		
		GlobalFiler Express	120		
2018 [18]	ANDE 6C	FlexPlex™	100	240	87.5%
	RapidHIT 200	GlobalFiler® Express	60		
	RapidHIT™ ID	GlobalFiler® Express	80		

21.3 Applications

The potential applications of Rapid DNA technology are quite expansive and, while originally intended for use outside of the laboratory (i.e. portable field use), it also offers numerous advantages to being implemented within the laboratory environment.

21.3.1 Booking Station Use

A major goal of the development of Rapid DNA technologies was for the systems to be portable for use outside of the laboratory in a decentralized environment (e.g. booking stations, border control, etc.). The implementation of Rapid DNA technology in booking stations offers immense potential to expedite arrestee identification, and even link the arrestee to prior crimes (Figure 21.15).

When an arrestee is booked in at a booking station, numerous levels of information are collected/searched, including fingerprints, criminal history information, etc. With the advent of Rapid DNA technology however, even further levels of information can now be collected immediately by taking a known reference buccal sample from the arrestee, generating a CODIS searchable DNA profile onsite, and uploading the resulting profile to search for linkages to previous crimes. This allows the law enforcement agency to search unsolved crimes of special concern while an arrestee is still in custody, and before they are released in non-charging states.

This high-speed processing and interpretation could significantly increase the rate at which prior and current crimes are solved. By identifying arrestees that are wanted in connection with other crimes (e.g. murders, rapes, burglaries, etc.), there is potential to identify repeat criminals and prevent them from repeating their crimes again. In addition, potentially millions of arrestee DNA profiles could be generated continually with Rapid DNA instruments running 24/7. This would also positively impact the current backlog of reference sample processing experienced nationwide in accredited crime laboratories.

With the passing of the Rapid DNA Act in 2017, US law enforcement booking stations were now able to implement Rapid DNA technologies to type single-source reference buccal samples. This offers a rapid turnaround time

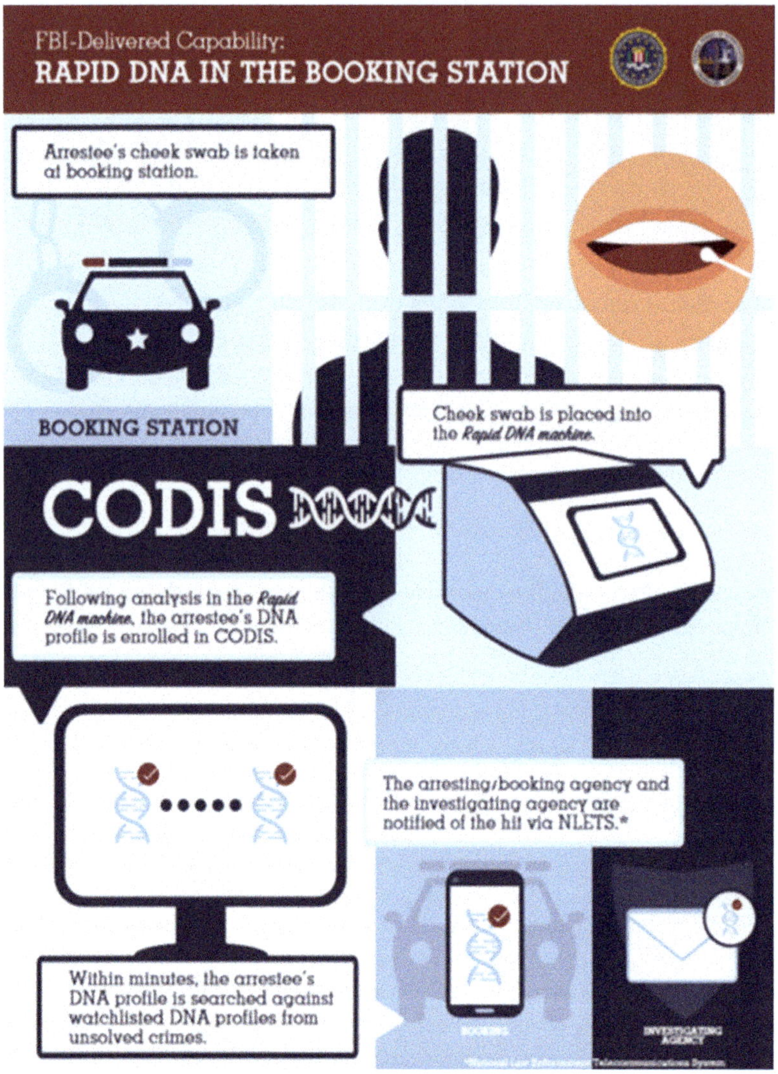

Figure 21.15 Workflow of Rapid DNA in the booking station. *Source*: www.fbi.gov.

and can enhance an investigation by offering quick elimination of potential suspects, or indeed, by identifying perpetrators of previous crimes who had remained unknown.

Currently (March 2020), within the United States, there are five pilot sites utilizing Rapid DNA technologies in booking station environments. The implementation is not as simple as placing a Rapid DNA instrument into a booking station. Within the last two years, the FBI have been making updates to their IT infrastructure which includes new CODIS software and other necessary CODIS interfaces in order to allow the data generated with Rapid DNA technologies in booking station environments to communicate with CODIS. This includes the integration of criminal history information.

Further, there are a number of prerequisites for implementation into the booking station environment. In order for federal, state, or local booking agencies to be granted approval to introduce Rapid DNA technologies in their booking stations, they must meet the following criteria [19];

- The state must have implemented an arrestee DNA collection law that authorizes DNA analysis at the time of arrest. Federal booking agencies already meet this prerequisite.
- They must have Electronic Fingerprint (Live Scan) integration during the booking process for obtaining State Identification numbers (SID) (Universal Control Number [UCN] for federal booking agencies) from the State Identification Bureau (FBI for federal) in near real time.
- The booking agency must have network connectivity with the State Identification Bureau (SIB)/Criminal Justice Information Services (CJIS) Systems Agency (CSA).

It will be critical for booking agencies to work with their State CODIS Administrator to ensure all requirements are met for implementation of Rapid DNA technologies in their agencies.

21.3.2 Accredited DNA Laboratory Use

On 4 June 2018, the FBI granted ANDE Rapid DNA Systems with NDIS approval for the ANDE 6C system. This allows NDIS-accredited crime laboratories to implement ANDE Rapid DNA technology to process DNA samples and also to upload the resulting DNA profiles to CODIS. The traditional DNA profiling process performed in an accredited DNA laboratory is a multistep process that requires numerous forms of instrumentation, multiple individual lab spaces for each step, and highly skilled personnel. The DNA profiling process within an accredited laboratory can take days, weeks, and even months to obtain results. The advent and success of DNA profiling over the last 30 years, together with the positive outcomes it has brought to investigations, have encouraged investigators to submit more and more samples to the lab for testing and therefore causing higher demand for DNA testing to be performed, which can result in sizeable backlogs of samples to be processed.

While the analysis of forensic samples by a Rapid DNA system is not compliant with the FBI Director's QAS for Forensic DNA Testing Laboratories, Rapid DNA technology could alleviate some of the backlogs currently experienced. Within accredited DNA laboratories, Rapid DNA technologies could be used to process a wide range of sample types, some of which include:

- Known reference buccal samples submitted to the laboratory.
- Blood, other forensically relevant body fluids, and touch DNA sampled from evidence items examined in the laboratory.
- Sexual Assault Kits (SAKs) that are currently backlogged and remain untested.

Not only would the implementation of Rapid DNA technologies in accredited crime laboratories reduce the time it takes to produce results and therefore provide investigative leads quicker, but it would also reduce the amount of "hands on work" required by the lab staff, thereby allowing them time to work on other crimes.

21.3.2.1 Known Reference Buccal Samples

Rapid DNA technologies have reported much success with the various platforms for use with single-source reference buccal samples. Reference buccal samples are collected from known individuals, including victims and suspects, from which an STR DNA profile is obtained and can be uploaded to a searchable database for matching. The reference buccal STR DNA profile can also be used to be compared against resulting DNA profiles from biological evidence recovered from evidence items, i.e. Questioned versus Known DNA profile comparison.

Much of the early studies investigating Rapid DNA technologies focused solely on their use with reference buccal samples, as the goal was to implement this technology in booking stations for arrestee identification purposes, and also in accredited DNA laboratories to alleviate the backlog of reference sample processing. Reference buccal samples typically contain high quantities of DNA and are not required to be quantified prior to STR Amplification, which is required for all other forensic samples by the FBI's DNA Advisory Board (DAB) QAS (Standard 9.3). The recommended best practice is the two-swab method: one swab for Rapid DNA profiling and one swab preserved for traditional DNA profiling.

Several studies have thoroughly validated the use of the current Rapid DNA systems with known reference buccal samples to assess limits of detection, species specificity, mixture implications, and susceptibility to PCR inhibition [6, 16]. Importantly, species specificity included interrogation of oral specific bacteria, which demonstrated limited cross reactivity.

21.3.2.2 Forensic Evidence Samples

21.3.2.2.1 Bloodstains and Other Forensically Relevant Body Fluids

Witnessing the success of the RapidHIT System with reference buccal samples, IntegenX created a new protocol, called "Run Other Samples" to investigate the use of the system with samples other than reference buccal samples, such as blood and saliva (from cigarettes and drink containers), while using the same cartridges as those used for reference buccal sample processing. Both blood and saliva samples demonstrated 88% success rate, by yielding DNA profiles of sufficient quality for upload to a database. Furthermore, the researchers noted it was possible to retrieve the blood and saliva samples from the chambers after the full processing run, which were suitable for retesting using standard/traditional forensic STR profiling methods [20].

The ANDE 6C Rapid DNA System was also investigated for use with forensic evidence samples where the analysis of 1705 casework samples was found to be comparable to traditional laboratory methodologies [8]. Forensically relevant samples typically encountered in real forensic investigations were included in the study to interrogate the effectiveness of the ANDE 6C Rapid DNA System to recover full and concordant DNA profiles that are compliant with the FBI's QAS and the NDIS operational procedures (although QAS standards require DNA quantitation for forensic samples that Rapid DNA systems do not perform). Sample types included dried blood on tile, blood on FTA paper, oral epithelial samples from drinking containers, saliva on non-FTA paper, saliva on FTA paper, neat semen, semen on fabric, blood recovered from a knife, blood on fabric, chewing gum, bone, and other Disaster Victim Identification (DVI) sample types. The results of the study demonstrated that the ANDE 6C Rapid DNA System and accompanying expert system produced reliable, reproducible, and robust results, demonstrating the capability of the system for automatic processing and interpretation of forensically relevant casework samples.

While the ANDE Rapid DNA System currently has NDIS approval for reference buccal sample processing, forensically relevant crime scene samples do not have approval for upload and searching in CODIS. However, until approval is granted, individual accredited forensic science laboratories can validate individual Rapid DNA technologies for use with casework samples comparison and searches within local and state databases (e.g. the state of Kentucky using Rapid DNA for Sexual Assault Kits).

21.3.2.2.2 Touch DNA/Low-DNA Content (LDC) Samples

Touch DNA is DNA left behind when a person touches a surface or object and leaves behind sloughed epithelial skin cells. This type of "trace" evidence is often low in abundance of DNA and depends upon the surface characteristics and shedder status of the donor. The exact origin of touch DNA is not always certain but could be: shed keratinoycytes or their constituent parts from outer layers, nucleated epithelial cells from other body fluids or body parts, cell-free DNA reserves endogenous to the hands (e.g. sweat), or cell-free DNA transferred from body fluids (e.g. saliva). The quantity of DNA recovered from touch DNA samples is highly variable, ranging from <6 to 100 ng. For context, 6 pg is approximately the quantity of genomic DNA in a single human cell. With regard to touch DNA samples and traditional STR profiling methods, considerable efforts have been made to improve the yield of probative information from these samples with regard to their collection methods, DNA extraction methods, and STR amplification methods.

In 2016, the ANDE platform was modified for use with low-DNA content (LDC) samples [21]. A modified BioChipSet (BCS) was produced, while the remainder of the platform and analysis were maintained (Figure 21.16).

The modification involved incorporating a microfluidic ultrafiltration (UF) module downstream from the purification, which essentially concentrates the extracted DNA following purification in order to maximize yield. The

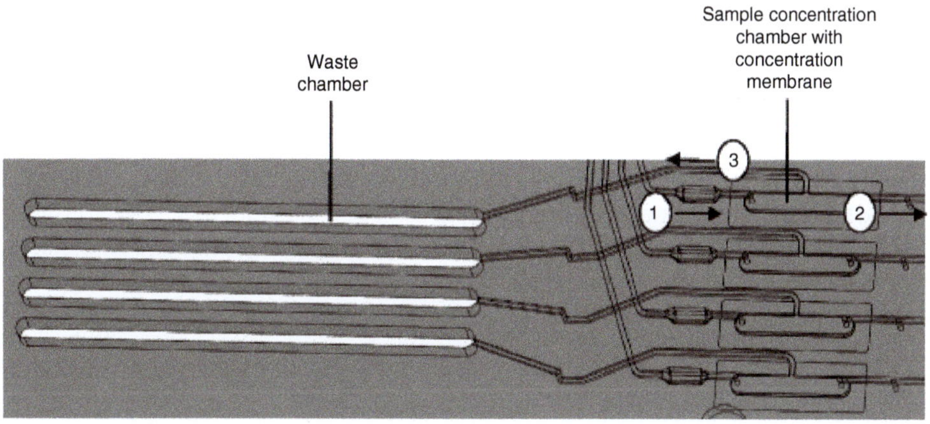

Figure 21.16 BioChipSet concentration channels. *Source*: Turingan et al. [21], 2020 Springer Nature. Licensed under CC BY 4.0.

Smart Cartridge accepts four LDC samples and contains the liquid reagents required for sample processing. Fluids are transported throughout the biochip by pneumatic pressure – there are no moving parts. Essentially, using pneumatic pressure, the purified DNA moves toward a semipermeable UF membrane, which prevents the passage of the DNA fragments, but allows the elution solution to flow through to a waste chamber, resulting in concentrated DNA on the membrane.

The four concentration channels can be seen in Figure 18, one for each independent sample. Purified DNA solution flows into the concentration chamber (*arrow 1*) from the purification module. Concentrated DNA solution flows out of the concentration chamber (*arrow 2*) to the PCR chamber. The permeate is collected in the waste chamber (*arrow 3*). This modified method was tested with a large variety of sample types, including, oral epithelial cells from drinking containers, blood on FTA paper, blood on untreated paper, buccal cells on FTA paper, buccal cells on untreated paper, dried blood on ceramic tile, dried blood on clothing, dried semen on clothing, chewing gum, cigarette butts, swabs of cellphones, and bone fragments. This modified BCS for LDC analysis has a runtime of 102 minutes. The BCS utilizes the PowerPlex 16 HS assay.

21.3.2.2.3 Tissue Samples

Exposure to harsh environmental insults can result in substantial degradation of human remains. With decomposed human remains, sometimes soft tissue samples are still viable for DNA testing. In other cases, the optimal sample types for DNA recovery are bones and teeth. In 2019, a study investigated the use of the ANDE 6C Rapid DNA System for use with buccal samples, teeth, bone samples, muscle tissue, liver tissue, and brain tissue [22]. The samples were exposed to aboveground decomposition at the Anthropology Research Facility ("Body Farm") at the University of Tennessee. Some human donors were placed above ground immediately after collection, while another group were refrigerated in a morgue cooler for up to three months. The results of this study reported that exposed and refrigerated samples consistently produced partial or full concordant DNA profiles throughout the decomposition period that lasted one year. The results revealed buccal samples to be the "sample of choice" from remains which were exposed for up to 11 days, until maggot infestation interferes. Bone and tooth samples were the sample of choice for the one-year duration of the study. Muscle tissue produced full or partial profiles up to day 6 of exposure and, as expected, the profile generated deteriorated as time increased. It was observed that bicep samples and quadriceps produced similar results. Tissue samples produced full concordant profiles from the various tissue samples recovered from remains which had been refrigerated at the morgue over the three-month testing period. This study provides informative data to assist crime scene responders and medical examiners in their selection of samples which would likely yield identity information if Rapid DNA technologies are to be employed.

21.3.2.3 Sexual Assault Kits (SAKs)

The forensic investigation of sexual assault crimes can be challenging even in the most well-equipped crime laboratory. The analysis of Sexual Assault Evidence Kits or Sexual Assault Kits (SAKs) is time-consuming and labor intensive. The reality is that there are hundreds of thousands of backlogged SAKs spanning many years, across the globe, that have yet to be examined. Not only is this detrimental to the victims' right to justice, but it means the perpetrator of the crime remains free with the opportunity to reoffend.

The State of Kentucky in the United States was the first to adopt Rapid DNA technology for use in sexual assault investigations. In April 2019, Governor Matt Bevin announced the implementation of Rapid DNA technology, specifically, the ANDE 6C system, within the Kentucky State Police Forensics Laboratories and performed rigorous evaluation of the system. At the time a SAK is collected from a victim, all conventional samples are collected for traditional processing, however additional Rapid DNA swabs are collected also. Upon receipt of the SAK to the crime laboratory, the Rapid DNA swabs are processed immediately, while the remainder of the SAK is stored. The resulting DNA data generated within two hours are then searched against the local and state criminal databases. If a potential suspect is identified by the database, and apprehended by law enforcement within a short time period, it could also allow for the collection of trace evidence/biological fluids/DNA from the suspect, which perhaps would have been lost if it had taken weeks/months/years to identify the suspect. This would further strengthen the investigation and eventual outcome of the case.

By eliminating the backlog and testing all SAKs submitted to a crime laboratory, it will be possible to identify serial rapists, in both acquaintance and stranger rape cases. In addition, testing of SAKs could link different crimes. Sexual assault offenders typically engage in other criminal behavior such as burglary and homicide. Further to this, if a suspect/perpetrator is presented with DNA evidence linking them to a crime, the perpetrator often pleads out, resulting in earlier resolution of the case, saving the expense of a trial, and brings justice to the victim, which aids in their recovery. One in five women in America will experience rape or attempted rape at some point in their life. Currently, only 6 in 1000 sexual assault kits result in prosecution. The implementation of Rapid DNA technology has the ability to change that statistic.

Benefits of Testing all Sexual Assault Kits with Rapid DNA

- Quick apprehension eliminates opportunity to reoffend.
- Quick apprehension increases likelihood of recovery of forensic evidence from offender.
- Presenting the offender with DNA evidence linking them to the crime often leads them to confess or plead out.
- Identifies serial rapists.
- Identifies linkages to other crimes, e.g. burglary, homicide, etc.
- Knowledge of success of Rapid DNA may discourage offenders from offending.
- Knowledge that all kits will be tested will encourage victims to report these crimes.

21.3.3 Field Use

Outside of its use in booking stations and within an accredited crime laboratory, one of the major aims of the development of Rapid DNA technologies was to create ruggedized instruments suitable for field use in a wide variety of environments, e.g. mass disaster/mass casualty sites, battlefields, and border control stations, etc. This requires the instrument to be small in footprint, require no maintenance or optical calibration following transport, and the ability to operate using a standard power supply. Although traditional forensic DNA methods have been utilized for analysis of samples recovered from mass disaster/casualty events (including the 9/11 terrorist attacks [23–25], other terrorist attacks [26], natural disasters [27–29], and airplane crashes [30–32]), it can take many years (even decades) for all of the remains to be processed within an accredited crime laboratory. Victim remains from the 9/11 terrorist attacks in 2001 are still being analyzed by the Office of the Chief Medical Examiner (OCME) in New York, almost 20 years later. Therefore, there is great need and a wide scope of applications for Rapid DNA technology to be utilized to field uses.

21.3.3.1 Border Control, Immigration, and Human Trafficking Investigations

In May 2019, DHS announced the beginning of a pilot program to test the DNA of migrants arriving at the US border from Mexico who are suspected of fraudulent parentage or kinship claims. It was stated that the aim of this program was to utilize Rapid DNA technology as an investigative tool at the border to identify fraudulent cases of migrants claiming parentage or kinship with children that they are traveling with. It has been suggested that some migrants may be trying to take advantage of loopholes within the US Immigration System, whereby individuals arriving at the border with a child will avoid detention and can enter the asylum process more swiftly [33]. By identifying this type of fraudulent case, vulnerable children can be removed from potentially dangerous situations. Traditionally, the identification of this kind of fraudulent case has been performed using document screening and other investigative methods. Rapid DNA technology however offers a faster, more accurate method for parentage and kinship determination. ANDE Corporation has been contracted to perform the testing, at the border crossing, with results obtained within two hours. Homeland Security investigators from US Immigration and Customs Enforcement (ICE) can request to collect reference buccal samples from individuals, if they suspect a parent–child relationship is not legitimate. The resulting DNA profiles produced do not need to be uploaded to federal databases, reducing privacy and data storage concerns. Instead, within the FAIRS software package, parent and kinship verification can be performed.

There are some obvious limitations to this method, such as cases whereby parentage is through adoption, or indeed a parent may not be aware they are not the biological parent of the child. Other cases may include a family member, e.g. cousin, uncle, or aunt, traveling with the child. Nevertheless, if employed correctly and in tandem with other investigative methods to evaluate the full circumstances, the use of Rapid DNA technology at border crossings to stem fraudulent cases could prove very effective and reduce risk to vulnerable children.

While the use of Rapid DNA technology is being piloted in the United States only at this time, there is significant interest from other countries, particularly Eastern European and Asian countries, for the use of Rapid DNA technology at their border crossings to identify cases of human trafficking. Traditional DNA profiling has already been applied to combat human trafficking through victim identification, family reunification, and prosecution of human traffickers [34, 35]. In 2004, the DNA-PROKIDS program was established with the aim of combatting human trafficking using genetic identification of victims and their families, especially children. The program has successfully reunited illegally adopted and/or missing children with their families, and continues to do so. The program aims to establish a worldwide database of victims.

In 2014, a unique study was performed whereby covert investigators collected DNA samples using ANDE/NetBio sample swabs from individuals suspected of being victims of human trafficking [36] in Nepal and Costa Rica. Trained investigative personnel obtained samples during ongoing undercover investigative efforts. In circumstances where the collection of reference buccal samples was not practical to obtain, objects known to be in contact with the suspected human trafficking victim were sampled, e.g. cigarette butts, drink containers, etc. Objects located within the environment (e.g. brothels, locations suspected to house or transport victims) were also sampled, e.g. condoms and tissues. Objects located in public environments documented as areas where traffickers would congregate or engage were also sampled, e.g. cigarette butts and drink containers. All samples were processed using the DNAscan™/ANDE 4C Rapid DNA Identification System, with the modified low-DNA content (LDC) BioChipSet cassettes. A total of 50 samples were collected during the investigation. Nineteen of those samples were collected from known reference buccal samples of suspected human trafficking victims and yielded 95% success rate for obtaining full DNA profiles. The remaining 31 samples were collected from objects within the environment or in public places and produced 71% success rate for obtaining informative DNA profiles, with 23% full, 42% partial, and 6% mixture profiles. The results of this study provide evidence of immense support for the establishment of teams to implement the databasing of DNA profiles of suspected victims and perpetrators of human trafficking in an effort to combat the epidemic.

The identification of human trafficking remains to be a considerable challenge. The creation of a global database of human trafficking victims would offer effective evidence to support the identification of human trafficking victims, family reunification, and prosecution of human traffickers. Traditional methods of utilized within human trafficking investigations involve surveillance, use of informants, and interview and interrogation of both victims and perpetrators. These methods used in tandem with DNA profiling methods have proven useful to date. In this technological age, the sophistication of human trafficking crimes has evolved, and so must the investigative modes utilized to combat them.

21.3.3.2 Disaster Victim Identification (DVI)

There are numerous scenarios encountered in forensic casework that may result in mass fatalities, including bombings/explosions, natural disasters, fires, terrorist attacks, war conflicts, aviation crashes, and other mass transit accidents (e.g. high-speed passenger trains, subways). Human remains recovered in these situations often are severely damaged, dismembered, fragmented, commingled, in varying states of decomposition, or even skeletonized. Although many national, state, and regional crime laboratories exist throughout the country with extensive expertise and training in forensic DNA identification, these fixed "brick-and-mortar" laboratories are not amenable to field-based operations and testing. Hence, although these laboratories are willing partners in DVI efforts, the logistics of coordinating subcontracts and transporting biological samples long distances for genetic testing is tedious. Additionally, traditional DNA casework methods are both labor- and time-consuming, increasing the amount of time that families of decedents must wait for a positive identification.

Rapid DNA technology offers a potential solution to meet the high-throughput and onsite testing demand of mass fatality events. The ANDE 6C Rapid DNA System was specifically engineered for field deployment and onsite testing operations. In addition to its relatively small size and weight (as mentioned previously), it is operational at altitudes as high as 3048 m (10,000 ft) and was ruggedized to account for vibration and shock during transportation, meeting the strict requirements of US Military Standard 810G (MIL-STD-810G) [7]. Moreover, reagents are stable at room temperature for up to six months [8]. Complementing the ANDE 6C Rapid DNA instrument's compact size, fast runtimes, and temperature-stable reagents is kinship analysis software. This software, called FAIRS, has two secure modules which compare DNA profiles obtained from unidentified victims to direct reference samples or family member exemplars: (i) the Claimed Relationships module, and (ii) the Familial Search module. When algorithms embedded in the software determine a match, a formal report containing the decedent's genetic profile and the associated match statistic calculations are automatically generated for the case file. Hence, this comprehensive human identification system is particularly well-suited for location-based testing at accident sites, in disaster zones, and/or in temporary buildings or tents constructed by Disaster Mortuary Operational Response Teams (DMORTs).

DHS Science and Technology (S&T) Directorate is currently exploring the possibility of implementing Rapid DNA technology in mass fatality responses, especially since it is already being used to support immigration investigations and to combat transnational human trafficking [2, 6, 33]. DHS S&T has developed numerous drills and mock disaster victim identification exercises in order to: (i) assess if the instruments can be shipped in an expeditious and efficient manner for disaster response operations, and (ii) to identify and mitigate potential logistical challenges that might be encountered during complex deployments. Additionally, in 2019, the American Society of Crime Laboratory Directors (ASCLD) formed a Rapid DNA DVI Subcommittee, the goal of which is to determine best practices and develop policies for mass fatality applications (www.ascld.org). Although the SWGDAM has not yet developed formal guidelines for Rapid DNA analysis of DVI casework samples, the FBI has formed a working group to facilitate discussions on this issue.

Recent publicized successes using Rapid DNA technology in DVI cases likely will expedite discussions among the DHS S&T, ASCLD, SWGDAM, and FBI working groups, and may facilitate formal implementation of the technology in mass fatality response efforts. One of the aforementioned high-profile cases involved the 2018 "Camp Fire" in Paradise, California, which claimed the lives of 85 people and is the deadliest wildfire in California history. In collaboration with Butte County's Sheriff's Office and the California Department of Justice (DOJ),

ANDE Corporation deployed a team to assist with identification efforts. Remains from 69 victims were recovered; however, due to the intensity and duration of the fire, traditional methods of identification (e.g. fingerprints, odontology, surgical hardware) were only feasible for 22 of the decedents. Using a mobile (recreational) vehicle and three ANDE 6C Rapid DNA instruments, a variety of sample types from the victims' remains (e.g. bone, muscle, brain tissue, liver) and approximately 300 family reference samples were tested. DNA profiles were obtained from 62 of the 69 victims recovered (89.9% of samples tested). Of the 62 victims for whom DNA profiles were successfully generated, 58 victims (93.5%) were able to be positively identified via comparison to exemplars from biological relatives [37–45]. In 2019, less than a year after the wildfires, another tragedy struck in California. Thirty-three passengers and one crew member aboard the dive boat *Conception* perished when it caught fire off the coast of Santa Cruz Island. All 34 victims were identified within 10 days using ANDE Rapid DNA technology, much faster than the identification effort would have taken using traditional laboratory-based testing [22, 46].

In addition to the human remains from the California wildfires and the *Conception* boat fire that were successfully processed for DNA, the ANDE Rapid DNA platform has been subjected to extensive research with a broad spectrum of biological samples. During developmental validation, a total of 101 samples across seven different DVI-relevant tissue types were processed for accuracy and concordance. Included in the study were 18 bones (femora, humeri, ribs), 3 M teeth, 24 liver samples, 34 skeletal muscles, 2 brain cortices, 11 lung tissues, and 9 kidney samplings [8]. In another study, 10 deceased (donated) human subjects were exposed to the environment (surface deposition) at an approved outdoor research facility, and various tissue samplings were collected at regular time intervals over the course of a one-year study period. In general, successful DNA typing of soft tissues (brain, muscle, liver) was limited to relatively short postmortem time intervals (1–11 days), attributed to accelerated decomposition during environmental exposure compared to hard tissues. The majority of the bone/tooth samples tested yielded useful DNA profiles from the first day of collection through the end of the one-year experimental period [22]. Both of these research studies – as well as the casework performed during the California wildfires and the *Conception* boat fire – demonstrate proof-of-concept for DVI applications and support the use of this technology as a way to expedite human identification and family reunification efforts.

ANDE Corporation's Rapid DNA platform has already demonstrated robust and reliable forensic genetic testing of human remains with relatively short postmortem intervals (i.e. recent deaths ranging from one day to one year). However, human remains from mass fatality events are not always located and recovered within such short time frames. Victims of homicide, oppressive regimes, genocide, and war conflicts, for example, often are buried in clandestine graves and may not be discovered for many years (sometimes decades). These types of samples – which have been buried and have naturally skeletonized over time due to environmental insult – will pose additional challenges for rapid DNA platforms. Buried remains present an entirely different set of challenges compared to "fresh" bones that are recovered immediately or shortly after a fatality event has occurred. As the structural components of bone (i.e. calcium hydroxyapatite, collagen) decompose in a process called diagenesis, endogenous DNA molecules become far more susceptible to damage. Thus, buried bones typically yield much lower DNA quantities and present templates with a multitude of problematic lesions that can prevent successful profiling. Perhaps future research efforts could expand to include buried human remains. Successful application of Rapid DNA technology to naturally skeletonized remains which have been exposed to extended burial conditions or surface deposition across a broad range of climates and environments could dramatically expedite and change the landscape of forensic human identification casework.

21.3.3.3 Enhancing Forensic Capabilities in Underdeveloped Countries

Rapid DNA technologies may represent an alternative mode of operation for the development of forensic science laboratories in underdeveloped countries. There are underdeveloped countries that do not have DNA profiling capabilities, and therefore could benefit from systems such as Rapid DNA technology whereby a more cost-effective, yet information-yielding approach could be employed. Within the United States, operational crime laboratories are a huge expenditure. They require significant investment to build an entire DNA laboratory. The average DNA crime laboratory requires:

- Large equipment and infrastructure investments.
- Highly skilled technical and scientific personnel.
- Instrument maintenance and calibration.
- Freezer and cold storage.

For example, the DNA section of the Washington D.C. DNA crime laboratory in the United States endured initial setup costs of approximately USD $22 million. They incur an annual DNA laboratory operating budget of about USD $4 million. Other underdeveloped countries do not have this level of funding or budgetary capabilities. Rapid DNA technology, however, could represent an opportunity to apply DNA profiling to their investigations with a much-reduced initial setup cost and operational expenditure in underdeveloped countries.

21.4 Limitations and Important Considerations

There are some drawbacks and known limitations to this technology, and these should be considered prior to use or implementation.

21.4.1 Minute/Precious Samples

With multiple different Rapid DNA technologies available and ready for use, law enforcement agencies are eager to implement these technologies. With reference buccal samples and crime scene samples (e.g. blood, and other body fluids etc.), the recommended best practice is the two-swab method, whereby one swab is processed using Rapid DNA technology and the other is preserved for traditional DNA profiling in a laboratory. This ensures that if the Rapid DNA profiling fails for some reason, there is another available sample for testing. However, biological samples recovered from crime scenes or evidence items can often be minute in sample volume, i.e. trace biological evidence. Therefore, investigators must make the decision whether or not to consume the available sample using Rapid DNA profiling and retrieve results faster, or to preserve the sample and submit to the laboratory for traditional DNA profiling. In addition, when dealing with very minute samples within the DNA laboratory, standard operating procedures to enhance the recovery and yield of DNA from the samples are performed, in addition to methods to increase amplification of the PCR products. While systems such as the ANDE 6C Rapid DNA System with the I-Chip have added a concentration within their BioChipSet to increase yield, this may not be as effective as in-laboratory processes. Ultimately, there is a risk that potentially useful information could be lost due to the consumption of a trace sample when using Rapid DNA technologies, which might be prevented if submitted directly to the laboratory for processing. Because of this, it is recommended that minute or trace biological samples are preserved for traditional DNA processing.

21.4.2 Mixtures

With the current/most recent Rapid DNA technologies, the ANDE 6C Rapid DNA System, and the RapidHIT ID System, mixture resolution and interpretation within the systems are not possible. Both systems, however, will flag a sample run as possibly containing a mixture of DNA profiles.

DNA mixtures are profiles that are produced when a sample contains DNA contributions from two or more individuals. Mixed DNA profiles are failed by the system under one or more of the following conditions:

- Two or more heterozygous loci with three alleles.
- One or more heterozygous loci with four alleles.
- One or more hemizygous loci with two alleles.

In 2019, Carney et al. showed that the ANDE 6C Rapid DNA System could correctly identify mixed DNA samples from reference buccal samples that were created using a five different ratios of two-person mixed samples (19 : 1, 5 : 1, 1 : 1, 1 : 5, and 1 : 19) [6]. With the 19 : 1 and 1 : 19 samples however, the ANDE Expert System software ignores the minor contributors, which if reviewed manually by DNA analysts, in some cases, could yield probative information. In a separate study, the RapidHIT ID System correctly flagged mixture profiles that were processed using four different ratios of two-person mixed samples (1 : 1, 1 : 2.5, 1 : 4, and 1 : 9). Similar to ANDE, the RapidHIT systems are designed to flag DNA mixtures, however only once the data have been exported to GeneMapper™ or GeneMarker™.

The ability to identify mixture samples quickly using Rapid DNA technology can aid an investigation by providing feedback to crime scene investigators who can then select other areas or stains to sample and improve or focus their collection. The ability to provide this type of strategy information is often lost, as it takes weeks or months to profile the sample using conventional methods.

21.5 Future Considerations and Conclusions

Smaller and faster: As with all field-based instrumentation, there is a continuous push to make the instruments smaller, and to produce results faster. The current Rapid DNA instruments are comparable in size to a large printer and produce results within approximately 90 minutes; however, this time frame includes a maximum of five samples per run. With future platforms, the time required to perform a run may not be amenable to faster run times, although they may be amenable to more samples per run.

Cost: As the technology advances and the demand increases, the cost of Rapid DNA technologies will be driven down. This cost will reduce not only the purchase of the instrument itself, but also the cost-per-sample to run.

Standardization: As there are just two major competing companies that produce Rapid DNA technologies at the moment, there is potential for other smaller companies to begin producing their own instruments, based on the success of the current two platforms. Inversely, there is also potential for the two major competing companies to join forces and merge to produce one super provider.

Best practices: Rapid DNA technologies were originally developed with the aim of using them in booking stations for arrestee profiling, requiring no technical expertise. The technology however has now expanded into more complex applications using crime scene and DVI samples, which can be low in DNA content and highly degraded. Therefore, best practices need to be developed to address the processing of individual evidence types, along with adequate training for these more complex sample types as they require technical expertise.

Policies and legislation: While the Rapid DNA Act of 2017 provided the springboard from which Rapid DNA technologies could truly be implemented into operational casework, it is crucial that new policies and legislation are continually being created and revised in order to keep up with technological advancements and as the methods are more broadly implemented. In addition, there needs to be buy-in from all parties involved where clear communication and agreement are maintained. This includes the instrument manufacturers, the law enforcement agencies that bear the responsibility for the evidence handling, and the state prosecutors who will handle the cases in the courtroom and have to defend the use of these methods.

As technology advances, it is certain that further improvements will be made to existing platforms. Indeed, more platform providers may appear, or some existing providers may merge. The range of applications for both field-based and laboratory-based use will continue to grow as researchers refine the methodologies involved. Industry demand will undoubtedly grow as further successes are reported. Continued research and development will drive the technology forward and may ultimately lead to Rapid DNA technologies becoming a staple of crime laboratories across the world.

A. Appendix

A.1 Acronyms

A-Chip	Arrestee Chip
ANDE	Accelerated Nuclear DNA Equipment
ASCLD	American Society of Crime Lab Directors
BCS	BioChipSet
CCD	Charge Coupled Device
CE	Capillary Electrophoresis
CJIS	Criminal Justice Information Services
CODIS	Combined DNA Index System
CSA	CJIS Systems Agency
DAB	DNA Advisory Board
DHS	Department of Homeland Security
DHS S&T	Department of Homeland Security Science & Technology Directorate
DOD	Department of Defense
DOJ	Department of Justice
DMORT	Disaster Mortuary Operational Response Team
DVI	Disaster Victim Identification
EPG	Electropherogram
ESS	European Standard Set
FBI	Federal Bureau of Identification
FTA	Flinders Technology Associates
HID	Human Identification
ICE	Immigration and Customs Enforcement
ID	Identification
I-Chip	Investigative Chip
ILS	Internal Lane Standard
IT	Information Technology
LDC	Low DNA Content
mRDA	modified Rapid DNA analysis
NDIS	National DNA Index System
ng	Nanogram
NIJ	National Institute of Justice
NIST	National Institute of Standards and Technology

OCME	Office of the Chief Medical Examiner
PCR	Polymerase Chain Reaction
pg	Picogram
QAS	Quality Assurance Standards
RDA	Rapid DNA analysis
RFID	Radio Frequency Identification
SAK	Sexual Assault Kit
SIB	State Identification Bureau
SID	State Identification (number)
STR	Short Tandem Repeat
SWGDAM	Scientific Working Group on DNA Analysis Methods
UCN	Universal Control Number
UF	Ultra-Filtration
USACIL	United States Army Criminal Investigation Laboratory

References

1 FBI.gov (2020). *FBI Webpage* [Internet]. https://www.fbi.gov/services/laboratory/biometric-analysis/codis/rapid-dna (accessed 1 April 2020).

2 Congress.gov (2017). *H. Rept. 115-117 – Rapid DNA Act of 2017* [Internet]. https://www.congress.gov/congressional-report/115th-congress/house-report/117/1 (accessed 30 March 2020).

3 French, J.L., Turingan, R.S., Hogan, C., and Selden, R.F. (2016). Developmental validation of the DNAscan™ Rapid DNA analysis™ instrument and expert system for reference sample processing. *Forensic Science International Genetics* 25: 145–156.

4 Tan, E., Turingan, R.S., Hogan, C. et al. (2013). Fully integrated, fully automated generation of short tandem repeat profiles. *Investigative Genetics* 4 (1): 16.

5 Moreno, L.I., Brown, A.L., and Callaghan, T.F. (2017). Internal validation of the DNAscan/ANDE™ Rapid DNA analysis™ platform and its associated PowerPlex® 16 high content DNA biochip cassette for use as an expert system with reference buccal swabs. *Forensic Science International. Genetics* 29: 100–108.

6 Carney, C., Whitney, S., Vaidyanathan, J. et al. (2019). Developmental validation of the ANDE™ rapid DNA system with FlexPlex™ assay for arrestee and reference buccal swab processing and database searching. *Forensic Science International. Genetics* 40: 120–130.

7 U.S. Department of Defense (2008). *Department Of Defense Test Method Standard: Environmental Engineering Considerations And Laboratory Tests Title* [Internet]. https://www.atec.army.mil/publications/Mil-Std-810G/Mil-Std-810G.pdf (accessed 12 April 2020).

8 Turingan, R.S., Tan, E., Jiang, H. et al. (2020). Developmental validation of the ANDE 6C system for rapid DNA analysis of forensic casework and DVI samples. *Journal of Forensic Sciences* 65 (4): 1056.

9 Grover, R., Jiang, H., Turingan, R.S. et al. (2017). FlexPlex27 – highly multiplexed rapid DNA identification for law enforcement, kinship, and military applications. *International Journal of Legal Medicine* 131 (6): 1489–1501.

10 Hennessy, L.K., Franklin, H., Li, Y. et al. (2013). Developmental validation studies on the RapidHIT™ human DNA identification system. *Forensic Science International: Genetics Supplement Series* 4: e7–e8.

11 Jovanovich, S., Bogdan, G., Belcinski, R. et al. (2015). Developmental validation of a fully integrated sample-to-profile rapid human identification system for processing single-source reference buccal samples. *Forensic Science International: Genetics* 16: 181–194.

12 Hennessy, L.K., Mehendale, N., Chear, K. et al. (2014). Developmental validation of the GlobalFiler® express kit, a 24-marker STR assay, on the RapidHIT® system. *Forensic Science International: Genetics* 13: 247–258.

13 Vallone, P.M. (2013). *Biometric Consortium Meeting, BCC, Tampa, FL, Sepember* 2013. NIST Rapid-DNA Interlaboratory Study [Internet]. https://strbase.nist.gov//pub_pres/Vallone_BCC_Talk_Sept2013.pdf (accessed 27 March 2020).

14 Larue, B.L., Moore, A., King, J.L. et al. (2014). An evaluation of the RapidHIT® system for reliably genotyping reference samples. *Forensic Science International: Genetics* 13: 104–111.

15 Holland, M. and Wendt, F. (2015). Evaluation of the RapidHIT™ 200, an automated human identification system for STR analysis of single source samples. *Forensic Science International: Genetics* 14: 76–85.

16 Salceda, S., Barican, A., Buscaino, J. et al. (2017). Validation of a rapid DNA process with the RapidHIT® ID system using GlobalFiler® express chemistry, a platform optimized for decentralized testing environments. *Forensic Science International: Genetics* 28: 21–34.

17 Romsos, E.L., Lembirick, S., and Vallone, P.M. (2015). Rapid DNA maturity assessment. *Forensic Science International: Genetics Supplement Series* 5: e1–e2.

18 Romsos, E.L., French, J.L., Smith, M. et al. (2020). Results of the 2018 rapid DNA maturity assessment. *Journal of Forensic Sciences* 65 (3): 953–959.

19 FBI.gov. *Rapid DNA* [Internet]. https://www.fbi.gov/services/laboratory/biometric-analysis/codis/rapid-dna (accessed 3 April 2020).

20 Gangano, S., Elliott, K., Anoruo, K. et al. (2013). DNA investigative lead development from blood and saliva samples in less than two hours using the RapidHIT™ human DNA identification system. *Forensic Science International: Genetics Supplement Series* 4: e43–e44.

21 Turingan, R.S., Vasantgadkar, S., Palombo, L. et al. (2016). Rapid DNA analysis for automated processing and interpretation of low DNA content samples. *Investigative Genetics* 7: 2.

22 Turingan, R.S., Brown, J., Kaplun, L. et al. (2019). Identification of human remains using Rapid DNA analysis. *International Journal of Legal Medicine* 134: 863–874.

23 Biesecker, L.G., Bailey-Wilson, J.E., Ballantyne, J. et al. (2005). DNA identifications after the 9/11 world trade Center attack. *Science* 310 (5751): 1122–1123.

24 Mundorff, A.Z., Bartelink, E.J., and Mar-Cash, E. (2009). DNA preservation in skeletal elements from the world trade center disaster: recommendations for mass fatality management. *Journal of Forensic Sciences* 54 (4): 739–745.

25 Holland, M.M., Cave, C.A., Holland, C.A., and Bille, T.W. (2003). Development of a quality, high throughput DNA analysis procedure for skeletal samples to assist with the identification of victims from the world trade Center attacks. *Croatian Medical Journal* 44 (3): 264–272.

26 Sudoyo, H., Widodo, P.T., Suryadi, H. et al. (2008). DNA analysis in perpetrator identification of terrorism-related disaster: suicide bombing of the Australian embassy in Jakarta 2004. *Forensic Science International: Genetics* 2: 231–237.

27 Prinz, M., Carracedo, A., Mayr, W.R. et al. (2007). DNA Commission of the International Society for forensic genetics (ISFG): recommendations regarding the role of forensic genetics for disaster victim identification (DVI). *Forensic Science International: Genetics* 1 (1): 3–12.

28 Deng, Y.J., Li, Y.Z., Yu, X.G. et al. (2005). Preliminary DNA identification for the tsunami victims in Thailand. *Genomics, Proteomics & Bioinformatics* 3 (3): 143–157.

29 Schou, M.P. and Knudsen, P.J.T. (2012). The Danish disaster victim identification effort in the Thai Tsunami: organisation and results. *Forensic Science, Medicine, and Pathology* 8 (2): 125–130.

30 Hsu, C.M., Huang, N.E., Tsai, L.C. et al. (1999). Identification of victims of the 1998 Taoyuan airbus crash accident using DNA analysis. *International Journal of Legal Medicine* 113 (1): 43–46.

31 Leclair, B., Frégeau, C.J., Bowen, K.L., and Fourney, R.M. (2004). Enhanced kinship analysis and STR-based DNA typing for human identification in mass fatality incidents: the Swissair flight 111 disaster. *Journal of Forensic Sciences* 49 (5): 939–953.

32 Piccinini, A., Betti, F., Capra, M., and Cattaneo, C. (2004). The identification of the victims of the Linate air crash by DNA analysis. *International Congress Series* 1261: 39.

33 Miroff, N. (2019). Homeland Security to test DNA of families at border in cases of suspected fraud. *Washington Post* [Internet]. https://www.washingtonpost.com/immigration/homeland-security-to-test-dna-of-families-at-border-in-cases-of-suspected-fraud/2019/05/01/8e8c042a-6c46-11e9-a66d-a82d3f3d96d5_story.html (accessed 3 April 2020).

34 Katsanis, S. and Kim, J. (2014). DNA in immigration and human trafficking. In: *Forensic DNA Applications: An interdisciplinary perspective*, Section 4, Chapter 22. Boca Raton: CRC Press, Taylor and Francis Group https://www.routledge.com/Forensic-DNA-Applications-An-Interdisciplinary-Perspective/Primorac-Schanfield/p/book/9781466580220.

35 Katsanis, S.H., Kim, J., Minear, M.A. et al. (2014). Preliminary perspectives on dna collection in anti-human trafficking efforts. *Recent Advances in DNA Gene Sequences* 8: 78–90.

36 Palmbach, T., Blom, J., Hoynes, E. et al. (2014). Utilizing DNA analysis to combat the world wide plague of present day slavery – trafficking in persons. *Croatian Medical Journal* 55 (1): 3–8.

37 Gin, K., Tovar, J., Bartelink, E.J. et al. (2020). The 2018 California wildfires: integration of Rapid DNA to dramatically accelerate victim identification. *Journal of Forensic Sciences* 65 (3): 791–799.

38 Department of Homeland Security (2019). Snapshot: S&T's Rapid DNA technology identified victims of California Wildfire. *DHS Science and Technology News* [Internet]. https://www.dhs.gov/science-and-technology/news/2019/04/23/snapshot-st-rapid-dna-technology-identified-victims (accessed 4 May 2020).

39 Border, G. (2018). California Taps War-Zone DNA Specialists After Wildfire. *Reuters* [Internet]. https://www.reuters.com/article/us-california-wildfires-remains/california-taps-war-zone-dna-specialists-after-wildfire-idUSKCN1NK1EI (accessed 3 April 2020).

40 Brown, K. (2018). California Turns to War-Zone DNA Test to ID Fire Remains. *Bloomberg* [Internet]. https://www.bloomberg.com/news/articles/2018-11-19/dna-testing-company-gets-call-to-help-id-california-fire-victims (accessed 3 April 2020).

41 Druga, M. (2019). No Rapid DNA Technology Identified Victims of California Wildfires. *Homeland Preparedness News* [Internet]. https://homelandprepnews.com/stories/33583-rapid-dna-technology-identified-victims-of-california-wildfires (accessed 3 April 2020).

42 Gin, K. (2019). California Wildfires: Rapid DNA and the Future of Mass Fatality Identification. *Forensic Science Executive* [Internet]. www.forensicscienceexecutive.org (accessed 3 April 2020).

43 Evan Sernoffsky (2018). DNA Technology Helps Identify Camp Fire Victims; Challenge is Finding Family. *San Francisco Chronicle* [Internet]. https://www.sfchronicle.com/california-wildfires/article/DNA-technology-helps-identify-Camp-Fire-victims-13406853.php (accessed 3 April 2020).

44 Waitt, T. (2019). Rapid DNA. ID's deadly 'Camp Fire' Disaster Victims. *American Security Today* [Internet]. https://americansecuritytoday.com/rapid-dna-ids-deadly-camp-fire-disaster-victims-see-how-video (accessed 3 April 2020).

45 Zimmer, K. (2019). Rapid DNA Analysis Steps in to Identify Remains of Wildfire Victims. *The Scientist* [Internet]. https://www.the-scientist.com/news-opinion/rapid-dna-analysis-steps-in-to-identify-remains-of-wildfire-victims-65156 (accessed 3 April 2020).

46 Woods, A. (2019). Officials Used "Rapid DNA" Tech to Identify California Boat Fire Victims. *The New York* Post [Internet]. https://nypost.com/2019/09/05/officials-used-rapid-dna-tech-to-identify-california-boat-fire-victims (accessed 3 April 2020).

22

Portable Biological Spectroscopy: Field Applications

Brian Damit and Miquel Antoine

Johns Hopkins Applied Physics Laboratory, Laurel, MD, USA

22.1 Introduction

Field uses for portable biological spectroscopy systems are highly diverse, ranging from biological defense to medical diagnostics, and even forensics. For some of these disciplines, application of portable spectroscopy is established, and represents either a primary or complementary tool for practitioners to analyze samples. In a field environment, where resources such as power, access to consumables, and analysis time can be limited, spectroscopy exhibits distinct advantages over other approaches (e.g. molecular assays) including near-instant feedback, simple operation and interpretation of data, and minimal or no need for consumables. Thus, demand for greater inclusion of portable spectroscopy into existing user workflows, especially in challenging field settings, is well deserved. Advances in fundamental spectroscopy research, ubiquity of smartphones, and world events (e.g. September 11) over the last several decades have expanded field applications for portable spectroscopy, irrevocably shaping the landscape of the discipline. Commercial off-the-shelf (COTS) portable spectroscopy technologies, which are systems users can deploy directly into the field, have evolved from requiring users to possess substantial technical background, to now exhibiting features like touchscreens, automated data interpretation, and requiring minimal user training. Separately, both academia and industry alike have reported headway in exploiting the smartphone, repurposing its onboard camera optics and software to perform spectroscopy; see the chapters by Scheeline in Volume 1 and Peveler and Algar in Volume 2. The promise of a "spectrometer in your pocket" may indeed represent an ideal embodiment of a portable system, further broadening the range of field applications for spectroscopic analysis in the future.

Table 22.1 provides an overview of this chapter, summarizing the range of field applications for portable biological spectroscopy. Key disciplines employing spectroscopy include biological defense, medical and healthcare settings, food and agriculture safety, environmental pollution monitoring, and forensics. The role of spectroscopy in these disciplines varies considerably, but retains the common theme of applying light, usually ultraviolet (UV), visible, and/or infrared, to probe for molecular structure or chemical composition, thereby supplying users with data regarding features of a biological sample. Figure 22.1 displays photos of commercial systems that have been designed for the specific purpose of field use. Within biodefense, portable spectroscopy systems maintain a critical function in early warning monitoring for an airborne biological attack, and detection of biothreat materials in solid-phase or liquid samples [1, 2]. Medical applications of portable spectroscopy are varied, spanning from analysis of human tissue to disease diagnostics [3, 4]. The use of portable spectroscopy systems in food and agriculture settings is similarly diverse, involving in situ use of spectroscopic techniques to rapidly identify biological contamination and analyze the welfare of crops. Growing environmental concerns in recent years have motivated on-site monitoring of biological pollutants in water, soil, and air [5, 6]. Finally, portable spectroscopy systems are

Portable Spectroscopy and Spectrometry 1: Technologies and Instrumentation, First Edition.
Edited by Richard A. Crocombe, Pauline E. Leary and Brooke W. Kammrath.

Table 22.1 General overview of field applications of portable biological spectroscopy technologies.

Application area	Description	Spectroscopy types	Example usage
Biodefense	Detection of biological threats (solid, liquid, air phase)	Fluorescence, Raman, breakdown spectroscopy (LIBS)	Distribute portable detectors in area to monitor for airborne biological threats
Medical and healthcare	Assessment of tissue health and disease biomarkers	Infrared absorption, fluorescence, Raman	Attach infrared spectrometers to human subjects to measure tissue oxygenation
Food and agriculture	Monitoring of food contamination and welfare of crops and livestock	Fluorescence, visible, infrared, Raman	Measure the fluorescence spectra from meat surfaces to infer bacterial loading
Environmental pollution	Measurement of biological pollutants in water, soil, and air	Fluorescence, infrared absorption and reflectance	Assess water potability by measuring fluorescence from coliforms using a smartphone
Forensics	Identification of the presence of body fluids, bruises, and wounds	Fluorescence, visible absorption	Use portable light systems to locate blood at a crime scene to assist in sample collection by investigators

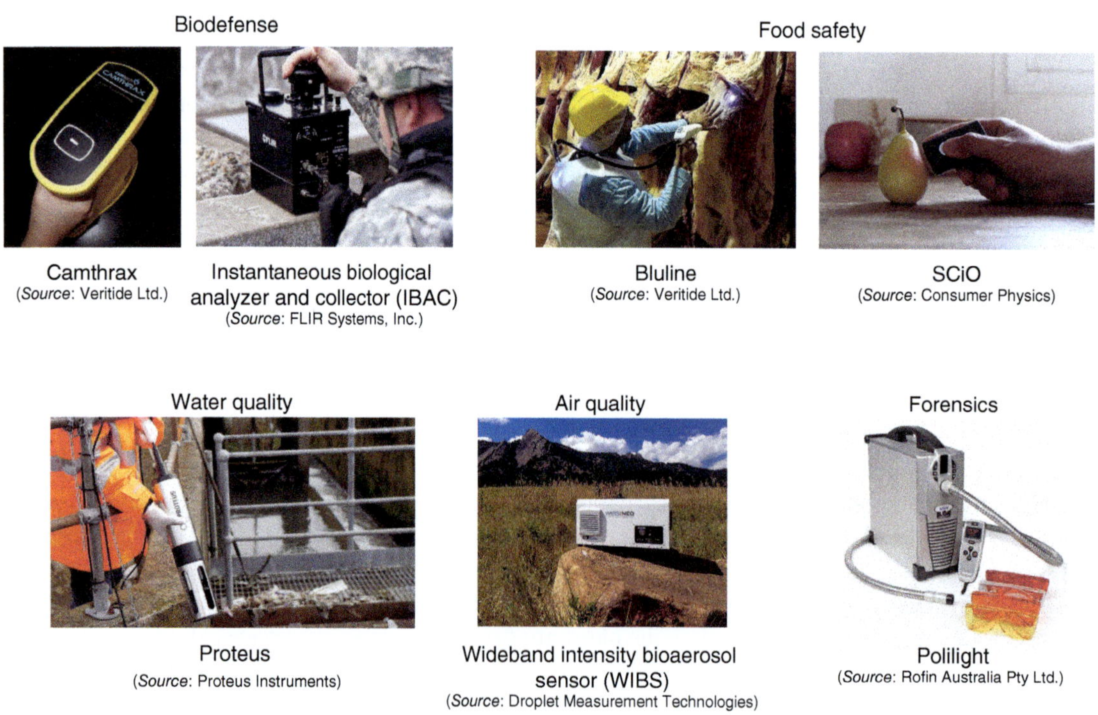

Figure 22.1 Photos of commercial portable biological spectroscopy systems developed for field use.

well established in the field of forensics, assisting crime scene investigators in locating and classifying biological samples like blood, other body fluids, and fingerprints present at a crime scene [7, 8]. This chapter reviews these disciplines, describing the use of portable biological spectroscopy from a field application perspective and providing practical considerations. To keep this chapter tractable, the scope of the discussions is constrained to reagentless spectroscopic detection approaches. This review also emphasizes technologies and concepts that are more mature, specifically those techniques that have been translated into practice, have demonstrated use in the field, and are integrated into the workflow of practitioners. However, less established, but notable advances that have potential to be transformative are referenced as well. This chapter concludes with a summary discussing challenges and outlook for portable biological spectroscopy in the coming years.

22.2 Organization of This Chapter

This chapter first describes the essential considerations for field-portable systems, namely the features that distinguish fieldable systems from their benchtop counterparts. The following discussion then summarizes specific field applications for portable spectroscopy: biological defense, medical and clinical, food and agriculture, environmental sciences, and forensics. These sections provide an overview of the field, and discuss specific applications, benefits, and limitations of portable spectroscopy in the respective field applications. The sections also introduce and describe several existing spectroscopy systems, as well as past systems to provide historical context to the reader. In addition to discussion of conventional technologies, this chapter includes discussion of other types of portable spectroscopy (e.g. smartphone), and their potential in future field applications.

22.3 Attributes of Field-Portable Spectroscopy Systems

Regardless of the specific field application, portable spectroscopy systems exhibit several important attributes that distinguish them from their benchtop counterparts. Paramount among these, portable systems must be transportable into their respective field setting, usually by a single user, which both limits the weight and form factor of a system. This limitation oftentimes results in sacrifice of features, reliability, and/or signal-to-noise ratio. Although "man portability" does not have a strict definition, portable systems are usually limited to a weight and size on the order of 14 kg and backpack sized (65 000 cm^3) to be comfortably transported by a user. For many of the field scenarios described in this chapter, portable systems must include the capability to be battery powered, as users may not have access to wall or generator power. Ideally, field-portable spectroscopy systems do not consume reagents or other consumables since this requires users to transport additional supplies into the field. Depending on the specific field usage, other attributes of portable systems include onboard data analysis, robust wireless communications, minimal maintenance and calibration, and reliable operation in a range of environmental conditions (e.g. precipitation, wind, temperature, relative humidity, etc.). For example, portable spectroscopy systems used for environmental monitoring purposes may need to operate outdoors for extended durations and transmit data remotely to a user. Therefore, the properties of a portable spectroscopy system are dictated by its intended usage in a field setting.

For many field applications, a key consideration for a portable spectroscopy system is the user experience. User experience is a user's perception or overall attitude toward a system – more precisely it refers to usability (ease-of-use), convenience, and even aesthetic of its hardware design or software interface. Although a largely subjective metric, a positive user experience is essential and dictates widespread adoption or rejection by practitioners. In challenging field environments, users prefer that interaction with the system be simple and intuitive, especially when users do not have expertise in the respective discipline. In general, it is desirable that these systems are accessible to a variety of users possessing both technical and nontechnical backgrounds. This characteristic empowers nonexperts like technicians, military personnel, law enforcement, volunteers, and citizen scientists alike to confidently collect and analyze samples. Regardless of their background, users usually prefer simple formats like integrated touchscreens or connection to tablets or smartphones to operate a spectrometer in field environments. Single-button operation, where a user simply presses a button to begin analysis, acquire data, interpret data, and display results is also ideal. For technical users desiring in-depth details, for instance, those users wishing to view a spectral plot, some systems include an advanced-user mode that can be accessed. For these advanced users, impromptu customization of spectral libraries or adjustment of other data acquisition settings is also highly desirable. Another notable element of user experience is the analysis time or time-to-detect – essentially the time required for a system to analyze a sample. Optimally, analysis time is near instantaneous rather than on the scale of hours, allowing users to efficiently obtain results.

With regard to data collection in field settings, some spectroscopy systems require initial setup and then sample autonomously, whereas other systems require users to actively collect and manipulate samples. In the latter case, sample introduction workflow is a critical aspect of user experience, with complex, multistep procedures reflecting poorly on usability. In an ideal operation, a user simply points the system toward a sample, initiates a measurement, and then obtains results. A similar, but a more involved process necessitates users to physically collect a solid, liquid, or tissue sample, and transport the sample to a system for subsequent analysis. In general, fewer sample preparation steps and manipulations by a user yield fewer user errors and biases. Several field scenarios require users to be in personnel protective equipment (PPE) while performing analysis, thus for some systems, operation must be streamlined and accommodate users with limited finger dexterity and range of motion. After sampling is complete, contamination of the system, also referred to as carryover, is preferably minimal or easily removed either automatically or by a user.

22.4 Field Applications

22.4.1 Biological Defense

Application of portable spectroscopy systems in biological defense is well established, especially in comparison to other fields, with a range of technologies available for analyzing solid, liquid, and air-phase samples for biological threats. Commercial systems are widely adopted by militaries and biodefense agencies across the world; moreover, growing concerns about bioterrorism and new biological threats have precipitated development of new technologies. The following sections provide a summary of historical and existing portable biodefense spectroscopy systems with emphasis on higher maturity technologies.

22.4.1.1 Biological Solid- and Liquid-Phase Detection

The 2001 anthrax letters incident (Amerithrax) motivated development of man-portable systems that can rapidly interrogate suspicious powders to detect biological materials without need for wet-chemistry preparation [9]. Marketed toward first responders, Hazmat teams, and military personnel, these handheld systems allow a user to load a powder sample into the system, press a button, and acquire results. Examples of such systems are the Camthrax and Ceeker (Veritide Ltd.). The Ceeker operates by measuring the fluorescence signature of dipicolinic acid (DPA), a compound uniquely present in *Bacillus* and *Clostridium* bacterial spores and responsible for the hardiness of the organisms. Due to the weakly fluorescing nature of DPA, the Ceeker applies a two-step analysis scheme: (i) 250 nm photodisassociation of DPA to facilitate conversion of DPA to more fluorescent picolinic acid, and (ii) fluorescence measurement of picolinic acid with 350 nm excitation. After a 10-minute measurement and analysis period, the system alerts the user to the results by displaying a simple "yes" or "no" answer on a screen located on the instrument. Since the system does not provide species-level identification, detection is deemed presumptive as molecular biology techniques (e.g. polymerase chain reaction [PCR]) must confirm that the spore is a biological agent. Although the Ceeker targets bacterial spores, development efforts have investigated extending these principles to detection of ricin and botulinum toxin – two proteins often implicated as biological weapons. A now defunct portable system, the Handheld Raman Imager (ChemImage) used Raman chemical imaging for the detection of both biological and chemical threats. The system combined molecular spectroscopy with digital imaging to analyze both powder/liquid samples and identify biological materials, especially those present in complex samples. In general, advances in assay technologies, particularly lateral flow assay (LFA) have largely superseded portable spectroscopy for field analysis of suspicious powder or liquid samples. These LFAs are simple nitrocellulose test strips in a pregnancy test-style format. In addition to offering species-level presumptive identification, LFA methods are less expensive, smaller, and provide more rapid results than portable spectroscopy systems.

22.4.1.2 Biological Aerosol Detection

Portable spectroscopy systems are used extensively in the detection of biological aerosols (bioaerosols). Bioaerosols, defined as airborne particles of biological origin, can originate from natural sources or be released intentionally for the purpose of biowarfare. Since the late 1990s, biodefense programs have integrated portable spectrometers into their biodetection strategies to support early warning alarming of an airborne bioattack [1]. Operationally, users distribute these portable systems throughout a high-risk location, and the systems monitor air continuously for anomalous aerosols – these locations can include subway stations, sporting venues, tourist attractions, military bases, or other densely populated areas. Most systems use some variant of laser-induced fluorescence (LIF) spectroscopy as the detection mechanism, although several systems use other spectroscopic modalities (e.g. Raman scattering) to supplement or instead of LIF. LIF systems detect bioaerosols by sampling ambient aerosols, aerodynamically focusing these aerosols into a single file, and then interrogating each aerosol with laser light to measure its fluorescence signature; bioaerosols, which intrinsically contain fluorescent molecules, produce autofluorescence whereas nonbiological particles (e.g. road dust) yield minimal or no fluorescence signature. Thus, measurement of an abundance of fluorescent aerosols can indicate an airborne biological attack has occurred. As air-phase particles are analyzed in real time, these systems can automatically provide a warning alarm, promoting nearby personnel to seek immediate protective measures. A key caveat for LIF systems is that these technologies cannot discriminate between biological threats and more benign biological particles (e.g. pollen), and thus molecular assays are needed to confirm the bioaerosols indeed contain threat organisms. LIF systems range in dimensions from ruggedized handheld units ($18 \times 15 \times 32$ cm) to larger units ($90 \times 50 \times 75$ cm) that require vehicles for transport to a field site. For biodefense applications, these systems often contain wireless communications to enable networking and transmission of data and alarming status to users stationed at a command center. Some systems feature integrated aerosol collection modules that can be autonomously initiated to collect a bioaerosol sample. Within the biodefense community, portable real-time spectroscopy systems are often referred to as "triggers," since they are operationally implemented to trigger aerosol collector systems that obtain an aerosol sample for subsequent processing with a molecular biology assay.

Portable LIF detectors are designed to target molecules naturally present in microorganisms and their growth media; the systems' excitation (λ_{ex}) and emission (λ_{em}) wavelengths overlap with the fluorescence spectra of high-intensity endogenous fluorophores in prokaryotic organisms. An exhaustive review of fluorophores and associated spectra for a variety of organisms is presented in Pöhlker et al. [10] Molecules specifically targeted by LIF systems include several amino acids (tryptophan, tyrosine, and phenylalanine), which have excitation and emission maxima of 260–280 nm and 280–350 nm respectively, and enzymes and coenzymes like nicotinamide adenine dinucleotide (NADH), which have excitation and emission of 290–450 nm and 440–535 nm, respectively [11]. Components of the cell wall, various pigments, and vitamins can also produce weak or moderate fluorescence signal. Nonvolatile constituents of cell-culture media, aerosolized with microorganisms, exist as residue on aerosols after a wet-dissemination, yielding exogenous autofluorescence. Lysogeny broth, for example, is comprised of tryptone and yeast extract, which contain fluorescent molecules like peptides and vitamins added to enhance microbial growth; indeed, these nonvolatile components can contribute significant fluorescence that is detectable by LIF systems.

Interest in portable LIF bioaerosol spectrometers for biodefense applications began in earnest with the Fluorescence Aerosol Particle Sensor (FLAPS, TSI Inc.), a design originally conceived by the Canadian Department of National Defence. Eventually adopted by the US military and Canadian biodefense programs, the FLAPS, later termed the Ultraviolet Aerodynamic Particle Sizer (UV-APS), employed laser excitation in the 340–360 nm range [1]. Nearly simultaneous investigations by the US Army Research Laboratory, White Sands Range, and Yale University were equally influential, establishing notable technological advancements and characterizing the spectroscopic properties of bioaerosols. Key contributions include experimental measurements of isotropy, yield, and degradation of bioaerosol fluorescence as well as introduction of sub-300 nm excitation laser configurations. Reduction in the form factor of portable LIF systems has benefited from semiconductors and light-emitting diode (LED) research supported through Defense Advanced Research Projects Agency (DARPA) funding [1].

As of 2020, a multitude of portable LIF bioaerosol systems exist, each possessing different features, optical configurations, and levels of maturity. Table 22.2 summarizes commercial detection systems and their distinguishing properties. Although these systems are available from different vendors, they fundamentally share an identical design – all systems contain some element for aerosol focusing, excitation lasers, photodetectors, signal-processing circuitry, algorithm, and user interface. The current trend in development of new LIF systems involves adding multiple excitation wavelengths and/or multichannel photodetectors to obtain spectrally resolved aerosol fluorescence signatures. These optical configurations enable real-time measurement of the excitation–emission matrix of single aerosols. Coupling these measurements with techniques in machine learning and pattern recognition to interpret the data can permit higher fidelity classification of aerosols, leading to better discrimination of threats to background aerosols [2]. Examples of such multichannel systems include two instruments developed by Droplet Measurement Technologies (DMT): the Wideband Integrated Bioaerosol Sensor (WIBS) and the Spectral Intensity Bioaersol Sampler (SIBS). The WIBS is the first commercial aerosol LIF system to feature dual-wavelength excitation (280 and 370 nm), which is generated from two separate xenon flash lamps in the WIBS. The single-aerosol fluorescence is collected, filtered, and then detected by two detection channels, channel 1 (310–400 nm) and channel 2 (420–650 nm). Like the WIBS, the SIBS applies two excitation wavelengths, 285 and 370 nm, but it possesses greater spectral resolution, recording fluorescence emission over 16 channels in the range of 302–721 nm [12]. Both DMT systems are marketed for academic and environmental sampling, but will likely attract increasing attention from biodefense customers in coming years.

Besides LIF, other spectroscopic techniques like Raman, breakdown spectroscopy, and elastic light polarization have demonstrated potential in classifying threat bioaerosols. The Resource Effective Bioidentification System (REBS), introduced by Battelle, is a man-portable Raman spectrometer that provides autonomous identification of bacteria, spores, toxins, and viruses within 15 minutes or less. The REBS deposits aerosols onto a moving metallic collection tape, and then excites the aerosol sample using 643 nm beam that is shaped into a line geometry. A charged coupled device (CCD) measures the Raman spectra from positions on the line in approximately 1 μm increments – afterward, the line is stepped across the tape to capture the spectra from all deposited particles on the tape, in a process referred to as line scanning. The REBS software algorithm then compares the measured spectra to known threat signatures for determination of a threat. Another portable system, the Rapid Agent Aerosol Detector (RAAD), combines LIF measurements with laser-induced breakdown spectroscopy (LIBS) to improve classification of an aerosol threat. RAAD employs three lasers: an 808 nm cueing laser, a 355 nm laser for elastic scattering and fluorescence measurement, and a 266 nm excitation laser for fluorescence measurement [2]. The system triggers LIBS on demand to obtain the elemental composition of single aerosols. Given the number of separate features measured during the analysis – two elastic scatter, two fluorescence, and a LIBS spectrum – data from the RAAD are well-suited to undergo statistical analysis to enhance discrimination of threat aerosols. An entirely non-LIF bioaerosol detector, the Polaron (ATI), relies on light polarization signatures of aerosols. The Polaron measures light depolarization upon interaction with an aerosol. After scattering, the degree of depolarization is measured by detection optics, and the data undergo analysis and comparison to a library of threat signatures. The Polaron is able to classify aerosols because different aerosol types have distinct chemical compositions and therefore uniquely

Table 22.2 Summary of portable biological spectroscopy used primarily for aerosol biodefense applications.

Name	Developer/Vendor	Modality	Detection properties	Status (as of 2020)
Ultraviolet Aerodynamic Particle Sizer (UV-APS)	TSI Inc.	Aerodynamic sizing Fluorescence	Ex: 355 nm Em: 420–575 nm	Discontinued
Wideband Integrated Bioaerosol Sensor (WIBS)	Droplet Measurement Technologies	Optical sizing Fluorescence	Ex: 280 and 370 nm Em: 310–400 nm and 420–650 nm	Available
Spectral Intensity Bioaerosol Spectrometer (SIBS)		Optical sizing Fluorescence	Ex: 285 and 370 nm Em: 302–721 nm (16 channel resolution)	Available
Instantaneous Biological Analyzer and Collector (IBAC)	FLIR	Optical sizing Fluorescence	Ex: 405 nm	Available
TACBIO® Gen II	Chemical Biological Center	Optical sizing Fluorescence	Ex: 270 nm Em: 317–700 nm	Available
BioScout	Environics	Optical sizing Fluorescence	Ex: 405 nm	Available
Rapid Aerosol Agent Detector (RAAD)	MIT Lincoln Laboratory	Optical sizing Fluorescence Breakdown spectroscopy	Scattering: 808 nm Ex: 355 and 266 nm	Under development
Instantaneous Microbial Detection (IMD-A)	BioVigilant	Fluorescence	Ex: 405 nm	Available
Biological-Agent Warning Sensor (BAWS)	MIT Lincoln Laboratory	Fluorescence	Ex: 260 nm	Discontinued
Resource Effective Bioidentification System (REBS)	Battelle	Raman	Ex: 643 nm	Available
VeroTect	Biral	Optical sizing Particle asymmetry Fluorescence	Ex: 280 nm Em: 330–350 nm and 420–650 nm	Discontinued

This list represents only a fraction of the technologies that have been developed.

modify the polarization state of scattered light. Generally speaking, portable non-LIF systems are less mature than their LIF counterparts, but these approaches yield additional information about an aerosol, and can assist in better threat classification. Overall, wider adoption of these alternative spectroscopic techniques, combined with more spectrally resolved LIF measurements, holds promise in improving real-time detection capability in biodefense.

22.4.2 Medical and Healthcare

Portable spectroscopy technologies can potentially transform diagnostic approaches in the medical and healthcare fields. For practitioners, these technologies can potentially yield timely acquisition of quantitative, reliable, and objective data, supporting rapid assessment of patient health and well-being. Within the healthcare field, portable spectroscopy systems, especially those used for on-site patient diagnosis and monitoring, are often referred to as point-of-care (POC) technologies [13]. As a whole, however, contemporary portable spectroscopy technologies are not widely commercialized or adopted in clinical settings due to their relative immaturity, requirement to meet stringent medical-device approvals, and challenges in educating medical professionals. The sections below detail the medical and healthcare areas that portable spectroscopy can contribute toward, offering researchers and healthcare workers unique tools that can complement their current approaches.

22.4.2.1 Human Tissue

Portable spectroscopy has several benefits in the analysis of human tissue – the techniques give rapid output, are noninvasive to patients, reagentless, and provide a wealth of diagnostic information. As spectroscopy yields information on a molecular level, the data are invaluable in elucidating fundamental properties and composition, oxygenation, and differentiation and maturation of human tissue [14]. The near-infrared (NIR) region (650–1100 nm), often referred to as the "optical window" due to its minimal absorption by water and hemoglobin compared to other wavelengths, is applied for acquiring spectral measurements at the skin surface as well as penetration depths into tissue [15]. Specific applications of NIR spectroscopy in human tissue analysis include in vivo measurement of oxygen in muscle, study of skeletal muscle metabolism, and disease diagnosis. Using NIR light, users can measure an absorption spectrum and interpret peaks to infer details about the composition of a tissue sample [14]. Comparison of a measured NIR spectrum to the spectra from a normal sample provides information on the state of health of the tissue, and can guide further diagnostic decision-making.

In clinical settings, using portable NIR spectroscopy to measure the oxygenated and deoxygenated hemoglobin (Hb) and myoglobin (Mb) is commonplace. Contemporary portable NIR devices are smartphone-sized modules that attach to a human subject, collect spectra, and either store or transmit the data wirelessly [16]. These devices consist of an LED, photodiode, battery, microprocessor, and wireless networking module. In vivo measurements are achievable during human activity (i.e. locomotion), making NIR techniques especially attractive for studies in sports and exercise sciences. To calculate useful metrics (e.g. oxygenated Hb/Mb), raw data are processed with the Beer–Lambert equation [16]. Besides NIR spectroscopy, Raman spectroscopy and fluorescence spectroscopy of human tissue have been explored, but these techniques have not been translated into portable systems for clinical use. Although their prospects are promising, additional research is needed before adoption of these techniques to supplement conventional human tissue analysis.

22.4.2.2 Disease Diagnosis

Spectroscopic approaches offer means to diagnose diseases in a POC manner by detecting the causative agent (i.e. pathogenic organism) or a biomarker of disease. On its own, spectroscopy largely does not have the fidelity to resolve specific identification of a target pathogen, protein, or other biomolecule. To gain specificity, a spectroscopy technique can be paired with a molecular biology assay that applies spectroscopy as means to report the detection of the target of interest. Among all the field applications discussed in this chapter, smartphone biosensing arguably holds the greatest potential, primarily due to both the ubiquity of the smartphone among the general public and widespread need for simple and rapid disease diagnosis. Indeed, academic literature contains numerous articles describing smartphone POC analysis, but these approaches remain immature and have not translated from basic research to practice [13]. Usually, these efforts have involved several key activities: (i) design of a custom smartphone cradle to house a sample, (ii) novel repurposing of smartphone optics, and (iii) development of software apps to process data and output a result; see the chapter in Volume 2 by Peveler and Algar. Proof-of-concept is then demonstrated by preparing and adding sample into the cradle, followed by recording readout from the smartphone setup [3, 4].

Examples of POC smartphone spectroscopy include detection of red and white blood cells, cholesterol, markers for hepatitis B and human immunodeficiency virus (HIV), syphilis antigens, and others [3, 4, 17, 18]. In many of these efforts, assays were immunoassay based, and used some form of colorimetric change or fluorescence spectroscopy to detect binding of antibodies to the antigen target. In all studies, researchers used the optics of popular smartphones like the Apple iPhone or Samsung Galaxy iterations. Some concepts were moreover facilitated by microfluidics to reduce reagent consumption and further reduce the form factor of the smartphone setup to improve portability. Generally, future technical maturation of the referenced techniques, coupled with validation in clinical trials, will promote greater use of smartphone spectroscopy in disease diagnosis. As the scope of this chapter is related to reagentless spectroscopic detection, the reader is referred to the cited sources to gain greater perspective on pairing of portable spectroscopic techniques with biological assays, as this deserves a thorough treatment in its own right [19].

22.4.3 Food and Agriculture

Portable spectroscopy systems serve several roles in the food and agriculture industry: evaluation of product quality, determination of food safety, assessment of plant welfare, and others. The first section focuses on use of portable spectroscopy to monitor for microbial contamination of food. The following section briefly summarizes applications in spectroscopic characterization of crops and livestock. Use of these portable technologies directly at a field site (e.g. food processing line or farm settings) affords unique advantages including rapid turnaround in results and ease-of-use over conventional assay techniques, thereby improving decision-making by users.

22.4.3.1 Food Safety and Hygiene

At food processing plants, the sanitation monitoring of food products, including meat, poultry, fruits, and vegetables is essential. Traditional techniques for analyzing foods for microbial contamination consist of coarse visual and odor inspection, and/or swabbing of the product and assaying the swab sample with plate counts or adenosine triphosphate (ATP) bioluminescence measurement [20]. Plate count or total viable count (TVC), which is reported in units of colony-forming units (CFU) per mL, g, or cm^2, indicates the bacterial loading in the food, whereas ATP reflects the quantity of biomass and thus cleanliness – although ATP can also originate from the food as a nonmicrobial source. Performing conventional analysis, however, is time-consuming, laborious, and necessitates skilled workers to obtain a swab sample. As such, portable spectroscopic systems have emerged as alternatives to monitor spoilage, wherein the food product undergoes analysis through optical techniques during processing. Academic literature contains a wealth of studies documenting development and application of fluorescence, Raman, NIR, and visible spectroscopic approaches to monitor for food spoilage [21–28]. A variety of foods have been investigated: beef, fish, chicken, pork, milk, apple juice, ham, spinach, cabbage, and others. In these studies, researchers built portable systems, and then determined the correlation between measured spectral signatures and bacterial loading (e.g. CFU/g) on the food product. The spectral signature of interest depends on spectroscopic approach; for example, fluorescence-based systems usually probe for the presence of microbial tryptophan and or other markers, whereas Raman systems scan for characteristic features in a measured Raman spectrum. The raw data are then processed with an appropriate data analysis method like partial least square (PLS) regression. As many foods exhibit heterogeneous compositions that interfere with discrimination, Raman spectra, which are inherently richer with information, must undergo data processing with more advanced multivariate techniques such as principal component analysis (PCA), neural networks, or other methods. In general, the studies have demonstrated that spectroscopy systems are indeed able to predict food contamination with reasonable accuracy – for instance, fluorescence techniques can achieve an R^2 value that exceeds 0.80 for fluorescence signal vs. bacterial loading. Raman approaches may offer improved accuracy over fluorescence spectroscopy in estimating microbial contaimination [27]. To further improve specificity, numerous authors have described the combination of biological assays with portable spectroscopy (e.g. smartphones); these approaches are described elsewhere [29].

In practice, several commercial systems are currently available for detection of microbial contamination across the food processing supply chain. The Bluline fecal detector (Veritide), a handheld system used for measuring contamination of beef, applies a fluorescence technique to detect chlorophyll molecules and associated metabolites on meat surfaces. Chlorophyll serves as an indirect marker of contamination because fecal matter of ruminants, like cattle and sheep, characteristically contains an abundance of chlorophyll [30]. The Bluline enables a user to simply point-and-shoot at a meat target and receive results, facilitating rapid analysis of samples. Another handheld point-and-shoot system, the BFD-100 (FreshDetect GmbH), also based on fluorescence spectroscopy, measures spectra from a meat surface, analyzes data, and reports results within five seconds. An NIR-based device, the SCiO (Consumer Physics), records the NIR reflectance from a sample in the wavelength range of 760–1150 nm. Comparison of the NIR spectra of fresh food products to their aged counterparts allows a user to infer potential microbial spoilage of the product [31]. While it is unlikely that these devices will eclipse the gold standard of TVC in the near future, the convenience of these approaches is undeniable and deserving of continued attention within the food industry.

22.4.3.2 Agriculture Applications

For farmers and ranchers, portable spectroscopy holds utility for in situ characterization of the crops and livestock during the product life cycle. In farming, portable systems can be deployed into a field to determine properties like plant growth, species and weed content. The resulting data can guide site-specific dispersal of fertilizer, water, or even herbicide to destroy the growth of undesirable species. Several studies have described the use of portable visible and NIR systems to differentiate crop from weeds by measuring the absorbance or reflectance spectra and analyzing data with PCA or other techniques [32, 33]. In these efforts, spectra were obtained using custom-built systems or commercial handheld spectroradiometers; these systems measure signatures originating from chloroplast-derived pigments, intercellular spaces, water, and cellulose. The accuracy of the classification by these techniques can depend strongly on data interpretation, but it is largely promising, demonstrating successful classification at a rate over 70% [33]. Another application for such technologies is in rapid extraction of information regarding the welfare of crops. For example, visible-light portable spectroscopy systems can detect chlorophyll content in leaves [34]. As chlorophyll is a marker of nutritional status and photosynthetic capacity, its detection can aid in quality assessment, and inform crop management. Uses of portable systems also extend to on-site quantification of biological contaminants like molds and mycotoxin on crops including maize, wheat, and peanuts [35, 36].

Within livestock science, portable spectroscopy has a unique role in profiling of livestock stool to predict forage quality of free-range animals. Characterizing the feces of the animals allows users to access diet and nutritional data in a rapid and noninvasive manner [37]. First introduced by Lyons and Stuth [32], and later adopted in numerous studies, this process consists of collecting fecal samples, performing an NIR reflectance measurement, and interpreting the data for dietary nutritional quality parameters like crude protein, digestible organic matter, and fiber [38]. Predictive models are developed beforehand by feeding animals known diets, recording the fecal spectra, and creating calibration equations; only 8–10 NIR wavelengths account for roughly 85–95% variation in diet [37]. The NIR spectra can reveal marked differences in fecal composition between animals possessing low-quality and high-quality diets for a variety of domesticated and even wildlife species. Continued progress in this area, especially in development of predictive models, will further improve the maturity of these techniques, enabling users to better monitor the health of livestock.

22.4.4 Environmental Pollution Monitoring

Portable spectroscopy systems can detect biological pollution levels for a variety of environmental monitoring field applications. Microbial contamination of water, soil, and air can produce adverse effects to human health, animals, and agriculture, and so rapid biological pollutant monitoring through spectroscopic tools offers an attractive alternative to laborious culturing or and/or molecular assays. A key advantage of portable spectroscopy in microbial pollutant detection is that users can readily bring these systems to a site to analyze samples, or even install a system at the site for in situ and autonomous monitoring. These tools can then inform and facilitate immediate remedial activities to reduce the environmental impact. The sections below describe various applications of portable spectroscopy in biological pollutant monitoring for water, soil, and air.

22.4.4.1 Water

Portable spectroscopy systems can assess microbial quality of wastewater and potability of water sources for rapid on-site analysis. Although Raman and other techniques have been investigated, academic literature has primarily emphasized fluorescence spectroscopy for measuring microbial properties of water [39]. Fluorescence originating from 280 nm excitation, referred to as tryptophan-like fluorescence (TLF) by the water quality community, is a marker of pathogenic microorganisms and anthropogenic organic matter, allowing a user to infer that the water is biologically contaminated. Numerous studies have described the development, characterization, or evaluation of compact handheld fluorimeters that measure TLF emission as a practical tool for pre-screening water to complement confirmatory culture and PCR techniques [40]. In general, these portable fluorimeters are technically simple, consisting of a water sample introduction port, UV–visible light source like an LED, light collection optics, photodetector, and associated software. Operation of the systems by a user is similarly straightforward, involving procurement of a water sample, insertion of the sample into the system, and interpretation of fluorescence results. This ease of operation is especially valuable as it is envisioned that these systems can be distributed to nonexperts in resource-limited settings for analysis of water quality. An example of a commercialized TLF detector is the Proteus Multiparameter Water Quality Meter (Proteus Instruments). Weighing 4.5 kg, the device measures fluorescence at 285 nm excitation, and outputs a real-time readout of coliform quantity and other water quality properties such as biochemical oxygen demand and total organic carbon.

Literature has established correlations of TLF intensity to coliform concentration (CFU/mL), obtaining log-linear correlation – shown over several logs – between the two types of measurements [41, 42]. This relationship was demonstrated for environmental water samples from multiple sites across several locales (United Kingdom, America, Africa, and China), although the correlation displays variability within a body of water, and TLF may overpredict coliform concentration compared to culturing measurements. Regardless, fluorescence measurements have potential use in predicting risk of microbial contamination, and improving prioritization of water samples that should undergo additional analysis [5]. At the moment, fluorimetry is not an established method for coliform analysis, but efforts to formalize it according to the World Health Organization (WHO) or other regulations are ongoing [43].

22.4.4.2 Soil

Compared to water quality applications, development and use of portable spectroscopy systems in soil sciences remain less explored. Microbial biomass influences the biogeochemical equilibrium of soil, responsible for 80–90% of biotransformations – through organic matter decomposition and nutrient cycles – and consequently plant productivity and ecological sustainability [44]. Therefore, assessment of the microbial populations and their activity in soil is of key importance. Given that soil microbial biomass comprises only 5% or less of total organic carbon, spectroscopy cannot directly resolve its contribution to the soil spectra. Thus, researchers focus on indirect markers of microbial biomass, applying visible–NIR spectroscopy to analyze mineralization of soil organic matter (SOM) and soil organic carbon (SOC) since these components strongly absorb in these wavelengths.

Although traditional analysis includes extraction of soil core samples and analysis in the laboratory, mobile proximal soil sensing (PSS) has emerged as a concept to rapidly acquire spectral data of soil biota [45]. PSS involves affixing compact spectroscopy systems or other sensors to vehicles (e.g. tractor, all-terrain vehicles, etc.) so that the system is positioned adjacent to the soil surface. As the vehicle traverses the field, these systems record real-time measurements from soil immediately beneath the vehicle. Termed "on-the-go" analysis, this type of platform, when coupled with a global positioning system, provides spatial resolution mapping of the soil properties of an area of interest. Academic literature has reported on use of on-the-go technologies based on visible–NIR diffuse reflectance to measure SOM and SOC. [45] Setup and operation of such systems in the field requires consideration of distance from the sensor window to the soil surface, and presence of outlier materials like gravel. Notably, these approaches have demonstrated moderate successes in showing a correlation between measured reflectance and SOM/SOC in the soil. Further field demonstrations and improvements to both hardware and data processing will facilitate broader adoption of these approaches by the soil science community.

22.4.4.3 Air

Portable fluorescence-based spectroscopy systems, which were pioneered for military and biodefense purposes, have found select applications in civilian air quality monitoring of environmental airborne microorganisms, often referred to as primary biological aerosol particles (PBAPs) within the academic air quality community. As detailed in Section 22.4.1.2, these systems use LIF to excite biogenic fluorophores present in microorganisms to discriminate bioaerosols from nonbiological aerosols. Academic literature contains numerous field test campaign studies describing the use these portable systems to measure the presence of pollutant bioaerosols in a diverse set of locales: Central Europe, the Amazon rainforest, China, and others [6, 46, 47]. In these campaigns, researchers installed systems at a field site, sampled ambient aerosols over timescales ranging from days to weeks, and interpreted the spatial, temporal, and spectral fluorescence patterns of the sampled bioaerosols. Early LIF studies, specifically those predating the development of the WIBS, focused on PBAPs exhibiting fluorescence emission of biogenic metabolic molecules (355 nm excitation), but advances in LIF bioaerosol technologies have enabled investigation into the multichannel fluorescence signatures of PBAPs. Greater spectral resolution in the fluorescence emission has allowed researchers to develop classification algorithms for PBAPs using statistical methods and machine learning approaches. These studies have demonstrated successes in combining multichannel LIF data with machine learning to classify PBAPs as bacteria, fungi, or other biogenic material, allowing moderate characterization of a biologically diverse air sample in real time [48]. Forthcoming LIF systems, which support even greater spectral resolution (see Section 22.4.1.2) and perhaps improved classification accuracy, can further establish the use of portable spectroscopy systems for monitoring of environmental bioaerosols.

22.4.5 Forensics

Portable biological spectroscopy systems have expanded into the field of criminal forensics, providing investigators with compact tools to more thoroughly examine crime scenes and gather vital physiological evidence [7]. Spectroscopic approaches possess several essential attributes desired by the forensic community – the techniques are nondestructive, rapid, are require no reagents or sample preparation, and involve a simple workflow that can be performed on site by a nonscientist; see the chapter by Kammrath et al. in Volume 2 of this book. In a field setting, such as a crime scene, portable spectroscopy systems can serve to locate the presence of secreted body fluids and stains that may not be visible to the naked eye, assisting investigators in collecting samples for subsequent extraction of genetic information from the sample. Spectroscopic analysis is presumptive, incapable of identification or connecting a body fluid sample to a specific individual. Several established uses of portable systems within forensics are described in the sections below.

22.4.5.1 Detection of Body Fluids

Generally, the workflow for spectroscopic analysis at a crime scene setting consists of an initial assessment and proper documentation of the site. Afterward, investigators survey the area for biological evidence that includes body fluids like blood, saliva, sweat, semen, vaginal fluids, and urine. Some human and animal body fluids naturally fluoresce as these biological substances are comprised of fluorescent amino acids; semen for example is excited at 300–500 nm (UV–violet), emitting visible-light fluorescence in the 460–520 nm range [49, 50]. Within forensics, fluorescence-based spectroscopic searches, often referred to as Alternative Light Sources (ALSs), colloquially termed blacklights or forensic light sources, involve illumination of areas suspected of containing biological fluids with an intense light source followed by visual observation of fluorescence emission [51]. Investigators are equipped with goggles to filter out scattered light and ambient light, improving their discrimination of fluorescence signal. ALS systems are highly portable, usually weighing less than 2 kg and can be carried with one hand. ALS systems are simple technologies containing only four major components: a broadband light source (e.g. xenon lamp), filters, light-shaping optics, and a flexible fiber-optic cable to allow the user to scan light across an area of interest. For some systems, excitation wavelength is selectable by the user; for example, Polilight (Rofin Forensic) systems generate light in 12 narrow wavelength bands ranging in color from UV to red, with the 415 nm band termed the "blood filter."

Portable ALS systems, which have proven use in the detection of body fluids, feature coarse selectivity between biological fluid types. Proper selection of ALS wavelength and goggles by an investigator can improve contrast of a body fluid and/or stain, although the substrate containing the sample can also affect imaging. In one study, it was demonstrated that 450 nm excitation light combined with long-pass goggles enabled a user to visibly distinguish between dried urine, blood, and other fluids [51]. Compared to reference chemical-based screening tests like acid phosphatase and Phadebas paper, ALS systems have similar sensitivity to these assays, displaying a low incidence of false-negative results [51]. On the other hand, false-positive results, defined as ALS detection with a corresponding negative chemical test, are common due to the fluorescence properties of many nonbiological materials. Besides the use of fluorescence emission for locating body fluids at crime scenes, other approaches have been proposed for greater specificity in body fluid identification. This has mostly involved Raman spectroscopy due to its potential to identify chemical functional groups, but currently none of these approaches have been translated into portable systems for field use by crime scene investigators [52].

22.4.5.2 Detection of Latent Fingerprints

Latent fingerprints, fingerprints that are undetectable by the naked eye, can also be located through use of spectroscopic approaches. Latent fingerprints are formed from residue of oil or perspiration deposited by an individual touching a solid surface with their finger. Several portable techniques have been proposed for detection and analysis of latent fingerprints: ALS, hyperspectral imaging, LIBS, and Raman [53, 54]. These spectroscopic approaches have considerable advantages over traditional methods for lifting fingerprints (i.e. dusting for fingerprints) since they can be performed rapidly and do not contaminate the fingerprint for further potential analysis. With ALS, investigators can scan surfaces, applying different light filters and optionally powders, and/or dyes to enhance imaging. Fingerprint evidence is then photographed and documented.

22.4.5.3 Analysis of Bruises, Bite Marks, and Other Wounds

ALS approaches can afford enhanced visualization of bruises and other wounds, revealing features to forensic investigators that are otherwise unobservable with natural light. Bruises and similar injuries, which develop from blunt force trauma that causes dermal blood to leak into tissue, strongly absorb violet and blue light, producing high contrast when illuminated with these colors. Fluorescence from bruises is also observable though not as identifiable [8]. Spectrally, fresh blood exhibits two absorption peaks in the visible region: the Hb Soret peak at 415 nm, and a less intense peak between 520 and 590 nm attributed to oxyhemoglobin. Controlled experimental studies are largely lacking, but several laboratory investigations have reported on the effectiveness of ALS for forensic injury analysis, describing it as either vastly improving or no better than visualization with natural light [55–57]. Clinical studies have described benefits of portable ALS in revealing injuries in strangulation and possible child abuse victims [58, 59]. ALS can be useful in discerning fine features of bruises, deeper injuries, and enhancing contrast for subjects with darker complexions, although it does not appear to aid in determining the age of a bruise [60, 61].

22.5 Summary, Challenges, and Outlook

This chapter presented only a brief overview of field applications of portable biological spectroscopy. With future advances in basic spectroscopy research, smartphones, as well miniaturized technologies, field applications, and users of portable spectroscopy will only continue to expand. As witnessed from the last several decades, unexpected world events and increasing concerns about the environment, healthcare, disease transmission, or other areas, can prompt the growth of new and exciting analysis tools. The main obstacle for biological spectroscopy remains its inferior specificity compared to molecular biology assays, and in the context of its field applications, this concern is no different. In a field setting, spectroscopic techniques are limited to presumptive detection of a biological material, and generally lack species-level specificity to unequivocally identify or characterize a target. More robust or confirmatory techniques like PCR, immunoassay, or genome sequencing must complement spectroscopic analysis, requiring users to bring additional equipment, supplies, or consumables into the field. Developments in fundamental spectroscopy approaches, machine learning pattern recognition, and better integration with assays can improve specificity, accelerating greater adoption of portable spectroscopy within biology. Another facet for portable spectroscopy, which can be viewed as an opportunity, is the continued pursuit of miniaturization of optical components, control circuitry, and other hardware in order to reduce the form factor of these systems while preserving signal-to-noise ratio and user experience. Current developments in optics, materials science, and related disciplines can contribute toward achieving this goal.

While this chapter focused on several disciplines that maintain established uses for portable biological spectroscopy, these techniques may have unique utility in other nonconventional areas including astrobiology, anthropology, art history, and others. In these disciplines, portable biological spectroscopy can offer novel means to detect and classify biological materials in samples through a nondestructive process without need for wet chemistry. For example, several astrobiology studies have proposed a miniature Raman spectrometer to noninvasively probe for the existence of life, since Raman scattering is able to differentiate between spectral signatures from a sample containing both biological and geological constitutes [62]. This type of system can be outfitted to a robotic lander or perhaps supplied to astronauts for analysis of samples; see the chapter in Volume 2 by Edwards et al. for more details. Within anthropology, archeologists have used portable systems to perform on-site identification of biomaterials present at historical burial sites; studies have reported on the Raman spectra of human and animal hair and skins cells dating back 4000 years ago found an Egyptian burial site [63]. Undoubtedly, portable biological spectroscopy will continue to find other meaningful applications as technologies improve and researchers desire to gather molecular description of their samples.

As stated in this chapter, smartphones, which already contain many of the necessary components for spectroscopic analysis, hold potential in transforming portable spectroscopy and, due to their ubiquity, even democratizing spectroscopy for citizen scientists (see the chapter in this Volume by Scheeline for more details). Clearly, as smartphone technologies undergo improvements to camera optics, their potential to perform a variety of biological spectroscopic analyses also grows. In the near future, smartphones can possibly serve as a generic spectrometer in any field application, with specific functionality being added by installing a relevant smartphone app. Furthermore, the existing ability of smartphones to upload real-time data that are rich in metadata like timestamps makes them well-suited to support the development of spectroscopic databases possessing spatial and temporal information. This capability can impact fields like environmental pollutant monitoring, where any individual with a smartphone can contribute high-fidelity environmental data at their specific location. With these advancements, the outlook for field-based portable spectroscopy is excellent, filled with exciting new possibilities and opportunities to improve practical biological detection.

Acknowledgements

The authors acknowledge the Johns Hopkins Applied Physics Laboratory Janney Program for the funding support. The authors recognize Ms. Leah Carol and Dr. Jody Proescher for compiling references and reviewing this chapter, respectively.

List of Acronyms

ALS	alternative light source
ARL	Army Research Laboratory
ATP	adenosine triphosphate
CCD	charged coupled device
CFU	colony-forming unit
COTS	commercial off-the-shelf
DARPA	Defense Advanced Research Projects Agency
DMT	Droplet Measurement Technologies
DPA	dipicolinic acid
FLAPS	fluorescence aerodynamic particle sizer
Hb	hemoglobin
LED	light-emitting diode
LFA	lateral flow assay
LIBS	laser-induced breakdown spectroscopy
LIF	laser- (or light)-induced fluorescence
NADH	nicotinamide adenine dinucleotide
Mb	myoglobin
NIR	near-infrared
PBAB	primary biological aerosol particles
PCA	principal component analysis
PCR	polymerase chain reaction
PLS	partial least squares
POC	point of care
PPE	personal protective equipment
PSS	proximal soil sensing
RAAD	Rapid Agent Aerosol Detector
REBS	Resource Effective Bioidentification System
SIBS	Spectral Intensity Bioaerosol Spectrometer
SOC	soil organic carbon
SOM	soil organic matter
TLF	tryptophan-like fluorescence
TVC	total viable counts
UV	ultraviolet
UV-APS	Ultraviolet Aerodynamic Particle Sizer
WIBS	Wideband Integrated Bioaerosol Spectrometer
WHO	World Health Organization

References

1 DeFreez, R. (2009). LIF bio-aerosol threat triggers: then and now. *Opt. Biol. Chem. Detect Def. V.* 7484 (September): 74840H. https://doi.org/10.1117/12.835088.

2 Huffman, J.A., Perring, A.E., Savage, N.J. et al. (2019). Real-time sensing of bioaerosols: review and current perspectives. *Aerosol Sci. Tech.* 54 (5): 1–31. https://doi.org/10.1080/02786826.2019.1664724.

3 Laksanasopin, T., Guo, T.W., Nayak, S. et al. (2015). A smartphone dongle for diagnosis of infectious diseases at the point of care. *Sci. Transl. Med.* 7 (273): 273re1. https://doi.org/10.1126/scitranslmed.aaa0056.

4 Giavazzi, F., Salina, M., Ceccarello, E. et al. (2014). A fast and simple label-free immunoassay based on a smartphone. *Biosens. Bioelectron.* 58: 395–402. https://doi.org/10.1016/j.bios.2014.02.077.

5 Sorensen, J.P.R., Lapworth, D.J., Marchant, B.P. et al. (2015). In-situ tryptophan-like fluorescence: a real-time indicator of faecal contamination in drinking water supplies. *Water Res.* 81: 38–46. https://doi.org/10.1016/j.watres.2015.05.035.

6 Yue, S., Ren, H., Fan, S. et al. (2017). High abundance of fluorescent biological aerosol particles in winter in Beijing, China. *ACS Earth Sp Chem.* 1 (8): 493–502. https://doi.org/10.1021/acsearthspacechem.7b00062.

7 Wee-chuen, L. and Bee-Ee, K. (2010). Forensic light sources for detection of biological evidences in crime scene investigation: a review. *Malaysian J. Forensic. Sci.* 1 (1): 20–30.

8 Lombardi, M., Canter, J., Patrick, P.A., and Altman, R. (2015). Is fluorescence under an alternate light source sufficient to accurately diagnose subclinical bruising? *J. Forensic. Sci.* 60 (2): 444–449. https://doi.org/10.1111/1556-4029.12698.

9 National Research Council (2011). *Review of the Scientific Approaches Used during the FBI's Investigation of the 2001 Anthrax Letters*. Washington, DC: The National Academies Press https://doi.org/10.17226/13098.

10 Pöhlker, C., Huffman, J.A., and Pöschl, U. (2012). Autofluorescence of atmospheric bioaerosols – fluorescent biomolecules and potential interferences. *Atmos. Meas. Tech.* 5: 37–71. https://doi.org/10.5194/amt-5-37-2012.

11 Ammor, M.S. (2007). Recent advances in the use of intrinsic fluorescence for bacterial identification and characterization. *J. Fluoresc.* 17 (5): 455–459. https://doi.org/10.1007/s10895-007-0180-6.

12 Könemann, T., Savage, N., Klimach, T. et al. (2019). Spectral intensity bioaerosol sensor (SIBS): an instrument for spectrally resolved fluorescence detection of single particles in real time. *Atmos. Meas. Tech.* 12 (2): 1337–1363. https://doi.org/10.5194/amt-12-1337-2019.

13 McCracken, K.E. and Yoon, J.Y. (2016). Recent approaches for optical smartphone sensing in resource-limited settings: a brief review. *Anal. Methods* 8 (36): 6591–6601. https://doi.org/10.1039/c6ay01575a.

14 Chiriboga, L., Xie, P., Yee, H. et al. (1998). Infrared spectroscopy of human tissue. I. Differentiation and maturation of epithelial cells in the human cervix. *Biospectroscopy* 4 (1): 47–53. https://doi.org/10.1002/(sici)1520-6343(1998)4:1<47::aid-bspy5>3.3.co;2-1.

15 Sakudo, A. (2016). Near-infrared spectroscopy for medical applications: current status and future perspectives. *Clin. Chim. Acta* 455: 181–188. https://doi.org/10.1016/j.cca.2016.02.009.

16 Hamaoka, T., McCully, K.K., Niwayama, M., and Chance, B. (2011). The use of muscle near-infrared spectroscopy in sport, health and medical sciences: recent developments. *Philos. Trans. R. Soc. A Math. Phys. Eng. Sci.* 369 (1955): 4591–4604. https://doi.org/10.1098/rsta.2011.0298.

17 Zhu, H., Sencan, I., Wong, J. et al. (2013). Cost-effective and rapid blood analysis on a cell-phone. *Lab Chip* 13 (7): 1282–1288. https://doi.org/10.1039/c3lc41408f.

18 Oncescu, V., Mancuso, M., and Erickson, D. (2014). Cholesterol testing on a smartphone. *Lab Chip* 14 (4): 759–763. https://doi.org/10.1039/c3lc51194d.

19 Kanchi, S., Sabela, M.I., Mdluli, P.S., and Inamuddin, B.K. (2018). Smartphone based bioanalytical and diagnosis applications: a review. *Biosens. Bioelectron.* 102 (August): 136–149. https://doi.org/10.1016/j.bios.2017.11.021.

20 Oto, N., Oshita, S., Makino, Y. et al. (2013). Non-destructive evaluation of ATP content and plate count on pork meat surface by fluorescence spectroscopy. *Meat Sci.* 93 (3): 579–585. https://doi.org/10.1016/j.meatsci.2012.11.010.

21 Schmidt, H., Sowoidnich, K., and Kronfeldt, H.D. (2010). A prototype hand-held Raman sensor for the in situ characterization of meat quality. *Appl. Spectrosc.* 64 (8): 888–894. https://doi.org/10.1366/000370210792081028.

22 Ye, X., Iino, K., and Zhang, S. (2016). Monitoring of bacterial contamination on chicken meat surface using a novel narrowband spectral index derived from hyperspectral imagery data. *Meat Sci.* 122: 25–31. https://doi.org/10.1016/j.meatsci.2016.07.015.

23 Saranwong, S. and Kawano, S. (2008). System design for non-destructive near infrared analyses of chemical components and total aerobic bacteria count of raw milk. *J. Near Infrared Spectrosc.* 16 (4): 389–398. https://doi.org/10.1255/jnirs.807.

24 Luo, B.S. and Lin, M. (2008). A portable Raman system for the identification of foodborne pathogenic bacteria. *J. Rapid. Methods Autom. Microbiol.* 16 (3): 238–255. https://doi.org/10.1111/j.1745-4581.2008.00131.x.

25 Sowoidnich, K., Schmidt, H., Kronfeldt, H.D., and Schwägele, F. (2012). A portable 671 nm Raman sensor system for rapid meat spoilage identification. *Vib. Spectrosc.* 62: 70–76. https://doi.org/10.1016/j.vibspec.2012.04.002.

26 Aït-Kaddour, A., Boubellouta, T., and Chevallier, I. (2011). Development of a portable spectrofluorimeter for measuring the microbial spoilage of minced beef. *Meat Sci.* 88 (4): 675–681. https://doi.org/10.1016/j.meatsci.2011.02.027.

27 He, H.J. and Sun, D.W. (2015). Microbial evaluation of raw and processed food products by visible/infrared, Raman and fluorescence spectroscopy. *Trends Food Sci. Technol.* 46 (2): 199–210. https://doi.org/10.1016/j.tifs.2015.10.004.

28 Liang, P.S., Park, T.S., and Yoon, J.Y. (2014). Rapid and reagentless detection of microbial contamination within meat utilizing a smartphone-based biosensor. *Sci. Rep.* 4: 4–11. https://doi.org/10.1038/srep05953.

29 Rateni, G., Dario, P., and Cavallo, F. (2017). Smartphone-based food diagnostic technologies: a review. *Sensors (Switzerland)* 17 (6) https://doi.org/10.3390/s17061453.

30 Oh, M., Lee, H., Cho, H. et al. (2016). Detection of fecal contamination on beef meat surfaces using handheld fluorescence imaging device (HFID). *Sens. Agric. Food Qual Saf. VIII.* 9864 (May): 986411. https://doi.org/10.1117/12.2227184.

31 Lee, S., Noh, T.G., Choi, J.H. et al. (2017). NIR spectroscopic sensing for point-of-need freshness assessment of meat, fish, vegetables and fruits. *Sens. Agric. Food Qual Saf. IX.* 10217 (May): 1021708. https://doi.org/10.1117/12.2261803.

32 Zhang, Y. and He, Y. (2006). Crop/weed discrimination using near-infrared reflectance spectroscopy (NIRS). *Fourth Int. Conf. Photon. Imaging Biol. Med.* 6047 (October): 60472G. https://doi.org/10.1117/12.710957.

33 Borregaard, T., Nielsen, H., Nørgaard, L., and Have, H. (2000). Crop-weed discrimination by line imaging spectroscopy. *J. Agric. Eng. Res.* 75 (4): 389–400. https://doi.org/10.1006/jaer.1999.0519.

34 Liu, B., Yue, Y.M., Li, R. et al. (2014). Plant leaf chlorophyll content retrieval based on a field imaging spectroscopy system. *Sensors (Switzerland)* 14 (10): 19910–19925. https://doi.org/10.3390/s141019910.

35 Sieger, M., Kos, G., Sulyok, M. et al. (2017). Portable infrared laser spectroscopy for on-site mycotoxin analysis. *Sci. Rep.* 7: 1–6. https://doi.org/10.1038/srep44028.

36 Levasseur-Garcia, C. (2012). *Infrared spectroscopy applied to identification and detection of microorganisms and their metabolites on cereals (corn, wheat, and barley).* Agricultural Science, Godwin Aflakpui, IntechOpen. doi: 10.5772/34762. https://www.intechopen.com/books/agricultural-science/infrared-spectroscopy-applied-to-identification-and-detection-of-microorganisms-and-their-metabolite.

37 Stuth, J., Jama, A., and Tolleson, D. (2003). Direct and indirect means of predicting forage quality through near infrared reflectance spectroscopy. *F Crop Res.* 84 (1–2): 45–56. https://doi.org/10.1016/S0378-4290(03)00140-0.

38 Lyons, R.K. and Stuth, J.W. (1992). Fecal NIRS equations for predicting diet quality of free-ranging cattle. *J. Range Manage.* 45 (3): 238. https://doi.org/10.2307/4002970.

39 Carstea, E.M., Bridgeman, J., Baker, A., and Reynolds, D.M. (2016). Fluorescence spectroscopy for wastewater monitoring: a review. *Water Res.* 95: 205–219. https://doi.org/10.1016/j.watres.2016.03.021.

40 Cumberland, S., Bridgeman, J., Baker, A. et al. (2012). Fluorescence spectroscopy as a tool for determining microbial quality in potable water applications. *Environ. Technol.* 33 (6): 687–693. https://doi.org/10.1080/09593330.2011.588401.

41 Baker, A., Cumberland, S.A., Bradley, C. et al. (2015). To what extent can portable fluorescence spectroscopy be used in the real-time assessment of microbial water quality? *Sci. Total Environ.* 532: 14–19. https://doi.org/10.1016/j.scitotenv.2015.05.114.

42 Baker, A., Ward, D., Lieten, S.H. et al. (2004). Measurement of protein-like fluorescence in river and waste water using a handheld spectrophotometer. *Water Res.* 38 (12): 2934–2938. https://doi.org/10.1016/j.watres.2004.04.023.

43 Nowicki, S., Lapworth, D.J., Ward, J.S.T. et al. (2019). Tryptophan-like fluorescence as a measure of microbial contamination risk in groundwater. *Sci. Total Environ.* 646: 782–791. https://doi.org/10.1016/j.scitotenv.2018.07.274.

44 Nannipieri, P., Ascher, J., Ceccherini, M.T. et al. (2003). Microbial diversity and soil functions. *Eur. J. Soil Sci.* 54: 655–670. https://doi.org/10.1046/j.1365-2389.2003.00556.x.

45 Dhawale, N.M., Adamchuk, V.I., Prasher, S.O. et al. (2015). Proximal soil sensing of soil texture and organic matter with a prototype portable mid-infrared spectrometer. *Eur. J. Soil Sci.* 66 (4): 661–669. https://doi.org/10.1111/ejss.12265.

46 Huffman, J.A., Sinha, B., Garland, R.M. et al. (2012). Size distributions and temporal variations of biological aerosol particles in the Amazon rainforest characterized by microscopy and real-time UV-APS fluorescence techniques during AMAZE-08. *Atmos Chem. Phys.* 12 (24): 11997–12019. https://doi.org/10.5194/acp-12-11997-2012.

47 Toprak, E. and Schnaiter, M. (2013). Fluorescent biological aerosol particles measured with the waveband integrated bioaerosol sensor WIBS-4: laboratory tests combined with a one year field study. *Atmos Chem. Phys.* 13 (1): 225–243. https://doi.org/10.5194/acp-13-225-2013.

48 Ruske, S., Topping, D.O., Foot, V.E. et al. (2018). Machine learning for improved data analysis of biological aerosol using the WIBS. *Atmos. Meas. Tech.* 11 (11): 6203–6230. https://doi.org/10.5194/amt-11-6203-2018.

49 Stoilovic, M. (1991). Detection of semen and blood stains using polilight as a light source. *Forensic Sci. Int.* 51 (2): 289–296. https://doi.org/10.1016/0379-0738(91)90194-N.

50 Nanda, K.D.S., Ranganathan, K., Umadevi, K.M., and Joshua, E. (2011). A rapid and noninvasive method to detect dried saliva stains from human skin using fluorescent spectroscopy. *J. Oral Maxillofac Pathol.* 15 (1): 22–25. https://doi.org/10.4103/0973-029X.80033.

51 Vandenberg, N. and Van Oorschot, R.A.H. (2006). The use of Polilights® in the detection of seminal fluid, saliva, and bloodstains and comparison with conventional chemical-based screening tests. *J. Forensic Sci.* 51 (2): 361–370. https://doi.org/10.1111/j.1556-4029.2006.00065.x.

52 Virkler, K. and Lednev, I.K. (2010). Raman spectroscopic signature of blood and its potential application to forensic body fluid identification. *Anal. Bioanal. Chem.* 396 (1): 525–534. https://doi.org/10.1007/s00216-009-3207-9.

53 Godwal, Y., Taschuk, M.T., Lui, S.L. et al. (2008). Development of laser-induced breakdown spectroscopy for microanalysis applications. *Laser Part Beams.* 26 (1): 95–103. https://doi.org/10.1017/S0263034608000128.

54 Nakamura, A., Okuda, H., Nagaoka, T. et al. (2015). Portable hyperspectral imager with continuous wave green laser for identification and detection of untreated latent fingerprints on walls. *Forensic Sci. Int.* 254: 100–105. https://doi.org/10.1016/j.forsciint.2015.06.031.

55 Trefan, L., Harris, C., Evans, S. et al. (2018). A comparison of four different imaging modalities – conventional, cross polarized, infra-red and ultra-violet in the assessment of childhood bruising. *J. Forensic Leg. Med.* 59 (April): 30–35. https://doi.org/10.1016/j.jflm.2018.07.015.

56 Nijs, H.G.T., De Groot, R., Van Velthoven, M.F.A.M., and Stoel, R.D. (2019). Is the visibility of standardized inflicted bruises improved by using an alternate ('forensic') light source? *Forensic Sci. Int.* 294: 34–38. https://doi.org/10.1016/j.forsciint.2018.10.029.

57 Olds, K., Byard, R.W., Winskog, C., and Langlois, N.E.I. (2016). Validation of ultraviolet, infrared, and narrow band light alternate light sources for detection of bruises in a pigskin model. *Forensic Sci. Med. Pathol.* 12 (4): 435–443. https://doi.org/10.1007/s12024-016-9813-x.

58 Holbrook, D.S. and Jackson, M.C. (2013). Use of an alternative light source to assess strangulation victims. *J. Forensic Nurs.* 9 (3): 140–145. https://doi.org/10.1097/JFN.0b013e31829beb1e.

59 Mimasaka, S., Ohtani, M., Kuroda, N., and Tsunenari, S. (2010). Spectrophotometric evaluation of the age of bruises in children: measuring changes in bruise color as an indicator of child physical abuse. *Tohoku J. Exp. Med.* 220 (2): 171–175. https://doi.org/10.1620/tjem.220.171.

60 Hughes, V.K., Ellis, P.S., and Langlois, N.E.I. (2006). Alternative light source (polilight®) illumination with digital image analysis does not assist in determining the age of bruises. *Forensic Sci. Int.* 158 (2–3): 104–107. https://doi.org/10.1016/j.forsciint.2005.04.042.

61 Owens, L., Warfield, T., Macdonald, R., and Krenzischek, E. (2018). Using alternative light source technology to enhance visual inspection of the skin. *J. Wound Ostomy Cont. Nurs.* 45 (4): 356–358. https://doi.org/10.1097/WON.0000000000000448.

62 Villar, S.E.J. and Edwards, H.G.M. (2006). Raman spectroscopy in astrobiology. *Anal. Bioanal. Chem.* 384 (1): 100–113. https://doi.org/10.1007/s00216-005-0029-2.

63 Edwards, H.G.M. and Munshi, T. (2005). Diagnostic Raman spectroscopy for the forensic detection of biomaterials and the preservation of cultural heritage. *Anal. Bioanal. Chem.* 382 (6): 1398–1406. https://doi.org/10.1007/s00216-005-3271-8.

Index

Portable Spectroscopy and Spectrometry 1: Technologies and Instrumentation, First Edition.
Edited by Richard A. Crocombe, Pauline E. Leary and Brooke W. Kammrath.

C

d

e

h

i

n

o

S

y

z